ANIMAL BREEDING: PRINCIPLES AND APPLICATIONS

ANIMAL BREEDING: PRINCIPLES AND APPLICATIONS

GURVINDER SINGH BRAH
Formerly Professor Cum Director
School of Animal Biotechnology
Guru Angad Dev Veterinary & Animal Sciences University,
Ludhiana, Punjab (India)

KALYANI PUBLISHERS
LUDHIANA - NEW DELHI - NOIDA (U.P.) - HYDERABAD
CHENNAI - KOLKATA - CUTTACK - GUWAHATI - KOCHI - BANGALURU

KALYANI PUBLISHERS

Head Office

B-1/1292, Rajinder Nagar, **Ludhiana-**141 008 • Ph : 0161-2760031, 2745756
E-mail : kalyanibooks@yahoo.co.in **Website :** www.kalyanipublishers.co.in

Administration Office

4779/23, Ansari Road, Daryaganj, New Delhi-110 002
Ph : 011-23271469, 23274393 **E-mail :** kalyani_delhi@yahoo.co.in

Works

B-16, Sector-8, **NOIDA** (U.P.)

Branches

1, Mahalakshmi Street, T. Nagar, **Chennai-**600 017 • Ph : 044-24344684
Gopabandhu Lane, Behind Govt. Bus Stand, Badambadi,
Cuttack-753 012 (Odisha) • Ph : 0671-2311391
3-5-1108, Narayanaguda, **Hyderabad-**500 029 • Ph : 040-24750368
10/2B, Ramanath Mazumdar Street, **Kolkata-**700 009 • Ph : 033-22416024
Arunalaya, 1st Floor, Saraswati Road, Pan Bazar, **Guwahati-**781 001 • Ph : 0361-2731274
Koratti Parambil House, Convent Road, **Kochi-**682 035 • Ph : 0484-2367189
No. 24 & 25, 1st Floor, Hameed Shah Complex, Cubbonpet Main Road, **Bengaluru-**560 002

KPP M 21227 05 P-30830 05 S-35687 05

Reprinted, 2020

Reprinted, 2022

Datalink Computer
(ANI-BR16)

ISBN 978-93-272-6976-5

PRINTED IN INDIA
at : B.B Press, A-37 Sector 67, Noida-201301

Preface

This book is the outcome of my teaching of Animal Breeding courses to undergraduate and postgraduate students for over 35 years. During this period, as a teacher, I realized that most students lack interest because of abstract nature of the subject and the mathematical derivations that carry little biological meaning to beginners. This book attempts to introduce the subject with barest possible statistical treatment and with greater emphasis on conceptual understanding and applications of Animal Breeding.

The book has primarily been designed as a textbook for undergraduate course in animal breeding for BVSc and AH students as per the syllabi prescribed by the Veterinary Council of India. Nonetheless, the book should be equally useful for postgraduate students as an overview of Animal Breeding, and as a basic reference book for the teachers.

The main features of the book are:

- **Background framework and sequential presentation:** Each chapter begins with background information pertinent to the topic followed by a statement of the relevant definitions illustrated with examples wherever required. Each topic is covered under a number of structured titles and sub-titles which facilitate readability and easy flow of information.
- **Solved and supplementary problems:**Animal breeding involves lot of computational work, and this is the area where students of biological, veterinary and applied sciences experience major difficulty primarily for want of sufficient practice in problem solving. Inferences, wherever required, are highlighted for better appreciation of the computations. A number of unsolved problems along with answers are provided at the end of chapter. These supplementary problems serve as a review of the topic and are designed to test mastery of computational aspects.
- **Boxed material:** In most chapters short topics have been put in “Boxes” which are additional relevant materials of the chapter. These boxed materials contain several historical aspects and anecdotes pertaining to animal breeding and quantitative genetics and are designed to induce reader’s interest. These boxes can be read independent of remaining text as they are not a part of the syllabi. Advanced students would find such information useful for broadening their vision of the subject.

Acknowledgments: Sincere thanks are to:

Dr ML Chaudhary, Formerly Head Department of AGB, GADVASU, a long-time associate and research collaborator, for sharing greater research responsibilities during writing of the book, constant support and stimulation, thoroughly going through all the chapters, checking calculations and offering suggestions and changes. His help and support has been invaluable and to be cherished forever. Dr Chandra Shekhar Mukhopadhyay and Dr GVPPS Ravi Kumar who read and cross-checked all the solved and unsolved problems on measurement of inbreeding and relationship coefficients. Dr Chandra also deserves special mention for cross checking of calculations of many other chapters and also for pushing me to complete the book.Dr Puneet Malhotra for proof reading and checking of most chapters including some of the cumbersome ones, which involve mathematical symbols and computations. Dr A K Thiruvenkadan of TNVASU deserves special thanks for providing the definition and description of sheep and horse traits, and for some queries on sheep and goat. Dr Ratan Choudhary for the subtle push to wind up the project during the freshly-brewed coffee sessions with him, and for reading the preface. Drs. P.P. Dubey and S.K. Dash for proof reading chapter no. 5 and 7. Dr. J.S. Arora for providing library support. My former MVSc and PhD students, Drs Hitesh Pawar (currently at Texas University, USA) and Pranav Mathur (currently at University of Utah, USA), deserve special appreciation for checking of numerical problems and in inserting the repeated correction made by me in the hard copy.

The writing of the book, though started several years ago, was completed after my formal retirement while being Adjunct Professor in the School of Animal Biotechnology. Special thanks are to Dr Ramneek, Director of School of Animal Biotechnology, GADVASU, for providing a conducive working environment, office facilities and ever willing support. Thanks are to Datalink Computer for bearing with my ever increasing demands of setting of the figures, diagrams and general format of the book.

Finally this book would not have been written but for the whole-hearted support, encouragement and forbearance of my family. My wife Amrit and son Gurkanwal had been instrumental in pushing me to complete this project when I had thought of abandoning it due to greater administrative responsibilities.

A book of this length and breadth, involving mathematical expressions, statistical formulas and computations, which for printing are converted to 'PageMaker', is likely to have typographical and other errors, despite best efforts for which the sole responsibility rests with me. Readers may point out these so that they are rectified in the next edition. The book is the outcome of my notes in which I used some diagrams and figures but the source from which these were derived or based has been lost during this long period. A number of photographs and figures that are taken from public domain contain possible citations. However, inability to trace the original source of few diagrams and photographs and copyrighted material, if any, may be brought to notice for rectification in the next edition.

August, 2016

Gurvinder Singh Brah

Email: gsbrah@gmail.com

Content

CH.NO.	CHAPTER	PAGES

SECTION 1: HISTORICAL PERSPECTIVE

SECTION 2: QUANTITATIVE TRAITS

CH.NO.	CHAPTER	PAGES

SECTION 3: SELECTION

SECTION 5: GERMPLASM RESOURCES

SECTION 7: BIOTECHNOLOGY

CONTENTS OF BOXES

SHORT TOPICS

SECTION - 1

HISTORICAL PERSPECTIVE

1

Animal Breeding: History and Overview

History:

"Yet to admire our successes as if they had no past, would make a caricature of knowledge":
Bronowski (1973)

Animal Heritage:

"By sharing our rich, multifaceted heritage with students, we become prisms that analyze the colorful episodes and recombine them into an integral picture that surrounds humans and the animals chosen to be domesticated. Our heritage joins the art and science of husbandry to the totality of human experience............ . Brilliant insights into creating the future come from those who are prepared to learn from the distilled wisdom of the ages."

(Willham)

Animal breeding, which is a sub discipline of the broad field of animal science, focuses on deliberate manipulation of animal's heredity for developing new stocks for use in the society. Animal breeding as a human activity has its origin in antiquity, and is perhaps as old as humanity itself. Iconographic documents give some idea as to how farm animals have evolved since domestication, particularly for external features.

This introductory chapter is devoted to a brief overview of animal breeding including a brief history of its development from art through science to an industry, its scope and accomplishments.

Domestication

Domestication is the process by which a population of wild animals/plants is brought under human provision and control. It involves partial or complete separation of a breeding stock from its wild ancestors. Animals have been domesticated for a wide variety of reasons viz; to produce food (milk, meat, eggs, etc.) or other valuable commodities (fiber, wool, hides, etc.), for draft and transportation, for protection of man and livestock, and for recreation and companionship.

Characteristics Associated with Domesticated Animals

Domestication implies several factors though no one feature defines it completely. Generally, the following four broad criteria are associated with the animals that are considered for domestication:

- **Ability to breed (reproduce) in captivity:** The ability of the wild-captured animals to reproduce (breed) freely in captivity is one of the most important criteria used for classifying a species as domesticated. Animals that are reluctant to reproduce in captivity are unable to produce offspring and cannot be domesticated (e.g., the pandas and cheetahs usually fail to reproduce in captivity).
- **Tameness and pleasant disposition:** Animals that are aggressive towards humans are difficult or even dangerous to be domesticated. For example, the African buffalo has an unpredictable nature and can be highly dangerous to humans. Similarly, animals with nervous disposition are difficult to be kept in captivity as they attempt to flee from confinement when scared. Domestic sheep, for example, still have a tendency to panic but their flocking instinct makes them stay together even when scared. Similarly, species of birds, which are flighty, are also difficult to be domesticated.
- **Modifiable social structure**: Wild animals that live in social groups are easy to be domesticated as their social structure makes them compatible with humans. Further, the animals, which recognize a social hierarchy of dominance, can be raised to recognize human as their leader. Animals with flocking instinct are more amenable to be herded not only by humans but by dogs as well.
- **Flexible diet and early sexual maturity:** Domesticated animals need to adapt to the type of the feed that is offered to them by humans which may be different from what the animals were used to in wild. Animals which can consume a wide variety of food are less expensive to be kept in captivity whereas carnivores (i.e. flesh eating mammals) require huge expenditure. Animals that are fast maturing as compared to human lifespan are more suited for domestication.

Some of the researchers associated with study of evolution of domesticated animals include not only the most common species of livestock but also a vast array of animals such as guinea pig, rabbits, hamsters, carps, and fishes, edible frogs, tortoise, crocodiles, silkworm etc as domesticated. However, most authorities include only the productive species such as cattle, buffaloes, yak, mithun (bovine), sheep (ovine), goat (caprine), and swine (porcine), horse (equine), poultry (avian), dogs (canine) and other pets as domesticated animals. A list of the major food and fiber producing and recreational and companion species of domesticated animals is given in Table 1.1

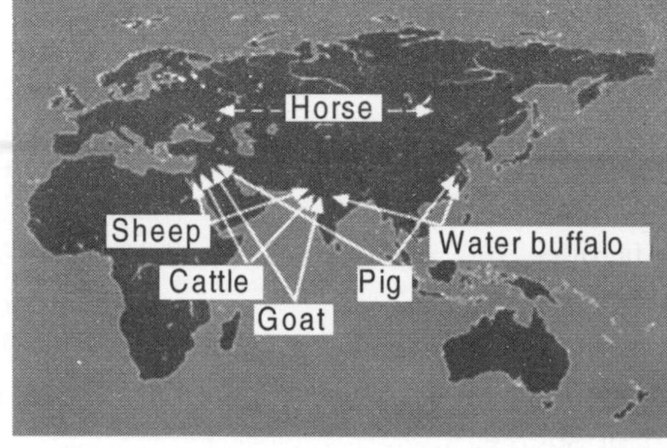

Fig. 1.1. Centres of origin of domesticated animals

Table 1.1 Major food and fiber producing and recreational/companion species of animals.

Species	Scientific name	Primary utility (purpose)
Cattle	Indigenous: *Bos indicus* Exotic: *Bos taurus*	Milk, meat and draft
Buffaloes	*Bubalus bubalis*	Milk, meat and draft
Pig	*Sus scrofa*	Meat
Sheep	*Ovis aries*	Meat and wool
Goat	*Capra hircus*	Meat and milk
Horse	*Equus caballus*	Draft, riding, racing
Fishes	Many species	Food, aesthetic
Poultry:		
Chicken	*Gallus gallus domesticus*	Meat and eggs
Ducks	*Anas platyrhyncha*	Meat and eggs
Turkey	*Meleagris gallopavo*	Meat and eggs
Geese	*Anser anser*	Meat and eggs
Japanese quail	*Coturnix coturnix japonica*	Meat and eggs
Dog	*Canis familiaris*	Companion
Cat	*Felis catus*	Recreational

Domestication of animals is believed to have started about 14000 years ago and ever since agriculture invented. Dog was the first animal species to be domesticated while sheep, goat, cattle, pigs and horses followed. Domestication of different species may have taken place at different times and probably at several different places (regions of world) independently. New species continue to be domesticated, as there are varying degrees of domestication. The domestication of fish for food and ornamental purposes are relatively recent compared to land animals.

Attempts at domestication continue as humans try to breed animals in captivity to prevent their extinction. The approximate locations of suggested domestication centers for sheep, goat, cattle pig and water buffalo are shown in Figure 1.1. Domestication of horse was probably a more diffuse and dispersed process both in time and space.

Forces in Domestication: How has Domestication Occurred?

Little documented information is available about how domestication was first accomplished. It is speculated that hunters brought home a few young as pet and the most gentle and cooperative were kept. Prior to domestication, natural selection chose the animals, which were most fit under the prevailing environment. The animals, which were most adapted to their environment, survived the most and contributed more offspring to future generations.

Domestication of animals and plants gradually changed the human society from early nomadic tribal to a stationary culture. The capture and controlled breeding of the animals were a part of such a transformation. Domestication, in most cases, resulted in large changes in behavior, morphology or physiology. The animals that were best adapted to human controlled environment must have been preferred by man as parents (breeders). Such animals contributed a larger share of their gene pool to next generations leading to genetic changes in behavior and other features, which suited human needs. Man must have practiced 'selective breeding' (i.e. artificial selection) as a part of domestication to suit his needs. The hypothesis of artificial selection ('selective breeding') as a force in domestication has been corroborated by the experiment of domestication of wild silver foxes started by Belyaev in 1959. After 40 years of selection for quiet temperament, silver foxes, that are otherwise aggressive to humans in wild, became as friendly as dog. Artificial selection for the desired attributes has simultaneously brought about some changes in physical appearance also. For example, some bones of the skull of dogs have been modified when compared to their ancestors (wolves).

Natural selection, however, did not completely cease with domestication. Thus both natural as well as artificial selection have played varied roles at different stages of domestication. And both these categories of selection are not mutually exclusive.

Consequences of Domestication

The major consequences of domestication were:

- Genetic changes in behavior, morphology and other features including changes in physiology of animals.
- Domestication led to mating of animals within a region of human settlement due to restricted movement of domesticated animals. This led to increased genetic relatedness among animals of an area. This subsequently helped in formation different breeds suited to specific needs or environment niche.
- Movement of humans from place to place, sometimes too far away places, along with their domesticated animals resulted in mating of genetically unrelated animals. This led to creation of genetic variability which may have helped in formation of new breeds.

Species that become completely domesticated are often unable to survive when reintroduced into the wild. This is so because selection (artificial selection) that drives domestication makes the animals lose their natural instinct of adaptive features that are critical for survival in the wild, but undesirable according to the needs of humans.

Domesticated animal have continually been selected as per new demands, both economic and aesthetic thereby gradually moving domesticated animals farther away (genetically, morphologically and physiologically) from their earlier counterparts. For example, chickens were first reared on free range which then shifted to confinement in sheds on floor.In this process some significant behavior changes, coupled with changes in production, were inevitable. Further, intensification of poultry led to cage rearing in which some more behavioral changes occurred.

Early Developments in Animal Breeding

The practice of animal breeding can be traced to the time of beginning of domestication. Domestication involved a control on mating and reproduction of captive animals. The animals, which suited human's needs, were probably chosen to produce offspring. The most important trait considered in choosing the animals was their behavior (temperament), and it was most conspicuous for dog than with other species. Some cave paintings do indicate that selection was also applied for some morphological traits (e.g. coat color, presence or absence of horns etc.) It is, however, difficult to trace the history of the actual methods used by man to change the domesticated animals. Indeed, animal breeding has for a long time been a collection of experiences which were transmitted orally from one generation to the next and thus accumulated.

The need of humans for food, draft or fiber might have had some effect on consideration of production traits in choosing the animals for reproduction. Written documents dating back to 4000 years indicate that humans:

- Appreciated the significance of resemblance between family members in making decisions about mating.
- Recognized the dangers of mating between relatives (inbreeding).
- Used specific husbandry practices such as castration to prevent reproduction of undesirable males.

Because of some degree of selection and controlled mating, the performance of domesticated animals might have also changed. The improvement through these breeding practices was very slow, and may have partly resulted from adaptation of animals to their surroundings.

Beginning of Systematic Improvement of Livestock

Distinct breeds of cattle and sheep, mainly because of natural selection and geographic isolation, had existed long ago. But the practice of pedigree recording with the aim of systematic stock improvement dates back only to mid 18th century in England. An English stockman, Bakewell is generally given the credit for initiating systematic system of improvement. Nonetheless, other stockmen of his time also conducted breeding on the basis of pedigree.

Robert Bakewell (1725-1795)

Robert Bakewell, an English stockman and agriculturist, is widely acclaimed to be the founder or father[1] of systematic practical animal breeding. Bakewell preceded Mendel by over a century but the soundness (genetic) of the techniques employed by him for developing improved stocks, strains or breeds has survived to present day. Perhaps it is for this legacy that he is referred to as the founder of practical animal breeding.

Bakewell's father had a farm of about 400 acres, at Dishley in Leicestershire County (England), of which 25% was suitable for cultivation while the remainder was pasture. As a young man, Bakewell travelled around the country on horse. He learned agricultural techniques, and studied the anatomy of animals when this field was relatively unknown.

At the age of 35, he assumed the charge of his father's Dishley estate. He put into practice the techniques of agriculture that he had learnt during his travel viz; water management, crop rotation, land manuring etc. The livestock in his time consisted of local types and non-descript. Salient features of his work on stock improvement are listed below:

Development of breeds[2] : Bakewell carried out the breeding of sheep, cattle and horse on his estate and laid the foundation of Leicester breed of sheep, Longhorn breed of cattle (for meat) and Shire horse. Selection in Leicester breed which was developed from the old Lincolnshire breed was applied for improvement of mutton yield because of which the animals of this breed reached market age a year earlier than the usual age of 3-4 years. The Leicester breed had good fatty shoulder (popular at that time) and had good quality fleece.

---

[1]Lush in his book states it big title to refer to him (Bakewell) as founder or father of practical animal breeding.

[2]Breed is a recognized group of interbreeding animals within a domesticed species. Animals of a breed usually have common origin, and are fairly uniform for some genetically controlled morphological traits.

Performance and pedigree recording: Bakewell's success as a breeder, in addition to the cohesive use of certain concepts (genetic), is also attributed to his *care in record keeping, of both performance and parentage.* He made comparisons of growth and feed intake of animals for making selection decisions and used measurements like preserved specimens to measure his progress in stock improvement.

Breeding Procedures Used by Bakewell

Bakewell used both selection and inbreeding for producing animals of the 'desired' type:

(i) **Use of selection:** He purchased animals possessing the 'desired' traits from different places and selected the ones which fulfilled his requirement (i.e. the ones most suitable for his aim). These selected animals were used as parents of the next generation.

He was the first to have conducted *systematic progeny testing of rams and bulls, which* was accomplished as follow:

- *Legal contract*: He signed legal contracts with farmers and hired-out his prized rams to them for a nominal fee. However, he had his terms of "*how the progeny of selected and hired sires were to be reared*". For example, he had the condition that only homegrown fodder was to be fed to all the animals. The idea was to provide uniform and standard management practices so that the progeny of different sires are exposed to more or less similar environment. This provision of good environment also allowed for fullest expression of their genetic potential.
- *Inspection of progeny*: He retained his right to inspect progeny of the hired sires. The sires producing the best progeny according to the trait(s) of his interest, were used by him in his own herd. This procedure led to genetic improvement not only in his own herd but also in the herds of farmers who used his sires (i.e. transmission of genetic improvement).

(ii) **Use of mating between relatives:** At Bakewell's time, there was a general prejudice against mating between close relatives (inbreeding) which was considered incest in humans. Bakewell went against all the preconceived notions and ideas, and made close matings for several generations. The work on inbreeding by Bakewell is said to have been conducted with more or less secrecy so far the public was concerned. Mating between close relatives demonstrated the power of inbreeding in '*fixing*' the desired traits in progeny and perpetuation of these specific traits generation after generation.

Bakewell also promoted some concepts that still have meaning today, such as :'like begets like' which means superior parents are more likely to produce superior progeny than are inferior parents, **'prepotency'**[3] is associated with inbreeding', 'breed the best to best' ; and 'on an average, and not every time' (indicating the role of chance).

Establishment of Breed Associations

As time passed the number of stockmen who were involved in stock improvement increased. It became difficult to remember the genetic relationship between animals far back in the pedigree. There thus arose a need for recoding pedigree on paper. This led to need for breed associations. In 1783, Bakewell established an association of sheep breeders, the 'Dishley Society' to protect the 'purity' of his stock. The society laid down rules for letting of rams for Leicester breeders in England, and proved to be forerunner of today's pedigree breed associations.

---

[3]Prepotency is the ability of an animal to stamp down it's trait(s) on its progeny.

The flock masters of the entire kingdom took note of his success and his farm became famous for breeding and management. His annual animal auctions created great attention:

- One celebrated ram, 'Two Pounder', is said to have earned 800 guineas in a single season (equivalent to today's lakhs of pounds sterling).
- Several of his Longhorn bulls (at a sale in Oxfordshire in 1791), fetched above 200 guineas each.
- A bull fetched 400 guineas at a sale in 1793.

King George III honored Bakewell for his "new discovery in stock breeding". A society named as 'New Dishley Society' exists in England to promote the memory of Bakewell and his students. Its aim is to disseminate knowledge of his work and appreciation of his pioneering legacy in breeding of improved farm livestock.

Blue Plaque for Bakewell placed by Charnwood Borough Council, Loughborough (UK)

A painting of Bakewell on horse

Robert Bakewell (1725-1795)

Bakewell's success in stock breeding attracted the attention of several English stockmen who came to his estate to learn breeding methods. **Charles Colling,** a young man was one of such pupils (in 1783). Colling also conducted intensive inbreeding for producing outstanding cattle. He along with his brother founded the Shorthorn breed, which became the most famous breed of cattle during 19th century. This breed was used in the formation of a number of other breeds. *The Shorthorn breed was one of the first organized breeds of livestock.* Since its founders practiced intense inbreeding, the entire breed was highly related to one or two bulls which were heavily advertised. Soon after the development of Shorthorn, many other breeds of cattle, sheep, swine and horses were also developed. Thomas Bates (1775-1849) studied Colling's methods, and initiated record keeping of feed consumption of Cattle.

Darwin on breeding

Darwin himself was impressed by the achievements of farmers, and artificial selection was a source of inspiration for his theory of evolution:

"We cannot suppose that all the breeds were suddenly produced as perfect and as useful as we see now them; indeed, in several cases, we know that this has not been their history. The key is man's power of accumulative selection: nature gives successive variations; man adds them up in certain directions useful to him. In this sense he may be said to make for himself useful breeds". DARWIN on origin of species (1859, p. 30)

Herd Books

The term **'herd book'** is used worldwide to designate the procedure by which lineage of livestock is recorded in a book. Herd books were published by respective breed organizations. The single most important function of herd books was identification of ancestry of an animal and its progeny. Herd books that contained information about pedigree of animals were started in 18th and 19th centuries:

- Herd book for English thoroughbred horses, referred to as '*An Introduction to General Stud Book*' was started (in 1791) to record pedigree of *'performers'*(meaning those that raced),and was published in 1808.
- The first herd book for cattle was for Shorthorn breed, called '*Coates Herd Book*' (1822). It emphasized only pedigree information, and not the performance.
- Later, herd books were published for different breeds of horses in France (1826), Germany (1827) and Australia (1847), and for cattle in late 19th century (viz; for Ayershire, Jersey, Holstein, Guernsey and Brown Swiss breeds in USA between 1863 and 1880).
- The first swine herd book was started by American Berkshire Association (1876).

At the beginning, the herd books were kept open for registration of outstanding animals though their parents had not been registered in that herd book. As time passed most of the herd books became closed which meant that only those animals who had parents in herd book were eligible for registration in that herd book. The animals whose parents were registered in herd or studbook of a breed were called '**purebreds**'.

Drawback: The biggest lacuna in herd books was that performance of animals was not required as the criterion for registration once the herd book was closed. Since both pedigree and performance are important for evaluation of genetic merit (breeding values), pedigree without performance served little purpose.

Box 1.1: Breed Associations.

Breed associations (societies) have been an integral component of animal husbandry traditions of many countries. In USA, the Dairy breed organizations, which publish their summaries, were started in the latter half of 19th century. These associations which are non-profit membership corporations (to be run on business lines) are meant for breeders of a particular breed. Their governance is through a Board of Directors who are elected by members of the association. The primary purpose of such organizations has been to maintain correct records of ancestry (pedigree) of animals and thus to ensure breed purity.

Other purposes of breed associations include promotion of that breed through variety of activities, to encourage the breed improvement programs, and to provide information for performance specifications, sales, standards, exports and imports etc. One of the most successful breed associations is the U.S. Holstein Association which enjoys a worldwide status. It also maintains an international herd book.

Breeding for Type

The word **'type'** in animal husbandry is used to refer to appearance (morphology) of animals (e.g. features of head, ear, color or other such characteristics). It is also used more or less synonymously with the word **'conformation'**. Following the introduction of herd books, it became a practice to fix a **'standard type'** for each breed. Animals of a specific breed were required to conform to this 'standard-type' as closely as possible to be registered in the herd book. The choice of animals for breeding on the basis of morphological features (e.g. horn shape, curvature, coat color etc.) has been called **'breeding for type'**.

During early phases, breeders of 'type' were of the view that producing a herd of uniform external appearance would also lead to uniformity in growth and milk production. To the extent, these morphological features were associated with productivity; it was desirable to breed for type. But formalism of attention to unnecessary details of type definitely retarded rate of genetic improvement for production traits.

Economic pressures on animal production after World War II have reduced the importance of breeding for type *per se*. But still breed organizations, though primarily improving their herds for productivity, try to maintain 'type traits' of that breed. The models used for recording of 'type traits' over the years have, however, undergone changes. For example, the models of 'ideal type' or 'true type' of Holstein breed in USA (developed in early 1920's) in terms of physical traits have progressed from simplest form of type grades to complex scoring procedures that are quite objective and related to productivity.

Gordon Dickerson on early breeding practices

"Perhaps we should not be too critical of the emphasis placed by earlier breeders on pleasing color patterns, horns, and beauty of conformation. Such intangible aesthetic rewards for the breeders' efforts are not as easily snatched away by the unfeeling laws of supply and demand, as are the returns from increased efficiency of production. However, to the extent that our role is to help reduce costs of animal products to consumers, we do need to focus on those genetic changes in form and function that are most relevant" ***(Dickerson, 1986)***

Production Recording

Recording of conformation and production, including milk yield of animals, began during latter half of 19th century. Some of the salient events pertaining to development of methods for recording of performance were:

- Method for rapid determination of fat content of milk (1889-93).
- Recording of growth rate, feed consumption and carcass quality of pigs which began in Denmark and Sweden in 1907 and 1923 respectively.
- Trapnesting of hens, which is used for recording of egg production, was introduced in Austria in 1879. Originally, the mechanical trapnest was developed as a means of identifying the broody hens. It is rather interesting to note that a device developed for one purpose would be later utilized so extensively in poultry breeding through for a different purpose. It has made a tremendous impact on genetic improvement for egg production.

In dairy-developed countries, breed organizations have played a leading role in recording the production performance of animals. The breed organizations either conduct production recording or provide services to assure reliability of performance records. Outstanding example of value of performance records, coupled with record of pedigree in evaluating the genetic worth (breeding value), is the progress shown by cows of Dairy Herd Improvement Associations of USA.

Importance of production recording: Systematic and comprehensive recording of performance is one of the most important requirements for assessing the genetic worth (breeding value) of animals. In fact, *the development of procedures for objective recording of production is thought to be the most notable event in animal breeding* after Bakewell's pioneering work. For example, the method for assessing of back fat thickness in pigs by use of a scalpel probe (by Hazel in 1950's), and subsequently the ultrasonic method for direct measurement of back fat has made tremendous impact on genetically reducing fat in pig carcass. Similarly, the techniques for measurement of many physiological and biochemical traits viz; enzymes, hormones etc. have been standardized.

Development of objective recording procedures for newly emerging traits of importance in livestock is a continual process for their potential use in selection. A few examples include assessing egg shell-quality (shell-strength), bone density in laying hens and abdominal fat in broiler chicken.

Box 1.2: Some definitions relevant to history and overview of animal breeding.

- **Animal husbandry:** The art of managing farm animals (sometimes it is also referred to animal agriculture) while animal science is the science that deals with study, breeding, feeding (nutrition) and care (management) of farm animals.
- **Breed society (association) :** An organization, formed by registered members of a breed, to record the ancestry and performance of animals.
- **Breeding goal (objective) :** The traits which are intended to be genetically improved constitute 'breeding goal (objective)'. The breeding goals/objectives are a statement of the relative values of genetic change in all of the desired traits that are included in a breeding plan.
- **Domestication vs. taming:** The term 'domesticated' refers to an entire species whereas the term, 'tame' may refer to just one or few animals of a species. Humans have been capable of 'taming' animals of many species though these species have never been truly domesticated (e.g. elephant, giraffes, bear etc.).
- **Economic value** (of a trait): The profit resulting from a unit genetic change in a trait with no change in other traits influencing the profit.
- **Genetic evaluation:** The term is used for prediction of genetic worth (breeding value).
- **Livestock or farm animals:** The animals which are kept to serve some economic purpose to man viz; food, fiber or work. The term livestock (used as singular or plural) usually includes domesticated animals, which are intentionally kept in an agricultural set-up for producing food, fiber etc. In this context, poultry and farmed fish are not included as livestock. In a broader context, however, poultry is also included in livestock.
- **Nucleus (elite) breeders or nucleus herds :** Breeders (herds, flocks) at the top of improvement pyramid and which supply breeding stocks usually purebred, to breeders at the lower strata of the pyramid.
- **Pets :** The animals which are domesticated (kept) for home companionship or recreation.
- **Performance:** The word 'performance' in this book is used interchangeably with the term 'phenotype' or 'phenotypic value'.
- **Pedigree:** A record of ancestry of an animal.
- **Pedigreed (registered) animal :** An animal whose ancestry is known, and usually, whose ancestry is recorded by a breed society (association).
- **Pure-bred/straight-bred:** Animal with both parents of the same breed.
- **Pure breeding or straight breeding:** The practice of mating between animals of the same breed.

Box 1.3: Dairy Herd Improvement Association (DHIA) of USA.

Dairy associations in USA were started in different states in the first decade of 20th century. By 1929, all of the 48 US states (of that time) had their associations which were originally called '*Cow Testing Associations*'.

The American Dairy Science Association developed uniform guidelines for the operation of these associations. In 1927, these organizations were renamed as "*Dairy Herd Improvement Associations*' (DHIA). These associations were organized under co-operative laws of the state, and were incorporated under legal statute. The associations employed 'supervisors' who were involved in recording production of cows including their feed records. The National DHIA started functioning in 1965.The DHIA's program is operated jointly by United States Department of Agriculture (USDA) and the colleges of land-grant universities.

Recording of milk production of cows has been an important role of DHIA. The records of cows are now computerized and maintained as a data bank. Identification of individual animals, which is important for evaluation of their breeding values, was accomplished by ear tagging system (started in 1935). The systematic recording of production traits has made the performance records collected by DHIA of considerable reliability. The combination of DHIA records, computer evaluation of breeding values, and the dissemination of genes of best bulls by artificial insemination has led to sizeable genetic improvement. The 'DHIA' production records have been extensively utilized in formal training of scientific workers by many of the US Universities. The earliest example is of Dickerson who used data of DHIA for his PhD (1937) at the University of Wisconsin.

Beginning of Science of Animal Breeding

General Background

Following the rediscovery of Mendelian principles, intensive work was started to study the inheritance of economic traits of farm animals (e.g. egg production, milk production, etc.) through Mendelian analysis but these efforts were unsuccessful.

Animal breeding as a scientific discipline (of biology) was not influenced by Mendelism until 1930 when Wright, Fisher and Haldane bridged the gap between Mendelism and Biometry. These pioneers developed the basic concepts for inheritance of quantitative traits. They introduced the terms that are now basic in today's vocabulary of population and quantitative genetics; for example, average effect of an allele, average effect of an allele substitution, covariance between relatives, additive genetic variance, variance due to dominance deviations, non-additive genetic variance, inbreeding coefficients, path coefficients.

The terms used in quantitative genetic analyses were based on statistical models and methods like regression of phenotype on genotype, analysis of variance to partition phenotypic variation. Consequently, skills in mathematics and statistics were necessary to comprehend the basic concepts presented by Fisher, Wright, and Haldane. Although there were differences among Fisher, Wright, and Haldane on the relative importance of specific genetic effects (*e.g.* epistasis, dominance, genetic drift) in evolution but these three individuals provided the basic framework for determining the relative importance of different genetic effects on variation in quantitative traits.

The work of Fisher, Wright and Haldane led to development of quantitative genetics as a distinct branch of genetics. This work also served as the basis for development of procedures for the genetic improvement of economic traits of livestock by Lush of Iowa State University (USA), and his contemporaries.

Box 1.4: Beginning of applied genetics: University of Wisconlin.

L.E. Cole (1877-1948)

Following the rediscovery of Mendelian principles at the turn of 20th century, the science of genetics developed in England, mainly due to Bateson and Punnett. Though Genetics started a little later in the US but it progressed very fast under Morgan at Columbia University in the department of Zoology. The application of genetics to plant and animal improvement however happened late. One of the earliest initiatives in this direction was made at University of Wisconsin. The Dean of the Agriculture College at Wisconsin, realizing the potential of application of genetics for improvement of plants and animals, established the 'Department of Experimental Breeding' (later named 'Department of Genetics'). It was thought that establishment of such a new department would stimulate application of genetics to plants and animals instead of scattering genetic work among departments such as animal, dairy, poultry, agronomy, horticulture, entomology etc.

Such an approach required a dynamic person who could lead and establish working relations with separate departments. This challenge was put on young genius Dr. Cole(1877-1948), and he demonstrated his administrative and scientific abilities fully as evidenced from his legacy

Cole obtained his PhD (1906) from Harvard, and by virtue of his talent and training, was appointed the youngest (29-yr) Chief of the 'Animal Breeding and Pathology' Division at the Experiment Station of University of Rhode Island. Later he served as Instructor in Zoology at Yale University for three years. In 1910, Dr Cole was brought in to Head the new department of genetics (then named the Department of Experimental Breeding).He thus became earliest to facilitate the application of genetics, (when it was still in its infancy), to the genetic improvement of animals and plants in USA. Roots of many of the pioneer animal breeders of 20th century (including Lush, Dickerson etc) go back to him.

His original area of research was on genetic traits of pigeons, which later included poultry as well. Subsequently, he initiated basic research on genetic study of contrasting traits in crosses of Angus with Jersey and Holstein breeds of cattle. The results of the crosses indicated the need for a much larger experiments. His work later expanded to include all species of farm animals as well as mink, laboratory mouse, etc.

Student guidance and training: Dr. Cole initially supervised postgraduate students in both plant and animal genetics but in 1919, he added a coworker who took over plant genetics. With the expansion of the Department, a broad spectrum of students came for graduate studies in genetics, which led to joint research programs with other departments. The areas of research included genetic improvement of corn, potatoes, clover, bees, pigeons, fur animals, poultry, dairy cattle, beef cattle, pigs and sheep.

Cole was not himself trained in quantitative genetics but he was aware of its potential in improvement of economic traits of animals and plants. His earlier association with renowned geneticist's like Jennings, Pearl, Castle and others was helpful in developing a diverse training and research program. He encouraged his students to study the work of

Wright, and, later, of Lush (who had also been his PhD student) and others in this emerging field. Course programs for candidates in animal genetics area were broad, including in-depth study of genetics and statistics, plus biochemistry, nutrition, physiology, endocrinology and economics.

Many of the later leaders in genetics, and particularly in animal breeding were among the *123 students (MS and PhD)*, trained under Cole's leadership. The big names in the field of animal breeding included **Lush** (father of scientific animal breeding); **Craft** (Director of Regional Swine Breeding Laboratory), **Johansson** (of Sweden who wrote one of the popular book on animal breeding),**Warwick** (author of another earliest book on animal breeding), and **Chapman** and **Dickerson** (both of whom became pioneers in animal breeding).

In 1923, he on the special request of US Agriculture Secretary served as chief of the Animal husbandry division.

Legacy:

- Establishment of a strong Department of Genetics at University of Wisconsin that attracted talented faculty and got international recognition in research and teaching. Later the department had some Nobel Laureates on faculty. Prof Crow, a world-renowned population geneticist, also spent his entire career in the department.
- Preparation of the first diverse and core group of well-trained geneticists (USA) who later on took over the breeding research with different livestock and plant species for training of the next generation of geneticists.
- As Chief of Animal Husbandry Division in the US Bureau of Animal Husbandry (1923), Cole stimulated fundamental research at branch stations of the Division, added key staff in poultry and nutrition; and in animal breeding, which led to expansion of basic work of Sewell Wright in quantitative genetics and animal breeding. He thus contributed in establishing strong scientific research program of United States Department of Agriculture.

Lush School of Animal Breeding

Animal husbandry had been taught in Land grant universities in the US from the beginning of 20th century (or earlier) but animal breeding did not find a place little more than a component of reproductive biology. Even after rediscovery of Mendelism, (as per Lush who was introduced to genetics in 1914), animal breeding then taught in universities went as follow: "the first principle was that 'like produces like', and the second principle was that 'like does not always produces like". Lush and many others who were later became leaders in animal and plant breeding, were trained at university of Wisconsin in 1920-30's (Box 1.4). These PhD's were then employed for teaching, research and extension in universities and in the newly established Regional Breeding Laboratories of various species. The scientific developments in animal breeding have occurred at two broad levels:

- Development of a comprehensive body of theory of population and quantitative genetics that is applicable to economic traits of farm animals.
- Translation of this theory into operational procedures which are relevant to genetic improvement of different species of farm animals.

Jay L. Lush
(Circa 1969)

Lush explaining a breeding concept through path analysis.

Lush's original desk.

Lush (1896-1982)

Jay L Lush is universally recognized as father of scientific animal breeding. In fact, these are the words used in his 'Wolf Prize' citation of 1979. He is credited for melding the theory of population and quantitative genetics (developed by Fisher, Wright and Haldane) into *a comprehensive, cohesive and logical science of animal breeding* (and genetics).

Research: His research contributions are too numerous to be listed but these can broadly be grouped into three:

- Development of a general theory and procedures for study of inheritance of economic traits of livestock viz; heritability, repeatability and genetic, environmental and phenotypic correlations. Though Lush did not coin the word 'heritability' but he defined heritability in broad- and narrow-sense and elucidated its methods of estimation.
- Development of procedures for assessing the genetic worth (breeding values) of animals using information from different types of relatives combined optimally.
- Demonstration of scientific effectiveness of selection, inbreeding and crossbreeding for genetic improvement of different species of livestock.

With these broad developments, it became possible to determine which selection and mating system could be expected to yield best results for the ganetic improvement of a given trait. Consequently, animal breeding became an applied science whereby genetic improvement could be conducted on scientific lines.

Books: His textbook **'Animal Breeding Plans'** first published in 1937, had enormous impact on scientific breeding principles and procedures. The book was translated into many different languages, and is still readable after 70 years of its first publication. The conceptual framework of the subject coupled with the analogies used to elucidate breeding concepts are highlights of the book.

His **'Mimeographed Notes in Animal Breeding'** entitled **'Genetics of Population'** (1945), written for postgraduate students, provided a documentation of the application of theory of population and quantitative genetics to genetic improvement of farm animals. These notes were the outcome of his teaching of two postgraduate courses. These notes had been published in the form of a book (1994) by two of his students and colleagues.

Student training: Professor Lush nurtured a generation of animal breeders after 2nd World War who could be found in most research institutes of the world. Training of students was well organized with a balance of statistics, genetics and physiology besides animal breeding. A total of 279 students graduated under his guidance, in collaboration with his colleague, Dr Hazel. These students came from almost all states of USA and from more than 32 foreign countries. Training of hundreds of postgraduate students in animal breeding, who in turn trained many-fold more students across the world, is an important impact of Lush's school on genetic improvement of farm animals.

His contributions have been honored through awards, symposia, nine honorary doctoral degrees besides membership to prestigious National Academy of Science (USA) and Royal Society of Edinburgh (UK). An auditorium in animal science building (Kildee Hall) at Iowa State University is named after him. His students and family members through contribution of over one million dollars have established 'Lush Endowed Chair' which is occupied by one of his professional grand chldren.

Hazel (1911-1992)

Hazel (Ph.D. in 1941 under Lush) was strongly trained in statistics and is ranked highly among early pioneers in animal breeding. Following Ph.D., he worked as a research scientist with United States Department of Agriculture (USDA) in the newly organized Regional Swine and Sheep Breeding laboratories. During this job, he applied many of his and Lush's techniques for genetic improvement of these species. He joined Iowa State University in 1947 as a Professor where he made significant contributions through research and training of postgraduate students.

His landmark contribution to animal (and plant breeding) was his Ph.D. research work on "*Development of method of 'selection index' for simultaneous genetic improvement of several traits*". This was a fundamental work on the theory and application of selection index for simultaneous genetic improvement of several traits. Through this work, he introduced the concept of **'economic value'** in genetic improvement which has since then been refined and expanded. In fact, the principle of construction of selection index continues to be mandatory part of training of postgraduate students throughout the world till date. The method of selection index has made tremendous impact on genetic improvement of all species of farm animals.

His work on 'methodology of estimation of components of variance and covariance' was also a landmark in animal breeding, and was forerunner of its elaboration by many of his and Lush's students. His three famous papers were published within a span of only three years (1942 to 1944). He was also the inventor of "mechanical measurement of carcass value (in particular measurement of back fat thickness) on live hogs" using a back fat probe.

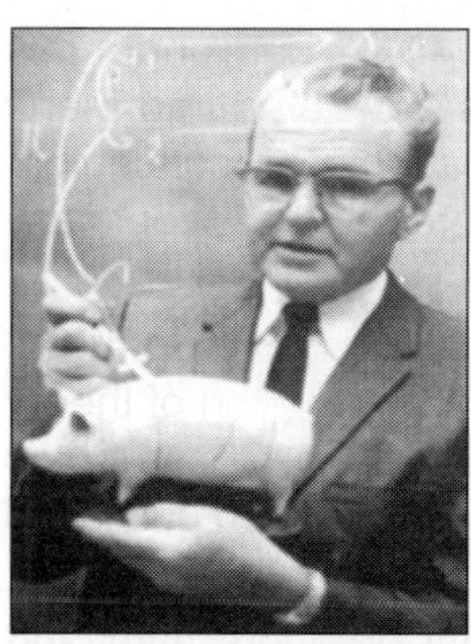

Hazel
(1911-1992)

Henderson
(1911-1989)

Dickerson
(1912-2000)

Harvey
(1919-2015)

Dickerson (1912-2000)

Gordon Dickerson was also an early leader in the application of quantitative genetics to livestock improvement. For his PhD (1937) under Prof Cole at University of Wisconsin, he analyzed production records of cows enrolled in 'Dairy Herd Improvement Association' of Wisconsin state (USA) with primitive computational equipment. This work later became the basis for National Dairy Sire Summary. Dickerson worked in more varied capacities and organizations than any other breeder of 20th century, and with almost all species of livestock including laboratory species like mice. During 1940's, when he was employed in Regional Swine breeding Laboratory in Iowa, he was an active member of the **'elite group of animal breeders'** of Iowa State University. Dickerson's notable contributions (within the scope of this book) include:

- Elucidation of the basic formula for predicting rate of genetic improvement through selection by taking into account the contribution of both male and female parents. It was the first attempt at manipulating components of genetic gain to maximize annual genetic progress.
- Explained the reasons for lack of agreement between predicted and actual genetic gains due to selection to which he referred to as **'genetic slippage in response to selection'.** This work, which he completed at the private poultry breeding firm, Kimber Farm utilizing data of the chicken stocks, is ranked among the classical papers of quantitative genetics.
- His manuscript **"Techniques for Research in Quantitative Animal Genetics"** (1959) has been one of his major contributions to research and teaching of animal breeding.
- Introduced and defined the concept of economic and biological efficiency in breeding programs involving whole system approach.

A number of students who later became leaders in animal breeding were trained under him while he was a professor at University of Missouri, and later during his last years at the University of Nebraska.

Henderson (1911-1989)

C R Henderson, after obtaining his MS in animal nutrition (1935) worked in various capacities including US Army's Medical Nutrition laboratory wherein he demonstrated a strong skill for statistical analysis. This statistical skill attracted him to do his Ph.D. in animal breeding under Lush and Hazel (Hazel was in fact chairperson of his PhD comprehensive examination and of his dissertation's reading committee). His strong training both in statistics and genetics, under some of pioneers of these fields at Iowa State University, proved to be a great strength in his later work. Following Ph.D. (1949), he joined Cornell University as Professor where he remained until his retirement in 1976 but continued his professional work until his death.

His notable contributions, which primarily relate to statistical methodology of animal breeding, include:

- Development of statistical methods (referred to as **Henderson's Methods** I, II and III) for estimation of variance and covariance components for complex but realistic data obtained from farmer's herds or from experimental herds in which many non-genetic sources of variation existed.
- Development of procedures for evaluating the genetic worth (breeding value) of dairy sires which were being used through A.I. under varied types of field conditions (method of sire evaluation). The **'Herd-mate method of sire evaluation'**, which was initiated

by him in New York State (1953), was later adopted by USDA in the 'US National Sire Evaluation Program'. The method of breeding value estimation (genetic evaluation), expanded by him, led to development of **"Best Linear Unbiased Prediction" (BLUP).** The application of the method of BLUP has been made possible by artificial insemination, and in turn, it has brought about a revolution in genetic improvement of dairy cattle.

Henderson's methods, since their development in 1950s, have been used in estimation of breeding values for virtually every class of farm animals including as diverse species as fishes. His work culminated in a book titled "Applications of Linear Models in Animal Breeding" (1982) which he completed after his retirement. The methodology outlined in this book is used not only by animal breeders but also by researchers in many branches of animal science as well as in other fields like plant breeding, forestry, fisheries and even in social sciences.

Harvey (1919-2015)

Walter Harvey (MS and PhD under Lush in 1947 and 1949 respectively), worked with Biometrical Division of USDA as a Biometrician, and then joined Ohio State University (1964) where he remained until his retirement. Harvey is best known for *Development and Application of least-squares techniques to the analysis of data from animal experiments'.* His least-squares programs (LSML), which are PC compatible, are widely used throughout the world. He has taught short courses on data analysis using method of least squares throughout the world. He has developed procedures to estimate genetic parameters from multiple-trait selection experiments.

Among others, the early development of principles of animal breeding and their application to genetic improvement of livestock is also credited to **Robertson** (1920-1989) in UK. The work of **Falconer** (1913-2004, UK) in quantitative genetics using mice as model species, and in particular his book **'An introduction to quantitative genetics'** has been a landmark. This book continues to be extensively used in training of post-graduate students of animal and plant breeding, and by evolutionary geneticists throughout the world.

Since the development of quantitative genetic theories, these have been amply verified by actual experiments with laboratory species viz; *Tribolium* (flour beetle), *Drosophila,* mice (as mammalian model), Japanese quails (as poultry model) etc. Among food and fiber producing animal species, chickens have been most extensively used for verification of the efficacy and efficiency of alternative breeding procedures (i.e. selection and mating systems).

Box 1.5: Animal Breeding at Iowa State University (ISU): Lush and Hazel group.

In 1930, Iowa State University hired Dr Lush for its teaching and research program in animal breeding. At the same time Regional Swine Breeding Laboratory, which was first of its kind, was established in Iowa. Dr Craft, having had his PhD from university of Wisconsin was its founder director. This made a good combination (of Lash and Craft) for research in animal breeding. Later Hazel joined the department of as animal breeder. At the same time, the department also hired a poultry breeder, Dr Nordskog, to augment teaching and research.

A well-established statistics department at Iowa already had statistical authorities like Snedecor and Cochran on its faculty. Later Prof Kempthorne, who later also became a statistical geneticist, besides being a statistician also joined the department. Similarly, in crop sciences some eminent plant breeders (like Sprague) had also joined.

Lush was pioneer in developing animal breeding, logically most of the early developments happened around him mainly through his students who later on dispersed to other US universities.

Lush's teaching, student training and research resulted in making ISU an internationally recognized center of animal breeding and genetics. It gained worldwide recognition and attracted the best of students from not only USA but from all over the world. The tradition of excellence in animal breeding at Iowa still continues.

J.L. Lush

J.L. Lush referred to as father of modern animal breeding, is credited with the application of principles of population and quantitative genetics to improvement of livestock species.

Education: Lush obtained B.S. (1916) in agriculture with major in animal husbandry, and MS (1919) in genetics and animal husbandry, both from Kansas State Agricultural College. He served as a teacher in a high school, and in US army and Air Force wing for about two years. In 1922, he obtained Ph.D. in animal husbandry and genetics under the renowned applied geneticist Prof. Cole at University of Wisconsin. Though his Ph.D. research was on "Possibility of sex control by artificial insemination with centrifuged spermatozoa" but his statistical and quantitative leaning was evident in the thesis where he identified the sources of variability, computed correlation coefficients, fitted normal curves to distribution of sperm-head and length measurements, and tested these hypotheses for goodness of fit.

Research scientist at Texas: Immediately after Ph.D., he was employed as a research scientist at Texas Agricultural Experiment Station where he worked for more than eight years doing research in animal husbandry. Most of that research pertained to animal breeding but some of it was in other areas of animal production. The necessities of research drew him further in biometry, and during this period, he must have studied the papers of Wright.

University Professor: In January 1930, Lush joined Iowa State University as Professor in the Department of Animal Husbandry to do research and teaching in Animal Breeding. All of his work revolved around finding *ways to apply genetics more efficiently for improving animals.* For these purposes, he used many biometrical tools developed by Wright and Fisher. Most of his developments grew out of some actual problem. In 1931, he made frequent trips to the University of Chicago to attend Wright's course on 'statistical genetics' and to consult with him. According to Lush, the 10 weeks he spent studying under Wright were the most fruitful time he ever had. He also invited Fisher to lecture at Iowa State during summers of 1931 and 1936.

Student training: Between 1930 and 1966, he was either major professor or on the committee of 279 post graduate students in animal breeding who came from not only North America (42 states) but from 32 different countries as well.

His success as a teacher and director of graduate students is reflected by statements of some of his students:

- "As an advisor of graduate students, Dr. Lush was highly respected and admired. He was firm without being threatening and he got his points and message across without raising his voice. He was a warm and friendly person with a tremendous respect and tolerance for students".
- "Lush could explain a principle based on statistical expression in simple words, and always drew a figure that made the principle crystal clear to his students. This ability coupled with his research, resulted in a center for study in animal breeding at Iowa State University that had an international reputation from 1930 to the present."

- "He started with the fundamental principles of genetics; then, by invoking a deductive argument, he gave them effect through use of tools of population and biometrical (quantitative) genetics. He made his students aware of the danger of accepting conclusions without taking into account the role played by chance, the frequent need to make simplifying assumptions and the errors likely to be made by semantic arguments that are not supported by experimental evidence or by quantitatively evaluated, deductive arguments".

Legacy: A former student captured the essence of Lush's legacy in the following statement: "His students and academic 'grandchildren' and 'great-grandchildren' and so on hold a large fraction of responsible animal breeding positions the world over. In effect, the field of Animal Breeding is a program of intellectual 'line breeding' to Lush. He has contributed more than any other individual, directly and indirectly through his many students, toward the continuing evolution of Animal Breeding from an art into a science throughout the world."

Lush first retired at the age of 65(1961), that was the limit fixed by the university. He was invited by Ohio State University for its animal breeding teaching and research where he spent around a year. But ISU again brought him back to officially retire at 70 years of age (1966).After his formal retirement, he continued to come to university for work and discussions till his death.

L.N. Hazel

Hazel has been usually ranked next to Lush through his research and training of students. He earned a B.S. in animal husbandry from Texas Technology College (1933). After spending few years with dairy operations, he went to Texas A&M College for M.S. in genetics (1938). This was followed by an instructor's position where he was introduced to a new book, "*Animal Breeding Plans"* (1937) by Lush. This book attracted him but there were parts of the book that he could not understand. This he confided to his biometry professor. The professor advised him to go to Iowa to study under Lush. He joined PhD under Lush in 1939, which he completed in 1941.

Hazell had a great aptitude for statistics, and was influenced by two towering statisticians of Iowa state university. His thesis was a fundamental work on simultaneous selection of several traits which is still considered a landmark. Hazel later recalled

> "*The subject of selecting simultaneously for several traits arose at an informal coffee session of J. L. Lush with several of his graduate students in 1940 when Lush suggested that this subject would make a good thesis topic.*"

Before that time, Lush had spent several summers at Kimber Farms in Niles, California, working with Lamoreux on problems of breeding for improvement of many different traits of poultry, including age at sexual maturity, rate of egg production, egg size, shell strength, internal egg quality and viability under differing exposures to disease vectors.

Anecdote of PhD: During the development of 'selection index method' for his PhD,as the anecdote goes :

> "*One day Hazel rushed into Lush's office and enthusiastically presented his ideas. Lush raised the green gambler's eyeshade (which he wore while at his desk), and looked perplexed. Hazel went back to his desk rather deflated. Later Hazel organized his ideas, and over a series of presentations, gave Lush the insights into the problem of selection for several traits and its solution*".

Later Lush is stated to have said:

"Hazel was the first student who wrote a thesis I couldn't altogether understand".

Whether it was true or not, but it made a good story that was often repeated. After PhD, Hazel occupied three varied positions in succession stated, which gave him opportunities to apply his new knowledge into practical research: on swine and sheep in the Regional Breeding Laboratories of USDA and at Kimber Farm on poultry breeding. In 1947, Hazel was invited by Lush to return to Iowa as Professor of Animal Breeding.

The association between Lush and Hazel was fully synergistic which led to developments that changed both the practices of animal breeding and the scientific methodologies of the breeding research. Hazel in a personal communication to one his colleagues (1990, personal communication) stated that :

"We [he and Lush] divided graduate students according to their major species interest. Lush always felt a major responsibility for all the MS and PhD students and that the division of students was somewhat a paper ruse to quiet a graduate dean who felt one man (Lush) could not guide the work of such a large number of students".

Students specializing in dairy cattle breeding were formally led by Lush, and those in beef and swine breeding by Hazel. During their association, they guided 279 graduate students for M.S. or Ph.D. in animal breeding who came from 42 states of USA and 32 foreign countries. Late in 1950's when Dr Freeman joined the department though Lush was advisor of some students on paper but he had them under Freeman for research in dairy cattle, and never took credit in research publication. One example is of an Indian student, D. Sundaresan whose official guide was Lush but he did research under Freeman. Majority of the graduate students of "Lush and Hazel pair" became leaders of animal breeding either at breeding corporations or at research or educational institutions throughout the world.

Although Hazel was recognized for exceptional ability with statistical concerns, his enthusiasm was greater for biological aspects of breeding wherein he also initiated some work on tissue culture. Often he would pass on laborious but routine tasks to students while he explored novel and innovative aspects. Once he found an answer to a problem, he quickly moved to other ideas of concern to him.

In 1968, Hazel reluctantly shifted his focus from direct teaching and research and became an administrator, Head of the large and diverse Animal Science Department at Iowa State University. But he took early retirement in 1973 (at 62yrs) to spend life on a large farm house with livestock and other agricultural activities.He did these activities as intensely as he did science. He rarely, if ever, seen in conferences. This was just the opposite of Dr Lush as he continued to come to the department until his death at the age of 86 years.

The excelence set by Lush and Hazel in animal breeding at ISU has continued. In 1957, the university hired a 'professional grandson' of Lush, Dr Freeman, (a student of Henderson). In 1960's the university added one more accomplished animal breeder Dr Willham who had trained under Lush and Hazel. Currently the university maintains its excellence in animal breeding which is itself a legacy of Lush and Hazel.

Box 1.6: Edinburgh School of Quantitative Genetics and Animal Breeding.

The science of animal breeding is built on principles of population and quantitative genetics. Two of the pioneers in formulating the theories of these branches were based in UK; Fisher (at Cambridge) and Haldane (London's university college).

By the 1940s, partly because of the war, Britain had fallen far behind the United States in quantitative genetics and its application to livestock improvement. In United States, Lush's '*Animal Breeding Plans*' had been published in 1937, but the principles set forth in it were virtually unknown in the UK. This despite the presence of Haldane and Fisher who had provided much of the mathematical background for the theory of quantitative genetics. There were only a few geneticists then in the UK, and they were regarded as 'eccentrics pursuing an incomprehensible subject'. Consequently, little or no genetics was taught in undergraduate courses.

Animal Breeding and Genetics Research Organization: Food rationing in 2nd world war brought about a realization in Britain for a need to improve agriculture. Britain's Agricultural Research Council (ARC), realizing need for government-funded research on animal breeding to ameliorate its dependence on import, (in 1945) set up, the Animal Breeding and Genetics Research Organization (ABGRO). At that time, the University of Edinburgh was having a department of animal genetics led by pioneer classical geneticist Professor Crew. The Genetics Section of ARC which had been in temporary quarters in London, moved to Edinburgh in 1947, and worked as a unit with the genetics department of university of Edinburgh.

Researchers: The ABGRO hired a group of young scientists to start its breeding work. These included Robertson, Rendel, Reeve, Falconer and many others. But most of them had little background in genetics, and some did not have a biological background like Reeve who was a Mathematician and Robertson was a Physical Chemist (though he did not complete his Ph.D. in Physical Chemistry). The only one who appeared close to quantitative genetics and breeding was Rendel who had his PhD under Haldane at London's University College. However, Rendel left in 1951 to join CSRI in Australia.

Falconer had PhD in Zoology from Cambridge (1947) with research work on earthworms. He had little background in genetics, as recalled by him "the nearest thing to genetics in my zoology course was the curious fact that *Ascaris* sheds most of its chromatin when it makes somatic cells".

Training: In preparation for joining the breeding unit, Falconer was sent to be trained under Fisher. He spent 18 months with Fisher in order to learn about mouse genetics. Since Fisher at that time was mainly interested in linkage, Falconer's time with Fisher was not entirely satisfactory. Falconer recalled in an anecdote:

"Fisher was engaged in a complicated experiment involving a large number of mouse genes, and also involving inbreeding, that is, brother by sister mating. Inbreeding is well known to reduce fertility, and I (Falconer) thought it would be interesting to see how much the fertility would be reduced in the next two or three generations, and see whether he would have sufficient animals to choose suitable pairs to mate, and I found out that he would not, that he would find that many of his lines would become extinct because there was not a suitable pair to continue. I thought I ought to tell him of this finding and see what he thought about it, so I did one day at tea. He listened to what I said and then without uttering a word, he turned round and walked out of the room. The next day he came to me and said: "*I think you should arrange to go to Edinburgh as soon as you can.*"

None the less this young and enthusiast group of ABGRO developed their own strengths through reading or visits to other places of learning, or through interaction with eminent visitors:

- Robertson spent nine months with Wright and Lush the two who had done most to develop quantitative genetics in its application to animal breeding. Consequently, he became much better informed about quantitative genetics and animal breeding than other members of the (his) group.
- In 1948 Lerner, a well-known teacher and researcher of genetics and poultry breeding of University of California, was a visiting scientist at Edinburgh. He made the group aware of the new techniques from the United States. He 'unfolded' the mystery of Wright's path coefficients, which were being used for deducing the necessary theoretical parameters for animal breeding as UK geneticist were more aware about Fisher's analysis of variance approach. Lerner wrote most of his **'Population Genetics and Animal Improvement'** book, (published in 1950) while he was at Edinburgh.
- In 1949, Wright spent a sabbatical year in Edinburgh to lecture. But most geneticists of Edinburgh at that time did not have sufficient background to understand much of what he taught. Nonetheless, it was a great stimulus for the group.

Research: The original intention of ABBRO's unit on quantitative genetics at Edinburgh was to conduct research on breeding on rabbits, mice, and *Drosophila* as the unit did not have farm facilities. The idea was to do basic quantitative genetics research with *Drosophila* because it was cheap and fast to work on this insect. However, the results from *Drosophila* could not be applied directly to farm animals because *Drosophila* is physiologically different from farm animals in a number of ways. The breeding methods first tried on *Drosophila* were then tested on mice.

One of he important early contributions of the group to animal breeding was development of method of **'contemporary comparison'**, which was formulated by Robertson and Rendel for genetic improvement of dairy cattle. This procedure enabled selection of dairy bulls for use in artificial insemination. Later this method revolutionized dairy cattle breeding.

Robertson (1920-89)

Robertson became an authority in quantitative genetics, and eventually earned a DSc from Edinburgh. Robertson's contributions to quantitative genetics in relation to animal breeding are well known to advanced students of animal breeding. He guided a number of PhD students who became leaders in their field (One Indian student was Dr. Prem Narain who became Director of Indian Agricultural Statistics Research Institute).

One of his students, Hill, was appointed to University staff of Edinbwgh in 1965, and worked in close association with the breeding unit. He made landmark contributions in quantitative genetics in its application to animal breeding. Hill, despite his formal retirement, continues to publish, and is considered an authority on theoretical quantitative genetics, particularly in relation to selection and the estimation of genetic parameters.

Another student of Robertson,MacKay, who became a renowned geneticist (at North Carolina State University, USA) continues to be in forefront of population and quantitative genetics. She later revised Falconer's book as a co-author.

Diploma and degrees: In 1940's little was formally taught about genetics in UK. In 1949, a postgraduate Diploma in Genetics was started by the university of Edinburgh, admitting students with little previous training in genetics to embark on Ph.D. studies. Then, in 1975, an MSc. in animal breeding was started in collaboration with the Agriculture Department and was run by Hill who had worked on animal breeding and had become a leading theorist in quantitative and population genetics. This was the only course of its kind in the United Kingdom. It attracted students from all over the world, many of whom went on to take PhD in Edinburgh and then went home to start quantitative genetics or animal breeding in their own countries, notably in Australia and New Zealand.

Introduction to Quantitative Genetics: Falconer (1913 – 2004) after joining the unit started research which fell into three broad areas: the formal genetics of mouse, quantitative genetics, and the genetics of human diseases:

Falconer (1913-2004)

- His work on mouse mutants and gene linkage was a major contribution to mapping mouse genome.
- His research in quantitative genetics produced theoretical and practical findings that have been of great benefit to animal breeders like asymmetry of selection response, selection limits, concept of genotype-environment interaction in terms of genetic correlation.
- His short and long-term selection experiments on mouse were of immense use in empirical evaluation of quantitative selection theory.
- In human genetics, he made the first attempt to understanding of traits of complex inheritance where he gave a procedure for estimation of heritability for liability to diseases (the **'proband's method'**).

However, his greatest contribution is considered his book. The origin of the book had its roots in diploma offered by University of Edinburgh.This diploma had one course in quantitative genetics. Teaching of this course in early 1950's was assigned to Falconer. The existing books of that time (Li's, Lerner's, Mather's and others) were not much relevant to this teaching as they were meant for a different readership. An earlier book by Kempthorne (1957) had rigorous mathematical treatment of theory for topics related to quantitative genetics without proofs.

He taught the course through the notes he developed from existing books supplemented by his own experimental work on mice. These notes finally took the shape of now considered classic book on quantitative genetics: 'Introduction to Quantitative Genetics' published in 1960. This book since its publication has had four editions in 36 years. The book is used as a standard text book worldwide in application of quantitative genetics to evolution, animal breeding and plant breeding.

Overview of Animal Breeding

Animal Breeding and Its Scope

Animal breeding is a deliberate effort by humans (breeders) to modify heredity of animals in a directed manner. Lush defined animal breeding as "the means available for improving the heredity of farm animals". Since it involves modifying the heredity, the changes made are relatively permanent and transmittable. This objective of changing the heredity is dictated by a desire of humans to improve certain traits of animals to meet needs of humans. Consequently, the term 'animal breeding' is often used synonymously with "animal genetic improvement".The word 'genetic improvement' implies that something better is expected from animals with reference to human needs. Animal breeding has been defined by some leaders in the field as follow:

- Animal breeding is the application of scientific knowledge to genetic improvement of animals (Legates).

- The science of animal breeding is concerned with the application of population and quantitative genetics to improvement of domestic animals (Willham).
- Animal breeding means a planned genetic alteration of the population so that the animals can better fulfill the demands dictated by production requirements (Johansson and Rendel).

The common element in these definitions is 'genetic improvement' which means that successive generations are to be genetically superior to previous ones. Genetic improvement of any animal species in the simplest possible terms is a process of replacing an existing population with another which is superior in net genetic worth. A renowned poultry breeder, Nordskog, referred to animal breeding as a 'man-made evolution' or 'micro evolution'. The concept of genetic improvement is thus central to animal breeding.

The definition by Johansson and Rendel is more explicit and explains different aspects involved in genetic improvement. For the purpose of beginner student, we redraft the definition of animal breding as follow:

"Animal Breeding is the (art and) science of changing the genetic constitution of animal populations in a planned way to improve productivity that is dictated by demands of the market and production system".

Some important elements of the animal breeding process need qualification:

Animal breeding as an art: The art and practice of animal breeding had been going on since beginning of domestication. In fact, many breeds of livestock were developed much prior to the formal birth of science of genetics. Animal breeding in those times was based on the experience gained by stockmen while working with animals. The stockmen in turn handed down the accumulated experiences to their sons/ disciples.

Animal breeding as an art involved an element of 'trial and error'. That is, the methods employed by early breeders sometimes worked and sometimes failed but it was not exactly known 'why they succeeded or failed'. For instance, Colling's herd suffered from reduced fertility but the genetic basis for it was not known. In some cases, the success was due to sheer luck.

The discovery of Mendelian principles at the turn of 20th century did not have much impact on genetic improvement of livestock. The main reason was that the traits of importance in livestock such as growth rate, milk yield, egg production etc did not fit into two or three classes but were of continuous nature. Until, the first two or three decades of 20th century, livestock improvement continued to be practiced more as an art or empirically.

Box 1.7: Animal Breeding : Art to science to industry.

Until the first two or three decades of 20th century, livestock improvement was considered and practiced as an art by stockmen who called them breeders. Attitude of breeders of that time as was given by Sanders, the Editor of Breeders' Gazette in 1925 as follow:

> *"What then, should be our estimate of the work of one who has first to conceive the figure in his brain; whose only tools are the laws of heredity ...., whose only materials are flesh and blood, ...; who battles ever against the stubborn forces of '**reversion**' to ancestral forms; who seeks, and succeeds in producing a creature pulsating with life .... a creation that serves the highest utilitarian purpose? The breeder of animals (or plants) directs the spark of life itself. The possibilities of his art are almost infinite".*

Livestock breeding was thus thought of as a highly creative activity of great pride as breeders thought themselves to be like a sculptor who creates a statue which he has conceived. Breeders relished this type of self-praise of being artist that gave them satisfaction.

Such views of practitioner of 'art of animal breeding' were rather difficult to be modified. Traces of such attitudes can be seen in pure line breeders (seed stock breeders) of some classes of farm animals.

As Science: Prof. Crew (University of Edinburgh, UK), in his book on 'Animal Genetics'(1925), though not undermining the 'art', opened the way for scientific applications in animal breeding while stating:

"The stock-breeder is concerned with the maintenance of desirable qualities of his stock and with the improvement of these qualities generation by generation. He has drawn upon his own standards of perfection, often in ignorance of scientific principles which undoubtedly underlie his practice, he has done wonderful things. Down through the ages has come a traditional craft of animal-breeding based on the accumulated experience of countless generations of husbandmen,and out of his methods of 'trial and error' the breeder has developed his wonderful art. But in this country(UK) at least the unit in animal breeding is no longer inexpensive, stock-breeding is now besieged with difficulties and the breeder is turning more and more to scientifically trained men for help in elucidation of his many and varied problems".

In 1936's '*Yearbook of Agriculture*' (of USDA), efforts made in genetic improvements in USA for different species were reviewed. The need for change in animal breeding was advocated as follows:

"The methods and practices of the past have accomplished a great deal, giving us fine breeds of livestock we have today, but these methods and practices have taken us about as far as they can, and the most we can expect to do, if we continue to follow them, is to hold the gains that have been made. Breeding in these classes of livestock, in other words, is likely to become a frozen and static art. There are indications, however, that livestock breeding may be at a turning point in its long history" (The credit for this as a footnote is made to Lush).

During subsequent years, there has been considerable scientific progress in animal breeding. The development of scientifically sound theory of quantitative genetics and animal breeding has made tremendous advances in bringing about genetic improvement.

As Industry: Early in 1940's, following the success of hybrid maize, poultry breeding industry started developing on lines of the maize industry. As competition among private breeders increased, more attention was laid on profit related traits for both layers and broilers. The swine industry has very much followed the same pattern as poultry breeding though about 20 years later. Dairy cattle industry is also well developed. Beef and sheep are slowly moving in the same direction.

In nutshell, animal breeding has progressed from art to science and eventually it has become an industry where profit is the main objective. Nonetheless the training of scientific manpower continues to be vested with public funded organizations (universities and institutes)

Animal breeding as a Science: Animal breeding in empirical sense is quite old. But it was only after the developments made by Lush and others (from late 1930's onwards) that it became a branch of science that could be conducted on sound scientific principles (like any of other branches of biological sciences).

The methods used for bringing about genetic improvement in animal populations heavily rely on the knowledge of population and quantitative genetics, and statistics. With the development of theory of animal breeding, it has been possible to predict the outcome of using a breeding program (selection and mating system). The theoretical predictions have been empirically verified using laboratory animals. Although logical but animal breeding is an abstract science, (like mathematics and physics), and therefore it requires mathematical and statistical bent of mind. Much of the early scientific work on genetic improvement of different classes of livestock had been conducted in public funded universities/institutes. The universities had also been closely associated with supply of superior stocks to farmers.

Animal breeding is an applied science; therefore, like other applied fields of science, 'intuition, skill, and judgment' of animal breeders are important to success of livestock breeding. This is what some workers have referred as **"breeder's eye"**. These attributes are still relevant in modern day scientific animal breeding because ultimately these are the animals that are to be chosen and bred. In fact, it is possible for different breeders to use same set of tools (selection and mating) to address the same kind of problem with different results, due in part to differences in skill and experience of working with animals.

Lush (1973) on working with animals

Lush's experience of working for eight years on the research farm of Texas Agricultural and Mechanical College sums up the importance of working with animals:

"Immediately on completing my Ph.D. degree, I did research for more than 8 years, with almost no interruption for teaching. I'm glad it happened that way. If I had taught the same course as much as three times in succession I would surely have come to believe those things myself so firmly that the errors among them could scarcely have been corrected by any amount of subsequent experience. ***As it was, the cattle and sheep and goats talked back to me.*** *Having no papers to grade or class rolls to call, I listened.* ***Usually the animals were saying something like:***

Most of the things you think you know may be true in principle but you have many of them out of all proportion to their actual importance.

When you draw a conclusion, you often overlook circumstances which, if you considered them properly, would upset your recommendations badly.

Lush despite his busy teaching and research schedule would be present while actually selecting the animals (bulls) as if using his 'breeder's eye' recalled one of his Indian students. Dr. Acharya

Livestock breeding in developed countries has now become an organized industry. The degree of implementation of scientific findings to industry varies from one species to another primarily due to differences in their reproductive biology. In some species, like poultry and to some extent in swine, private industries have taken over work of genetic improvement and its dissemination to end users. Nonetheless, the training of scientific manpower required for industries continues to rest with public funded universities/institutes.

Changing genetic constitution (structure) : Genetic make-up of a population is determined by gene and genotypic frequencies which can be systematically changed through selection and a suitable system of mating (e.g. inbreeding, crossbreeding, etc.). Through selection, breeders change the genetic structure of populations in a way that performance of population improves. Unplanned (random) changes in genetic structure of a population are of no consequence to livestock breeders.

Improving productivity: Genetic improvement sought by breeders could be to increase, decrease or for some traits to maintain level of performance for a trait depending upon its nature and importance. For example, decreasing age at first calving but increasing milk production constitute improvement. For some traits, improved productivity may mean achieving an optimal level of production (i.e. an intermediate value). For example, increase in egg weight in chickens beyond a certain level may fetch no extra economic returns as eggs are sold on grade and per-dozen basis. On the other, hand if eggs are too small (called 'peeve' eggs), they are not marketable. Hatchability of eggs (a measure of reproduction) bears a curvilinear relation with egg weight such that hatchability is maximum at intermediate egg weight. The overall productivity of a livestock population is dependent on several traits, therefore, improving productivity means changing each trait in the desired direction, and to the desired level.

Market or production requirements: In most traditional food and fiber producing species, the users are commercial producers (e.g. dairy, poultry, swine and sheep farmers). A **commercial producer** is the one whose primary product is meant for public consumption. Commercial producers may have small- or large-scale operations; they may have extensive or intensive system of production but their products (milk, meat, eggs, wool, etc) are meant for public consumption and utility. While bringing about genetic improvement, therefore, needs of the commercial producers as well as those of consumers have to be considered.

The demands of markets or production systems may vary not only from time to time but also from place to place (region/country), as the examples cited below illustrate:

- At one time, fat in beef, pigs and broiler chickens was not a matter of concern as a large segment of human population was engaged in physical work. Today, in most developed countries/societies, the objective has become to produce leaner pigs, beef and broilers because of sedentary life style (e.g. the cholesterol scare).
- When food was in shortage, the objective was primarily to increase level of production but with passage of time, the quality and composition of product (meat, milk, wool, eggs etc) have also become important in most countries.

Gianola and Rosa on 'One hundred years of statistical developments in animal breeding' (2015)

"Statistical methods have been an important feature of the World Congress on Genetics Applied to Livestock Production, the first of which was held in Madrid (1974) and the tenth of which was held in Vancouver in August 2014. Also, statistics has had a visible place in the International Conference on Quantitative Genetics, which was held first in Ames, Iowa (1976), and most recently held in Edinburgh (2012) and will take place next in Madison, Wisconsin (2016). Further, many statistical approaches grew out of collective efforts and informal scientific exchanges made in animal breeding workshops and scientific meetings and even in coffee or beer breaks. Examples of hot spots of free scientific give and take of ideas are : the Edinburgh and Cornell coffee breaks, the Iowa State QTL Lounge, and the Wisconsin breakfast".

These authors also point out to the increasing use of statistical methods and modeling in using the genomic information in selection, and in its integration with phenotypic performance records.

- At one time, the American homemakers preferred yellow-skinned broilers whereas British housewives preferred white-skinned ones. It was, therefore, necessary to incorporate the suitable allele for skin color (white or yellow) to cater to each of these two markets. Similarly, in some markets there is a preference for brown (tinted) eggs.
- Newer issues like animal welfare etc are becoming important to be considered in breeding programs in developed countries. Accordingly breeding efforts are being directed towards it.

With increasing input costs, efficiency of production (input to output ratio), has become increasingly important. Additional constraints on breeding are imposed by societal and political issues of the country. Genetic improvement of farm animals (i.e. animal breeding) is thus a continuous and long-term process because of constantly changing market demands.

General Goals of Animal Breeders

Animal breeding revolves around genetic improvement. The basic question to beginner is "What needs to be improved". The trait(s) that are intended to be genetically improved constitute the breeding goals or objectives. The breeding goals (objectives) are a statement of the relative values of genetic change in all of the desired traits that are included in a breeding plan.

The commercial breeders of all the economically important species of farm animals are concerned with the profitability of their enterprises. The three most important aspects (categories of traits) associated with profitability from farm animals are:

- *Productivity:* to increase the product output per animal.
- *Efficiency:* to increase the efficiency of producing the product (input-output ratio)
- *Quality*: to improve the quality of product

These broad goals apply to most species that are reared for economic purposes though the relative emphasis placed on different traits or category of traits may vary. Improvement of growth, production (milk, egg, meat, and fiber), reproduction and survival; feed efficiency and quality (milk, meat, eggs and fiber) are all the facets of these three broad goals. In companion and pet animals, breeding goals deviate from these traits, as aesthetic traits (e.g. type and breed traits like shape, size, color, etc.) attain greater importance. Because of the economics of animal production, modern breeders realize that inclusion of many morphological features that are not directly associated with productivity slows down the rate of genetic improvement for main economic traits.

Livestock Breeding Industry

Lush in his book stated: **"animal breeding is a business"**. Therefore, as an activity of a commercial breeder it entails making profits. It is generally agreed that the ultimate objective of livestock breeding is more efficient production and animal products which are more suited to market conditions. In this scenario, all the stakeholders from basic breeder, multiplier to consumer need to be benefitted.

Livestock production in many industrialized countries has developed from an unorganized sector into a well-organized industry. Consequently, animal breeding in these countries has also acquired the status of industry. The transition of breeding into an industry started in early 1940's following the success of hybrid maize. Poultry breeding industry started developing on the lines of plant industry. Commencement of crossbreeding in layer and broiler chicken attracted larger industry enterprises and increased competition led to more attention being laid on profit related traits for both layers and broilers. Large-scale corporate breeders had been employing qualified (PhD's) animal/poultry breeders, who have incorporated vast performance traits of individual animals as well as of their relatives in genetic improvement programs. The swine industry has followed the same pattern as poultry breeding though about 20 years later. Dairy cattle industry is also well developed. Beef and sheep are slowly moving in the same direction.

Understanding the genetic structure of breeding populations (breeds, strains etc) is helpful as to know how the genetic improvement achieved in base flocks is disseminated further, and in identifying the ways to fasten the rate of transmission of improved stocks to producers. To understand the implications of genetic improvement, it is important to have a broad picture of breeding industry and its structure.

Breeding pyramid: Breeding sector in most livestock species has a triangular structure, usually termed a **'breeding pyramid'**, consisting of usually a three-tier system (Fig. 1.2):

- *Nucleus, elite, primary or seed stock breeders*: The breeders who produce animals of high genetic merit and supply their genetic material to the next strata (multiplier) of pyramid in the form of live animals, semen, embryos (germplasm), constitute the first strata of pyramid. The term **'seed stock'** refers to a breeding stock which is meant for producing the progeny to be sold to other breeders or producers. Historically, seed stock breeders have been breeders of pure breeds though currently it is not so in some species (e.g. swine and beef).

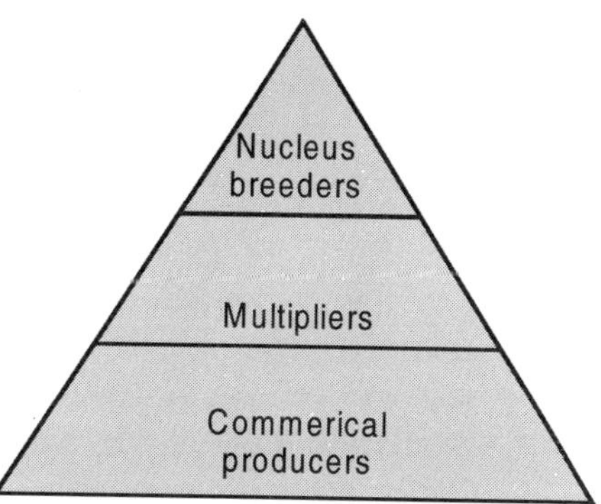

Fig. 1.2 Pyramidal structure of breeding industry

- *Multipliers*: The multipliers are those who expand the germplasm obtained from elite breeder by multiplying them in larger numbers.
- *Commercial producers*: The multipliers of most food and fiber producing species, in turn, sell their stock to commercial producers who sell these products for public consumption or utility.

Sometimes there may be more or less than three-tiers, and various combinations of strata are possible. In general, the flow of germplasm occurs from the top to lower strata (i.e. from seed-stock breeders to commercial producers).

A pyramid characteristic of pig and poultry breeding structures is composed of three tiers that direct genetic improvement. Actual genetic improvement takes place at the top of the pyramid (nucleus), followed by multiplication of that genetic improvement to produce larger number of purebreds and/or crossbreds which are then sold to commercial farms. In this structure, there is one way transfer of genetic material: from top to lower strata (called closed nucleus structure). An alternative to the pig or poultry breeding pyramid exists in dairy cattle in which animals (in this case females) from lower tiers can be brought into an upper tier by generating sons that are chosen as if they were bred in the nucleus. This is the so-called open nucleus structure. In beef breeding industry, the structure is a mixer where there is no clear differentiation between nucleus and multiplier tiers. In general, industry is at a greater advantage when genetic improvement can be concentrated in nucleus herds, especially when a breeder must select animals from another herd to introduce outside inheritance without loss in overall genetic superiority.

Time scale of genetic improvement and transmission: There is always a substantial time lag between a breeding decision and the transmission of genetic improvement to commercial producer or consumer. The time period varies depending on the species and its industrial structure. In poultry and swine, which have well-established industries and high reproductive rate, it takes about 5 years before genetic improvement gets transferred to commercial producers. In cattle, the time-scale may extend up to 20 years. Therefore, the breeder has to keep in touch with the consumer preferences, and be able to cope up with changes in market demands.

Branches of Science Useful in Animal Breeding

An animal breeder besides a requiring a general background in animal sciences, also needs sound background in the following branches :

1. **Population and quantitative genetics:** Genetic improvement involves altering the genetic structure of populations. Since the economic traits are quantitative in nature, the breeders need to have a sound training in population and quantitative genetics. In fact, the study of quantitative genetics is an integral component of training in animal breeding. In most universities in western countries in particular in North America, quantitative genetics is taught through the departments who have qualified quantitative geneticists independent of the college of the university where they are placed. For example, in the US, quantitative genetic courses are available in animal science, crop (plant) science, biology, genetics and statistics departments, and many a times cross-listed across related departments.
2. **Statistics:** One of the most important aspect of animal breeding is identification and hence selection of genetically superior animals (i.e. those superior in their breeding values). The computation of breeding values utilizes performance (phenotype) of the individual and or its relatives, and it requires application of regression analysis. The estimation of genetic parameters (heritability, repeatability, genetic correlation) required for formulation of a breeding plan employs the technique of analysis of variance and covariance. A sound training in statistics is, therefore, also an important part of training in animal breeding. The method of 'Best Linear Unbiased Prediction' (BLUP) is the standard method of genetic evaluation. BLUP requires variance components (genetic and environmental) for predicting genetic values simultaneously accounting for genetic relationship between animals, and taking care of differences in non-genetic factors which different animals experience. With the advent molecular genetic markers, the role of statistics has further increased, as large body of information (phenotypic and genomic) has to be integrated optimally to assess the genetic worth of animals (See Box. 1.8 and 1.9).
3. **Computer application:** Advances in computational technology have facilitated the application of quantitative genetic and statistical methodologies in genetic improvement. Animal breeders maintain and process performance records of a very large number of animals as well as of several traits of each individual. The number of such records run into several thousand.
   - Computations of breeding values, estimation of genetic parameters and construction of selection indices all require fast and accurate computations.
   - The voluminous performance records collected on all the animals for a large number of traits have to be processed as early as possible to make timely selection before reproductive rate of selected parents starts declining.

- Implementation of most of the advanced methods of genetic evaluation (breeding value calculation) like herd-mate comparison, best linear unbiased prediction (BLUP), REML etc has become possible due to advances in computational technology.

Computers have therefore become invaluable in animal breeding (Box. 1.8).

Specialized computer packages (software's) for analysis of animal breeding data, beginning with Harvey's LSML package, have been developed and are being extensively used on personal computers. In fact, no other branch of animal science has so far used the computer technology as extensively as animal breeders have.

Box 1.8: Computers in Animal Breeding.

Prior to development of computers, many of the tasks of animal breeders were performed by traditional hand-operated machines. Hazel solved his dissertation problem of selection index using such a traditional hand operated machine. Learning of matrix algebra thus was a routine in those times as it facilitated solving of simultaneous equations fast. Still computations were long and took hours for solution to very basic problems. For example, inversion of a 10-by-10 matrix by best mathematicians would have taken a day or so as compared to about 7 minutes taken by the computers of late 1950's. Dr Searle, a mathematician and a authority on application of matrix algebra in biological sciences recalled:

"My Ph.D. thesis, which involved inter-related regressions for a new method of estimating the effect of a young cow's age on the amount of milk she produced, involved calculations that would have been horrendous without a computer for the 3,000 or so cows involved; but it involved 'only' two or three 8-hour nights using the computer, which by that time (1958) the Animal Science Dept. had acquired, an IBM 650."

In those times one had to book time for use of computers beforehand. And most of time, the computer slot allotted for animal breeding students was after 11 at night till morning.

The development of computers has made the application of many of the now standard methods of animal breeding very easy. Throughout the period from late 1970s to mid-1990s, evolution in computer technology was a key for sophisticated genetic evaluation. Most of the improvements made in statistical models used in animal breeding are linked to the computational power being made available at increasing speed but decreasing costs. For example, the first computer used by researchers at Colorado State University (USA) in early 1980s was the Cyber 205 from Control Data Corporation that cost more than $12 million. It had a nominal memory capacity of 16 megabytes of volatile memory and 2 gigabytes of disk storage. A typical genetic evaluation for beef cattle breeding data involving about 200,000 records which resulted in about one million equations, used to take more than 5 hours of computing time to obtain solution to the linear system. Current desktop computers can complete the same task in a matter of minutes.

In addition to these branches, reproductive technologies (e.g. artificial insemination, semen freezing) have greatly aided the rate of genetic improvement through sire path. Multiple ovulation embryo transfer technologies (MOET) offer further opportunities to increase the rate of genetic improvement. New molecular genetic approaches (marker- assisted selection, genomics etc.), which are being developed, will not only help in better understanding of genetic basis of economic traits but are also expected to complement the existing breeding procedures though not replace them. Animal breedes have therefore to be trained in molecular genetic techniques as well.

Box 1.9: Shortage of quantitative-genetics and statistics trained animal breeders.

Dr Misztal of the Department of Animal Science of University of Georgia (USA) in the Editorial of the Journal of Animal Breeding and Genetics (2007) states as follow regarding the state of quantitative geneticists in animal breeding:

*"**More and more I receive phone calls from various breeding companies looking to hire a PhD in quantitative genetics.** They inquire if I know of a graduate versed in quantitative genetics and mixed models, with some programming skills, who can speak and write passable English, has a general understanding of markers and molecular genetics, can run and troubleshoot a genetic evaluation, and in general be a problem solver. I do not know of anyone available, I reply. There were many of them 10–15 years ago, but now they are rare. If they show up, they usually have very good offers well before graduation. My colleagues outside the USA are telling me of similar problems, although the severity of the PhD shortage is country dependent."*

Dr Misztal and Dr Eisen (of North Carolina State University have attributed this shortage mainly to two factors:

- "Reallocation of research funds from federal agencies in the US from quantitative to molecular genetics has been one of the most important factors in the change of profile of trained professional breeders. Animal breeders working in universities are competing for research grants from public funded organizations with new generation of molecular biologists, and in general are falling behind them in garnering these. The developing countries though lagging far behind the advanced countries in enhancing livestock productivity are equally affected by this change".
- "There is a general glamour associated with molecular genetics mainly originating from breakthroughs in human molecular genetics and genomics. There has therefore been a greater allure for postgraduate students of finding genes than of assuming 'black box' genetics, especially as the latter requires extensive training in quantitative genetics, statistics, and programming. The graduate students instead of undergoing a rigorous training in quantitative genetics, statistics and programming, prefer to opt for molecular genetics, partly because of availability of big grants in this area, and with the unreasonable thinking that everything possible through quantitative genetics has been accomplished, and that molecular genetics will be a panacea for livestock genetic improvement."

Dr Misztal in the editorial concluded as follow:

*"The future is bright for well-trained quantitative geneticist's who can adapt to new challenges; however, in many countries they are disappearing. There is still time for funding bodies to recognize an impending shortage, tap into the slowly disappearing infrastructure for training, and take action. **Countries without an adequate supply of quantitative geneticists will be forced to outsource their genetic evaluation and perhaps even their breeding programs. They also may be denied some of the potential payoffs of genomic selection.** Perhaps, the final product of genomics will be a model of an animal involving at least a few hundred genes that describe dynamics and complex relationships among traits within the animal. Perhaps computer simulations will be able to look at changes in the phenotype under pressure from a variety of environments and challenges, and find genotypes useful for each environment. Perhaps, we will find a machine that creates an embryo given the genotype. Such a scenario is possible, though not anytime soon, and **I bet that such a model will involve quantitatively-trained geneticists**".*

Similar concerns have been raised by many well-known animal breeders in my (author of this book) interaction with them in U.S. In fact sheep breeder indicated that U.S. sheep breeders are outsourcing their genetic evaluation work to Australia. Though there will definitely be a role of molecular genetics and genomics in animal breeding but integration of molecular genetic information with performance records and pedigree will require a much stronger training in quantitative genetics, statistics, bioinformatics etc.

A few universities in Europe like Wageningen University of Netherlands are maintaining a balance of training in quantitative genetics and statistics along with genomics. The MS and PhD program in animal breeding at this university combines expertise in different domains (quantitative genetics, statistics and machine learning, bioinformatics, genomics).

A few universities in the US are attempting to impart balanced training in various fields related to animal breeding. For example Iowa State University (ISU) offers PhD with a traditional emphasis on quantitative genetics and statistics though the students are provided a general training in a broad spectrum of related disciplines, including molecular genetics, immunogenetics, and bioinformatics. The PhD research (dissertation) program of students has the option for specialization either in quantitative genetics or molecular genetics and immunogenetics. Despite specialization in either quantitative genetics or molecular genetics, PhD students have to take larger number of credit hours in quantitative genetics and statistics. Both these specializations put substantial emphasis on integration of quantitative and molecular genetics including bioinformatics in student's training program.

Due to shortage of quantitative-genetics and statistics trained university professors in US for offering MS and PhD programs in animal breeding, some universities have formed a group to offer joint online courses in population and quantitative genetics, statistics and molecular genetics for students of the participating universities.

Currently the shortage of quantitative-genetics and statistics trained animal breeders in developing countries like India is no different, if not worse, from that in the US. But neither efforts are being made to handle the issue nor any concerns been raised. For example, one of the earliest US trained group of animal breeders in 1960-70's was located at Punjab Agricultural and Haryana Agriculture universities. The genetics and statistics department at these universities were equally strong to offer courses relevant to training of breeders. Later, similar competence groups developed at other institutions. A good number of animal breeders trained at these organizations later dispersed to other universities (institutes) across the country.

Beginning this century, their numbers in India have consistently been declining, and currently majority of them have retired, and with no replacements being made. The glamour for molecular genetics continues with both the researchers and students with the hope of getting research grants, and publication of research in highly ranked journals. The problem has precipitated due to concentration of courses of biostatistics, molecular genetics including immunogenetics and cytogenetics, and population and quantitative genetics in a single department of 'Animal Genetics and Breeding'(AGB) rather than the earlier mode where several of these courses were taught in the departments of 'Statistics and Genetics'.

Contribution of Animal Breeding to Livestock Productivity Enhancement

Animal husbandry aims at improving the productivity of farm animals through interventions in breeding (genetics), nutrition, management and health coverage. During the past 5-6 decades, tremendous increase in productivity has taken place in most species of farm animals. Accomplishments of animal breeding are numerous; these can be better appreciated in the last few chapters of this book which deal with genetic improvement of livestock species. Broadly, there has been improvement in yield, product composition and quality and efficiency of feed utilization.

Three illustrative examples are cited here to emphasize the improvements accomplished, one each from poultry and swine (high reproductive rate) and the other from dairy cattle (low reproductive rate)

Dairy cattle:

- Milk yield per cow in US Holstein and Jersey breeds has more than doubled from 1957 to 2013. Fat yield and protein yield have also been consistently improved. (Fig 1.3).
- Somatic cell score, a measure of degree of mastitis, has also declined (improved). Daughter pregnancy rate which was declining until 1999-2000, has started improving as selection for it has been incorporated in the genetic improvement program.

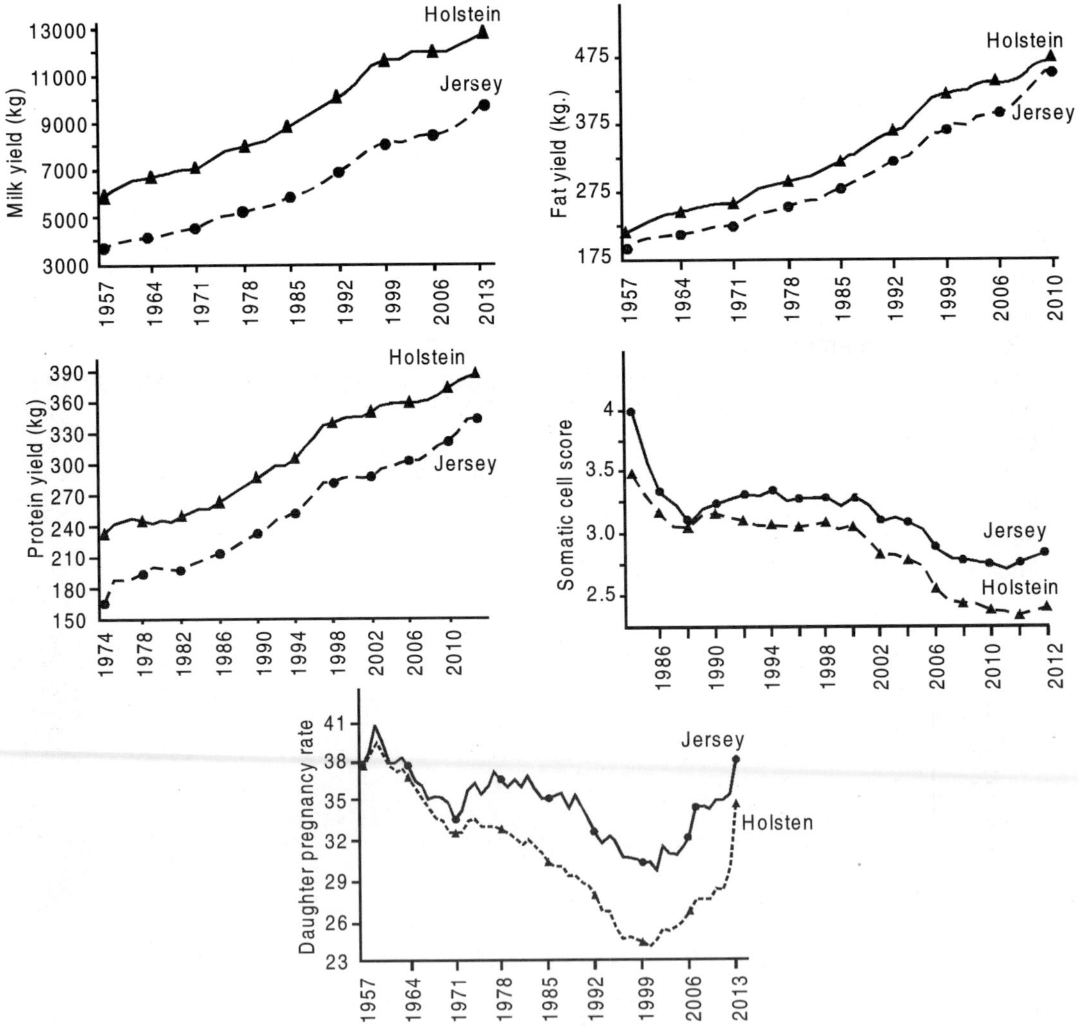

Fig 1.3 Changes in milk traits, somatic cell score and pregnancy rate of US registered Holstein and Jersey.

As a result of the tremendous improvement attained in all the important components of productivity in the US, the number of cows in dairy operations had been declining while milk yield per cow continues to improve. (Fig. 1.4)

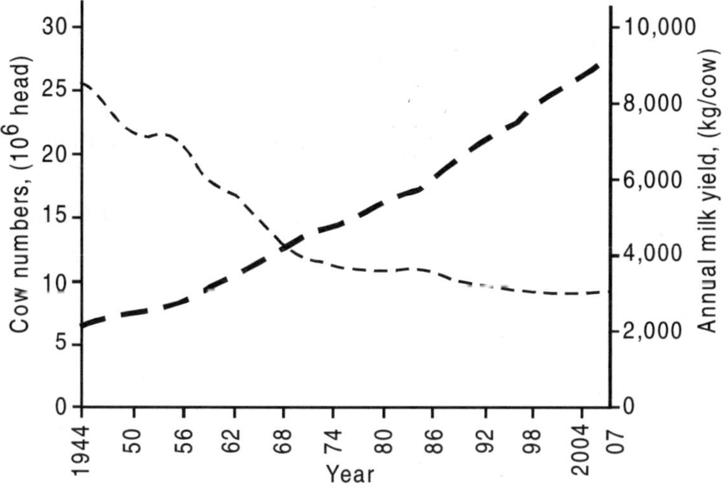

Fig. 1.4 Changes in per cow productivity (- - - -) and cow numbers (- - - -) in USA.

Poultry:

- Broiler chickens attained body weight of 1.6 kg at 12 weeks of age about 50 years ago. But now they attain the same weight by 4 to 5 weeks of age. Feed efficiency (defined as the amount of feed consumed per unit gain in body weight) has improved from 4.0 (1957) to about 1.6 (2001). Not only this the yield of body components has also been modified in the desirable direction. For example, body weight has improved by over five times from 1957 to 2001 (Fig. 1.5) with a concomitant increase in breast yield. Due the positive relation between carcass fat and body weight, the carcass fat increased from 1957 to 1991 as no selection was exerted against it. But subsequent selection against it has checked it's further increase though body weight continues to improve.

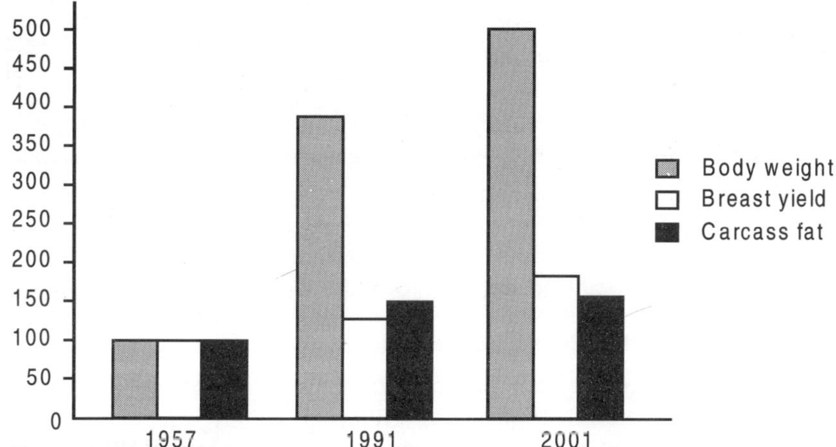

Fig. 1.5. Genetic changes in some broiler traits over time(Source: Hill 2008).

- Egg production (per-hen per-annum) has increased from about 100 (in 1920) to around 330 (in 2000). Concomitantly egg weight has also been improved. The commercial laying stocks available in the market now a days, during their first year, produce about 22 kg of egg mass (i.e. product of egg number and average egg weight) which comes out to be 59 g egg mass per-hen per-day. In 1950, on the average, a hen produced only 34 g egg mass per day

Swine

American Yorkshire club has accumulated one of the largest data base which has been subjected to genetic evaluation by researcher to assess the actual genetic change achieved from 1986 to 1996. As an illustration, genetic changes in maternal line index which is an index of reproduction expressed in terms of earning per litter (US dollars per litter) are presented in Fig. 1.6.

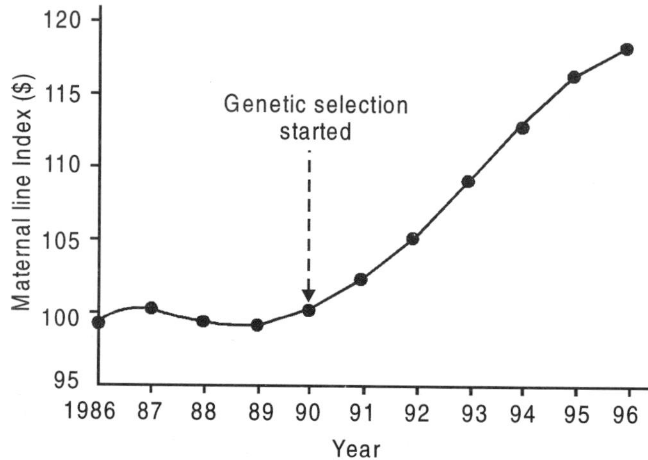

Fig 1.6. Genetic trend for maternal line index for American Yorkshire(in units of US $).

The units for the index are dollars per litter from the base value of one hundred. Evidently there has been a dramatic genetic increase in combined performance, virtually all of it has occured after 1990 when the first sound genetic evaluation program (across-herd genetic evaluation) was initiated. The cumulative genetic improvement has been US$ 18.0. If we assess it from 1990, it amounts to about $ 3.0 per-litter per-annum which shows the strength of a sound genetic improvement (selection) program in enhancing productivity.

Similarly, tremendous improvement in productivity has occurred in other species of farm animals. The increase in productivity observed in most food-and fiber-producing species has led to much better utilization of feed resources and, in turn, the cost of production has dramatically been reduced. Some of the animal products which at one time were a delicacy (e.g. broiler chicken) have come within the reach of common person. Looking at the level of productivity of the modern breeds (strains) of farm animals, it seems unimaginable today to even think of using wild stocks for production of animal products (e.g. milk, meat, eggs, wool etc.).

Genetic versus Environmental Improvement

Improvement in productivity has resulted from genetic as well as nutritional, managemental and health interventions. Estimates of improvement attributable to genetics are available, some of which are listed in Chapters on species improvement. (Chapters 18 to 22). For example, it has been estimated that genetic component has accounted for about 60% of the improvement achieved in milk yield of Holstein between 1980 and 2000. More than 80% increase in broiler body weight between 1957 and 2001 has been due to genetic component. Similarly more than 60% improvement in layer chickens is attributed to genetic component.

The following observations further highlight the importance of breeding and genetics in enhancing productivity:

- Improvement attained through breeding (in particular due to selection) is of rather permanent nature and hence it continues to pay long-term dividends. Genetic improvement is a cumulative process wherein current improvement adds to past improvement, and so on.
- Small genetic improvement for each individual (animal) of the population and for each trait adds to large genetic gains for the population as a whole for overall productivity. Typical rates of genetic improvement for many livestock species are in the order of 0.7 to 3.0 % of the mean per annum. Such rates can run to substantial cumulative changes over long-term.
- Better nutrition, management practices, health coverage can exploit the performance potential fully provided the animal has the required genetic make-up.

Genetic improvement is relatively a long-term process and hence it takes a few generations before its impact becomes evident. On the other hand, feeding, management, disease prevention etc. have immediate and noticeable impact on improvement productivity. ***Breeding has been equated with 'capital assets' while nutrition, management and health coverage with 'operating investments'***. Genetic and non-genetic (feeding, housing, health) factors have to move together to achieve the highest possible productivity.

The cost-benefit ratios of genetic improvement programs in farm animals have been estimated to be between 1:5 and 1:50, depending on the species. Despite the long history of genetic improvement in most farm animals, their performance continues to improve further for a vast number of traits despite the existence of antagonistic relation among traits.

Genetic improvement in its simplest form is a two-step process, as operationally it involves:

- *Choice of parents (selection)*: The choice of individuals to become parents of next generation, called selection, is the first and most critical step in genetic improvement. The choice of parents is based on standards laid down by breeder (the breeding goals). Selection involves increasing the frequency of the 'desirable' alleles/genes; and hence the change brought about by selection is of rather permanent nature.
- *Choice of mates (mating system):* The second step in genetic improvement involves a system by which the selected individuals (belonging either to one or more than one breed or population) are mated. The objective of a mating system is to produce superior genotypes (i.e. gene combinations). The change brought about by a mating system is of transient nature as gene combinations are broken down in the next generation. A mating system primarily utilizes the non-additive component of genetic variance.

In a genetic improvement program, selection is the most critical element, therefore it occupies a wider coverage in animal breeding. A schematic representation of genetic improvement of animals within a population (i.e. through selection) is given below in Fig. 1.7.

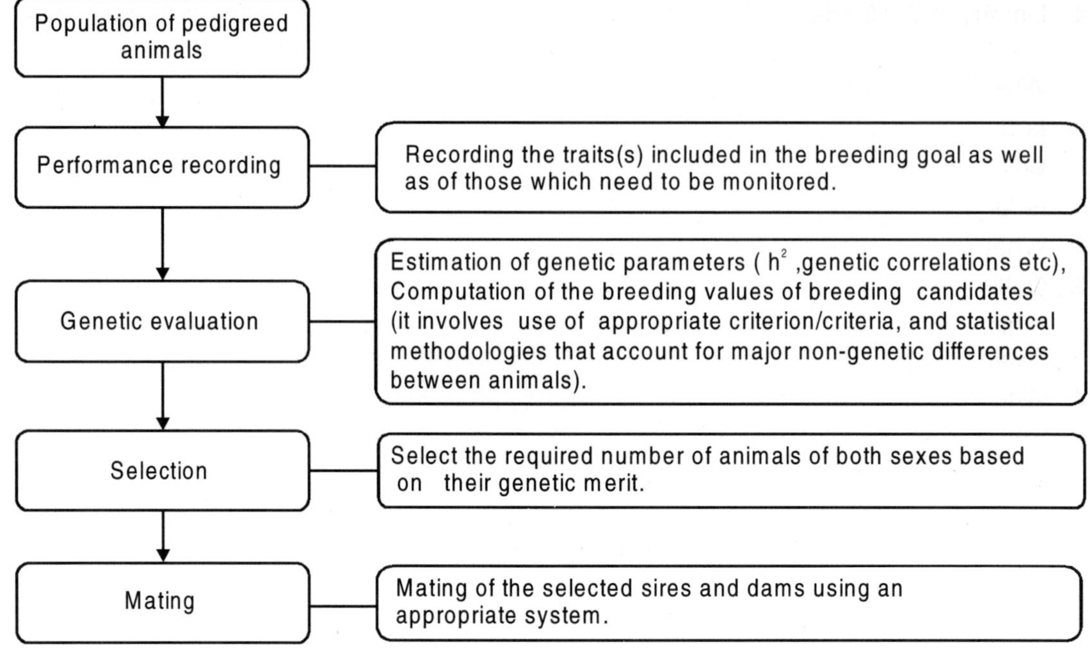

Fig 1.7. Flow chart of a genetic improvement program (within population selection).

Review Questions

1. Define/explain the following as used in this chapter:
   (*a*) Domestication
   (*b*) Livestock and farm animals
   (*c*) Breed associations/societies
2. What attributes are associated with domesticated animals? Explain.
3. What forces have been responsible for domestication?
4. Has domestication resulted in genetic changes in different traits ? Which trait seems to have changed the most through domestication?
5. Outline the contributions of Robert Bakewell in stock improvement.
6. Name the breeds whose development is attributed to Bakewell.
7. Explain the breeding procedure used by Bakewell.
8. What is a herd book? What information did the herd books of 18th and 19th century contain? What was the greatest lacuna in these herd books?
9. What is meant by 'breeding for type'?
10. List two traits for which performance-recording procedures were developed in 19th century.
11. What is the importance of production recording in animal breeding?
12. Who is called the father of scientific animal breeding? List his notable contributions.
13. List the most notable contribution(s) of :
    (*a*) L.N. Hazel
    (*b*) C R Henderson
    (*c*) Walter Harvey
    (*d*) Gordon Dickerson
    (*e*) D S Falconer

14. List the major food-and fiber-producing species of farm animals including their zoological names and the purpose(s) for which they are reared.
15. What is animal breeding? Outline its scope.
16. Is animal breeding an art or science? Explain.
17. What does the word 'improving productivity' means in relation to animal breeding?
18. Is 'animal breeding' a dynamic or a static process? Explain.
19. What are the general or broad goals of breeders of livestock species?
20. Outline the general structure of livestock breeding industry.
21. Which branches of science have contributed the most to developments in animal breeding? Explain.
22. What have been the accomplishments of animal breeding in enhancing livestock productivity? Cite one example from each poultry and dairy cattle.

SECTION - 2

QUANTITATIVE TRAITS

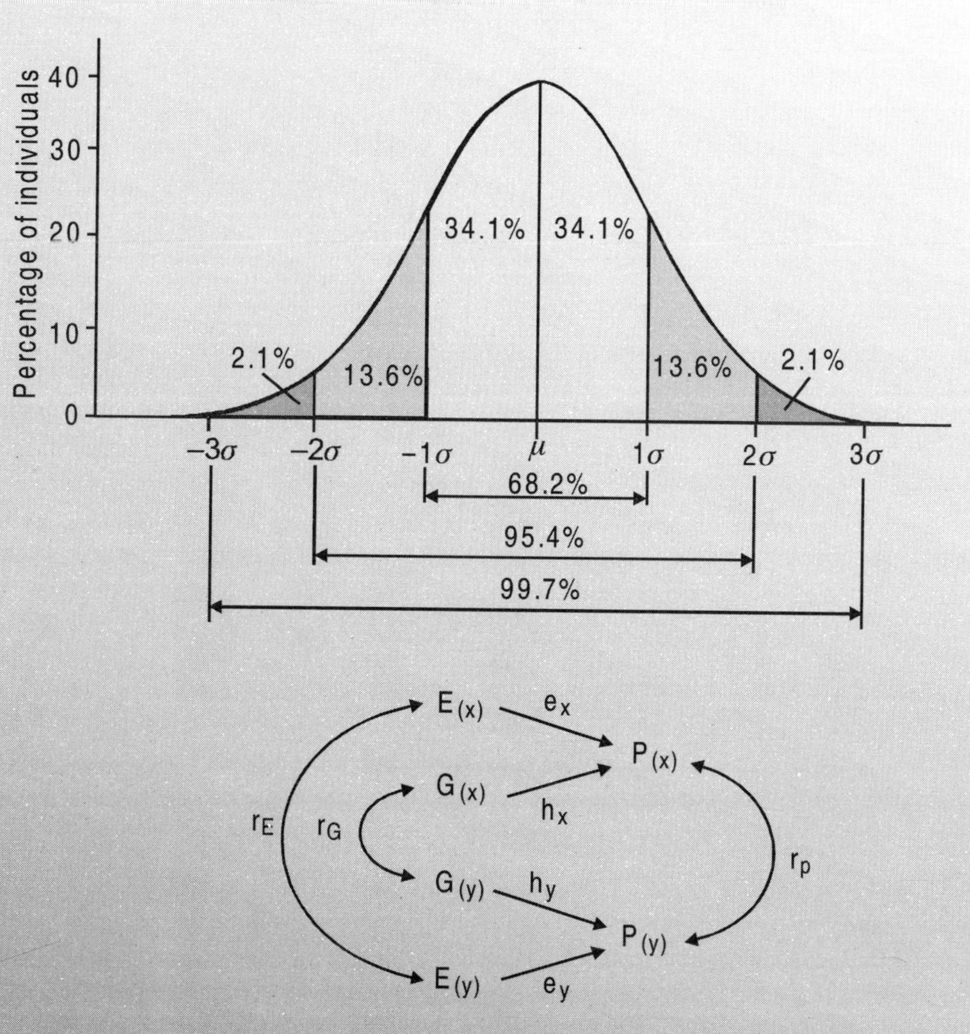

2

Economic Traits and their Standardization

Animal breeding entails bringing about genetic improvement in the traits of interest to breeders of the species/population concerned. Therefore, it is important to study different categories of economically important traits, and the methods used to bring the performance records of different animals to a common basis such that they are comparable. This chapter deals with economic traits, their standardization, and finally with the commonly used statistical methods to describe these traits.

Economic Traits

Farm animals are usually described in terms of appearance (e.g. body shape, size, color and other morphological features), performance (e.g. body weight, milk production, egg production etc.) or a combination of both appearance and performance. The term 'trait' or 'character' (used interchangeably) is applied to describe any morphological or performance feature of an animal. A **trait** is any observable or measurable feature of an animal. Examples of observable features are coat color, shape of horns, comb shape, etc, while those of measurable features include body weight, milk yield, egg weight, wool weight etc. The level to which a trait is expressed in an individual is referred to its **phenotype** or **phenotypic value** (e.g. actual body weight of an animal or amount of milk produced by a dairy cow etc).

Qualitative versus Quantitative Traits

Traits of farm animals can be classified by many criteria. From genetic standpoint, traits are classified as qualitative (monogenic or simply inherited) or quantitative (polygenic or multi–factorial). Salient features of qualitative and quantitative traits are listed in Table 2.1.

Table 2.1 Features of qualitative and quantitative traits.

Qualitative traits	Quantitative traits
1. Controlled by one, two or few genes/loci.	1. Controlled by large number of genes/loci.
2 Each allele of the gene has a large effect on phenotype.	2. Each allele of the gene has small effect but the cumulative effect of all the genes on phenotype is large.
3. Environment has little or no effect on phenotypic expression.	3. Environment usually has large effect on their phenotypic expression.
4. Simple and direct correspondence between genotype and phenotype (P=G).	4. Relation between genotype and phenotype is complex: (P = G + E).
5. Differences between individuals are of kind, therefore, such traits can be classified into distinct phenotypic classes.	5. Differences between individuals are of degree, therefore, such traits are difficult to be classified into distinct phenotypic classes.
6. Exhibit discontinuous variation, represented by binomial distribution.	6. Exhibit continuous variation, represented by normal distribution.
7. Studied by making crosses and from pedigrees by counting the number of progeny.	7. Studied by descriptive statistical methods like mean, variance, analysis of variance etc.
8. Examples include morphological traits like coat color, presence or absence of horns, blood groups and enzymes.	8. Examples include most of the economic traits like production, growth rate, disease resistance etc.

Threshold Traits

Most polygenic traits exhibit normal distribution of phenotypic values(Fig 2.1).But there is a special category of polygenic traits that show discrete phenotypic classes (like those of qualitative traits) though they are controlled by many genes and also modified by environmental factors. The traits which are controlled by several genes and also modified by environmental factors but exhibit few phenotypic classes are referred to as '**threshold**', '**all–or–none**', '**categorical**' or '**quasi–continuous**' traits. Tne threshold traits are presumed to have an 'underlying' scale which is continuous but unobservable (referred often as a '**scale of liability**', Fig 2.1).

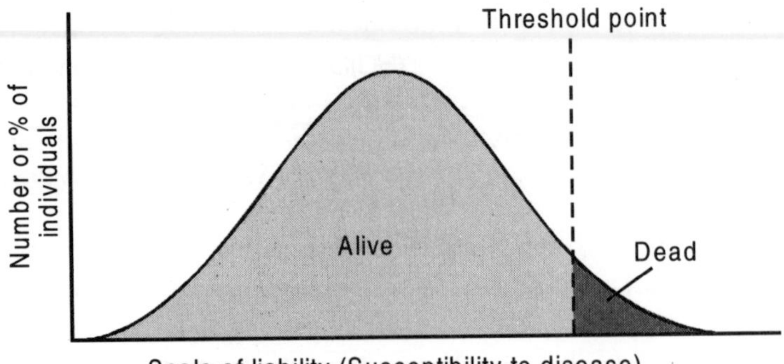

Fig 2.1. Illustration of an 'all or none' trait on scale of liability.

The underlying scale can be conceptually thought to be made–up of genetic as well as environmental values. The point on the liability scale above which animals exhibit one phenotype and below which they exhibit another phenotype is referred to as **'threshold point'** (Fig 2.1). When the phenotype of an individual is below the threshold point, the phenotype falls into one category (i.e. the individual is alive). When an individual crosses the threshold point, it expresses the phenotype of the other category (i.e. the individual dies). Threshold traits pose some special problems for their analysis, and genetic improvement.

A number of threshold traits are known in different species of livestock or farm animals. Some of these may be of interest to livestock producers and breeders. Survival of an animal is one such trait where there are only two versions (survival or death).Some features and examples of threshold traits are listed in Table 2.2.

Table 2.2 Features and examples of threshold traits.

Features

- Controlled by many genes
- Influenced by environmental factors
- Exhibit categorical (two or few) phenotypic classes on an individual–animal basis and hence show discontinuous distribution.
- The model for describing the phenotypic value of an individual for a threshold trait is same as that for a continuously distributed quantitative trait (i.e. P = G + E)
- Relatively more difficult for genetic analysis

Examples

- *All species*: Survival (livability) or mortality of an animal up to a specific age (e.g. an animal is either 'alive' or 'dead' (these two categories are respectively coded as 'one' and 'zero' on the measurement scale for subsequent genetic analysis).
- *Mammals:* Conception rate/fertility (as measured by success or failure to conceive)
- *Cattle:*
  - o Dystocia (i.e. difficulty in giving birth/calving): – Assisted or unassisted calving
  - o Twining: A cow either does or does not give birth to twins
  - o Cryptorchidism (i.e. failure of one or both testicles to descend into the scrotum): Whether or not the testes descend into scrotum.
- *Pig, sheep* : Litter size(i.e. number of offspring born in a single calving/lambing): one, two, three etc
- *Poultry:* Leg defects; either a bird is normal or affected (i.e. exhibits a leg defect).
- *Guinea pig*: Polydactylism– Presence of four toes on hand/ feet; the normal number of toes is three.
- *Mice:* Number of lumbar vertebrae (5 or 6)

On an individual–animal basis, threshold traits can be categorized into two or a few discrete classes (i.e. alive or dead) but for a group of animals (e.g. breed/herd/flock/family), the phenotypic potential can be expressed as the arithmetic mean. Then the arithmetic means for different groups can be compared.

Illustrative example: In a family of chickens, 10 chicks were hatched. Three of these died before reaching sexual maturity. The phenotypic value of the family for mortality would be

30% (i.e. (3/10) ×100 =30%). Similarly, another family may exhibit 50% mortality. These two averages can be compared, and if a selection decision is to be made, the one with the low mortality can be selected. On a single bird basis, the trait mortality can have only two classes (usually coded as '0' and 1).

There are some threshold traits that can be classified into more than two categories (but still a limited number of classes) and this aids in their analysis. An example is hip dysplasia, spelling commonly seen in Border Collies breed of dogs, which can be visually classified into few categories (e.g. normal, mildly affected or severely affected). Based on radiographic features, however, hip dysplasia can be scored into a larger number of classes.

Component Traits

Some quantitative traits can be biologically partitioned into two or more contributing traits. The traits that are mathematically or biologically function of two or more contributing traits are called **component traits**. Some examples of component traits are:

- Weaning weight in beef cattle and pigs can be partitioned into birth weight and gain in weight from birth to weaning age.
- Yearling weight is determined by weaning weight and post weaning gain.
- Annual egg production in chickens can be partitioned into two components: part–year and residual egg production.

In the above examples, the contributions of components of the trait are additive. In some cases, the component trait may be multiplicative function of the contributing traits:

- Feed efficiency in meat animals is usually defined as amount of feed consumed per unit gain in body weight:

$$\text{Feed efficiency} = \frac{\text{Amount of feed consumed}}{\text{Gain in body weight}}$$

Feed efficiency is a component trait whose contributing traits are multiplicative function of each other (i.e. feed consumed times the reciprocal of gain in body weight). Similarly, the total egg mass output per hen is a multiplicative function of number of eggs and the average weight of eggs. Similarly, economic efficiency of an animal can be calculated as the ratio of cost (input) of animal rearing to output from that animal in terms of animal product (e.g. meat, milk, wool, eggs, etc.).

Categories of Economic Traits

Livestock producers strive to increase profit from their enterprise that in turn is dependent on a number of traits. Traits that have monetary value in the production of farm animals (livestock) are referred to as **economic traits**. The monetary value may be direct or indirect.

Most of economically important traits in farm animals are quantitative in nature but some qualitative (single–gene) traits may also have some importance in some situations or market conditions. For example, coat color/pattern usually fetches premium in pet animals. Similarly, skin color and plumage color in broiler chickens may influence market returns. For example, colored broilers are preferred in some tribal regions of India. Sometimes breeder's fads act to create demand for unusual type traits simply for 'show' purposes (usually such attributes relate to morphological features). Some important qualitative traits (single–gene) are listed in Table 2.3.

Table 2.3 List of some Mendelian traits of economic/aesthetic importance.

Cattle:

- *Coat color:* A number of loci control coat color. But in many breeds, there is no variation at some of these loci or it is masked by variation at other loci. Therefore, coat color is effectively controlled by a single locus.
- *Polled vs. horned condition*: In exotic breeds of cattle (*Bos taurus*) it is controlled by 'P' locus ('P' for polled and 'p' for horned). In *Bos indicus* breeds, a different (but single) locus controls polledness.
- *Beta–casein milk protein*: β– casein protein in milk is controlled by a single locus with co–dominant alleles, the most common are A1 and A2 types. The A1 allele is thought to be undesirable for humans for a number of reasons, in particular digestion.
- *Double muscling*, also called muscular hypertrophy ('mh' locus): It is controlled by a recessive gene (myostatin). Double–muscle phenotype is due to increase in number of muscle fibers (hyperplasia) rather than the enlargement of individual muscle fibers (hypertrophy). Cattle with double muscling phenotype have 80% lean meat as compared to 65% for conventional– type meat. Double muscling phenotype is common in Belgian blue cattle breed, and is useful for production of lean muscle and tender meat with higher percentage of most desirable cuts of meat.

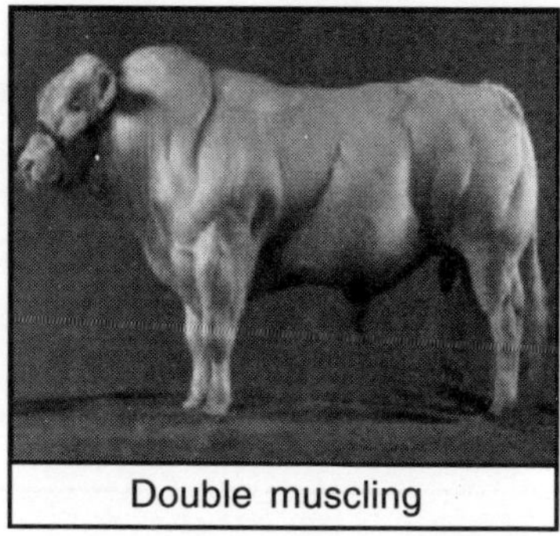

Double muscling

- A number of genetic disorders/diseases in cattle (e.g. 'amputated' lethal or 'bull–dog' calf' syndrome, and 'bovine leukocyte adhesion deficiency' syndrome (BLAD, an autosomal recessive disease) are controlled by single genes and have some economic value.

Swine:

- *Porcine stress syndrome (*PSS), controlled by a single gene is associated with sudden death due to stress; pale soft exudative meat (PSE) and sensitivity to halothane induced malignant hyperthermia.

Sheep:

- *Booroola gene*: In Australian Merino sheep it influences fecundity (by increasing ovulation rate and litter size).
- *Inverdale* gene in Romney sheep influences fertility (through increased litter size).
- *'Spider syndrome'* (caused by an autosomal recessive gene) is associated with abnormal bone and cartilage growth that leads to an outward deviation of front legs; the affected lambs are unable to stand.
- *Susceptibility to scrapie* (a fatal neurodegenerative disease with clinical symptoms of excitability, trembling etc.) is governed by a single locus with multiple alleles.

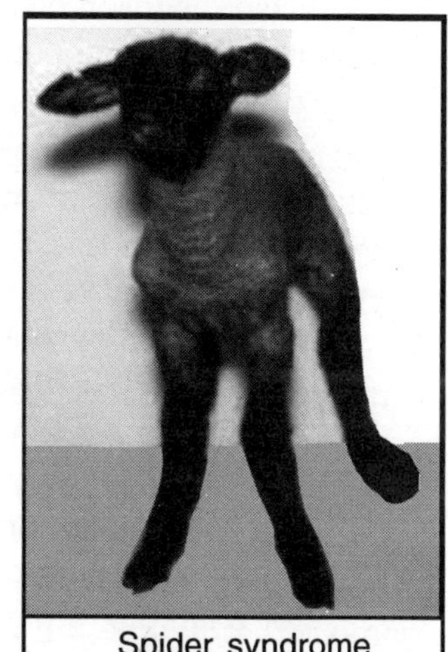

Spider syndrome

Poultry:

- *Plumage color* (important for fanciers).

- *Skin color* (genes), though consumer liking varies from one place/country to another.
- *Rate of feathering*, controlled by sex–linked genes, is used in day–old chick sexing.
- *Dwarfism, caused* by sex–linked gene, has been commercially exploited in broilers for reducing cost of chick production.

Dog:

- *Coat color/pattern* (e.g. in Labrador, Dalmatian breeds)
- *Hair length*
- *Hairlessness* in American terriers (caused by a recessive gene)

Horse:

- *Coat color/pattern*: Palomino: CC: Chestnut (reddish); $c^{cr}c^{cr}$: Cremello (Whitish); Cc^{cr}. Palomino pattern (golden/ yellow body with white mane and tail is preferred by horse breeders.
- *Hyperkalemic periodic paralysis (HYPP)*: A muscular disorder associated with uncontrolled muscle twitching during attack, causing intense weakness. This defect disrupts sodium ion channel's ability to open or close. It is a semi lethal defect which occurs in Quarter horses (has been traced to a mutation in a single quarter horse named 'Impressive').

Fish (Ornamental):

- *Body color*
- *Fin and body shape*

Majority of the economically important traits of farm animals, however, are quantitative in nature. The list of such traits in different species is large but they fall under four or five general categories:

1. **Production (including growth) traits**: These traits usually have direct impact on economic returns from livestock production enterprise. These include milk yield in dairy cattle; growth rate, feed efficiency and body weight in meat–producing animals (swine, broiler chickens, turkeys, beef cattle, mutton–type sheep, fish etc); wool production in sheep; racing or pulling performance in horses etc. These traits need to be defined with respect to age of animal at which they are recorded or expressed (e.g. body weight at six–week in broiler chicken, milk yield of a cow in a complete lactation).
2. **Fitness (reproduction and viability) traits**: The traits that are concerned with reproductive and survival ability of a species/population constitute the fitness traits. The reproductive and survival ability are defined in terms of number of offspring produced and those surviving to a given age. The list includes conception rate(fertility) and calving interval in cattle; litter–size in swine; prolificacy in sheep ; fertility in poultry; foaling rate in horse, and survivable up to market or reproductive age in all species of livestock.

   The contribution of fitness traits to economics of farm animals is generally indirect. In some species or class of animals, however, these may exert direct influence. For example, a pig breeder who sells piglets to commercial producers would have considerable interest in litter–size and the number of litters produced (farrowing) per year.

3. **Quality and composition traits**: The traits that determine the quality or composition (physical or chemical) of animal product constitute quality traits. These include carcass composition and quality in meat–type animals; milk constituents (fat, protein, and solids–not–fat) in dairy animals; egg quality and composition in egg–type poultry; and wool curliness, staple length, and fiber diameter in sheep. The economic value of quality traits varies according to market demands. These traits are becoming increasingly important as health consciousness grows.
4. **Type / Aesthetic traits**: These traits are usually features of aesthetic nature and involve personal preference/judgment of the producer/keeper. Traits like coat–color/ pattern; physical appearance, teat placement and udder shape, horn shape and several other aspects of conformation are examples of type/ aesthetic traits. Features related to structural soundness are also included in this category. During early stages of development of breed societies/organizations, type traits in dairy animals received considerable importance. With mounting economic pressure on livestock enterprise, the type traits which did not have direct bearing on productivity have became lesser important. However, for breeders of companion and pet animals, these traits continue to get attention.
5. **Behavioral traits:** This category of traits is also gaining importance in some species of farm animals from welfare standpoint. These include general temperament, docility in most species of farm animals, herding ability of sheep dog, reaction to stress etc in most species.

Measurement of Economic Traits

Genetic improvement involves recording the trait(s) of interest on all the animals of a population at the same age, and then selecting the ones with high performance (at genetic level). For selection to be effective, the trait(s) under consideration must be carefully defined and measured so that small differences between animals can be detected. Some traits can only be assessed by subjective measures while others can be measured in quantitative terms:

- **Subjective assessment**: Subjective measures involve sensory (visual and tactile) observation and are usually recorded without the aid of measuring devices. Such traits are usually scored/graded (e.g. a hedonic scale where judges score the extent of liking or disliking of a food product). Temperament, behavior, type, conformation, and some attributes of carcass are given subjective ratings. Skills, developed through training and practice, of the evaluator is closely associated with accuracy of subjective estimates (scores/grades). Nevertheless, the subjective estimates are seldom as accurate as the methods that employ measuring devices. However, the cost–benefit ratios indicate that for some traits (like those that are of lesser importance in a breeding program); the subjective estimates may be optimum. For example, body conformation in broiler chicken (particularly related to breast region) can be measured by breast angle meter but most breeders rely on trained persons for its assessment. Similarly, traits such as wool fiber strength, wool color etc in sheep are subjectively assessed.
- **Quantitative measurement**: Most of the traits observed on live animals (e.g. body weight, milk production, egg production, wool yield, feed efficiency etc.) can be measured

on a continuous scale and hence are little affected by measurement errors. Reproductive traits can also be quantified to a certain degree. Some quality traits (e.g. fat and protein percentage, egg composition etc.) can be measured quantitatively while others are assessed using subjective scoring.

Genetic improvement of a trait becomes easier if the trait can be recorded quantitatively. Indeed, the development of methods for measuring and recording of production (milk, fat, egg production) was one of the most notable events in animal breeding during the latter half of the nineteenth century. Since then methods have been developed for measurement of growth rate, feed efficiency and quality in swine, cattle, sheep, poultry and other species of farm animals. Considerable efforts have been made in development of instruments for measuring the traits on a fine scale. Development of methods for precise and accurate recording of traits, particularly the newly emerging ones, is a continuing process as the examples cited below illustrate:

- Back fat thickness in pigs can now be determined accurately on live animals by an ultrasonic technique without causing any harm. Earlier it was recorded by using a scalpel to cut the skin and then thrusting a ruler (or a ruled probe) into the fat until it stopped by contact with the muscle fibers of the body.
- Indirect methods of recording of abdominal fat in broilers have been developed and standardized.

From the standpoint of genetic improvement, it is valuable if quality, type and behavior traits could be measured objectively. In retrospect, genetic improvement has been most successful in those species or traits in which performance could be measured objectively on a continuous scale.

Economic Importance of Traits

Although economic returns from farm animals depend on a number of traits but all the traits do not have equal economic importance. The importance placed upon a particular trait depends on its **economic value**, which is defined as the profit resulting from a unit genetic change in the trait when other traits influencing the profit are kept constant.

- Production traits generally have largest economic value as income is directly dependent on these. They are also easier to be recorded.
- Reproductive traits are also important as these determine the survival of the population. High emphasis is thus placed on reproductive traits (e.g. litter size and survivability to weaning age) in litter–bearing species (e.g. swine). For a swine breeder/producer who is involved in selling of pigs, reproductive traits like the number of healthy pigs weaned and the number of litters per year per sow attains the highest economic importance.
- Quality traits usually have lesser economic value because the price of livestock products for a wide range of quality scores remains static. However, some exceptional scores for quality may provide large bonus or loss.

Similarly, type/ aesthetic traits directly have little economic significance but for fanciers these may be of considerable importance. The relative economic value of aesthetic traits is usually a function of 'eye' of the breeder/producer. In general, the relative economic importance of different traits is likely to vary from breeder to breeder depending on the segment of market being catered. A summary of the five broad categories of traits is given in Table 2.4.

Table 2.4. Summary of different categories of economic traits.

Trait:	Definition	Measurement	Economic value	Examples
Production:	Related to production and growth aspects	Quantitative	Usually direct and high	Milk production, growth rate, wool production, egg production,racing ability etc.
Reproduction:	Related to ability to produce offspring	Usually quantitative	Usually indirect and low	Calving interval, conception rate, litter–size, foaling rate, survivability etc.
Quality :	Related to quality and composition of product	Quantitative as well as subjective	Varies according to market (usually low in developing countries)	Milk constituents, egg quality & composition, meat–quality & composition, wool quality etc.
Type:	Related to appearance	Subjective (scoring)	Little or very low	Body color; body shape, conformation etc.
Behavior :	Related to welfare aspects	Subjective (scoring)	Little or very low	Docility, herding ability, reaction to stress etc.

List of Economic Traits of Farm Animal Species

The traits of economic importance vary not only from one species to another but also from one class to another class within the same species (e.g. egg–type vs. broiler chicken; beef versus dairy cattle; mutton–type vs. wool–type sheep etc.).Traits also vary in importance at different periods. Listed below are some of the economically important traits in different species. The list is extensive but not exhaustive because the importance of traits varies between countries/regions depending upon the market conditions and the industry structure. Details about description and measurement of traits of some important livestock species are given in the chapters dealing with their breeding.

(a) Cattle and Buffaloes

Production:

- Milk yield/production: Quantity of milk produced per lactation (complete lactation or 305–d standardized lactation or per year).
- Peak yield: Highest milk yield on a single day during lactation.
- Lactation length: Period for which the animal is in milk (day /month).

Milk constituents:

- Fat % or total yield of fat in the lactation.
- Protein % or total yield of protein in the lactation.
- Solids–not–fat % or total yield of solids–not–fat in the lactation.

Reproductive:

- Age at first calving: Age when the animal calves for the first time(day /month).

- Calving interval: Period between consecutive calvings (also called inter–calving interval).
- Conception rate: Number of services per conception (natural or artificial insemination).
- Dry period (days dry): Number of days for which the animal is not in milk (between consecutive calvings).
- Service period(days open): Number of days from calving to next conception.
- Dystocia (in 1st calving): Difficult parturition i.e. parturition requiring assistance.

Other features:

- Aspects of type like body shape, udder shape and placement.
- Somatic cell count (indicator of mastitis).
- Temperament.

(b) Swine

Production/growth :

- Birth weight: Weight at birth (either on per piglet or per litter basis).
- Weaning weight: Weight at weaning (usually at 8 weeks of age); it can be expressed on per–pig or per-litter basis.
- Market weight: Weight at the time of marketing. In India, pigs are marketed at around 5–6 months of age when they usually weigh more than 60 Kg.
- Feed efficiency, also called feed conversion ratio (FCR): Amount of feed consumed per unit gain in weight (from weaning to marketing).

Reproductive and survival :

- Litter size: Number of piglets born.
- Number of farrowings/ litters per year.
- Mothering ability, which includes not only milk yield of sow but also its ability to take care of piglets.
- Viability of piglets (percentage).

Carcass :

- Back–fat thickness: It is used as a measure of amount of fat in the animal.
- Dressing percentage: Dressed weight expressed as percentage of live weight.
- Other features of carcass desirability: These include a number of features like body length, area of loin–eye muscle; percentage of lean cuts, muscle quality etc. These are considered for inclusion in selection program in some western countries where pig industry is well developed.

Others features:

- Type and conformation score .

(c) Sheep

Mutton– type

Growth :

- Birth weight (of lamb).

- Weaning weight : Weight of lamb at weaning (usually 90 days of age).
- Post–weaning weight gain: Measured as difference in weight from weaning to 9 months of age which is the usual marketing age for mutton purpose.
- Market weight: Weight at the time of marketing (normally 8-9 month of age when dressing % is high and mutton is lean).
- Feed efficiency (during post–weaning period): Feed consumed per unit gain in weight between weaning and marketing.

Carcass :

- Dressing/killing–out percentage: Dressed weight expressed as % of live weight.
- Muscling quality grade.
- Fat depth between muscles.
- Meat–fat–bone proportions.

Wool– type

Fleece :

- Greasy fleece weight: Quantity of wool including impurities likes grease, contaminants such as vegetable matter and bacterial, fungal and water stains. It is recorded either on per–shearing or on per-annum basis as sheep are normally sheared twice).
- Clean fleece weight: Total quantity of clean wool i.e. excluding impurities (either on per–shearing or on per-annum basis).
- Number of shearings per year: Sheep are usually shorn twice a year but high wool-yielding breeds/strains can be shorn more than twice.

Wool quality :

- Fiber diameter
- Staple length: Length of the fiber/staple
- Medullation: Measured as proportion or percentage of medulated fibers in wool.

Maternal (mothering) ability:

- Milk production.
- Lamb survivability (percentage of lambs surviving to a specific age).

Reproduction (both wool and meat type sheep):

- Age at 1st breeding/mating.
- Numbers born(singles and twins).
- Number of litters per year.

(d) Goat

In India goat are reared for both meat and milk. Their meat traits are quite similar to meat–type sheep.

Growth :

- Birth weight.
- Weaning weight : Weight at weaning (usually 90 days of age).

- Market weight: Weight at the time of marketing (normally 9 to 12 month depending on breed).

Carcass traits:

- Dressing percentage: Dressed weight expressed as percentage of live weight.
- Meat–bone proportion.

Maternal/mothering ability:

- Milk yield.

Reproduction and survival:

- Age at 1st breeding/mating.
- Numbers born(singles and twins).
- Survivability (percentage of kids surviving to a specific age).

(e) Chickens

Chickens are reared for egg and meat. Those reared for eggs are called egg–type or layer chickens while those reared for meat are called meat–type or broilers.

Egg– type (Layer) :

Egg production traits:

- Annual egg production/number: Number of eggs laid up to 72 weeks of age.
- Part–year egg production/number: Number of eggs laid in the part–year (usually up to 40 weeks of age)
- Age at 1st egg (age at sexual maturity) :Age at the laying of first egg
- Rate of lay (percentage): Number of eggs laid in a specified period divided by the number of days for which the bird was in laying and expressed as percentage.

Egg quality / composition traits:

- Egg weight: Weight of eggs at a specific age (usually recorded as average of 6–7 consecutive eggs per bird).
- Albumen quality: Firmness of albumen and is measured as Haugh Unit Score.
- Shell thickness: A measure of egg shell strength.
- Blood and meat spots: Incidence of blood and meat spots in eggs.

Meat–type (Broiler):

Growth traits:

- Body weight at marketing age (usually 5–7 weeks of age).
- Feed efficiency: Amount of feed consumed from hatching to marketing divided by the gain in body weight from hatching to marketing age.

Carcass traits:

- Dressing percentage: Dressed weight expressed as percentage of live weight.
- Conformation: Measured as breast angle, it is indicator of flesh on the rib cage (breast region).

Reproductive & viability traits *(Both egg– and meat-type chicken)*

- Fertility percentage: Number of eggs fertile divided by the number of eggs set in the incubator, and expressed as percentage.
- Hatchability percentage: Number of chicks hatched divided by the number fertile eggs, and expressed as percentage.
- Mortality percentage: From hatching to 5–7 weeks of age in broiler chicken, and during the laying period in egg–type chicken.

Other:

- Incidence of leg defects: Modern–day heavy broilers exhibit a number of skeletal defects, the most common being defects of the leg.

(f) Horse

Performance traits:

Horses are bred for many different purposes viz; racing, jumping, draft, sporting etc. The performance has to be measured for the specific attribute for which they are bred.

Racing: Horse racing being an equestrian sport is now a major international industry. Racing's economic importance lies in the gambling associated with it. Two common types of horse racing are :

- *Flat racing*: It is prevalent throughout the world. In flat racing the horses are not required to jump over obstacles (e.g. hurdles or fences).It is run over a predetermined distance (generally 1000–2500 m). In Indian race courses the two different distances are: 1200m (sprinting) and 2400 m (stayers). Flat racing is a test of speed and stamina.
- *Jump racing*: It is racing over jumps, and is run over long distances, usually from two miles (3200 m) up to four and a half miles (7200 m) and includes national hunt–racing (hurdling), steeple chasing and point–to–point racing. It is popular in European countries.

In addition to these two types, harness racing is also practised where horses trot or pace while pulling a driver in a small, light cart known as a sulky.

Measure of racing performance: The different measures of racing performance are broadly classified into three categories but the two important ones for the purpose of this book are:

- Racing time viz; time to finish, also called winning time, indicates the time taken to complete a race. Other measures of racing time are average time, which is the average of all the horse's times over a given distance; best time which is the horse's fastest time over a given distance.
- Earnings (like earnings, average earning index, annual earnings etc.).

Gait: It refers to any of the various foot movements of a horse (as a trot, pace, canter and gallop). All horses move naturally with four basic gaits:

- Four–beat walk, which averages 6.4 kilometers per hour (4.0 mph).
- Two–beat trot (a gait in which legs are lifted in alternating diagonal pairs) or jog at 13 to 19 kilometers per hour (8.1 to 12 mph).
- Canter or lope, a three–beat gait that is 19 to 24 kilometers per hour (12 to 15 mph).

- Gallop which averages 40 to 48 kilometers per hour (25 to 30 mph) but the world record for a horse galloping over a short, sprint distance is 88 kilometers per hour (55 mph).

Sport: Horses are trained to be ridden or driven in a variety of sporting competitions such as:

- Show jumping(where the horse and rider jump a set course of obstacles usually within a time period).
- Dressage (a sport that uses classical training comprising of movements that have been developed over centuries).
- Cross–country jumping(where the jumping horses and riders jump a series of obstacles over varied terrain).
- Three–day eventing.
- Competitive driving.
- Endurance riding
- Rodeos etc.

Reproduction:

- Foaling rate: Measured as the number of foals born divided by the total mares covered, and expressed as percentage.

Conformation:

- *Canon bone circumference*: Cannon bone is the bone of lower leg, (it stretches from knee joint to fetlock joint), which is a weight bearing bone. Circumference of the cannon just under the knee is a guide to horse's ability to bear weight and do hard work.
- *Wither height:* Height of horse is measured at the highest point of withers where neck meets the back. This point has been chosen as it is a stable point of the anatomy, unlike the head or neck, which moves up and down.

(g) Fish (Indian major carps):

Growth:

- Body weight: Weight at marketing age which for Indian major carps is between 7 and 8 month of age (when they attain body weight of more than 500 g).
- Feed conversion ratio: Feed consumed from fry or fingerling to market age per unit gain in weight during this period.

Carcass quality/composition:

- Dressing percentage: Dressed weight (i.e. the live weight minus the weight of head, tail, scales, and viscera) expressed as percentage of live weight.
- Fat percentage
- Protein percentage

Reproductive and survival:

- Age at sexual maturity (called spawning).
- Survival percentage during a specified period,usually from fry/fingerling to marketing age.

Standardization of Performance Records

Selection which is the most important aspect of genetic improvement, involves making comparisons between one animal and another for the trait(s) of economic importance. It is, therefore, important that performance values/records of the animals are made under as similar environmental conditions (e.g. management, nutrition and hygiene) as possible. Uniformity of environmental conditions helps in identifying the animals, which are superior because of their genes, and not because they have been well-fed and –managed. This, however, is not always possible because there are a number of non–genetic factors (both physiological and environmental) which are beyond the control of breeder.

Examples of Non–genetic Factors

Some of the commonly encountered non–genetic factors are:

- **Sex of animal:** Sex of animal is an important factor that influences the performance, particularly the growth traits. For example, body weight and growth rate of farm animals are greatly influenced by the sex of the animal: the males are generally heavier than females.
- **Season/year of calving:** Milk production of a cow is greatly influenced by the season in which she calves: those calving in the season when environmental factors such as ambient temperature, humidity etc are moderate produce more milk than those calving in unfavorable season. Same holds true for the year of calving/birth, herd–management conditions etc. Season also exerts great influence on several traits of sheep, swine, goat, poultry and other farm species.
- **Age of animal:** It influences the performance for a number of traits (e.g. body weight, milk production, wool yield, egg production etc.).
- **Age of dam:** Young mothers usually have low body weight, therefore, they produce lesser milk, and in turn, their offspring grow slower.
- **Parity** (i.e. order of birth/calving): Cows produce lower milk in their 1st and 2nd calving than in subsequent calvings. Similarly, heifers experience higher dystocia (difficult calving) than older cows.
- **Litter size:** Young ones from large litters grow slowly than those from smaller litters.

These are examples of some of the common types of non–genetic factors that influence performance across all species of farm animals. On the other hand, there are a number other non–genetic factors which are specific to some species. For example, length of lactation, frequency of milking, length of dry period, service period, calving interval etc all influence the milk production traits in dairy animals. Season of hatching in poultry influences age at sexual maturity and egg production. Though we can not determine the 'exact' effect which any one of such factors has on the performance of any single animal but we *can estimate (quantitatively) the average effect of these factors on the performance of a group of animals.* Based on their average effect, we can adjust the performance records of different animals to a common basis so that the comparison of performance of different animals becomes valid. One of the illustrative examples of adjustment of performance records is in dairy cattle in USA where Dairy Herd Improvement Association (DHIA) adjusts the lactation milk yield of all cows to the following standards:

- *305 day lactation length*: Lactation milk yield of cows which were in milk for a longer period is adjusted back to 305 day i.e. what would have been the estimated milk yield if the cow would have been in milk for 305 day?. This adjustment reduces the variation in yield caused by varying lactation length and also reduces the effect of pregnancy on milk production.Also lactation length of 305 days is desirable as ideally cows should calve each year and have a dry period of about two month.

- *Twice a day milking*: In most cases, cows are milked twice a day. But the cows that are milked thrice a day produce about 10–20 % more milk than if they are milked twice a day. Hence milk yield of cows which were milked thrice a day is converted back to 'twice a day' milking.
- *Mature age*: Lactation milk yield increases from first lactation until the animals are 6–8 years of age, and then starts declining. Therefore, milk yield records of cows of different ages are adjusted to their expected values had they been mature (called *mature equivalent*).

In this way, performance records of different animals are converted to a common basis. This permits a fair comparison of performance of different animals. However the value of such adjustment depends on the magnitude of the environmental effects and the accuracy with which they can be estimated.

Methods of Adjustment of Performance Records: Adjustment Factors

There are two general types of adjustment or correction factors: additive or multiplicative:

Additive Correction/Adjustment Factors: Performance record of an animal is adjusted by adding a certain value to its actual record depending on the particular class to which the animal belongs (e.g. all those cows which calve in a particular season constitute one class while all those which calve in a different season constitute another class). Alternately, we can subtract a certain amount from the actual performance records of animals that belong to a certain class.

Illustrative example: Weaning weight of lambs is dependent on whether they were born as singles or twins. In a large flock of sheep, the following data were available for weaning weight of lambs:

Mean weaning weight (per lamb in kg) for :	
Single birth	Twin birth
22	20

It can be seen that the average weaning weight of single–born is 2 kg more than those born as twins. The difference in average weight of single– and twin–born lambs serves as adjustment/correction factor: we can either add 2 kg to the average weight of those born as twins or subtract 2 kg from the average weight of those born single. Suppose we have two lambs whose weaning weight and birth pattern are as follow:

ID number	Birth pattern	Weaning weight(kg)
1	Single	25
2	Twin	24

The question is "which one of the two lambs is genetically superior for weaning body weight". In order to make a valid comparison of body weight of the two lambs, we have to adjust their body weights for the birth pattern i.e. single vs. twin. If we wish to compare both lambs on single–born basis, the weight of second lamb is adjusted to single birth basis as follow:

ID number	Birth pattern	Weaning weight(kg):	
		Original	Adjusted to single–birth basis:
1	Single	25	25
2	Twin	24	24 + 2 = 26

Similarly, we know that males are heavier than females of the same age. We can convert the body weight of males to a female–basis or convert the female body weight to a male–basis. If the birth weight of a female is to be corrected to male basis, the adjustment factor is added to the body weight of all the female calves. Alternately, subtracting the correction factor from body weight of male calves, converts their body weight to female basis.

Additive correction factors are widely used across all species of farm animals. These are easy to use and do not make any assumption about the average genetic make–up (breeding value) of animals of different classes.

Multiplicative Correction/Adjustment Factors: Multiplicative correction factors involve adjusting the performance record of an animal by multiplying it with the correction/adjustment factor. That is, the actual performance record of animal is multiplied by a certain factor to obtain its adjusted record.

Illustrative example: Suppose average birth weights of male and female lambs in a flock are 22 and 20 Kg respectively. We can see that the birth weight of males is 10% higher than of female lambs (i.e. $[(22 - 20) / 20] \times 100 = 10\%$).Therefore, we can convert the female weight to male–basis by multiplying the female weight by 1.10. Suppose we have two lambs: one a male with weight of 25 Kg and the second a female with body weight of 24 Kg. We are to adjust the body weight of female lamb to a male–basis which can be done by multiplying the female weight by factor of 1.10. The adjusted body weight of female lamb is calculated as follow:

$$\text{Adjusted weight} = \text{Original weight} \times \text{Adjustment factor}$$
$$= 24 \times 1.10 = 26.4 \text{ Kg}$$

A realistic example of multiplicative adjustment factors used by Holstein Friesian breeders of USA (breed association) for adjusting milk yield to 305–day lactation length is given in Table 2.5.

Table 2.5. Multiplicative factors for adjustment of milk yield for lactation length to 305–d basis.

Lactation length (days)	Adjustment factor
305–308	1.0
309–312	0.99
333–336	0.93
337–340	0.92
342–344	0.91
361–364	0.86

Multiplicative correction factors are especially useful when there is a relationship between the mean performance and standard deviation of different groups. If, for example, females not only have lower body weight than males but also have lower standard deviation, the multiplicative correction factor would be more suitable. The multiplicative correction factor would not only elevate the mean of females to the equivalent of males but would also increase the variation. Therefore multiplicative correction factors can be used across herds/flocks which have different levels of performance and different magnitude of variability.

Limitations of adjustment factors: In general, both types of adjustment (correction) factors have certain limitations:

- Performance records of a large number of animals, which have been exposed to a wide variety of environmental conditions, are required to work out the correction factors.
- The adjustment/correction factors are specific to particular herds, flocks or breed from which these have been developed.
- In some circumstances they may over– or under–adjust performance values because some of the assumptions involved in their computations may not have been fulfilled.

Adjustment factors have been developed in western countries for a wide variety of traits in different species of livestock. These are developed from large sets of data that were obtained under a wide range of environmental conditions. Mostly this work is handled by respective breed associations.

Use of Contemporary Groups for Adjustment of Performance Records

In addition to the above two methods of adjusting the performance of animals for known causes of non-genetic factors, another method involves the use of contemporary groups. The animals which have been exposed to similar conditions and treated in a similar way are called **contemporaries** (e.g. animals born over a relatively short period of time ; on the same farm/ location; and fed and managed similarly), and the groups to which they belong is called a **contemporary group**. In order to bring the performance records of different animals to a comparable basis, the performance record of each animal is expressed as a deviation from the mean of its contemporary group (usually in standard deviation units).

Illustrative example: Average birth weights of calves of both sexes born in two seasons (winter, summer) of a year in a population are tabulated below:

Season	Sex	Average birth wt.(kg)	Standard deviation (kg)
Winter	Male	30	6
	Female	25	5
Summer	Male	28	7
	Female	24	6

Suppose we wish to compare the body weight of two calves born in this population for selection purposes: a male calf with body weight of 35 kg, born in winter, and a female calf weighing 31 kg, born in winter. Performance (birth weight) of the calves can be expressed as deviation from its contemporary group (i.e. from the mean birth weight of the same sex- and-season):

Adjusted record of male calf (35 kg) born in winter

= 35 – 30 = + 5 Kg (above contemporary group mean).

Adjusted record of female calf (31 kg) born in winter

= 31 – 25 = + 6 Kg (above contemporary group mean).

This example shows how the use of performance of contemporary group has helped to standardize the performance records (birth weight) such that these have become independent of season and sex, and have become directly comparable. If one of these two calves were to be selected, it would be the female calf as she is 6 kg better than her contemporary group while the male calf is only 5 kg better.

Performance records can be standardized and expressed in standard deviations units as follow:

Adjusted record of male calf born in winter:

$$= \frac{35 - 30}{6} = \frac{5}{6} = 0.83 \text{ standard deviation units (above the class mean).}$$

Adjusted record of female calf born in winter:

$$= \frac{31 - 25}{5} = \frac{6}{5} = 1.2 \text{ standard deviation units (above the class mean).}$$

In this case, the standardized records (birth weight) have not only become independent of season and sex but have also been adjusted for differences in standard deviations (variances), and have, therefore, become more directly comparable than when expressed simply as deviations from the contemporary group. These records can be better appreciated and interpreted.

Breeders of some classes of livestock like beef cattle sometimes prefer to express adjusted record as a percentage of its contemporaries (called the *trait ratio*). The adjusted body weight (trait ratio) of the female calf born in winter whose own weight is 31 kg would be: (31/25) = 1.24. This means that the female calf born in winter had 24 % higher weight than the average body weight of its contemporaries.

The advantage of use of contemporaries for adjustment is that it does not require prior information on the effect of different non–genetic factors on performance; therefore, it can be used in a wide range of herds/flocks. The main disadvantage is that it assumes that the average genetic make–up of different contemporary groups is same.

General Remarks: Calculation of proper adjustment/correction factors is not as easy as has been made out in the discussion above. These factors are applicable to the specific populations (breed) from which they have been calculated, and are computed from large volume of data that have accumulated in large breeding programs. In western countries, breed associations are involved in developing the adjustment factors which are appropriate for valid comparison of animals of that breed only.

Statistical Measures and Methods in Animal Breeding

Most of the economically important traits, which are quantitative in nature, can be measured on a continuous scale. Statistical measures like mean, variance, standard deviation and correlation and regression coefficients are used to describe such traits. The summarization of the data by a single number constitutes a **descriptive measure.** The various statistical measures (descriptive) are broadly classified as measures of:

- Central tendency (mean, mode, median)
- Dispersion (range, variance etc.)
- Covariation (correlation and regression)

The science of animal breeding, which involves bringing about genetic improvement in the performance of population, uses a number of statistical concepts. This selection briefly deals with some of the most commonly used statistical measures and methods.

Elementary StatisticalTerminology

Population and Sample: In genetic terminology, population is an inter–breeding group of individuals. Since a population is usually very large (theoretically infinite), the entire population cannot be observed / studied due to cost, time and logistic involved. Therefore, the researcher tries to study a small group from a population, which is representative of the entire population. Such a group, which is subset of the population, is called a **sample**. If accurate inferences about a population are to be drawn, the sample should be obtained in such a way that it reflects the characteristics of the population from which it is drawn. Statistical procedures for obtaining samples typically assume that the samples are obtained in a random fashion. A **random sample** is one in which each member of the population has an equal and independent chance of being included. That is, not only must each individual in the population have an equal chance of being chosen as a member of the sample but also the selection of any member of the population must not influence the selection of any other member. The word 'sample' in this book will imply 'random sample'.

Variable: If we observe a trait, we will find it can take on different values for different animals, we say it is a variable. The values which are recorded on individuals are called **variables**. A variable can be discrete or continuous:

- *Discrete variable*: A variable which can have a limited number of values, usually whole number, between two fixed points(e.g. 1,2,4,5, 7, ... etc)
- *Continuous variable*: A variable that can have any number of values (which can be non–integer) between two fixed values (e.g. 1, 1.2, 1.3, 1.4, ..., 10.0). In animal breeding we mostly deal with continuous variables

Parameter and Statistic: Several measures help to describe a population or sample (i.e. measures of central tendency, measures of dispersion etc). A quantity such as a measure of central tendency (i.e. mean) or a measure of dispersion (i.e. variance) when it describes a population is called a **parameter.** Since the study of the entire population is seldom possible, one has to rely on a sample to arrive at conclusions about the population. An estimate of the population parameter obtained from a sample is called a **statistic** (use of the terms 'parameter' and 'statistic' was defined by R A Fisher in 1925).

In summary, a descriptive measure (mean, variance, standard deviation etc) computed from data of a sample is called a statistic while that computed from data of a population is called a parameter. It is a statistical convention to represent parameters by Greek letters and statistic by Latin letters as shown below:

	Parameter	Statistic
Mean	u	$\overline{X}$
Variance	σ^2	s^2

The statistic we calculate will vary from one sample to the other taken from the same population. Since we use sample statistics (e.g. $\overline{x}$, s^2 etc) as estimates of the population parameters (e.g. u, σ^2), we strive to arrive at the **'best estimate'** possible. Of the various criteria used for choosing an estimate as 'best' in animal breeding are the '**unbiasedness**' and '**precision**'. An estimate is said to be 'unbiased' if the average of indefinitely large number of samples taken from a population equals the 'parameter'. The 'precision' refers to as "how close the statistic obtained from a single sample is to the parameter being estimated". It is quantified by the variance of the statistic. These two criteria (unbiasedness and precision) are generally used together to provide the definition of a 'best' estimator.

Mean

The most commonly used measure of central tendency is the arithmetic mean. It is a single value that is representative of the population/ group/ sample as a whole. The adjective, 'arithmetic' distinguishes this mean from other types of mean (e.g. geometric mean, harmonic mean). In animal breeding, we are mostly concerned with arithmetic mean; therefore, we shall refer to arithmetic mean simply as the mean unless it is specified otherwise.

Arithmetic mean is notated as μ (pronounced as 'mu') for a population and as $\bar{x}$ or $\bar{P}$ (X– or P– bar) for a sample. Mean (or phenotypic mean) for a trait is calculated by obtaining the sum all the observations / values (phenotypic values) in a population/ sample, and dividing it by the number of observations / values. In mathematical notation, the mean is represented as follow:

$$\bar{P} = \frac{1}{N} \sum P_i \quad \text{or} \quad \bar{X} = \frac{1}{N} \sum X_i$$

Where:

$\bar{P}$ or $\bar{X}$ represents the mean

P_i or X_i (read as 'P sub i' or 'X sub i') refers to phenotypic value of the ith animal

N is the number of animals/ observations in the group/ sample

Σ represents sum of all the values of a variable that vary from first to the last

Arithmetic mean is easy to compute, easily understood by non–technical persons and has desirable statistical properties. Since each value of the set enters into the computation of mean, it is affected by each value. Extreme values can influence the mean, and in some cases, these can distort it. One of the most important algebraic properties of arithmetic mean is that; "sum of deviations of a set of values (observations) from their arithmetic mean equals zero, that is:

$$\Sigma(X_i - \bar{X}) = 0$$

When a large number of observations/values (X_i) are grouped into a frequency table (f_i) the arithmetic mean is obtained as follow:

$$\bar{X} = \frac{\sum f_i X_i}{\sum f_i}$$

Where: $N = \Sigma f_i$ which is the total frequency (i.e. the total numbers of observations.)

In animal breeding, mean (also called average) is useful for describing a breed, strain, herd or a management group. The arithmetic mean is appropriate to describe most of the traits of economic importance in livestock.

Other types of means, which may be used for some traits, are the geometric and harmonic means:

- ***Geometric mean*** **(GM)** of a set of positive values/observations is defined as the Nth root of the product of values/ observations of all 'N' individuals in a group/ sample:

$$\text{Geometric mean} = \sqrt[N]{X_1 \times X_2 \times X_3 \text{ ....} \times X_N}$$

Calculation of geometric mean thus uses the product of individual values (as opposed to the arithmetic mean, which uses their sum). For instance, the geometric mean of two numbers, say 2 and 8, is just the square root of their product; that is square root of multiple of 2 and 8 equals four.Similarly geometric mean of numbers 2, 4 and 8 is the cube root of their product:

$$GM = \sqrt[3]{2 \times 4 \times 8} = 4.0$$

The geometric mean applies only to numbers of the same sign. It is often used for a set of numbers whose values are meant to be multiplied together or are exponential in nature, such as data on growth of human population etc. In animal breeding, there are some traits which exhibit multiplicative properties for which statistical analysis uses geometric mean. (e.g. somatic cell count which is an indicator trait of mastitis; parasitic fecal egg count in sheep, 'Ct' values obtained in Real Time PCR in genomic studies).

- ***Harmonic mean (HM)*** is defined as reciprocal of the arithmetic mean of the reciprocal of values (observations) of all 'N' individuals in the group/ population:

Harmonic mean: $$HM = \frac{1}{\frac{1}{N}\sum X_i} = \frac{N}{\sum \frac{1}{X_i}}$$

For example, HM of the numbers 2, 4 and 8 is:

$$HM = \frac{3}{\frac{1}{2}+\frac{1}{4}+\frac{1}{8}} = 3.43$$

Harmonic mean is less sensitive to a few large values than are the arithmetic or geometric means. So it is sometimes used for distribution which are highly skewed.In animal breeding, harmonic mean is used in few situations like working out the effective population size(Ne).For example, say the population size was 10, 100, 50, 80, 20, 500 over six generations. Then the effective population size is the harmonic mean of these, giving:

$$\frac{1}{N_e} = \frac{\frac{1}{10}+\frac{1}{100}+\frac{1}{50}+\frac{1}{80}+\frac{1}{20}+\frac{1}{500}}{6} = \frac{0.1945}{6} = 0.032416667$$

$$\therefore \quad N_e = 30.8$$

Other measures of central tendency include median (the central value) and mode (the class with the highest frequency).

Variation

Mean is representative of the group/ population from which it has been computed but it reveals nothing about how far the individuals/ observations deviate from the mean (i.e. how the individuals are scattered around the mean). Variation in a sample/ population is commonly described and measured by variance.

Variance: Variance is defined as the average of the squared deviation of observations (values) from the mean. Variance, symbolized as σ^2 or σ_X^2 or σ_P^2 for a population is calculated as:

$$\sigma^2 = \frac{\sum (X_i - \mu)^2}{N}$$

Where: σ^2 = population variance, X_i = values of the trait/variable, and u = population mean.

Variance carries the squared unit in which the trait is measured. In computing the variance, the deviations are squared, therefore, a phenotypic value or an observation that is farthest from the mean contributes more to the variance than the one closer to it.

For a sample, the variance (s^2) is calculated as:

$$s^2 = \frac{\sum (X_i - \overline{X})^2}{N - 1}$$

Variance can be more easily calculated from the sum of squared observations and sum of observations squared, rather than calculating, the square of each deviation from the mean, as follow:

$$\text{Sample variance : } \quad s^2 = \frac{\sum X_i^2 - \frac{(\sum X_i)^2}{N}}{N - 1}$$

The following points need to be noted:

- The denominator in the expression (i.e. N–1) is referred to as 'degrees of freedom', and is one less than the total number of observations in the sample. An appropriate question is "Why do we use 'N–1' in the denominator for sample". The reason for this is the concept of degrees of freedom. It can be noted that the calculation of variance requires information on mean such that there are only 'N–1' independent observations once the mean is known. For example, the mean of measurements on five animals is 3.6, such that if measurements on the first four animals, X_1, X_2, X_3, and X_4, are known, then the value of 5th observation must be equal to : $5(3.6) - (X_1 + X_2 + X_3 + X_4)$.
- Size of the sample in most of the animal breeding problems runs into several hundreds or thousands, therefore, we frequently use the notation of σ^2_X or σ^2_P (instead of s^2).

  Variance has some useful statistical properties, but unlike mean, it is difficult to be conceptualized.

Standard Deviation: The square root of variance is called 'standard deviation' (designated as σ or σ_P):

$$\sigma_P = \sqrt{\text{variance}} = \sqrt{\sigma^2}$$

Standard deviation indicates by how much the value in a distribution typically varies from the mean: larger the standard deviation, more the values would be spread about the mean and *vice versa*. Since standard deviation carries the unit of mean (i.e. the unit in which the trait/ variable has been measured), it is more commonly used than the variance to describe the variability in a trait.

Standard Error: Standard deviation of the sample means is referred to as standard error (SE). It measures the accuracy with which the sample represents population mean and is calculated by dividing the standard deviation by square root of the number of observations:

$$SE = \sigma / \sqrt{N}$$

Illustrative example: The concept of SE can be better appreciated by an example. Suppose we wish to obtain mean body weight of the entire adult buffalo population of a state. Obviously, we cannot weigh all the animals; therefore, we draw a sample from this population to estimate its mean (parametric). However, our sample mean will not be exactly equal to the parametric mean. Therefore, we would like to have an idea of how close our sample mean is likely to be. If our sample size is small, our estimate of the mean will not be a good estimate as compared to one based on a larger sample size. Standard error is a measure the reliability of sample as an estimate of population mean, and for this reason standard error is defined as standard deviation of the many random samples from a population. The smaller the standard error, the more representative the sample will be of the overall population mean. Standard error is inversely proportional to sample size: larger the sample size, smaller the standard error because the statistic will approach the actual value.

Standard error along with mean $(\overline{X} \pm SE)$ describes the reliability of sample mean as an estimate of the true/population mean.

Coefficient of Variation: Another measure used to describe variability in a sample/ population is the coefficient of variation (CV).It is defined as the standard deviation expressed as percentage of the arithmetic mean.

$$\text{C.V.} = \frac{\text{Standard deviation}}{\text{Mean}} \times 100$$

Since mean and standard deviation are expressed in the same unit, the units of measurement cancel out in computing the coefficient of variation. Consequently, it makes the variance independent of mean, and therefore allows the comparison of variability between traits within a population or between populations for the same trait.

Note: Variance, standard deviation, standard error, coefficient of variation etc are measures of a population/ sample and not of an individual animal.

Normal Distribution

Most of the economic traits, which are quantitative in nature, can be measured on a continuous scale, and hence show continuous variation. When data on large number of individuals from a population are recorded, and then plotted on a graph paper with abscissa (X–axis) denoting the phenotypic values and the ordinate (Y–axis) denoting the numbers of individual or the frequency/percentage, a "bell–shaped curve", referred to as **normal curve**, is obtained. The distribution of such a curve is called a **normal distribution**. It is also referred to as **'Gaussian distribution'** to recognize the contributions of its originator, the mathematician Carl Fried Gauss (1777–1855).

Fig 2.2b shows the distribution of 6 week body weight in broiler chickens with mean of 1500 g and phenotypic standard deviation of 150g while Fig. 2.2 a shows a general example. The height of the curve at a particular point represents the frequency / percentage of individuals having that phenotypic value.

The normal curve, regardless of the trait or unit of measurement, has some important properties. Given the properties of normal distribution and knowledge of mean and standard deviation of a population, certain predictions about the distribution of observations for the population can be made. The important properties (features) are:

- The curve is symmetrical about its mean (u), that is the mean lies in the middle of the distribution, and half the records (50%) are above the mean, and half below it. That is the right and left halves of the curve are mirror image of one another.
- About 68% of the records lie in the range of $-1\sigma_P$ to $+1\sigma_P$ from mean. For example, the standard deviation of 6–week body weight in chickens is 150g with mean of 1500 g. In this case, about 68% of broiler chickens are expected to fall in the range

of 1350 to 1650 g (the mean minus one standard deviation is 1350 g, and mean plus one standard deviation is 1650 g). This is illustrated in Fig.2.2b.

- About 95% of the records lie in the range of $-2\sigma_P$ to $+2\sigma_P$ from the mean (for the example of 6–week body weight considered above, we expect 95% broilers to fall between 1200 and 1800g.
- Over 99% of the records lie in the range of $-3\sigma_P$ to $+3\sigma_P$ from the mean (for the example of body weight, we expect 99% broilers to fall between 1050 and 1950 g.)

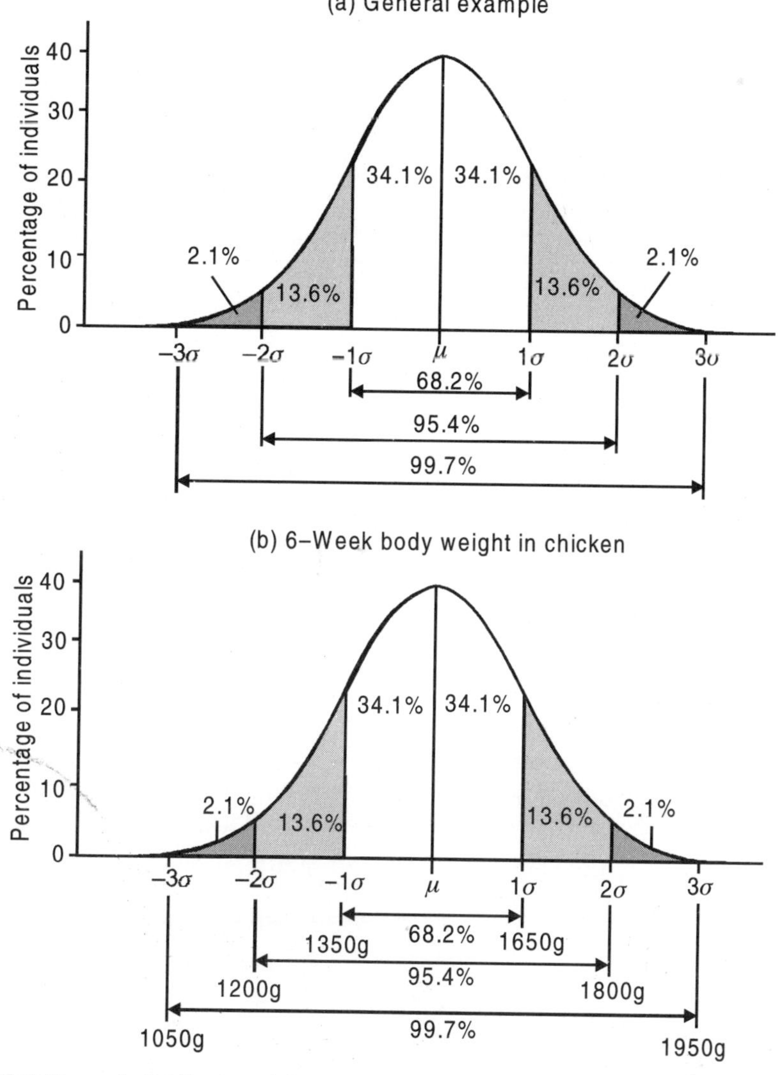

Fig 2.2 Normal distribution: (a) general example (b) example of six–week body weight in chickens with mean of 1500g and standard deviation of 150g.

Note: The curve of normal distribution, as shown in the plot above, is an idealized one. For most of the traits in farm animal populations, the plot is likely to be less regular, particularly when the number of individuals measured is small.

Normal distribution is the most important distribution encountered in biological data. Its importance, in particular in animal breeding, arises from the fact that:

- Most of the economically important traits in farm animals have a normal distribution.
- Many other distributions (like Binomial and Poisson) tend towards normal distribution as the sample size increases

Its main feature of general utility arises from the **central limit theorem.** This theorem states that the distribution of a sample drawn from a given distribution (with mean = u and variance = σ^2) tends towards normal distribution with mean = u and variance = σ^2/n as the value of 'n' approaches infinity. In more simple terms it means that given a sufficiently large sample size drawn from a population (with a finite level of variance where mean of all samples from the same population will be approximately equal to the mean of the population) all of the samples will follow an approximate normal distribution, regardless of the underlying distribution.

Standard Normal Distribution : Normal distribution is completely determined by two parameters: mean (u) and standard deviation(σ). This means that a different normal distribution is specified for each different value of u and σ. In other words, one normal distribution is distinguished from another normal distribution on the basis of the values of u and σ. In order to alleviate the difficulty of a large number of distributions, the curve is standardized in such a way that it has a mean of zero and a standard deviation of one; such a curve/distribution is called **standard normal distribution/curve**. A standard normal variate, Z, is obtained as follow:

$$Z = \frac{X_i - u}{\sigma}$$

For any value of X_i from a standard normal population with mean u, and standard deviation σ, the value of 'Z' tells us "how many standard deviations from mean the X_i value is located". This feature of standard normal curve is used in animal breeding to work out selection intensity when individuals above a particular value are selected (under truncation selection).

Covariation

Covariation refers to as "how the values recorded on the same animals for two traits vary together in a population/sample". Covariation which refers to the joint distribution of two traits/variables, measured as **'covariance'**, quantifies the degree to which the values for the two traits vary together in the sample/population. Covariance between two traits/variables in a sample, denoted by 'Cov (x,y)' or σ_{xy}, is calculated as follows:

$$\text{Cov (X,Y) or } \sigma_{XY} = \frac{\Sigma XY - \frac{(\Sigma X \Sigma Y)}{N}}{N-1}$$

The numerator of the covariance formula is the sum of cross products for the two traits/variables. Each individual cross product can be positive, zero or negative. Therefore, the sign of the sum of cross products, and of covariance can be negative, zero or positive.

Covariance carries the units of both the variables. Technically, the computation of covariance is similar to that of variance. The variance of 'X' is equivalent to covariance of 'X' with 'X' (that is covariance of 'X' with itself)

The nature of covariation between two traits can be better visualized by plotting the two traits/ variables on Y– and X–axis, and with vertical and horizontal lines drawn at the mean value of each trait. Fig 2.3 provides an example of back fat thickness and growth rate in swine. In such a case, the graph can be divided into four areas (quadrants) as shown in Fig 2.3 a,b,c. The scatter of points in four quadrants indicates the direction of association between the two traits:

- If majority of the data points lie in the bottom–left and top–right quadrants, then the trait on Y–axis increases with the increase in the value of the trait on the X–axis, and the traits are said to be positively correlated (Fig 2.3a).
- If majority of the data points lie in the top–left and the bottom–right quadrants, then the trait on Y–axis decreases as the trait on X–axis increases, and the traits are said to be negatively correlated (Fig 2.3b).
- If the data points are equally distributed in the four quadrants, the variables are said to be unrelated. (Fig 2.3c)

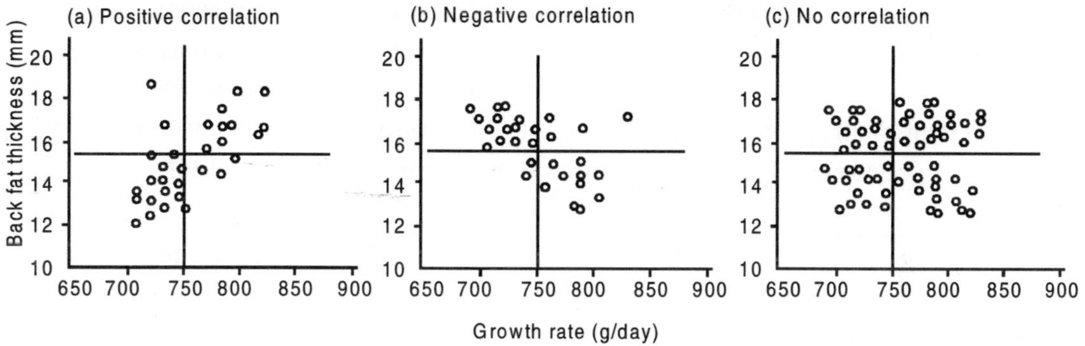

Fig. 2.3. Illustration of : (a) positive correlation (b) negative correlation and (c) no correlation.

Covariance is useful from statistical standpoint but it has little conceptual appeal as the numerical value of covariance is dependent on units of measurements of both the traits. For example, if covariance between body weight (kg) and egg weight (g) is 32 kg × g, the value of '32' by itself does not reveal much about the association except that the direction of association between the two traits is positive.

On the other hand, correlation and regression coefficients allow a better visualization of the degree of association between the two traits.

Correlation

Correlation coefficient, symbolized as 'r' or 'r_{xy}', is a measure of degree of association between two traits. The correlation between two traits/variables, which are expressed on pair-basis, is calculated as:

$$r_{xy} = \frac{\text{Cov}(x, y)}{\sqrt{V_X \times V_Y}} = \frac{\Sigma XY - \dfrac{\Sigma X \Sigma Y}{N}}{\sqrt{\left(\Sigma X^2 - \dfrac{(\Sigma X)^2}{N}\right)\left(\Sigma Y^2 - \dfrac{(\Sigma Y)^2}{N}\right)}}$$

The correlation coefficient is thus covariance divided by the geometric mean of variances of the two traits (i.e. covariance per unit of average variance). The type of the correlation discussed above is more specifically called **linear correlation coefficient** as it measures the linear relationship between two variables). The correlation coefficient is easily interpretable as:

- It is independent of the variability in the two traits (as it has been 'standardized').
- It is unitless as the units in numerator (of covariance) get cancelled with the units in the denominator (variances)

Since variances are always positive, the sign of covariance determines the sign of correlation coefficient. Some useful properties of correlation coefficient are:

- Correlation coefficient is mathematically bound to vary between –1.0 (perfect negative correlation) to +1.0 (perfect positive correlation).
- Correlation and variances are related as they represent two different ways of examining the variation. Square of the correlation coefficient, r^2, is the proportion of variation that is common to the two variables/traits. In other words, r^2 is the proportion of the variation in one variable/trait that is accounted for by its linear association with the other variable/trait.
- Correlation is a valid measure of only the linear association between traits. If the relationship between variables is not linear (e.g. curvilinear), the correlation coefficient is not appropriate to represent the relationship. When the linear correlation coefficient is zero or close to it, one may have to examine for the existence of non–linear type of correlation.
- Existence of correlation between two traits (variables) does not imply or assume any "cause and effect" relationship between them. (i.e. the correlation coefficient does not tell as to which variable is the 'cause' and which is the 'effect'). It only quantifies the observed association between the two. Evidence about the cause and effect comes from the study of the biological relationship between the traits/variables involved.

Regression

Regression coefficient is another measure that is commonly used to study the linear association between two traits, particularly when the relationship involves 'cause and effect'.In such a case, one is a dependent variable(Y) and the other an independent variable(X).For example, the amount of feed consumed by an animal is dependent on its body weight. Therefore, we call feed consumed as dependent and body weight as independent variable. Like correlation coefficient, it is more appropriately called simple linear regression coefficient.

Regression coefficient is a measure of the amount of change in the dependent variable per unit change in the independent variable (trait). It is most commonly used for predicting the value of one variable/trait for a given the value of the second variable/trait. Regression of trait 'Y' on trait 'X', designated as $b_{YX,}$ is calculated as:

$$b_{YX} = \frac{\text{Cov(x,y)}}{V_X} = \frac{\sigma_{XY}}{\sigma_x^2} = \frac{\Sigma XY - \dfrac{(\Sigma X \Sigma Y)}{N}}{\Sigma X^2 - \dfrac{(\Sigma X)^2}{N}}$$

Numerator of the expression carries the units of both the 'Y' and 'X' traits/variables while denominator carries the squared unit of trait 'X' only. The regression coefficient, therefore, carries the unit of 'Y' per unit of 'X'.Regression coefficient can be negative, zero, or positive, depending upon the sign of covariance.

When data on two variables/traits are plotted on a graph paper, with Y on vertical- and X on horizontal-axis, the relationship between two traits can be represented by drawing a line, called the **regression line**. The regression coefficient represents the slope (gradient) of the linear regression line.

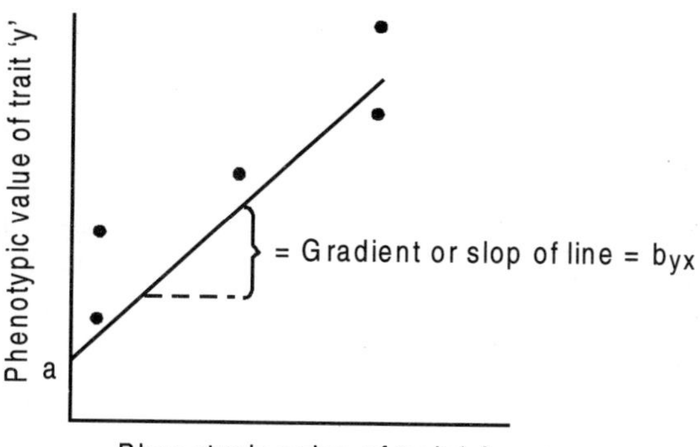

Fig 2.4. Linear regression of 'Y' on 'X' (when b_{yx} is positive)

The linear regression equation (line) has the following mathematical form:

$$Y_i = a + b_{yx} X_i \quad ...(2.1)$$

Where:

Y_i is the predicted value of Y for the 'i'th individual

X_i is the measured value of X for the 'i'th individual

b_{yx} is the regression coefficient of Y on X (slope of regression line)

'a' is the intercept (constant) which refers to the value of Y when the value of variable X equals zero.

The regression line, when plotted on a graph, passes through the means of both the traits (i.e. through $\overline{Y}$ and $\overline{X}$). Therefore, the estimate of intercept, 'a', is obtained as follow:

$$a = \overline{Y} - (b_{YX} \times \overline{X}) \quad ...(2.2)$$

Putting the value of 'a' (as in 2.2) in equation 2.1:

$$Y_i = (\overline{Y} - b_{YX} \times \overline{X}) + (b_{YX} \times X_i)$$

Or

$$Y_i = [\overline{Y} + b_{YX}(X_i - \overline{X})] \quad ...(2.3)$$

Equations 2.1 and 2.3 can be interchangeably used for predicting the value of 'Y' for a given value of 'X'. However, the prediction of 'Y' from 'X' is strictly valid only for the range of values of the Y and X studied.

One of the most important uses of regression in animal breeding is the prediction of one variable from the other variable: the variable to be predicted is usually the breeding value (A) while the variable from which it is predicted is the phenotypic value (P).

Correlation versus Regression

The sign of the correlation and the regression coefficients is same since numerator of both 'r' and 'b_{YX}' is the covariance, which carries the sign (positive or negative). Correlation and regression coefficients are mathematically related. Algebraically:

$$b_{YX} = r_{XY} \times \frac{\sigma_Y}{\sigma_X}$$

One of the two statistics can thus be computed as a function of the other provided the value of σ_y and σ_x are known. It can be seen that the correlation and regression coefficients would be equal, if the variances of the two variables are equal ($\sigma_X^2 = \sigma_Y^2$). Salient points of comparison between the correlation and regression coefficients are listed in Table 2.6.

Table 2.6 Correlation versus regression coefficient.

Correlation	Regression
1. Measure of degree of association between two traits.	1. Measure of change in one trait per unit change in the other.
2. Pure number with no unit of measurement.	2. Carries the actual unit of measurement of dependent variable per unit of independent variable.
3. Mathematically varies between –1.0 to +1.0.	3. Can take any numerical value (negative to positive).
4. No cause and effect relationship between the variables is assumed, and 'r' is bidirectional: $r_{xy} = r_{yx}$.	4. Cause and effect relation (dependent and independent) is assumed (conventionally 'Y' is designated the 'effect' and X the 'cause') and $b_{xy} \neq b_{yx}$.
5. Correlation is a general–purpose statistic that is useful to quantify the linear association between two variables.	5. Regression is used specifically to quantify the effect of one variable (X) on the other (Y), and to predict the numerical value of one variable from the value of the second variable.
6. Used when cause and effect relationships are not known or are not of interest.	6. Used where cause and effect relation is known.

A summary of notations, units and ranges for the different statistical measures used for description of economic traits is given in Table 2.7.

Table 2.7 Notations, units and numerical values for various statistics using back-fat thickness, Y (mm), and body weight, X (Kg) in swine.

Statistic	Notation	Unit for variable :		Numerical value
		Y	X	
Mean	X or ī	mm	kg	Any value: +ve (but for some variables it may can have a negative value, e.g. below freezing temperature)
Variance	σ^2 or s^2	mm^2	kg^2	Any value ; always +ve
Standard deviation	σ or s	mm	kg	Any value ; always +ve
Coefficient of variation	C.V.	%	%	Any value between (0–100), taken +ve
Covariance	Cov (x,y)	mm x kg		Any value: + ve or – ve
Correlation	r_{xy}	Unitless		–1.0 to +1.0
Regression	b_{YX}	mm per kg		Any numerical value: + ve or –ve

Analysis of Variance

Analysis of variance, introduced by R.A. Fisher, is a statistical technique of partitioning the observed variation into its components (i.e. into identified causes). For example, the total observed variation among a number of animals, which are categorized into groups (like treatments), can be partitioned into:

- Between–group variation (i.e. average variation between the groups)
- Within–group variation (i.e. average variation within the group)

The method (abbreviated as 'ANOVA') is extensively used in animal breeding for partitioning the total observed variation (total phenotypic variance) into genetic and environmental components. The use of 'ANOVA' for partitioning the total variation into its components (i.e. estimation of variance components involving random effects), and then for estimation of genetic parameters is discussed in Chapter 3. Other uses of ANOVA include estimation of fixed effects, and testing of hypothesis about sources of variation. Though there are different uses of 'ANOVA' but operationally it makes little difference since the basic computations remain same.

Intra–class Correlation

Sometimes a correlation coefficient is desired among members of a group for a single trait . In such a situation the traditional correlation coefficient (i.e. the product–moment correlation coefficient) cannot be computed as the values/ observations are not paired. For example, when the same trait is recorded on members of a family (i.e. full or half sib family), the product–moment correlation cannot be calculated. In such a situation, intra-class correlation is used to describe how closely members in the same group resemble each other.

Intra-class correlation, unlike product moment correlation, is obtained from the calculation of variance 'between–classes' (σ_B^2), and 'within–classes' (σ_W^2), as follows:

$$\text{Intra-class correlation:} \quad t = \frac{\text{Between-class component of variance}}{\text{(Between-class variance + Within-class variance)}}$$

$$t = \frac{\sigma_B^2}{\sigma_B^2 + \sigma_W^2}$$

Estimates of the between– and within–class components of variance are obtained from ANOVA. In animal breeding, intra–class correlation is used for estimation of genetic parameters viz; heritability and repeatability. It may be noted that intra–class correlation is viewed as a type of correlation but unlike most other types of correlations it operates on data structured as groups, rather than data structured as paired observations. In biological studies it is used for assessment of consistency or reproducibility of quantitative measurements made by different observers measuring the same quantity.

Review Questions

1. Define/explain the following with examples wherever necessary :

   (a) Trait
   (b) Phenotype
   (c) Threshold traits
   (d) Component traits
   (e) Economic traits
   (f) Descriptive statistics/measures
   (g) Random sample
   (h) Variable

(i) Variance
(j) Standard deviation
(k) Coefficient of variation
(l) Normal distribution
(m) Standard normal distribution
(n) Correlation coefficient
(o) Regression coefficient
(p) Covariation
(q) Regression prediction equation
(r) Analysis of variance
(s) Intra-class correlation

2. Differentiate between the following :
   (a) Quantitative and qualitative traits
   (b) Population and sample
   (c) Parameter and statistic
   (d) Correlation and regression coefficients
   (e) Trait and phenotype
   (f) Discrete and continuous variables
   (g) Product moment correlation and intra class correlation
3. List the salient features alongwith examples of the threshold traits.
4. List at least two examples of Mendelian traits of economic/aesthetic importance for each of the following species: cattle, sheep, poultry, horse and swine.
5. What are the four general categories of economic traits in farm animals? Explain with examples.
6. What are the two broad ways of measurement of economic traits? Which one the two is most appropriate in animal breeding? Explain.
7. Why is the development of accurate methods of recording of economic traits a continuous process? Explain with an example.
8. Which general category of traits have highest and which the lowest economic value?
9. List and define at least two economic traits of importance for cattle/buffaloes, swine, sheep, chickens, horse, and fish for each the following categories: production, growth, reproductive and quality.
10. What are different types of means/averages? Which of these is most commonly used for description of quantitative traits ?
11. List and define with appropriate formula the various measures of variation.
12. What is the unit in which standard deviation is expressed?
13. Name the most common type of distribution exhibited by economic traits. What are its importance features?
14. How can the distribution of a quantitative trait be graphically illustrated?
15. Why do most economic traits of farm animals exhibit normal distribution?
16. What statistical measures are used to describe a quantitative trait?
17. What is meant by covariation? How is it measured? What are its units?
18. How can the covariation between two traits be depicted diagrammatically?
19. What purpose does the correlation coefficient serve over covariance?

20. Does the existence correlation between two traits means the cause and effect relationship between them? Explain.
21. What does the square of correlation coefficient between two traits implies?
22. What numerical values can the correlation coefficient assume?
23. What information is conveyed by the correlation coefficient?
24. What is a regression coefficient? What are its uses in animal breeding?
25. What information is conveyed by the regression coefficient?
26. Explain the meaning of the terms dependent variable and independent variable?
27. Diagrammatically show the relationship between a dependant and an independent variable in terms of regression coefficient and regression line.
28. Describe the elements of a simple (linear) prediction equation.
29. What is the mathematical relationship between the correlation and regression coefficients?
30. What is intra–class correlation coefficient? What is its use in animal breeding?
31. Which of the following parameter(s) always have(has) a positive value :
    (a) Mean ....
    (b) Standard deviation ....
    (c) Coefficient of variation ....
    (d) Regression coefficient ....
32. What is analysis of variance? What is its most important use in animal breeding?
33. Why do breeders of livestock species attempt to reduce the environmental variation by providing uniform environmental conditions and through adjustment of performance records?
34. List some of common types of non–genetic factors which influence the performance of farm animals?
35. Why are performance records adjusted for the effect of non–genetic factors? Give three examples of adjustment of records in dairy cattle.
36. For what factors are the milk yield records commonly adjusted?
37. What are the two general methods used for adjustment of performance records? Explain with suitable examples.
38. What are the general limitations of adjustment factors?
39. What is meant by contemporaries? How can these be used for reducing the variation caused by known non–genetic factors?
40. What are the merits and demerits of the use of contemporaries for adjustment of performance records?

3

Genetic Parameters

Quantitative traits being governed by several genes and modified by environmental factors, exhibit continuous variation. Consequently, individuals in a population can not be classified into distinct phenotypic classes because the differences between consecutive classes are very small. Such traits can not be studied by Mendelian methods of analysis. Phenotype of an individual for a quantitative trait is represented by a numerical value (referred to as phenotypic value); therefore, statistical methods that are appropriate for continuous variation are employed for their description and analysis (e.g. mean, variance, correlations, regression, analysis of variance etc.). The parameters which are used for describing the variation and covariation in quantitative traits are collectively referred to as **genetic parameters.** These are heritability, repeatability, and genetic and phenotypic correlations.

This chapter deals with the concept, importance and methods of estimation of the various genetic parameters.

Heritability

Individuals within a population usually differ from one another for quantitative traits. These phenotypic differences are caused by genetic as well as environmental effects. The total phenotypic variation (σ_P^2) can be partitioned into genetic (σ_G^2) and environmental (σ_E^2) components:

$$\sigma_P^2 = \sigma_G^2 + \sigma_E^2$$

The relative role of heredity (genetic make-up) and environment to the phenotypic variation of a trait is studied by the **coefficient of heritability** or simply **heritability**. Heritability is the ratio of genetic variance to phenotypic variance, and is mathematically expressed as:

$$\text{Heritability} = \frac{\sigma_G^2}{\sigma_P^2} = \frac{\sigma_G^2}{\sigma_G^2 + \sigma_E^2}$$

Unit and Range

Both the genetic and phenotypic variances (for a trait) carry the same units. Heritability, being ratio of two types of variances, does not carry any unit (unitless). Since phenotypic and genetic variances can not be negative; heritability always has a positive value. If all of the phenotypic variability for a trait is genetic in nature (i.e. environmental factors do not contribute to the observed variability: $\sigma^2_E = 0$), then heritability equals one ($\sigma^2_G / \sigma^2_G = 1$). This holds true for most Mendelian (single-gene) traits (e.g. blood types, coat color etc.) as they are not affected by environmental factors. On the other hand, if all the individuals of a population have identical genetic make–up (as is true for a genetically homozygous population), then whole of phenotypic variability would be environmental in nature (i.e. there would be no genetic variability: $\sigma^2_G = 0$), and heritability would be zero. Theoretically, the true heritability of a trait is mathematically bound to vary between zero and one:

$$\text{Heritability} : \geq 0 \text{ and } \leq 1$$

In general, however, heritability of quantitative traits is always greater than zero but less than one because phenotypic variation for such traits in most populations is caused by both genetic and environmental factors. Heritability can also be expressed on a percentage scale (zero to 100%).

Two Ways of Expression of Heritability

There are two ways of expressing the heritability (as pointed out by Lush, 1937).

Heritability in broad–sense (symbolized as H^2 or $h^2_{B.S.}$): Heritability in the broad – sense is the ratio of genetic variance to total phenotypic variance.

$$H^2 = \frac{\sigma^2_G}{\sigma^2_P} = \frac{\sigma^2_G}{\sigma^2_G + \sigma^2_E}$$

Stated in words, broad-sense heritability is the proportion of total phenotypic variance that is caused by genetic differences among individuals in a population. Since genetic variance is made-up of additive, dominance and epistatic components (i.e. $\sigma^2_G = \sigma^2_A + \sigma^2_D + \sigma^2_I$), heritability in the broad–sense can alternately be expressed as:

$$H^2 = \frac{\sigma^2_A + \sigma^2_D + \sigma^2_I}{\sigma^2_A + \sigma^2_D + \sigma^2_I + \sigma^2_E}$$

The broad-sense heritability measures the influence of all the genetic effects (additive and non– additive), and is hence useful in studying the relative influence of heredity and environment to the phenotypic variation of a trait.

Genotypes as such are not transmitted from parents to the progeny; therefore, any superiority or inferiority of an individual due to dominance and epistatic effects is broken down at the time of gamete formation. Consequently, the broad–sense heritability is not useful in studying the degree to which differences among parents in a population are transmitted to their offspring. Broad–sense heritability thus has a limited practical use as it can not be used for predicting the performance of progeny generation. The performance of progeny generation can be predicted with the knowledge of heritability in the narrow–sense

Heritability in narrow-sense (symbolized as h^2 or $h^2_{N.S.}$): Breeders are usually concerned with the portion of phenotypic variation that is expected to be recovered in the progeny generation. Of the total genetic variance, it is only the additive genetic fraction that is transmitted to offspring because dominance and epistatic effects are not stably transmitted from parental to progeny generation. Heritability in narrow–sense is the ratio of additive genetic variance to total phenotypic variance:

$$h^2 = \frac{\sigma_A^2}{\sigma_P^2} = \frac{\sigma_A^2}{\sigma_A^2 + \sigma_D^2 + \sigma_I^2 + \sigma_E^2}$$

Stated in words, narrow–sense heritability is the proportion of total phenotypic variance that is due to differences in breeding values of individuals in a population. For a trait with high heritability, the differences in performance among animals are largely due differences in their breeding values; and the non-additive genetic effects (dominance and epistatic) and environmental factors contribute less to the phenotypic variation. On the other hand, if a trait is lowly heritable, the differences among individuals are caused more by differences in non–additive genetic and environmental effects, and less by differences in their breeding values. A comparison of the expressions for heritability in broad-and narrow-sense is given below:

$$H^2 = \frac{\sigma_A^2 + \sigma_D^2 + \sigma_I^2}{\sigma_A^2 + \sigma_D^2 + \sigma_I^2 + \sigma_E^2} \quad ; \quad h^2 = \frac{\sigma_A^2}{\sigma_A^2 + \sigma_D^2 + \sigma_I^2 + \sigma_E^2}$$

The denominator for the expressions of both H^2 and h^2 is same but the numerator is different. The numerator for h^2 is only the additive genetic variance (σ_A^2) while for H^2 it is the total genetic variance (σ_G^2) which includes additive, dominance and epistatic variances. The broad-sense heritability is, theoretically, expected to be greater than the narrow–sense heritability. However, if the non-additive genetic variance is absent ($\sigma_D^2 + \sigma_I^2 = 0$), then heritability in broad-sense equals heritability in narrow-sense.

Both the symbols, H^2 and h^2, stand for 'heritability' and not for their squares. The exponent '2' on letters H and h (i.e. the square sign) is assigned to signify variances (rather than absolute values). Narrow-sense heritability is of more practical utility in animal breeding, therefore, any reference or statement about heritability unless otherwise specified, implies heritability in the narrow-sense.

Alternative Definitions of Heritability

In addition to the definition given above (ratio of additive genetic to the phenotypic variance), there are some alternative ways of defining h^2. These are useful in understanding the concept of h^2 and its applications:

Heritability as correlation coefficient: The correlation between breeding values (A) and phenotypic values (P), designated as r_{AP}, equals the square root of heritability :

$$r_{AP} = \sqrt{h^2} = h$$

Correlation coefficient (in statistical terms) is a measure of the degree of association between two variables. Therefore, the square root of heritability determines the degree of association between the phenotypic value and breeding value. The correlation between breeding value and phenotypic value (as defined in context of heritability) varies between zero and one (as contrasted to the general range of correlation coefficient from -1.0 to +1.0).

Alternatively, heritability (h^2) equals the square of the correlation between breeding values and phenotypic values:

$$h^2 = r_{AP}^2$$

The square of correlation coefficient between two variables, in statistical sense, is an estimate of the proportion of variation that is common to two variables. In genetic context as defined here, therefore, the square of correlation between phenotypic and breeding values (r^2_{AP}) represents the proportion of variation in phenotypic values that is accounted for by variation in breeding values.

Illustrative example: Suppose heritability of a trait is 0.6, it would mean that 36% (i.e. 0.6 × 0.6 = 0.36, expressed as percentage) of the variation in phenotypic values is accounted for by the variation in breeding values.

Heritability as a regression coefficient: Heritability (h^2) can also be defined as the regression of breeding value on phenotypic value:

$$b_{AP} = h^2$$

Heritability as a regression coefficient is a measure of the expected change in breeding values per unit change in phenotypic values. It is therefore used for prediction of breeding values from phenotypic values.

Illustrative example: Figure 3.1 depicts the regression lines for three hypothetical situations viz; a trait with perfect heritability (h^2=1); a trait with zero heritability (h^2 = 0) and a trait of 0.5 heritability (h^2 =0.5). Slope of regression line (of breeding value on phenotypic value) equals heritability of the trait. A trait with h^2=1.0, has a perfect relation between breeding value and phenotypic value (i.e. a unit change in phenotypic value in the population is associated with a unit change in breeding value). For a trait with zero h^2, a change in phenotypic value is not associated with any change in breeding value (Fig 3.1b). For heritability of 0.5, each unit change in phenotypic value is associated with 0.5 unit change in breeding value (Fig 3.1c). Stated in other words, if the h^2 of a trait is high an individual who is phenotypically above population average is likely to be above population average for its breeding value and vice versa. Conversely, if a trait has low heritability, an individual which is phenotypically above the population average may be just average for its breeding value.

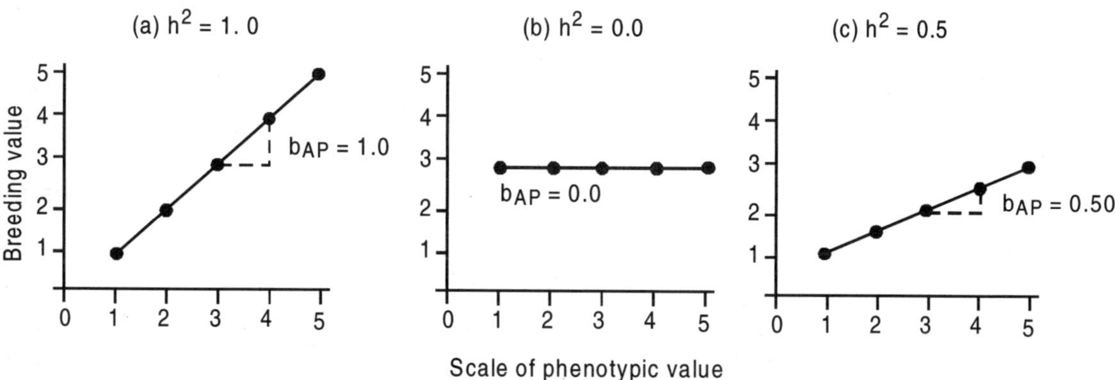

Fig. 3.1 Illustration of regression of breeding value on phenotypic value for three hypothetical characters: (a) h^2=1.0 (b) h^2=0.0(b) h^2=0.5. Note the slope of regression line equals the value of heritability

Heritability as a fraction of phenotypic superiority of parents: Heritability can also be defined as the proportion of phenotypic superiority of parents over the population average which, on the average, is passed on to its offspring. Therefore heritability measures the degree of association between the performance of parents and progeny. In this respect, heritability can be used for predicting the performance of progeny if performance of parents is known. If h^2 of a trait is high, a large fraction of superiority of parents over population average is passed on to the offspring. If heritability of trait is low, then only a small fraction of superiority of parents is passed on to its offspring.

Illustrative example: If the selected parents are superior over population average by 'X', then "Xh^2" fraction of superiority of parents over population average is likely to be transmitted to offspring. For example, h^2 of body weight in chickens is 0.4. A breeder selects birds which are better than the flock average by 100g, then we expect that "40 g" of the superiority of selected birds would be transmitted to the offspring.

Usefulness of Heritability

The concept of h^2 is central to quantitative genetics, and animal and plant breeding. The knowledge of heritability is important in a number of respects:

(i) **Determining the relative role of heredity and environment** (*Nature vs. Nurture*)**:** Broad-sense heritability measures the relative influence of heredity (nature) verses environment (nurture) to the phenotypic variation of a trait. Magnitude of broad-sense heritability is useful to both the geneticists (in particular, human geneticists) as well as to breeders of animal and plant species. There are a number of human traits (e.g. stature, intelligence, susceptibility to certain diseases like diabetes, cardiac ailments and behavioral attributes like alcoholism, depression etc.) for which the physicians and social scientists are interested to know whether ***nature*** (genotype) or ***nurture*** (environment) has more influence on their expression. The degree to which such traits are determined by 'genetic' verses 'environmental' effects is important so that preventive measures and genetic counseling strategies can be developed. In human genetics, the broad sense heritability is also referred to as **coefficient of genetic determination**.

For animal breeders, the value of H^2 indicates the potential scope for genetic improvement of a trait. A trait with high H^2 can be more easily improved by genetic means.Changes in environmental factors shall have relatively lesser influence on the modification of expression of such a trait. On the other hand, a trait having very low broad–sense heritability can be more easily improved by improvement in environment conditions (viz. management, nutrition and health care etc.). Nonetheless, it does not imply that traits with low H^2 can not be modified genetically. Rather it means that it may be more economical to improve such traits by improving the environmental conditions.

Illustrative example: H^2 of egg weight in chickens is more than 0.6 while that of viability (disease resistance) is less 0.1. This means that egg weight can be more easily improved genetically while viability can be improved easily through improvement in environmental conditions (like hygiene, management etc).

(ii) **Type of gene action v*is a vis* choice of breeding plan:** Different traits of farm animals may be influenced to different extent by additive and non–additive genetic effects, and by environmental factors. A comparison of values of broad-sense and narrow-sense heritabilities indicates the relative importance of additive and non-additive genetic variances. A high value of narrow–sense heritability indicates the importance of additive genetic variance. If additive genetic variation is more important, selection within a population **(**referred to as simply **'selection'** or **'intra-population selection'** or **'within-population' selection)** shall be more effective in genetic improvement. If

H^2 is considerably higher than h^2, it implies that non-additive genetic variance exerts relatively more influence than additive genetic variance. When non-additive genetic variance is more important, some forms of crossing (**mating systems**) that make use of non-additive genetic variance (like inbreeding, crossbreeding etc.) shall be more effective for genetic improvement. Magnitude of the two types of heritabilities thus helps in assessing the type of gene action governing a trait, and consequently in choice of the breeding plan (i.e. selection versus crossing) to be used for its genetic improvement.

(iii) **Choice of basis of selection:** Value of heritability (h^2) is required for choice of the most effective basis of selection to be used for genetic improvement of a trait. Principal objective of selection is to choose the animals having high breeding values so as to be used as parents of the next generation. However, it is only the phenotypic value (performance) that is available for selection of genetically superior animals.

The correlation between the breeding value (unobserved) and henotypic value (observed), referred to as **accuracy of selection**, equals the square root of heritability and may be represented as follows:

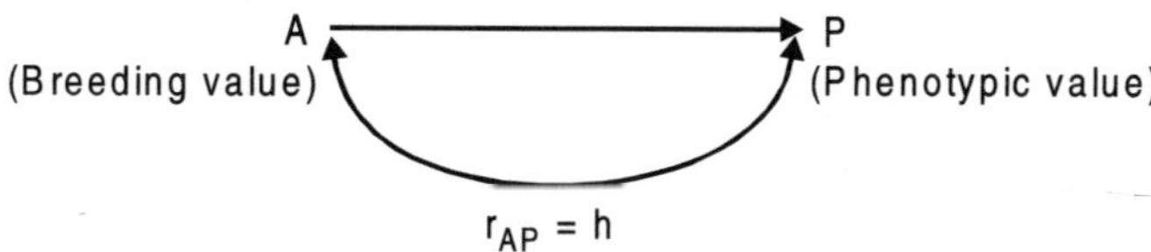

Strength of the relationship between phenotypic value and breeding value is thus directly proportional to heritability of the trait. Stated in other words, the square root of heritability is a measure of the degree of closeness of phenotypic value to the breeding value. In artificial selection programs h^2 thus determines the precision with which breeding value can be predicted from phenotypic information (performance):

- If heritability of a trait is high, phenotypic value of an animal, on the average, is a reliable indicator of its breeding value. Therefore, for a trait of high h^2, and which is measurable on the individual, an effective selection scheme would be to select individuals on the basis of their own phenotypic value. Selection of individuals on the basis of their own phenotypic value (performance), referred to as **individual or mass selection,** is one of the simplest basis of selection and is recommenced for high heritability traits.
- On the other hand, if the heritability of a trait is low, phenotype of an animal would be a poor indicator of its breeding value. Consequently, selection on the basis of animal's own performance shall be less effective. For low heritability traits, therefore, selection on the basis of performance of relatives (e.g. 'family' or 'progeny' performance) is generally more effective.

The magnitude of heritability, therefore, dictates the choice of most effective criterion of selection (**basis of selection**) to be used for genetic improvement of a trait. In general, it is easier to improve high than low heritability traits.

Illustrative example: h^2 of egg production in chickens is 0.2 while that of body weight is 0.4. Individual selection is used for improvement of body weight whereas family selection is employed for improvement of egg production.

(iv) **Prediction of genetic gain (response to selection):** Selection is the most important means for genetic improvement of a population. It involves retaining a certain number (fraction) of phenotypically superior animals from a population and using them as parents for producing the progeny generation. The magnitude of genetic improvement expected due to selection for a quantitative trait (referred to as '**genetic gain**' or '**response to selection**') is predicted with the help of heritability as follows:

Genetic gain = Heritability × Selection differential

Stated in words, the magnitude of genetic gain for a trait equals the product of heritability and the selection differential (phenotypic superiority of selected individuals over population mean). If equal selection differential can be applied for two different traits, the trait with high heritability shall produce higher genetic gain than the one with low heritability. Hence h^2 is central to predicting the response to selection. This topic is considered in detail in Chapter 5.

(v) **Prediction of breeding value:** An individual's phenotype for a quantitative trait (called phenotypic value) is composed of genotypic value and the environmental value (P= G+E). The genotypic value, in turn, is composed of breeding value (additive genetic effects) and gene combination value (non–additive genetic effects). However, these are only the additive genetic effects that are transmitted from parents to progeny, therefore, prediction of breeding value (called 'expected breeding value' EBV or 'predicted breeding value' PBV) is of prime importance to animal breeder. Breeding value of an animal (which is not known) can be predicted from its phenotypic value (which is measurable) as follows :

$$\text{Expected /Predicted breeding value: EBV / PBV} = \bar{P} + h^2\ (P_i - \bar{P})$$

Where: $\bar{P}$ = population average

P_i = phenotypic value of the individual (animal's own performance)

h^2 = heritability of the trait

The uses of heritability, as discussed above, show that h^2 is the most important genetic parameter that is useful in all aspects of selection. Most of the calculations involved in selection require the value of heritability of the trait. The concept of heritability, therefore, has become a corner stone in animal breeding because the scope for genetic improvement is directly linked to the value of heritability.

A summary of comparison of broad- and narro-sense heritability is given in Table 3.1.

Table 3.1 Broad-sense vs. narrow-sense heritability.

Broad-sense	Narrow-sense
1. Ratio of genetic to phenotypic variance : $H^2 = \sigma^2_G / \sigma^2_P$.	1. Ratio of additive genetic to phenotypic variance: $h^2 = \sigma^2_A / \sigma^2_P$.
2. Usually has a higher value than narrow-sense heritability as it includes the contribution of additive as well as non – additive genetic effects (H^2 is the upper limit of h^2).	2. Usually has a lower value than the broad–sense heritability because it includes the contribution of additive genetic effects only.
3. Useful in determining relative influence of heredity and environment to phenotypic variation.	3. Useful in choice of most effective basis of selection, and in prediction of genetic gain and breeding values.
4. Primarily of theoretical interest and has limited practical use.	4. Wide application in genetic improvement cf farm animals.

General Features of Heritability

Heritability is the ratio of (additive) genetic variance to total phenotypic variance. Hence the magnitude of heritability can vary due to changes in genetic variance, the environmental variance or due to both.

Genetic variance depends on allelic frequencies, gene effects and the type of gene action. In general, genetic variance is highest at intermediate level of gene frequencies. Environmental variance arises from differences in environmental conditions encountered by different individuals. Variable conditions of management, nutrition, hygiene etc. increase the environmental variance while uniform conditions decrease it. Thus a population raised under highly variable conditions of environment is expected to yield low heritability. The same population when raised in a carefully controlled uniform environment will yield higher heritability although the genetic structure of the population remains same.

- *Heritability may vary from one trait to another in a population*: Firstly, traits may differ for the number of genes, average effect of genes, gene frequencies and the type of gene action. Secondly, traits may also differ in their susceptibility to environmental conditions. Some traits are more susceptible to changes in environmental conditions than others. In general, traits that are more susceptible to changes in environmental conditions have low heritabilities than those which are less susceptible to changes in environmental conditions.
- *Heritability of a trait may also differ between populations*: This is because of differences in their genetic structures, the environmental conditions experienced or due to both. For example, different populations (say different strains or breeds of a species) reared under similar environmental conditions are likely to have different heritabilities for the same trait. On the other hand, the same population when measured in different environments (e.g. say two different geographic locations) may exhibit different heritabilities for the same trait.
- *Heritability of a trait may also change over time*: Heritability of a trait undergoing intense selection may change as selection, particularly if it is intense, progresses because of the changes (usually a decrease) in genetic variance.

Representative Values of Heritability

The value of heritability calculated for a given trait is strictly speaking applicable only to the specific population that has been reared under a given set of environmental conditions. However, the values of heritability of a trait obtained from different populations, reared under fairly similar environmental conditions, and based on large size of data, are usually of similar magnitude. For example h^2 of egg production obtained from different breeds/strains of chickens reared under different conditions falls between 0.15 and 0.25 while that for body weight is close to 0.4. As a general rule, heritabilities are classified as very low; low; moderate, high or very high. This categorization is only for the sake of convenience although the range of values for different categories is not clearly defined or accepted.

The following values are generally accepted for different categories:

Category:	Very low	Low	Moderate	High
Range of h^2:	< 0.10	0.10 –0.20	>0.2 to 0.4	>0.4

A number of studies over the last five to six decades have been conducted across various species of farm animals for estimation of heritability of a large number of traits. Though the estimates obtained from different studies have usually been different but certain typical ranges for heritability estimates have been observed. Tables 3.2 to 3.6 give the typical ranges of heritabilities for economic traits in different species of farm animals.

Table 3.2 Typical range of heritabilities for some economic traits of dairy cattle and buffaloes.

Traits	h^2
Milky yield:	
Per-lactation or -year	Low
Milk constituents:	
Fat % or yield	High
Protein % or yield	High
SNF% or yield	High
Reproductive:	
Age at first calving	Low
Calving interval	Very low
Conception rate	Very low
Dry period	Very low
Service period	Very low
Calving difficulty (dystocia)	Very low
Somatic cell score (a measure of mastitis)	Very low

Table 3.3 Typical range of heritabilities for some economic traits of sheep and goat.

Traits	h^2
SHEEP	
Meat –type:	
Growth traits:	
Birth weight	Moderate
Weaning weight	Moderate
Mature weight	Moderate
Age at market	Moderate
Feed efficiency	Low
Carcass traits:	
Muscling quality grade	Moderate - high
Dressing yield	Moderate - high
Loin eye area	Moderate - high
Wool type:	
Fleece traits:	
Fleece weight	Moderate
Fiber diameter	High
Staple length	High
Medullation %	Moderate
Reproduction (both types):	
Maternal ability	Very low - low
Age at 1st breeding	Very low - low
Lambing %	Very low - low
Litters/year	Very low - low
Lamb survival	Very low

GOAT	
Body weight	Moderate
Milk yield	Low
Fat %	High
Kidding rate	Very low

Table 3.4. Typical range of heritabilities for some economic traits of swine.

Traits	h^2
Growth traits:	
Weaning weight	Moderate
Market weight	Moderate
Age at market	Moderate
Feed efficiency	Low
Reproductive:	
Litter size	Very low
No. of farrowings /year	Very low
Viability Very low	
Carcass traits:	
Back fat thickness	High
Rib (loin) eye area	High
Dressing %	High
Carcass length	Very high
Leg length	Very high

Table 3.5 Typical range of heritabilities for some economic traits of poultry.

Traits	h^2
EGG TYPE:	
Egg production: part or annual	Low
Age at 1st egg	Low-moderate
Egg quality :	
Egg weight	High
Albumen quality	High
Shell thickness	Moderate
Blood and meat spots	Low
MEAT TYPE :	
Growth traits:	
Body weight (6 to 8 wks)	Moderate
Feed efficiency	Moderate
Carcass traits:	
Dressing %	Moderate
Breast angle (conformation)	High
Reproduction & viability (both types):	
Hatchability (%)	Very low
Fertility (%)	Very low
Mortality (%)	Very low

Table 3.6 Typical range of heritabilities for some economic traits of horse.

Traits	h^2
Performance traits:	
Racing/ walking speed	Low
Time to run a mile	Low
Time to trot one mile	Low
Pulling power	Low
Reproduction:	
Foaling rate	Very low
Conformation:	
Wither height	High
Canon bone circumference	Very high

An examination of Table 3.2 to 3.6 shows that there is a general connection between the nature of the trait and the magnitude of heritability across all the species as follows:

Type of trait	Range of heritability
Reproduction and viability	Very low to low
Production including growth	Low to moderate
Composition & quality	High
Morphological and skeletal measurements	Very high

It's evident that the traits closely related with reproduction and survival (fitness traits) have low heritabilities. Production traits (e.g. milk production, growth rate, wool yield, egg production, racing performance etc.) tend to have low to moderate heritabilities while quality and composition traits are highly heritable.

Heritability estimates have also been obtained for a very number of human traits. Most of these estimates pertain to heritability in the broad sense. Values for a few traits are listed in Table. 3.7. These should, however, be considered as broad guides only.

Table 3.7 Estimates of broad sense heritabilities for certain human traits.

Trait	Heritability
Stature	$\geq$ 0.90
Schizophrenia	0.80
Diabetes mellitus:	
Late onset	0.35
Early onset	0.70
Asthma	0.80
Congenital heart disease	0.35
Peptic ulcer	0.35
Depression	0.40

Commonly Held Misconceptions about Heritability

The concept of heritability is sometimes misinterpreted /misunderstood by beginners. The most commonly held misconceptions are:

- *Heritability of a trait is concerned with differences among individuals and not with their absolute values.* In other words, heritability tells us what proportion of the phenotypic differences is due to genetic differences. It does not tell us what portion of an individual's phenotypic value is due to its genotype and what portion is due to the environment experienced by it. For example, if the h^2 of body weight in the flock of broiler chickens is 0.5 (or 50%) and the average body weight of the flock is 1.0 kg. Heritability value of 0.5 does not mean that 50% of body weight (i.e. 500g) is due to heredity (genetic make-up) and the remaining 500g due to environment. It means that of all the observed differences among birds in that flock for body weight, approximately 50% is due differences in their genetic make-up and the remaining 50% due to environmental differences.
- *Heritability is a population measure (parameter) and not the property of an individual:* Heritability refers to a group of individuals (i.e. a population), and is not a value to be associated with an individual of the population. We state, for example, that heritability of body weight in a flock of chickens is 0.50 and heritability of milk production in a herd of cows of 0.15. It would be incorrect, for example, to say that heritability of a specific cow is low or high. This is because a single individual can not have any variance and hence the concept of heritability, like variances, is applicable to a group of individuals.

Methods of Estimation of Heritability

Heritability can be estimated by two broad approaches:

- Phenotypic resemblance among relatives
- Selection experiments

Of these two approaches, the phenotypic resemblance among relatives is most commonly used. This approach is based on the principle of resemblance among relatives (i.e. the individuals who are related to each other due to common ancestry): relatives resemble each other more than do the unrelated individuals (i.e. non-relatives).

The chief cause of resemblance among relatives (except identical twins) is the additive genetic variance (σ^2_A). The degree of resemblance among relatives, measured as covariance between relatives, provides an estimate of the additive genetic variance (Table 3.8). Heritability can thus be obtained from an estimate of either the regression of one relative on the other (b) or by computing the correlation coefficient between relatives (t). The composition of regression (b) or the correlation (t) in terms of h^2 or H^2 is also given in Table 3.8.

Theoretically, any types of relatives can be used for estimation of heritability but the following sets of relatives are frequently available in farm animals and humans:

- Twins (identical and fraternal)
- Offspring-parent
- Half-sibs (individuals with one common parent)
- Full-sibs (individuals with both parents common)

Table 3.8 Composition of covariance between relatives.

Relatives	Composition of covariance (a measure of resemblance)	't'(correlation coefficient) or 'b'(regression coefficient)
Monozygotic twins :	Cov (MZ) $= \sigma_G^2 = \sigma_A^2 + \sigma_D^2 + \sigma_I^2$	$t = \frac{\text{Cov (MZ)}}{\sigma_P^2} = H^2$
Offspring-parent:	Cov (O,P) $= \frac{1}{2}\sigma_A^2 + \frac{1}{4}\sigma_I^2$	$b_{OP} = \frac{Cov_{OP}}{\sigma_P^2} = \frac{\frac{1}{2}\sigma_A^2 + \frac{1}{4}\sigma_I^2}{\sigma_P^2}$ $= \frac{1}{2}h^2 + \frac{1}{4}$ epistatic effects
Full sibs*:	Cov (FS) $= \frac{1}{2}\sigma_A^2 + \frac{1}{4}\sigma_D^2 + \frac{1}{4}\sigma_I^2$	$t = \frac{Cov_{FS}}{\sigma_P^2} = \frac{\frac{1}{2}\sigma_A^2 + \frac{1}{4}\sigma_D^2 + \frac{1}{4}\sigma_I^2}{\sigma_P^2}$ $= \frac{1}{2}h^2$ + dominance & epistatic
Half sibs*:	Cov (HS) $= \frac{1}{4}\sigma_A^2 + \frac{1}{16}\sigma_I^2$	$t = \frac{Cov_{HS}}{\sigma_P^2} = \frac{\frac{1}{4}V_A + \frac{1}{16}V_I}{\sigma_P^2}$ $= \frac{1}{4}h^2 + \frac{1}{16}$ epistatic

*Full sibs: individuals having both parents common;*half sibs: individuals having one parent common.

Twin Analysis

Identical twins (one-egg or monozygotic) have exactly the same genetic make-up as they result from single fertilization. If identical twins are reared apart in different environments, then differences between such twins are environmental in nature. The correlation coefficient between the phenotypic values (performance) of identical twins is an estimate of the broad-sense heritability:

$$\text{Correlation between identical twins} = H^2$$

The essential requirement for estimation of heritability from identical twins is that they must be separated from each other early in life such that one member of the pair is reared in one environment while the other is reared in a different environment. Heritability for many traits in humans (e.g. height, weight, I.Q etc) has mostly been obtained from twin analysis. Similarly, if identical and fraternal (dizygous or dizygotic) twins are simultaneously available, heritability can be obtained from the comparison of correlation between identical and fraternal twins. The use of both types of twins simultaneously for heritability estimation is more complicated and beyond the scope of this book.

Limitations:

- Frequency of twining (monozygotic) in most species of farm animals is very low such that adequate numbers of twin pairs are not available. For example, frequency of twining in various breeds of cattle varies between one and four percent and further only about 10% of the same-sex twins are monozygous (identical). Therefore, considerable expenses are involved in diagnosing identical twins, and in accumulating the data on the required number of twin pairs.
- Twin analysis provides an estimate of the broad-sense heritability, therefore, the heritability estimated by this method can not be used for prediction of genetic gain and breeding values.

- Identical twins, in addition to genetic similarity, share the same intra-uterine environment and may also be exposed to same environment. The covariance between identical twins may thus also include the contribution of common environmental (maternal and neonatal) effects. Therefore, for traits influenced by maternal and neonatal effects (e.g. birth weight, weaning weight etc.), identical twins resemble more than expected due to their common genes. Consequently, the correlation between identical twins may overestimate the broad-sense heritability. The heritability obtained by this method is thus higher than that obtained by other methods.

Regression of Offspring on Parent(s)

The relationship between the offspring and parent(s) can be measured as either the correlation coefficient (r_{op}) or the regression of offspring on parent(s) (b_{op}). Both the correlation and the regression coefficients provide identical estimates of heritability if the phenotypic variance among parents is same as that among offspring (i.e. $\sigma^2_{Offspring} = \sigma^2_{Parents}$).This condition is fairly valid in humans. In livestock species this condition is usually not fulfilled because the parents are usually a selected group from a population undergoing selection. Phenotypic variance in parental group is lesser than phenotypic variance in offspring. When the variances between parental and offspring groups are unequal ($\sigma^2_{Offspring} \neq \sigma^2_{Parents}$), the regression of offspring on parent (s) provides a more valid estimate of heritability than based on correlation coefficient. This is due to the reason that regression coefficient is not affected by selection practiced in parents because selection in parents reduces the covariance to the same extent as it reduces the variance among parents. Offspring-parent(s) regression can be obtained using the performance of one parent or average performance of both the parents:

- *Regression of offspring (O) on one Parent (P)*: Regression of offspring on one parent (either on sire or dam) measures half of the narrow-sense heritability. Heritability thus equals twice the regression of offspring on one parent: $h^2 = 2b_{OP}$
- *Regression of offspring (O) on mid-parent ($\bar{P}$):* Regression of offspring on average performance of parents (mid-parent value), measures the narrow-sense heritability: $h^2 = b_{O\bar{P}}$ where 'O' refers to performance of offspring and '$\bar{P}$' refers to the average performance of the two parents.

Depending upon the availability of performance records of offspring and parent, heritability can be estimated from either of these two types of regressions. For example, if the trait is sex-limited (e.g. egg production, milk production etc.), the regression of daughters on dams can be used to calculate the heritability. On the other hand, if a trait can be measured in both parents (e.g. body weights, wool traits etc), the regression of offspring on the average of the two parents can be used. Similarly, if the trait is expressed in males only, (e.g. semen traits) the regression of sons on sires can be used for h^2 estimation.

Advantages:

- It provides a true estimate of narrow-sense heritability because the covariance between offspring and parents is primarily due to additive genetic variance only.
- The offspring-parent relationship is free of environmental covariance as the parents and offspring are measured over different time periods.

Limitations:

- The use of offspring-parent regression requires that performance records of parents(s) are available. In populations where parental performance is not available, this method cannot applied.

- Maternal effects, if present, may increase the covariance between offspring and dam. The heritability calculated from offspring-dam regression shall be overestimated if the trait is greatly influenced by maternal effects. Birth weight is one trait that is greatly influenced by the weight of the mother. The traits which are measured early in the life of an individual (young) are more influenced by maternal effects than those expressed late in life.

An appreciation of the use of offspring-parent regression in estimation of heritability can be gained by plotting the phenotypic values of offspring against the average phenotypic values of their parents (mid–parent value) as shown in Fig. 3.2.

Slope of the regression line (b_{OP}) is an estimate of narrow-sense heritability. Thus if, the phenotype of the offspring lies exactly midway between the phenotype of parents, it implies that all the variation among offspring groups is caused by variation in mid–parental values. In such a case, the slope of regression line would be unity and, therefore, h^2 would be one (Fig 3.1a). On the other hand, if there is no correspondence between the performance of offspring and their mid-parental values, the points would be randomly scattered on the graph (Fig. 3.1b). In this case, the slope of regression line would be zero (Fig 3.1b) and therefore, heritability would be zero because the variation among offspring groups is due to environment (and not due to their parents). And when there is no perfect association of offspring values with mid-parent values, then heritability will have a value more than zero but less than one(say, $h^2 = 0.5$ as shown in Fig 3.2c).

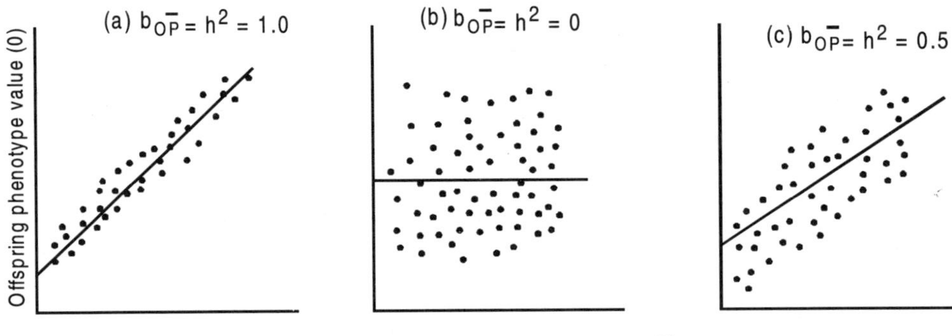

Fig. 3.2 Illustration of regression of offspring (o) on mid-parent value ($\bar{P}$).

Sib Analysis: Intra-class Correlation between Sibs (t)

The method of "Analysis of Variance" is used to partition the observed phenotypic variation into component parts, viz. genetic and environmental. In farm animals, individuals are grouped into families such that the total phenotypic variation can be separated into variation 'between families' and variation 'within families'. Two types of families are common: half-sib and full-sib. Depending upon the availability of the types of sibs, sib-analysis can be categorized as:

(i) half-sib analysis

(ii) full-sib analysis

The use of half-sib analysis for the estimation of heritability is detailed below. The method of full-sib analysis is given briefly as it is computationally more complicated and beyond the scope of the book.

Half-sib Analysis

Individuals having one common parent, referred to as half sibs, can of two types:

- Paternal half sibs –individuals having male (sire) as common parent.
- Maternal half sibs–individuals having mother (dam) as common parent.

Of the two types of sibs, paternal half sibs are more common in livestock species (e.g. cattle, buffaloes, horse etc.) because a sire can be mated to a number of dams resulting in a number of half sibs.

Mating Structure: Each sire is randomly mated with several dams and each dam in turn produces one offspring. The progeny group thus produced by a sire constitutes a paternal half -sib family. The mating-structure (family-structure) involved in the paternal half sib analysis is shown diagrammatically below:

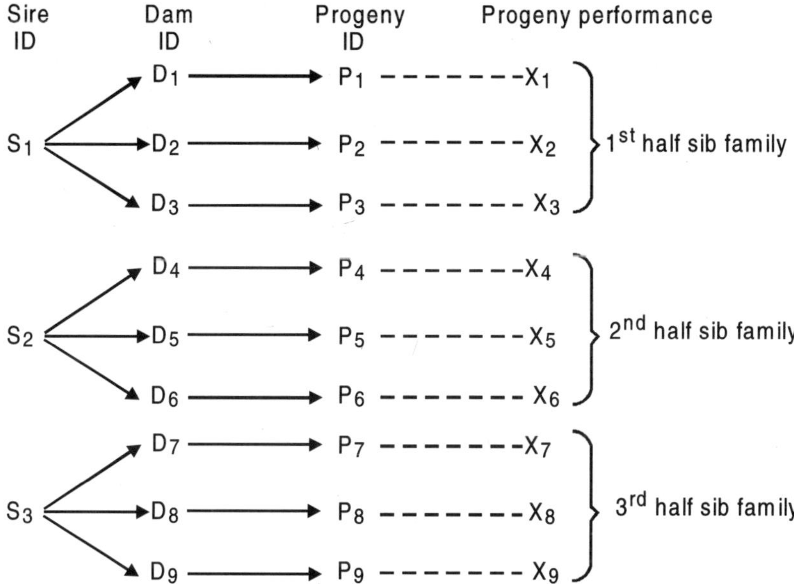

The total phenotypic variation observed among progeny is partitioned into two components:

- Variation between the progeny of different sires, referred to as 'between-sire (family)' variation
- Variation within progeny of sires, referred to as 'within-sire' (family) variation.

The partitioning of total variation involves obtaining the total sum of squares, designated as "SS_T" and its partitioning into: "between-sire sum of squares" (SS_S) and "within-sire sum of squares" (SS_W). The mean sum of squares (MS) are obtained by dividing the corresponding sum of squares by their respective degrees of freedom. The analysis of variance table used for half-sib analysis is shown below:

Source of variation	Degrees of freedom (d.f.)	Sum of squares (SS)	Mean squares (MS)	Composition of mean squares (EMS)
Between sire	S – 1	SSs	MS_s	$\sigma^2_W + K\ \sigma^2_S$
Within sires	N – S	SS_W	MS_W	σ^2_W
Total	N	SS_T		

Where: S = Total number of sires in the herd/flock

K = Number of progeny per sire (when the number of progeny varies from one sire to other, it refers to the average number of progeny per sire).

N = Total number of progeny in the herd/flock (N = S× K)

The composition of means squares (EMS) in terms of variance components is given in the right-hand column of the table of 'Analysis of Variance'. The 'sire-and within-sire' components of variance are obtained as:

Within-sire component of variance :

$$\sigma_W^2 = MS_W$$

Between-sire component of variance:

$$\sigma_S^2 = \frac{MS_S - MS_W}{K}$$

Sire-component of variance (σ_S^2) estimates ¼th of additive genetic variance:

$$\sigma_S^2 = \frac{1}{4}\sigma_A^2$$

'Within-sire component of variance' (σ_W^2) estimates the remaining genetic variance ($\frac{3}{4}\sigma_A^2 + \sigma_D^2 + \sigma_I^2$) and the environmental variance (σ_E^2):

$$\sigma_W^2 = \left[\frac{3}{4}\sigma_A^2 + \sigma_D^2 + \sigma_I^2 + \sigma_E^2\right]$$

Therefore, the total phenotypic variance equals:

$$\sigma_P^2 = \sigma_S^2 + \sigma_W^2 \text{ and } h^2 = \frac{4\sigma_S^2}{\sigma_S^2 + \sigma_W^2}$$

Alternatively: intra-class correlation between half sibs,

$$t = \frac{\sigma_S^2}{\sigma_S^2 + \sigma_W^2} = \frac{1}{4}h^2, \text{ and thus, } h^2 = 4t.$$

Advantages:

- Paternal half-sib analysis provides an unbiased estimate of heritability in the narrow sense because the resemblance between paternal half-sibs is primarily caused by additive genetic variance (σ_A^2) (Table 3.8). The heritability estimated from half sib correlation is thus free of non-additive genetic and environmental effects.
- Half-sib analysis is conducted on performance records of offspring; therefore, it does not require knowing the performance of parents. It requires only the parentage of offspring with respect to the sires.
- Most livestock populations (herds/flocks) provide data appropriate for paternal half-sib analysis as the information required for such analysis is available as a routine. Therefore, estimating heritability from such data is fairly inexpensive and cost/effective as well. Also fairly good numbers of paternal half sibs are available in farm species.

For all these reasons, paternal half-sib analysis is most widely used for estimation of heritability for most species of farm animals. Maternal half-sibs are used much less frequently than paternal half sibs mainly for two reasons:

- Number of maternal half-sibs per family is very small as most livestock species give birth to singles.
- Covariance between maternal half-sibs may be increased due to existence of maternal effects. This is due to the fact that maternal half-sibs are exposed to similar uterine and post-natal maternal environment provided by the same dam.

Full-sib Analysis

When mating of a sire to a dam produces several offspring, the progeny thus produced constitute a 'full-sib family'. Full-sib families are common in poultry and swine. The mating structure (family structure) involved in full-sib analysis where each progeny is identified with respect to dams and sires is as follow :

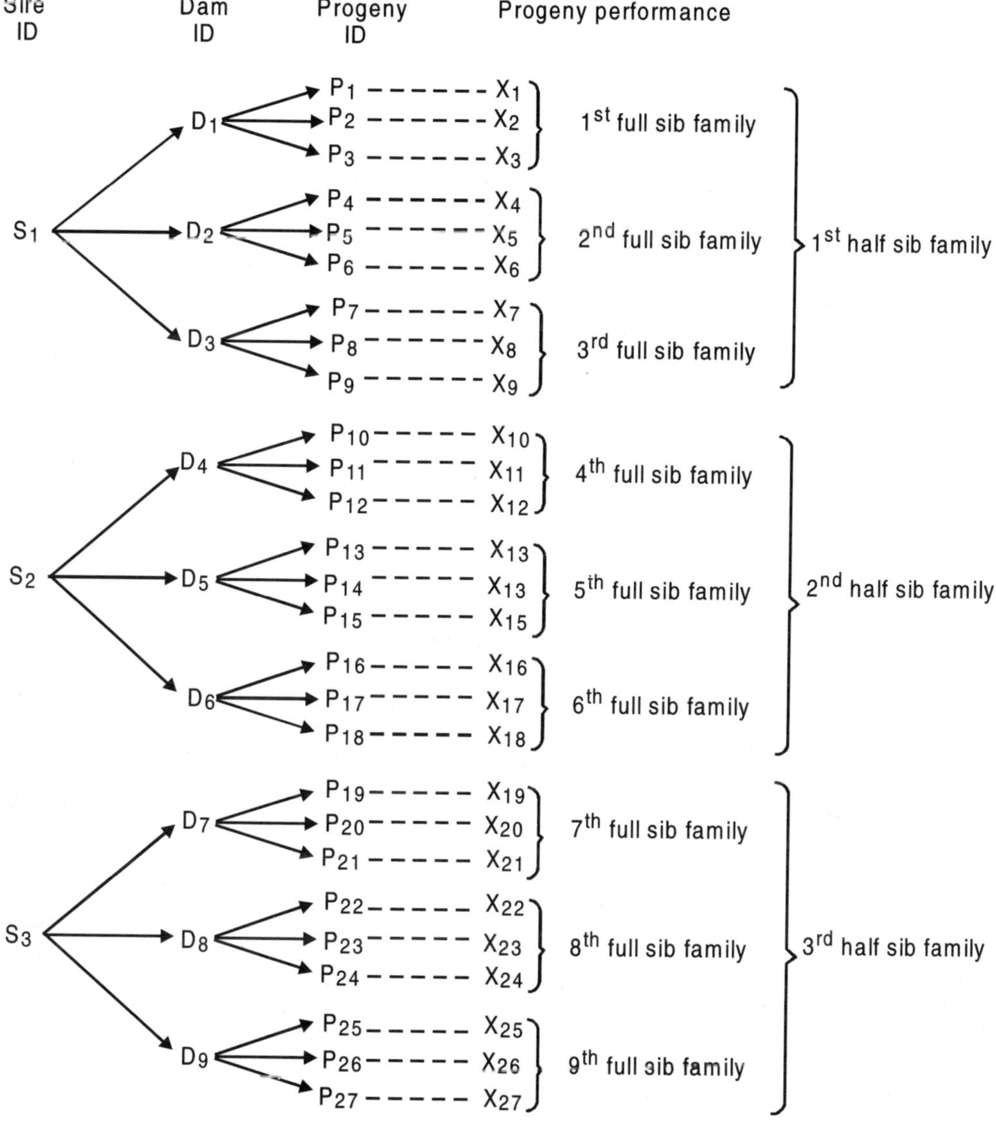

Under the family structure shown above, the offspring resulting from mating of a sire to a dam constitute a group of full-sibs as well as half-sibs. The total phenotypic variation can then be partitioned into the following components:

- Differences between progeny of different sires which constitute 'between-sire' component of variance (σ_S^2)
- Differences between progeny of dams mated to the same sire which constitute "between-dam" component of variance (σ_D^2).
- Differences between progeny of the same dam, which constitutes the "within-dam" component of variance (σ_W^2).

These three components of variance viz; sire-component σ_S^2, dam component σ_D^2 and within-dam component σ_W^2 are used for the estimation of heritability as follow:

Heritability from sire component: $$h_S^2 = \frac{4\sigma_S^2}{\sigma_S^2 + \sigma_D^2 + \sigma_W^2}$$

Heritability from dam component: $$h_D^2 = \frac{4\sigma_D^2}{\sigma_S^2 + \sigma_D^2 + \sigma_W^2}$$

Heritability from sire-plus-dam component: $$h_{S+D}^2 = \frac{2(\sigma_S^2 + \sigma_D^2)}{\sigma_S^2 + \sigma_D^2 + \sigma_W^2}$$

The computations involved in calculating the heritability from full-sib analysis are cumbersome, and beyond the scope of this book. The full-sib correlation method is not commonly used in most species of livestock because full-sib families of fairly large size are not available in single-birth species. Further, the heritability estimated from full-sib correlation method is an overestimate of the true h^2 because it includes the contribution of dominance and maternal effects.

Heritability can also be calculated from the results of the actual selection experiments conducted for the improvement of a trait. This method of h^2 estimation is dealt in (Chapter 6).

Which method of heritability estimation to use? The choice of the method of heritability estimation is usually dictated by the biology of organism (specifically the reproductive biology) and the type of data available (pedigree structure and performance records). For example, fewer full-sibs are available in cattle, buffalo, horse and other uniparous species; therefore, full-sib analysis cannot be used in such species. Parent-offspring relationship can not be used for the estimation of heritability of slaughter traits and also when parental performance records are not available. Each method has some advantages and some limitations that make them suitable for some species and for some traits; and not in other situations. However, in general, paternal half-sib correlation method is most commonly employed for the estimation of heritability in most species of livestock and poultry.

Repeatability[1]

The concept of repeatability alongwith its application in animal breeding was introduced by Lush (1937). Repeatability is a genetic parameter that is closely associated with heritability, and is applicable for traits that are expressed several times in the lifetime of an individual. For example, dairy cows complete many (four to five) lactations in their life time, sheep produce

[1]The concept of repeatability as discussed here pertains to repetition of a trait over time(called temporal repetition). The concept can be extended to the repetition of a trait over space like yield of fruits on different branches of the same tree (called spatial repetition).

many wool clips in their life. Similarly, there are several traits of interest in different species of farm animals which can be measured more than once in the life time of an animal. A few examples of such traits are given below:

- Cattle, Buffaloes : Milk yield, milk constituent's etc in different lactations.
- Sheep : Fleece characters like fleece weight, staple length, staple diameter etc in different shearing/ clippings.
- Goat : Milk production and milk constituents in different lactations.
- Swine : Litter size, litter weight in different farrowings.
- Horse : Racing, jumping and pulling performance in different competitions
- Poultry : Egg production, egg weight and other egg quality traits at different stages (periods) in laying cycle.

The assumption underlying the principle of repeatability is that performance of an animal during successive periods is determined by the same set of genes. **Repeatability** refers to the ability of an animal to repeat its performance over time, and is measured as the correlation between successive records (performance) of same animals. Suppose each cow in a herd has milk production records of two consecutive lactations. Repeatability then represents the correlation between milk yield of the 1st and the 2nd lactations :

$$\text{Repeatability} = r_{\text{1st, 2nd lactation}}$$

When the number of records on each animal is more than two, repeatability measures the average correlation between performances in different records.

Repeatability as a correlation coefficient is thus a measure of strength of association between repeated records. When repeatability of a trait is close to unity it indicates that the performance of animals between successive periods is highly correlated: the animals having high performance in first period are likely to have high performance in the second period, and *vice-versa*. On the other hand, when repeatability of a trait is close to zero, there would be no association between performance in consecutive periods. Consequently, the first record of the animal shall be of little use in predicting its performance in the second period.

Basis of Concept of Repeatability

Phenotypic value (P) of an individual can be split into genetic and environmental components:

$$P = G + E$$

The genotypic value (G), which is a made up of additive and non-additive effects of genes possessed by an individual, is determined at the time of fertilization and does not change during the lifetime of that individual. The genotypic value is therefore a permanent effect associated with an animal during its life time. The environmental component of performance for a repeatable trait can be further broken down into permanent (E_P) and temporary (E_T) environmental effects:

Permanent environmental effects (E_P): The environmental effects which stay with the animal for its life are known as permanent environmental effects. These non-genetic effects influence the performance of the animal throughout his/her lifetime. A few examples of permanent environmental effects are:

- *Udder damage due to mastitis*: Damage of a teat (or a quarter of an udder) of a heifer due to severe mastitis becomes a permanent effect associated with that cow and would influence her milk production in all the subsequent lactations. The damage of secretary tissues of udder thus becomes a permanent environmental effect with respect to that trait.

- *Calf hood nutrition:* Early nutrition of dairy heifer is known to have permanent effect on mammary gland development that influences milk production in all the lactations.
- *Training of young horses:* Early training of horses usually also exerts a permanent influence on racing ability throughout their life.

Similarly many events during fetal development and early growth phase may influence the performance of an animal throughout its life. Though the permanent environmental factors influence the performance of an animal throughout its life but, unlike genes, these are not transmitted to progeny.

Temporary environmental effects (E_T): The non-genetic factors that influence the performance of an animal in one period only (i.e. they vary from one record to other) constitute 'temporary environmental effects'. These effects are specific to a particular period only. For example, fodder supply, environmental temperature, humidity etc. vary from one season of the year to another. Milk yield of a cow may get temporarily depressed in a particular lactation if it happened to calve in a season when fodder supply was low. When the cow gets better feed/ fodder in a subsequent lactation, the previous effect would disappear and the performance of the cow improves. The 'quantity and quality feed/fodder' would thus be classified as a temporary environmental factor. Similarly, racing performance of a horse in a particular competition may be affected by the condition of the track (muddy, dry etc.), and has a specific effect on racing ability on that day only.

The contribution of environmental effects, as a function of permanent and temporary effects, is expressed as:

Environment = Permanent environment + Temporary environment

or,

$$E = E_P + E_T$$

The model to represent the phenotypic value of an animal for a repeatable trait (P) is:

$$P = G + E_P + E_T$$

Where:

P = phenotypic value (performance) of the animal.

G = genotypic value of the animal.

E_P = permanent environment effect(s) associated with the animal.

E_T = temporary environment effect (s) which the animal experiences in a specific period

All the components of the model (G, E_P, E_T) are assumed to be independent of each other. For example, a temporary effect like high ambient temperature has no relation with the genetic make-up of the cow or with the permanent environmental effects associated with that cow. Similarly, permanent environmental effects are usually independent of temporary environmental effects as well.

The total environmental variation (σ_E^2) for a repeatable trait can be partitioned into two components:

- Variation due to permanent environmental effects: σ_{EP}^2
- Variation due to temporary environmental effects : σ_{ET}^2

Total environmental variance can be expressed mathematically as:

$$\sigma_E^2 = \sigma_{EP}^2 + \sigma_{ET}^2$$

The model for describing the total phenotypic variation for a repeatable trait then becomes:

$$\sigma^2_P = \sigma^2_G + \sigma^2_{EP} + \sigma^2_{ET}$$

or Phenotypic variance = Genetic variance + Permanent environmental variance + Temporary environmental variance

Alternative Definitions of Repeatability

Repeatability which refers to ability of an animal to repeat its performance over time is measured as the correlation between successive (performance) records of the same animals. There are some other ways of describing the repeatability:

- **Repeatability as a ratio of variances:** Repeatability, (like heritability), can also be defined as the ratio of variances: ratio of genetic (σ^2_G) and permanent environmental variances (σ^2_{EP}) to the total phenotypic variance (σ^2_P) of the trait:

$$\text{Repeatability} = \frac{\sigma^2_G + \sigma^2_{EP}}{\sigma^2_G + \sigma^2_{EP} + \sigma^2_{ET}}$$

  Stated in words, repeatability is the proportion of observed (phenotypic) variation that is caused by permanent differences among individuals in a population. The magnitude of repeatability indicates the extent to which the observed differences in performance between animals are expected to be repeated in their future (subsequent) records.

- **Repeatability as regression of future on past performance:** Regression of future performance of the animal on its past performance equals the repeatability:

$$b_{F,P} = \text{repeatability}$$

  Where: F = expected future performance of the animal (in subsequent records)

  P = present (past) performance of the animal

  Repeatability as a regression coefficient can be used to predict future performance of animals on the basis of their past performance.

Range and unit of repeatability: Repeatability, like heritability, does not carry any unit, and varies between zero and one (between zero and 100 on percentage scale). If all the differences between animals are caused by genetic and permanent environmental effects (i.e. $\sigma^2_{ET} = 0$), then repeatability equals one. On the other hand, if all the differences are caused by temporary environmental factors (i.e. $\sigma^2_G + \sigma^2_{EP} = 0$), then repeatability equals zero.

Heritability vs. Repeatability: Repeatability is Upper Limit of Heritability

A comparison of the expressions of broad-sense heritability (H^2) and repeatability is given below:

$$H^2 = \frac{\sigma^2_G}{\sigma^2_P} \; ; \; \text{Repeatability} = \frac{\sigma^2_G + \sigma^2_{EP}}{\sigma^2_P}$$

The denominator in the expressions of both H^2 and repeatability is same but the numerator is different. The numerator for H^2 is genetic variance (σ^2_G) while that for repeatability is the sum of genetic (σ^2_G) and permanent environmental (σ^2_{EP}) variance. Repeatability is theoretically bound to be either equal to or greater than the broad–sense heritability. Board-sense heritability in turn may be greater than or equal to narrow-sense heritability:

$$h^2 \leq H^2 \leq \text{repeatability}$$

Because of this relationship, repeatability sets the upper limit to H^2. A comparison of some features of repeatability and heritability is made in Table 3.9.

Table 3.9 Repeatability vs. heritability.

Repeatability	Heritability
1. Ratio of genetic plus permanent environmental variance to the total phenotypic variance: $(\sigma_G^2 + \sigma_{EP}^2 / \sigma_P^2)$.	1. Ratio of genetic to total phenotypic variance: σ_G^2 / σ_P^2.
2. Sets an upper limit to the heritability of a trait.	2. Usually has a lower value than repeatability of a trait.
3. Most important use is computation of MPPA which is used for deciding whether or not an animal is to be kept in the herd.	3. Most important use of narrow-sense heritability is prediction of genetic gain, breeding value and in choice of the breeding and selection methods.
4. Calculated as the correlation between repeated records on the same animals.	4. Estimated from the phenotypic resemblance between relatives.
5. Estimation is relatively easier as it does not require pedigreed data.	5. Estimation is more complicated as it requires large volume of pedigreed data.

General Features of Repeatability

Repeatability, like h^2, is not a biological constant. Being a ratio, it can change due to changes in genetic, permanent environmental and temporary environmental variances:

- *Repeatability may vary from one trait to another within the same population* because of differences in genetic, permanent environmental and temporary environmental components of variances between traits. Damaged udder of a heifer, for example, will influence its future milk production but it may have no influence on other traits like calving interval, service period etc. Similarly, environmental factors that vary between periods (the temporary environmental factors) may exert more influence on one but lesser effect on the other trait. Different traits of animals may also exhibit different magnitude of genetic variances.
- *Repeatability may vary from one population to the other for the same trait* because of differences in genetic variance, and permanent and temporary environmental variances for different populations.
- *Repeatability, like heritability, is a population (herd/flock) measure,* and is thus not a value to be associated with an individual. We refer, for example, to the repeatability of racing performance in a population (stud) of horse or the repeatability of lactation yield in a herd of cattle. It is incorrect to refer to repeatability for racing performance of a particular horse.
- *Repeatability measures the relative ranking of animals in different periods:* High repeatability for a trait does not mean that the animals will have exactly the same performance in subsequent periods. It means that animals that were better than the herd average in first period would tend to come consistently at top in subsequent periods irrespective of herd average. Similarly, animals that were near the bottom in the herd would perform much below the herd average in subsequent periods.

Representative Values of Repeatability

Repeatability values are generally classified as: low (< 0.2), moderate (0.2 - 0.5) and high (> 0.5). A number of studies have been conducted to estimate repeatability of different traits in many different populations (breeds / strains etc.). Estimates obtained from different studies have usually been different due to differences in genetic and environmental variances. However certain typical ranges for different traits have been observed across different populations of the same species. Table 3.10 gives such typical ranges for some repeatable traits.

Table 3.10 Typical estimates of repeatability (r) for some traits in different species.

Trait	r	Trait	r
Cattle/Buffalo:		*Swine :*	
Lactation length	0.20–0.4	Litter size	0.15
Calving interval		Litter weight	
Service period		Back fat thickness	
Conception rate		*Horse:*	
Dry period		Wither height	>0.40
Milk yield	>0.50	Cannon bone circumference	
% Fat		Walking speed	0.02–0.3
% Protein		Time to trot one mile	
Sheep:		Time to run one mile	
Litter size	0.15	Pulling power	
Fleece weight	0.4-0.6	Temperament	
Fleece grade		*Poultry:*	
Staple length		Egg weight	> 0.50
Fiber diameter		Egg shape	
		Shell thickness	
		Shank length	
		Conformation	

As a general rule, reproductive traits (litter size, calving, internal, service period *etc.*) have lower repeatabilities than growth and production traits. The traits concerned with quality, composition and morphological features have highest repeatabilities.

Usefulness of Repeatability

Knowledge of repeatability of a trait is useful in the breeding of farm animals in a number of ways:

Computation of 'Most Probable Producing Ability' (MPPA): Repeatability of a trait is most commonly used for predicting the future performance of an animal in the herd during his/her lifetime. The expected future performance of an animal has been referred (by J. L. Lush) to as it's 'Most Probable Producing Ability'[2]. The term 'most probable producing ability' was first applied to milk production in dairy cattle but it is now commonly used in all species of livestock for repeatable traits. The term MPPA or simply as 'producing ability' represents the potential of an animal for milk production (cattle/buffaloes) or wool production (sheep) or litter size (swine) or racing ability (horse) etc.

[2]MPPA is sometimes also referred to as "future producing ability" or simply as 'producing ability'.

The ability of an animal to repeat its performance over time (in its life time) is dependent not only on its genes but also on permanent environmental factors associated with it. The MPPA is of practical importance to commercial producers who usually have fixed facilities for maintenance of animals in the herd. Since each year a new crop of animals is added to the herd/flock, the livestock producer is forced to cull a certain proportion of older animals so as to improve herd average in subsequent lactations/periods. The producer would like to keep only those animals that are expected to give high performance in their lifetime rather than those who produce the best progeny. Selection of animals to be kept in the herd for future performance is based on their 'most probable producing ability' (MPPA). The MPPA of an animal is computed as follow:

$$\text{MPPA} = \text{Herd average} + b\ (\text{Animal average} - \text{Herd average})$$

Where: b= 'weighting coefficient' assigned to the animal and is calculated as:

$$b = \frac{nr}{1 + (n-1)r}$$

Where: n = number of records available on that animal

r = repeatability coefficient

For a fixed value of repeatability, this weighting factor can vary between 'r' (for n = 1) and one (for n = ∝). The expression for MPPA of an animal thus becomes:

$$\text{MPPA} = \text{Herd average} + \frac{nr}{1 + (n-1)r}(\text{Animal average} - \text{Herd average})$$

The MPPA depends on the following factors:

- average performance of the herd.
- average performance of the animal based on 'n' records.
- Repeatability of the trait (r).
- Number of records available on the animal (n).

Illustrative example: Suppose average milk yield in a herd of cows is 1800 kg per lactation. Repeatability of milk yield is 0.3. Milk yield of two cows alongwith the number of records per cow, is given below alongwith the calculation of MPPA.

Cow no.	No. of records (n)	Milk yield in (lactation);				Cow avg.	Herd avg.	(b) *	MPPA
		1st	2nd	3rd	4th				
A	4	2110	2100	2060	2090	2100	1800	0.63	2289
B	1	2300	—	—	—	2400	1800	0.30	2250

$$*b = \frac{nr}{1 + (n-1)r}$$

Average performance of cow 'A' is lower than of cow 'B'. But the MPPA of cow 'A' is higher than of cow 'B' because its performance is based on four records while that of cow 'B' is based on a single record. Consequently, a low weighting coefficient (b=0.3) has been assigned to cow 'B'. If one of the two cows is to be kept in the herd for future performance, cow 'A' would be kept though its own performance was lower than of cow 'B'. The MPPA has brought the average performance of all the animals in the herd (with variable number of records) to a comparable basis, and provides a valid comparison of performance of different animals. If a cow did not have a performance record (like a heifer), its MPPA will be equal to herd average (since b=0).

Number of records required for selection: Magnitude of repeatability of a trait is an indicator of the extent to which an animal is likely to repeat its performance in subsequent periods. For example, repeatability of fat percentage in dairy cows is very high, fat percentage of a single lactation would thus be a reliable indicator of its potential in subsequent lactations. A cow can thus be selected on the basis of its fat percentage in first lactation. On the other hand, repeatability of milk yield is low to moderate; therefore, milk yield of first lactation would not be a very reliable indicator of its potential for milk production throughout its life.

When the repeatability of a trait is low, selection of animals has to be based on a number of records. The knowledge of repeatability of a trait is thus useful in deciding about the number of records required before an animal can be selected for producing the progeny. In general, more number of records would be required to determine a cow's potential for low repeatability traits (e.g. calving interval) than for high repeatability trait (e.g. fat percentage).

Idea about magnitude of heritability: Repeatability sets the upper limit for broadsense heritability as it includes genetic and permanent environmental effects which contribute to real differences between animals in a herd. Estimation of heritability is relatively cumbersome as it involves voluminous and pedigreed data while the estimation of repeatability is relatively easier. The magnitude of repeatability provides preliminary idea about the value of heritability of a trait.

In addition to the uses listed above, repeatability is also used in estimation of breeding values for repeatable traits and in calculating the gain in accuracy of selection to be expected from the use of repeated records. Estimation of repeatability is quite similar to that of heritability by half sib analysis. Salient features of MPPA and EBV are listed in Table 3.11.

Table 3.11 EBV vs. MPPA.

Breeding value (EBV)	Most probable producing ability (MPPA)
1. Value of an individual judged by the mean value of its progeny	1. Expected future performance of an animal during its lifetime.
2. Refers to genetic merit of an animal, half of which is transmitted to its offspring.	2. Refers to ability of an animal to perform in subsequent periods (and not to the genetic merit transmitted to the progeny).
3. Depends on the (average) effect of the genes possessed by the animal.	3. Depends on additive and non-additive effect of genes as well as on permanent environmental effects.

Phenotypic, Genetic and Environmental Correlations

The overall productivity in farm animals is dependent on a number of traits which are usually interrelated. For example, cattle heavier at birth are usually heavier at weaning; heavier hens usually produce large-sized eggs; high milk producing cows have lower fat percentage; heavier sheep produce more wool (fleece) and so on. Thus different traits recorded on the same animal may tend to be related to each other. These are the phenotypic associations between traits. There are three types of correlations which are commonly used in animal breeding: phenotypic correlations, genetic correlations and environmental correlations

Phenotypic correlation (r_P): Phenotypic correlation between two traits is a measure of the degree of association between their phenotypic values. Phenotypic correlation between two traits (say X and Y), designated as $r_{P(xy)}$, is a function of the phenotypic covariance between them, and their phenotypic standard deviations as follows:

$$r_{P(X,Y)} = \frac{Cov(P_X, P_Y)}{\sigma_{P(X)} \times \sigma_{P(Y)}}$$

Where: Cov (P_x, P_y) = covariance between phenotypic values of two traits

$\sigma_{P(x)}$ = phenotypic standard deviation of trait 'X'

$\sigma_{P(Y)}$ = phenotypic standard deviation of trait 'Y'

Genetic correlation (r_G): Genetic correlation between two traits is a measure of the degree of association between their breeding (genotypic) values. Genetic correlation, designated as $r_{G(XY)}$, is a function of the genetic covariance between them and their genetic standard deviations as follows:

$$r_{G(X,Y)} = \frac{Cov(A_X, A_Y)}{\sigma_{A(X)} \times \sigma_{A(Y)}}$$

Where: Cov (A_x, A_y) = genetic (additive) covariance between two traits

$\sigma_{A(x)}$ = additive genetic standard deviations of trait 'X'

$\sigma_{A(y)}$ = additive genetic standard deviations of trait 'Y'

Environmental correlation (r_E): Environmental correlation between two traits is a measure of the degree of association between environmental factors associated with the two traits. Environmental correlation, designated as $r_{E(XY)}$, is a function of environmental covariance between them and their environmental standard deviations as follows:

$$r_{E(X,Y)} = \frac{Cov_{(E_X, E_Y)}}{\sigma_{E(X)} \times \sigma_{E(Y)}}$$

Where: Cov (E_x, E_y) = covariance between environmental values of the two traits

$\sigma_{E(X)}$ = environmental standard deviation of trait 'X'

$\sigma_{E(Y)}$ = environmental standard deviation of trait 'Y'

Environmental correlations arise when same environmental factor(s) exert simultaneous influence on more than one trait. For example, r_E between gain in weight and back fat thickness in pigs is positive. This means that environmental conditions that cause pigs to attain high weight also tend to produce fatter pigs and *vice versa*. Similarly, diseased condition (illness) of animals influences food consumption, growth rate, sexual maturity, milk yield etc and therefore becomes an environmental factor to influence these traits. If an environmental factor causes improvement in performance of both the traits, the environmental correlation between these would be positive. On the other hand, if an environmental factor is favorable for one trait but depresses the other, the environmental correlation between these would be negative.

Environmental correlations are often useful for management purposes. For example, the r_E between body weight and back fat thickness in pigs is positive. If a producer wants to produce leaner pigs, he may have to consider a feeding regimen that would cause lower gain in weight. Similarly, feed restriction in broiler chicken before sexual maturity is employed as it reduces the gain in weight which in turn leads to increased egg production during laying period.

The relationship between two traits in terms of phenotypic, genetic and environmental correlations can be better understood diagrammatically as shown below:

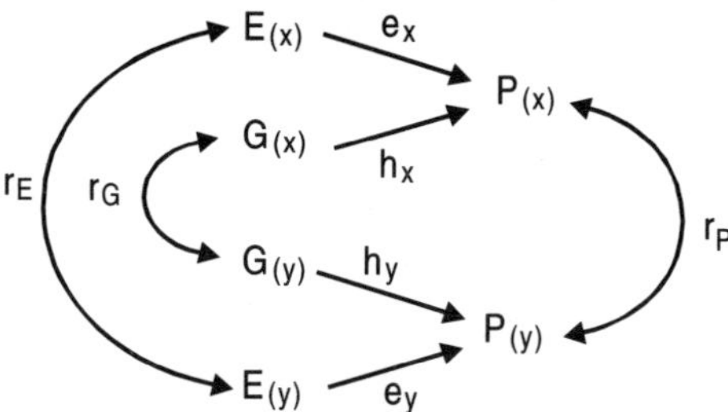

The diagrammatic representation shows that both genetic (r_G) and environmental (r_F) correlations contribute to phenotypic (r_P) correlations between traits. However, two points need to be noted for the relationship between the three types of correlations:

- Phenotypic correlation between two traits is not simply the sum of the genetic and environmental correlations between them: $r_{P(XY)} \neq r_{G(XY)} + r_{E(XY)}$
- Phenotypic correlation between two traits is not the numerical average of the genetic and environmental correlations between them: $r_{P(XY)} \neq \frac{r_{G(XY)} + r_{E(XY)}}{2}$

Phenotypic correlation between two traits may, theoretically, differ from the corresponding genetic and environmental correlations both in direction and magnitude. This means that magnitude and sign of phenotypic correlation between two traits may not be similar to the magnitude and sign of genetic correlation between them. In practice, however, phenotypic correlations between traits usually lie somewhere between the corresponding genetic and environmental correlations. The three types of correlations theoretically vary between – 1.0 to 1.0, and are unitless.

Causes of Genetic Correlations

Of the three types of correlations, genetic correlations are more important for breeders as they are interested in the genetic improvement of a number of economically important traits. Two possible causes of genetic correlations are linkage and pleiotropy.

- **Linkage:** Presence on the same chromosome of genes affecting more than one trait, referred to as linkage, tends to prevent different genes from segregating independently at gamete formation. Therefore, the chromosome carrying either both the dominant (or desirable) genes or both the recessive (or undesirable) genes affecting different traits tend to be transmitted together to the progeny causing genetic association between the traits. Since linked genes do not always stay together due to crossing over, linkage cannot be a cause of permanent genetic correlations. Genetic correlations resulting from linkage tend to decrease or disappear over generations due to crossing over. Furthermore, the genetic correlation between two traits caused by linkage may be expected to differ in magnitude and sign between different populations, except when the linkage between genes is very close. Genetic correlations caused by linkage are therefore transient and do not persist over generations.
- **Pleiotropy:** The phenomenon where one gene influences the expression of several traits is referred to as pleiotropy, and the gene that influences more than one trait is called pleiotropic gene. Pleiotropy is a common property of most genes considering

their biological functions. Genes control synthesis of amino acids, proteins and enzymes, and in turn control a number of biochemical reactions and physiological functions of the organism. It is therefore logical that a gene exerts effect on a number of traits of the organism. There are a number of examples of single-gene traits exhibiting pleiotropic effect. A few are listed below:

- *Sickle-cell anemia gene in humans:* Sickle-cell anemia allele (Hb_S) in homozygous form (Hb_S Hb_S) causes sickle shaped red-blood cells, reduced O_2 carrying capacity of blood, anemia, severe swelling and the associated pain.
- *Halothane gene in pigs:* The gene that causes malignant hyperthermia when pigs are exposed to halothane, also improves meatiness (lean meat yield), feed efficiency but decreases litter size, survival rate and meat quality.
- *Double muscling gene in cattle:* A single gene is responsible for double muscling observed in many cattle breeds. This gene is also associated with reduced fertility, lower survival of calves and reduced resistance to stress. Similarly, Merle coat color gene in dogs also causes eye defects.

Economic traits of livestock are controlled by a large number of genes each with small individual effect. Conceptually, a gene is likely to influence several traits of the animal because expression of such traits may be mediated through a series of common biochemical and physiological mechanisms. For example, a gene responsible for poor digestion or absorption of nutrients could affect growth, and production of milk or wool or eggs. Pleiotropy is the major reason for existence of genetic correlation between traits. Genetic correlations caused primarily by pleiotropy tend to persist over successive generations while those arising from linkage are transient in nature.

Usefulness of Genetic Correlations

Genetic correlations are important to breeders of farm animals in a number of ways:

- **Prediction of correlated response to selection**: When selection is practiced for the improvement of one trait, a change in the performance of other traits from parental to the progeny generation may also result which is referred to as **correlated response to selection**. Correlated response occurs due to existence of genetic correlation between traits. The direction and magnitude of the correlated response can be predicated from the knowledge of genetic correlation. For example, the genetic correlation between egg production and egg weight in poultry is negative; consequently selection for improvement of egg production would lead to decline in egg size. Knowing the magnitude and sign of the genetic correlation between two traits, the breeder can predict beforehand the likely changes to occur in correlated traits due to selection for the primary trait. Knowledge of the direction and magnitude of genetic correlation is, therefore, important in the formulation of efficient selection program for simultaneous improvement of several traits
- **Computation of selection indices:** Computation of selection indices, which are used for the simultaneous improvement of several traits, also requires the value of genetic correlation. In fact, genetic correlation coupled with heritability is useful in all aspects of selection of genetically superior animals. Many uses of genetic correlations will become clear in the chapters to follow.

Estimates of some of the commonly and consistently observed genetic correlations across different strains/breeds of some species are listed in Table 3.12.

Table 3.12 Values of a few consistently observed genetic correlations.

Species	Traits	r_G	Remarks
Cattle/Buffalo :	Milk yield and fat yield	0.80	Desirable
	Milk yield and fat percentage	–0.50	Undesirable
	Milk yield and total protein	0.90	Desirable
	Age at first calving and milk yield	–0.3	Desirable
Sheep :	Wool yield and fiber diameter	–0.3	Undesirable
	Clean fleece weight and staple length	0.3	Desirable
Swine :	Growth rate and feed efficiency	0.80	Desirable
	Growth rate and feed intake	0.40	Undesirable
	Body weight and back fat thickness	0.25	Undesirable
Poultry:	Age at first egg and egg number	–0.40	Desirable
	Egg production and egg weight	–0.40	Undesirable
	Egg weight and body weight	0.40	Desirable
	Shank length and body weight	0.40	Desirable
	Keel length and body weight	0.50	Desirable

Classification of Genetic Correlations

Genetic, phenotypic and environmental correlations can also be classified in a number of ways depending on:

(a) *Magnitude of correlation:* Genetic correlations may be high, moderate or low (though no uniform pattern is followed for classifying correlations as high, moderate or low.)

(b) *Direction of correlation:* Genetic correlations between traits may be positive, negative or close to zero. A positive genetic correlation implies that some of the genes which increase the performance for one trait also increase performance of the other trait (i.e. the same genes act synergistically to influence the two traits.).Some of the consistently observed positive genetic correlations are:

- Adult body weight with egg weight in poultry.
- Milk yield with total milk fat, total milk protein and lactation length in cattle/buffalo.

A negative genetic correlation implies that genes responsible for increased performance for one trait decrease the performance of second trait (i.e. genes influencing different traits act antagonistically.) Examples of negative genetic correlations are:

- Egg number with egg weight and age at sexual maturity in poultry.
- Milk yield with fat %, protein % and age at first calving in dairy cattle.
- Clean fleece weight with fiber diameter in sheep.

Absence of a correlation (zero) between two traits means that the traits are controlled by independent set of genes.

(c) *Favorable vs. unfavorable correlations:* The classification of correlations as favorable or unfavorable depends on the objectives of the breeder, nature of traits and market requirements. A favorable genetic correlation between two traits means that a desirable genetic change in one trait is associated with a desirable genetic change in the other while an undesirable correlation implies the reverse.

- A desirable genetic correlation can be negative or positive. For example, the genetic correlation between age at first egg and the number of eggs laid in poultry is negative. But it is a desirable correlation because early sexual maturity is desirable from producer's stand point as is the number of eggs laid. On the other hand, a negative genetic correlation between egg number and egg weight is undesirable because hens laying more eggs would be producing progeny with small-sized eggs. But a poultry farmer would like to have hens that produce not only more number of eggs but of heavier size as well.
- Further a positive correlation does not mean that it is a desirable correlation, and similarly a negative correlation does not mean that it is undesirable. It depends on the biological nature of the trait or its definition. For example, the genetic correlation between body weight gain and feed efficiency (in pigs, chickens) is desirable which means genetic improvement in body gain would lead to genetic improvement in feed efficiency. If the feed efficiency is expressed as gain in weight divided by amount of feed consumed, the correlation is positive and desirable. On the other hand if feed efficiency is expressed as amount of feed consumed-divided by gain in weight, then the genetic correlation would be negative but desirable.

Estimation of Correlations by Sib Analysis

The principle underlying the estimation of genetic correlations is same as that of estimation of heritability. In general, any method of estimating heritability can also be adapted to estimate genetic correlations. Half sib analysis of variance, the most commonly used method of heritability estimation, is also most commonly used for genetic, phenotypic and environmental correlations. Broadly it involves:

- Conducting analysis of variance and covariance.
- Computing the components of variances (between- and within-sire components, σ_S^2 and σ_W^2) and covariance (between- and within-sire components, $\sigma_{S(X,Y)}$ and $\sigma_{W(X,Y)}$).
- Computing the correlation from ratio of covariance and variances.

Use of ANOVA for calculation of sire- and within-sire components of variance was outlined in estimation of heritability by half sib analysis. The analysis of covariance table for calculating the different components of covariances from half-sib analysis is given below:

Analysis of Covariance Table

Source	d.f.	Sum of cross products (SCP)	Mean cross products (MCP)	Composition of MCP
Between sires	S – 1	SCP_S	MCP_S	$Cov_W + K\ Cov_S$
Within sires	N – S	SCP_W	MCP_W	Cov_W
Total	N – 1	SCP_T		

Where: S = Number of sires

N = Total number of paired observations (i.e. the total number of animals)

K = Average number of progeny per sire

The computations involved in obtaining the sum of cross products (SCP) and mean cross products (MCP) are similar to those involved in obtaining the sum of squares (as in case of ANOVA). The sire-component (Cov_S) and within-sire component (Cov_W) of covariance are calculated from SCP and MCP (the detailed calculations of SCP and MCP are given in next chapter) by equating MCP to their expected values given in the last column of analysis of covariance table and solving for Cov_S and Cov_W as shown below:

Within-sire component of covariance: $Cov_W = MCP_W$

Between-sire component of covariance: $Cov_S = \dfrac{MCP_S - MCP_W}{K}$

'Sire-component of covariance' (Cov_S) is an estimate of ¼th of additive genetic covariance:

$$Cov_S = \frac{1}{4} Cov_{A(X,Y)}$$

'Within-sire component of covariance' (Cov_W) is an estimate of remaining genetic covariance ($\frac{3}{4}Cov_{A(X,Y)} + Cov_{D(X,Y)}, + Cov_{I(X,Y)}$) and the environmental covariance ($Cov_{E(X,Y)}$):

$$Cov_W = \left[\frac{3}{4}(Cov_{A(X,Y)}) + Cov_{D(X,Y)} + Cov_{I(X,Y)} + Cov_{E(X,Y)}\right]$$

The sum of 'sire-component' and 'within-sire component' of covariance equals total phenotypic covariance:

$$Cov_S + Cov_W = \left[\left(\frac{1}{4}Cov_{A(X,Y)}\right) + \left(\frac{3}{4}(Cov_{A(X,Y)}) + Cov_{D(X,Y)} + Cov_{I(X,Y)} + Cov_{E(X,Y)}\right)\right]$$

$$= \left[\left(Cov_{A(X,Y)} + Cov_{D(X,Y)} + Cov_{I(X,Y)} + Cov_{E(X,Y)}\right)\right] = Cov_{P(X,Y)}$$

Estimates of phenotypic, genetic and environmental correlations are obtained from Cov_S and Cov_W as follow:

Phenotypic correlation (r_P):

$$r_{P(XY)} = \frac{Cov_{P(X,Y)}}{\sqrt{\sigma^2_{P(X)} \times \sigma^2_{P(Y)}}} = \frac{Cov_{S(X,Y)} + Cov_{W(X,Y)}}{\sqrt{(\sigma^2_{S(X)} + \sigma^2_{W(X)})(\sigma^2_{S(Y)} + \sigma^2_{W(Y)})}}$$

Genetic correlation (r_G):

$$r_{G(XY)} = \frac{Cov_{S(X,Y)}}{\sqrt{(\sigma^2_{S(X)} \times (\sigma^2_{S(Y)})}} = \frac{Cov_{A(X,Y)}}{\sqrt{\sigma^2_{A(X)} \times \sigma^2_{A(Y)}}}$$

Note: Though we call the genetic association between the two traits as genetic correlation but it should we kept in mind that actually it is correlation between the breeding values of the two traits.

Environmental correlation (r_E):

$$r_{E(XY)} = \frac{Cov_{W(X,Y)}}{\sqrt{\sigma^2_{W(X)} \times \sigma^2_{W(Y)}}}$$

Precision of Genetic Parameters

Estimates of genetic parameters are obtained from a limited number of animals in a herd/ flock that are really a sample of an infinitely large population. Hence these are sample statistics like mean, variance etc. and hence subject to sampling variation. Precision (reliability) of any sample statistics is provided by its standard error. For example, precision of arithmetic mean is given by its standard error. Similarly, standard errors of heritability, repeatability and genetic correlations can be calculated using appropriate formulas which are given in standard text books of quantitative genetics and animal breeding. These expressions are beyond the scope of this book. However, a few general observations about the magnitude of standard errors of genetic parameters are important to note:

- Standard errors are dependent not only on the total number of animals available for analysis but also on the mating design used viz; number of families (number of sires) and family size.
- Repeatability and heritability as compared to genetic correlations are less subject to sampling. That is for a given flock/herd with same number of animals, same mating structure, standard errors of repeatability and heritability will be lower than of genetic correlations.

The estimates of genetic parameters should therefore be considered in relation to the standard errors associated with them.

Review Questions

1. Explain / define the following :

I. Heritability as ratio of variances
II. Heritability as correlation coefficient
III. Heritability as regression coefficient IV. Accuracy of selection
V. Repeatability as correlation coefficient
VI. Repeatability as ratio of variances VII. Permanent environmental effects
VIII. Temporary environmental effects IX. Most probable producing ability
X. Genetic correlation XI. Environmental correlation
XII. Phenotypic correlation XIII. Pleiotropy

2. Which genetic parameter provides information about the relative role of heredity and environment to the phenotypic variation of a trait? Explain

3. Name the two broad causes of phenotypic variation among individuals within a population.

4. What are the different components of genetic variance? Which of these and to what extent is/are transmitted to next generation?

5. Explain the range and unit of heritability.

6. What will be the magnitude of heritability?

I. If all the differences among individuals in a population are caused by environmental factors.
II. If all the differences among individuals in a population are caused by genetic factors.

7. Can a quantitative trait have zero heritability? Explain.

8. Can a quantitative trait have complete heritability? Explain.

9. Can the heritability of a trait be negative or more than one? Explain.

10. What are the two ways of expression of heritability? Explain.

11. Which of the two is more useful to livestock breeders: narrow-sense or broad-sense heritability?

12. Can a trait be affected by both additive and non-additive gene action? Explain.
13. Why is accuracy of selection generally higher for high heritability traits than for low heritability traits?
14. Explain the different uses of heritability in animal breeding.
15. How is heritability useful for prediction of breeding value?
16. How can the knowledge of heritability be used to determine if a trait is determined by additive gene action, non- additive gene action or both?
17. How does heritability help us in determining the choice of the most appropriate breeding plan for genetic improvement?
18. How does heritability estimate help in choice of the most effective basis of selection?
19. Narrow- sense heritability of body weight in a population is 0.7? How much of the variation in body weight in this population is caused by variation in breeding values?
20. Explain how the heritability of the same character may differ between different populations?
21. When does broad-sense heritability equal narrow-sense heritability?
22. Can the heritability of a trait change over time? Explain.
23. Can the heritability of one trait differ from the heritability of another trait within the same population? Explain.
24. Is heritability the property of an individual? Explain.
25. Is heritability a constant? Explain.
26. What are the two most commonly held misconceptions about heritability?
27. What is wrong with the following statement: "Heritability of a buffalo for milk production is high". How can the statement be re-drafted to convey the actual meaning of heritability?
28. Differentiate between the broad-sense and narrow-sense heritabilities.
29. Categorize the following types of traits into 'very low', 'low to moderate', 'high' and 'very high' heritability: reproductive, growth and production, quality, morphological.
30. (a) List four methods of estimation of heritability.
    (b) Which of these methods is most commonly employed for estimation of heritability in:
    (i) cattle and buffalo (ii) humans (iii) poultry
31. List the composition of covariance between :
    I. Mono-zygotic twins
    II. Offspring and parent
    III. Full sibs
    IV. Half sibs
32. What is the principle underlying the estimation of heritability from resemblance between relatives?
33. What fraction of phenotypic superiority of parents over the population mean is transmitted to their offspring?
34. What fraction genetic variation is transmitted from parents to offspring?
35. Which type of heritability (broad-sense or narrow-sense) is more accurate in predicting response to selection? Explain.
36. Which is more accurate in predicting response to selection, a heritability estimated from twins or one estimated from resemblance between parents and offspring? Explain.

37. Explain the 'twin analysis' method of estimation of heritability. What are its limitations?
38. What are the advantages and limitations of regression of offspring on parent for estimation of heritability?
39. Heritability can be determined from correlation between offspring and parent or by regression of offspring on parent. Which of these two is the most appropriate method of estimation of heritability in livestock species? Explain.
40. Write a short note on 'half- sib' analysis method of estimation of heritability including the mating structure and analysis of variance.
41. What are the advantages of 'half-sib' analysis for estimation of heritability?
42. Outline the mating structure of 'full-sib' analysis of variance for heritability estimation.
43. List one repeatable trait for each of the following animal species specifying the period over which it is repeated: cattle / buffalo, sheep, swine, horse, poultry.
44. Write the model for describing phenotypic value and phenotypic variation for a repeatable trait.
45. Repeatability is the upper limit of heritability. Explain.
46. Differentiate between repeatability and heritability.
47. Which of the two can change over the lifetime of an individual, phenotype or genotype? Explain.
48. Cite at least two examples of permanent environmental factors.
49. What is the unit and range of repeatability? Explain.
50. Can the repeatability of a trait vary from one population to the other? Explain.
51. Repeatability can vary from one trait to another in the same population. Explain.
52. Repeatability measures the relative ranking of animals over different periods. Explain.
53. Cite at least two examples for each of the following: (i) low repeatability traits (ii) high heritability traits.
54. Write a short note on usefulness of repeatability in animal breeding.
55. What factors determine most probable producing ability of an animal?
56. What are the different types of correlations in animal breeding? Give mathematical expressions for these. Which of these is most important in animal breeding?
57. Diagrammatically show how common genetic and environmental factors (effects) contribute to phenotypic, genetic and environmental associations between two traits.
58. What are the possible causes of genetic correlations? Explain.
59. What is the usefulness of genetic correlations in animal breeding?
60. What does a negative genetic correlation between two traits indicate?
61. What does a positive genetic correlation between two traits indicate?
62. Give at least two examples of each of the positive and negative genetic correlations.
63. Differentiate between favorable and unfavorable genetic correlations citing suitable examples.
64. Outline the procedure for estimation of genetic correlations.

●●●

4

Estimation of Genetic Parameters

Genetic and phenotypic parameters viz. heritability, repeatability and genetic and phenotypic correlations are used for describing the variation and covariation in quantitative traits. These are used for designing effective genetic improvement programs for economic traits of farm animals.

The theory relevant to heritability, repeatability and different types of correlations including their application in animal breeding has been presented in the last chapter. The present chapter covers the computation of these genetic/phenotypic parameters including the relevant inferences.

Concept of Heritability and Inferences

4.1. What will be the heritability of a trait if : (a) whole of the phenotypic variability is genetic in nature, (b) whole of the phenotypic variability is environmental in nature, (c) half of the phenotypic variability is due to genetic effects.

Solution :

(a) When whole of the phenotypic variability for a trait is genetic in nature, it implies that environmental factors do not contribute to the variation of that trait, therefore, $\sigma^2_E = 0$ and thus, $\sigma^2_G = 1.0$

$$H^2 = \frac{\sigma^2_G}{\sigma^2_P} = \frac{1.0}{1.0} = 1.0$$

- ***Inference :*** If there is no environmental variation for a trait, then heritability equals one.

(b) When whole of the phenotypic variability for a trait is environmental in nature, it implies that genetic effects do not contribute to the variability of that trait, therefore, $\sigma^2_G = 0$, and thus $\sigma^2_E = 1.0$ and $\sigma^2_P = 1.0$

$$H^2 = \frac{\sigma^2_G}{\sigma^2_P} = \frac{0.0}{1.0} = 0.0$$

- ***Inference*** **:** If genetic variability for a trait is absent, then heritability equals zero.

(c) Let, $\sigma_P^2 = X$, then, $\sigma_G^2 = ½\ X$ and $\sigma_E^2 = ½\ X$

$$\sigma_P^2 = \sigma_G^2 + \sigma_E^2 = ½\ X + ½\ X = X$$

$$H^2 = \frac{\sigma_G^2}{\sigma_P^2} = \frac{1/2\ X}{X} = 0.5$$

- ***General Inference***: Heritability varies between zero and one.

4.2. Total genetic variance for protein yield in a herd of dairy cattle is 750 Kg^2. Environmental variance is 850 kg^2 and variances due to dominance and epistatic effects are 25 and 25 kg^2, respectively.

(a) Calculate the heritability of protein yield in narrow- and broad-sense.

(b) What inference can be drawn about the type of gene action and the breeding programme (whether selection or crossing) to be followed for the genetic improvement of protein yield?

Solution :

(a) Phenotypic variance:

$$\sigma_P^2 = \sigma_G^2 + \sigma_E^2 = 750 + 850 = 1600 \text{ kg}^2$$

$$\sigma_A^2 = (\sigma_G^2) - (\sigma_D^2 + \sigma_I^2) = 750 - (25 + 25) = 700 \text{ kg}^2$$

$$h^2 = \frac{\sigma_A^2}{\sigma_P^2} = \frac{700}{1600} = 0.438$$

$$H^2 = \frac{\sigma_G^2}{\sigma_P^2} = \frac{750}{1600} = 0.469$$

(b) A comparison of the broad- and narrow-sense heritabilities provides information about the relative role of additive and non-additive gene action. The broad-sense heritability of protein yield is marginally higher than the narrow-sense heritability (0.47 vs. 0.44). Therefore, the trait is primarily governed by additive gene action. Selection (*i.e.* intra-population selection) can be effectively used for the genetic improvement of protein yield.

4.3. In a flock of White Leghorn chicken, the following information on egg weight was obtained: Mean = 60g, Phenotypic standard deviation = 3g, Coefficient of variation = 5%, Narrow-sense heritability = 0.50, Non-additive genetic variance = 2.0 g^2.

(a) Calculate the phenotypic variance, additive genetic variance and heritability in broad-sense

(b) What does a comparison of broad- and narrow-sense heritability show?

Solution :

(a) Given: Standard deviation: $\sigma_P = 3$g.

Therefore, phenotypic variance: $\sigma_P^2 = 9\text{g}^2$

Since $h^2 = \sigma_A^2 / \sigma_P^2$, therefore, $\sigma_A^2 = h^2 \times \sigma_P^2 = 0.5 \times 9 = 4.5\text{g}^2$

Non- additive genetic variance: $(\sigma_D^2 + \sigma_I^2) = 2.0\text{g}^2$ (given)

Total genetic variance: $\sigma_G^2 = (\sigma_A^2) + (\sigma_I^2 + \sigma_D^2) = 4.5 + 2.0 = 6.5 \text{ g}^2$

$$H^2 = \frac{\sigma_G^2}{\sigma_P^2} = \frac{6.5}{9.0} = 0.72$$

(b) H^2 is considerably more than h^2 ,therefore, non-additive gene action is also important for the trait (in addition to the additive gene action).

4.4. The different components of variance for litter-size (number of piglets born) in a flock of Large White Yorkshire breed of swine were as follows:

Additive genetic variance (σ_A^2) = 0.8 pig^2, dominance variance (σ_D^2) = 0.6 pig^2

epistatic variance (σ_I^2) = 0.6 pig^2, environmental variance (σ_E^2) = 0.6 pig^2

(a) Calculate the broad- and narrow–sense heritabilities.

(b) What is the percentage contribution of genetic and environmental effects to the variation of litter size?

(c) What percentage of the total genetic variance is caused by additive- and non–additive genetic effects?

Solution :

Given: $\sigma_A^2 = 0.8$ pig^2; $\sigma_D^2 = 0.6$ pig^2; $\sigma_I^2 = 0.6$ pig^2 and $\sigma_E^2 = 0.6$ pig^2

(a) $\sigma_G^2 = \sigma_A^2 + \sigma_D^2 + \sigma_I^2 = 0.8 + 0.6 + 0.6 = 2.0$ pig^2,

$\sigma_P^2 = \sigma_G^2 + \sigma_E^2 = 2.0 + 0.6 = 2.6$ pig^2,

$$h^2 = \frac{\sigma_A^2}{\sigma_P^2} = \frac{0.8}{2.6} = 0.307$$

$$H^2 = \frac{\sigma_G^2}{\sigma_P^2} = \frac{2.0}{2.6} = 0.769$$

Percentage contribution of genetic effects $= \frac{\sigma_G^2}{\sigma_P^2} \times 100 = \frac{2.0}{2.6} \times 100 = 76.9$

Percentage contribution of environmental effects $= \frac{\sigma_E^2}{\sigma_P^2} \times 100 = \frac{0.6}{2.6} \times 100 = 23.1$

(c) Percentage of total genetic variance caused by:

Additive genetic effects $= \frac{\sigma_A^2}{\sigma_G^2} \times 100 = \frac{0.8}{2.0} \times 100 = 40$

Non-additive genetic effects $= \frac{(\sigma_D^2 + \sigma_I^2)}{\sigma_G^2} \times 100 = \frac{1.2}{2.0} \times 100 = 60$

4.5. In a large flock of White Leghorn variety of chicken, egg number up to 72 weeks of age and egg weight were measured. The different types of variances calculated for the two traits are tabulated below:

Type of variance	Egg number (eggs^2)	Egg weight (g^2)
Phenotypic :	625	16.0
Genetic :	205	10.0
Environmental :	420	6.0
Additive genetic:	95	9.6
Dominance :	50	0.2
Epistatic :	60	0.2

(a) Calculate the broad- and narrow-sense heritabilities for the two traits.

(b) Which of the two traits is expected to respond to selection more and why?

(c) How do these two traits differ for the type of gene action?

Solution :

(a)

Genetic parameter	Egg number	Egg weight
$h^2 = \sigma^2_A / \sigma^2_P$:	95 / 625 = 0.15	9.6 / 16 = 0.60
$H^2 = \sigma^2_G / \sigma^2_P$:	205 / 625 = 0.32	10.0 / 16 = 0.63

(b) h^2 of egg weight is higher than of egg number, therefore, egg weight should respond to selection more than egg number (provided similar magnitude of selection pressure is applied for the two traits).

(c) The broad-sense heritability of egg number is twice as large as its narrow-sense heritability which means that non-additive gene action is also important for egg number. However, for egg weight the gene action is primarily additive because the difference between h^2 and H^2 for this trait is very small.

Methods of Estimation of Heritability

General note : Estimation of heritability requires data on large number of pedigreed animals (usually in thousands) involving several sires and dams. Computation of h^2 has been greatly facilitated by computational software. However, for learning the computational steps involved, a very small-sized data have been used here. The students should keep in mind that the actual data involved are very large. The theory behind the use of different methods has been discussed in the previous chapter.

I. Regression of Offspring on Parent(s)

(a) Regression of offspring on one parent

4.6. Given below are data on daily milk yield(kg) of ten dams (X) and their daughters (Y) :

Dam's ID no. :	1	2	3	4	5	6	7	8	9	10
Dam's milk yield (X) :	9	6	9	6	7	8	7	7	8	9
Daughter's milk yield (Y):	8	6	7	8	8	7	7	9	9	8

(a) Calculate the regression of daughter on dam.
(b) Estimate the heritability of milk yield from daughter-dam regression and daughter-dam
(c) Compare the heritabilities calculated by regression and correlation methods, and explain 'why do they differ'?

Solution :

(a) The calculations involved in computing the variances and covariances are shown below:

Milk yield of:		Cross product	Squares:	
Dam (X)	Daughter (Y)	(XY)	X^2	Y^2
9	8	72	81	64
6	6	36	36	36
9	7	63	81	49
6	8	48	36	64
7	8	56	49	64
8	7	56	64	49
7	7	49	49	49
7	9	63	49	81
8	9	72	64	81
9	8	72	81	64
$\Sigma X = 76$	$\Sigma Y = 77$	$\Sigma XY = 587$	$\Sigma X^2 = 590$	$\Sigma Y^2 = 601$

Variance of dam's yield : $\sigma^2_X = \dfrac{\Sigma X^2 - \dfrac{(\Sigma X)^2}{N}}{N-1} = \dfrac{590 - \dfrac{(76)^2}{10}}{N-1} = \dfrac{590-577.6}{9}$

$= 1.37\ Kg^2$

Variance of daughter's yield : $\sigma^2_Y = \dfrac{\Sigma Y^2 - \dfrac{(\Sigma Y)^2}{N}}{N-1} = \dfrac{601 - \dfrac{(77)^2}{10}}{N-1} = \dfrac{592.9}{9}$

$= 0.9\ Kg^2$

Covariance (daughter-dam) : Cov (X,Y)

$$= \frac{\Sigma XY - \dfrac{(\Sigma X \Sigma Y)}{N}}{N-1} = \frac{587 - \dfrac{(76 \times 77)}{10}}{9} = \frac{587-585.2}{9} = 0.2$$

Regression of daughter on dam : $b_{yx} = \dfrac{Cov(XY)}{Variance(X)} = \dfrac{0.2}{1.37} = 0.145$

(b) Correlation between daughter and dam(r_{XY}) :

$$r_{XY} = \frac{Cov(XY)}{\sqrt{VarianceY}\ X\sqrt{VarianceY}} = \frac{0.2}{\sqrt{1.37}\sqrt{0.9}} = 0.18$$

h^2 from regression coefficient $= 2 \times b_{yx} = 2 \times 0.145 = 0.29$

h^2 from correlation coefficient $= 2 \times r_{xy} = 2 \times 0.18 = 0.36$

(c) The heritabilities calculated from regression and correlation coefficients are different because of the difference in the variance of daughters and variance of dams. Had the daughter and dam variances been equal, the heritabilities by correlation and regression methods would have been same.

- ***Inference***: Heritability can be calculated by both the correlation and the regression methods. However, the regression method provides a better estimate of h^2 than the correlation method because regression is not influenced by selection practiced in dams.

(b) Regression of offspring on average of parents

4.7. Fleece weight (lbs) was measured in a sample from a sheep flock. The data listed below are fleece weight of parents (average of both parents and their offspring:

Mid-parent (X) :	11.8	8.4	9.5	10.0	7.6	10.8	8.5	11.8	10.5
Offspring (Y) :	7.7	5.7	7.2	7.3	5.4	7.2	5.6	8.4	7.0

(a) Calculate the regression of offspring on mid-parent, and estimate the h^2 of fleece weight (from this small sample).

(b) Interpret the value of the regression coefficient.

(c) Plot the data on a graph paper and draw the regression line.

(d) What would be the expected performance of offspring whose parents averaged 8.0 lbs of fleece?

(e) If two mid-parental groups differed by 4 lbs, what would be the expected difference between the performance of their offspring groups ?

Solution :

(a) The computations involved in estimation of variance and co-variances are shown given below :

Mid-parent (X)	Offspring (Y)	Squares: X^2	Y^2	Cross product XY
11.8	7.7	139.24	59.29	90.86
8.4	5.7	70.56	32.49	47.88
9.5	7.2	90.25	51.84	68.40
10.0	7.3	100.00	53.29	73.00
7.6	5.4	57.76	29.16	41.04
10.8	7.2	116.64	51.84	77.76
8.5	5.6	72.25	31.36	47.60
11.8	8.4	139.24	70.56	99.12
10.5	7.0	110.25	49.00	73.50
$\Sigma X = 88.9$	$\Sigma Y = 61.5$	$\Sigma X^2 = 896.19$	$\Sigma Y^2 = 428.83$	$\Sigma XY = 619.16$

$$\overline{X} = \frac{\Sigma X}{N} = \frac{88.9}{9} = 9.87 \text{ lbs}$$

$$\overline{Y} = \frac{\Sigma Y}{N} = \frac{61.5}{9} = 6.83 \text{lbs}$$

Regression of offspring (Y) on mid-parent (X), b_{yx} is calculated as follow :

$$b_{yx} = \frac{\Sigma XY - \frac{(\Sigma X \Sigma Y)}{N}}{\Sigma X^2 - \frac{(\Sigma X)^2}{N}} = \frac{619.16 - \frac{88.9 \times 61.5}{9}}{819.16 - \frac{(88.9)^2}{9}}$$

$$= \frac{619.16 - 607.5}{896.16 - 878.13} = \frac{11.67}{18.06} = 0.65$$

h^2 = Regression of offspring on mid-parent = 0.65

(b) Regression coefficient of 0.65 means that for each pound increase in parental fleece weight (mid-parent) values, the performance of progeny is expected to increase by 0.65 lbs.

(c) For plotting the regression line, we need the value of intercept, 'a', which is obtained by rearranging the regression equation as shown below :

$$a = \overline{Y} + b_{yx} \times \overline{X} = 6.833 - (0.65 \times 9.87) = 6.833 - 6.387 = 0.41 \text{ lbs}$$

The regression line is :

$$Y_i = a + b_{yx} \times X_i = 0.41 + 0.65X_i$$

The performance of daughters (Y) plotted against mid-parent values (X) along with the regression line (shown as a straight line) is shown below :

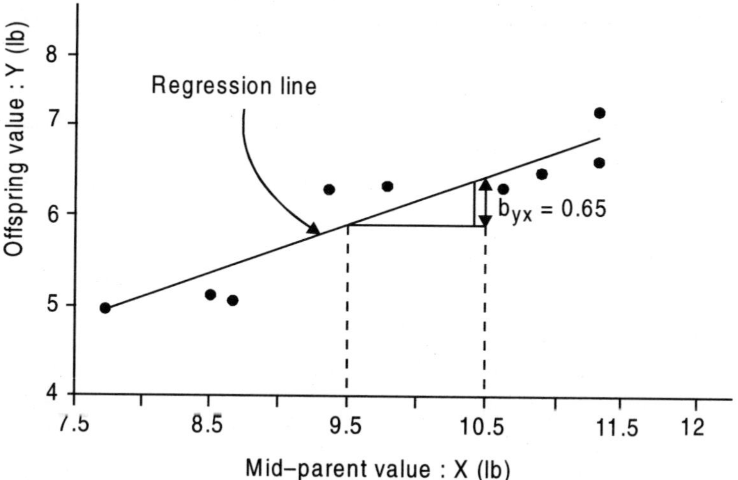

The plot shows that as the average performance of parental groups' increases, the performance of their progeny also increases.

(d) The expected performance of progeny group whose parents had average fleece weight of 8 lb is calculated by putting the value of 'X' (as '8.0') in the regression equation:

$$Y = 0.41 + 0.65 \times 8.0 = 5.61 \text{lb}$$

(e) Let : X_1 be the mid-parent value for 1st group of offspring (Y_1).
X_2 be the mid-parent value for 2nd group offspring (Y_2).

Since X_2 differs from X_1 by 4 lb, therefore, $X_2 = X_1 + 4$

Average performance of 1st group of offspring (Y_1) is expected to be :

$$Y_1 = a + b_{yx} \times X_1 = 0.41 + 0.65X_1 \qquad ...(4.1)$$

Average performance of 2nd group of offspring (Y_2) is expected to be:

$$Y_2 = a + b_{yx} \times X_2 \qquad ...(4.2)$$

Putting the value X_2 in terms of X_1 (i.e. $X_2 = X_1 + 4$) in equation 4.2

$$Y_2 = a + b_{yx} \times X_2 = 0.41 + [(0.65 \times (X_1 + 4)] = 0.41 + (0.65\ X_1) + 2.6$$

$$Y_2 = 3.01 + 0.65\ X_1 \qquad ...(4.3)$$

The difference between Y_2 and Y_1 is obtained from the difference between equation 4.3 and 4.1 :

$$Y_2 - Y_1 = (3.01 + 0.65\ X_1) - (0.41 + 0.65\ X_1) = 3.01 + 0.41 = 2.6 \text{ lb}$$

Therefore, progeny of the parents which differed from each other by 4 lb are expected to differ by 2.6 lb.

- ***Inference :*** The regression of offspring on mid-parental value, which equals h^2, multiplied by the difference between the performance of parental groups (*i.e.* $X_1 - X_2$ or $X_2 - X_1$), gives the expected difference between their progeny groups:

$$Y_2 - Y_1 = b_{yx}(X_2 - X_1) = 0.65 \times 4 = 2.6 \text{ lb}$$

Note : Prediction of progeny performance and of genetic gain are two important uses of heritability. The uses of h^2 for prediction of progeny performance and genetic gain are discussed in Chapter VI.

II. Sib Analysis: Intra-class Correlation between Sibs

(a) Half-sib Analysis

4.8. Thirty dairy sires were each mated to five dams, and each mating produced one daughter. The daily milk yield (kg) of daughters was recorded. The sum of squares obtained from the analysis of variance conducted on the milk yield records are given below:

Source of variation	Sum of squares
Between sires :	1830
Within sires :	4800

(a) Calculate the phenotypic variance, additive genetic variance and the h^2 of milk yield.

(b) Calculate the intra-class correlation (t), and show that : $h^2 = 4t$.

Solution :

(a) Let : S = number of sires = 30

K = number of daughters per sire = 5

N = total number of daughters (N = K × S) = 30 × 5 = 150

The completed ANOVA:

Source	d.f.	Sum of squares	Mean squares	Composition of mean squares
Between sires	S–1 = 29	1830	MSs = 1830/29 = 63.1	$\sigma_W^2 + K\ \sigma_S^2$
Within sires	N–S = 120	4800	MSw = 4800/120 = 40.0	σ_W^2
Total	N–1 = 149	6630	—	—

'Within-sire' component of variance : $\sigma_W^2 = 40$

'Between-sire' component of variance:

$$\sigma_S^2 = \frac{MS_S - MS_W}{K} = \frac{63.1 - 40}{5} = 4.62$$

'Sire-component' of variance is a measure of one-fourth of the additive genatic variance:

$$\sigma_S^2 = \frac{1}{4}\sigma_A^2$$

Therefore, additive genetic variance: $\sigma_A^2 = 4\sigma_S^2 = 4 \times 4.62 = 18.68$

Total phenotypic variance: $\sigma_P^2 = \sigma_S^2 + \sigma_W^2 = 4.62 + 40 = 44.62$

$$h^2 = \frac{4\sigma_S^2}{\sigma_S^2 + \sigma_W^2} = 0.41$$

Alternately, $$h^2 = \frac{\sigma_A^2}{\sigma_P^2} = \frac{18.68}{44.62} = 0.41$$

(b) The intra-class correlation between half-sibs, (t), is calculated as the ratio of 'between-sire' component to 'total variance' as:

$$t = \frac{\sigma_S^2}{\sigma_S^2 + \sigma_W^2} = \frac{4.62}{4.62 + 40.0} = 0.1035$$

$$\therefore \quad h^2 = 4 \times t = 4 \times 0.1035 = 0.41$$

4.9. As a simplified example, suppose four sires were each mated to five dams and each mating produced one female offspring (daughter). Age at first calving (months) was recorded and is tabulated below for the daughters of the four sires:

Age at 1st calving (months) for sire number :			
1	2	3	4
28	31	31	27
30	30	29	28
30	29	29	29
28	29	28	30
29	30	29	29

Calculate the additive genetic variance, phenotypic variance and the h^2 of age at first calving by half-sib analysis of variance.

Solution :

Notations: The notations that are commonly used in analysis of animal breeding data are:

'X' refers to the phenotypic value (age at first calving) of an individual (progeny),

Subscripts on 'X' are used to designate the identity of the sire and its offspring. For example;

X_{11} = age at 1st calving of 1st daughter of 1st sire;

X_{12} = age at 1st calving of 2nd daughter of 1st sire, and so on.

The notation, '$X_{1.}$' (*read as 'X one dot'*) refers to sum of age at first calving of daughters of 1st sire and so on,

The notation, 'X..' (read as 'X dot dot') refers to sum of age at first calving of daughters of all the sires.

The table given in the problem can be re-arranged by writing the appropriate notation as:

Sire number:			
1	2	3	4
X_{11} = 28	X_{21} = 31	X_{31} = 31	X_{41} = 27
X_{12} = 30	X_{22} = 30	X_{32} = 29	X_{42} = 28
X_{13} = 30	X_{23} = 29	X_{33} = 29	X_{43} = 29
X_{14} = 28	X_{24} = 29	X_{34} = 28	X_{44} = 30
X_{15} = 29	X_{25} = 30	X_{35} = 29	X_{45} = 29
Sum for 1st sire: $X_{1.}$ = 145	Sum for 2nd sire: $X_{2.}$ = 149	Sum for 3rd sire: $X_{3.}$ = 146	Sum for 4th sire: $X_{4.}$ = 143
Grand total : $X_{..} = X_{1.} + X_{2.} + X_{3.} + X_{4.} = 583$			

Total number of progeny : N = 20

Correction term/factor : CT or CF $= \frac{(\text{Grand total})^2}{N} = \frac{(583)^2}{20} = 16994.45$

Total sum of squares (corrected) : SS_T :

$$SS_T = (X_{11})^2 + (X_{12})^2 + X_{13})^2 + (X_{14})^2 + \ldots. + (X_{43})^2 + (X_{44})^2 + (X_{45})^2 - CT$$

$$= (28^2 + 30^2 + 30^2 + 28^2 + 29^2 + \ldots + 27^2 + 28^2 + 29^2 + 30^2 + 29^2) - 16994.45$$

$$= 17015 - 16994.45 = 20.55$$

Between-sire sum of squares (corrected) : SS_S :

$$SS_S = \frac{(\text{Sum for 1st sire})^2}{\text{No. of progeny for sire1}} + \frac{(\text{Sum for 2nd sire})^2}{\text{No. of progeny for sire 2}} + \ldots + \frac{(\text{Sum for 4th sire })^2}{\text{No. of progeny for sire 4}} - CT$$

$$= \frac{(X_{1.})^2}{K} + \frac{(X_{2.})^2}{K} + \frac{(X_{3.})^2}{K} + \frac{(X_{4.})^2}{K} - CT = \frac{(X_{1.})^2 + (X_{2.})^2 + (X_{3.})^2 + (X_{4.})^2}{K} - CT$$

$$= \frac{(145)^2 + (149)^2 + (146)^2 + (143)^2}{5} - 16994.45$$

$$= 16998.2 - 16994.45 = 3.75$$

Within-sire sum of squares:

$$SS_W = SS_T - SS_S = 20.55 - 3.75 = 16.8$$

Analysis of Variance Table

Source	d.f.	SS	MS	Composition of MS
Between sires	4 − 1 = 3	3.75	$MS_S = 1.25$	$\sigma^2_W + K\sigma^2_S$
Within sires	20 − 4 = 16	16.8	$MS_W = 1.05$	σ^2_W
Total	N − 1 = 19	20.55		

$$\sigma^2_W = MS_W = 1.05$$

$$\sigma^2_S = \frac{MS_S - MS_W}{K} = \frac{1.25 - 1.05}{5} = 0.04$$

Total (phenotypic) variance :

$$\sigma^2_P = (\sigma^2_s + \sigma^2_w) = (0.04 + 1.05) = 1.09$$

Additive genetic variance:

$$\sigma^2_A = (4 \times \sigma^2_s) = 4 \times 0.04 = 0.16$$

Intra-class correlation (t) $= \frac{\sigma^2_s}{\sigma^2_s + \sigma^2_w} = 0.0366$

$$h^2 = 4 \times t = 0.146$$

Alternately, h^2 can be calculated as :

$$h^2 = \frac{4\sigma^2_s}{\sigma^2_s + \sigma^2_w} = \frac{4 \times 0.04}{0.04 + 1.05} = 0.146$$

Alternately, $h^2 = \dfrac{\sigma_A^2}{\sigma_P^2} = \dfrac{0.16}{1.09} = 0.146$

(b) Full-sib Analysis

4.10. Twenty poultry sires were each mated to five dams, and each dam produced 10 offspring. The trait was measured on all the progeny. The analysis of variance for the trait is given below:

Source	d.f.	SS	MS	Composition of MS
Between sires	S–1=19	6,650	MS_S=350	$\sigma_w^2 + n\sigma_D^2 + nd\sigma_s^2$
Between dams/sires	S(d–1)=80	12,000	MS_D=150	$\sigma_w^2 + n\sigma_D^2$
Between progeny/ dam/sire	900	90,000	MS_W=100	σ_w^2
Total	N–1= 999	1,08,650	—	—

(a) Calculate the sire, dam and sire-plus-dam components of variance, and the heritability from the different components.

(b) What genetic inference (about the type of gene action) can be drawn from comparison of h^2 obtained from sire and dam components?

Solution :

(a) Given : No. of sires: S = 50

No. of dams mated to each sire: d = 5

No. of progeny per dam: n = 10

The mean squares are equated to their respective expectations composition and solved for individual components as follow:

$$\sigma_w^2 = MS_W = 100$$

$$\sigma_D^2 = \frac{MS_D - MS_W}{n} = \frac{150-100}{10} = 5.0$$

$$\sigma_s^2 = \frac{MS_S - MS_D}{nd} = \frac{350-150}{10 \times 5} = 4.0$$

$$\sigma_P^2 = MS_S + MS_D + MS_W = 100 + 5.0 + 4.0 = 109.0$$

Heritability from sire-component:

$$h_S^2 = \frac{4\sigma_S^2}{\sigma_S^2 + \sigma_D^2 + \sigma_W^2} = \frac{4\sigma_S^2}{\sigma_P^2} = \frac{4 \times 4.0}{109.0} = 0.146$$

Heritability from dam-component:

$$h_D^2 = \frac{4\sigma_S{}^2}{\sigma_S{}^2 + \sigma_D{}^2 + \sigma_W{}^2} = \frac{4\sigma_S{}^2}{\sigma_P{}^2} = \frac{4 \times 4.0}{109.0} = 0.146$$

Heritability from sire-plus-dam component:

$$h_{S+D}^2 = \frac{2(\sigma_S^2 + \sigma_D^2)}{\sigma_S^2 + \sigma_D^2 + \sigma_W^2} = \frac{4\sigma_S^2}{\sigma_P^2} = \frac{2(4.0 + 5.0)}{109.0} = 0.165$$

(b) Comparison of h^2 from sire and dam components indicates the role of dominance and maternal effects. Since h_D^2 is only marginally higher than h_S^2, therefore, dominance and maternal effects are relatively unimportant for the trait.

4.11. Given below is the wool yield of 36 sheep resulting from mating of four sires, each sire mated to three dams, and each dam in turn producing three progeny:

Sire ID	Dam ID	Progeny ID	Wool yield (lb) of progeny	Sire ID	Dam ID	Progeny ID	Wool yield (lb) of progeny
1		1	X_{111} = 19	3		19	X_{311} = 09
	1	2	X_{112} = 15		7	20	X_{312} = 14
		3	X_{113} = 13			21	X_{313} = 14
		4	X_{121} = 14			22	X_{321} = 12
	2	5	X_{122} = 06		8	23	X_{322} = 16
		6	X_{123} = 11			24	X_{323} = 12
		7	X_{131} = 07			25	X_{331} = 16
	3	8	X_{132} = 12		9	26	X_{332} = 15
		9	X_{133} = 10			27	X_{333} = 14
2		10	X_{211} = 12	4		28	X_{411} = 19
	4	11	X_{212} = 14		10	29	X_{412} = 14
		12	X_{213} = 18			30	X_{413} = 14
		13	X_{221} = 10			31	X_{421} = 17
	5	14	X_{222} = 16		11	32	X_{422} = 17
		15	X_{223} = 18			33	X_{423} = 13
		16	X_{231} = 18			34	X_{431} = 14
	6	17	X_{232} = 14		12	35	X_{432} = 11
		18	X_{233} = 16			36	X_{433} = 13

(a) Calculate the heritability of wool yield from sire, dam and sire-plus-dam components

(b) What inference(s) can be drawn about the type of gene action from comparison of sire- and dam-component heritabilities?

Solution :

(a) Given : No. of sires: S= 4

No. of dams mated to each sire: d = 3

No. of progeny per dam: n = 3

Notations used :

X_{111} refers to 1st progeny of 1st sire and 1st dam.

X_{112} refers to 2nd progeny of 1st sire and 1st dam, and so on.

$X_{1..}$ = sum of values of the trait for 1st sire.

Similarly, $X_{2..}$, $X_{3..}$, $X_{4..}$ refer to sum of observations for 2nd, 3rd and 4th sire, respectively.

$X_{11.}$ = sum of observations of 1st dam of 1st sire.

Similarly, $X_{12.}$, $X_{13.}$, ——, $X_{43.}$ refer to dam totals for different sires

$X_{...}$ = Grand total = sum of all the observations (values of the trait)

Computations:

Sum of all the observations (sum of wool yield of all the progeny): GT = X... = 497

Sum of observations for sires:

Sum for sire 1: $X_{1..} = 107$ Sum for sire 2: $X_{2..} = 136$
Sum for sire 3: $X_{3..} = 122$ Sum for sire 4: $X_{4..} = 132$

Sum of observations for dams:

Sum for dam 1: $X_{11.} = 47$ Sum for dam 2: $X_{12..} = 31$
Sum for dam 3: $X_{13.} = 29$ Sum for dam 4: $X_{21..} = 44$
Sum for dam 5: $X_{22.} = 44$ Sum for dam 6: $X_{23..} = 48$
Sum for dam 7: $X_{31.} = 37$ Sum for dam 8: $X_{32..} = 40$
Sum for dam 9: $X_{33.} = 45$ Sum for dam 10: $X_{41.} = 47$
Sum for dam 11: $X_{42.} = 47$ Sum for dam 12: $X_{43.} = 38$

Calculation of sum of squares:

$$\text{Correction Term : CT} = \frac{(\text{Grand total})^2}{N} = \frac{(X..)^2}{N} = \frac{497^2}{36} = 6861.36$$

Total sum of squares (corrected):

$$SS_T = SS_T \text{ (uncorrected)} - CT$$
$$= (X_{111})^2 + (X_{112})^2 + \ldots + (X_{431})^2 + (X_{432})^2 + (X_{433})^2 - CT$$
$$= 7201 - 6861.36 = 339.64$$

Between-sire sum of squares (corrected):

$$SS_S = \frac{(X_{1.})^2 + (X_{2.})^2 + (X_{3.})^2 + (X_{4.})^2}{nd} - CT$$
$$= \frac{(107)^2 + (136)^2 + (122)^2 + (132)^2}{9} - CT$$
$$= 6917.0 - 6861.36 = 55.64$$

Between-dam sum of squares (corrected): SS_D

$$= \frac{(X_{11.})^2 + (X_{12.})^2 + ----+(X_{43.})^2}{n} - SS_S \text{ (uncorrected)}$$
$$= \frac{(47)^2 + (31)^2 + (29)^2 + ----+(38)^2}{3} - SS_S \text{ (uncorrected)}$$
$$= 7014.33 - 6917.0 = 97.33$$

Between-progeny sum of squares is obtained from the difference between 'total sum of squares' and 'sum of squares due to sires plus sum of squares due to dams' as follow:

$$SS_W = SS_T - (SS_S + SS_D) = 186.67$$

Analysis of Variance Table

Source	d.f.	SS	MS
Sires	S −1 = 4 −1 = 3	55.64	MS_S = 18.55
Dam/sires	(d − S) = 12− 4 = 8	97.33	MS_D = 12.17
Progeny/dam/sire	24	186.67	MS_W = 7.78
Total	N −1 = 35	339.64	

Note : The correction term, CT, for dams is the SS_S (uncorrected)

Calculation of components of variance :

$$\sigma_W^2 = MS_W = 7.78$$

$$\sigma_D^2 = \frac{MS_D - MS_W}{n} = \frac{12.17 - 7.78}{3} = 1.46$$

$$\sigma_S^2 = \frac{MS_S - MS_D}{nd} = \frac{18.55 - 12.17}{9} = 0.71$$

$$\sigma_P^2 = \sigma_S^2 + \sigma_D^2 + \sigma_W^2 = 0.71 + 1.46 + 7.78 = 9.95$$

Heritability from sire-component:

$$h_S^2 = \frac{4\sigma_S^2}{\sigma_P^2} = \frac{4 \times 0.71}{9.95} = 0.285$$

Heritability from dam-component:

$$h_D^2 = \frac{4\sigma_D^2}{\sigma_P^2} = \frac{4 \times 1.46}{9.95} = 0.587$$

Heritability from sire-plus-dam component:

$$h_{S+D}^2 = \frac{2(\sigma_S^2 + \sigma_D^2)}{(\sigma_S^2 + \sigma_D^2 + \sigma_W^2)} = \frac{2(0.71 + 1.46)}{9.95} = 0.436$$

(*b*) Heritability from dam-component is about twice as that from sire-component which indicates that the trait is also influenced by dominance and maternal effects.

Concept of Repeatability and Inferences

4.12. Total genetic variance of protein yield (in 305-days) is 1,800 kg^2. The environmental variance is 5,400 kg^2. The variances due to dominance and epistatic genetic effects are 20 and 20 kg^2, respectively. Permanent environmental variance is 1800 kg^2.

(a) Calculate the heritability in the narrow- and broad-sense.

(b) Calculate the repeatability.

(c) Calculate the relative contribution (%) of different components of variance to the total phenotypic variance.

Solution :

Given: $\sigma^2_G = 1800\ kg^2$, $\sigma^2_D = 20\ kg^2$, $\sigma^2_I = 20\ kg^2$, $\sigma^2_E = 5400\ kg^2$

and $\sigma^2_{EP} = 1800\ kg^2$.

(a) The additive genetic variance (σ^2_A), temporary environmental variance (σ^2_{ET}) and the total phenotypic variance (σ^2_P) are obtained as follow:

$$\sigma^2_A = \sigma^2_G - (\sigma^2_D + \sigma^2_I) = 1800 - 40 = 1760\ Kg^2$$

$$\sigma^2_P = \sigma^2_G + \sigma^2_E = 1800 + 5400 = 7200\ Kg^2$$

$$\sigma^2_{ET} = \sigma^2_P - \sigma^2_{EP} = 5400 - 1800 = 3600\ Kg^2$$

$$h^2 = \frac{\sigma^2_A}{\sigma^2_P} = \frac{1760}{7200} = 0.244$$

$$H^2 = \frac{\sigma^2_G}{\sigma^2_P} = \frac{1800}{7200} = 0.25$$

(b) Repeatability $= \dfrac{\sigma^2_G + \sigma^2_{EP}}{\sigma^2_P} = \dfrac{1800+1800}{7200} = 0.50$

(c) Relative contribution (%) of different components to total (phenotypic) variance:

$$\text{Genetic variance} = \frac{1800}{7200} \times 100 = 25\%$$

$$\text{Environmental variance} = \frac{5400}{7200} \times 100 = 75\%$$

$$\text{Additive genetic variance} = \frac{1760}{7200} \times 100 = 24.4\%$$

$$\text{Non-additive genetic variance} = \frac{40}{7200} \times 100 = 0.55\%$$

$$\text{Permanent environmental variance} = \frac{1800}{7200} \times 100 = 25\%$$

$$\text{Temporary environmental variance} = \frac{3600}{7200} \times 100 = 50\%$$

4.13. Given below are the components of genetic and environmental variances for calving interval in cattle :

Phenotypic variance = 6400 day^2
Genetic variance = 800 day^2
Environmental variance = 5600 day^2
Additive genetic variance = 100 day^2
Non-additive genetic variance = 700 day^2
Temporary environmental variance = 4200 day^2

(a) Calculate the broad-sense heritability and repeatability of calving interval

(b) What does the comparison of repeatability and broad-sense heritability indicate?

Solution :

(a) $$H^2 = \frac{\sigma^2_G}{\sigma^2_P} = \frac{800}{6400} = 0.125$$

Permanent environnemental variance:

$$\sigma^2_{EP} = \sigma^2_E - (\sigma^2_{ET}) = 5600 - 4200 = 1400 \text{ day}^2$$

$$\text{Repeatability} = \frac{\sigma^2_G + \sigma_{EP}^2}{\sigma^2_P} = \frac{800+1400}{6400} = \frac{2200}{6400} = 0.344$$

(b) Repeatability is considerably higher than broad-sense heritability (H^2) which means that temporary environmental factors are quite important in influencing the variability in calving interval over successive calvings.

Methods of Estimation of Repeatability

There are two methods for the estimation of repeatability; both involve calculating the correlation between repeated records on the same animals.

I. Inter-class Correlation (Product–moment Correlation)

When each animal has two records, repeatability can be estimated as the product-moment correlation (inter-class correlation) between the two records.

Let, 'X' and 'Y' represent, the first and second performance records respectively, then r_{XY} is an estimate of repeatability:

$$r_{XY} = \frac{\text{Cov (1}^{st} \text{ and } 2^{nd} \text{ record)}}{\sqrt{\text{Var } 1^{st} \times \text{Var } 2^{nd}}} = \frac{\sigma_{XY}}{\sqrt{\sigma^2_X \times \sigma^2_Y}} = \frac{\sigma_{XY}}{\sigma_X \times \sigma_Y}$$

Repeatability can also be obtained as the regression of second record (Y) on the first record (X):

$$b_{XY} = \frac{\text{Cov (1}^{st} \text{ and } 2^{nd} \text{ record)}}{\text{Variance of } 1^{st} \text{ record}} = \frac{\sigma_{XY}}{\sigma^2_X}$$

If variances of first and second records are equal ($\sigma^2_X = \sigma^2_Y$), then the correlation and regression coefficients yield similar values of repeatability.

4.14. The numbers of pig born per sow (litter size) in each of the two consecutive litters (farrowing) for a highly prolific breed of swine are listed below:

Sow ID no.:	1	2	3	4	5	6	7	8	9	10
1st litter size :	9	7	11	8	7	6	8	9	8	11
2nd litter size :	8	10	10	8	6	4	10	7	10	10

(a) Calculate the repeatability of litter size from the correlation between 1st and 2nd records, and from the regression of 2nd on 1st litter size (record).

(b) Why does the repeatability of litter size differ by the two methods?

Solution :

(a) Let: 'X' be the 1st litter size and 'Y' be the 2nd litter size

The computations required for estimation of correlation and regression coefficients are shown below:

Sow	Litter size in :		Squares:		Cross products:
ID	1st (X)	2nd (Y)	X^2	Y^2	(XY)
1	9	8	81	64	72
2	7	10	49	100	70
3	11	10	121	100	110
4	8	8	64	64	64
5	7	6	49	36	42
6	6	5	36	25	30
7	8	10	64	100	80
8	9	7	81	49	63
9	8	10	64	100	80
10	11	10	121	100	110
Sums:	$\Sigma X = 84$	$\Sigma Y = 84$	$\Sigma X^2 = 730$	$\Sigma Y^2 = 738$	$\Sigma XY = 721$

$$(\Sigma X)^2 = 84^2 = 7056, \quad (\Sigma Y)^2 = 84^2 = 7056$$

$$(\Sigma X)(\Sigma Y) = 84 \times 84 = 7056$$

$$\text{Var (X)} = \frac{\Sigma X^2 - \frac{(\Sigma X)^2}{N}}{N-1} = \frac{730 - \frac{(7056)}{10}}{(10-1)} = \frac{24.4}{9} = 2.711$$

$$\text{Var (Y)} = \frac{\Sigma Y^2 - \frac{(\Sigma Y)^2}{N}}{N-1} = \frac{738 - \frac{(7056)}{10}}{(10-1)} = \frac{32.4}{9} = 3.6$$

$$\text{Cov (X, Y)} = \frac{\Sigma(XY) - \frac{(\Sigma X)(\Sigma Y)}{N}}{N-1} = \frac{721 - \frac{(7056)}{10}}{(10-1)} = \frac{15.4}{9} = 1.711$$

Corrélation between 1st and 2nd record :

$$r_{XY} = \frac{\text{Cov}(XY)}{\sqrt{\text{Var}(X) \times \text{Var}(Y)}} = \frac{1.711}{\sqrt{2.711 \times 3.6}} = 0.548$$

Régression of 2nd on 1st record :

$$b_{YX} = \frac{\text{Cov}(XY)}{\text{Var}(X)} = \frac{1.978}{2.711} = 0.631$$

Estimates of repeatability by the two methods differ because the variance of 1st litter size is different from the variance of 2nd litter size. (i.e. $\sigma^2_X \neq \sigma^2_Y$)

II. Intra-class Correlation

When the numbers of records per animal are more than two, the calculation of repeatability by product- moment correlation becomes cumbersome since there will be a number of combination of records. For example, if there are three records per animal, the product moment correlations among all the possible pairs of records (between 1_{st} & 2_{nd}; between 1_{st} & 3_{rd}; between 2_{nd} & 3_{rd} etc.) have to be calculated. Furthermore when the numbers of records vary from one animal to the other (unequal number of records per animal), the product-moment correlation cannot be computed.

Under such situations, which are common in farm animals, repeatability can be more conveniently estimated as the intra-class correlation between repeated records using the 'Analysis of Variance' technique. The intra-class correlation estimates the overall correlation between repeated records as a ratio of variances. The procedure for calculation of repeatability, through analysis of variance, is computationally similar to that for estimation of heritability from half-sib analysis. The total phenotypic variation for repeated records is partitioned into two components:

- Variation due to differences between animals, and estimated as σ^2_B
- Variation due to differences between repeated records on the same animal, and estimated as σ^2_W.

Intra-class correlation is defined as ratio of 'between-group' component of variance to the total observed variance. Therefore, repeatability (r), estimated as intra-class correlation, is the ratio of 'between-animal' component of variance (σ^2_B) to the total phenotypic variance (σ^2_P):

$$\text{Repeatability : } r - \frac{\sigma^2_B}{\sigma^2_P} = \frac{\sigma^2_B}{\sigma^2_B + \sigma^2_W}$$

Computation of repeatability as ratio of variances is based on the principle that the 'within-animal' component of variance (σ^2_W) is entirely environmental in nature, and is an estimate of the temporary environmental variation (σ^2_{TE}). The 'between-animal' component of variance (σ^2_B) is composed of genetic variance (σ^2_G) and the permanent environmental variance (σ^2_{EP}). The ratio of σ^2_B to σ^2_P is therefore an estimate of repeatability :

$$r = \frac{\sigma^2_B}{\sigma^2_P} = \frac{\sigma^2_B}{\sigma^2_B + \sigma^2_W} = \frac{\sigma^2_G + \sigma^2_{EP}}{(\sigma^2_G + \sigma^2_{EP}) + \sigma^2_{ET}}$$

The analysis of variance table for partitioning the observed variation into 'between-animal' and 'within-animal' components is shown below:

Source	d.f.	Sum of squares	Mean squares	Composition of MS
Between animals	L – 1	SS_B	MS_B	$\sigma^2_W + K\sigma^2_B$
Within animals	N – L	SS_W	MS_W	σ^2_W
Total	N – 1	SS_T		

Where : L = number of animals

K = number of repeated records per animal

N = total number of records in the group (N = K × L)

The steps involved in obtaining the sum of squares (SS_B, SS_W and SS_T) are the same as those followed for estimation of h^2 from half-sib analysis, and hence not outlined here. The different components of variance are obtained from 'ANOVA' as follow:

'Within-animal' component of variance: $\sigma^2_W = MS_W$

'Between-animal' component of variance:

$$\sigma^2_B = \frac{MS_B - MS_W}{K}$$

$$\text{Repeatability} = \frac{\sigma^2_B}{\sigma^2_B + \sigma^2_W}$$

4.15. Fat percentage of four cows of Holstein Friesian breed was recorded for eight consecutive days and is tabulated below:

Cow	Fat % on day:							
ID	1st	2nd	3rd	4th	5th	6th	7th	8th
1	4	4	5	6	7	5	3	4
2	7	9	8	6	8	6	5	7
3	10	8	9	7	6	7	8	9
4	3	4	5	6	4	3	5	6

Calculate the repeatability of fat percentage.

Solution :

Let : L = number of cows (animals) = 4

K = number of repeated records per cow = 8

N = total number of records in the sample (N = K x L) = 32

Grand total: GT: Sum of all the values = X.. = 194

Correction term: $CT = \frac{(GT)^2}{N} = \frac{(194)^2}{32} = 1176.125$

Total sum of squares (corrected):

$$SS_T = (4^2 + 4^2 + \ldots\ldots + 5^2 + 6^2) - CT$$
$$= 1292 - 1176.125 = 115.875$$

Between-cow sum of squares (corrected): SS_B

$$= [(\text{Sum for cow 1})^2 + \ldots.. + (\text{Sum for cow 4})^2]/\ 8 - CT$$
$$= 1246.5 - 1176.125 = 70.375$$

(iv) Within-cow sum of squares (corrected):

$$SS_W = SS_T - SS_B = 115.875 - 70.375 = 45.5$$

Analysis of Variance Table

Source	d.f	Sum of squares	Mean squares
Between cows(animals)	L – 1 = 3	70.375	23.458
Within cows(animals)	N – L = 28	45.5	1.625
Total	N – 1 = 31	115.875	—

$$\sigma^2_W = MS_W = 1.625$$

$$\sigma^2_B = \frac{MS_B - MS_W}{K} = \frac{23.458 - 1.625}{8} = \frac{21.838}{8} = 2.729$$

$$\text{Repeatability} = \frac{\sigma^2_B}{\sigma^2_B + \sigma^2_W} = \frac{2.729}{2.729 + 1.625} = \frac{2.729}{4.354} = 0.627$$

Computation of Most Probable Producing Ability (MPPA)

4.16. A dairy farmer has three cows whose milk yield records (liters) for different lactations are tabulated below:

Cow ID	Milk yield in lactation :			
	1st	2nd	3rd	4th
A	2070	2130	—	—
B	2110	2100	2060	2090
C	2120	—	—	—

Herd average = 2000 liters. Assume repeatability of lactation yield to be 0.3.

(a) Calculate the MPPA of the three cows and indicate which two of the three cows should be retained in the herd for future lactations?

(b) If repeatability of milk yield were 0.8, calculate the MPPA

(c) Compare the ranking of the cows for the two values (levels) of repeatability ($r = 0.3$ and $r = 0.8$) and draw inferences about the relative ranking.

Solution :

(a) MPPA = Herd average + *b* (Cow average – Herd average)

Where, '*b*' is the weighting coefficient which is computed by the following expression

$$b = \frac{(nr)}{[1+(n-1)r]}$$

Repeatability (*r*) = 0.3

The computations for MPPA are shown below:

Cow ID	No. of records (n)	Milk yield in lactation : 1st	2nd	3rd	4th	Cow avg.	Herd avg.	Weighting coefficient (b)*	MPPA (liters)	Rank
A	2	2070	2130	—	—	2100	2000	0.462	2046.1	2nd
B	4	2110	2100	2060	2090	2090	2000	0.632	2056.8	1st
C	1	2120	—	—	—	2120	2000	0.300	2036.0	3rd

Cows 'B' and 'A' should be retained as their MPPA is higher than of 'C'.

Note: Cow 'C' has the highest milk yield but it has not been retained (selected) because its milk yield is based on a single record, and hence it got lowest weighting coefficient.

(b) For repeatability (*r*) = 0.8, the computations for MPPA are shown below:

Cow ID	(n)	Cow avg.	Herd avg.	Weighting coefficient (b)	MPPA (liters)
A	2	2100	2000	0.889	2088.8
B	4	2090	2000	0.941	2084.7
C	1	2120	2000	0.800	2096.0

(c) Comparison of relative rankings of cows for the two values(levels) of repeatability:

Cow ID	Weighting coefficient (b) for: r = 0.3	r = 0.8	MPPA for: r = 0.3	r = 0.8	Ranking for: r = 0.3	r = 0.8
A	0.462	0.889	2046.2	2088.9	2nd	2nd
B	0.632	0.941	2056.8	2084.7	1st	3rd
C	0.300	0.800	2036.0	2096.0	3rd	1st

With the change in the value (level) of repeatability, the relative ranking of cows has changed. Cow 'B' which had the highest MPPA (2056.8) at low level of repeatability (*r* = 0.3) has the lowest MPPA (2048.7) at high value of repeatability (*r* = 0.8). The change in ranking has occurred because of change in weighting coefficients assigned to the cow's average performance. For example, for 'r' = 0.3, cow 'B' got about twice the weighting coefficient as compared to that for cow 'C', whereas, for r = 0.8, cow 'B' got only slightly higher weighing coefficient as compared to cow 'C'.

- ***Inference:*** Increase in the number of records per cow (animal) is more useful when the repeatability of the trait is low.

Computation of Genetic and Phenotypic Correlations

4.17. Given below is the information for egg number (up to 72 weeks of age) and egg weight in a strain of White Leghorn chickens:

Parameter	Egg number	Egg weight
Mean :	300 eggs	55 g
Phenotypic variance:	1600 eggs^2	20 g^2
Heritability (h^2):	0.20	0.60
Genetic covariance :	– 20.0	
Phenotypic covariance :	– 15.0	

Calculate the genetic and phenotypic correlations between egg number and egg weight and interpret their values.

Solution :

Computation of genetic (r_G) and phenotypic (r_P) correlations between two traits, X and Y, requires the following information:

Genetic covariance: $\sigma_{A\ (XY)}$, Phenotypic covariance: $\sigma_{P\ (XY)}$

Additive genetic variances: $\sigma^2_{A(x)}$, $\sigma^2_{A(Y)}$, and phenotypic variances: $\sigma^2_{P\ (X)}$, $\sigma^2_{P(Y)}$

Since the additive genetic variances for the two traits are not given, these can be calculated from h^2 and phenotypic variances as follows:

$$\sigma^2_{A\ (X)} = h^2 \times \sigma^2_{P(X)} = 0.20 \times 1600 = 320 \text{ egg}^2$$

$$\sigma^2_{A\ (Y)} = h^2 \times \sigma^2_{P(Y)} = 0.60 \times 20.0 = 12 \text{ g}^2$$

The genetic [$r_{G\ (XY)}$] and phenotypic [$r_{P(xy)}$] correlations between egg number and egg weight are calculated as follows:

$$r_{G\ (XY)} = \frac{\text{Genetic cov}(X, Y)}{\sqrt{\text{Genetic Var}(X) \times \text{Genetic Var}(Y)}}$$

$$= \frac{\sigma_{A(X,Y)}}{\sqrt{\sigma^2_{A(X)} \times \sigma^2_{A(Y)}}} = \frac{-20.0}{\sqrt{320 \times 12.0}} = -0.320$$

$$r_{P\ (XY)} = \frac{\text{Phenotypic cov}(X, Y)}{\sqrt{\text{Phenotypic Var}(X) \times \text{Phenotypic Var}(Y)}}$$

$$= \frac{\sigma_{P(X,Y)}}{\sqrt{\sigma^2_{P(X)} \times \sigma^2_{P(Y)}}} = \frac{-15.0}{\sqrt{1600 \times 20.0}} = -0.083$$

Interpretation:

- Egg number and egg weight are negatively correlated at the genetic level. This means that selection for the improvement of egg number would lead to decline in egg weight of the progeny generation. Alternatively selection for increase of egg weight would lead to decline in egg number.
- The phenotypic correlation between egg number and egg weight is also negative but very low in magnitude. The negative phenotypic correlation means that, within the flock, the hens producing more number of eggs tend to have relatively lower egg weight & *vice-versa*.

4.18. From the information given below calculate the :

(a) heritability of milk and protein yield

(b) genetic and phenotypic correlations between them

Parameter	Milk yield	Protein yield
Phenotypic variance:	30,00,000 Kg^2	1800 Kg^2
Genetic variance(additive):	7,00,000 Kg^2	600 Kg^2
Genetic covariance :	18,000	
Phenotypic covariance :	60,000	

Solution :

(a) h^2 of milk yield $= \dfrac{7,00,000}{30,00,000} = 0.23$

h^2 of protein yield $= \dfrac{600}{1800} = 0.33$

$$r_{G\ (XY)} = \frac{\text{Genetic cov. (X,Y)}}{\sqrt{\text{Genetic Var(X)} \times \text{Genetic Var(Y)}}} = \frac{18000}{\sqrt{7,00,000 \times 600}} = 0.878$$

$$r_{P\ (XY)} = \frac{\text{Phenotypic cov.(X,Y)}}{\sqrt{\text{Phenotypic Var(X)} \times \text{Genetic Var(Y)}}} = \frac{60,000}{\sqrt{30,00,000 \times 1800}} = 0.817$$

Estimation of Correlations by Half-Sib Analysis of Variance and Covariance

The principle underlying the estimation of genetic correlations is same as that used for estimation of heritability. Broadly it involves:

- Conducting the analysis of variance and covariance
- Computing the components of variances(between- and within- sire components: σ^2_S and σ^2_W) and covariance (between-and within- sire components: $\sigma_{S(X,Y)}$ and $\sigma_{W(X,Y)}$)
- Computing the correlations from ratio of covariance and variances.

The use of ANOVA for calculation of sire- and within-sire components of variance has been outlined in an earlier section of this chapter. The analysis of covariance table for calculating the different components of covariances is given below:

Analysis of Covariance Table

Source	d.f.	Sum of cross products (SCP)	Mean cross products (MCP)	Composition of MCP
Between sires	S – 1	SCP_S	MCP_S	Cov_W + K Cov_S
Within sires	N – S	SCP_W	MCP_W	Cov_W
Total	N – 1	SCP_T	—	—

Where:

S = Number of sires

N = Total number of paired observations (i.e. the total number of animals)

K = Number of progeny per sire

Cov_S = Between-sire component of covariance

Cov_W = Within-sire component of covariance

The computations involved in obtaining the sum of cross products(SCP) and mean cross products (MCP) are similar to those involved in obtaining the sum of squares(as in case of ANOVA). The sire-component (Cov_S) and within-sire component (Cov_W) are calculated from SCP and MCP (the detailed calculations of SCP and MCP are given in the next Problem) by equating the SCP and MCP to their expected values (*i.e.* the composition of MCP given in the last column of analysis of covariance table) and solving for Cov_S and Cov_W as shown below :

Within-sire component of covariance:

$$COV_W = MCP_W$$

Between-sire component of covariance:

$$COV_S = \frac{MCP_S - MCP_W}{K}$$

'Sire-component of covariance' (Cov_S) is an estimate of ¼th of the additive genetic covariance:

$$COV_S = \frac{1}{4} Cov_{A(X,Y)}$$

'Within-sire component of covariance' (Cov_w) is an estimate of the remaining genetic covariance (¾ $Cov_{A(X,Y)}$ + $Cov_{D(X,Y)}$, + $Cov_{I(X,Y)}$) and the environmental covariance ($Cov_{E(X,Y)}$):

$$Cov_W = \left[\frac{3}{4} Cov_{A(X,Y)} + Cov_{D(X,Y)} + Cov_{I(X,Y)} + Cov_{E(X,Y)}\right]$$

The sum of 'sire-component' and 'within-sire component' of covariance equals the total phenotypic covariance:

$$Cov_S + Cov_w = \left[\left(\frac{1}{4} Cov_{A(X,Y)}\right) + \left(\frac{3}{4}(Cov_{A(X,Y)}) + Cov_{D(X,Y)} + Cov_{I(X,Y)} + Cov_{E(X,Y)}\right)\right]$$

$$Cov_S + Cov_w = \left[\left((Cov_{A(X,Y)}) + Cov_{D(X,Y)} + Cov_{I(X,Y)} + Cov_{E(X,Y)}\right)\right] = Cov_{P(X,Y)}$$

The estimates of phenotypic, genetic and environmental correlations are obtained as follow :

Phenotypic correlation (r_P):

$$r_{P(XY)} = \frac{Cov_{P(X,Y)}}{\sqrt{\sigma^2_{P(X)} \times \sigma^2_{P(Y)}}} = \frac{Cov_{S(X,Y)} + Cov_{W(X,Y)}}{\sqrt{(\sigma^2_{W(X)} + \sigma^2_{S(W)}) \times (\sigma^2_{W(Y)} + \sigma^2_{S(Y)})}}$$

Genetic correlation (r_G):

$$r_{G(XY)} = \frac{Cov_{S(X,Y)}}{\sqrt{(\sigma^2_{S(X)} \times (\sigma^2_{S(Y)})}} = \frac{¼ Cov_{A(X,Y)}}{\sqrt{¼\sigma^2_{A(X)} \times ¼\sigma^2_{A(Y)}}} = \frac{Cov_{A(X,Y)}}{\sqrt{\sigma^2_{A(X)} \times \sigma^2_{A(Y)}}}$$

Note: Though we call the genetic association between the two traits as genetic correlation but it should be kept in mind that actually it is the correlation between breeding values of the two traits.

Environmental correlation (r_E):

$$r_{E(XY)} = \frac{Cov_{W(X,Y)}}{\sqrt{\sigma^2_{W(X)} \times \sigma^2_{W(Y)}}}$$

4.19. Three dairy sires were each mated to three dams, and each dam in turn produced one daughter. Milk yield per day (Kg) and fat % recorded on daughters are tabulated below:

Sire 1		Sire 2		Sire 3	
MY/d (X)	Fat % (Y)	MY/d (X)	Fat % (Y)	MY/d (X)	Fat % (Y)
23.9 (X_{11})	6.2 (Y_{11})	20.0 (X_{21})	6.0 (Y_{21})	25.5 (X_{31})	5.5 (Y_{31})
22.0 (X_{12})	6.0 (Y_{12})	21.9 (X_{22})	6.0 (Y_{22})	23.3 (X_{32})	5.8 (Y_{32})
25.0 (X_{13})	5.5 (Y_{13})	23.8 (X_{23})	5.8 (Y_{23})	23.2 (X_{33})	5.6 (Y_{33})
$X_{1.}$ = 70.9	$Y_{1.}$ = 17.7	$X_{2.}$ = 65.7	$Y_{2.}$= 17.8	$X_{3.}$= 72.0	$Y_{3.}$=16.9

Estimate the genetic and phenotypic correlations between milk yield and fat %, and comment on their values.

Solution :

Steps in the calculation of genetic and phenotypic covariances:

(i) Grand total of X: GT_x or X.. = $X_{1.} + X_{2.} + X_{3.}$ = 70.9 + 65.7 + 72.0 = 208.6

Grand total of Y: GT_y or Y.. = $Y_{1.} + Y_{2.} + Y_{3.}$ = 17.7 + 17.8 + 16.9 = 52.4

Alternately, the GT's can be obtained by summing all the values for a trait as follow:

$$GT_X \text{ or } X.. = X_{11} + X_{12} + \ldots\ldots\ldots\ldots + X_{33} = 208.6$$

$$GT_Y \text{ or } Y.. = Y_{11} + Y_{12} + \ldots\ldots\ldots\ldots + Y_{33} = 52.4$$

(ii) Correction term (factor) :

$$CT_{XY} = \frac{(X..)\times(Y..)}{N} = \frac{208.6\times52.4}{9} = 1214.516$$

Where 'N' is number of paired observations

(iii) Total sum of cross products (uncorrected) : SCP_T

$$= (X_{11} \times Y_{11}) + (X_{12} \times Y_{12}) + \ldots\ldots\ldots\ldots + (X_{33} \times Y_{33})$$

$$= (23.9 \times 6.2) + (22.0 \times 6.0) + \ldots\ldots\ldots + (23.2 \times 5.6)$$

$$= 1212.43$$

Total sum of cross products (corrected)

$$= SCP_T - CT_{XY} = 1212.43 - 1214.516 = -2.086$$

(iv) Between-sire sum of cross products (corrected): (SCP_S)

$$= \frac{(X_{1.} \times Y_{1.}) + (X_{2.} \times Y_{2.}) + (X_{3.} \times Y_{3.})}{K} - CT_{XY}$$

$$= \frac{(70.9\times17.7)+(65.7\times17.8)+(72.0\times16.9)}{3} - 1214.516$$

$$= \frac{3641.191}{3} - 1214.516 = 1213.73 - 1214.516 = -0.7856$$

(v) Within-sire sum of cross products:

$$SCP_W = (SCP_T - SCP_S)$$

$$= -2.086 - (-0.7856) = -1.3$$

Analysis of Covariance Table

Source	d.f.	Sum of cross products (SCP)	Mean cross products (MCP)	Composition of MCP
Between sires	2	– 0.7856	– 0.3928	$Cov_W + K.Cov_S$
Within sires	6	– 1.3	– 0.2167	Cov_W
Total	8	– 2.086	—	—

Where, K = number of progeny per sire = Total number of progeny/number of sires = 9/3 = 3

$$Cov_W = MCP_W = -0.2167$$

$$Cov_S = \frac{MCP_S - MCP_W}{K} = \frac{-0.3928-(-0.2167)}{3} = -0.0587$$

$$Cov_T = COV_W + COV_S = -0.2167 - 0.0587 = -0.2753$$

For obtaining genetic and phenotypic correlations, we also need the values of genetic (additive) and phenotypic variances. These are obtained from the ANOVA conducted on the two traits (X and Y) .The 'ANOVA' for milk yield and fat % is shown below:

Source	d.f.	Milk yield (X)		Fat % (Y)	
		SS	MS	SS	MS
Between sires	2	7.549	3.774	0.162	0.081
Within sires	6	15.207	2.534	0.333	0.056
Total	8	22.756	—	0.496	—

Computation of components of variance for milk yield (X):

$$\sigma^2_{W(X)} = 2.534 \text{ and } \sigma^2_{S(X)} = \frac{3.774-2.534}{3} = 0.413$$

$$\sigma^2_{P(X)} = \sigma^2_{S(X)} + \sigma^2_{W(X)} = 0.413 + 2.534 = 2.947$$

Computation of components variance for fat % (Y):

$$\sigma^2_{W(Y)} = 0.056 \text{ and } \sigma^2_{S(Y)} = \frac{0.081-0.056}{3} = 0.0085$$

$$\sigma^2_{P(Y)} = \sigma^2_{S(Y)} + \sigma^2_{W(Y)} = 0.056 + 0.0085 = 0.0645$$

Computation of correlations:

$$r_{G\,(XY)} = \frac{\text{Genetic cov}(X,Y)}{\sqrt{\text{Genetic Var}(X) \times \text{Genetic Var}(Y)}} = \frac{Cov_{S(XY)}}{\sqrt{\sigma^2_{S(X)} \times \sigma^2_{S(Y)}}}$$

$$r_{G\,(XY)} = \frac{-0.0587}{\sqrt{0.413 \times 0.0085}} = \frac{-0.0587}{0.0592} = -0.992$$

$$r_{P\,(XY)} = \frac{\text{Phenotypic cov}(X,Y)}{\sqrt{\text{Phenotypic Var}(X) \times \text{Phenotypic Var}(Y)}} = \frac{\sigma_{P(XY)}}{\sqrt{\sigma^2_{P(X)} \times \sigma^2_{P(Y)}}}$$

$$r_{P\,(XY)} = \frac{-0.2753}{\sqrt{2.947 \times 0.0645)}} = \frac{-0.2753}{0.4359} = -0.632$$

- ***Inference***: Genetic and phenotypic correlations between milk yield and fat % are negative. The negative genetic correlation means that selection for the improvement of milk yield would lead to decrease in fat % of the progeny generation and *vice versa*. The negative phenotypic correlation between these traits means that cows with high milk yield tend to have lesser fat % in their milk and *vice-versa*. Hence these correlations are undesirable.

Supplementary Problems

4.20. Given: σ_G^2 = 15 and σ_P^2 = 25, calculate the heritability and interpret its value.

4.21. Given: H^2 = 0.9 and σ_G^2 = 45, what does σ_P^2 equal ?

4.22. From the information given below calculate the dominance and environmental variances:

Heritability =0.3 , phenotypic variance = 200 lb^2, genetic variance= 100 lb^2 , epistatic variance = 0.0

4.23. Yearly wool yields (lb) of 10 sheep of a Merino flock were: 11.8, 8.4, 9.5, 10.0, 10.9, 7.8, 10.8, 8.5, 11.8, and 10.5. Suppose the phenotypic and additive genetic variances for wool yield are 0.60 and 2.0 lb^2, what is the heritability of wool yield in this flock?

4.24. A non-breeder states that the narrow–sense heritability for human stature (height) is 0.7 whereas the broad-sense heritability is 0.3. Is the statement correct? Explain.

4.25. If environmental variance for a trait is three times as large as the genetic variance, what will be the broad-sense heritability?

4.26. A very large herd of Sahiwal cattle grown under farmer's field in loose-housing system had daily milk yield of 10 Kg with standard deviation of 2 Kg^2. The same herd when shifted to highly uniform conditions of management, nutrition etc. had daily milk yield of 12 Kg. In which of the two cases higher heritability is expected? Explain

4.27. The components of variance for two traits of sheep are shown below:

Variance component	Weaning wt. (kg^2)	Staple length (cm^2)
Phenotypic :	20.0	1.00
Total genetic :	16.0	0.35
Environmental :	4.0	0.65
Additive genetic:	8.0	0.30

(a) Calculate the non–additive genetic component of variance, and the broad-and narrow-sense heritabilities for the two traits.

(b) What inferences can be drawn about the type of gene action and the choice of basis of selection to be used for the genetic improvement of each trait?

4.28. Staple length (cm) of lambs and of their sires and dams at first shearing/clipping were recorded in a flock of sheep, and are given below. Calculate the h^2 of staple length using the regression of offspring on average of parents.

Sire:	5	9	8	6	7	5	10	7	10	9
Dam:	9	11	8	6	7	8	8	7	8	7
Lamb:	6	6	8	5	10	7	11	7	9	7

4.29. Given below are the gain in weights of sheep (lb) recorded on sires and their progeny.

Sire :	2	7	9	9	10	10	11	11	12	12	13	13	13	14	14	14	15
Progeny :	12	14	14	15	16	15	15	15	14	15	15	16	14	14	17	15	15

Sum of values for: sires = 189; dams = 251

Which type of heritability (broad- or narrow-sense) can be computed from this data? What will be its value?

4.30. What would be the heritability estimated from each of the following correlations or regressions, assuming that resemblance between relatives due to environment or dominance is absent?

(a) Regression of offspring on father = 0.21

(b) Regression of offspring on mother = 0.27

(c) Regression of offspring on mean of parents = 0.32

(d) Correlation between half sibs = 0.02

4.31. In a flock of broiler chicken, ten sires were each mated to five dams, and each dam in turn produced one progeny. Body weight at 5 weeks was recorded on all the progeny. The corrected sums of squares, obtained from the ANOVA, were as follows:

Sum of squares due to sires = 400

Sum of squares within sires = 800

(a) What type of sib analysis (full- or half-sib) can be used to conduct the ANOVA, and to estimate the heritability?

(b) Set up the analysis of variance table giving the degrees of freedom, sum of squares and mean sum of squares.

(c) Compute the 'sire' and 'within-sire' components of the variance.

(d) Calculate the intra-class correlation between half sibs and heritability of the trait.

4.32. In a flock of desi swine, three sires were each mated to five dams and each dam produced one female progeny. Data recorded for three traits on the progeny of the three sires are given below:

Weaning weight (kg)			Yearling weight (kg)			Feed conversion ratio		
Sire 1	Sire 2	Sire 3	Sire 1	Sire 2	Sire 3	Sire 1	Sire 2	Sire 3
12	12	12	90	90	65	3.4	3.8	4.0
16	12	11	70	60	70	4.2	3.6	4.6
14	16	10	80	75	60	4.6	4.0	4.2
18	14	13	70	70	70	4.2	4.2	4.4
16	12	16	80	80	80	3.0	3.8	4.4

(a) Construct the analysis of variance table for calculation of heritability.

(b) Calculate the intra-class correlation between sibs (t) and the heritability.

4.33. Body weight gain from day-old to five weeks of age in chicken was analyzed using full sib analysis. The mating structure involved equal number of dams mated to each sire and each dam producing equal number of offspring. The ANOVA is given below:

Source	d.f.	MS	Composition of MS
Between sires	16	34.1	$\sigma^2_W + n\sigma^2_D + nd\,\sigma^2_S$
Between dams/sires	51	17.6	$\sigma^2_W + n\sigma^2_D$
Between progeny/dams/sire	136	9.3	σ^2_W

(a) List the number of sires, number of dams per sire and number of progeny per dam that were used in the analysis.

(b) Calculate the total variance and the components of variance for gain in weight.

4.34. Six-week body weight of 24 birds of a White plumage strain of Japanese quail maintained at GADVASU, Ludhiana poultry farm, resulting from mating of four sires each mated to two dams, and each dam producing three progeny, are given below:

Sire ID	Dam ID	Progeny ID	6-week body wt(g)	Sire ID	Dam ID	Progeny ID	6-week body wt(g)
		1	280			13	260
	1	2	290		5	14	260
1		3	270	3		15	280
		4	300			16	270
	2	5	330		6	17	290
		6	310			18	290
		7	280			19	270
	3	8	280		7	20	280
2		9	250	4		21	270
		10	300			22	250
	4	11	310		8	23	210
		12	280			24	320

(a) Which type of sib analysis (full- or half-sib) can be used to calculate the heritability?

(b) Calculate the heritability of six week body weight (from this small data).

(c) What genetic inferences can be drawn from the h^2 obtained from different components

4.35. Given below are the litter size (number of offspring born per litter) of five consecutive litters produced by four female rabbits.

Rabbit	Litter number :				
ID	1st	2nd	3rd	4th	5th
1	3	4	2	3	3
2	2	1	2	4	2
3	3	4	3	3	4
4	3	3	4	2	1

(a) Calculate the between- and within-animal components of variances.

(b) Calculate the repeatability of litter size.

4.36. The following information on milk yield (kg) was available on five cows from their first five lactations in each case:

Total sum of squares (corrected) = 9200 Kg^2

Between-cow sum of squares (corrected) = 4350 Kg^2

Calculate the:

(a) Components of variances required for calculation of repeatability.

(b) Repeatability of milk yield.

4.37. What would be the most probable producing ability of a heifer cow being kept in a herd averaging 2100 kg of milk in a lactation ?

4.38. Repeatability of milk yield and calving interval are 40 and 15 per cent, respectively. How many number of records per animal are required for calving interval so as to have almost as much confidence/ reliability for culling for calving interval as one could have for a single record of milk yield ?

4.39. Given below are the wool yield records of eight sheep in three consecutive clippings/ shearings:

Sheep	Shearing no.:		
ID	1st	2nd	3rd
1	28	29	27
2	30	33	31
3	28	28	25
4	30	31	28
5	26	26	28
6	27	29	29
7	27	28	27
8	25	21	32

(a) Set up the ANOVA table.

(b) Calculate the between- and within-sheep components of variance.

(c) Calculate the repeatability of wool yield.

4.40. Repeatability of milk yield in cattle is 30 per cent. In a herd that averaged 2400 kg, cow 'A' with four lactations was superior over herd average by 600 kg and cow 'B' with two lactation records was superior over herd average by 700 kg. Which one of the two cows is superior for keeping in the herd for future lactation? Explain

4.41. Racing performance of six thoroughbred horses, measured as time to run 2 km (in seconds) at two competitions was recorded. The horses participated in the first competition held at Bangalore race course; and in the second competition (6 months later), at Calcutta race course. The racing performance records are tabulated below:

Horse	Time to run 2 km (seconds)	
ID	Bangalore (1st)	Calcutta (2nd)
1	200	190
2	145	143
3	190	185
4	140	138
5	150	175
6	155	160

(a) Set up the analysis of variance table for computation of repeatability showing the source of variation, degrees of freedom, sum of squares and mean squares.

(b) Calculate the repeatability of racing performance.

(c) Is the performance in 1st race a good indicator of performance in the 2nd race? Explain.

4.42. Fleece weights in different shearings of 50 sheep with equal number of clipping per sheep were recorded. The ANOVA conducted on these records is given below:

Source	d.f.	Mean squares
Between sheep	49	24.0
Within sheep	100	3.0

(a) How many shearings per animal were made?

(b) Calculate the repeatability of fleece weight.

(c) What would be your guess about the magnitude of broad-sense heritability of fleece weight?

4.43. Estimate the genetic and phenotypic correlations between egg number (EN) and feed conversion ratio (FCR) in a poultry flock from the following information obtained from six sires each having fifteen daughters:

Components of variance & Covariance	Trait	
	EN	FCR
Between-sire component of variance : σ^2_S =	1974.0	1374.0
Within-sire component of variance : σ^2_W =	16102.8	1023.1
Between-sire component of covariance: $\sigma_{S(X,Y)}$ = 263.0		
Within-sire component of covariance : $\sigma_{W(X,Y)}$ = 1142.4		

4.44. In a flock of swine, body weight at birth and marketing-age were recorded on 200 offspring which resulted from mating of 20 sires each mated to 10 dams with each dam producing one progeny. The analysis of variance and covariance table is given below:

Source	d.f.	Sum of squares for :		Sum of
		Birth wt.	Marketing wt.	cross products
Between sires	19	300	260	220
Within sires	190	1600	1200	900
Total	209	1900	1460	1120

(a) Calculate the components of variances for birth and marketing weight.

(b) Calculate components of covariance between birth and marketing weight.

(c) Calculate the genetic and phenotypic correlations between birth weight and marketing-age weight.

(d) If the breeder selects for the genetic improvement of birth weight, what will be its genetic effect on marketing weight (correlated response)?

Answers to Supplementary Problems

4.20. Heritability = 0.6. Heritability is very high, therefore, genetic effects contribute much more to phenotypic variability than do the environmental effects.

4.21. $\sigma_P^2 = 50$

4.22. $\sigma_D^2 = 40$; $\sigma_E^2 = 100$

4.23. Heritability = 0.3

4.24. No, because heritability in the narrow-sense can't be more than the broad-sense heritability.

4.25. Heritability = 0.25

4.26. Higher h^2 is expected in second case because environmental variance would be lesser under uniform environmental conditions.

4.27. (a)

Parameter	Weaning wt. (Kg^2)	Staple length (cm^2)
Non-additive genetic variance :	8.00	0.05
Broad-sense heritability :	0.80	0.35
Narrow-sense heritability:	0.40	0.30

(b) ***Inferences:***

- Weaning weight is influenced both by additive and non-additive gene action whereas staple length is influenced primarily by additive gene action.
- Narrow-sense heritability (h^2) of both the traits is moderately high, therefore, individual selection (i.e. selection based on individuals' own performance) can be effectively used for their genetic improvement

4.28. $h^2 = 0.51$

4.29. Narrow-sense heritability; $h^2 = 0.42$

4.30. (a) 0.42 (b) 0.54 (c) 0.32 (d) 0.08.

4.31. (a) Half-sib analysis.

(b) ANOVA table:

Source	d.f.	SS	MS
Between sires	10–1=9	400	44.44
Within sires	50–10=40	800	20.0
Total	49	1200	

(c) $\sigma_S^2 = 4.8$, $\sigma_W^2 = 20.0$

(d) t = 0.198 , $h^2 = 0.774$.

4.32. (a) Analysis of variance Table

Source	d.f.	Weaning weight		Yearling weight		Feed conversion ratio	
		SS	MS	SS	MS	SS	MS
Between sires	2	20.8	10.4	210	105	0.646	0.323
Within sires	12	54.8	4.566	1000	83.33	2.144	0.1768
Total	14	75.6	-	1210	-	2.790	-

(b)

	Weaning weight	Yearling weight	Feed conversion ratio
Intra-class correlation (t) :	0.203	0.049	0.139
Heritability :	0.814	0.197	0.555

4.33. (a) No. of sires = 17, No. dams per sire = 4, No. of progeny per dam = 3

b) $\sigma_P^2 = 13.442$, $\sigma_S^2 = 1.375$, $\sigma_D^2 = 2.77$, $\sigma_W^2 = 9.3$

4.34. (a) Full-sib mating design

(b) Sum of squares: Between sires = 2945.8

Between dams/sires = 3416.7

Between progeny = 8533.0

$\sigma_W^2 = 533$, $\sigma_D^2 = 107.06$ and $\sigma_S^2 = 21.28$

$h_S^2 = 0.1298$, $h_D^2 = 0.6473$, $h_{(S+D)}^2 = 0.3886$

(c) Heritability from dam-component is much higher than from sire-component which means that dominance and maternal effects are also important for body weight.

4.35. (a) $\sigma_B^2 = 0.1016$, $\sigma_W^2 = 0.825$ (b) Repeatability = 0.109

4.36. (a) $\sigma_B^2 = 169.0$, $\sigma_W^2 = 242.5$ (b) Repeatability = 0.410

4.37. MPPA = 2100 kg

4.38. Four records (rounded for whole number)

4.39. (a) Analysis of variance table:

Source	d.f	Sum of squares	Mean squares
Between sheep	7	63.625	9.089
Within sheep	16	85.333	5.333
Total	23	148.958	—

(b) $\sigma_B^2 = 1.252$ $\sigma_W^2 = 5.333$ (c) Repeatability = 0.190.

4.40. Cow 'A' is superior as its MPPA is higher.

4.41. (a) Analysis of variance table:

Source of variation	d.f.	Sum of squares	Mean squares
Between horses	5	5164.75	1032.95
Within horses	6	391.5	65.25
Total	11	5556.25	—

(b) 0.872.

(c) Yes, because repeatability of racing performance is very high. Therefore, the horses would tend to perform in similar order/rank in the second race.

4.42. (a) 3 (b) 0.7

(c) less than 0.7 because bread-sense heritability is generally lower than repeatibility.

4.43. $r_G = 0.159$, $r_P = 0.213$

4.44. (a) Birth weight : $\sigma_S^2 = 0.6901$, $\sigma_W^2 = 8.888$;

Marketing weight : $\sigma_S^2 = 0.7018$, $\sigma_W^2 = 6.666$

(b) Cov_S = Sire-component of covariance = 0.6578, Cov_W = Within-sire component of covariance = 5.0

(c) $r_G = 0.945$, $r_P = 0.6759$

(d) Selection for birth weight will also increase the marketing–age weight since the genetic correlation is positive and very high.

●●●

SECTION - 3

SELECTION

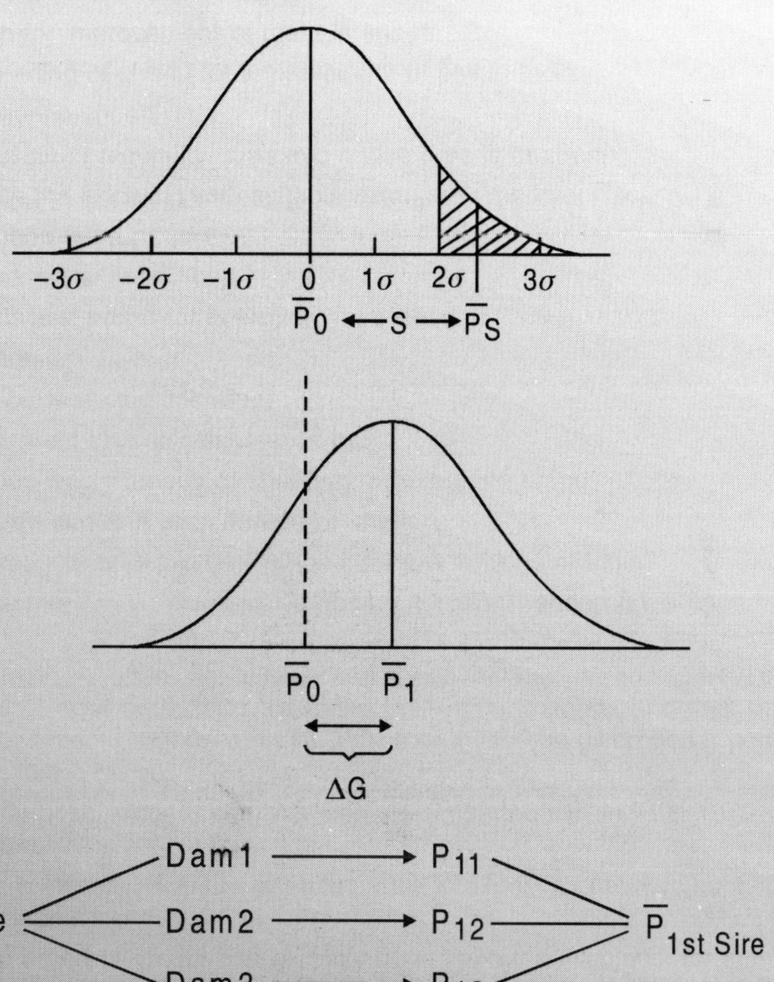

11. What are the main differences in skin structure of exotic fine wool breeds and indigenous breeds of sheep?
12. Categorize the native sheep breeds of four regions into meat or wool.
13. Why is fiber diameter of wool measured at constant temperature and humidity conditions?
14. How is staple length measured?
15. What are the two most important traits contributing to income from wool-type sheep? How are these genetically related to each other (the direction and magnitude of the r_G)? Can these two traits be simultaneously improved in a breeding program? If yes, which selection procedures can be employed for their simultaneous improvement? Is there any evidence to show that these two traits can be simultaneously improved?
16. How is staple length genetically related to fleece yield? Is the r_G between these favorable or unfavorable? Explain
17. List three desirable types of genetic correlations in sheep.
18. At what age are the native sheep of India weaned?
19. Give the range of the following traits of Indian improved breeds of sheep:
    (a) [illegible] wool [illegible] (b) [illegible]
    (c) [illegible] (d) annual greasy fleece weight
20. What is the most [illegible] for the genetic improvement of meat-type sheep? What is the magnitude of its heritability?
21. List the two important components of reproductive performance in sheep. What is the magnitude of [illegible] traits?
22. List [illegible] traits of sheep which can be measured on group basis.
23. List the [illegible] following traits; [illegible] choice of the [illegible]

5

Response to Selection

The principal objective of animal breeder is to bring about a change in genetic composition of population such that the progeny generation is genetically better than the parental population(s).There are two broad ways to bring about genetic improvement: selection (choice of individuals to be used as parents) and mating/crossing system (the ways by which the parents or populations are mated/crossed to produce progeny). Of these two broad ways available for genetic improvement, selection is most important for directed genetic improvement of a population[1], and has consequently been termed keystone in the arch of genetic improvement. This chapter deals with response to selection and the various means of improving response to selection.

Selection and Response to Selection

The fundamental genetic effect of selection, as pointed out earlier, is to change the allelic frequencies: it increases the frequency of desirable alleles at the cost of decrease in the frequency of the undesirable alleles. The change in allelic frequencies in turn leads to change in genetic mean of the population. Since the quantitative traits are controlled by a large number of genes, and further modified by environmental factors, the changes in gene frequencies for such traits cannot be observed. Effectiveness of selection for quantitative traits has, therefore, to be studied at the level of phenotype. Nonetheless, the fundamental genetic effect of selection for quantitative traits is same as that for qualitative traits i.e. increasing the frequencies of favorable alleles/ genes. The effect of changes in the frequencies of favorable alleles at the concerned loci is reflected as an improvement in the mean of the population from one generation to the next.

> *"The key is man's power of accumulative selection : nature gives successive variations; man adds them up in certain directions useful to him".* **Darwin**

[1]Selection practiced within a population is also referred to as intra-population selection.

Natural vs. Artificial Selection

Selection involves differential survival and reproduction of different individuals (genotypes) in the population. It may be mediated by the forces of nature (**natural selection**) or man (**artificial selection**). In nature, the main force responsible for selection is the survival of the fittest: the animals that are best fit in their environment have higher survival rate and produce more number of progeny. The cumulative effect of natural selection over time is to make the population well adapted to its prevailing environment. **Artificial selection** is the process of choosing the individuals to become parents of the next generation. The choice is based on the standards laid down by breeders for a trait or a set of traits. Artificial selection, as compared to natural selection, is a fast process, and can bring about a considerable change in the genetic composition of a population over a short period (Table 5.1).

Table 5.1 Natural vs. artificial selection.

Natural selection	Artificial selection
1. Selection effected by the forces of nature.	1. Selection effected by the efforts of man / breeder.
2. 'Survival of fittest' is the main force in natural selection.	2. Standards laid down by the man/breeder, which are usually economic in nature, are the determining factors.
3. A very slow process and usually takes hundreds/ thousands of generations to produce meaningful effects.	3. Relatively a fast process, and may bring about considerable change in the genetic composition of a population over short periods.
4. Can occur at any stage from zygote formation to adulthood.	4. Can be conducted only when the trait of interest has been recorded.
5. Examples include origin and development of new species of animals and plants.	5. Examples include development of new varieties/breeds of animals and plants suited for economic use.

Natural selection may occur simultaneously with artificial selection. But it may not be as forceful as in nature because the domesticated animals are usually not exposed to the same harsh environmental conditions as the wild animals. When intense artificial selection for a trait is practiced consistantly in one direction over successive generations, natural selection usually comes to oppose it through reduced viability and reproduction. One of the illustrative examples of the conflict between natural and artificial selection is observed in chickens. Artificial selection in broilers for increased body weight over the last three to four decades has led to severe decline in their viability and reproduction. In this chapter we shall deal with artificial selection.

Three Forms of Selection

Artificial selection can be performed in three broad ways: stabilizing, diversifying and directional (Fig. 5.1).

- **Stabilizing selection** involves selecting the individuals which are around the population mean and discarding the individuals with extreme phenotypes (Fig 5.1a). Stabilizing selection aims at keeping the mean of population fairly stable over generations but to decrease its variability. This type of selection is performed when the population has intermediate optima for the trait (e.g. egg weight optima in poultry for hatchability; fat content optima in fishes for filet).

- **Diversifying selection (**also called **disruptive selection**) involves selecting individuals from the two extremes of the distribution (Fig 5.1b). If the selected parents (from the high and low groups) are mated between them, little change in average performance is expected but the variability in the next generation may be increased. But if the selected parents of each extreme group are mated within its own group eventually two distinct subpopulations could emerge. This type of selection is rarely applied in farm animals. The only exception being in laboratory species for specific experimental purposes (e.g. bi-directional selection for testing the validity of quantitative genetic theory).
- **Directional selection** involves choosing the individuals (high performing) from one direction only. If high phenotypic value for a trait is desirable, then we select the animals with high values (e.g. those with high growth rate, milk yield, wool yield etc). If low value is desirable, then we select from the lower extreme side of distribution (e.g. back-fat thickness in swine, time to run a race in horse, age at sexual maturity in laying chicken etc). This is the most common form of selection that is applied in farm animals, and results in improvement of mean performance of the population in succeeding generations (Fig 5.1c). We shall henceforth deal with directional selection.

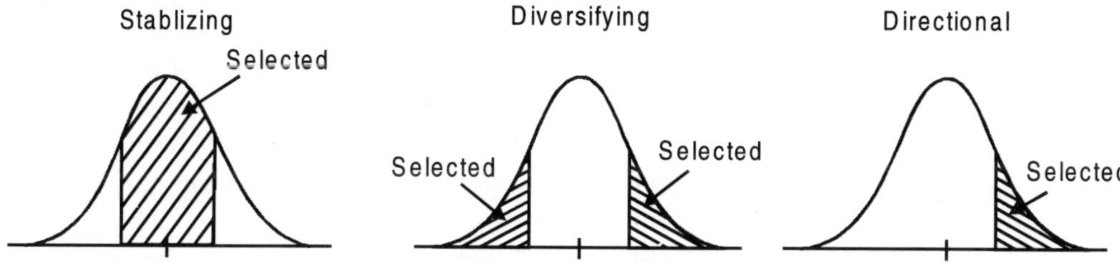

Fig. 5.1 Three forms of selection

Directional selection involves recording the performance of all the individuals (potential parents) of the population, ranking them in order of phenotypic values (merit) and then retaining the required number of individuals. The individuals chosen to become parents of the next generations are superior over the population average and constitute the **selected group** (Fig. 5.2). The remaining animals of the population (excluding the selected ones) are the **culled animals**. These animals are not used to produce the progeny but may be used for commercial purposes (e.g. meat, milk, eggs etc). The process of selection of superior individuals is, therefore, accompanied by the reverse process of culling of undesirable animals.

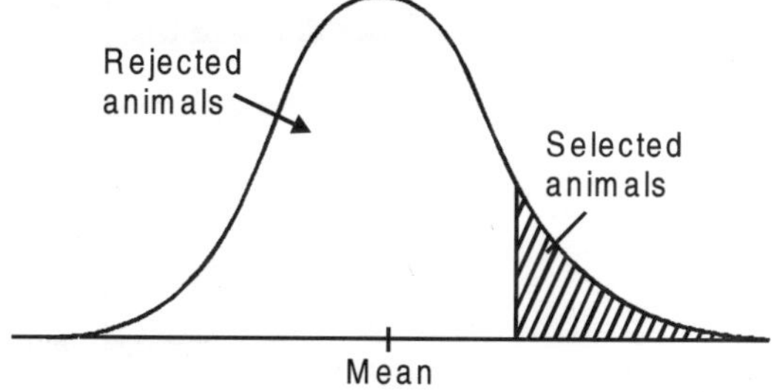

Fig. 5.2 Illustration of directional selection for a quantitative trait showing selected and rejected animals

Response to Selection: Genetic Gain

Consider directional selection for improvement of a quantitative trait by selecting individuals on their own performance.Let, $\bar{P}_0$ be mean of population and $\bar{P}_S$ be mean of animals selected as parents. Naturally, the average performance of selected individuals is expected to be better than the population average. The difference between the mean of selected individuals and the mean of the population before selection is termed **selection differential**[2], and designated as 'S':

Selection differential = Mean of selected animals – Mean of population

Symbolically: $S = \bar{P}_S - \bar{P}_0$

Selection differential is thus a measure of phenotypic superiority of selected individuals (parents) over population average (before selection). Selection differential may be positive or negative depending on whether the trait of interest is to be increased or decreased. For example, if selection is practiced for increasing milk production, selection differential would be positive. On the other hand, if selection is practiced for reducing age at first calving, selection differential would be negative. Directional selection for a normally distributed trait is diagrammatically shown in Fig 5.3.

Random mating among selected animals (males and females) would produce the offspring generation with mean of $\bar{P}_1$. Assuming that the environmental conditions remain same from parental to progeny generation, the difference in average performance between the parental and progeny generation, $\bar{P}_1 - \bar{P}_0$, represents the response achieved through selection.

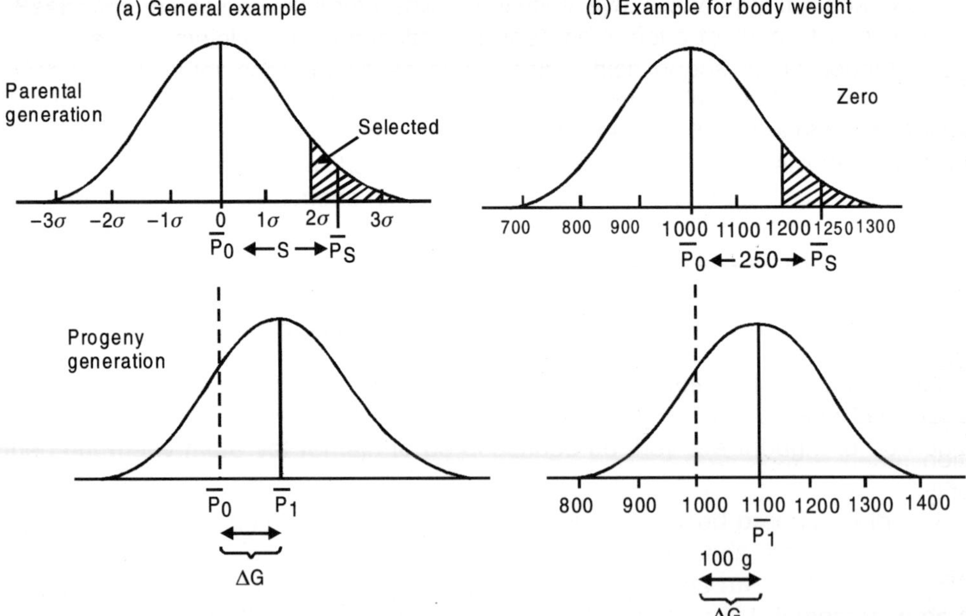

Fig 5.3 Illustration of response to selection (a) General example where the horizontal axis is in units of phenotypic standard deviation. (b) Example of 6-week body weight in chickens (mean = 1000 g and σ_P = 100g; h^2 = 0.4). The Y-axis in both the cases represents the number of individuals or the frequency (%) of individuals.

[2]Selection differential is sometimes also referred to as 'reach' or 'attempted gain'.

The change in mean performance from parental to progeny generation brought about by selection is referred to as **'response to selection'** (R) or **'genetic gain'** (ΔG)[3] per generation:

Genetic gain/generation = $\bar{P}_S$ mean of progeny generation – mean of parental generation.

Symbolically, $\Delta G = \bar{P}_S - \bar{P}_0$ or $R = \bar{P}_1 - \bar{P}_0$

When selection is continued over successive generations, the mean of population would continue to improve generation after generation as shown in Fig 5.4. Each generation of selection leads to accumulation of genetic gain as every generation of selection begins from the performance level that was achieved in the previous generation. Directional selection thus produces rather permanent or long-lasting, results. For example, performance level of generation five is based on accumulation of genetic gains from all the previous generations.

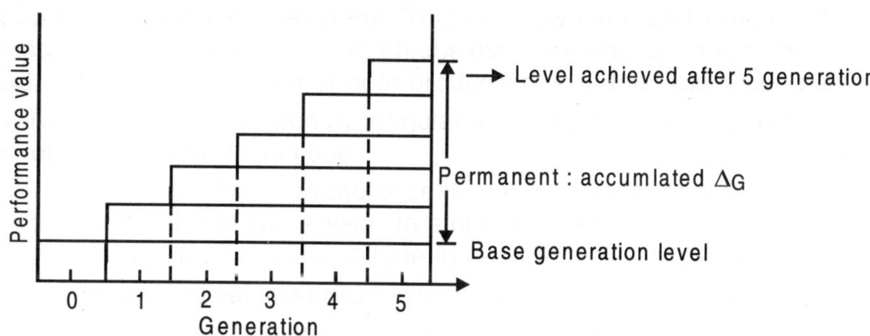

Fig 5.4 Illustration of how selection brings about permanent and accumulated changes in performance over generations

Prediction of Genetic Gain

Genetic gain expected due to one generation of selection is a function of heritability and selection differential, and can be predicted as follow:

$$\Delta G/\text{generation} = h^2 \times S$$

Therefore, higher the selection differential, higher will be the genetic gain and *vice versa*. If all the individuals of a population are retained for breeding (absence of selection), selection differential would be zero and in turn genetic gain would also be zero.

Why is selection differential multiplied with heritability to predict genetic gain? Selection for a quantitative trait is based on phenotypic values of the individuals because their genetic (breeding) values are not known. The phenotypic superiority of selected parents over population mean is thus partly made up of genetic superiority and partly of environmental superiority:

Phenotypic superiority = genetic superiority + environmental superiority

The genetic superiority is further composed of superiority due to additive, dominance and epistatic effects of genes. Neither environmental nor the non-additive genetic effects are transmitted from parents to progeny ; it is only the additive genetic effects that are transmitted. As defined earlier, heritability is the fraction of superiority of parents over population mean that is expected to be recovered in their progeny. Therefore, multiplying the selection differential by heritability

[3]Alternative terms used for 'response' and 'genetic gain' include 'genetic progress' or 'genetic advance'

gives the expected genetic superiority of progeny over parental generation (i.e. genetic gain). In simpler words, heritability is the weighting coefficient assigned to selection differential for predicting genetic gain.

Illustration of genetic gain as product of h^2 and selection differential: The concept of prediction of genetic gain as product of heritability and selection differential (ΔG/generation $= h^2 \times S$) can be better understood by assuming three hypothetical values of heritability: $h^2 = 0.0$; $h^2 = 1.0$ and $h^2 = 0.5$ (Fig. 5.5).

- **h^2 = 0.0**: If heritability of a trait is zero, it implies that all the observed differences among individuals in the population are caused by environmental factors. Phenotypic superiority of selected parents over population mean (selection differential) would then be purely environmental in nature. Consequently, the phenotypic superiority will not be transmitted to offspring. The mean of progeny generation would be same as mean of parental generation (i.e. $\bar{P}_1 = \bar{P}_0$), hence $\Delta G = 0$ (Fig. 5.4a). Thus, when heritability of a trait is zero, selection would fail to increase mean performance of population.
- **h^2 = 1.0:** If h^2 of a trait is unity, it implies that all the observed differences among individuals are due to differences in their breeding values. Selection of parents on the basis of their phenotypic values is thus equivalent to selection on the basis of their breeding values. The phenotypic superiority of selected parents over population mean would be solely due to superiority in breeding values and consequently, whole of it would be passed on to their offspring. Mean of progeny generation would then equal mean of selected parents (i.e. $\bar{P}_1 = \bar{P}_S$), and genetic gain would be maximum ($\Delta G = S$). In other words, the "attempted gain" would be fully realized in the offspring. Successive generations of selection would result in rapid genetic progress.
- **h^2 = 0.5:** Both the above considered situations are unrealistic because expression of a quantitative trait is governed by genetic (both additive and non–additive) as well as environmental effects (i.e. $h^2 > 0$ but <1.0). Therefore, only a fraction of phenotypic superiority of selected parents over population mean (selection differential) is transmitted to progeny. Selection differential is multiplied by heritability to account for the fraction of phenotypic superiority that is transmittable to next generation (Figure 5.5c).

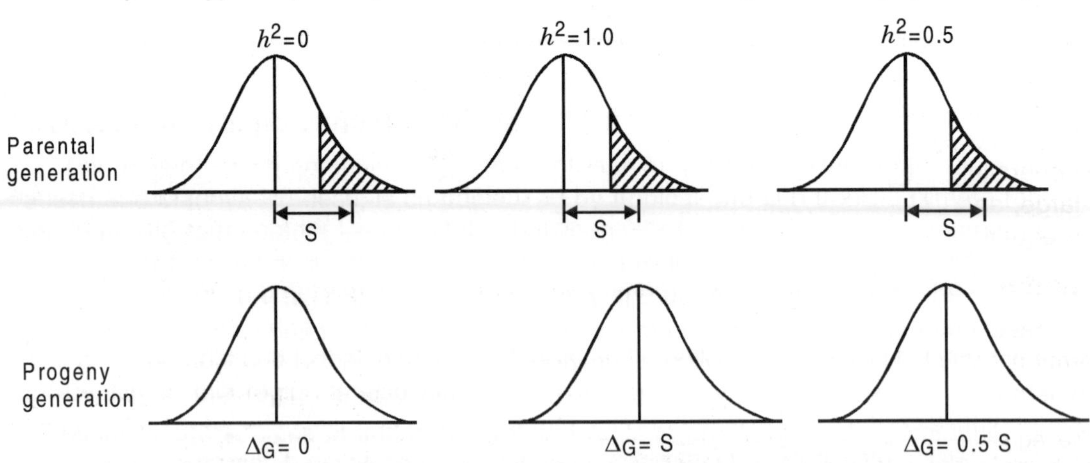

Fig 5.5 Illustration of effect of selection at three levels of heritability: (a) $h^2 = 0.0$; (b) $h^2 = 1.0$ and (c) $h^2 = 0.5$

The mean performance of progeny generation ($\bar{P}_1$) following one generation of selection can be predicted from mean of parental generation ($\bar{P}_0$) if we know the selection differential (S) and heritability of the trait (h^2). Assuming that there is no environmental change from parental to progeny generation, then the mean of progeny generation is given by the expression:

Mean of progeny generation = Mean of parental generation + Genetic gain

$$\bar{P}_1 = \bar{P}_0 + h^2 \times S \quad \text{Or} \quad \bar{P}_1 = \bar{P}_0 + h^2(\bar{P}_S - \bar{P}_0)^4$$

Genetic Gain per Year

We have expressed genetic gain on per-generation basis. Since length of generation turn over varies between species, it is more practical to express genetic gain on per-year basis. Genetic gain can be expressed on annual (yearly) basis if we know the generation interval in the flock/ herd.

Generation interval (symbolized as 'L'.) is defined as the average age of parents when their offspring, who are to become parents of next generation, are born. It may be noted that generation interval is not simply the age of parents at birth of their offspring but it is the age at the time of birth of those that are likely to replace the parents for producing the next generation. For example, if the parents produce the progeny at 2 years of age but the progeny are sold for commercial purposes (e.g. meat, milk, egg, wool etc). And the progeny are kept for breeding purposes when the parents are four- year old, the generation interval would be four years (and not two years). Generation interval is thus the time interval between the birth of parents and the birth of their replacements.

Expected genetic gain can be expressed on per-year basis by dividing genetic gain per-generation by the generation interval (in years):

$$\Delta G/\text{year} = \frac{\text{Heritabilty} \times \text{Selection differential}}{\text{Generation interval (yr)}} = \frac{h^2 \times S}{L}$$

Genetic gain per year in a population undergoing selection is thus dependent on three factors:

- heritability of the trait (h^2)
- selection differential (S)
- generation interval (L)

The above equation relating genetic gain to three factors viz; h^2, selection differential and generation interval, is often called the **'key equation for genetic gain' or 'breeders equation'**.[5] This equation which is the core of a breeder's selection program, states that rate of genetic gain is directly proportional to the heritability and selection differential, and is inversely proportional to generation interval. Genetic gain per annum is maximal when h^2 is high, selection differential is large, and the generation interval is short. If any of the factor that appears in numerator of key equation (heritability or selection differential) is zero, genetic gain would be zero.

Standardization of Selection Differential: Selection Intensity

Selection differential carries the unit of measurement of the trait and hence it does not permit a comparison of magnitude of selection pressure applied between traits/species. This problem is overcome by expressing selection differential in units of standard deviation (σ_P).

[4]The equation for predicting mean of progeny generation is a simple linear regression equation in which the value of regression coefficient equals h^2 of the trait.

[5]Ollivier (a Lush student) in a symposium held in honor of Lush in 2016 has suggested the breeder equation to be referred to as Lush equation.

Selection differential divided by phenotypic standard deviation is referred to as **'selection Intensity'**(symbolized as 'i'):

$$\text{Selection intensity} = \frac{\text{Selection differential}}{\text{Phenotypic standard deviation}} \quad \text{or} \quad i = \frac{S}{\sigma_P}$$

Standardized selection differential, 'i', expresses the number of phenotypic standard deviations by which the selected parents exceed the average of population. Since selection intensity is independent of unit of measurement of the trait, it permits a valid comparison of the selection pressure applied for different traits/populations/species etc.

Computation of selection intensity by the above expression ($i = S/\sigma_P$) requires the value of 'S' and 'σ_P'. But for a normally distributed trait, selection intensity can be more easily obtained from the properties of normal distribution. Consider directional selection for a normally distributed trait for which all the individuals above a certain value on the phenotypic scale are selected and those below this point are culled (Fig 5.6). The point that separates the selected and rejected animals on the phenotypic scale is referred to as **point of truncation (t).** Selection of this type in which all the individuals above a certain point on phenotypic scale are selected while those below that point are rejected is referred to as **truncation selection**. Selection intensity (i) under truncation selection is given by the ratio of the height of ordinate (z) at the point of truncation (t) to the proportion selected (p):

$$\text{Selection intensity} = \frac{\text{height of ordinate at the truncation point}}{\text{proporatoin selected}} \quad \text{or} \quad i = \frac{z}{p}$$

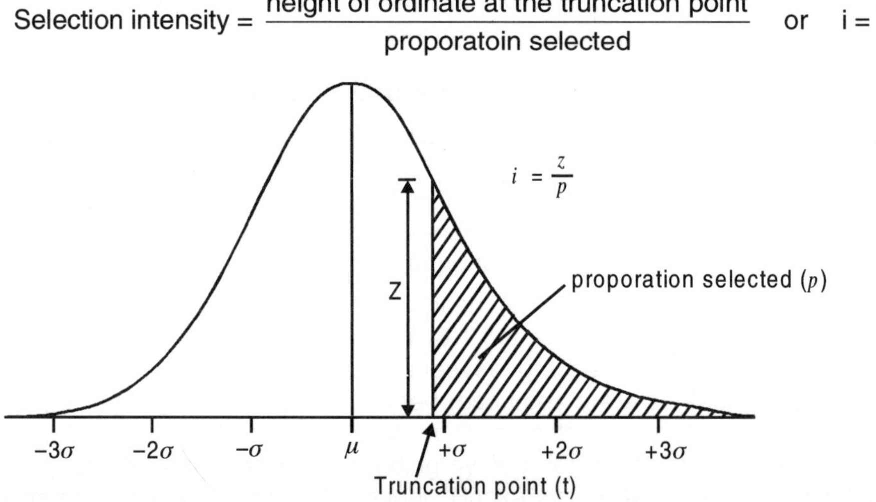

Fig. 5.6 Illustration of selection intensity for a normally distributed trait under truncation selection where t = truncation point (in phenotypic standard deviation units), p = proportion selected for breeding, z = height of ordinate at the point of truncation, and i = selection intensity, and equals z/p

For a normally distributed trait, there is a simple relationship between the proportion of population retained for breeding and selection intensity. The value of 'i' can be conveniently obtained from the Tables of areas of the normal curve when the proportion saved (p) is known. Fig. 5.7 shows the relation between values of selection intensity and percentage of animals being selected.The actual values of intensity for varying percentage selected are given in Chapter 6.

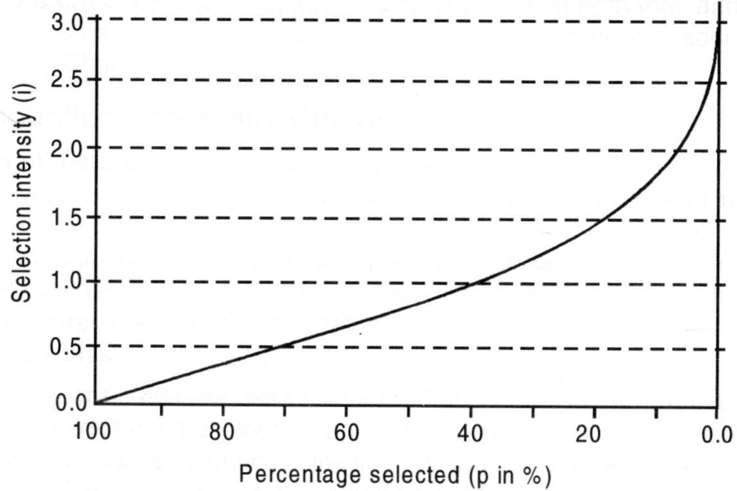

Fig. 5.7 Relation between percentage selected (p) and selection intensity (i)

A given value of selection intensity gives the number of phenotypic standard deviations by which the selected animals are better than the population average. For example, if fraction saved is 6%, the intensity of selection is 1.99, which means that the selected individuals are better than population average by 1.99 σ_P. Selection intensity is inversely proportional to proportion (percentage) of the animals retained for breeding: lower the proportion selected, higher the selection intensity and *vice versa.* If all the animals are kept for producing the next generation (p = 1.0), selection intensity (differential) would be zero.

The relationship $i = S/\sigma_P$ can be rearranged to calculate the selection differential from the proportion selected and standard deviation as: $S = i \times \sigma_P$.

Selection intensity offers two advantages:

- It permits comparison of selection pressure applied between traits/species or between alternative selection programmes.
- Computation of selection intensity does not require information about the performance of the population and the selected animals. Hence it can be used to compute selection differential before actually doing selection.

The selection intensity obtained from the tables of normal distribution requires that the trait is normally distributed, and the selection is strictly of truncation type.

Alternative Equation for Prediction of Genetic Gain: Generalized Equation

To be able to predict genetic change using equation : $\Delta G = (h^2 \times S)/L$, we need to know which animals are actually selected, and also their phenotypic values(performance) so as to calculate the selection differential. When planning a selection program it would, however, be useful to be able to predict the genetic gain before actually selecting the animals. The key equation for prediction of genetic change can be expressed in terms of selection intensity as follow:

$$\Delta G/\text{year} = \frac{h \times i \times \sigma_A}{L}$$

B

Since 'h' has earlier been defined as the correlation between the breeding value and phenotypic value, (r_{AP}), replacing 'h' with 'r_{AP}' in the above equation gives:

$$\Delta G/\text{generation} = r_{AP} \times i \times \sigma_A \quad \text{and} \quad \Delta G/\text{year} = \frac{r_{AP} \times i \times \sigma_A}{L}$$

$$\text{Genetic gain/year} = \frac{\text{accuracy of selection} \times \text{selection intensity} \times \text{genetic standard deviation}}{\text{generation interval}}$$

Genetic gain /year (expressed in terms of selection intensity) is, therefore, dependent on four factors:

- *Accuracy of selection (r_{AP}):* Correlation between breeding value and phenotypic value is referred to as accuracy of selection. Accuracy of selection equals 'h' if individuals are selected on the basis of their own phenotypic values (**individual selection**). In general terms, however, accuracy of selection is defined as the correlation between breeding value and criterion of selection. The criterion of selection could be individual's own performance or the performance of its sibs (family), ancestors or progeny.
- *Selection intensity (i)*: Standardized selection differential and equals S/σ_P.
- *Genetic standard deviation* (σ_A): Additive genetic standard deviation which can be calculated as : $h \times \sigma_P$ (since $h = \sigma_A/\sigma_P$)
- *Generation interval (L) :* Length of generation turn over (years)

Partitioning of Genetic Gain into Male and Female Paths

In the prediction of genetic gain so far, we have not separated the contribution of male and female parents. It was assumed that different components of genetic gain viz; accuracy of selection, selection intensity, genetic standard deviation and generation interval were similar for male and female parents. In practice, however, some of these factors, particularly accuracy of selection and selection intensity, differ between the sexes.

The key equation for genetic gain, therefore, needs to be modified to account for unequal contribution of male and female parents to genetic gain: the male path and the female path:

$$\Delta G/\text{year} = \frac{1}{2}[\Delta G/\text{year (male selection)}] + \frac{1}{2}[\Delta G/\text{year (female selection)}]$$

Generalized equation genetic gain through male and female paths becomes

$$\Delta G/\text{year} = \frac{1}{2}\left[\frac{r_{AP(M)} \times i_M \times \sigma_A}{L_M}\right] + \frac{1}{2}\left[\frac{r_{AP(F)} \times i_F \times \sigma_A}{L_F}\right]$$

Where $r_{AP(M)}$ = accuracy of selection for males,

$r_{AP(F)}$ = accuracy of selection for females

i_M = selection intensity for males,

iF = selection intensity for females

σ_A = additive genetic standard deviation.

Improving Response to Selection

Genetic improvement through selection in a population is expressed as an increase in its average performance over successive generations. The magnitude of genetic gain per year depends on four factors: accuracy of selection, selection intensity, genetic standard deviation and generation interval. Since the first three terms appear in numerator of the key equation while generation interval appears in denominator, the rate of genetic gain can, theoretically, be increased by increasing the accuracy of selection, selection intensity and genetic standard deviation, and by decreasing the generation interval. All the four factors can be modified to varying degree to enhance genetic gain.

Increasing Accuracy of selection

Accuracy of selection is a measure of the degree of association between breeding value (A) and criterion of selection (P_i). The criterion of selection may be individual's own performance (individual selection) or performance of its relatives (family, ancestors, and progeny). When selection is based on individual's own phenotypic value, the accuracy of selection (r_{AP}) equals the square root of heritability of the trait:

$$r_{AP} = h$$

Accuracy of selection is thus directly proportional to heritability of the trait: higher the heritability higher will be the correlation between the phenotypic value and breeding value. Because of the direct relationship between heritability and accuracy, selection on the basis of animals own performance is more effective for moderate to high heritability traits (e.g. growth and quality) than for low heritability traits (e.g. reproductive traits). Accuracy of selection, numerically takes a value between zero and one, and can vary depending on the criteria of selection used. It varies for different criteria of selection depending on heritability and the number and types of relatives whose performance is available for selection.

Accuracy of selection is the most important factor for comparing the efficiency of different bases of selection because other factors (intensity, generation interval, genetic standard deviation) are less amenable to large modifications. Improving the accuracy of selection is, therefore, the principal means of increasing genetic gain. Since accuracy is directly proportional to heritability, all those means which can increase heritability, also improve accuracy. There are two principal means of increasing heritability and in turn accuracy of selection: increasing genetic variability (σ_A^2) and decreasing environmental variation (σ_E^2).

The magnitude of additive genetic variance is population and trait specific, the breeder has little control on systematic modification of additive genetic variance within a ***closed population***[7] undergoing selection. The only practical and effective means of increasing heritability (and in turn increasing accuracy) is by reducing the environmental variance.

Reducing Environmental Variation: Environment, from genetic standpoint, includes all the non-genetic influences which an individual experiences from fertilization until the time of expression of the trait. Factors like fetal and post-natal nutrition, environmental temperature, humidity, housing; hygienic conditions and a large number of such known and unknown factors constitute the environment. Environmental factors, for the purpose of present discussion, can be classified into two categories: systematic/tangible and random/intangible.The variation caused by environmental factors can't be completely eliminated but can be reduced through certain measures:

[7]A population into which introduction of outside animals is not allowed.

- **Reducing variation caused by tangible or systematic sources:** The recognizable causes of environmental variation include factors like level of feeding, type of management (open vs. barns for cattle; floor vs. cages for hens), month/season of calving, sex of animal etc. The variation caused by known sources can be reduced by providing uniform environmental conditions and by statistical adjustment of performance records:
  - o *Providing uniform environment:* One of the important ways of reducing the environmental variation is to rear all the animals of the herd/flock under as similar and as consistent conditions of environment as possible. This can be partially achieved by following standard norms with respect to nutrition, management and health practices for all animals of the herd.
  - o *Statistical adjustment of performance records*: There are certain sources of non-genetic variation which influence the performance of animals in a consistent way. Performance records of animals for such factors can be statistically adjusted so as to bring records of all animals to a comparable basis (as discussed in Chapter 2). Typical examples include adjustment of milk yield records of cattle to a standard lactation length of 305 days and to twice-a-day milking; adjustment of body weight and gain in weight for sex of animal; adjustment of milk yield and composition for age at first calving, parity etc. The adjustment (correction) factors reduce the environmental variation such that a larger portion of remaining variation is genetic in nature. Nonetheless adjustment factors have certain limitations.
- **Reducing variation caused by random sources:** Random environmental factors include a wide variety of factors such as random errors in measurement (recording), random variations in temperature, feeding and management, and differences due to maternal effects, and many other non-specific reasons. Since the effects of these factors are difficult to be estimated or quantified it is difficult to control the variation caused by these. Some of the means of reducing the variation caused by random sources are:
  - o *Use of repeated records*: Use of repeated records on the same animal tends to reduce or cancel-out the variation caused by temporary environmental factors. The resultant reduction in environmental variation increases the accuracy of selection.
  - o *Use of records of relatives*: Use of records of relatives (e.g. full- and –half sibs, progeny), if based on a fairly good numbers, reduces the environmental variation leading to increased accuracy of selection.
  - o *Accurate measurement of trait*: Random errors in recording of the trait may also be reflected in environmental variation. Some traits can be recorded on a continuous scale (e.g. body weight, milk yield, fat % etc.). The 'error of measurement' variance for such traits is small. There are others which may only be graded into a limited number of classes (e.g. carcass quality and behavior traits). For such traits, error of measurement variance may be large. Recording of trait be as accurate as possible, and it should preferably be made on a continuous scale.

Increasing Selection Differential (Intensity)

Intensity of selection (or selection differential) is inversely proportional to the fraction selected: smaller the proportion selected, higher the selection intensity (differential) and *vice-versa*. For example, genetic gain will be more if we select the top 10% of the animals than say the top 30%. And, if all the animals are kept as parents, selection intensity will be zero. The proportion selected (p) is a function of number of animals selected and the total number of animals that are available for selection:

$$\text{Proportion selected} = \frac{\text{Number of animals selected}}{\text{Total number of animals available}}$$

High selection intensity can theoretically be achieved by either selecting fewer parents from fixed total number of animals available in the herd/flock or by increasing the number of animals reared and thus selecting parents from a larger group of potential parents. Theoretically both the means appear easier. But in practice there are a number of constraints in applying high selection differential which are discussed below.

Factors Determining Magnitude of Selection Differential (Intensity)

The size of selection differential (intensity) is dependent on a number of factors:

1. **Sex:** Livestock species being polygamous, a single male can produce larger number of offspring compared to a female. Therefore, fewer males than females are needed for (as replacements) and as a general rule, higher selection differential can be applied in males than in females. Usually, the best 1-10% males can be retained as parents. On the other hand, a higher percentage (usually 20–60%,) of the females are retained to maintain a herd/flock of constant size. In some herds, if the total number of available females is less, every available female replacement may have to be retained for breeding to keep a constant herd size. In such a situation, selection differential in females may be virtually absent, and genetic improvement may only come from male selection (i.e. the male path).

   Furthermore, with the use of artificial insemination (AI) the number of males required is much lesser as compared to natural service. Through AI, a bull can, for example, serve up to 20,000 to 30,000 cows per year whereas in natural service, a bull might serve 30 to 50 cows only. As a result, considerably fewer males are needed under AI than with natural service.

2. **Reproductive and survival rate:** The number of offspring that can be produced from a fixed number of parents in a specified period is dependent on reproductive and survival rate of the species/population. Higher the reproductive rate more will be the number of offspring produced from a fixed number of parents. Alternately, higher the reproductive rate of a herd/flock, fewer will be the parents required to produce the same number of offspring. Therefore, higher selection intensity is possible in herds/flocks having high reproductive and survival rate.

   Illustrative example: Consider selection of female replacement in two flocks of sheep which differ in their reproductive rate, as measured by lambing percentage. Assume that a fixed number of female replacements (60) are required to be selected in each flock to maintain a constant flock size. The effect of reproductive rate (that differs between two flocks) on selection intensity is shown below:

	Flock 1	Flock 2
No. of ewes available :	300	300
Lambing percentage (% lambs born per100 ewes):	100	130
No. of offspring born	300	390
No. of females born (1:1 sex ratio):	150	195
No of females required to be selected:	60	60
Percentage female selected*	40	30.7
Selection intensity (i):	0.97	1.16

*Calculated as : (number selected divided by the total number available × 100)

It can be seen from the above calculations that the number of females available for selection is more in flock 2 (*i.e.* the flock having higher reproductive rate), therefore, higher selection intensity is possible in this flock. Thus, higher the reproductive rate of a flock/ herd, higher the selection intensity that can be applied.

Reproductive rate varies between species: As a general rule, higher selection differential can be applied in highly prolific species (e.g. poultry, swine, and other litter-bearing species) because selection can be made from a larger number of animals than in single-birth species (e.g. buffalo, cattle, horse etc). The natural reproductive rate of a species may thus become a limiting factor in applying high selection intensity. Within a species reproductive rate may also vary between breeds/ strains or class of animals (e.g. meat-type vs. egg-type chickens). For example, indigenous cattle have lower reproductive rate than exotic cattle. Similarly, reproductive rate is lower in broiler than in layer chickens as high body weight of broiler parents interferes with their natural mating ability leading to reduced fertility.

Like reproductive rate, survival rate also influences the selection intensity/differential. Higher the survivability of a population, more will be the number of animals available for selection, and in turn higher will be the selection differential. In a population that experiences heavy mortality, selection intensity will be lower as lesser number of animals will be available for selection.

Interventions in nutrition, management and health practices can improve reproductive as well as survival rate of a flock/herd. Reproductive technologies (e.g. semen freezing, artificial insemination, super ovulation, embryo transfer etc.) have helped in enhancing reproductive rate of many species of livestock. In poultry, artificial incubation of eggs using modern incubators has tremendously improved reproductive rate thereby permitting very high selection intensity.

The maximal selection differential possible for a species is largely determined by its typical reproductive rate. Table 5.2 gives some typical values for the percentage of males and females to be kept as replacements for maintaining a herd of constant size (with natural service).

Table 5.2 Typical percentages of parents required for breeding (replacement rate).

Species	Percentage retained*	
	Males	Females
Dairy cattle:		
Exotic	4 – 5	50 – 60
Indigenous	5 – 10	>70
Buffalo	5 – 10	>70
Horse	3 – 5	30 – 50
Sheep and goat	2 – 4	40 – 50
Swine	1 – 2	10 – 20
Poultry: Chicken	0.5 – 2.0	10 – 15

* Values are only broad guides as the percentage retained depends on a number of factors.

3. **Herd size and replacement rate:** In most breeding herds/flocks, the facilities available for rearing and recording of animals are fixed. Further a certain number of animals have to be kept to maintain a constant herd size. In order to maintain a constant herd size, a certain number of female replacements have to be kept for reproduction. Therefore, it may not always be possible to increase flock size and to decrease the number of replacements below a certain limit. If a herd / flock is expanding in numbers, a larger number of female replacements have to kept, and this will lower the selectiondifferential. Thus both herd size and replacement rate may restrict high selection intensity.
4. **Number of traits considered for selection:** Intensity of selection for any single trait decreases as the number of traits considered for selection increases. As an example, consider that selection in a dairy herd is applied only for increasing milk production and, say, the best 30% of the cows, based on their milk production, are selected. On the other hand, consider a second situation in which milk yield is to be improved along with some improvement in fat percentage. The selection intensity for milk yield will decrease because some of the cows selected on the basis of their high milk yield may have to be rejected from the selected group because of their low fat percentage. And some cows with a slightly lower milk yield but with a higher fat percentage may have to be included in the selected group. Thus if selection is applied for several traits, the selection differential for each individual trait decreases. Selection should, therefore, be directed towards the traits that have immediate and real economic importance. Otherwise simultaneous consideration of several traits in selection program would decrease the selection differential for any single trait.
5. **Accurate measurement of trait:** Measurement of the trait to be improved should be as accurate as possible because measurement errors can inflate selection differential. If the trait is not carefully and accurately recorded, some of the individuals selected may become superior because of recording errors (e.g. divisions of weighing balance, weighing of animals vis a vis full-fed or off feed for the same duration). But the actual (realized) selection differential will be lowered.

Increasing Genetic Variability

Genetic variability is a pre-requisite for selection to operate. If all the animals were genetically alike, there would be no scope for selection. On the other hand, if genetic variation is large, the selected animals are likely to be genetically far better than the average of population. For this reason, variability (genetic) is referred to as the raw material for the breeder.

Genetic variance is trait and population specific as it depends on factors like gene action, gene frequencies, number of loci influencing the trait. Some traits exhibit low while others exhibit high genetic variance: as a general rule reproductive traits have lower genetic variability than production and growth traits. Since genetic variance is population specific as well, it is usually difficult to modify the genetic variability systematically once a population has been closed to outside inheritance (**closed population[8]**).

Genetic variance is maximal at intermediate gene frequencies. Before launching selection in a population, it is advisable to infuse genetic variability through introduction of genes/ (inheritance) from other populations. However introduction of genes from other populations needs to be made cautiously. If the donor population (or its individual's which are being

[8]A population in which introduction of animals from other breeding populations is not permitted is called a closed population. i.e. a population that is closed to genetic material from outside.

introduced) is genetically inferior, introduction of its inheritance into an improved population, may cause greater loss in performance than the potential increase in genetic variability. For these reasons, the scope for increasing genetic gain through systematic increase in genetic variation is rather limited.

Shortening Generation Interval

Shorter the generation turnover, higher will be the genetic gain per annum. Some typical values of generation interval (in years) for some important of species are given in Table 5.3. Generation interval, in general, can be decreased to a certain extent through better reproductive and management practices.

Table 5.3 Typical values for generation interval in different species.

Species	Generation interval (Years)	
	Male	Female
Horse	8 – 14	8-12
Buffalo	3 – 4.0	5-7.0
Cattle:		
Exotic	2 – 4.0	3.0 – 5.0
Indigenous	4.0 – 5.0	4.0 – 6.0
Sheep	2 – 3.0	4 – 4.5
Goat	2 – 3.0	4.0
Swine	1.5 – 2.0	1.5 – 2.0
Dog	2.0	3.0
Poultry:		
Chickens:	1.0	1.0
Meat type	<1.0	<1.0
Egg type	1.0 – 1.5	1.0 – 1.5
Japanese quails	0.3 – 0.4	0.3 – 0.4

Factors Affecting Generation Interval

Generation interval depends upon a number of factors viz. species, sex, time of measurement of trait, basis of selection applied etc.

1. **Species:** Generation interval varies considerably between species (Table 5.4) primarily due to differences in age at sexual maturity and reproductive rate, and to a certain extent due to differences in gestation length. For example, generation interval in humans, relative to other species, is very long (approximately 25 years) due to their late puberty. On the other hand, mice reproduce within a few months of birth; Japanese quails by six weeks, fruit flies by 14 days. Faster generation turnover is one of the important reasons for the use of fruit fly, mice, and quails as model laboratory animals for genetic studies of farm animals.

   Generation interval, as a general rule, is shortest for species having early sexual maturity, higher reproductive rate and short gestation/incubation length (e.g. poultry, swine, sheep and goat). All these factors collectively allow for early replacement of parents in poultry and swine than in cattle, buffalo and horse. Genetic improvement per annum thus is higher in poultry and swine than in cattle, buffaloes, horse etc.

Within a species, generation interval may also vary across herds, breeds depending on age at sexual maturity and reproductive rate. Age at sexual maturity is the earliest age at which an animal can be used for reproduction. Improved nutritional, managemental and health practices by reducing age at sexual maturity can decrease the generation interval. Improvement of reproductive rate, which is a function of fertility and embryo/fetal mortality, through better management practices can make fast generation turnover because the required number of offspring can be obtained in a shorter period.

2. **Sex:** Generation interval frequently differs between males and females. Males usually have shorter generation interval than females for two reasons:
   - Firstly, males can produce several offspring in a short time whereas females produce a limited number of offspring in a given period. For example, a bull can serve 50 to 60 cows in a year under natural service while a cow can at the most produce one offspring in a year. Therefore, a bull can produce the required number of offspring in a much shorter period. Artificial insemination has further widened the gap in generation interval between males and females.
   - Secondly, males can be used for reproduction once they start semen production. Females, besides production of ova, need to have proper development of reproductive tract so that they can properly nurture the fetus.

   If the males are selected on the basis of pedigree performance (e.g. selection of bulls for milk yield and other sex-limited traits), their generation interval is shorter than of females. However, when males are selected on the basis of their progeny performance (e.g. selection of dairy bulls for milk production), then their generation interval is doubled.

3. **Time of measurement of trait:** Age of animals when the trait to be selected becomes available affects the length of generation interval. Selection for traits, which are manifested late in life, extends the generation interval as compared to selection for traits which are manifested early in life (e.g. longevity).

   Generation interval may also be limited by the time taken to obtain the performance record on which selection is to be based. Short-term records (i.e. recording of egg production up to 40 weeks of age rather than up to 72 weeks of age in chickens) are commonly used for improvement of full–year egg production as these shortens generation interval. Similarly, part-lactation milk records are commonly used in dairy cattle for improvement of complete lactation milk yield. It should, however, be emphasized that though a trait may be recorded at birth or before sexual maturity (e.g. 6-week body weight in broiler chicken) but offspring can only be produced when the animals attain sexual maturity. Therefore generation interval for a species cannot be shorter than its age at sexual maturity.

4. **Basis of selection:** The criterion of selection employed for improvement also affects generation interval. Pedigree selection (i.e. selection on the basis of ancestral performance), for example, has the shortest generation interval while selection on the basis of progeny performance (i.e. progeny testing) has the longest generation interval. Similarly, use of repeated records also prolongs generation interval.

From the discussion of the factors determining genetic gain, it is apparent that genetic gain can primarily be increased by increasing the accuracy of selection, intensity of selection, and shortening the generation interval. Of these three factors, it is the accuracy of selection that can be more easily modified for increasing rate of genetic gain. Most of the attention of animal breeder is, therefore, focused on improving the accuracy of selection.

Optimizing Genetic Gain

Genetic gain per year can be maximized by increasing the accuracy of selection, intensity of selection and genetic variability, and by decreasing the generation interval. Genetic variance within a closed population cannot be systematically modified. Therefore, we need to manipulate selection accuracy, intensity and generation interval in such a way that "$(r_{AP} \times i)/L$" is maximized. However, some of the components of genetic gain are interrelated which means improving one component may lead to deterioration of another. There are many potential interrelationships between the three components but the most common ones are:

- **Increasing the accuracy of selection may prolong generation interval:** Accuracy of selection can be increased if the individuals are selected on the basis of their progeny performance (i.e. progeny testing). Further, the accuracy of progeny testing can be increased by increasing the number of progeny per sire. This means that the breeder has to wait for a long period for the progeny test to become available, and this prolongs the generation interval. Use of repeated records is another way of increasing the accuracy of selection, but it also increases the generation interval because longer the parents are kept for breeding, the greater will be their average age when their offspring are born. On the other hand, pedigree selection has the shortest generation interval but it gives low accuracy.
- **Increasing the accuracy of selection may lower selection intensity:** Testing facilities for evaluating the performance of animals on farm are limited/fixed. The accuracy of progeny testing can be increased by increasing the number of daughters per sire. But then the total number of sires that can be progeny tested with the fixed testing facility decreases. Consequently selection intensity is reduced.
- **Shortening the generation interval may decrease selection intensity:** Early use of parents decreases the generation interval but it also decreases the selection intensity as the total number of individuals available for selection at an early age may also be less.

There are some other undesirable genetic effects of modifying some of the key factors in the desirable direction. For example, selection intensity can be increased by reducing the number of parents but it leads to increased inbreeding.

The conclusion from the above discussion is that improving one of the components of genetic gain does not necessarily lead to higher genetic improvement. The breeder has therefore to make a balance between different factors so that genetic gain is optimized. Designing of a selection program in fact involves making a compromise among the three factors (r_{AP}, i and L). The best selection program for a particular situation is the one that has the optimum combination of these three factors.

The decision about the choice of a selection programme is generally made by predicting the magnitude of response per year for different programs, and then choosing the one that gives the highest response per year. Finally, the economics of running a selection program may also have to be taken into account.

Realized Genetic Change and its Measurement

Selection programs are carried out by selecting superior individuals, mating them at random, and producing the progeny generation. When a selection program has been in operation for some generations, it should be checked for "what genetic changes have been achieved". The actual amount of genetic improvement achieved is referred to as **realized genetic gain** (ΔG_R). It should be noted that realized genetic gain may not always be same as predicted genetic gain.

Prediction of genetic gain helps us to optimize a selection program whereas estimating the realized genetic change helps us to find out whether the selection program has been effective in achieving the genetic gain to the extent it was predicted. One should not assume that since one applies selection it would be realized. The essence of this being that realized genetic gain needs to be monitored at regular intervals. There can be a number of causes between the discrepancy between the predicted and realized genetic gain[9].

Genetic gain is measured as the difference between mean performance of progeny and parental generations on the assumption that environment remains unchanged between generations. But the assumption of constancy of environment between generations is rarely fulfilled in farm animals. Temperature, humidity, day-length, nutrition, hygienic conditions and a host of micro-climatic conditions invariably change from one generation to the other. Even in populations of laboratory animals (e.g. Fruit fly, *Tribolium*, Mice etc) which are kept in controlled conditions (e.g. diet, temperature etc); fluctuations in mean performance between generations due to environment are common. The diference in average performance between successive generations due to changes in environmental conditions is termed as **non-genetic change** or **environmental trend**. Changes in performance of selected population from parental to the progeny generation are, thus due both to genetic (ΔG) as well as the environmental change (ΔE). In other words, the change between generations is phenotypic in nature (ΔP) and equals the sum of genetic and the environmental change:

Phenotypic change = Genetic change + Environmental change

Symbolically: $$\Delta P = \Delta G + \Delta E$$

Therefore, measurement of realized genetic gain requires partitioning the phenotypic change into genetic and environmental components. There are a number of ways by which realized (actual) genetic change can be measured. One of the most common and effective method, which is widely used in poultry and pig breeding, is the use of a '**genetic control population**'.

A genetic control population is an unselected population which is contemporarily maintained with the selected population to measure genetic gain. The control line is reproduced without any intentional selection (i.e. by randomly choosing the potential parents) such that its genetic constitution remains unchanged between generations. The genetic mean of the control population thus remains stable over generations (i.e. $\Delta_G = 0$). Any change that occurs in average performance of control between generations is a measure of the environmental change (ΔE). On the other hand, the change that occurs in the selected population from one generation to next is a measure of phenotypic change (ΔP). The difference between the phenotypic and the environmental change provides an estimate of realized genetic gain:

$$\Delta G = \Delta P - \Delta E$$

---

[9]Dickerson, a renowned animal breeder, has used the term 'genetic slippage' when realized genetic gain does not equal the predicted genetic gain.

For the control population to be effective in measuring the environmental change, it must fulfill one important criterion: the control must be reproduced at the same time and raised in the same environment as the selected population.

Random bred control lines are widely used in assessing the effectiveness of selection programs in poultry, swine and laboratory species. The most serious limitation of the use of control populations in livestock is its cost. Maintaining an unselected control line, which otherwise serves no economic purpose, involves huge cost and resource investment

Illustrative example: Consider a selection program conducted for improvement of six-week body weight in chickens. The mean body weights of the selected and control lines in two consecutive generations along with the changes are given below:

Population:	Generation:		Change (Δ)
	Parental	Progeny	
Selected	1000	1200	$\Delta P = 200$
Control	1000	1100	$\Delta E = 100$

It is apparent that mean body weight of control population has increased by 100g. Since control is presumed to remain genetically unchanged, the change of 100g is due to environment. The change of 200g observed in the selected line is due to both genetic and environmental factors. Therefore, $\Delta G = \Delta P - \Delta E = 200 - 100 = 100$g

Had no control population been contemporarily maintained, the change in mean of selected population from parental to progeny generation would have only provided an estimate of phenotypic change.

Realized Heritability

When a selection experiment (program) has been carried out, the actual selection differential applied, called realized selection differential (S_R), and the realized (actual) genetic gain (ΔG_R) can be used to estimate heritability. Heritability calculated from realized genetic gain and selection differential obtained from an actual selection experiment is referred to as **realized or effective heritability** (h_R^2) and is computed as follows:

$$\text{Realized heritability } (h_R^2) = \frac{\Delta G_R}{S_R}$$

Realized h^2 can only be calculated if actual selection has been conducted and the performance of progeny generation measured. Since actual genetic gain can only be measured when the offspring reach the age equivalent to that at which their parents were measured, therefore, the estimation of realized heritability takes at least one generation to be obtained. Heritability calculated from selection experiments is usually considered true estimate of heritability of that population. But the practical application of this method of estimation of heritability in livestock is limited for two reasons:

- It requires large facilities and is costly as it involves actually conducting a selection experiment.
- It is difficult to control environmental fluctuations between generations. The difference between the mean of parental and progeny generation may not be an estimate of actual genetic gain but only of the phenotypic change (provided a control population is not maintained).

Correlated Response to Selection

When actual selection is practiced a change in mean performance of progeny generation for the trait being selected is naturally expected. The genetic gain in the trait undergoing selection is referred to as **'direct response'** or **'direct genetic gain'** and the trait being selected is referred to as **'primary trait'**. If the trait being selected is designated as 'X' (called primary trait), then direct genetic gain, designated as $\Delta_{G(X)}$ which equals the difference in mean between parental and progeny generation:

$$\Delta G_X = \bar{P}_{1(X)} - \bar{P}_{0(X)}$$

Where: $\bar{P}_{1(X)}$ = mean of progeny generation for trait X

$\bar{P}_{0(X)}$ = mean of parental generation for trait X

Many a times selection for a given trait simultaneously leads to changes in other traits. The change in performance from parent to progeny generation for the trait which is not being directly selected is referred to as **'correlated response** or **correlated genetic change'**. The trait which is not being directly selected is referred to as **secondary trait** or **correlated trait.** The correlated genetic change occurs due to existence of genetic correlation. Correlated response in trait 'Y' (unselected trait) due to direct selection for trait 'X' (selected trait), designated as $\Delta_{GY/X}$, and is given by the following expression :

$$\Delta_{GY/X} = r_{G(XY)} \times h_X \times h_Y \times i_X \times \sigma_{P(Y)}$$

The expression shows that :

- Direction of correlated response is solely dependent on the sign of genetic correlation between the selected and unselected traits.
- Unit of correlated response is that of the correlated trait.

Correlated response will be high if both the traits have high heritabilities, and the genetic correlation between traits is also high.

Livestock productivity is dependent on a number of traits, which are usually genetically correlated. Prior to start of a selection program, it is useful to predict the genetic change not only in the trait to be selected (direct response) but also in other correlated traits (correlated response). Prediction of correlated response is important in formulation of a selection program for bringing desired changes in several traits. For example, selection for increased egg production in chickens decreases egg weight as the genetic correlation between them is negative. Therefore, prediction of correlated response in egg weight due to direct selection for improving egg production would indicate the extent to which egg weight is likely to decline if selection is conducted for improvement of egg production only. One can then apply a suitable method of selection for simultaneous improvement of both egg number and egg weight. There are number of well-documented examples of correlated genetic changes when direct selection is applied for a trait:

- Chicken: selection for increased egg production reduces egg weight (undesirable).
- Sheep: selection for increased fleece weight increases fiber diameter (undesirable).
- Cattle: selection for increased milk yield decreases fat percentage (undesirable).
- Swine and chicken: selection for improvement of growth rate improves feed efficiency (desirable).

Review Questions

1. What are the two broad means (ways) of bringing about genetic improvement?
2. Why the effectiveness of selection for quantitative traits cannot be studied in terms of gene and genotypic frequencies?
3. Define and differentiate between :
   (a) Natural selection and artificial selection
   (b) Selection differential and selection intensity
4. Define/explain the following :
   I. Stabilizing selection
   II. Disruptive (diversifying) selection
   III. Directional selection
   IV. Selection differential
   V. Response to selection
   VI. Generation interval
   VII. Accuracy of selection
   VIII. Key equation for genetic gain
   IX. Truncation selection
   X. Selection intensity
   XI. Correlated response to selection
   XII. Realized heritability
   XIII. Genetic control population
5. Is artificial selection for a trait in one direction favored or opposed by natural selection? Explain.
6. Diagrammatically show the response to selection from one generation to the next for a normally distributed trait including the appropriate terms used.
7. What factors determine the response to selection from one generation of selection?
8. Discuss how can genetic gain through selection be increased?
9. Why is selection differential multiplied by heritability to predict genetic gain?
10. What will be the magnitude of genetic gain when: (a) h^2 = 0.0 (b) h^2 = 1.0 (c) h^2=0.6. Assume that the selection differential is 'S'.
11. Why whole of the phenotypic superiority of selected parents over population average not transmitted to offspring?
12. What fraction of superiority of selected parents over herd/flock average is transmitted to progeny?
13. Give the expression for prediction of mean performance of progeny generation in terms of population mean and mean of selected parents.
14. What is the key equation of genetic gain? Explain its components.
15. Why is selection differential standardized for phenotypic variability?
16. What are the advantages of computation of selection intensity over selection differential?
17. Diagrammatically show the response to selection in terms of section intensity, truncation point, height of ordinate at truncation point and proportion saved.
18. Higher the proportion selected lower the selection differential. Explain.
19. Give the equation of genetic gain due to variable selection in two sexes in terms of accuracy of selection, selection intensity, genetic standard deviation and generation interval.

20. Accuracy of selection is directly proportional to heritability of the trait. Justify.

21. What constitutes environment from a breeder's view point ?

22. What are the different ways of reducing environmental variation for increasing accuracy of selection?

23. Why are performance records statistically adjusted?

24. What factors determine the magnitude of selection differential? Explain.

25. Why higher selection differential can be applied in males as compared to females?

26. Why higher selection differential is possible in males under artificial insemination compared to natural service?

27. Higher the reproductive rate, higher will be the selection differential. Justify.

28. How do herd size and replacement rate influence selection differential?

29. Larger the number of traits considered for selection, lower will be the selection differential for any single trait. Explain.

30. How does accurate measurement of a trait influence the selection differential?

31. Why can't genetic variability be increased within a closed population?

32. Introduction of outside animals (genes) into a herd/flock is suggested as a means of increasing the genetic variability. What precautions should be taken into consideration while making such an introduction?

33. List the factors affecting generation interval. Which of these can be modified within a herd/flock to increase genetic gain per annum?

34. Higher genetic gain per annum is possible in poultry than in cattle. Justify.

35. Why do males usually have a shorter generation interval than females?

36. Why are short-term or part-year records used in animal breeding?

37. Explain how an increase in each of the following increases the rate of genetic change: (i) accuracy of selection (ii) selection intensity (iii) genetic variation, (iv) generation interval.

38. What is a control population? How does it help in measurement of genetic change?

39. What are the limitations of selection experiments to estimate heritability?

40. List the factors on which correlated response to selection depends.

41. Cite one example of correlated genetic changes for each of the following: chickens, sheep, swine and cattle.

●●●

6

Computation of Selection Differential, Generation Interval and Genetic Gain

Response to selection per generation for a quantitative trait is dependent on selection differential/intensity of selection, heritability/accuracy of selection, genetic variability and generation interval. The theory pertaining to response to selection including the factors determining it has been outlined in the last chapter. In the present chapter, the computations involved in selection differential, selection intensity, generation interval, realized heritability and correlated response to selection alongwith appropriate interpretations and inferences are outlined.

Computation of Selection Differential

Selection differential, symbolized as S, is computed as the difference between mean of selected parents ($\bar{P}_S$) and mean of population prior to section ($\bar{P}_0$) and carries the unit of measurement of the trait. When selection is applied in both sexes, selection differential can be calculated separately for each of the sexes , and then averaged over the two sexes :

$$\text{Average } S = \frac{1}{2}\left[S(\text{males}) + S(\text{females})\right]$$

6.1. Average 20-week body weight in a flock of sheep was 50 kg. The males and females selected as breeders averaged 70 and 60 kg, respectively. Calculate the selection differential for males, females and the average selection differential for the two sexes.

Solution :

Population mean : $\bar{P}_0 = 50$ kg

Mean of selected males : $\bar{P}_{SM} = 70$ kg

Mean of selected females : $\bar{P}_{SF} = 60$ kg

Selection differential for :

Males : $S_M = \left(\bar{P}_{SM} - \bar{P}_0\right) = 70 - 50 = 20$ kg

Females : $S_F = \left(\bar{P}_{SF} - \bar{P}_0\right) = 60 - 50 = 10$ kg

Average : $S = \dfrac{S_M + S_F}{2} = \dfrac{20 + 10}{2} = 15$ kg

Selection in One Sex

Sometimes the trait to be improved is expressed only in one sex (sex-limited) or it may have been recorded in one sex only. Under such a situation, selection may not be possible in one sex (usually the males). Absence of selection in males does not mean that males are not used for reproduction. When selection is not applied in one sex (*i.e.* absence of selection), the individuals of that sex are randomly chosen from the population without any consideration to their performance (it may be appropriate to refer to such a procedure as "random selection"). Average performance of randomly chosen parents is then expected to equal the population mean and therefore selection differential for that sex would be zero. Consequently, average selection differential for the population will be reduced.

6.2. Average annual egg production in a flock of Rhode Island Red chicken was 250 eggs. The females (hens) selected on the basis of their own production had average egg production of 300 eggs. The cocks were randomly chosen to become parents. Calculate the selection differential for dams (hens), sires (cocks), and the average selection differential for two sexes.

Solution :

Population mean : $\bar{P}_O$ = 250 eggs

Mean of selected females : $\bar{P}_{SF}$ = 300 eggs

Mean of breeding males : $\bar{P}_{SM}$ = 250 eggs (since males are randomly chosen)

$$S_M = 250 - 250 = 0 \text{ eggs and } S_F = 300 - 250 = 50 \text{ eggs}$$

$$\text{Average } S = \frac{S_M + S_F}{2} = \frac{0+50}{2} = 25 \text{ eggs}$$

- ***Inference :*** In the absence of selection in one sex, the overall selection differential in the flock is reduced.

Computation of Selection Intensity

Selection intensity (i) can be calculated from the selection differential and phenotypic standard deviation as : $i = S/\sigma_P$

Alternately, selection intensity can be obtained from the Table of normal distribution (Table 6.1) provided the proportion (or percentage) retained for breeding is known, and the distribution of the trait is normal.

Table 6.1 Selection intensity values (i) under truncation selection when different fraction (percentage) are retained for breeding.

% or fraction selected :		Selection Intensity	% or fraction selected :		Selection intensity	% or fraction selected :		Selection intensity
%	p	*i*	%	p	*i*	%	p	*i*
0.5	0.005	2.89	11	0.11	1.71	30	0.30	1.16
1	0.01	2.67	12	0.12	1.67	35	0.35	1.06
2	0.02	2.42	13	0.13	1.63	40	0.40	0.97
3	0.03	2.27	14	0.14	1.59	45	0.45	0.88
4	0.04	2.15	15	0.15	1.55	50	0.50	0.80
5	0.05	2.06	16	0.16	1.52	60	0.60	0.64
6	0.06	1.99	17	0.17	1.49	70	0.70	0.50
7	0.07	1.92	18	0.18	1.46	80	0.80	0.35
8	0.08	1.86	19	0.19	1.43	90	0.90	0.20
9	0.09	1.80	20	0.20	1.40	100	1.00	0.00
10	0.10	1.76	25	0.25	1.27			

Selection intensity, like selection differential, can also be calculated separately for the two sexes as well as on sex-pooled basis.

6.3. Examine the values of selection intensity for various proportion (or percentage) of animals kept for breeding as given in Table 6.1. What type of association between the proportion (and percentage) selected and the value of selection intensity is evident?

Solution : There is an inverse relationship between the proportion selected (p) and selection intensity (i) : smaller the fraction retained for breeding, higher the selection intensity and *vice versa*.

6.4. In a flock of broiler chicken with mean 6-week body weight of 1200 g and standard deviation of 120 g, the selected parents were superior over flock average by 240 g. Calculate the selection intensity and interpret its value in terms of percentage (of the parents) retained for breeding.

Solution :

Selection differential : S = 240 g

$$\text{Selection intensity :} \quad i = \frac{S}{\sigma_P} = \frac{240}{120} = 2.0$$

Selection intensity value of 2.0 implies that the selected birds were better than the flock average by 2.0 standard deviation units. The value of 'i' = 2.0 (in Table 6.1) corresponds, approximately, to a proportion selected of 0.06 (p = 0.06) which means that the top 6% of the birds had been kept for breeding.

6.5. Following information is available for two populations of White Plymouth breed of chicken for six-week body weight :

Population	Mean (g)	σ_P(g)
A :	1250	142
B :	1300	188

The selected birds in each of the two populations exceeded the average of the respective population by 150 g. In which population selection is more intense and why?

Solution : Since the means and standard deviations of the two populations differ from each other, comparison of selection differentials of the two populations is not a valid comparison of the amount of selection applied. The values of selection intensity can, however, permit comparison of selection pressure applied in two populations :

$$i_A = \frac{S}{\sigma_P} = \frac{150}{142} = 1.056 \quad \text{and} \quad i_B = \frac{S}{\sigma_P} = \frac{150}{188} = 0.798$$

The selected birds in population 'A' exceeded the population average by 1.056 standard deviation units while those in population 'B' exceeded the population average by 0.798 standard deviation units. Selection is thus more intense in population 'A' because the value of 'i' is higher in this population.

- ***Inference :*** Selection intensity can be used for comparison of selection pressure between different populations for a given trait or between two or more than two traits within a population.

6.6. Four-month body weight in a flock of swine was normally distributed and had mean of 70 kg with phenotypic standard deviation of 15 kg. The best 20% boars and 40% gilts were retained for breeding.

(*a*) Calculate the selection intensity for boars, gilts and the average for both sexes using the information given above.

(*b*) Calculate the selection differential for boars, gilts and the average for both the sexes.

Solution :

Flock mean : $\bar{P}_0$ = 70 kg ; Standard deviation : σ_P = 15 kg.

Proportion saved (*p*) for boars = 0.20 and for gilts = 0.40

(*a*) Values of selection intensity for '*p*' = 0.20 and '*p*' = 0.40 are 1.40 and 0.97, respectively (Table 6.1) :

i_M = 1.40 and i_F = 0.97

$$\text{Average selection intensity for both sexes} = \frac{1.40 + 0.97}{2} = 1.18$$

(*b*) Selection differential can be calculated from value of selection intensity and phenotypic standard deviation as : $S = i \times \sigma_P$

$$S_M = i_M \times \sigma_P = 1.40 \times 15 = 21 \text{ kg}$$

$$S_F = i_F \times \sigma_P = 0.97 \times 15 = 14.6 \text{ kg}$$

$$\text{Average selection differential} = \frac{S_M + S_F}{2} = \frac{21 + 14.6}{2} = 17.8 \text{ kg}$$

Computation of Generation Interval

Generation interval, designated as L, is computed as the time interval between the birth of parents and birth of their offspring, and is usually expressed in years. If the male and female parents vary in the age when their offspring are born, the generation interval is computed separately for the two sexes, and can then be averaged :

$$\text{Average L} = \frac{(L_M + L_F)}{2}$$

6.7. Given below are the months of birth of sires and dams of the pigs born in January 2010.

Month & Year of Birth of :	Pig ID number :				
	1	2	3	4	5
Sire :	July, 2006	Nov., 2006	May 2007	Nov., 2006	Jan., 2006
Dam :	Nov., 2006	Feb., 2007	Jan., 2007	Nov., 2006	Feb., 2007

Calculate the generation interval for sires, dams and the average generation interval for the two sexes.

Solution : The information given above can be rearranged as follows :

Pig ID	Month & year of birth of :			Parent's age at birth of offspring (month):	
	Offspring	Sire	Dam	Sire	Dam
1	Jan., 2010	July, 2006	Nov., 2006	42	38
2	Jan., 2010	Nov., 2006	Feb., 2007	38	35
3	Jan., 2010	May, 2007	Jan., 2007	32	36
4	Jan., 2010	Nov., 2006	Nov., 2006	38	38
5	Jan., 2010	Jan., 2006	Feb., 2007	48	35

$$\text{Generation interval for sires} = \frac{(42+38+32+38+48)}{5} = 39.6 \text{ month}$$

$$\text{Generation interval for dams} = \frac{(38+35+36+38+35)}{5} = 36.4 \text{ month}$$

$$\text{Average generation interval} = \frac{(39.6+36.4)}{2} = 38.0 \text{ month}$$

- ***Comment :*** Generation interval can vary between sexes, and in the present case it is slightly more for sires than dams.

Generation Interval for Variable Age-structure of Herd

In highly prolific species (*e.g.* poultry), a large number of offspring can be obtained (or are born) at the same time. Consequently, parents at the time of selection are more or less of the same age. But in other species of livestock (*e.g.* buffaloes, cattle, sheep, goat, pigs etc.), the situation is more complicated as all the individual are not born at the same time and therefore, the selected parents are of different ages. Further, the fertility of parents of different age groups may also differ, therefore, different proportion of animals become available for different age groups.

Both these factors have to be taken into account for calculating the generation interval. The examples below illustrate this point.

6.8. In a flock of sheep, 50% two-year old and 50% three-year old parents were selected for producing the progeny. Calculate the generation interval.

Solution : Average age of selected parents would be the simple average of the two age–groups since both the age groups are represented by equal number/proportion of parents:

$$L = \frac{(2+3)}{2} = 2.5 \text{ years}$$

6.9. A herd of cattle is managed in such a way that a typical cow produces a total of six calves beginning when she is two years old, and then calving every year while a typical bull produces three calves beginning when he is two year old and then producing one calf every year. Calculate the generation interval for sires, dams, and average for both sexes.

Solution : Generation interval for:

Sire : $$L_M = \frac{(2+3+4)}{3} = 3 \text{ years}$$

Dam : $$L_F = \frac{(2+3+4+5+6+7)}{6} = 4.7 \text{ years}$$

Pooled (Overall) : $$L = \frac{(L_M+L_F)}{2} = \frac{(3+4.7)}{2} = 3.85 \text{ years}$$

6.10. In a flock of sheep, the selected parents at the birth of their offspring had the following structure : 60% were two-year old and 40% were three-year old. Calculate the generation interval.

Solution : The average age in the flock would be the weighted average of the two age-group parents :

$$L = (0.6 \times 2.0) + (0.4 \times 3.0) = 2.4 \text{ years.}$$

6.11. A flock of sheep consisted of 100 ewes and 50 rams of different age groups. Sixty and forty per cent of all the ewes and rams respectively were selected for producing the next generation. The age-group of selected ewes and rams, their numbers, and the number of lambs born for each age-group are given below :

Age at lambing	For ewes :		For rams :	
	No. of ewes in each age-group	No. of lambs born from each age-group	No. of rams in each age-group	No. of lambs born from each age-group
2	15	12	10	40
3	15	13	10	60
4	15	30	–	–
6	15	15	–	–
Total :	60	70	20	100

Calculate the generation interval for ewes, rams and average of two sexes.

Solution :

Ewes :

Age of ewes at lambing	No. of ewes in each age-group	No. of lambs born from each group	Proportion of lambs contributed by each age group	Proportional contribution of ewes to their age group
2	15	12	12/70 = 0.17	0.17 × 2 = 0.34
3	15	13	13/70 = 0.19	0.19 × 3 = 0.57
4	15	30	30/70 = 0.43	0.43 × 4 = 1.72
6	15	15	15/70 = 0.21	0.21 × 6 = 1.26
Total :	60	70	1.00	3.89

Rams :

Age of rams at lambing	No. of rams in each age-group	No. of lambs born from each group	Proportion of lambs contributed by each age group	Proportional contribution of rams to their age group
2	10	40	40/100 = 0.40	0.40 × 2 = 0.80
3	10	60	60/100 = 0.60	0.60 × 3 = 1.80
Total :	20	100	1.00	2.60

Average age of ewes at the birth of their offspring : $L_F = 3.89$

Average age of rams at the birth of their offspring : $L_M = 2.60$

$$\text{Average generation interval} = \frac{(3.89 + 2.6)}{2} = 3.25 \text{ year}$$

Computation of Expected Genetic Gain and Progeny Performance

(A) Equal selection differential and generation interval in both sexes :

Expected genetic gain on per-generation or per-year basis is obtained as follows :

ΔG/generation = Heritability × Selection differential = $h^2 \times S$

$$\Delta G/\text{year} = \frac{\text{Heritability} \times \text{Selection differential}}{\text{Generation interval}} = \frac{h^2 \times S}{L}$$

Alternately, expected genetic gain on per-generation and per-year basis can be calculated as :

$$\Delta G/\text{generation} = h \times i \times \sigma_A \text{ and } \Delta G/\text{year} = \frac{h \times i \times \sigma_A}{L}$$

Average performance of progeny generation $(\bar{P}_1)$ is obtained by adding the expected genetic gain (ΔG) the mean of parental generation $(\bar{P}_0)$:

$$\bar{P}_1 = \bar{P}_0 + \Delta G$$

6.12. Staple length in a flock of sheep averaged 6 cm with phenotypic standard deviation of 0.5 cm. The parents retained for breeding had average staple length of 6.5 cm. The selected rams and ewes at the time of birth of their offspring averaged to be three years. The h^2 of staple length is 0.5.

(*a*) Calculate the expected genetic gain on per-generation and per-year basis using the formula of genetic gain in terms of selection differential as well as in terms of selection intensity.

(*b*) What would be the average staple length of the progeny generation?

(*c*) What would be the average staple length after 12 years if the same magnitude of selection is applied every generation ?

Solution :

(*a*) Flock average : $\bar{P}_0$ = 6 cm, Standard deviation : σ_P = 0.5 cm,
Generation interval : L = 3 years

Mean of selected parents : $\bar{P}_s$ = 6.5 cm

∴ Selection differential : $S = \left(\bar{P}_s - \bar{P}_0\right) = 6.5 - 6.0 = 0.5$ cm

Selection intensity : $i = \frac{S}{\sigma_P} = \frac{0.5}{0.5} = 1.0$

The information required for use of both the formulas of expected genetic gain is available therefore the computations are shown by both the expressions (formulas).

- **Using formula in terms of selection differential**

  ΔG/generation = $h^2 \times S = 0.5 \times 0.5 = 0.25$ cm

  $$\Delta G/\text{year} = \frac{h^2 \times S}{L} = \frac{0.5 \times 0.5}{3} = 0.083 \text{ cm}$$

- **Using formula in terms of selection intensity**

  For the computation of genetic gain by the second formula, the value of additive genetic standard deviation (σ_A) is required which is calculated from heritability and phenotypic standard deviation as follow :

  $$h = \frac{\sigma_A}{\sigma_P},$$

Therefore, $\sigma_A = h \times \sigma_P = \sqrt{0.5} \times 0.5 = 0.707 \times 0.5 = 0.3535$ cm

$\therefore$ $\Delta G/\text{generation} = h \times i \times \sigma_A = 0.707 \times 1.0 \times 0.3535 = 0.25$ cm

$$\Delta G/\text{year} = \frac{h \times i \times \sigma_A}{L} = \frac{0.25}{3} = 0.083 \text{ cm}$$

- **Inference :** Both the formulas yield the same value of genetic gain. Therefore, any of these two expressions, depending upon the type of information available, can be used for calculation of genetic gain, For example :
  - If the mean of selected parents is given but the value of selection intensity (*i*) is not given, then genetic gain can only be calculated by the first expression :

    $$\Delta G = h^2 \times S$$

  - If the proportion selected and σ_P are given but the mean of selected parents Is not given, then genetic gain can be calculated by the second expression:

    $$\Delta G = h \times i \times \sigma_A$$

(*b*) Average staple length of next generation :

$$\bar{P}_1 = \bar{P}_0 + \Delta G = 6.0 \text{ cm} + 0.25 \text{ cm} = 6.25 \text{ cm}$$

(*c*) Number of generations possible in 12 years = Number of years/generation interval (years) = 12/3 = 4 yrs.

$\therefore$ Average staple length after four generations

$$= \bar{P}_0 + (\Delta G/\text{generation}) \times 4 = 6.0 \text{ cm} + (0.25 \times 4) = 7.0\text{cm}$$

(B) Unequal selection differential and generation interval in two sexes

Since fewer males than females are required for breeding, selection differential usually varies between sexes. Similarly, generation interval may also vary between sires and dams. Expected genetic gain can be calculated using the average selection differential and average generation interval of the two sexes.

Sometimes, the breeder may be interested to calculate the genetic gain attributable to selection of sires and dams separately. The total genetic gain is then obtained by adding the genetic gain due to selection of sires and due to selection of dams. Further, expected genetic gain can be calculated using either the formula involving selection differential or involving selection intensity, provided the information required for use of both the formulas is available.

Considering variable selection differential/intensity and different generation intervals for the two sexes, ΔG can be predicted by using the following formula :

$$\Delta G/\text{yr} = \frac{1}{2}\left(\frac{h^2 \times S_M}{L_M}\right) + \frac{1}{2}\left(\frac{h^2 \times S_F}{L_F}\right)$$

$$\Delta G/\text{yr} = \frac{1}{2}\left(\frac{h \times \sigma_A \times i_M}{L_M}\right) + \frac{1}{2}\left(\frac{h \times \sigma_A \times i_F}{L_F}\right)$$

6.13. In a flock of Japanese quails, 4-week body weight was normally distributed and averaged to be 200g with phenotypic standard deviation of 20g. The best 10% males and 20% females were kept for producing the next generation. The h^2 of body weight is 0.3. One generation of quails can be obtained in four months.

(*a*) Calculate the expected genetic gain on per-generation and per-year basis

(*b*) What would be the average 4-week body weight in the next generation and after five years if same intensity of selection were applied every generation?

Solution : Flock average : $\bar{P}_0$ = 200g, h^2 = 0.3, σ_P = 20 g

Additive genetic standard deviation : $\sigma_A = \sigma_P \times h$

$$= 20 \times \sqrt{0.30} = 20 \times 0.548 = 10.95 \text{ g}$$

The value of selection intensity (i) for 10 and 20 % selected, as obtained from Table 6.1, is:

For p = 0.10, i_M = 1.76

For p = 0.20, i_F = 1.40

(*a*) Since only the values of selection intensities are available, we use the following expression of genetic gain :

$$\Delta G/\text{gen.} = \frac{1}{2}\left[h \times i_M \times \sigma_A\right] + \frac{1}{2}\left[h \times i_M \times \sigma_A\right]$$

$$\Delta G/\text{gen.} = \frac{1}{2}\left[\sqrt{0.3} \times 1.76 \times 10.95\right] + \frac{1}{2}\left[\sqrt{0.3} \times 1.40 \times 10.95\right]$$

$$= 5.28 + 4.20 = 9.48 \text{ g}$$

$\therefore$ ΔG/generation = 9.48g

Generation interval = 4/12 = 0.33 yrs

Predicted ΔG/year = 9.48/0.33 = 28.7 g

(*b*) Expected mean 4-week body in the next generation :

$$\bar{P}_1 = \bar{P}_0 + \Delta G/\text{generation} = 200 + 9.48 = 209.5 \text{ g}$$

Flock average after five years of selection will be :

$$\bar{P}_{5yrs} = \left[\bar{P}_0 + (\Delta G/\text{year} \times 5)\right] = 200 + (28.7 \times 5) = 343.5 \text{ g}$$

Alternate solution for (*a*) :

Genetic gain can also be calculated by using the average selection intensity of two sexes as follows :

$$\text{Average selection intensity : } i = \left(\frac{i_M + i_F}{2}\right) = \left(\frac{1.76 + 1.40}{2}\right) = 1.58$$

$$\Delta G/\text{generation} = (h \times i \times \sigma_A) = (0.548 \times 1.58 \times 10.95) = 9.48 \text{ g}$$

- ***Inference :*** Genetic gain can be calculated separately due to selection of sires and dams or it can be calculated using the average selection differential (or selection intensity) of the two sexes.

6.14. Yearling weights (kg) of 40 lambs (20 of each sex) along with their identification numbers (ID) obtained from a flock of sheep are tabulated below :

Males				Females			
ID no	Body wt.	ID No	Body wt.	ID no	Body wt.	ID no	Body wt.
1	88	11	61	21	85	31	57
2	76	12	91	22	71	32	87
3	82	13	54	23	82	33	51
4	99	14	77	24	96	34	73
5	77	15	96	25	72	35	92
6	64	16	73	26	59	36	68
7	86	17	78	27	84	37	93
8	80	18	62	28	76	38	56
9	95	19	84	29	90	39	75
10	59	20	56	30	54	40	49

Mean and phenotypic standard deviation for this sample (of both sexes) are 75.2 and 14.4 kg, respectively. Selection in this flock is to be conducted for genetic improvement of body weight. (***Note :*** *A small flock of 40 animals has been chosen only for ease of computation, the actual size of breeding flock usually runs into hundreds or thousands*). The best 20% and 40% of the rams and ewes, respectively, are retained for breeding on the basis of their own body weight. The h^2 of yearling body weight is 0.3. The average age of rams and ewes at the birth of their progeny is 4.0 and 2.7 years, respectively.

(*a*) Calculate the selection differentials for the two sexes separately and the average selection differential (sex-pooled).

(*b*) Calculate the selection intensities for the two sexes separately and the average for two sexes.

(*c*) Compare the calculated value of selection intensity (obtained from S and σ_P) with the value given in Table 6.1 and draw appropriate inferences.

(*d*) Calculate the expected genetic gain due to one generation of selection of male and female candidates separately, and on combined-sex-basis, using both the formulas : (*i*) in terms of selection differential and (*ii*) in terms of selection intensity.

(*e*) What percentage of the total genetic gain per-generation has been contributed by rams and ewes? Comment on their relative contribution.

(*f*) Calculate the expected genetic gain per-year due to selection of males and females and on combined sex-basis.

(*g*) What percentage of the total genetic gain per-year has been contributed by rams and ewes ? Comment on their relative contribution.

(*h*) Calculate the average yearling weight of the next generation.

Solution : The first step in selection involves ranking of animals in descending order on the basis of their body weight. The ranking is done within each sex separately.

Selection of the best 20% males and 40% females implies that best four males (0.20 × 20 = 4.0) and best eight females (0.40 × 20 = 8) are to be retained for breeding.

Ranking of the required number of selected animals in descending order yields the following table :

Selected	ID no	4	15	9	12				
males :	Body wt	99	96	95	91				
Selected	ID no	24	37	35	29	32	21	27	23
females :	Body wt	96	93	92	90	87	85	84	82

(*a*) Calculation of selection differential :

$$S_M = \bar{P}_{SM} - \bar{P}_0 = 95.3 - 75.2 = 20.1 \text{ kg}$$

$$S_F = \bar{P}_{SF} - \bar{P}_0 = 88.6 - 75.2 = 13.4 \text{ kg}$$

Average $$S = \frac{1}{2}(S_M + S_F) = \frac{1}{2}(20.1 + 13.4) = \frac{1}{2} \times 33.5 = 16.75 \text{ kg}$$

- ***Inference :*** Selection differential for males is higher than for females because fewer number of males than females had been selected.

(*b*) Calculation of selection intensity :

$$i_M = \frac{S_M}{\sigma_P} = \frac{20.1}{14.4} = 1.395 = 1.4$$

$$i_F = \frac{S_F}{\sigma_P} = \frac{13.4}{14.4} = 0.93$$

Note : Since phenotypic standard deviation is given for the whole flock (both sexes) it is presumed that standard deviation is same in both sexes. Therefore, same value of σ_P has been used for both sexes. Had σ_P been given for males and females separately, then σ_P of that sex would have been used for computation of selection intensity.

$$i_{Average} = \frac{\frac{1}{2}(S_M + S_F)}{\sigma_P} = \frac{16.75}{14.4} = 1.163$$

Average selection intensity can also be obtained by averaging the selection intensity of males and females :

$$i_{Average} = \frac{i_F + i_M}{2} = \frac{1.40 + 0.93}{2} = 1.165$$

(*c*) Comparison of tabulated and calculated selection intensities :

For 20 and 40% of the males and females retained for breeding, the tabulated selection intensity values (Table 6.1) respectively are 1.40 and 0.97. The values of '*i*' calculated from the selection differential and phenotypic standard deviation are 1.40 and 0.93 for males and females respectively. The tabulated and calculated values for males are exactly the same but for females the calculated value ($i = 0.93$) is slightly lower than the tabulated value which could be due to the non-normality of the distribution of female body weights (as the use of tabulated value assumes normal distribution of the trait).

(*d*) Calculation of expected genetic gain per-generation :

(*i*) *Using formula in terms of selection differential :*

$$\Delta G/\text{generation} = \frac{1}{2}\left(h^2 \times S_M\right) + \frac{1}{2}\left(h^2 \times S_F\right) = \frac{1}{2}(0.3 \times 20.1) + \frac{1}{2}(0.3 \times 13.4)$$

$$= 3.02 + 2.01 = 5.03 \text{ Kg}$$

$\therefore \Delta G$ due to male selection = 3.02 Kg and due to female selection = 2.01 kg.

ΔG/generation on combined sex-basis = $h^2 \times S_{average} = 0.3 \times 16.75 = 5.03$ kg.

(*ii*) *Using formula in terms of selection intensity (we use the value of 'i' as obtained from S and σ_P)*

Additive genetic standard deviation required for the formula is calculated as :

$$\sigma_A = h \times \sigma_p = 0.548 \times 14.4 = 7.89$$

$$\Delta G/\text{generation} = \frac{1}{2}(h \times i_M \times \sigma_A) + \frac{1}{2}(h \times i_F \times \sigma_A)$$

$$= \frac{1}{2}(0.548 \times 1.40 \times 7.89) + \frac{1}{2}(0.548 \times 0.93 \times 7.89)$$

$$= 3.02 + 2.01 = 5.03 \text{ Kg}$$

ΔG/generation (*combined sex–basis*) = $h \times i \times \sigma p$

$$= 0.548 \times 1.163 \times 7.89 = 5.03 \text{ kg}$$

Comment : Both the formulas yield the same value of genetic gain therefore any one of the two can be used depending upon the information available.

(*e*) Percentage of the total genetic gain contributed by each of the two sexes separately:

$$\text{Males} = \left(\frac{\Delta\text{G due to male selection}}{\text{Total genetic gain}}\right)\times 100 = \frac{3.02}{5.03}\times 100 = 60\%$$

$$\text{Females} = \left(\frac{\Delta\text{G due to female selection}}{\text{Total genetic gain}}\right)\times 100 = \frac{2.01}{5.03}\times 100 = 40\%$$

Comment : Males contribute more to genetic gain because of the higher selection differential applied/possible.

(*f*) Computation of ΔG/generation :

Using formula in terms of selection differential :

$$\Delta\text{G/generation} = \frac{1}{2}\left(\frac{h^2 \times S_M}{L_M}\right)+\frac{1}{2}\left(\frac{h^2 \times S_F}{L_F}\right)$$

$$\Delta\text{G/generation} = \frac{1}{2}\left(\frac{0.3\times 20.1}{4}\right)+\frac{1}{2}\left(\frac{0.3\times 13.4}{2.7}\right)$$

$$= 0.75 + 0.74 = 1.49 \text{ kg}$$

Using formula in terms of selection intensity :

Additive genetic standard deviation required for the formula is calculated as :

$$\sigma_A = h \times \sigma_p = 0.548 \times 14.4 = 7.89 \text{ kg}$$

$$\Delta\text{G/generation} = \frac{1}{2}\left(\frac{h\times i_M \times \sigma_A}{L_M}\right)+\frac{1}{2}\left(\frac{h\times i_F \times \sigma_A}{L_F}\right)$$

$$= \frac{1}{2}\left(\frac{0.548\times 1.40\times 7.89}{4}\right)+\frac{1}{2}\left(\frac{0.548\times 0.93\times 7.89}{2.7}\right)$$

$$= 0.75 + 0.74 = 1.49 \text{ kg}$$

Total ΔG/year (both sexes) = 1.49 kg

ΔG/year due to male selection = 0.75 kg

ΔG/year due to female selection = 0.74 kg

- ***Inference :*** Both formulas yield the same value of genetic gain/year.

(*g*) Percentage of the total genetic gain contributed by each sex on per-year basis :

Genetic gain per-year due to male selection (0.75 kg) equals the genetic gain per-year due to female selection (0.74 kg) which means that each sex contributes about 50% to the genetic gain on per-year basis though genetic gain on per-generation basis was higher due to male selection. The shorter generation interval of females (2.7yrs) has compensated for their low selection intensity, and, in turn, has made the genetic gain of the two sexes equal.

(*h*) Average yearling weight of the next generation :

$$\bar{P}_1 = \bar{P}_0 + (\Delta\text{G/generation})$$

$$= 78.2 + 5.03 = 83.2 \text{ kg}$$

Computation of Realized Heritability

The actual or realized genetic gain (ΔG_R) and the actual/realized selection differential (S_R) can be used for estimating the heritability, referred to as 'realized heritability' (h_R^2) as follows:

$$\text{Realized heritability : } h_R^2 = \frac{\Delta G_R}{S_R} = \frac{\bar{P}_1 - \bar{P}_0}{\bar{P}_S - \bar{P}_0}$$

6.15. Average wool yield in a flock of sheep was 4 Kg. The parents selected from this flock for breeding had mean of 5 kg. The progeny produced from mating among selected parents had wool yield averaging 4.5 kg. Calculate the realized heritability of wool yield.

Solution :

Flock average in parental generation : $\bar{P}_0 = 4.0$ kg

Flock average in progeny generation : $\bar{P}_1 = 4.5$ kg

∴ Realized genetic gain : $\Delta G_R = \bar{P}_1 - \bar{P}_0 = 4.5 - 4.0 = 0.5$ kg

Actual (realized) selection differential : $S_R = \bar{P}_S - \bar{P}_0 = 5.0 - 4.0 = 1.0$ kg

$$\text{Realized heritability : } h_R^2 = \frac{\Delta G_R}{S_R} = \frac{0.5}{1.0} = 0.5$$

6.16. Selection for body weight in a population of broiler chicken increased the average body weight of progeny generation by half the amount of selection differential applied. Calculate the realized h^2 of body weight.

Solution :

Given : Realized genetic gain equals half the selection differential :

$$\Delta G_R = \frac{1}{2} S_R$$

$$\text{Realized heritability : } h_R^2 = \frac{\Delta G_R}{S_R} = \frac{\frac{1}{2} S_R}{S_R} = 0.5$$

Computation of Realized Heritability from Selection over Generations

Selection, being a continuous process, is carried over several generations. When selection has been carried over several generations, realized heritability can be calculated for each generation, and can also be calculated on pooled-generation basis (*i.e.* average across generations).

A pooled estimate of realized heritability is calculated by using following formula :

Pooled realized heritability :

$$h_R^2 = \frac{\Delta G_1 + \Delta G_2 + \dots\dots\dots + \Delta G_n}{S_1 + S_2 + \dots\dots\dots + S_n} = \frac{\sum G_i}{\sum S_i}$$

Where :

$\Delta G_1, \Delta G_2 \dots\dots\dots \Delta G_n$ refer to genetic gain for each of the 'n' generations

$S_1, S_2 \dots\dots\dots S_n$ refer to selection differential for each of the 'n' generations

$\sum G_i$ = Cumulative genetic gain over 'n' generations

$\sum S_i$ = Cumulative selection differential over 'n' generations

6.17. Given below are mean 4-week body weights of a Japanese quail strain undergoing selection. The means of selected parents for each generation are also listed :

	Generation :				
	0	1	2	3	4
Population mean ($\bar{P}$) :	120	130	142	152	165
Mean of selected parents $\bar{P}_s$	135	150	160	179	Selection terminated

Calculate the realized h^2 for different generations as well as on generation- pooled basis.

Solution : Realized selection differential and genetic gain for each generation are calculated as shown below :

	Generations :				
	0	1	2	3	4
$\bar{P}$	120	130	142	152	165
$\bar{P}_s$	135	150	160	179	Selection terminated
S_R	135–120 = 15	150–130 = 20	160–142 = 18	179–152 = 27	—
ΔG_R	–	130–120 = 10	142–130 = 12	152–142 = 10	165–152 = 13
h^2_R		10/15 = 0.667	12/20 = 0.60	10/18 = 0.56	13/17 = 0.765

$$\text{Pooled } h_R^2 = \frac{\Delta G_1 + \Delta G_2 + \dots\dots\dots + \Delta G_n}{S_1 + S_2 + \dots\dots\dots + S_n} = \frac{10+12+10+13}{15+20+18+27} = \frac{45}{80} = 0.562$$

Computation of Correlated Response to Selection

Correlated response to direct selection for a trait is calculated by the following expression:

$$\Delta G_{Y/X} = h_X \times h_Y \times i_X \times \sigma_{P(Y)} \times r_{G(XY)}$$

Where :

$\Delta G_{Y/X}$ = expected correlated response in trait 'Y' when direct selection is applied for trait 'X' (primary trait)

h_X = square root of heritability of the trait being selected (primary trait)

h_Y = square root of heritability of the correlated trait (*i.e.* the trait being not directly selected)

i_X = selection intensity for trait 'X' (*i.e.* for primary trait under selection)

$\sigma_{P(Y)}$ = phenotypic standard deviation for trait 'Y' (the correlated trait)

$r_{G(XY)}$ = genetic correlation between traits 'X' and 'Y'.

6.18. Given below is the information on different parameters of body weight and wool yield in a flock of a sheep :

Trait	Mean	σ_P	h^2	r_G
Body weight (kg) :	40	6.0	0.25	+ 0.30
Wool yield (kg) :	6.0	1.5	0.16	

Direct selection is conducted for body weight with the best 20% sheep being retained (selected) for breeding.

(*a*) Calculate the expected direct response in body weight.

(*b*) Calculate the expected average body weight of the next generation.

(*c*) Calculate the correlated response in wool yield.

(*d*) Calculate the expected average wool yield of the next generation.

(*e*) Comment on the direction of correlated response.

Solution :

Selection intensity (i) for proportion selected of 0.20 = 1.40 (from Table 6.1)

σ_A for body weight : $\sigma_{A(X)} = \sigma_P \times h = 6.0 \times \sqrt{0.25} = 3.0$ kg

(*a*) Direct ΔG_X (for body weight) = $h_X \times i_X \times \sigma_{A(X)} = \sqrt{0.25} \times 1.40 \times 3.0$ kg = 2.10 kg

(*b*) Average body weight in next generation :

$$\bar{P}_1 = \bar{P}_0 + \Delta G = 40 + 2.10 = 42.1 \text{ kg}$$

(*c*) Correlated response (genetic change) in wool yield (Y) :

$$\Delta G_{Y/X} = (h_X \times h_Y \times i_X \times \sigma_{P(Y)} \times r_{G(XY)})$$

$$= (0.5 \times 0.4 \times 1.40 \times 1.5 \times 0.30) = 0.13 \text{ kg}$$

(*d*) Expected wool yield in the next generation :

$$\bar{P}_1 = \bar{P}_0 + \Delta G(Y) = 6.0 \text{ Kg} + 0.13 \text{ Kg} = 6.13 \text{ kg}$$

(*e*) Direct selection for body weight is expected to improve wool yield also because the genetic correlation between body weight and wool yield is positive. Therefore, the correlated response is desirable.

Supplementary Problems

Selection Differential and Generation Interval

6.19. Six-week body weight in a flock of meat-type chicken averaged 1200g with phenotypic standard deviation of 180g. The males and females retained as breeders averaged 1500 and 1400g respectively.

(*a*) Calculate the selection differential for male and female parents separately and the average selection differential of both sexes.

(*b*) Suppose the males had been chosen randomly ('random selection') from the population, what would be the average selection differential ?

6.20. Average wither height in a breed of horse was 200 cm with phenotypic standard deviation of 20 cm. The selected male and female candidates exceeded the population mean by 40 and 30 cm respectively.

(*a*) Calculate the selection intensity for males, females, and on sex-pooled basis.

(*b*) What percentage of male and female replacements had been retained for breeding?

6.21. A flock of swine consisted of 300 gilts and 250 boars. Best 75 gilts and 25 boars, based on their own 4-month body weight, are selected for producing the next generation.

(*a*) Calculate the selection intensity for gilts and boars and on sex-pooled basis

(*b*) Suppose body weight in the flock averaged 50 kg with phenotypic standard deviation of 5 kg. What would be the selection differential for males and females ?

(*c*) What would be the mean of selected males and females ?

6.22. Given below is the information for two flocks of sheep :

	Flock 1	Flock 2
Number of offspring born :	250	250
Survival rate up to sexual maturity (%) :	70	80
Number selected :	70	70

(*a*) In which flock selection has been more intense ?

(*b*) Comment on the effect of survival rate on selection intensity.

6.23. Given below is information obtained from selection in two flocks of chickens and in one flock of swine :

Chickens			Swine	
Trait	Flock		Trait	
Body weight (g) :	A	B	Body weight (kg) :	
Population mean :	1500	1400	Population mean :	40
Standard deviation :	150	200	Standard deviation :	10
Mean of selected :			Mean of selected :	
males	1800	1710	male	60
females	1710	1654	temale	54
Egg production(no.) :			Litter size(no.)	
Population mean :	250	250	Population mean :	6.0
Standard deviation :	50	76	Standard deviation :	1.5
Mean of selected :			Mean of selected :	
males	No selection		males	nil
females	300	300	females	9.0

Compare the selection pressure applied between :

(*a*) the two populations of chicken for body weight.

(*b*) the two populations of chickens for egg production.

(*c*) body weight and egg production in flock 'A' of chicken.

(*d*) body weight and litter size in swine.

(*e*) body weight of flock 'A' in chicken and with body weight of swine.

6.24. Given below are the month and year of birth of parents of the calves born in December 2000 in a herd of cattle :

Calf ID	Month and year of birth of : Sire	Dam	Calf ID	Month and year of birth of : Sire	Dam
1.	Jan,1995	Sept, 1995	6.	Jan, 1996	April, 1995
2.	Dec,1995	Nov, 1996	7.	Feb, 1996	Sept, 1994
3.	Feb,1995	Jan, 1994	8.	July, 1997	Nov, 1995
4.	April, 1993	July, 1995	9.	June, 1997	Jan, 1997
5.	Dec, 1996	June, 1996	10.	Aug, 1995	March, 1995

Calculate the generation interval for sires, dams and the average for both sexes.

6.25. Calculate the generation interval for a flock of sheep for which three age-group parents were used for reproduction as follows : 25% were 3-year, 60% were 4-year and 15% were 5-year old at the time of birth of their offspring.

Expected Genetic Gain

(A) Equal selection pressure in both sexes

6.26. Average time taken to trot one mile in a population of horse is 140 seconds. Animals selected as parents to produce the next generation had average of 110 seconds. Assume heritability of 0.25 for the time to trot. What would be the :

(*a*) selection differential

(*b*) genetic change

(*c*) expected average time to trot in the progeny of selected parents ?

6.27. Narrow-sense heritability of body weight in a flock of White Plymouth Rock breed of chicken is 0.3. The selected parents are heavier than the average of population by 100g. Generation interval is nine months. What would be the expected genetic improvement on per-generation and per-year basis ?

6.28. The following information was gathered from French horse racing competitions for earning index : mean annual earning = 6.9 ; standard deviation = 1.26 and h^2 = 0.15. The breed association claims to have been applying selection intensity of 0.9. What would be the genetic change in earning index ?

6.29. In Landrace breed of swine, four-month body weight averaged 75 Kg with standard deviation of 7.5 kg. Heritability of body weight is 0.5. Average weight of selected breeders is 90 kg. Calculate the :

(*a*) selection differential and selection intensity

(*b*) expected genetic gain using both the formulae *i.e.* in terms of selection intensity and selection differential

(*c*) expected average weight of progeny generation

6.30. Average back fat thickness in a breed of swine is 3 cm and the average of individuals selected from this population to be parents of the next generation is 2 cm. What would be the average back fat thickness in next generation ? Assume h^2 to be 60%.

6.31. Average 5-week body weight in a strain of chicken is 1200 g. The birds selected from this population for breeding exceeded the population average by 300 g. Heritability of body weight is 0.3. Calculate the :

(*a*) selection differential

(*b*) expected genetic gain

(*c*) expected average 5-week body weight of progeny generation

6.32. Mangalarga is the oldest horse breed of Brazil. The following phenotypic and genetic parameters were obtained for cannon bone circumference in this breed : h^2 = 0.4 ; mean = 19.3 cm ; coefficient of variation = 4.5%. If selection intensity of 0.5 standard deviation were applied for the improvement of cannon bone circumference, calculate the expected genetic improvement through one generation of selection.

6.33. Average body length in a stock of channel catfish at 48 weeks of age is 300 mm with a standard deviation of 30 mm and the narrow-sense heritability of 0.6. A male measuring

320 mm was mated to two females in individual spawning pens. Average body length of the two females mated to the male was 310 mm. The mating produced large number of offspring. What would be expected average body length of their offspring at 48 weeks of age ?

6.34. A selection differential of 20 g per-generation was applied in an experiment to select for increased 4-week body weight in a strain of Japanese quail. The narrow-sense heritability of body weight was estimated to be 0.25. The mean 4-week body weight was 180 g at the onset of the experiment. Selection was practiced for 10 generations with the same magnitude of selection differential applied in each generation. Calculate :

(*a*) total genetic gain after 10 generations

(*b*) mean body weight after 10 generations

6.35. In a population of Pashmina goat, average hair yield is 110 g and the narrow-sense heritability is 0.20.

(*a*) What would be the average hair yield of offspring if parents with a mean of 140 g were chosen as replacements ?

(*b*) What should have been the average hair yield of selected parents in order to obtain average hair yield of 125 g in their offspring ?

(B) Unequal selection pressure in two sexes

6.36. In a flock of swine, average weaning weight was 5 kg with standard deviation of 0.5 kg. The top 50% females and 10% males were kept for breeding. Assume heritability of weaning weight to be 0.3.

(*a*) Calculate the : (*i*) selection differential for sires, dams and the average for both sexes. (*ii*) average weaning weight of the selected parents. (*iii*) expected genetic gain

(*b*) What percentage of the total genetic gain is contributed by male and female parents separately ?

(*c*) If generation interval is 2 year for males and 1.5 year for females, calculate the genetic gain per annum.

6.37. In a flock of sheep, the average weaning weight of rams and ewes selected as parents was 24 and 20 Kg, respectively, and the average for the entire flock was 18 Kg. The heritability for weaning weight was 0.3 and the average generation interval was 3.5 year for male and 2.5 year for female replacements.

(*a*) Calculate the expected genetic gain on per-generation and per-year basis.

(*b*) What would be the average weaning weight after nine years of selection?

6.38. Average staple length in a flock of sheep is 6.0 cm and the heritability of staple length is 0.50. Rams and ewes selected for breeding had average staple length of 8.5 and 7.5 cm respectively.

(*a*) Calculate the expected genetic gain due to selection of rams and ewes separately and due to selection in both sexes.

(*b*) What would be the expected mean performance of the progeny generation.

6.39. The narrow-sense heritability of daily weight gain in a population of Danish Landrace pigs is 0.4 and the mean gain in the population is 0.5 kg. Three males with daily weight gain of 0.9, 1.0 and 1.1 kg were selected and mated to six females with daily gain of 0.8, 0.8, 1.0, 0.9, 1.0 and 0.9 kg. Age of the three selected males at the birth of their offspring was 4, 5 and 6 year, while the corresponding age of the 6 female replacements at the birth of their offspring was 3, 3.5, 4, 5, 3.5 and 5 years. A large number of progeny resulting from random mating among the parents were measured for daily gain in weight.

(*a*) Calculate the genetic gain per-generation due to selection of males and females.

(*b*) Calculate the genetic gain per-year due to selection of males and females.

(*c*) What would be the average daily gain in weight in the progeny generation?

6.40. In a flock of 400 Japanese quail (equal number of both sexes), four week body weight averaged 200g with standard deviation of 20g. The best 5 and 20% of the males and females, respectively were retained for breeding. The h^2 of body weight is 0.3. One generation of quails can be obtained in 6 months.

(*a*) Calculate the : (*i*) means of selected males and females (*ii*) selection differential for males and females (*iii*) expected genetic gain on per-generation and per-year basis (*iv*) mean 4-week body weight in next generation.

(*b*) If constant selection intensity is applied over successive generations, what would be the mean performance of the population after 5 years ?

6.41. In a flock of sheep, containing equal number of males and females, average annual yield of wool (greasy fleece weight) for ewes is 5 kg. with σ_P = 0.50 kg. and for rams is 6 kg. with σ_P = 0.64. Heritability of wool yield in the flock (including males and females) is 0.36. The breeder retains the best 5% and 40% of the rams and ewes respectively for producing the next generation.

(*a*) Calculate the : (*i*) selection differentials for two sexes (*ii*) average wool yield of selected rams and ewes.

(*b*) Calculate the expected genetic gain due to selection of rams and ewes separately, and the total genetic gain (due to selection in both sexes).

(*c*) What percentage of the total genetic gain is contributed by rams and ewes ?

(*d*) What inference would you draw about the relative contribution of rams and ewes for genetic improvement ?

(*e*) Assuming that selection was conducted in ewes only, and the rams were not selected but randomly retained from the population, what will be the expected genetic gain ?

6.42. In a swine selection programe, four boars are selected that averaged 0.6 cm less back fat than the average of the flock while the 10 female replacements averaged 0.4 cm less back fat than the flock average. The selected males and females were mated at random to produce the progeny. The four selected boars were 9, 10, 11 and 12 months old respectively when they produce their first litter while the average age of the female replacements is 14 months when their progeny were produced. The h^2 of back fat thickness is 0.4 and σ_P is 0.4 cm.

(*a*) Calculate the generation interval for sires and dams.

(*b*) Calculate the total genetic change per-generation and -year.

(*c*) What percentage of the males and females replacements has been saved under truncation selection assuming normal distribution of back fat thickness ?

(*d*) Is reducing back fat thickness desirable ?

(c) Selection in one sex only

6.43. Given normal distribution of annual egg production in layer chickens with mean of 260 eggs and standard deviation of 30 eggs and h^2 of 0.15. The best 50% of the hens are kept as breeders with no selection in cocks. Calculate the :

(*a*) selection differential applied in the flock for males and females.

(*b*) average egg production of the selected females.

(*c*) expected genetic gain per-generation.

6.44. Average lactation milk yield in a large herd of indigenous cattle is 2000 liters with phenotypic variance of 1,60,000 liter^2. The h^2 of milk yield is 0.20 and generation interval for females is 5 yrs. Suppose 30% worst females are culled each generation with no culling in bulls. Calculate the :

(*a*) average selection intensity and selection differential applied in the herd.

(*b*) expected genetic gain on per-generation and per-year basis.

(*c*) herd average after 30 years if same selection intensity is applied every generation.

6.45. A poultry breeder is selecting for early age at sexual maturity in a flock of White Leghorn chicken in which the mean age at first egg is 170 days with phenotypic standard deviation of 25 days. Thirty per cent early maturing hens are selected as parents while the males are chosen at random to produce the next generation. The narrow-and broad-sense heritabilities of age at sexual maturity are 0.3 and 0.5 respectively.

(*a*) Calculate the average selection differential applied in the flock.

(*b*) What would be the average age at sexual maturity of the selected hens ?

(*c*) Calculate the expected genetic gain per-generation.

(*d*) What would be the average age at first egg of the hens in the next generation?

(*e*) Has selection been effective in modifying the trait in the desirable direction?

(*f*) Could a higher genetic gain been obtained had the males been also selected?

6.46. A breeder wishes to decrease calving interval in a herd of cattle in which average calving interval is 450 days and standard deviation is 75 days. The females chosen by him for reproduction had average calving interval of 400 days. The bulls were randomly chosen to be used as parents. The heritability for calving interval is 0.15. With selection applied for calving interval, the generation interval is 5 years.

(*a*) What will be the : (*i*) the genetic gain through one generation of selection (*ii*) average calving interval in the next generation (*iii*) average calving interval after 15 years if same selection pressure were applied every generation ?

(*b*) If selection is discontinued after three generations and the population is allowed to mate at random, what will be the calving interval after 30 years ?

Realized Heritability

6.47. In a population of mouse, the average plasma cholesterol level is 2.5 mg/100ml. Six mice (3 males and 3 females) with an average cholesterol level of 1.5 mg/100ml are selected and interbred to produce progeny. The mean cholesterol in their progeny is 2.0 mg/100 ml. What is the realized heritability of cholesterol level ?

6.48. In a flock of sheep, the average yearling weight of all animals was 60 kg. with standard deviation of 7 kg. The best 5% and 25% rams and ewes respectively were kept for breeding. The progeny of selected parents had an average yearling weight of 65 Kg. What is the realized heritability of the trait ?

6.49. A flock of chickens had average body weight of 1800 g, with standard deviation of 200 g. The best 40% parents were retained for breeding. The offspring generation had a mean of 1850 g. Estimate the heritability of body weight in this flock.

6.50. A fish breeder selects for increased body length at 6 weeks after hatching. The mean body length of 6-week old fingerling stock is 12 cm with standard deviation of 1.2 cm. The fish that had a mean length of 15 cm at 6-weeks of age were used to produce the new generation of fingerlings. The mean length of progeny of selected parents was 13.5 cm. Estimate the realized heritability of fingerling length.

6.51. If the selected parents transmitted whole of their phenotypic superiority over population mean for fleece weight to the next generation, what would be the realized h^2 ?

6.52. The sires and dams selected for high weaning weight in sheep failed to increase average weaning weight of their offspring. What is the realized h^2 of weaning weight ?

6.53. Self-confidence (defined as ability to react to new situations) in German shepherd dogs is an important component of behavior, and several dog breeding clubs select for improvement of this trait. The Swiss German shepherd dog breeding clubs score this trait on a scale of one(most confident) to five(least confident). Average self confidence score in one generation was 2.8. The males selected had average score of 2.0 while the selected females had an average score of 2.4. In the offspring generation which consisted of equal number of male and female, the average score was 2.5 for males and 2.7 for females. Estimate the realized heritability of self-confidence.

6.54. Average lactation milk yield in a herd of Holstein cows is 4000 kg. while that of selected females is 4500 kg. The sires are chosen at random. Average milk yield of offspring generation is 4100 kg. Estimate the realized heritability of milk yield in this herd.

6.55. Average greasy fleece weight in a population of sheep is 3 kg. with standard deviation of 0.3 kg. The best 10% rams and 30% ewes were chosen to produce the offspring. The progeny resulting from random mating among selected parents had average fleece weight of 3.2 kg. Calculate the realized heritability.

Correlated Response to Selection

5.56. A breeder is selecting for increasing body weight in a flock chicken with the best 20% and 40% of the males and females, respectively being retained for reproduction. Mean body weight of the flock at beginning of experiment is 1500 g with standard deviation of 150 g and heritability of 0.3. Mean egg weight of flock is 55 g with h^2 of 0.6 and σ_p = 3 g. The genetic correlation between egg weight and body weight is 0.5. Calculate the:

(*a*) genetic change in body weight and egg weight due to one generation of selection for body weight.

(*b*) mean body weight and egg weight after 5 generations of selection for body weight.

6.57. Given below are the genetic parameters for daily gain in weight and feed efficiency (measured as amount of feed consumed per unit gain in weight) in pigs :

	Daily gain(g)	Feed efficiency
h^2	0.5	0.36
σ_A	19.5	0.10
r_G between gain and efficiency = – 0.80		

Assume that the best 10% boars and 20% sows are selected on the basis of their own gain in weight.

(*a*) Compute the expected direct genetic change in weight gain and correlated genetic change in feed efficiency.

(*b*) Had direct selection been conducted for feed efficiency (instead of gain in weight) with the same amount of selection intensity as was applied for gain in weight, what would be the direct genetic change in feed efficiency, and the correlated genetic change in daily gain. Direct selection for feed efficiency would mean selecting for those having lower values of feed efficiency (*i.e.* negative selection differential).

(*c*) Which of the two selection programs (direct selection for weight gain vs direct selection for feed efficiency) would you recommend if the objective is to improve both the traits ?

6.58. Given below is the information for two traits in sheep :

Parameter	Fleece weight(kg)	Staple length(Cm)
Mean :	3.0	4.0
σ_P :	0.30	0.80
H^2 :	0.30	0.50
h^2 :	0.20	0.40
	r_G = + 0.2	
	r_P = + 0.3	

The best 10% males and 20% females, on the basis of their fleece weight, are selected to produce the next generation. Calculate the :

(*a*) selection differential for fleece weight.

(*b*) genetic gain for fleece weight and expected mean fleece weight of progeny generation.

(*c*) expected genetic change in staple length and expected mean staple length of the progeny.

6.59. Consider direct selection for egg number with h^2 of 0.20, mean of 100 eggs and standard deviation (σ_P) of 12 eggs. The best 20% of the females are selected with no selection in males. Egg weight in the flock averaged 56 g with standard deviation of 2.8 g and heritability of 0.6. Calculate the :

(*a*) expected genetic gain in egg number, and the mean egg number of next generation.

(*b*) expected genetic change in egg weight due to direct selection for egg number, given genetic correlation between egg number and egg weight is –0.4.

(*c*) average egg weight of the next generation.

6.60. Given below are the genetic parameters of two important economic traits of swine :

Trait	h^2	σ_P	% selected :	
			male	female
Litter size (no.) :	0.22	2.0	Random selection	25
Body weight (g) :	0.35	600	10	25

Genetic correlation between litter size and body weight is 0.8.

(*a*) Consider two selection programs :

(*i*) Direct selection is applied for increased litter size : calculate the direct response in litter size and correlated response in body weight.

(*ii*) Direct selection is applied for increased body weight : calculate the direct response in body weight and correlated response in litter size.

(*b*) Which of the two selection programs should be used for genetic improvement of litter size and why ?

6.61. Complete the following table pertaining to four selection scenarios :

	Selection scenario			
	A	B	C	D
ΔG/generation :	7.26	5.16	–	– 6.456
Standard deviation:	10.0	20.0	–	14.0
Selection differential:	24.2	–	–	–
Selection intensity:	–	–	1.99	–
Heritability :	–	0.2	0.3	0.2
ΔG/year :	–	1.29	1.99	–1.2712
% saved :	–	4	–	–
Generation interval (yrs.) :	3	–	2	5

6.62. Which one of the following two selection programmes would result in greater annual rate of genetic improvement ?

	Program A	Program B
Male selection:		
Accuracy of selection	0.30	0.90
Proportion selected	0.02	0.10
Generation interval	2 years	7 years
Female selection :		
Accuracy of selection	0.60	0.65
Proportion selected	0.90	0.90
Generation interval	4 years	4 years

Answers to Supplementary Problems

6.19 (a) 300g for males, 200g for females, 250g for average of two sexes

(b) 100g

6.20 (a) i = 2.0 (males), 1.5 (females) and 1.75 (pooled)

(b) Approximately, males best 6% and females best 17% (on the assumption of normal distribution).

6.21 (a) i (females) = 1.27 , i (male) = 1.76 , i (pooled) = 1.51

(b) S for male = 8.8 kg and for female = 6.35 kg

(c) mean of selected males = 58.8 kg and of selected females = 56.4 kg.

6.22 (a) Selection has been more intense in Flock 2 (Hint: The number of potential parents available at sexual maturity would be :

Flock 1 : No. at birth X survival rate = 250 X 0.70 = 175

Flock 2 : 250 X 0.80 = 200

Per cent selected for : flock 1 = (70/175)X100 = 40%,

flock 2 = (70/200)X100 = 35%

(b) Survival rate influences the number of animals available at the time of selection. The calculations show that the numbers available in the two flocks respectively were 175 and 200. With the same number of animals to be kept for breeding (i.e. 70), more number of animals become available at the time of selection in the flock in which survival rate is higher (i.e. flock 2). Therefore, the percentage retained for breeding in such a flock (flock 2) would be less and therefore, selection intensity will be more.

6.23 For chickens (a,b,c): The required computations are presented below :

	Flock 'A'	Flock 'B'
Body weight :	S_M = 300 ; S_F = 210	S_M = 310 ; S_F = 254
	S_{avg} = 255 ; i = 1.7	S_{avg} = 282 ; i = 1.141
Egg production :	S_M = 0 ; S_F = 50	S_M = 0 ; S_F = 50
	S_{avg} = 25 ; i = 0.50	S_{avg} = 25 ; i = 0.33

(a) Selection is more intense in flock 'A' as selection intensity (i) in this flock is higher (1.70) than in flock 'B'(1.41).

(b) Selection is more intense in flock 'A'(i = 0.50)

(c) In flock 'A', selection is more intense for body weight (i = 1.7) than for egg production (i = 0.50).

For swine (d&e)

Body Weight	Litter Size
S_M = 20 ; S_F = 16	S_M = 0 ; S_F = 3.0
$S_{avg.}$ = 18 ; i = 1.8	$S_{avg.}$ = 1.5 ; i = 1.0

(d) Selection is more intense for body weight than for litter size.

(e) Selection is more intense for body weight in swine (i = 1.8) than for body weight in flock 'A' of chicken(i =1.7)

6.24

Calf ID	Sire	Dam	Calf ID	Sire	Dam
1	5yr 11month	5yr 3month	6	4yr 11month	5yr 8month
2	5yr	4yr 1month	7	4yr 10month	6yr 3month
3	5yr 10month	6yr 11month	8	3yr 5month	5yr 1month
4	7yr 8month	5yr 5 month	9	3yr 6month	3yr 11month
5	4yr	4yr 6month	10	5yr 4month	5yr 9month

Generation interval for sires = 5yr 15 day, for dams = 5 yr 3 month 12 day, and average for population = 5 yr, 1 month 29 days.

6.25 Generation interval would be the weighted average of the different age structures: $(0.25 \times 3 + 0.60 \times 4 + 0.15 \times 5)$ = 3.9 years = 3yr 10 month 24 days

6.26 (a) S = –30 seconds

(b) Genetic gain = –7.5 seconds

(c) Expected average time to trot in progeny = 132.5 seconds

6.27 ΔG/generation = 30 g ; ΔG/year = 40 g ;

6.28 0.1701

6.29 (a) S = 15 Kg and i = 2.0 (b) 7.5 Kg (c) 82.5 Kg.

6.30 2.4cm

6.31 (a) S = 300g (b) Δ_G = 90g (c) 1290g.

6.32 0.1737 cm ; 309 mm

6.34 (a) 50g (b) 230g

6.35 (a) 116 g (b) 185g

6.36 (a) (i) S (females) = 0.4 Kg, S (males) = 0.88 kg, S (average) = 0.64 Kg (ii) 5.64 kg. (iii) expected genetic gain = 0.192 kg

(b) by males = 69% and by females = 31%

(c) = 0.110 kg.

6.37 (a) ΔG/generation = 1.2 Kg ; ΔG/year = 0.40 kg

(b) Average weaning weight after 9 years of selection = 21.6 kg

6.38 (a) ΔG(rams) = 0.625, ΔG(ewes) = 0.375, ΔG (total) = 1 cm

(b) 7 cm

6.39 (a) ΔG/generation (males) = 0.2 kg and (females) = 0.18 kg; ΔG (total) = 0.38 kg.

(b) ΔG/year (males) = 0.04 kg and for females = 0.045 kg (c) 0.88 kg.

6.40 (a) (i) males=241.2g; females = 228g, (ii) S for males= 41.2g and for females = 28g. (iii) ΔG/generation = 10.4g and ΔG/year = 20.8g (iv) 210.4g

(b) 304g

6.41 (a) (i) S for rams = 1.318 and for ewes = 0.485 kg (ii) average wool yield of rams = 7.32 kg and of ewes = 5.485 kg

(b) ΔG from ram selection = 0.2372 kg and from ewe selection = 0.0873 kg ; total ΔG = 0.3245 kg.

(c) Percentage of total genetic gain contributed by rams = 73.1% and by ewes = 26.9%

(d) Since the males (rams) have been more intensely selected, therefore, genetic improvement due to male selection is about 3 times as compared to female (ewe) selection

(e) ΔG = 0.0873 kg.

6.42 (a) 10.5 months for sires and 14 months for dams

(b) ΔG/generation = –0.2 cm; ΔG/year = –0.196 cm

(c) selection intensity for males = 1.5 and for females = 1.0. These selection intensity values correspond approximately to the best 15% and 40% of the male and female parents retained for breeding.

(d) Yes, because back fat thickness is a useful measure of amount of fat in pig carcass, and consumers prefer lower fat.

6.43 (a) 24 eggs for females and zero (nil) for males

(b) 284 eggs

(c) 1.8 eggs

6.44 (a) i = 0.25, S = 100 liters

(b) ΔG/generation = 20 liters and ΔG/year = 4 liters

(c) 2120 liters

6.45 (a) –14.5 days (b) 155.5 days (c) –4.35 days (d) 165.6 days

(e) Yes, because age at first egg has been reduced, which means birds start laying at an early age and will, therefore, produce more number of eggs up to a specified age.

(f) Yes

6.46 (a) (i) –3.75days (ii) 446.25days (iii) 438.75days

(b) 438.75 days because the change brought about by selection for 15 years (3 generations) would be permanent, and therefore, the mean would remain same in subsequent years.

6.47 0.5

6.48 0.429

6.49 0.258

6.50 0.5

6.51 one

6.52 Zero

6.53 0.33

6.54 0.4

6.55 0.45

6.56 (a) ΔG for body weight = 53.3 g and ΔG for egg weight = 0.75 g

(b) mean body weight = 1766.7 and egg wieght = 58.8 g.

6.57 (a) ΔG for daily gain = 21.8 g, ΔG for feed efficiency = –0.090

(b) ΔG for feed efficiency = –0.095, ΔG for daily gain 14.8 g

(c) Direct selection for body weight (1st selection programme) as it results in hihger improvement in body weight but at the same time it results in almost same improvement in feed efficiency as would have been possible with direct selection for feed efficiency.

6.58 (a) 0.474 kg (b) 0.0948 kg and 3.1 kg (c) 0.071 cm and 4.07 cm

6.59 (a) 1.7 eggs and 101.7 eggs (b) –0.28 g (c) 55.75 g

6.60 (a) (i) 0.279 pigs, 84.5 g (ii) 318 g, 0.672 pigs

(b) the second one, because it results in higher genetic gain in both the traits as it permits selection in both the sexes.

6.61

	A	B	C	D
ΔG/generation	7.26	5.16	3.98	–6.456
Standard deviation	10.0	20.0	6.66	14.0
Selection differential	24.2	25.8	13.2	–32.48
Selection intensity	2.42	1.29	1.99	2.3
Heritability	0.3	0.2	0.3	0.2
ΔG/year	2.42	1.29	1.99	–1.29
% saved	2.0	4.0	6.0	3.0
Generation interval (yrs.)	3	4.0	2	5

6.62 Genetic gain for 'A' = 0.316 σ_A units and for 'B' = 0.225 σ_A units.

7

Bases of Selection

Selection of genetically superior individuals is the most important tool available to breeder for genetic improvement of a population. Since genes of an individual (not its genotype) are transmitted to its progeny, it is necessary to have some information about the types of genes possessed by the individual. The sum of average (additive) effect of the genes possessed by an individual is referred to its **breeding value**[1]. The breeder would like to identify the animals with superior breeding values and then use them as parents of the next generation. However, for quantitative traits neither the number of genes nor their effects on phenotypic expression of the trait are known. Assessment of breeding value, therefore, is always based on the phenotypic value (performance) which is measurable. The various sources or criteria of information used for estimating the breeding value of an individual or a group of individuals are referred to as **bases of selection.**

An individual may be selected on the basis of its own performance or/and the performance of its relatives. The closeness of the criterion of selection (phenotypic value) to the breeding value, referred to as accuracy of selection, directly determines the genetic improvement possible through selection.

Classification of Basis of Selection

Broadly, there are three criteria or bases by which an individual can be selected:

- Individual's own performance
- Performance of it's relatives
- Combination of individual's own performance and the performance of it's relatives

Fig. 7.1 shows the biometrical relationship of an individual with its relatives alongwith the percentage of genes in common between the individual and its relatives. The word **'relative',**

[1]Breeding value is alternately also referred to as 'additive genetic value',' genetic worth' or 'genetic merit'

as used in animal breeding, is a general term to refer to all the individuals, which are related through common ancestry. The parents, grandparents, sibs, and progeny all constitute relatives.

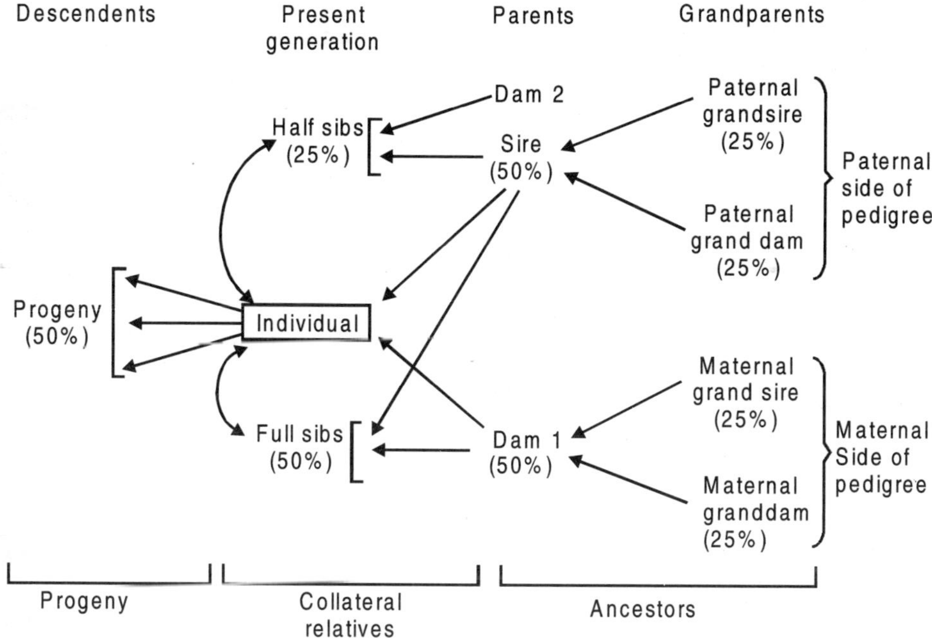

Fig. 7.1 Biometrical relationship of an individuals with different kinds of relatives and proportion of genes common between the individual and its relatives (shown within the parenthesis).

An individual receives a random half of its genes from one parent, and the other random half from its second parent (with the exception of the genes located on sex chromosomes). Similarly, an individual receives one fourth of its genes from each grandparent. Full sibs (i.e. full-brothers and-sisters) of the individual share one-half of their genes while half sibs (i.e. half-brothers and-sisters) share one fourth of their genes. Similarly, progeny receive a random half of their genes from the individual. However "which sample of genes is shared between an individual and its relative" is determined entirely by chance due to sampling nature of inheritance. The various bases or criteria used for selection of an individual (or a group of individuals) are:

- Individual's own performance (Individual selection)
- Pedigree performance (Pedigree selection)
- Family performance (Family or Sib selection)
- Progeny performance (Progeny testing)
- Combination of performance of individual and its relatives (Combined selection)

The individuals of a herd/flock from which selection is to be made are referred to as **'breeding candidates'**

Individual Selection (Mass Selection)

Selection based on individual's own phenotypic value (performance) is refered to as **individual selection**. The phenotype may be measured on a continuous scale (e.g. weight, milk production, egg production etc.) or it could be subjectively scored (e.g. body shape, conformation or other type characters). It is, however, desirable if the trait to be improved is recorded on a continuous scale so that small differences between breeding candidates can be detected. In plants and laboratory organisms (e.g. fruit fly, flour beetle etc.) selection based on individual's own performance is commonly referred to as **mass selection** because the selected individuals are usually put together ***'en mass'*** for mating (e.g. fruit fly and floor beetle in a vial). The term 'individual selection' is used more specifically in livestock species where mating are controlled. Measurement of performance of individuals for selection purposes is called performance testing.

Selection based on individual's own performance is one of the oldest and simplest procedures of selection. Livestock farmers had been using this procedure much before the scientific foundations of animal breeding were laid. Accuracy of individual selection, measured as correlation between the breeding value and phenotypic value (r_{AP}), equals the square root of heritability:

$$A \xrightarrow{r_{AP} = h} P$$

The genetic gain possible through individual selection is thus directly proportional to the h^2 of the trait. It is thus most effective for high heritability traits (e.g. growth, quality and composition traits in all species of farm animals etc). If employed for low heritability traits (e.g. reproduction, viability and some production traits) it will yield lower genetic gain and will be inefficient. Values of accuracy for different h^2 values are given in Table 7.1.

Table 7.1 Accuracy of individual selection at different values of heritability.

h^2 :	0.1	0.2	0.3	0.4	0.5	0.6	0.7	0.8	0.9	1.0
r_{AP}:	0.32	0.45	0.55	0.63	0.71	0.77	0.84	0.9	0.95	1.0

Theoretically the accuracy of individual selection can attain unity if the trait were 100% heritable (Table 7.1). In that case, phenotype of the individual would be identical to its breeding value. Since the quantitative traits are invariability influenced by non-additive genetic and environmental effects, the accuracy of individual selection in practice is never one.

Merits: Individual selection offers a number of advantages:

- *Operationally easy:* It is the easiest and quickest procedure of selection; all it requires is performance records of individuals in the herd/flock. Indeed much of the early genetic progress in livestock species has been brought about by individual selection. For high h^2 traits, it yields rapid response.
- *Possible in non-pedigreed herds:* Individual selection does not require knowing the pedigree of breeding candidates, and hence it is applicable in non-pedigreed herds. But avoidance of inbreeding in such herds becomes a problem. Over a long period, especially in small-sized herds, matings without knowing pedigree may lead to accumulation of inbreeding which in turn depresses herd performance. It is suggested that in non-pedigreed herds the number of breeding parents (particularly the sires) should be kept high (> 30 sires) so that lower inbreeding accumulates.

Constraints: There are certain traits and situations under which individual selection has certain limitations.

- *Not applicable for sex-limited traits:* Many economically important traits of farm animals (e.g. milk production in dairy animals, egg production in poultry, litter size in swine etc.) are only manifested in females. Selection of males for such traits cannot be based on their own performance. Similarly, individual selection is not possible for female selection if the traits are limited to males (e.g. semen quality etc.)

- *Slaughter or carcass traits*: A prerequisite for use of individual selection is that the trait to be improved is measurable on the breeding candidate itself while being alive. Consequently, it can not be applied for traits that need animal's slaughter for recording (e.g. carcass quality and composition in broiler chickens, pigs, beef cattle and mutton-type sheep) or for traits which can not be recorded on an individual (e.g. viability or mortality- an all-or-none trait).
- *Late-expressed traits*: Some economically important traits are expressed or recorded late in life. For example, longevity (i.e. the time for which the animal remains in the herd; lifetime milk production etc) is recorded late in life. The breeder has to wait until such traits are expressed which delays selection and increases generation interval. And by the time the trait is recorded, some of the animals may become too old to be used for reproduction or may die.
- *Low heritability traits:* Since accuracy of selection is directly proportional to heritability of the trait, the genetic progress possible through individual selection for lowly heritable traits would be slow. Individual selection is usually not recommended for traits with heritability of less than 0.25.

Despite some of its shortcomings, performance of an individual, if available, always receives some consideration in making selection decisions even though selection may primarily be based on some other criterion (e.g. performance of relatives). Usually the animals whose own performance is below the herd average are not selected regardless of performance of their relatives.

Aids to Individual Selection: Situations for their Use

Selection based on individual's own performance is one of the simplest techniques of selection. However, under certain circumstances, it may not be the most efficient procedure; therefore, some aids to selection (individual) may be necessary or desirable. These alternative bases of selection fall into two main classes:

- Individual selection based on repeated records and
- Selection based on performance of relatives or family

These alternative bases of selection are sometimes referred to as **'aids to individual selection'** or simply as **'aids to selection'**. There are three main reasons for use of various aids to individual selection:

- *To increase accuracy of selection:* When heritability of a trait is low, the accuracy of individual selection based on single record would also be low, therefore, individual selection will be inefficient. Use of repeated measurements on the same animal or use of records of relatives may be used to increase the accuracy of selection.
- *To achieve early selection:* Individual selection if applied for traits which are expressed or recorded late in life, increases the generation interval and in turn decreases annual genetic progress. Early selection based on short-term or part-records may be used to shorten the generation interval and in turn to increase annual genetic progress. Selection for late-expressed traits may be based on performance of ancestors which are available before birth of potential breeding candidates.

- *Inability to use individual selection:* Selection of males for sex-limited traits (e.g. milk or egg production) cannot be effected by individual selection. Similarly, females cannot be selected by individual selection for the traits which are limited to males only (e.g. semen traits). Also this selection procedure cannot be practiced for traits that require slaughter of animals. Selection for such traits may be based on performance of relatives.

Individual Selection based on Repeated Records

Selection of individuals for repeatable traits can be based on their average performance. For example, if P_1, P_2, P_3, ..., P_n are 'n' repeated records of a particular animal, then average performance of successive records of that animal, $\bar{P}$, is used as the criterion of selection:

$$\bar{P} = \frac{P_1 + P_2 + P_3 + \dots\dots\dots + P_n}{n} = \frac{\sum P_i}{n}$$

Averages based on repeated records are sometimes also referred to as **'long- term' or 'life-time' production averages.** Differences in performance between successive records of the same animal result from temporary environmental factors that vary from one period to other. For example, a cow calved during winter in one year and during summer in the following year; the influence of season of calving on milk yield would tend to cancel-out in its average based on two records.

The principle involved in use of repeated records is that: average based on repeated records is a better measure of breeding value of the animal because temporary environmental effects tend to cancel out in averages. Since the non-genetic influences on the same animals decrease, the breeding value is estimated more accurately than based on a single record.

Accuracy of selection based on repeated records, defined as the correlation between breeding value (A) and average of 'n' records ($\bar{P}_n$), designated as $r_{A\bar{P}_n}$, is given by the following expression:

$$r_{A\bar{P}n} = h\sqrt{\frac{n}{1 + (n-1)r}}$$

Where: h = square root of heritability of the trait

r = repeatability of the trait

n = average number of records per animal

It can be seen that for any value of repeatability ($r > 0.0$) and for any number of repeated records ($n > 1.0$), the value of '$\sqrt{n/1 + (n-1)r}$' exceeds unity. Therefore, the accuracy of repeated records is always more than accuracy of a single record. Values of accuracy of selection based on average of several records for various levels of repeatability for a fixed value of heritability ($h^2 = 0.1$) are given in Fig 7.2

Since each additional record of the individual provides some extra information with respect to breeding value, the accuracy of selection improves with increase in number of records (Fig 7.2). The increase in the number of records is however more useful for low repeatability traits (< 0.3) because for such traits temporary environmental factors make a large contribution, and such factors tend to cancel out more effectively as 'n' increases.

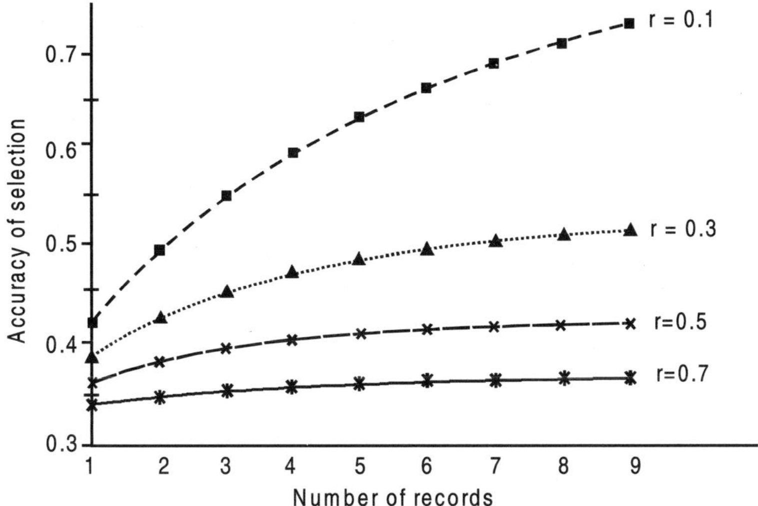

Fig 7.2 Accuracy of selection based on repeated records for different levels of repeatability

Repeated records are more commonly used for selection in cattle, buffaloes, horse etc. which have low reproductive rate. Since each additional record takes at least a year or so to complete, a breeder having fixed facilities for rearing of animals, can only afford to obtain a limited number of records. In most species, 4-5 records are considered sufficient for obtaining a reasonably good accuracy even when the trait has low repeatability (< 0.3).

The efficiency of selection based on repeated records as compared to that based on single record (called 'relative efficiency') is given by the ratio of accuracy of selection based on repeated records to that based on a single record as shown below :

$$\frac{h\sqrt{n/1 + (n-1)r}}{h} = \sqrt{\frac{n}{1 + (n-1)r}}$$

Values of relative efficiency for different values of repeatability are shown in Fig 7.3.It is apparent that the relative efficiency increases with decrease in values of repeatability (for any

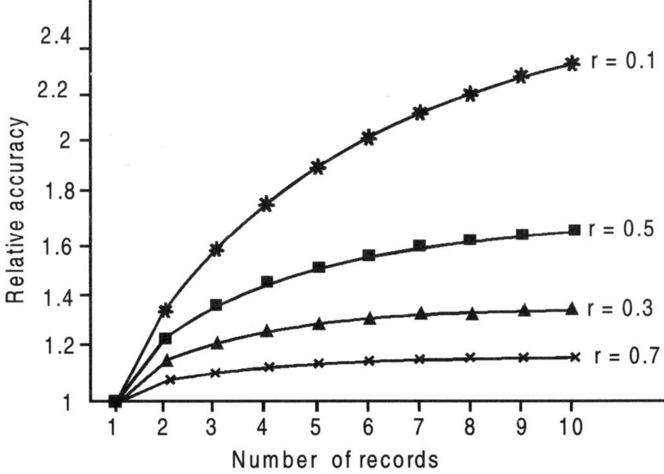

Fig 7.3 Relative accuracy of selection based on repeated records relative to a single record

given value of 'n'). This means that repeated records are more useful for low repeatability traits. The reason being a larger contribution of temporary environmental variation for such traits which tends to cancel out in the average.

Usefulness: Repeated records offer some advantages:

- *Increased accuracy*: Repeated records increase the accuracy of selection by decreasing the variation caused by temporary environmental factors.
- *Slightly higher selection intensity:* Repeated records may permit a slightly higher intensity of selection as waiting for additional records may lead to increase in total number of individuals available for selection.

Constraints/Limitations: Use of long-term production averages (repeated records) has certain disadvantages:

- *Increased generation interval*: Repeated records prolong generation interval since every new record takes time and hence annual genetic gain is reduced.
- *Increased cost:* Waiting for additional records before selection takes place increases both the cost of performance recording as well as the cost of rearing of animals particularly of less productive animals.

In summary, the use of repeated records though improves the accuracy of selection but it concomitantly prolongs generation interval. Therefore, the increase in accuracy of selection using repeated records needs to offset the prolonged generation interval to increase annual genetic progress. A balance of these two factors, based on annual genetic gain, need to be worked for evaluating the relative efficiency of repeated records over a single record.

Pedigree Selection

Pedigree of an animal is a genealogical record of its ancestry (i.e. a list of names of ancestors), and is a prerequisite for application of different bases of selection, (with the only exception of individual selection). Recording of parentage (pedigree) originally began for dairy cattle through herd book system but now the practice of pedigree keeping is common in all species of farm and companion animals. But mere recording of parentage by itself is of limited value unless the performance records of parents are maintained so that these can used for selection.

Selection of an individual on the basis of performance of its ancestors (parents, grandparents or remote ancestors) is refereed to as **pedigree selection**. Pedigree selection had been used by breeders of many species in older times, the most common being the selection of dairy bulls for milk production on the basis of their mother's milk yield.

The basic principle involved in pedigree selection is that an individual receives a certain proportion of genes from its ancestors: an individual receives 50% of its genes from its sire and the other 50% from its dam. The percentage of inheritance received by an animal (offspring) from its ancestors decreases as the ancestor becomes remote:

- Each parent contributes a random half of its genes to the animal: $(½)^1$.
- Each grandparent contributes a random one-quarter (25%) of its genes to the animal: $(½)^2$
- Each great-grandparent contributes one-eighth (12.5%) of its genes to the animal: $(½)^3$, and the remaining seven-eighth comes from all others ancestors.

The genetic contribution of each ancestor to the animal under consideration thus decreases with each receding ancestral generation. It can be obtained by raising "½" to the appropriate term: $(½)^n$, where 'n' is the number of generations separating the individual from its ancestor.

The value of pedigree record thus depends upon the closeness of relationship of ancestor to the animal under consideration (breeding candidate).

The accuracy of pedigree selection, measured as correlation between breeding value of the individual and performance of its ancestor(s), each with single record, is given by the following expressions:

Accuracy of selection based on one parent = 0.5 h

Accuracy of selection based on both parents = 0.7 h

Expressions have been derived for calculating the accuracy of pedigree selection based on performance of parents and grandparents; and on the combination of performance of parents, grandparents and great-grandparents. Numerical values for accuracy of different types of pedigree performance are given in Table 7.2.

Table 7.2 Accuracy of pedigree selection at different levels of heritability.

Accuracy of selection					
		Pedigree selection based on:			
h^2	Individual selection (h)	One parent (0.5h)	Both parents (0.7 h)	Parents plus grandparents (2P + 4GP)	Complete pedigree (2P+4GP+8GGP)
0.1	0.32	0.16	0.22	0.27	0.29
0.2	0.45	0.22	0.32	0.37	0.39
0.3	0.55	0.27	0.39	0.43	0.45
0.4	0.63	0.31	0.45	0.49	0.50
0.5	0.71	0.35	0.50	0.53	0.54
0.6	0.77	0.39	0.55	0.57	0.57
0.7	0.84	0.42	0.59	0.61	0.61
0.8	0.89	0.45	0.63	0.64	0.64
0.9	0.95	0.48	0.67	0.67	0.67
1.0	1.0	0.50	0.70	0.70	0.70

Salient conclusions from the Table are:

- Limit of accuracy of selection based on performance of one parent is 0.5, and that based on performance of any number of ancestors (e.g. parents, grandparents and great-grandparents) is 0.7. The reason for this is sampling nature of inheritance: an offspring receives a random sample of parental genes, and chance decides which of the two alleles at each locus will be transmitted.
- Accuracy of selection based on individual's own performance is higher than all types of ancestors and at all levels of heritability. Importance of individual's own performance relative to pedigree selection increases as heritability increases. That is, performance of individual itself becomes progressively more valuable than the performance of its ancestors as the heritability increases.
- Grandparents add little to accuracy of the selection, and that too when heritability of the trait is low. The reason for this is that if heritability of a trait is low, the trait is greatly influenced by environmental and non-additive genetic effects, which are not transmitted to offspring.

- Ancestors beyond grandparents (i.e. great-grandparents etc) contribute virtually nothing to accuracy except when heritability is low (< 0.20).
- Limit of accuracy of pedigree selection with complete pedigree (parents, four grandparents and eight great-grandparents) is 0.7. That is, pedigree selection of any type can never be 100% accurate in determining breeding value even when heritability of the trait is unity.

Advantages/Situations for Use: Pedigree selection can be employed under certain situations :

- *Shortest generation interval:* It has the shortest generation interval compared to all other bases of selection because selection decisions can be made even before birth of the individual (potential parent). For example, selection for milk production, egg production etc can be made even before the animal is born or the trait of interest has been expressed. In fact, pedigree provides the first piece of information about performance potential of an animal, and thus gives first opportunity for selection. A male can be selected on the basis of its pedigree performance, and can then be used for breeding once it becomes sexually mature. Generation interval for pedigree selection can be close to the average age at sexual maturity of the species.
- *Applicable for sex-limited and late-expressed traits* : Pedigree selection is particularly useful for sex-limited traits (e.g. milk yield in dairy cattle, egg production in poultry) and for traits which are expressed or measured late in life (e.g. longevity or lifetime milk production).
- *Useful for preliminary selection:* In order to save cost of rearing of a large number of animals, it is a common practice to undertake some preliminary selection on the basis of performance of parents. For example, young dairy bulls undergo preliminary selection on the basis of performance of their dams. Pedigree selection is later supplemented with progeny testing or family selection. Preliminary selection of females in dairy cattle is also based on performance of their dams.
- *Cost effective*: Pedigree selection is cost effective; all it needs is compilation of pedigree which now a days is routinely done.

Constraints/ Limitations: There are certain constraints and limitations :

- *Low accuracy:* Pedigree selection has low accuracy as compared to individual selection. If performance of the individual is known with precision, pedigree selection may not provide any additional information about its breeding value.
- *Lack of reliable records*: Reliable performance records of ancestors alongwith their contemporaries are usually not well maintained. This is particularly true for species like cattle, buffalo, horse etc, and in situations where performance recording is done at farmers' level. In such situations, the ancestors of different individuals may have experienced different environmental conditions as they might have been reared in different herds, years, seasons etc.
- *Genotypic-environment interaction*[2] : Performance records of ancestors might have been made several years ago when environmental conditions (like feeding, management and health regimes etc) may have been quite different from those existing today. This

[2]Genotype- environment interaction means differential response of different genotypes (i.e. parental vs progeny genotypes) to different environments (i.e. environment in the parental vs progeny generation).

might lead to strong 'genotype-environment' interactions, and hence bias evaluation of genetic merit of the individual.

- *Errors in pedigree*: Errors in recording of pedigree may occur particularly on male side. Records of some of the Breed Societies indicate that 5-10 % of pedigrees may be in error. Under extensive production system (i.e. the one in which animals are reared on pasture with free excess to movement), the chances of recording errors are more than under the intensive system. Some organizations use blood group typing to determine the accuracy of parentage.
- *Inability to discriminate among offspring of same ancestor:* Pedigree selection cannot be used for selection among offspring of the same ancestor as it provides no basis for their discrimination.

In addition to these limitations, history of selection of many species of farm animals is replete with examples where unwarranted or undue favoritism was given to some ancestors due to "fashion/fad", and not because of performance. Fortunately, such unwanted crazes are diminishing though breeders of companion animals continue to over-advertize their pedigree (e.g. horse and dog). Pedigree selection as the only procedure of selection is not much used in modern animal breeding.

Combining Information from more than one Ancestor

Frequently performance records of several types of ancestors (e.g. parents, grandparents etc.) may be available. A linear combination of performance records of different ancestors can then be made by assigning weighting coefficients to performance of different ancestors. The most important consideration in using combination of performance of different ancestors is: 'how should the performance records of different ancestors be weighed to obtain an estimate of its breeding value (EBV)'. The optimum weighting coefficients assigned to the performance of the parent and grandparents for different levels of heritability are given in Table 7.3.

Table 7.3 Accuracy of selection based on performance of parent and grandparent, and the relative weightage given to the parent versus to grandparent.

	Heritability:		
	0.10	0.20	0.40
Accuracy for one parent (0.5h) :	0.15	0.205	0.275
Accuracy for one grandparent (0.25h) :	0.071	0.091	0.103
Relative weightage (parent to grandparent) :	2.13	2.25	2.67

In general, the relative weights given to performance of parent to that of grandparent depend on:

- Degree of genetic relationship of the individual with the ancestor: More weightage is given to recent ancestor (parent) as compared to remote ancestors (e.g. grand parent or great grand parent).
- h^2 of the trait: Performance of parent receives increasingly more weightage than performance of grandparent as h^2 of the trait increases.That is, performance of grandparent is useful as a supplement to performance of the parent when h^2 of trait is low.

Selection Based on Performance of Collateral Relatives

Relatives of an individual can be grouped into two categories: direct and collateral. The relatives that are related to each other on gene donor-recipient basis are called **direct relatives** while the relatives which are neither direct ancestors nor direct descendants of an individual are referred to its **collateral relatives**. Full sibs, half sibs, cousins, uncles, aunts, nephews and nieces constitute collateral relatives. Since they are related to the individual because of common ancestry, the individual does not receive any genes directly from its collateral relatives. But due to common ancestry,the collateral relatives share a certain proportion of genes with the individual, and therefore they provide information about an individual's breeding value. Higher the degree of relatedness of an individual with its collateral relatives, the more valuable is the information they provide about the individual for selection.

In breeding of farm animals, information from only close collateral relatives (e.g. full- and half- sibs) is used for selection. The major advantage of selection based on collateral relatives is that it does not prolong the generation interval, especially when compared to selection based on progeny performance. There are three selection procedures that involve collateral relatives: family selection, sib selection and within-family selection.

Family Selection

The term 'family' was originally defined as all the animals coming from an unbroken line of females from an outstanding foundation female (like cow family in cattle,tail mares in horse). In scientific animal breeding, '**family**' refers to a group of individuals bearing common genetic relationship by virtue of having one or both parent's common (e.g. full- and half-sib families). Members of a family are only collaterally related to one another. In this usage, a family can include first cousins as well which are progeny from same grandparents. By virtue of common ancestry, members of a family have a certain fraction of the identical genes. This genetic similarity is measured as **coefficient of genetic relationship** (or simply as **relationship coefficient**), and designated as 'R'. According to collateral relationships, two types of families are common in livestock:

- **Full-sib families:** A group of individuals having both parents in common (sire and dam) constitutes a full-sib family. Genetic relationship coefficient (R) among members of a full-sib family is 0.5. Full-sib families are sometimes also termed as **dam-families** because each full-sib family results from mating of a sire to a dam, and the dam in turn producing a number of offspring. Full-sib families are common in highly prolific species (e.g. poultry, swine and other litter-bearing species like mice, rabbit etc.).
- **Half-sib families:** A group of individuals having one common parent constitutes a half-sib family. The common parent in most species of farm animals is male (sire). Relationship coefficient among members of a half-sib family is 0.25. A family having male as common parent is called a **paternal half-sib family**. When dam is the common parent, the family is called a **maternal half-sib family**. Since a sire can be mated to a number of dams, half-sib families are also referred to as **sire-families.** A sire-family is thus progeny of a given sire but from different dams. Paternal half-sib families are available in livestock species like cattle, buffaloes, horse etc. besides the highly prolific species.

Selection of a family on the basis of its average performance is referred to as **family selection.** The entire family, as a unit, is either selected or rejected on the basis of its average performance (phenotypic value) irrespective of performance of individual family members. The criterion of selection is thus the average performance of the family ($\bar{F}$), and the unit of selection is also the entire family and not any single individual though an individual contributes to the family average. Family selection is diagrammatically shown below :

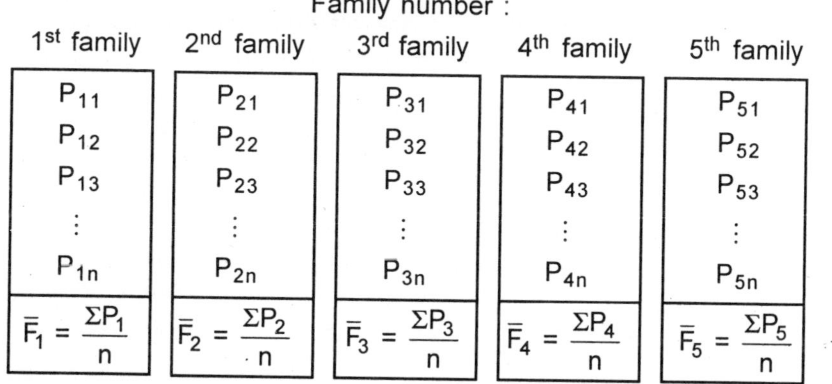

The families are ranked on the basis of their average performance (weighted for the number of individuals in the family and h^2), and the families with highest performance are retained for producing the next generation. Since selection on family performance discriminates between families; this type of family selection is more appropriately referred to as **between-family selection**. The families may be of full-sibs (in highly prolific species) or of half-sibs (in species with low reproductive rate). Families of more remote relationships (like cousins) are of little practical significance.

Accuracy of family selection, measured as correlation between average breeding value[1] ($\bar{A}$) and average phenotypic value of the family ($\bar{F}$), is given by following expression:

$$r_{\bar{A}\bar{F}} = h\sqrt{\frac{[\,1 + (n-1)R\,]}{n\,[\,1 + (n-1)t\,]}}$$

The accuracy of family selection thus depends on:

- *Heritability:* It is most useful for low heritability traits.
- *Family size (n)*: More the number of individuals per family, higher will be the accuracy as larger the family-size more will be the correspondence between family average and its (mean) breeding value. If each family contains a single individual, accuracy of family selection equals 'h' which is also the accuracy of individual selection. Family selection is thus more efficient for highly prolific species (e.g. poultry, swine, aquatic species like fish) where large sized families are easily available.
- *Genetic relationship coefficient among family members (*'type of family): Relationship coefficient (R) equals 0.5 for full-sib and 0.25 for half-sib families. With other factors being equal, full-sib families are more useful than half sib families.
- *Phenotypic coefficient of relationship among family members* (t): Phenotypic relationship coefficient among family members equals the product of heritability and genetic relationship coefficient: $t = Rh^2$. That is, 't' is a direct function of genetic relationship and heritability of the trait.

[3]The term breeding value has thus far been defined in context of an individual but it can be extended to the family as well.

The efficiency of family selection rests on the fact that phenotypic mean of family comes closer to its average breeding value because environmental and non-additive genetic effects tend to cancel out in mean of the family, provided mean is based on large number of individuals. Consequently, family selection is most efficient when environmental and non-additive genetic effects constitute a large part of the phenotypic variation, which means the trait has low h^2. Thus the *ideal conditions for use of family selection are : low* h^2 *and large family-size.*

Relative Efficiency of Family over Individual Selection: Relative efficiency of family selection over individual selection is measured as the ratio of accuracy of family selection to accuracy of individual selection:

$$\text{Relative efficiency} = \frac{\text{Accuracy of family selection}}{\text{Accuracy of individual selection}}$$

$$= \frac{h\sqrt{\dfrac{[1+(n-1)R]}{n[1+(n-1)t]}}}{h} = \sqrt{\frac{1+(n-1)R}{n[1+(n-1)t]}}$$

When the value of relative efficiency (relative accuracy) is less than one, selection on individual's own performance is more accurate than family selection; when the value is greater than one, family selection is more efficient than individual selection. As a general rule, family selection is more efficient than individual selection when 'R' is considerably more than 't'. It means that family selection is most effective when genetic relationship coefficient between family members (R) is much more than the phenotypic coefficient of relationship (t). That is, *family selection is most effective when family members resemble each other phenotypically the least* (see illustrative example below: 1st situation). This implies that family selection is more efficient when each family contains high as well as low performing individuals. Such is the situation when the heritability of a trait is low. Therefore, the chief circumstance for use of family selection is a large component of environmental variance relative to phenotypic variance (i.e. low h^2).

Illustrative example: Given below are two scenarios for application of family selection involving egg production to 72 weeks of age in chickens of two families. In the first situation, the two families overlap with respect to performance of its members. In the second situation, members of the two families have little overlapping.

Egg production	1st situation		2nd situation	
	1st	2nd	1st	2nd
93			✓	
95	✓	✓	✓	
97		✓	✓	
99	✓			✓
101	✓	✓		✓
105				✓

Family selection will be more efficient for the 1st scenario because performance of members of the two families overlaps considerably. That is, there is less variation between two families. On the other hand in 2nd scenario, there is more variation between families.

Advantages / Uses: Family selection is particularly suited for certain situations:

- *High accuracy at low heritability:* When heritability of a trait is low, phenotype of an animal is poor indicator of its breeding value because environment constitutes a larger part of phenotypic variance. For low heritability traits, family selection gives higher accuracy than does individual selection.
- *All or none traits*: Family selection is particularly useful for traits of 'all or none' nature (e.g. viability, fertility, defects or other such traits). For such traits, percentage value of a family can be used as selection criterion.
- *Low cost of recording* **:** The cost involved in measuring a trait for use in family selection can sometimes be low if *en mass* recording is possible. The best example of *en mass* recording is that of feed efficiency in most species of livestock. Recording of feed consumption and weight-gain of each animal is time consuming, less accurate and costly. On the other hand, group (family) recording is economically feasible. Family selection has found practical application in poultry and swine where large-sized families are available.

Limitations/ Constraints: There are some practical difficulties in application of family selection in farm animals **:**

- *Non-availability of large-sized families:* Large-sized families are usually not available in most species of livestock, the few exceptions being poultry, swine and fish.
- *Accumulation of inbreeding*: Family selection involves retaining all members of a family and hence, a limited number of families are selected for producing the next generation. The selected individuals of two sexes are more likely to come from a few families. Avoidance of inbreeding in family selection thus becomes a problem. Consequently inbreeding accumulates at faster rate as family selection progresses over generations.
- *Common environmental effects*: Family members, particularly if they are full sibs, usually share more similar environment as compared to randomly chosen individuals of the population. This increases the phenotypic resemblance between family members (t) which reduces the efficiency of family selection. For example, full sibs in swine (offspring of a litter) experience similar maternal environment (from conception to weaning), and hence they resemble each other phenotypically more than expected on the basis of genetic relationship and h^2. Such common environmental effects, referred to as 'C-' effects, decrease the efficiency of family selection.

Full-sib versus Half-sib Family Selection : An important question in application of family selection is whether to use full-sib or half-sib families. Important considerations in the comparison are:

- Full-sib families are much smaller though genetic relationship among full-sibs (R=0.5) is twice than among half-sibs (R=0.25).
- On the other hand, half-sib families are of much larger size but the genetic relationship coefficient among half-sibs is only half (R=0.25) than among full-sibs (R = 0.50).

- Full-sib families have the disadvantage of being exposed to common maternal or post-natal environment. In many species of farm animals, large-sized families of contemporarily reared half-sibs can be obtained easily.

The accuracy of half-sib family selection under practical situations is usually higher than those full-sib families.

Within-family Selection

This is a variant of family selection which is applied when there is a large proportion of environmental variation common to members of a family (large 'C-' or 'C^{2}'-effects). In such a situation, selection of whole families on the basis of their average performance would be inappropriate because differences between families would include not only genetic differences but also differences due to common environment. This problem can be overcome if selection is practiced within families. **Within-family selection** involves selecting the same proportion of individuals from each family and using them as parents of next generations. That is, the criterion of within-family selection is deviation of each individual from mean of its family. In a way, it is individual selection but practiced within each family.

As mentioned above, within-family selection is most useful when there is a large component of environmental variation that is common to members of a family. Examples of such situations are:

- *Early growth and survival traits* : Dams in swine and sheep provide common maternal environment to their offspring. Therefore differences in average performances between families for early growth and survival would include not only genetic differences but also include differences in maternal environment provided by dams. If early growth is to be genetically improved, then selection within-families should be practiced. Other examples of common maternal environment include passive transmission of antibodies by dams to all their offspring and the influence of egg size on chick weight in poultry.
- *Feed efficiency recording*: Sometimes, for ease of recording of a trait (e.g. feed efficiency in pigs and poultry), members of a family are housed together as one unit. If proper randomization of families in pens is not done, some families may inadvertently get differential environment. Differences between families are then largely due to differences in the environment in the sense that whole families tend to experience either 'good' or 'poor' environmental conditions.

Within-family selection cannot be effectively practiced in most species of farm animals. The most suited species for within-family selection are the ones with high prolificacy (e.g. poultry, fish and swine). Since large-sized families can be easily obtained in these species, a satisfactory level of selection intensity can be applied within each family.

A comparison of individual versus between-family with within-family selection is shown in Figure 7.4

Between-family versus within-family selection: It is important to note the difference between within-family and between-family selection. When selection is practiced on within-family basis, then every family gets represented in the next generation. But when selection is practiced between-families, then only a few families are represented in the next generation. Consequently, the level of inbreeding for within-family selection is quite low as compared to a rather high level of inbreeding under between-family selection. This is an additional advantage of within-family selection over between-family selection.

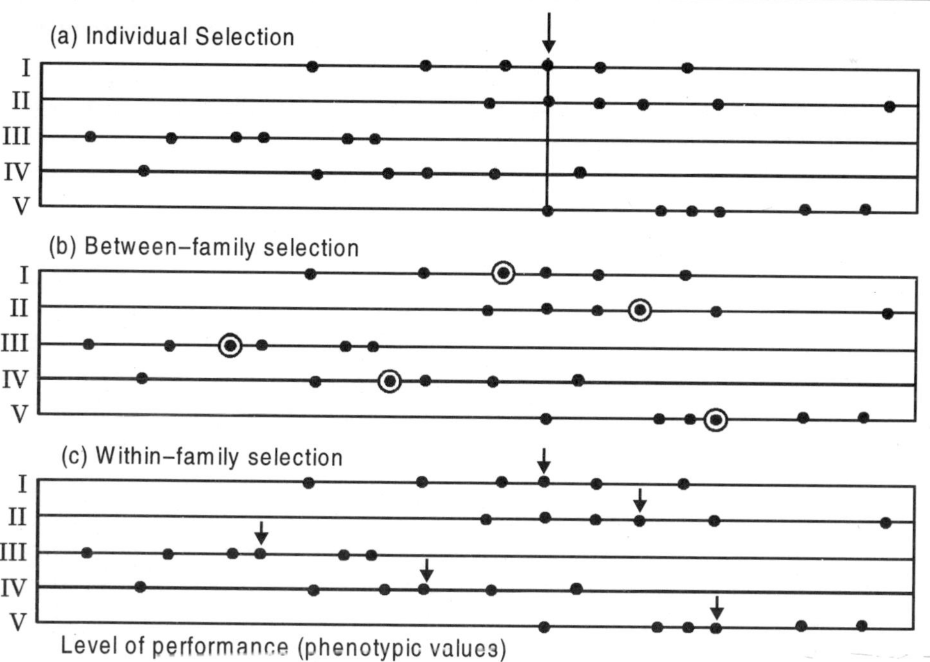

Fig 7.4 Illustration of comparison of individual vs between-family vs within-family selection. Each of the five families contains six individuals (each dot represents one individual).Fifty percentage of best breeding candidates are to be retained. Note: Under individual selection,all the individuals exceeding a defined level of performance (marked with arrow)are selected irrespective of the family from which they come; in between-family selection, three families having highest average performance are selected (family average is marked with a broad circle ◉); under within-family selection, the best fifty percent individuals from each of the five families are selected.

Sib Selection

The word 'sib' is acronym for 'brothers and sisters'. Selection of an individual on the basis of average performance of its sibs (full-or half-sibs) is referred to as **sib selection**. That is, an individual is kept or rejected on the basis of performance of its brothers and sisters. Sib selection is a variant of family selection but with two differences:

- The individual under consideration for selection does not contribute to family average to which it belongs.
- The unit of selection is an individual rather than entire family but the criterion of selection is family average.

Accuracy of sib selection, measured as correlation between breeding value of the individual and average performance of its sibs, is given by the following expression:

$$\text{Accuracy of sib selection} = Rh\sqrt{\frac{n}{1 + (n-1)\,t}}$$

Accuracy of sib selection, like that of family selection, depends on heritability, number of sibs in the family, degree of genetic relationship of individual with its sibs (R), and phenotypic coefficient of relationship among sibs (t). 'R' equals 0.5 for full sibs and 0.25 for half sibs, and 't' equals Rh^2 and has a value of $0.5h^2$ for full sibs and of $0.25h^2$ for half sibs.

The limiting value of accuracy of half-sib selection is 0.5 and that of full-sib selection is 0.7. Sib selection is, in general, inferior to individual selection except when heritability of the trait is very low and the number of sibs is very large. In fact, comparison of sib selection with individual selection is not justified as sib selection is usually employed due to force of circumstances rather than by choice.

Circumstances/Conditions for Use of Sib Selection:

- *Selection of males for sex-limited traits:* Sib selection is valuable for sex-limited traits. For example, cocks in poultry are frequently selected for improvement of egg production on the basis of average performance of their sisters. Similarly, it can be applied for selection of bulls for milk production but the requirement of large-sized families within a specific period limits its application in cattle, buffalo etc.
- *Slaughter traits:* Individual selection can not be practiced for traits requiring slaughter of animals (e.g. various aspects of carcass quality in pigs, poultry, mutton-type sheep, fish etc). Sib selection is particularly suited for such traits wherein a randomly chosen sample of sibs from each family is slaughtered for obtaining sib average. Based on average performance of slaughtered sibs, the surviving sibs can be selected and used for producing the next generation. Advent of artificial insemination and semen storage, however, has made sib selection less useful for selection of male parents. Semen of all the potential parents can be collected before slaughter, and later used for reproduction. Nonetheless, sibs selection continues to be used for improvement of carcass composition in pigs and broiler chickens.
- *All or none traits:* It can be used for 'all or none' traits like viability, fertility, incidence of defects etc.

In summary, sib selection, which is usually applied for the traits for which individual selection can not be applied because of the force of circumstances, is inferior to individual selection except when the heritability of the trait is very low and the number of sibs is very large.

Sib Selection versus Pedigree Selection: Genetic coefficient of relationship of each parent with its offspring is 0.5 which is exactly the same as the genetic relationship coefficient among the full-sibs. Therefore, a discerning reader would like to know which of the two is more efficient. The ratio of accuracy of full-sib selection to pedigree selection (based on one parent) is given by the following expression:

$$\frac{r_{AP\ (\text{Full sib selection})}}{r_{AP(\text{Pedigree selection})}} = \frac{0.5h\sqrt{\dfrac{n}{1+(n-1)\,t}}}{0.5h} = \sqrt{\frac{n}{1+(n-1)\,t}} = \sqrt{\frac{n}{1+(n-1)0.5h^2}}$$

It can be seen that for n = 1,the relative accuracy of sib selection space to pedigree selection equals unity which means that pedigree and full-sib selection are equally efficient. But for n>1.0, the ratio becomes greater than one, which means that accuracy of sib selection exceeds that of pedigree selection. The efficiency of selection based on full-sibs over pedigree selection (single parent) thus results only from availability of a large number of sibs. An individual can have only two parents but many full-sibs especially in multiple-birth species, which makes it more efficient.

Sib selection has an additional advantage over pedigree selection in that usually all the full-sibs are available at the same time. In pedigree selection, the ancestors are raised in different years, and hence the modifying influence of environment on performance of ancestors may be large.

Selection on Progeny Performance: Progeny Testing

Selection of an individual on the basis of average performance of its progeny is called **progeny testing**. Progeny testing was advocated by Varro, a Roman, over 2000 years ago, much before the laws of heredity were discovered. In 18th century, the pioneer English breeder Robert Bakewell successfully employed this procedure wherein he let-out bulls and rams on contract basis for use in farmers' herds with the right to inspect performance of their progeny. The bulls or rams that produced superior progeny were selected and used by Bakewell in his own herd for its genetic improvement. Denmark was the first country to have started systematic progeny testing through the 'Danish Progeny Testing Stations'

It has been said, *"Individuality tells what an animal looks to be, performance of his pedigree tells what he ought to be, but the performance of his progeny tells about his actual genetic potential*". This statement sums up the usefulness of progeny testing in determining the true breeding value of an animal. Since progeny testing has been most effectively used for the selection of dairy bulls, it has become synonymous with the selection of dairy sires for milk production. The procedure of progeny testing of sires is illustrated below though it is equally applicable for both sexes and for all the classes of animals.

The procedure of progeny testing involves mating of sires (potential parents) with a number of dams, and each dam in turn producing one progeny (daughters of a sire are half sisters with R = 0.25). The performance of daughters (or sons) of the potential parents is recorded and the average performance of offspring/daughters is used as the criterion for the selection of sires. The procedure of progeny testing is shown below:

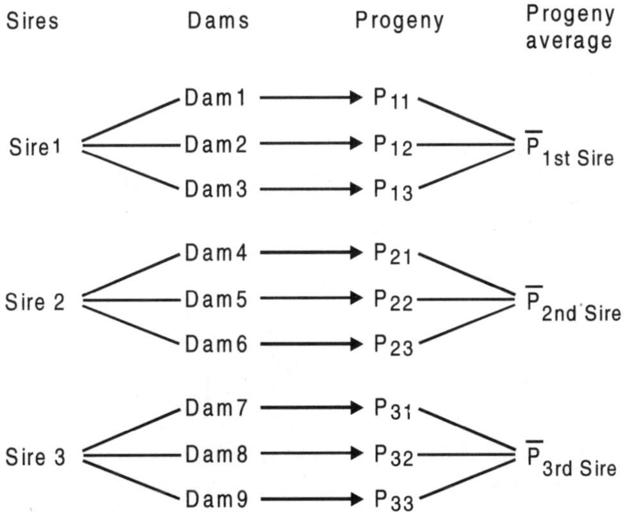

Principle of Progeny Testing

The principle of progeny testing rests on the sampling nature of inheritance (gene segregation and random union of gametes). Each offspring receives a random sample of half of the genes from each of its parents. Offspring produced by the same sire differ from one another because of random sampling of sire's genes. If the dams mated to a sire were a random sample of the dams taken from the population, then the average contribution of dams to the genetic merit of offspring would tend to cancel out in the progeny average. Thus, the Mendelian segregation and recombination, and the random environmental effects that influence the performance (phenotypic value) tend to cancel out in the average performance of offspring of a sire. The average performance of the offspring becomes a reliable measure of the breeding value of the sire. But a large number of offspring need to be measured for elimination of random environmental influences and the genetic-chance deviations (i.e. the possible errors due to genetic sampling and due to non-random environmental factors) in the progeny average. Larger the number of offspring per sire, more will be the reliability of the progeny test as a measure of true breeding value of the sire.

Accuracy of selection based on progeny average, measured as the correlation between breeding value and progeny average, $r_{\bar{A}\bar{P}}$, is given by the following expression:

$$r_{\bar{A}\bar{P}} = 0.5h\sqrt{\frac{n}{1 + (n-1)t}}$$

Where: h = square root of heritability

n = number of progeny per sire

t = phenotypic correlation among progeny of a sire and equals Rh^2. Since progeny of a sire constitutes a half-sib group, R = 0.25; therefore, 't' = $0.25h^2$

The accuracy of progeny testing, when progeny of a sire are half-sibs, can be written as:

$$r_{\bar{A}\bar{P}} = 0.5h\sqrt{\frac{n}{1 + (n-1)0.25\,h^2}}$$

This expression shows that the accuracy of progeny test depends on the heritability of the trait and number of progeny (daughters) available for progeny testing. The values of accuracy for different number of progeny and at different levels of h^2 are shown in Figure 7.5.

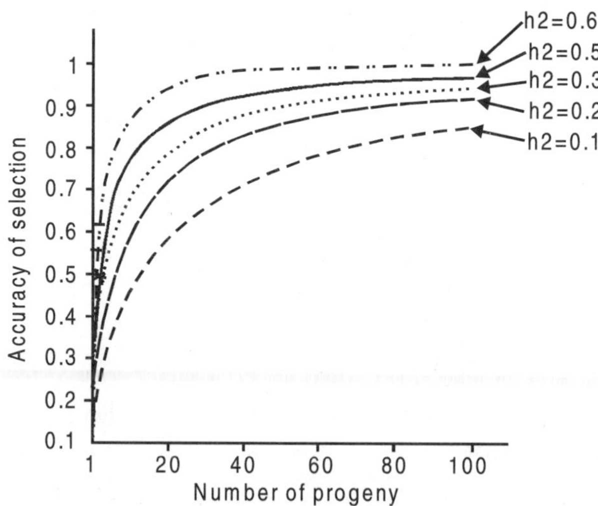

Fig. 7.5 Accuracy of progeny testing for different levels of heritability and variable number of progeny.

Salient conclusions from Fig 7.5 are:

- Accuracy of progeny testing based on a single progeny is same as that based on one parent or on one full-sib (i.e. 0.5h), and is only half of the accuracy of individual selection. This is so because an individual is genetically equally related to his parent and progeny (R=0.5).

- Accuracy of progeny testing increases with an increase in the number of progeny but the increase is more at low heritability. When h^2 is low, more number of progeny would be required to attain a high accuracy.
- As the number of progeny becomes very large (>100), the accuracy of progeny testing approaches unity even when h^2 of trait is low. An accuracy of unity implies that average performance of progeny provides a true estimate of breeding value of the sire being evaluated. Thus, progeny testing is the ideal basis of selection of genetically superior animals. In fact, we have defined breeding value of an individual in terms of the average performance of its progeny. It should, however, be underlined that the effectiveness of progeny test as a true measure of breeding value assumes that :
    - o dams mated to each test sire represent a random sample of dams taken from the population
    - o environmental factors act randomly on progeny of different sires.

Failure to satisfy any of these conditions may lead to errors in progeny testing.

Why are dams not progeny tested? : Progeny testing is widely applied for selection of males (sires) for sex-limited traits (e.g. milk production in dairy cattle, litter-size in swine and many other lowly heritable sex-limited traits). Nonetheless, it can be applied for selection of females (dams) as well. Since a dam can produce few progeny in a given time span, therefore, it is not possible to progeny test a dam with the same accuracy as a sire can be. Further, selection of dams on the basis of their own performance (particularly with use of repeated records) may yield higher accuracy than progeny testing based on a few progeny. For example, a progeny test based on at least five or more progeny is required to give the same accuracy as that given by individual selection even at low heritability. Therefore, for sex-limited traits like milk production (h^2 = 0.2), it is more feasible to select a cow based on her own performance.

Pedigree selection versus progeny testing: A question frequently crops up in the mind of beginners : "Why is performance of progeny of an individual a better indicator of its breeding value than the performance its parent (pedigree selection) though the genetic relationship coefficient of the individual being evaluated with its parent is same as it is with its progeny (R=0.5)"?. The reason is that an individual can have only two parents but it can have large number of progeny. And the large number of progeny tends to nullify the effect of Mendelian sampling and random environmental effects. For this reason, information on the breeding value of an individual based on several offspring is more reliable than the performance of a parent.

Common Environmental Effects in Progeny Testing

So far, we have assumed that phenotypic relationship between progeny of a sire (t) is exclusively determined by h^2 and coefficient of genetic relationship (R) among progeny of a sire : $t = Rh^2$. In other words, the genes transmitted by the parent cause all of the phenotypic similarity among progeny of a sire. However, if progeny of a sire are subjected to more similar environmental conditions than do unrelated individuals in the herd, then it creates an additional source of resemblance among progeny (i.e. environmental correlation between progeny or sibs). Such non-genetic causes of similarity, called C -effects reduce the accuracy of progeny testing. The correlation caused by C-effects is frequently designated as 'C^2' :

$$t = Rh^2 + C^2$$

Since 't' appears in denominator of the formulae of accuracy of selection, the presence of C- effects reduces the accuracy. Accuracy of progeny testing as influenced by the presence of common environmental effects of a very small magnitude (C^2 =1/16 = 0.06) for various numbers of daughters for a low heritability trait (h^2 = 0.25) is given in Table 7.4.

Table 7.4 Effect of C^2 effects on accuracy of progeny testing.

No. of progeny (n)	Accuracy of selection when common environmental effects are:	
	Absent $c^2 = 0.0$	Present $c^2 = 0.06$
1	0.25	0.25
3	0.41	0.39
10	0.62	0.54
50	0.88	0.66
100	0.93	0.69
∞	1.0	0.71

Apparently, a very small value of c^2 drastically reduces the accuracy of selection, and never allows it reach unity.

How to reduce common environmental effects: The following measures can overcome this problem:

- *Standardization of feeding, managemental and hygienic practices:* If standard feeding, managemental and hygienic practices are followed in a progeny-testing program, progeny of all the sires would receive more or less similar environment. Robert Bakewell, the pioneer practical breeder, might have realized the importance of this because he ensured that the farmers who used his rams/bulls must have their own ranch so that the progeny could be provided with minimum uniform standard of fodder and feed.
- *Random distribution of progeny:* If progeny of each sire are randomly distributed across all the environmental conditions (e.g. herd-management conditions), the contribution of common environmental effects to phenotypic resemblance will be very small.

Designing a Progeny Testing Program: Sources of Errors in Progeny Testing

Progeny testing is, theoretically, the ideal method for evaluating breeding values of sires for sex-limited traits. However, accuracy of progeny testing is dependent on fulfillment of certain conditions viz;

- *Comparable genetic merit of mates (dams):* The first practical difficulty encountered in progeny testing is that each offspring receives half of its inheritance from dam also. However, it is not known "what type of genes have been contributed by the dam to one particular offspring". Performance of an offspring could be 'good' or 'bad' because of the dam or sire. The best way of overcoming this difficulty consists of mating the test sires to randomly chosen group of dams from the population so that contribution of dams to the differences between performances of progeny of different sires is eliminated. Since the sires to be progeny tested (called 'test sires') in dairy cattle have usually been preliminarily selected on the basis of their dam's performance, it should be ensured that genetically 'superior' sires do not get mated with genetically 'superior' dams and *vice-versa*. If there is preferential allocation of a sire to dams, breeding value of the sire would be over- or under-estimated.

- *Uniformity of environmental conditions:* Progeny of different sires should be raised and compared under as uniform environmental conditions as possible (e.g. feeding, housing, hygiene etc). Since environmental influences vary between seasons, years and herds, progeny of different sires should be randomly distributed across test environments. The performance records, wherever possible, may also be adjusted for the systematic effects like season and year of calving, parity etc. using appropriate adjustment factors.
- *Large number of progeny:* Accuracy of progeny testing for a given level of h^2 is directly proportional to the number of progeny per sire. One of the important sources of bias or error in progeny testing is the contribution of randomly distributed sources (e.g. Mendelian segregation and random environmental effects). The contribution of these sources of variation can be reduced by increasing the number of offspring per sire. With the advent of artificial insemination and semen progeny, it has become possible to produce a large number of progeny per sire within a short period.

When these conditions are fulfilled then the differences among sire progeny averages provide an unbiased estimate of differences in breeding values among sires.

Merits :

- *High accuracy for low-heritability traits:* The greatest advantage of progeny testing is its high accuracy for low heritability traits, provided sufficient numbers of progeny per sire are recorded. With very large number of progeny, the accuracy of progeny testing can approach unity even when heritability of the trait is low (Fig 7.5).
- *Selection of sires for sex-limited traits:* Most of the traits of primary interest in dairy cattle (e.g. milk production) are only expressed in females. Since fewer males than females are required for breeding, the scope for genetic improvement is greatest through selection of males, especially when AI is used. Progeny testing has consequently become synonymous with selection of dairy bulls. During the first few decades of 20th century, progeny testing was also used for the selection of cocks for improvement of egg production. But it was eventually replaced by other bases of selection that offered high accuracy due to the availability of large number of contemporarily-reared sibs (e.g. sib selection and combined selection).
- *Detection of carriers of deleterious genes:* There are a number of single-gene traits that are lethal or semi-lethal in homozygous state. Since most of the deleterious genes are recessive, and hence they remain hidden in heterozygote individuals. Progeny testing offers an effective and practical means of detecting the carriers of such genes. Since males can produce a large number of progeny compared to females, progeny testing is specifically suited for detection of deleterious genes carried by male parents.

Drawbacks/ Limitations :

- *Prolonged generation interval:* The greatest limitation of progeny testing is its prolonged generation interval as compared to all other bases of selection. For example, a dairy bull may be 5-7 years of age before he has sufficient progeny to allow an accurate estimate of his breeding value for milk production. By the time the results of the progeny test become available, some of the sires under test may have died or become unfit for reproduction. The extended generation interval is particularly marked in buffaloes and Indian cattle because of their late sexual maturity. The increased generation interval reduces the rate of genetic progress per annum. Therefore, for the success of progeny testing, the increase in accuracy of progeny testing should offset the prolonged generation interval so that genetic progress per annum is not decreased. The techniques of semen freezing and artificial insemination have partially overcome this problem because they permit a sire to be used for breeding even after his death.

- *Low selection intensity:* Selection intensity that can be applied in a progeny-testing program is usually low because a limited number of sires can be progeny tested under fixed facilities. The available facilities have to be appropriately allocated between testing a large number of sires with fewer progeny versus testing lesser number of sires with larger number of progeny. In any case, the number of sires undergoing progeny test is limited, out of which a certain number of sires are eventually required for use in selection program.
- *High operational cost:* Progeny testing is costly in terms of time and facilities required for its implementation. Since a large number of daughters reared under varied field conditions are needed for an effective progeny-testing program, the cost of running such a program is very high. Further, the cost is also increased as a large number of waiting candidate bulls have to be maintained for long periods until the results of progeny test become available.

Use of Progeny Tested Sires

Progeny testing is one of the ideal bases for selection of sires for sex-limited traits in dairy cattle. The success of progeny testing program in bringing about genetic improvement is dependent on the extent to which the progeny tested sires are subsequently utilized in breed improvement program. That is, there must be ways of extensive utilization of superior bulls once they have been identified. Semen storage and artificial insemination have proved invaluable in extensive use of progeny tested sires. Through these reproductive techniques, a bull can be used not only throughout the state or region but globally as well. Progeny testing has therefore to be taken as a regional or national organized program.

A summary of accuracy of selection based on single record of individual verses single record of different categories of relatives is given in Table 7.5

Table 7.5 Accuracy of selection based on single record of the individual versus single record of different types of relatives.

Criterion of selection	Accuracy of selection
Individual :	h
One parent :	1/2 h
One grand parent :	1/4 h
One great-grandparent :	1/8 h
One identical twin :	h
One full-sib :	1/2 h
One half-sib :	1/4 h
One progeny :	1/2 h

The general inference from these values is that a single performance record of one relative is always less accurate than a single record of the individual (with the exception of identical twin). The accuracy of selection using any type of relative can only be increased by measuring large number of relatives.

Selection Using Several Sources of Information: Combined Selection

When several sources of information (i.e. performance records of several types of relatives) are simultaneously available, these can be compiled into a linear combination to arrive at a single value (called '**combined index value**'). This value is then used as the basis for selection. Selection based on combination of performance records of an individual and its relatives is referred to as **combined selection**. Different sources of information that are commonly available for selection of an individual are:

- individual's own performance
- performance of its family members (full- and half-sibs)
- performance of its ancestors and
- performance of its progeny

An individual's own performance and the average performance of its family (full- or half-sibs) are the most commonly used sources of information for estimating the breeding value of an individual through combined selection. The combined index value, which is a measure of breeding value (EBV), is calculated by assigning weighting coefficients to individual's own performance (b_1) and the performance of its family/sibs (b_2) as follow:

$$EBV = b_1 \text{ (Own performance)} + b_2 \text{ (Family Average)}$$

Estimate of the breeding value (EBV) obtained by combining the information from different sources, referred to as 'combined index value', is then used as the criterion/basis of selection.

The relative weighting coefficients ('b' values), and thus, the emphasis given to different sources of information (e.g. individual's own performance and family average) depend on following factors:

- *Heritability of the trait (h^2):* If heritability of a trait is high, individual's own performance receives more weightage than its family average. This is because performance of an individual is a good indicator of its breeding value when heritability of the trait is high. When h^2 of a trait is low, the correspondence between phenotypic value of an individual and its breeding value is low in which case family average gets more weightage than individual's own performance.
- *Family size (n):* For a given value of heritability, large-sized families receive more weightage as their average is more likely to be close to breeding value. This is so because Mendelian segregation and environmental effects are more likely to cancel out if size of families is large.
- *Type of family (R and t)*: With other factors being equal (i.e. h^2 and family size), full-sib families get more weightage than half-sib families. Since the number of half- sibs available within a flock is usually much more than that of full- sibs, a combination of family size (n) and type of family (R) determine the weightage assigned to a family.

Accuracy of selection based on combination of individual's own performance and the performance of its family is always higher than the accuracy of individual selection alone. Combined selection involving two sources of information (i.e. individual and its family/sibs) is one of the simplest forms of combined selection. Nevertheless, any number of sources (relatives) can be incorporated in combined selection provided reliable performance records of different types of relatives are available. For example, selection may be based on combination of individual's own performance (P_i), average performance of its full- sibs (FS), half- sibs (HS); performance of dam, and performance of progeny.

Combined selection has found widest application for genetic improvement of egg production in chickens. Selection of hens for egg production, for example, is usually based on the egg production of the hen and average egg production of her full-sisters and half- sisters as follow:

Combined index value of hen:

= b_1 (Hen's performance) + b_2 (Full-sister's performance) + b_3 (Half-sister's performance)

Egg production being sex-limited, the cocks are selected using combination of their full- and half- sister's egg production as follow:

Combined index value of cock = b_1 (Full- sisters average) + b_2 (Half-sisters average).

As indicated above combined selection has found most wide application for improvement of egg production. The reasons are two-fold:

- h^2 of egg production is low which makes selection on hen's own record (individual selection) inefficient.
- Large-sized families of contemporarily reared full- and half-sibs with reliable performance records are available in chicken.

In fact, combined selection has become synonymous with genetic improvement of egg-production in chickens. Similarly, in pigs, both male and female breeding candidates are often selected on the basis of combination of their own performance and performance of their sibs (depending on nature of trait).Combined selection can also be used in other species. For example, young bulls can be selected for milk production on the basis of a combination of milk yield of their dams and of their half-sisters. Similarly, cows can be genetically evaluated for milk production on a combination of their own performance and that of their dams and/ or their half-sisters.

Advantages/ Merits : Combined selection has a number of merits :

- *High accuracy of selection:* Combined selection yields high accuracy for low heritability traits. Its accuracy is much higher for low heritability traits as compared to accuracy of either individual or family selection.
- *Generation interval is not prolonged:* Combined selection does not increase generation interval as performance records of most classes of relatives are simultaneously available with that of the individual. The only exception being inclusion of progeny performance in combined selection.
- *No additional cost:* Use of combined selection does not involve any additional cost like maintenance and performance recording of animals over that required for individual selection.

Limitations/Constraints: Although combined selection makes optimal use of performance of different relatives, it has a few drawbacks:

- Performance records of different relatives (e.g. dam, sibs in cattle) may have been made in different environments and may introduce another source of variation.

- Computationally it is more laborious than other bases of selection but with the availability of high-speed computers, it is no more a limitation.

Indirect selection

When selection is actually practiced for a given trait, a change in its average performance from parent to progeny generation is referred to as **direct response,** and the selection that is applied directly for the improvement of the trait is called **direct selection.** Direct selection for a trait may concomitantly produce genetic change in another trait, which is genetically correlated with the trait under direct selection, the phenomenon called correlated response to selection. Under some situations, it may be more feasible to improve a trait through selection based on a correlated trait rather than to select directly for it. For example, if we want to improve trait 'Y', instead of selecting for 'Y', we may select for a correlated trait, 'X' and achieve higher genetic gain in trait 'Y'. This type of selection in which selection is based on a correlated trait with the objective of improving another trait is called **indirect selection.** The correlated trait when used for selection is also referred to as **indicator trait** (for the trait of primary interest).

The efficiency of indirect selection relative to direct selection is evaluated as the ratio of correlated response to direct response:

$$\text{Relative efficiency} = \frac{\text{Correlated response in trait Y due to selection for trait X}}{\text{Direct response to selection for trait Y}}$$

$$= \frac{h_x \times h_y \times i_x \times \sigma_{P(Y)}}{h_y^2 \times i_y \times \sigma_{P(Y)}} \times r_G \times \frac{GI_y}{GI_x} = \frac{h_x \times i_x}{h_y \times i_y} \times r_G \times \frac{GI_y}{GI_x}$$

When this ratio is greater than one, indirect selection is more efficient than direct selection. Assuming that equal selection intensities can be applied for both traits ($i_x = i_y$) and generation intervals for direct and indirect selection are equal ($GI_x = GI_y$), indirect selection will be more efficient than direct selection when " ($r_G \times h_x$)" > 'h_y'. This means that indirect selection is expected to be better than direct selection when:

- genetic correlation between the two traits is very high and
- the trait being directly selected (i.e. trait 'X') has substantially higher heritability than the trait to be improved (correlated trait).

Illustrative example: In a Fish breeding program, the aim is to improve body weight at marketing which can be accomplished through either: (a) direct selection for body weight or (b) indirect selection for body length. Genetic parameters of body weight and body length are given below : h^2 of body weight = 0.16 ; h^2 of body length = 0.36; r_G = 0.95. Assume that equal selection intensity can be applied for both the traits, and that generation interval for selection based on either of the two traits is same. We wish to see which of the two selection programs would be more efficient for improvement of body weight.

We calculate the relative efficiency of indirect selection (i.e. selection based on body length) with direct selection (for body weight):

$$\text{Relative efficiency} = \frac{\text{Correlated response in body weight due to selection for length}}{\text{Direct response to selection for body weight}}$$

$$= \frac{h_x}{h_y} \times r_G = \frac{\sqrt{0.36}}{\sqrt{0.16}} \times 0.95 = \frac{0.6 \times 0.95}{0.4} = 1.42$$

This means that selection based on body length will be 42% more efficient for improvement of body weight as compared to direct selection for body weight. This is because the h^2 of body length is considerably more than h^2 of body weight, and the genetic correlation between the two traits is very high (close to unity).

Illustrative example: Selection on part-year egg production (up to 40 weeks of age) is commonly employed for genetic improvement of annual egg production (up to 72 weeks of age). Assume heritabilities of part- and full-year records are 0.15 and 0.25 respectively, and the genetic correlation between these two traits is 0.90.The generation interval for part-year selection is one year while that for selection based on annual egg production is 1.9 year. Assuming equal selection intensities for the two criteria compare the efficiency of indirect selection on part-year egg production with direct selection for annual egg production.

$$\text{Relative efficiency} = \frac{\begin{array}{c}\text{Correlated response in annual egg production}\\ \text{due to selection for part record}\end{array}}{\text{Direct response to selection for annual egg production}}$$

$$= \frac{\sqrt{0.15}}{\sqrt{0.25}} \times 0.90 \times \frac{1.9}{1.0} = \frac{0.5814}{0.50} = 1.628$$

Thus, selection based on part-year egg production would be 1.628 times more efficient than selection on annual egg production.

Situations for Use of Indirect Selection

There are certain practical situations or considerations, which might make indirect selection preferable over direct selection:

- *When measurement of trait is difficult and expensive*: When the trait to be improved is costly and difficult to measure, then it may be economically better to select for an easily measured correlated trait. For example, recording of feed efficiency in swine, broiler chicken and fish is both expensive and difficult. In most species of farm animals, the genetic correlation between growth rate and feed efficiency is high and desirable. Consequently, selection for growth rate improves feed efficiency as a correlated response. The best example of indirect selection is observed in broiler chicken where selection for growth rate during the past 3-4 decades has concomitantly improved the efficiency of feed conversion.
- *Inability to apply high selection intensity*: If greater selection pressure can be exerted on correlated than the primary (desired) trait, indirect selection may become preferable. This is the case when primary trait is sex-limited or recorded very late in life by which time a certain proportion of breeding candidates may die.
- *When direct selection increases generation interval:* Indirect selection is advantageous if it results in substantial reduction in generation interval (e.g., correlated trait can be measured earlier, which results in short generation interval). For instance, selection based on part-year egg production (usually up to 40 weeks of age) in chickens that is used for improvement of annual egg production in poultry (usually up to 72 weeks of age), decreases the generation interval substantially. Other examples of use of part records as indicator traits for shortening generation interval are: part-year wool recording in sheep, part-lactation milk recording in buffalo and cattle, and recording of growth rate and feed efficiency for short periods.

Review Questions

1. Define/explain the following :

I. Breeding value	II. Bases or criteria of selection
III. Relative	IV. Breeding candidates
V. Individual selection	VI. Aids to (individual) selection
VII. Pedigree selection	VIII. Family selection
IX. Sib selection	X. Progeny testing
XI. Combined selection	XII. Indirect selection

2. Diagrammatically show the biometrical relationship of an individual with its relatives.
3. Which relatives of an individual are his direct relatives and which are collateral relatives?
4. Name the groups into which relatives of an animal can be classified.
5. What are the different relatives of an individual who can be considered for estimation of its breeding value?
6. What is the genetic basis of estimating the breeding value of an individual on the basis of performance of its relatives?
7. List the various criteria/bases of selection by which an individual or a group of individuals can be selected.
8. Differentiate between :
   I. Direct relatives and collateral relatives
   II. Full-sib family and half-sib family
   III. Sire-family and dam-family
   IV. Between-family selection and within-family selection
   V. Family selection and sib selection
9. Which of the two is more appropriate term for use in animal breeding: mass selection or individual selection? Explain.
10. What determines the correspondence between phenotypic value and breeding value?
11. Can the breeding value of an individual be completely determined from its phenotypic value if the trait were 100 percent heritable? Explain.
12. Name the most important genetic requirement for the effectiveness of individual selection.
13. What are the merits and demerits of individual selection?
14. What is the greatest danger in application of individual selection in non-pedigreed herds?
15. Why individual selection is not recommended for low heritability traits?
16. What are the situations for use of different aids to individual selection?
17. Under what situation, the estimation of breeding value on the basis of its own performance is not accurate enough.
18. Why does selection based on average of repeated records provide higher accuracy than that based on a single record?
19. What factors determine the accuracy of selection based on repeated records?
20. Why are repeated records more useful when repeatability of the trait is low?

21. What are advantages and disadvantages of repeated records for selection?

22. Why is the long-time (lifetime) production average a better criterion of selection as compared to performance based on a single record?

23. Why is knowledge of pedigree without knowing pedigree performance of little use in selection?

24. What is the basic principle involved in pedigree selection?

25. The value of pedigree record as an aid to selection depends on the closeness of genetic relationship of the ancestor to the individual. Justify.

26. Pedigree performance can never be 100% accurate in predicting the breeding value of an individual for a quantitative trait. Explain.

27. List the advantages and disadvantages (constraints) of selection on the basis of pedigree performance.

28. What factors determine the relative weightage assigned to performance of parent and to that of grandparents?

29. Which gives the greatest accuracy of selection: performance of an individual or combination of performance of its parents, grandparents and great-grandparents?

30. Why are performance records of grandparents less useful in selection than performance records of parents?

31. What is the principle underlying family selection?

32. What factors determine the accuracy of family selection?

33. Why family selection is must efficient when heritability of the trait is low?

34. List the advantages and limitations of family selection.

35. Name the situations under which family selection is more efficient than individual selection.

36. How does the number of family members (family size) affect the effectiveness of family selection?

37. Which is more useful for selection: full-sib or half-sib families?

38. What is the criterion of selection for within-family selection?

39. Cite two examples of the traits for which within-family selection can be practiced.

40. What factors determine the accuracy of sib selection?

41. Under what situations is sib selection used?

42. Which is more effective in determining the breeding value of an individual: selection on basis of performance of that individual or selection on the basis of performance of its full sibs?

43. Under what conditions is selection on the basis of half-sibs most accurate?

44. Which gives higher accuracy selection: selection based on performance of half-sibs or full-sibs? Explain.

45. What is the principle on which progeny testing rests?

46. Why are dams frequently not progeny tested?

47. Which is most crucial factor in determining the accuracy of progeny testing?

48. What is meant by 'common environmental effects'? How do these affect the accuracy of progeny testing? How can their contribution be reduced?

49. How can a sound progeny-testing program be designed? Explain.

50. What are the merits of progeny testing?

51. What are the major drawbacks of progeny testing?

52. Which gives the greatest accuracy of selection: selection on individual's own performance or performance of its single progeny? Explain.

53. What factors influence the accuracy of progeny testing?

54. Why are large number of offspring needed to obtain accurate estimate of breeding value of a sire?

55. What factors determine the relative weightage assigned to individual's own performance and the performance of its family?

56. Cite two examples of use of combined selection.

57. What are the merits and limitations of combined selection?

58. What factors determine the relative efficiency of indirect selection to direct selection?

59. What are the genetic situations under which indirect selection is more efficient than direct selection?

60. What are the practical situations under which indirect selection is useful?

8

Computation of Breeding Values

In the preceding chapter we dealt with the various criteria or the bases that are used for selection of genetically superior animals. Selection involves computation of breeding values of the breeding candidates, ranking them in order of their genetic merit (breeding values), and then retaining the required number of animals (parents).

In this chapter, we outline the procedure of computation of breeding values using different criteria. Though the computation of breeding values is an integral part of selection (dealt in the last chapter) but in this chapter we outline the computational steps only. Nonetheless the chapter has been designed such that it can be studied without necessarily going through the last chapter.

Why Compute Breeding Value ?

Selection of genetically superior individuals to become parents of next generation is the most important tool for bringing about genetic improvement. Since genes (and not the genotypes) are transmitted from parents to progeny, therefore, it is necessary to have some information about the genes possessed by the breeding candidates (potential parents). The sum of the average (additive) effects of all the genes possessed by an individual [for a trait] is referred to as **'breeding value'** or **'additive genetic value'** of the individual. Half of the breeding value of the parent is passed on to his/her offspring.

Quantitative traits are determined by large number of genes with small but cumulative effects. Such traits are also influenced by environmental factors. Neither the number of loci (genes) affecting a quantitative trait nor their effects on the trait [phenotype] are known, therefore, the above definition of breeding value is of a conceptual nature. It is only the phenotypic value (performance) which is available for making decisions about selection (*i.e.* for computing the breeding value). The performance record (phenotypic value) of an individual or its relatives provides the basis for the estimation of its breeding value, and in turn for making selection decisions.

Why is Breeding Value called Expected or Predicted Breeding Value ?

The true or actual breeding value of an animal for a quantitative trait is practically never known because of the large number of genes involved and the masking effects of the environment and gene-combinations (dominance and epistatic). Breeding value can only be estimated by using performance of the animal or its relatives. In fact, breeding value can only be predicted because it refers to the performance of the animal as breeding parent (the average performance of their progeny). The calculated breeding value of the individual is therefore referred to as 'predicted breeding value' (PBV) or sometimes also as 'estimated breeding value' (EBV); both the terms 'PBV' and 'EBV' are used interchangeably.

Different Criteria Used for Predicting the Breeding Value

Various sources or criteria of information used for computing the breeding value of an individual or a group of individuals are :

- Individual's own performance (individual selection).
- Pedigree performance (pedigree selection).
- Family performance (family selection).
- Progeny performance (progeny testing).
- Combination of performance of the individual and its relatives (combined selection).

Choice of the criteria for computing the breeding value and hence for selection of animals to serve as parents of next generation is dependent, besides genetic considerations, on :

- Reproductive rate (prolificacy) of the species.
- Nature of the trait (*i.e.* whether sex-limited or not).
- The time when the trait is expressed (*i.e.* early or late in life).

The solved problems outline the general procedure, irrespective of the species, for computation of breeding values. The examples have been chosen from the species where that specific criterion is more commonly used.

Computation of EBV or PBV[1]

8.1. Outline the generalized expression for computation of 'EBV or PBV' and explain the different components of the equation.

Solution : Breeding value of an individual, EBV_i, is calculated by the following general equation[2] (i^{th} individual) :

$$EBV_i = \bar{P} + b\,(P_i - \bar{P})$$

Where :

$\bar{P}$ = average performance of the herd/flock/population [in which the animal has been reared/performed].

P_i = phenotypic value (performance) of the concerned animal or its relatives (i^{th})

'b' = weighting coefficients[3] assigned to animal's performance (phenotypic value) or performance of its relatives.

1. The examples of computation of breeding values given in this chapter involve only a limited number of animals as the objective is to show the computational steps only. The actual breeding populations are of large size, usually involving thousands of breeding candidates for which computations are made through computers.
2. Which is a simple regression equation.
3. Statistically, weighting coefficient is the regression of breeding value (additive genetic value) on phenotypic value (performance).

Numerical value of the weighting coefficient varies for different criteria of selection depending on :

- h^2 of the trait
- number of relatives
- repeatability and the number of repeated records (if the trait is repeatable).

Theoretically, the weighting coefficient can vary between 0 and 2.0

$(P_i - \bar{P})$ = deviation of performance of the individual (or its relative) from the herd, flock or population mean.

The deviation can be negative (if the animal is below the population average), zero (if animal's performance equals population average) or positive (if animal is above population average). The deviation is thus a measure of the phenotypic superiority or inferiority of the individual over or below population average.

The equation for computation of breeding value can therefore be written as :

EBV = Population avg + (Weighting coefficient × phenotypic superiority or inferiority over population avg)

Or

EBV = Population avg + expected genetic superiority or inferiority over population avg.

The EBV calculated by the above expression is an absolute value and carries the unit of measurement of the trait. In many species it is a common practice to express the breeding value as deviation from population average as follows :

$$EBV_i = b\,(P_i - \bar{P})$$

Or

EBV_i = (Weighting coefficient x deviation of animal's performance from population avg)

The EBV expressed in deviation term can be positive, negative or zero.

Note : The formula for computing the breeding value in deviation term (*i.e.* $EBV_i = b\,(P_i - \bar{P})$) is equivalent to that used for predicting the genetic gain ($\Delta G = h^2 \times (\bar{P}_{\text{selected animals}} - \bar{P})$). Similarly, the formula of EBV as an absolute value is the same as for predicting the performance of progeny generation due to selection of parents. The only difference being that PBV or EBV refers to the individual animal whereas genetic gain refers to all the progeny born from the selected parents.

I. Using Individual's Own Performance (Individual Selection)

Individuals can be selected on the basis of their own performance. The performance may be based on a single record or average of several records of the individual. Further, the animals whose breeding values are to be calculated may come from a single herd or from different herds.

(A) Single Herd Evaluation

When all the breeding candidates are reared in the same herd/flock, it is reasonable to assume that they have been exposed to more or less similar types of environmental factors. The computation of breeding value of the animals from a single herd is easier than of those coming from different herds.

(i) Single Record

$$EBV_i = \bar{P} + b_{AP}\left(P_i - \bar{P}\right)$$

Where :

$\bar{P}$ = population/herd/flock average.

P_i = performance of the individual.

b_{AP} = weighting coefficient which is regression of breeding value on phenotypic value and equals heritability of the trait ($b_{AP} = h^2$).

$(P_i - \bar{P})$ = deviation of individual's performance from population/herd average.

The formula for EBV thus reduces to :

$$EBV_i = \bar{P} + h^2\,(P_i - \bar{P}) \qquad \text{Or}$$

EBV_i = Population avg + (h^2 × deviation of individual's performance from herd/population avg)

8.2. Given below are the first lactation milk yield (kg) records of five Sahiwal cows. Herd average is 2000 Kg and h^2 of milk yield is 0.25.

Cow ID :	1	2	3	4	5
Milk yield :	2200	2100	2000	1900	1800

(*a*) Calculate the breeding values, rank the cows on the basis of their EBV as well as on their own performance (milk yield) and comment on the breeding values (assuming that the five cows constitute the herd).

(*b*) What does the value of EBV/PBV of a cow (say cow 'A') mean ?

(*c*) Comment on the ranking by the two procedures.

(*d*) Why do EBV's tend to regress toward herd/population average ?

Solution :

(*a*) Computations required for EBV or PBV are shown below in steps :

Cow ID	Milk yield (P_i)	Herd average ($\bar{H}$)	Weighting co-efficient* (b)	Deviation ($P_i - \bar{H}$)	EBV (kg)		Rank based on	
					Absolute	Deviation	Milk yield	EBV for milk yield
A	2200	2000	0.25	200	2050	50.0	I	I
B	2100	2000	0.25	100	2025	25.0	II	II
C	2000	2000	0.25	0.0	2000	0.0	III	III
D	1900	2000	0.25	–100	1975	–25.0	IV	IV
E	1800	2000	0.25	–200	1950	–50.0	V	V

* Weighting coefficient, $b = h^2$

Comments on breeding values :

- Breeding values can be expressed as absolute values or as deviations from herd average, and carry the unit of measurement of the trait (*e.g.* Kg. in this example).
- As deviations, EBV's can have positive or negative values, or be equal to zero.
- The sign associated with EBV (in deviation term) indicates whether the animal is expected to be genetically above (+) or below (–) the average of the herd in which it was reared.

(b) What does the value EBV or PBV mean ?

- The value of EBV of a cow in absolute term means that if that cow (say cow 'A') is mated to a sire of average genetic merit (whose breeding value is equal to the average breeding value of the herd) and produced a large number of daughters which are reared under same environmental conditions as the cow under consideration had been, its daughter's average is expected to be equal to its EBV (*i.e.* equal to 2050 Kg).
- The EBV of a cow (say cow 'A') in deviation term means that daughters of that cow are expected to perform so many units higher or lower than the average performance of the herd in which its progeny performed (in case of cow 'A' it means 50 Kg). In essence, it means that "how much superiority or inferiority of the cow over or below the herd average is due to additive effect of genes."

(c) Ranking :

- Ranking of cows on the basis of their performance (milk yield) is exactly in same order as on the basis of their breeding values. This is so because the weighting coefficient, *b*, assigned to phenotypic superiority/inferiority over herd average is same for all the cows ($b = h^2$) since every animal has one performance record.

(d) Regression towards mean :

- Breeding values of cows have narrower range than their performance values (milk yield records). This is so because breeding values regress ('shrink') toward the herd average. The extent of 'regressing' or 'shrinking' is dependent on the heritability (*i.e.* the value of weighing coefficient) of the trait.
- Higher the heritability lower the proportion of environmental variation, and therefore, the less severe is the 'shrinking' or 'regression' towards average.

(ii) Multiple Records

When several records for a trait are available on the same animal, selection can be based on the average of these repeated records. Selection on repeated records is most commonly used in dairy cows. Breeding value based on repeated records is calculated by the following expression :

$$EBV_i = \bar{H} + b\,(\bar{P}_i - \bar{H})$$

Where :

$\bar{H}$ = herd/population average

$\bar{P}_i$ = average performance of ith cow based on '*n*' records

b = weighting coefficients assigned to average performance of the cow and equals

$$b = \frac{nh^2}{1+(n-1)r}$$

Where :

n = number of records available on the cow

r = repeatability of the trait

h^2 = heritability of the trait

$P_i - \bar{H}$ = deviation of average performance of cow from herd average.

Thus, EBV = Herd average + $\frac{nh^2}{1+(n-1)r}$ x (Cow average – Herd average)

8.3. Given below are lactation milk yield records of five cows reared in a herd averaging 2100 Kg.

Cow ID	Milk yield (kg) in different lactations				
	1st	2nd	3rd	4th	5th
A	2900	–	–	–	–
B	2800	2600	–	–	–
C	2750	2700	2500	–	–
D	2800	2700	2500	2400	–
E	2700	2600	2500	2400	2300

Assume, h^2 = 0.2 and repeatability = 0.3.

Calculate the breeding values of cows and rank them in order of genetic merit (breeding value). Compare the ranking of cows based on their EBV with that based on their average performances and draw inferences.

Solution : Computations for EBV are shown below :

Cow ID	No. of records (n)	Cow avg ($\bar{P}_i$)	Herd avg ($\bar{H}$)	Weighting coefficient* (b)	Deviation ($\bar{P}_i - \bar{H}$)	EBV (kg)	Rank on : Cow avg	EBV
A	1	2900	2100	0.200	800	2260.0	I	V
B	2	2700	2100	0.307	600	2284.2	II	III
C	3	2650	2100	0.375	550	2306.2	III	II
D	4	2600	2100	0.421	500	2310.5	IV	I
E	5	2500	2100	0.454	400	2281.6	V	IV

$$*b = \frac{nh^2}{1+(n-1)r}$$

Comments :

- The weighting coefficients (*b*) assigned to the average performance of cows are dependent on h^2, repeatability and the number of records available for that cow. Since h^2 and repeatability are population parameters, and are thus applicable to all the cows of the herd, the weighting coefficients are directly proportional to the number of records (*i.e.* the weighting coefficients increase with the increase in number of records.) For example, the weighting coefficient for a cow with a single record (Cow 'A') equals h^2 of the trait (*b* = 0.20). The cow with the minimum number of records (cow 'A') gets the lowest weighting coefficient while cow 'E' having maximum number of records (*n* = 5), gets the highest weighting coefficient (b = 0.454).

- Cow 'A' has the highest phenotypic superiority over the herd average ($\bar{P}_i - \bar{H} = 800$ kg) but its genetic (additive) superiority is lowest (160 Kg). Consequently, ranking of cows on the basis of EBV is different than that based on their average performance. Selection of cows should therefore be based on their EBV's.

- ***Inference :*** The EBV's of animals with repeated records, which are calculated by taking account of the number of records, are useful in identifying genetically superior animals (high genetic merit) for improving the average of next generation.

(B) Multi Herd Evaluation

In some situations, performance records are available on the animals which have been reared in different herds (*e.g.* in dairy cattle) or in different test stations (*e.g.* in swine and poultry). The environmental (managemental) conditions tend to vary among different herds or test stations, therefore, the differences in herd averages need to be taken into account for calculation of breeding values. On the assumption that the average genetic merit of different herds is equal, $(\bar{G}_{1st} = \bar{G}_{2nd} = \bar{G}_{n\,th})$, the differences in herd averages can be considered purely due to environmental differences between herds. Under such a situation, the EBV is calculated as follows :

$$EBV_i = \bar{P} + b\,(P_i - \bar{H})$$

Where :

$\bar{P}$ = population average (*i.e.* average of all the herds)

P_i = performance of the ith individual (cow)

$\bar{H}$ = average of the herd in which the individual performed

b = weighting coefficient = h^2

8.4. Given below are milk yield records (kg) of six cows along with their herd average and population average. Heritability of milk yield is 0.15.

	Herd 1			Herd 2		
Cow ID :	1	2	3	4	5	6
Milk yield $(\bar{P}_i)$	3100	3000	3200	2800	2900	2850
Herd avg $(\bar{H})$		3100			2850	
Population avg $(\bar{P})$			2975			

(*a*) Calculate the EBV's of the six cows by :

(i) taking account of herd averages

(ii) ignoring differences in herd averages (*i.e.* by assuming that averages of the two herds are equal).

(*b*) What inferences can be drawn from the comparison of ranking of cows by the two procedures ?

Solution :

(*a*) 1st herd average = $\bar{H}_1 = \dfrac{(3100 + 3000 + 3200)}{3}$ = 3100 Kg

2nd herd average = $\bar{H}_2 = \dfrac{(2800 + 2900 + 2850)}{3}$ = 2850 Kg

Population average = $\bar{P} = \dfrac{(3100 + 2850)}{2}$ = 2975 Kg

(i) Computations for EBV, taking account of herd average $(\bar{H})$, are shown below :

Cow ID	$EBV_i = \bar{P} + h^2(P_i - \bar{H}_i)$
1	2975 + 0.15 (3100 – 3100) = 2975 kg
2	2975 + 0.15 (3000 – 3100) = 2960 kg
3	2975 + 0.15 (3200 – 3100) = 2990 kg
4	2975 + 0.15 (2800 – 2850) = 2967.5 kg
5	2975 + 0.15 (2900 – 2850) = 2982.5 kg
6	2975 + 0.15 (3100 – 2850) = 2975 kg

(ii) Computations for EBV by ignoring average of herd (and thus using population average only) are shown below :

Cow ID	$EBV_i = \bar{P} + h^2(P_i - \bar{P})$
1	2975 + 0.15 (3100 – 2975) = 2993.75 kg
2	2975 + 0.15 (3000 – 2975) = 2978.75 kg
3	2975 + 0.15 (3200 – 2975) = 3008.75 kg
4	2975 + 0.15 (2800 – 2975) = 2948.75 kg
5	2975 + 0.15 (2900 – 2975) = 2963.75 kg
6	2975 + 0.15 (3100 – 2975) = 2956.25 kg

(*b*) Comparison of ranking by two procedures.

Cow	EBV's by :		Ranking by :	
ID	1st method	2nd method	1st method	2nd method
1	2975	2993.75	III	II
2	2960	2978.75	V	III
3	2990	3008.75	I	I
4	2967.5	2948.75	IV	VI
5	2982.5	2963.75	II	IV
6	2975	2956.25	III	V

- ***Inference :***
  - The ranking of cows on EBV's calculated by taking account of herd averages (1st method) is different than that calculated by ignoring herd average (2nd method).
  - The first method has adjusted for the difference in herd average, and has therefore identified the cows (cow ID 5 and 6) that would not have been identified to be genetically superior by the second method (*i.e.* ignoring herd differences).
  - When animals have been reared in different herds, it is appropriate to compute EBV's by adjusting for differences in herd averages.

II. Using Pedigree Performance (Pedigree Selection)

Selection of an individual for sex-limited traits (*e.g.* milk production) or for traits that are expressed late in life (*e.g.* longevity, life-time milk production etc.) can be done on the basis of performance of the dam (female). The dam may have a single record or multiple records.

(A) Use of Single Record of Dam

$$EBV_i = \bar{P} + b\,(P_i - \bar{P})$$

Where :

$\bar{P}$ = population average (in which the dams performed)

P_i = performance of the dam

b = weighting coefficient assigned to dam's performance : $b = 0.5h^2$

$(P_i - \bar{P})$ = deviation of dam's performance from population average

The equation for EBV can then be written as :

$$EBV_i = \bar{P} + 0.5h^2\,(P_i - \bar{P})$$

(B) Use of Repeated Records of Dam

$$EBV_i = \bar{P} + b\,(\bar{P}_i - \bar{P})$$

Where :

$\bar{P}$ = population average (in which the dam performed)

$\bar{P}_i$ = average performance of dam based on 'n' records

$(\bar{P}_i - \bar{P})$ = deviation of average performance of dam from population average

b = weighting coefficient assigned to dam's average performance and is calculated as : $b = \dfrac{0.5nh^2}{1+(n-1)r}$

where :

n = number of records of the dam

r = repeatability of the trait

h^2 = heritability of the trait

The expression for EBV can alternately be written as :

$$EBV_i = \bar{P} + \left[\frac{0.5nh^2}{1+(n-1)r}\right]\left(\bar{P}_i - \bar{P}\right)$$

8.5. Given below are the average milk yield records of dams of five young bulls along with the number of lactations on which average performance of their dams is based.

Bull ID	No. of records of dam (n)	Dam's avg milk yield (kg)
A	2	2575
B	3	2550
C	4	2525
D	2	2575
E	5	2500

Assume : Herd average = 2000 kg, h^2 = 0.2, repeatability = 0.3

(*a*) Calculate the breeding values of bulls.

(*b*) Rank the bulls on the basis of their EBV's as well as on the basis of their dam's simple average and comment on their relative ranking.

Solution :

(*a*) The computations for EBV's are shown below :

Bull ID	Dam's avg ($\bar{P}_i$)	No. of records of dam (n)	Weighting coefficient (b)*	Deviation ($\bar{P}_i - \bar{P}$)	EBV	Rank based on : Dam's avg	
						Dam's avg	EBV
A	2575	2	0.154	575	2088.5	I	IV
B	2550	3	0.188	550	2103.1	II	III
C	2525	4	0.211	525	2110.7	III	II
D	2575	2	0.154	575	2088.5	I	IV
E	2500	5	0.227	500	2113.5	IV	I

$$* \; b = \frac{0.5nh^2}{1+(n-1)r}$$

(*b*) The weighting coefficients assigned to dam's average increase with the increase in the number of records of the dam. The ranking of bulls based on their breeding values is different than that based on the simple average performance of dams. It is so because the computation of EBV involves giving weightage to the number of records of the dam as well to the h^2 of trait.

III. Using Family Performance (Family Selection)

Family selection involves selecting a family on the basis of its own average performance. Breeding value of a family is calculated by the following expression :

$$\text{EBV}_i = \bar{P} + b\,(\bar{F}_i - \bar{P})$$

Where :

$\bar{P}$ = population average

$\bar{F}_i$ = family average (of the ith family)

b = weighting coefficient assigned to family average and is calculated as :

$$b = h^2\left[\frac{1+(n-1)R}{1+(n-1)t}\right]$$

Where :

n = number of individuals in the family (family size).

R = coefficient of genetic relationship among family members (R = 0.5 for FS and 0.25 for HS families).

t = phenotypic correlation among family members : $t = Rh^2$.

The expression for EBV can then be written as :

$$\text{EBV}_i = \bar{P} + h^2\left[\frac{1+(n-1)R}{1+(n-1)t}\right](\bar{F}_i - \bar{P})$$

8.6. Given below are average annual egg production records of five best full-sib families along with the respective family size. Flock average = 300 eggs, h^2 = 0.10.

Family ID	Family size (n)	Family average
1	2	310
2	7	305
3	5	300
4	10	303

Compute the breeding values of the five families, and rank them in order of their genetic merit.

Solution :

Family ID	Family size (n)	Family avg. ($\bar{F}$)	Weighting coeff. (b)	Family deviation ($\bar{F}_i - \bar{P}$)	EBV of family	Rank on EBV
1	2	310	0.14	10	301.4	II
2	7	305	0.30	05	301.5	I
3	5	300	0.25	00	300.3	IV
4	10	303	0.37	03	301.1	III

IV. Using Performance of Sibs (Sib Selection)

Selection of an individual on the basis of average performance of its sibs, called sib selection, is commonly practiced for sex-limited traits, carcass traits, longevity etc. Breeding value of an individual is obtained as :

$$\text{EBV}_i = \bar{P} + b\,(\overline{FS}_i - \bar{P})$$

Where :

$\bar{P}$ = population average

$\overline{FS}_i$ = average performance of sibs of the ith individual

b = weighting coefficient assigned to sib average and is calculated as :

$$b = \left[\frac{nRh^2}{1+(n-1)t}\right]$$

Where :

n = number of sibs of the individual,

R = coefficient of genetic relationship of the individual with its sibs (R = 0.5 for full-sibs and 0.25 for half-sibs)

t = phenotypic correlation of the individual with its sibs : $t = Rh^2$

The expression for EBV can then be written as :

$$\text{EBV}_i = \bar{P} + \left[\frac{nRh^2}{1+(n-1)t}\right]\left(\overline{FS}_i - \bar{P}\right)$$

8.7. Given below are average egg production records of full sisters of five cocks :

Cock ID	Full sisters' average ($\overline{FS}$)	Number of sisters (n)
A	300	3
B	290	10
C	288	15
D	285	10
E	280	5

Flock average is 250 eggs and heritability of egg production is 0.10. Calculate the breeding values of the 10 cocks on the basis of their sister's performance and rank them on genetic merit (breeding value).

Solution :

Cock ID	Full sisters avgerage ($\overline{FS}$)	No. of sisters	Weighting coefficient (b)*	Family deviation ($\overline{FS}-\overline{P}$)	EBV	Rank based on : Full sister avg	Rank based on : EBV
A	300	3	0.136	50	256.8	I	IV
B	290	10	0.345	40	263.8	II	II
C	288	15	0.441	38	266.8	III	I
D	285	10	0.345	35	262.1	IV	III
E	280	5	0.208	30	256.3	V	V

$$* \; b = \frac{nRh^2}{1+(n-1)t} = \frac{n \times 0.5h^2}{1+(n-1)0.5h^2}$$

8.8. Given below is the information on 12 hens of four full-sib families for annual egg production. Each family has one cock (*i.e.* the total of four cocks). Fifty percent of the hens and 25% of the cocks need to be saved as parents. Two procedures are available for selection :

(*i*) selection on own performance (for hens only) and

(*ii*) selection on family performance.

Show which hens and cocks would be selected by the two selection strategies and comment on their efficacy. Comment on the problem of inbreeding in family selection assuming that one cock is to be mated with six hens.

Family ID	Annual egg production of different hens of each family : 200	220	240	260	280	300	320	340
A	A_1	A_2	–	–	–	A_3	–	–
B	B_1	–	–	B_2	–	–	–	B_3
C	–	C_1	–	–	C_2	–	C_3	–
D	D_1	–	–	–	D_2	D_3	–	–

Solution :

Procedure I (individual selection) : Based on the ranking of hens on their own egg production, the following six hens will be selected :

Hen ID :	B_3	C_3	A_3	D_3	C_2	D_2
Egg production :	340	320	300	300	280	280

Cocks do not have their own egg production records, therefore, they are randomly chosen. Alternatively, cocks can be selected by the second procedure (family selection).

Procedure II (Family Selection) : The EBV's of the four families need to be calculated for applying family selection. Since all the families have equal number of hens (n = 3), ranking of the families on the basis of their phenotypic average would be same as on the basis of their EBV's (the students can cross check it).

We can use average performance of each family for selection of cocks :

Family ID :	A	B	C	D
Family average	240.0	266.7	273.3	260.0

Family numbered 'C' has the highest average egg production, therefore, its cock will be selected.

On the basis of family average, families 'C' and 'B' will be selected *i.e.* hens with ID number C_1, C_2, C_3, B_1, B_2, and B_3 will be selected. The cock of family 'C' will be selected.

Comparison of Two Procedures : The ID of the hens selected by two procedures *viz.,* family selection and individual selection, is given below :

Family selection :	C_1	(C_2)	(C_3)	B_1	B_2	(B_3)
Individual selection :	A_3	(C_2)	(C_3)	(B_3)	D_2	D_3

It is apparent that some of the better hens are selected by either of the two procedures. Hens numbered C_2, C_3 and B_3 (marked by circle) get selected by individual selection as well as by family selection.

Inference : Individual selection and family selection, though conceptually different, are operationally overlapping (i.e. some of the common animals are selected by both the procedures).

Problem of Inbreeding : One of the important limitations of family selection is its inability to avoid inbreeding. It can be seen that by family selection we have selected the cock of family 'C'. Since one cock is to be mated with six hens, it is apparent that this cock will also get mated with its sisters (*i.e.* those with ID no. C_1, C_2 and C_3) which will lead to increased inbreeding.

V. Using Performance of Progeny (Progeny Testing)

Selection of an individual on the basis of performance of its progeny, referred to as progeny testing, uses the performance records of progeny for estimating the breeding value. Progeny testing is usually practiced for sex-limited traits and carcass characteristics. The EBV of the individual being progeny tested is calculated as follows :

$$EBV_i = \bar{P} + b\,(\bar{P}_i - \bar{P})$$

Where :

$\bar{P}$ = population average.

$\bar{P}_i$ = average performance of progeny of the individual being tested.

b = weighting coefficient assigned to average performance of progeny and is calculated as :

$$b = \left[\frac{0.5nh^2}{1+(n-1)t}\right]$$

Where :

n = number of progeny per sire

h^2 = heritability of the trait

t = phenotypic correlation among the progeny of a sire = 0.25 h^2

The expression for EBV then becomes :

$$\text{EBV}_i = \bar{P} + \left[\frac{0.5nh^2}{1+(n-1)0.25h^2}\right](\bar{P}_i - \bar{P})$$

8.9. Given below are lactation milk yield records of daughters of five bulls in a herd averaging 2000 Kg, h^2 = 0.25. Calculate the breeding values of the bulls; rank them in order of genetic merit (EBV) and comment

Bull ID	No. of daughters (n)	Average yield of daughters (kg)
A	1	2500
B	3	2450
C	5	2440
D	10	2450
E	20	2390

Solution : The computations for calculating the EBV are shown below :

Bull ID	No. of daughters (n)	Daughters' avg ($\bar{P}_i$)	Weighting coefficient (b)*	Daughter's deviation ($\bar{P}_i - \bar{P}$)	EBV	Rank on : EBV	Rank on : Simple daughter avg.
A	1	2500	0.125	500	2062.5	V	I
B	3	2450	0.333	450	2149.9	IV	II
C	5	2440	0.500	440	2220.0	III	IV
D	10	2450	0.800	450	2360.0	II	III
E	20	2390	1.143	390	2445.8	I	V

$$^*b = \left[\frac{0.5nh^2}{1+(n-1)0.25h^2}\right]$$

Comments :

- The weighting coefficients assigned to progeny performance increase as the number of progeny per sire increases *i.e.* the sire with maximum number of progeny receives the highest weightage and *vice-versa.*
- The ranking based on EBV is different from that based on simple daughter average as the computation of EBV uses different weighting coefficients for bulls having different number of daughters.

8.10. Examine the effect of number of progeny per sire on the weighting coefficient assigned to daughter's average for different number of progeny (one to 10, 20, 50, 100, 250, 500, 1000, 2000, 5000 and 10,000). What will be the EBV of a sire whose daughters had average yield of 2500 kg for different number of daughters (as listed above). Assume herd average = 2000 kg, h^2 = 0.2.

Solution : Information given : $h^2 = 0.2$; hence, $t = 0.25 \times 0.2 = 0.05$

$\bar{P}_i$ = average milk yield of progeny of the ith sire = 2500 Kg

$$\text{Weighting coefficient : } b = \left[\frac{0.5nh^2}{1+(n-1)t}\right]$$

Values of the weighting coefficient (b) for different number of progeny, along with EBV's of bulls, are listed below (h^2 = 0.2) :

No. of progeny (n)	Weighting coefficient (b)	EBV of bull :	No. of progeny (n)	Weighting coefficient (b)	EBV of bull :
1	0.10	50.00	20	1.03	512.82
2	0.19	95.24	50	1.45	724.64
3	0.27	136.36	100	1.68	840.34
4	0.35	173.91	250	1.86	929.37
5	0.42	208.33	500	1.93	963.39
6	0.48	240.00	1000	1.96	981.35
7	0.54	269.23	2000	1.98	990.59
8	0.59	296.30	5000	1.99	996.21
9	0.64	321.43	10000	2.00	998.10
10	0.69	344.83	—	—	—

Comments :

- Weighting coefficients increase with the increase in number of daughters.
- The limiting value of weighting coefficient is 2.0.
- With very large number of daughters (*e.g.* n =10,000), the EBV approaches a value of 1000 which is twice the deviation of daughters average from herd average. This is how we have defined breeding value (practical definition of breeding value) : "If an individual (sire) is mated to a large number of dams taken at random from the population, its breeding value is twice the deviation of its daughter's from the population/herd average".

VI. Using Combination of Performance of Individual and Family

Selection based on a linear combination of performance of an individual and its relatives is referred to as combined selection. Individual's own performance and the average performance of the family are commonly used sources of information for estimating the breeding value of the individual using combined selection. The procedure is most commonly used in selection for egg production in poultry, and swine. Breeding value of the individual is estimated by the following equation :

$$EBV_i = \bar{P} + b_1(\bar{P}_i - \bar{P}) + b_2(\bar{F}_i - \bar{P})$$

Where :

$\bar{P}$ = flock average

P_i = individual's own performance

$\bar{F}$ = family average

b_1 = weighing coefficient assigned to individual's own performance

b_2 = weighing coefficient assigned to family average (families could be of full-sibs or half-sibs)

The weighting coefficients, b_1 and b_2, are calculated by the following expression :

$$b_1 = h^2\left(\frac{1-R}{1-t}\right) \quad \text{and} \quad b_2 = h^2\left(\frac{1+(n-1)R}{1+(n-1)\,t}\right)$$

Where :

R = coefficient of genetic relationship among family members, (R = 0.5 for full-sib and 0.25 for half-sib family)

t = phenotypic correlation among family members

n = number of family members (family size)

8.11. Given below are the egg production records of five best hens along with their full-sib family average and family size. Flock average = 200 eggs, h^2 = 0.10.

Hen ID	Annual egg production of : Hen (P_i)	Family avg. ($\bar{F}_i$)	Family size (n)
1	270	220	5
2	260	230	5
3	250	260	3
4	245	260	7
5	235	270	2

(*a*) Compute the breeding values of the five hens using combined selection.

(*b*) Rank the hens on their EBV using combined selection, and compare this rank with that based on the EBV calculated using their own egg production (individual selection) and comment.

Solution :

(*a*) Steps for computation of breeding value using combined selection :

Hen ID	Family size (n)	Egg production: Own (P_i)	Family avg ($\bar{F}$)	Weighting coefficient for : Individual (b_1)	Family (b_2)	Individual deviation	Family deviation	EBV*
1	5	270	220	0.0526	0.25	70	20	208.7
2	5	260	230	0.0526	0.25	60	30	210.7
3	3	250	260	0.0526	0.18	50	60	213.6
4	7	245	260	0.0526	0.31	45	60	221.0
5	2	235	270	0.0526	0.14	35	70	211.9

*Calculated by the expression : $EBV_i = \bar{P} + b_1\,(P_i - \bar{P}) + b_2\,(\bar{F} - \bar{P})$

(*b*) Breeding values of the hens based on their own performance are calculated by the expression :

$$\text{EBV}_i = \bar{P} + h^2(P_i - \bar{P})$$

The EBV's calculated using individual's own performance alongwith those calculated by combination of individual and family performance (combined selection) are tabulated below :

Hen	EBV Calculated by :		Rank by :	
ID	Combined selection	Individual selection	Combined selection	Individual selection
1	208.7	207.0	V	I
2	210.7	206.0	IV	II
3	213.6	205.0	II	III
4	221.0	204.5	I	IV
5	211.9	203.5	III	V

General Comments :

- The EBV for combined selection is computed as weighted score of individual and family performance.
- The weighting coefficient assigned to individual's own performance is same for all the hens because each hen has one performance record.
- The weighting coefficient assigned to family average varies for different hens because of differences in their family sizes.
- Since h^2 of egg production is low, the weighting coefficient assigned to family average is 3 to 5 times higher than that assigned to individual's own performance.
- Ranking on EBV by combined selection is different from the EBV calculated using individual's own performance because combined selection uses weighted scores of both sources of information.

8.12. Given below is the information available on two Holstein Friesian bulls :

Milk yield of :	Bull 'A'	Bull 'B'
Dam based on average of 6 lactations :	6000 kg	–
Three daughters :	–	5900 kg

Suppose one of these two bulls is to be selected for use in the herd. Take h^2 and repeatability of milk yield to be 0.2 and 0.4, respectively, and breed/population average to be 5000 Kg. Which of the two bulls should be selected for genetic improvement of the herd ? Show all the calculations used for computation of breeding values.

Solution. We need to compute EBV's of the two bulls based on the information available. EBV of bull 'A' is calculated from its pedigree information (dam's milk production) as per the following expression :

$$\text{EBV}_A = \bar{P} + b\,(\bar{P}_i - \bar{P})$$

The value of weighting coefficient (*b*) is obtained as follows :

$$b = \frac{0.5nh^2}{1+(n-1)r} = \frac{0.5 \times 6 \times 0.2}{1+(6-1)0.4} = 0.2$$

$$EBV_A = 5000 + 0.2\ (6000 - 5000) = 5200.0 \text{ Kg}$$

EBV of bull 'B' is calculated from its daughter's milk yield as per the following expression:

$$EBV_B = \bar{P} + b\,(\bar{P}_i - \bar{P})$$

The value of the weighting coefficient (b) is obtained as per the following expression :

$$b = \left[\frac{0.5nh^2}{1+(n-1)r}\right] = \frac{0.5\times3\times0.2}{1+(3-1)0.05} = 0.272$$

$$EBV_B = 5000 + 0.272 \times (5900 - 5000) = 5244.8 \text{ Kg.}$$

As the EBV of bull 'B' is higher, it gets selected.

Box 8.1: Summary of weighting coefficients and accuracies of selection

Under practical situations variable type of information from different sources is available for the computation of EBV. A summary of the weighting coefficients for different sources of information along with the accuracies of selection is tabulated below :

Bases of selection	No. of records and related Information	Weighting coefficient (b)	Accuracy of selection (r_{AP})
Individual's own performance:	Single record:	h^2	h
	Multiple records: (n = number of records)	$\frac{nh^2}{1+(n-1)r}$	$h\sqrt{\frac{n}{1+(n-1)r}}$
Pedigree performance:	Single record of parent:	$0.5h^2$	$0.5h$
	Multiple records of parent : (n = no. of records per parent)	$\frac{0.5nh^2}{1+(n-1)r}$	$\frac{1}{2}h\sqrt{\frac{n}{1+(n-1)r}}$
	Single record of one grandparent	$0.25h^2$	$0.7h$
Family performance:	n = family size, R = 0.5 for F.S. & 0.25 for H.S. families, $t = Rh^2$	$\frac{[1+(n-1)R]h^2}{1+(n-1)t}$	$h\sqrt{\frac{[1+(n-1)R]}{n[1+(n-1)t]}}$
Sib performance:	n = no. of sibs of the individual, R = 0.5 for F.S. & 0.25 for HS $t = Rh^2$	$\frac{nRh^2}{1+(n-1)t}$	$Rh\sqrt{\frac{n}{1+(n-1)t}}$
Progeny testing:	n = number of progeny per sire t = 0.25h2	$\frac{0.5nh^2}{1+(n-1)t}$	$0.5h\sqrt{\frac{n}{1+(n-1)0.25h^2}}$
Combination of individual and family performance:	R = 0.5 for FS and = 0.25 for HS family $t = Rh^2$, n = family size	$b_1 = \frac{(1-R)h^2}{(1-t)}$	$\left[1+\frac{(R-t)^2}{1-t}\times\frac{n-1}{1+(n-1)t}\right]^{1/2}$
		$b_2 = \frac{[1+(n-1)R]h^2}{1+(n-1)t}$	

Supplementary Problems

8.13. A boar has average daily weight gain of 4 kg. Flock average = 3.5 kg. The boar is mated to randomly chosen females from the flock and produces a large number of offspring which are reared under the same flock conditions as the parents had been. Heritability of daily gain in weight is 0.3. Calculate the EBV of the boar for gain in weight.

8.14. A breeder selects a young ram having EBV of 2.0 kg for gain in weight based on his own performance. Subsequently the ram produces 30 progeny and its EBV based on progeny average is 1.5 Kg. Which of the two estimates of breeding value is more reliable ? Explain.

8.15. Average lactation milk yield of two cows, 'A' and 'B', based on two and five lactations respectively is 5200 and 5000 kg. Heritability of milk yield is 0.2, its repeatability is 0.3 and herd average is 4200 kg. Calculate the breeding values of the two cows.

8.16. A cow had four lactations with average milk yield of 200 kg above the herd average. What would be the genetic superiority of its daughters over the herd average ? Assume, h^2 = 0.2 and repeatability = 0.4.

8.17. Two thoroughbred horses, 'A' and 'B', owned by two renowned breeders are available for sale. Horse 'A' has competed in 5 races with average racing time of 10 seconds above average of its racing competitors. Horse 'B' has competed in 15 races with average racing time of 8 seconds above average of its racing competitors. Heritability of racing performance is 0.2 and repeatability is 0.3. Which of the two horses would you purchase for genetic improvement ? Explain.

8.18. Heritability of growth rate in swine is 0.3. Two boars, 'A' and 'B', are available for genetic evaluation. Boar A's sire had average daily gain of 4.0 kg and boar 'B' had a maternal grand sire whose average daily gain was 3.5 kg. The average daily gain in parental generation was 3.5 kg and in the grand parental generation was 3.0 kg. Which of the two boars is genetically superior ? Explain.

8.19. A cock has 5 half sisters whose average egg production is 250 eggs per annum. Flock average is 240 eggs; h^2 of egg production is 0.2. Calculate the breeding value of the cock.

8.20. Two full-sib families of chicken are available for selection. Information on annual egg production of hens of the two families is listed below :

Family ID	No. of full-sisters	Avg egg production
A	5	310
B	3	320

Each family has one cock. Flock average is 300 eggs and h^2 is 0.2. Calculate the breeding values of the cocks of the two families.

8.21. Which one of the following two bulls should be selected for genetic improvement of milk yield based on the following information?

Herd average = 5000 Kg; h^2 = 0.25.

Average of 50 daughters of bull 'A'= 5300 kg

Average of 10 daughters of bull 'B' = 5500 kg.

8.22. Given below is the information on two cows, 'A' and 'B', for milk yield :

Cow ID	Milk yield of :		No. of
	Cow	Half-sister's	half sisters
A	4980	4920	5
B	4970	4990	4

Herd average = 4500 kg and h^2 = 0.2. Which one of the two cows should be selected for genetic improvement ?

8.23. Information on four dairy bulls for milk yield is given below :

Bull ID	Source of information	No. of records	Average milk yield (kg)
A	Dam	1 record	5500
B	Daughter	10 daughters	4500
C	Half-sister	10 half-sisters	4600
D	Daughter	15 daughters	4400

Herd average = 4000 kg; h^2 = 0.2,

Calculate the breeding values of the four bulls, and rank them in order of their genetic merit.

8.24. Given below is the information on four cows for milk yield :

Cow ID	Performance	No. of records or relatives	Average milk yield (kg)
A	Own	1	5500
B	Own	4	5400
C	Progeny	2	5400
D	Half-sisters	50	5300

Herd average = 5100 kg; h^2 = 0.2 and repeatability = 0.4.

Calculate the breeding values of the four cows from the information given below, and rank them on genetic merit.

8.25. Average gain in weight in a flock of swine is 3.8 kg. Heritability of gain in weight is 0.35.Five boars with the following performance records are available for genetic evaluation.

Boar ID	Source of information	No. of records or relatives	Weight gain (kg)
A	Own performance	1	4.0
B	Sire's performance	1	4.3
C	Half-sib's average	10	3.9
D	Full-sibs' average	4	3.8
E	Progeny average	20	3.9

Calculate the breeding values of the boars and rank them in order of genetic merit.

8.26. Given below are the egg production records of three best hens along with full-sib average and family size. Flock average = 200 eggs; phenotypic standard deviation = 50 eggs; heritability = 0.10.

Hen ID	Hen's own avg.	Family avg.	Family size
I	250	260	3
II	245	260	7
III	335	270	8

(*a*) Compute the breeding values of the hens based on : (*i*) their own performance and rank them in order of their genetic merit (*ii*) combination of own performance and family average, and rank them in order of genetic merit.

(*b*) Why is there a difference in ranking by the two criteria ?

(*c*) Explain the factors 'b_1' and 'b_2' in the formula of computation of EBV by combined selection.

Answers to Supplementary Problems

8.13. 3.65 kg.

8.14. EBV of 1.5 kg is more reliable as it is based on progeny average.

8.15. EBV of cow 'A' = 4507.7 Kg and EBV of cow 'B' = 4563.6 kg.

8.16. 72.7 kg.

8.17. Horse 'B' because of its higher breeding value.

8.18. Boar 'A' is genetically superior as its EBV is higher (EBV = 3.57) than of boar 'B' (EBV = 3.037 kg).

8.19. 242.08 eggs.

8.20. EBV of cock 'A' is 304.3 eggs and of cock 'B' is 306.67 eggs.

8.21. Bull 'A' is selected as its EBV is 5461.54 Kg while that of 'B' is 5400.0 kg.

8.22. Cow 'A' is selected as its EBV is high.

8.23.

Bull ID	EBV (kg)	Rank
A	4150	IV
B	4344.8	II
C	4206.9	III
D	4352.9	I

8.24

Cow ID	EBV (kg)	Rank
A	5180	III
B	5209.1	II
C	5157.1	IV
D	5244.9	I

8.25

Boar ID	EBV (Kg)	Rank
A	3.87	III
B	3.89	II
C	3.85	IV
D	3.82	V
E	3.93	I

8.26 ('a' to 'b')

Hen ID	EBV based on :		Ranking based on :	
	Individual performance	Combination of individual and family performance	Individual EBV	Combination of relative's EBV
I	205.0	215.91	2	1
II	204.5	222.96	1	2
III	213.5	236.83	3	3

(c) Combined selection gives variable weightage to family average depending on family size.

(d) 'b_1' is the regression coefficient of hen's own performance on its own breeding value, and b_2 is the regression coefficient of family performance on the breeding value of the hen.

9

Sire Evaluation and Multi-Trait Selection

In an earlier chapter, we examined the various bases of selection of breeding candidates for a single trait at a time in which progeny testing was also discussed. Progeny testing though a general method of breeding value evaluation is most widely applied in selection of dairy sires. There are different procedures of evaluating the breeding value of sires through progeny testing. This chapter deals with most commonly used methods of dairy sire evaluation. The chapter also deals with the various methods of simultaneous selection for several traits.

Sire Evaluation Methods

Progeny testing implies evaluating the breeding value of an individual on the basis of average performance of its offspring. Such an evaluation can be done for both the sexes (males and females). Progeny testing is most commonly used for evaluating dairy sires for sex-limited traits like milk yield because a sire can produce a large number of progeny in a short time as compared to the dam. Progeny testing has become a standard practice in dairy industry to estimate breeding values of bulls for milk production and related traits. The estimate of the breeding value of a sire through progeny testing is known as **'sire index or sire proof'**. A sire index is an indicator of the breeding worth (value) of the sire. The objective of calculation of sire indices is two- fold:

- To provide an estimate of the breeding value (genetic worth) of the sire. That is, it is a measure of predicted performance of future daughters of a sire.
- To rank the sires (sire indexing) in order of their genetic merit for use in genetic improvement programs.

Progeny testing of dairy bulls is being practiced in most of the countries. Different methods of sire indexing have been proposed from time to time to minimize the variation between progeny groups due to other sources than due to differences in sire's breeding values.

Other sources of variation include genetic level of dams, randomly distributed sources and systematic sources like age, year, season etc. Many methods of dairy sire evaluation have been proposed, and different forms of the same method are also used in different countries. However, essentially there are three methods that have been used for indexing of sires.

Simple Daughter Average Index

This index, proposed by Edwards (1932), measures the breeding value of a sire in terms of average performance of its daughters.

Sire index: I = Daughter average ($\overline{D}$)

Symbolically: $I = \overline{D}$

It is one of the earliest methods of sire indexing and is simplest to compute. However theoretically it is more appropriate for predicting the performance of future daughters of the sire born to:

- dams of the same genetic level as those used for producing the current daughters and
- that the future daughters will be reared and tested under similar environmental conditions as the current daughters.

The simple daughter average index suffers from the following drawbacks:

- The variable numbers of daughters available for different sires are not taken into consideration.
- The method assumes that the average genetic level of the dams mated to different sires is same and is, therefore, subject to bias if the genetic production level of dams allotted to different sires were different.
- Assumes that the daughters of different sires have been reared and tested under similar herd-environmental conditions, which is rarely the case in dairy cattle and buffaloes, and other large livestock's species.
- It does not take into consideration heritability of the trait.

Despite its many shortcomings, simple daughter average when based on large number of daughters has been widely used in early genetic improvement schemes of sires.

Equi Parent or Intermediate Index or Daughter-dam Comparison Method

Equi-parent index, originally suggested by Hansson (1913), also referred to as '*Hanson-Yapp's index*', was first used for selection of cocks for egg production at Mount Hope farm (USA) in 1928, and in early years it was referred to as *'Mount Hope Index'*. The principle underlying this index is that progeny receive 50% of their inheritance from the sire and the other 50% from their dam.That is, both parents contribute equally towards the genetic make-up of the daughters. This is equivalent to saying that the average performance of progeny (daughters) should lie midway between the performance of the sire and dam(s), and hence the name equi-parent or intermediate index. Therefore, the average performance of the daughters of a sire mated to several dams can be expressed as follows:

$$\text{Daughter's average} = \frac{\text{Sire's breeding value + Dam's breeding value}}{2}$$

Since, we are interested in the part of the inheritance contributed by the sire, which can be calculated by rearranging the above formula as follows:

Sire's breeding value (Sire Index) = (2 × Daughter's average) – (Dam's average)

Where: 'Daughter's average' refers to average performance of all the daughters of the sire whose index is to be calculated.

'Dam's average' refers to the average performance of all the dams mated to the sire whose index is to be calculated.

The expression shows that this index will only be valid if the numbers of daughters of a sire are sufficiently large to eliminate the errors due to sampling nature of Mendelian inheritance and due to various environmental influences. The best advantage of this index is that it adjusts the daughter's average for varying production levels of the dams.

There are number of problems associated with the use of equi-parent index:

- The dams and daughters may be seldom used under similar environmental conditions. Most of the dams in dairy cattle make their performance 3-4 years before those of their daughters and are thus exposed to different environmental conditions.
- Equi-parent index is valid only if the trait is assumed to be completely heritable ($h^2 = 1.0$) which is not the case for most of the economic traits. Milk production has low to medium heritability and hence largely influenced by gene combination and environmental factors that are not accounted for in the index.
- The index also does not take into account the variable number of daughters available for different sires.

This method, also referred to as daughter-dam comparison, was adopted officially in 1935 by the US Department of Agriculture for dairy sire evaluation in the USA. Prior to its official adoption for US national sire evaluation, several indexes[1] were advocated/developed, most of them being variations of the basic method of daughter-dam comparison.

Before calculating this index, the records of both daughters and dams are corrected for some known sources of variation viz. age, lactation length, and times milked per day. If females had several previous lactations, their average is used in the calculations as well. The general assumption in all the indices was that the differences in environmental conditions between the years in which dams and daughters made records were negligible. In general, the ranking of bulls by various forms of daughter-dam comparison remained same though values of sire indices varied.

Herd-Mate Comparison Method (Contemporary Comparison)

The herd-mate comparison method of sire evaluation (also referred to as '*Contemporary daughter average index)* which is credited to Henderson's group (of Cornell University, USA), was adopted in USA in 1962. The herd-mate comparison method compares the *production of a bull's' daughters with their contemporaries in the same herd and same year.* In other words, the performance of daughters of a bull is compared with the performance of daughters of all other bulls (herd-mates) under similar condition of farm- and herd- management. The term 'herd mate' used above refers to all the daughters of all other sires which had calved in the same month or the preceding or succeeding two months as the daughters of the sire being evaluated (i.e. ± 2 months). That is, the average performance of the herd mates is based on five months moving average in the year. For example, the herd mate average for a sire's daughters calving

[1]A number of indices using dam–daughter comparison were proposed and used over years by various workers viz; Hanson,Yapp, Pearl, Gowen, Turner, Goodale, Wright, Norton, Allen, Rice, and Lush. Among various indexes, those of Hansson-Yapp,(recommended by Lush for intensive use in USA) and Rice were used the most.

in May would be obtained by averaging the production of daughters of all other sires calving in that herd from March to July in that year. This procedure helps in adjusting the performance of daughters of a bull with respect to variable environmental conditions (season, year and level of management of the herd).The sire index is calculated by using the following formula:

$$\text{Sire index} = \text{Herd av.} + b\ (\text{Daughters average} - \text{Herd mate average})$$

where $b = \dfrac{0.5nh^2}{1 + (n-1)\ 0.25h^2}$

and n = number of daughters for the sire under evaluation

h^2 = heritability of the trait.

$$\text{Thus, Sire index} = \text{Herd average} + \frac{0.5nh^2}{1 + (n-1)\ 0.25h^2}\ (\text{Daughter's avg.} - \text{Herd mate avg.})$$

The herd- mate comparison method takes care of many of the limitations of the first two methods of sire indexing viz;

- The 'b' value (weighting coefficient) assigned to phenotypic superiority or inferiority of daughters over their herd mates takes into account the number of daughters and h^2 of the trait
- It adjusts the production level of daughters' of the bull for variable environmental condition, referred to as systematic environmental differences between progeny groups, viz; season, year and level of herd-management.

In mid-1950's, Robertson and Rendel in UK introduced '**Contemporary Comparison**' method of sire evaluation under artificial insemination when *daughters of a bull were scattered in many herds.* This method is similar to US herd-mate comparison with the additional requirement that a herd- mate be of the same age. Thus, age correction factors are not required since daughters and their contemporaries start their first lactation at about the same age. This method is applicable when enough number of daughters per sire is available.

It is apparent that both the methods viz. herd-mate comparison and contemporary comparison are fundamentally similar but for differences in definition of contemporary groups. Many different countries use one or the other form of the basic herd-mate comparison method. Over the years, differences between these two testing procedures have diminished.

Several refinements in the basic herd-mate comparison method have been made depending upon specific conditions. For example in USA, the herd-mate comparison method was slightly modified in 1968, called '**modified herd-mate comparison**' after discussions with industry groups[2]. The modified herd- mate comparison method imposed certain conditions on the participating bulls viz; a bull must have 10 daughters with herd-mates for the first time; had semen marketed actively through an AI organization etc.

The use of frozen semen has made it possible to use bulls over wide areas (herds, regions, and even continents) and therefore made the problem of sire evaluation more complex. The discussion of these is beyond the scope of this publication.

---

[2]Prior to this modification, different breed associations and states were using different sire evaluation procedures.The multiple lists were a source of confusion in choosing sires. Consequently all dairy cattle breed associations discontinued their own sire evaluations which they had been doing using only a small fraction of the cows on Dairy Herd Improvement Association test

Methods of Selection for more than one Trait

Our discussion of selection procedures in Chapter 7 was restricted to improvement of a single quantitative trait at a time using information from different sources viz; Individual, its relatives or a combination of individual and relative's performance. Since economic returns from livestock enterprise depend on a number of traits, selection has to be usually directed at simultaneous improvement of several traits (multi-trait selection). For example:

- In poultry, the breeder needs to genetically increase egg production, egg weight as well improve the efficiency of feed conversion.
- Pigs need to be improved for not only growth rate and litter-size (number of piglets born) but also for reduced carcass fat.
- In sheep, it is often desirable to increase wool yield as well as wool fineness.
- In race horses, breeders strive to improve both speed and endurance (stamina).

These examples show that selection in practice is conducted for a number of economically important traits. But the different traits to be improved usually do not have equal economic importance. Economic importance of a trait is measured by its **economic value** that represents the contribution of that trait towards determining the monetary returns (profit).

When selection is simultaneously applied for several traits, the amount of selection pressure (selection intensity) applied for each individual trait decreases. Consequently, less genetic progress will be made for any single trait. Selection should thus be directed towards traits that have real economic importance. When selection is directed towards 'n' equally important and independent traits (the traits which are genetically unrelated, $r_G = 0.0$), the intensity of selection for any one trait is reduced and is only '$1/\sqrt{n}$' times as much as if selection were applied for a single trait.

Relative intensity of selection values for individual traits when different numbers of equally important and independent traits are selected are shown below:

Number of traits being selected (n) :	1	2	3	4	N
Relative intensity of selection :	1.00	0.71	0.58	0.50	$1/\sqrt{n}$

Obviously, inclusion of too many traits drastically reduces selection intensity for any single trait. When simultaneous genetic improvement of several traits is desired, three methods are available for selection viz; tandem selection; independent culling levels and total score or selection index

Tandem Selection

The literal meaning of 'tandem' is 'one after the other'. In this method, selection is practiced for one trait (say X_1) for a number of generations until a satisfactory level of performance for that trait has been achieved. Selection is then relaxed or stopped for that trait, and is directed towards the improvement of second trait (X_2) for a number of generations; and then for the third trait (X_3) and so on; until each has been improved to the desired level. For example, egg number and egg weight are economically important traits as they determine the income from laying chickens, both needs to be genetically increased. Tandem selection for these traits means that selection is conducted for increasing egg number for a number of generations until a certain level of egg number in the flock has been achieved. This is followed by selection for increased egg weight for a number of generations. Thus, selection at a time is for one trait, ignoring all other traits. Tandem selection in its pure form thus does not involve simultaneous selection for several traits at a time but involves selection in succession.

Effectiveness of tandem selection, largely depends on the direction and magnitude of genetic correlations between the traits to be improved:

- $r_G = 0.0$: If genetic correlation between the traits is zero, (i.e. the traits are independent) selection for one trait (X_1) will not cause any genetic change in the second trait (X_2). Consequently, selection for the second trait (X_2) would not be necessary for its genetic improvement.
- r_G *is desirable*: If genetic correlation between two traits is desirable, selection for the first trait would also lead to genetic improvement in the other trait. For example, genetic correlation between growth rate and feed efficiency in chickens and pigs is desirable. Selection for growth rate, which is easy to measure, also improves the efficiency of feed conversion. Selection for feed efficiency *per se* would not be required. In fact, both poultry and swine breeders have used selection for growth rate for concomitant genetic improvement of feed efficiency also.
- r_G *is undesirable*: If genetic correlation between traits is undesirable (and negative) then continued selection for the first trait would result in decline in the second trait. When selection is subsequently conducted for the second trait, the genetic improvement brought about in the first trait (through several generations of selection) would be neutralized. Two examples of antagonistic genetic relationships between economic traits are: egg production and egg weight in chickens; and clean wool weight and crimpiness (crimps per cm) in sheep.

In both the species, the genetic correlation between the two traits is negative and undesirable but from economic standpoint, both need to be improved.

Tandem selection is the simplest of the three methods as it involves measuring one trait at a time. The main problem in the application of tandem selection is to determine the number of generations for which selection be devoted for each trait. It seems reasonable that the trait with highest economic value but with moderate to high heritability should be selected first. The level of performance of different economic traits should be monitored every generation, and if a trait is found to have reached an unsatisfactory level, selection should then be directed towards that trait to counteract the decline.

In practice, this method is not commonly used in livestock because the time lag required for the genetic improvement of several traits is long. The method therefore becomes uneconomical for the time and effort expended by the breeder.

Independent Culling Level

In this method, a minimum level of performance (referred to as 'culling level') for each trait is fixed, and the animals that fall short of the minimum standard for a trait are culled regardless of their performance in other traits.

Illustrative example: Egg number to 40 weeks of age and egg weight (g) were recorded on a set of laying chickens. The values of egg weight (Y-axis) plotted against egg number (X-axis) are shown in Fig. 9.1. It can be noted that there is a negative correlation between the two traits i.e. the hens laying more number of eggs, in general, tend to produce small-sized eggs. From economic standpoint, the hens to be selected as parents need to be the ones that combine high egg number with high egg weight. Suppose culling levels of 88 eggs for egg umber and 54 g for egg weight have been fixed. Therefore, the animals with egg weight of

< 56 g and those with egg number of < 90 eggs (lower limit) will be rejected. Each dot in the figure below refers to one hen. It is apparent that a hen that produced 94 eggs but had egg weight of 52 g would get rejected because it failed to fulfill the minimum culling level for egg weight though it had highest egg number.

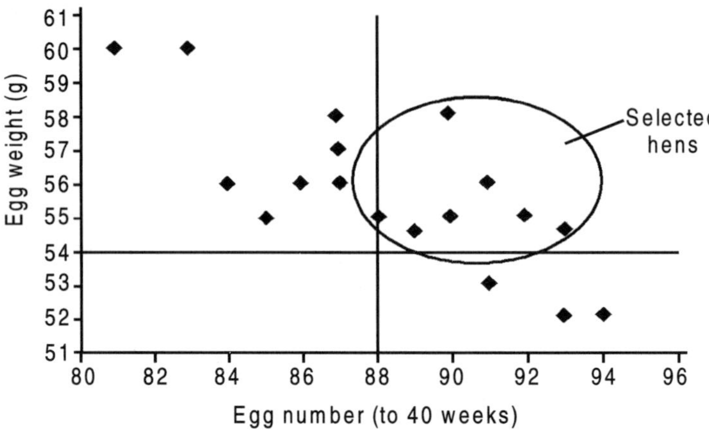

Fig 9.1 Illustration of use of culling level method of selection

The method of independent culling for improvement of several traits may be carried out at the same time provided all the traits are available at same time in life of the animals. Alternately, it may be applied in a sequential (multi-stage) scheme as and when different traits become available.

Illustrative example: Body weight at marketing (5 to 6 weeks of age) in meat-type chickens is the most important economic trait. The number of eggs produced by hens up to a fixed age (say up to 52 weeks of age) also needs to be improved so that more number of chicks can be obtained from each selected hen. That is, both the traits need to be improved (increased). Independent culling in a sequential fashion (multi-stage) is commonly practiced in breeding of broiler chickens:

- When broiler females reach marketing stage (5-6 week), 60% of the top birds (on body weight basis) could be selected based on body weight. These selected birds are reared up to 52 weeks of age and their egg production is recorded.
- From the lot selected on body weight, a certain proportion (say 20% best) is selected based on their egg number (high egg number). The net selected would be 12.0% ($0.60 \times 0.20 = 0.12$ or 12%). Such a sequential scheme using independent culling economizes the breeding operation as 40% low body weight birds are rejected at 5 or 6 weeks of age.

The relative emphasis to be placed on each character is adjusted depending on the economic importance of the trait, and the final proportion of animals to be used as breeders.

Suppose, selection is to be conducted for two uncorrelated traits viz; X_1 and X_2 in that order. Let, 80% of the flock be selected on trait X_1 alone; (i.e. proportion saved, $p_1 = 0.80$). Of those selected at the first stage, let 60% be selected on trait X_2 ($p_2 = 0.6$). Since selection is done independently for each of the two traits, the total fraction selected (p_t) is the product of fractions selected for each of the two traits: (i.e. $p_{total} = p_1 \times p_2$) and equals 0.48.

In practice, the total proportion of animal retained as parents is usually dictated by breeding (replacement) requirements. Accordingly, the culling level for each trait can be modified to meet the requirement of total number of animals to be selected. For example, if a fraction of 0.48 needs to be kept as breeders, proportion selected could be 0.8 for trait X_1, and 0.6 for trait X_2. Alternately, the proportion selected for each of the two traits could be 0.6 for X_1 and 0.8 for X_2.

The culling levels to be used for each trait should be decided with the knowledge of:

- h^2 of the trait
- economic importance of the trait
- number of animals available, and the total proportion to be selected.

The culling levels for each trait have to be fixed anew every generation depending on the average performance level of the herd/flock and the proportion to be saved for reproduction.

Merits: The main advantages of the independent culling method are its cost effectiveness and computational ease:

- *Cost effective*: In most species of livestock, performance records for different traits accumulate over a period as per biological development of the organism. The method of independent culling is especially suited for sequential (step-wise) selection as and when the trait of interest is expressed. For example, a certain fraction of poultry birds can be culled at hatching on the basis of their low chick weight; others can be culled at the time of sexual maturity; and still others on completion of laying period. Selection can thus be made in stages corresponding to biological age of the animals. Such a selection sequence reduces the number of animals that have to be maintained for the entire period. The cost of maintenance and recording of low producing animals is reduced. And, at the same time some income accrues from sale of animals discarded at different stages. It should, however, be ensured that the early-expressed traits do not get undue emphasis so that economically more important but late-expressed traits receive lower selection intensity.
- *Computational ease*: The method of independent culling is operationally easy and can be conveniently applied where computational facilities are limited, and the time-interval between trait measurement and the actual selection is short.

Demerits: There are a few lacunas in the method of independent culling level:

- *No compensation for exclusive merit in one trait:* This method involves fixing of culling levels for rejection of animals for each trait without regard to their merit for other traits. Consequently, it does not permit the superiority of an animal for one trait to be compensated for its deficiencies in other traits. Some of the animals which are extremely superior for one trait may thus be rejected for want of a minimum level of performance for the second trait.
- *Problems of fixing of culling levels and selection intensity:* The determination of optimum culling levels to be set for each trait is tedious, particularly when the traits to be improved are genetically correlated. Further difficulty lies in determining the amount of selection pressure to be applied for each trait. With computer facilities, it is possible, to some extent, to determine the optimum combination of selection intensities for each trait keeping the total proportion selected fixed. However, when the traits to be improved are correlated to each other, the choice of proportion selected becomes difficult.

The method of independent culling level has been extensively used in earlier times, particularly in selection of cattle for 'show-purpose' traits. Nevertheless, selection for such traits, unless their genetic correlation with the economic traits is favorable, is not justified. In practice, most breeders still use independent cullings in many situations either alone or in combination with the method of selection index. For example, the animals, which have been selected by selection index based on several traits but fall below an acceptable threshold level for traits like fertility, defects, disease resistance or temperament etc., are culled more or less independently.

Total Score or Selection Index Method

In this method, two or more traits of an individual are combined to obtain an 'index value' by assigning weights to each trait. The value so calculated is referred to as **'total score'** or **'selection index'** value, symbolized as 'I', and serves as the criterion of selection. The animals with the highest 'total score' or 'selection index' values are retained and used as parents of the next generation.

The selection index value can be mathematically written as:

$$I = b_1X_1 + b_2X_2 + ... + b_nX_n$$

Where: 'I' is the total score or selection index value

'b_1, b_2 ....., b_n' are weighting coefficients assigned to different traits in the index.

'X_1, X_2 ....., X_n' are performance records for various traits obtained on the animal whose index value is to be calculated.

The total score method thus involves assigning appropriate weighting coefficients to different traits such that the merits and demerits of an individual for different traits are combined to arrive at a single score or selection index value. The criterion of selection, the total score or index value (I) is a single phenotypic value but derived from several traits. The total score by itself does not have any biological meaning but is only used for ranking of animals for selection. Selection based on 'total score' leads to genetic improvement in all the traits included in the index in such a way that the total economic returns are maximized.

Box 9.1: Selection Index - Historical.

The theory of selection index for simultaneous genetic improvement of several economically important traits in farm animals was developed by Hazel for his PhD dissertation under Lush (in 1942). In his 1943's paper on selection index, Hazel stated :

"The idea of a yardstick or selection index for measuring the net merit of breeding animals is probably almost as old as the art of animal breeding itself. In practice, several or many traits influence an animal's practical value, although they do so to varying degrees.. . . .. These factors make wise selection a complicated and uncertain procedure; in addition fluctuating, vague, and sometimes erroneous ideals often cause the improvement resulting from selection to be much less than could be achieved if these obstacles were overcome."

Recalling the origin of the problem, Hazel stated "the subject of selecting simultaneously for several traits arose at an informal coffee session of Lush with several of his graduate students in 1940 when Lush suggested that this subject would make a good thesis topic"

Hazel chose it for his PhD thesis. Hazel also recalled about the development of concept of selection index:

"I rushed into the office of Lush and enthusiastically presented the ideas for his thesis. Lush raised the green gambler's eyeshade he wore while at his desk, and looked perplexed".

Hazel went back to his desk rather deflated. Slowly he organized his ideas and presented these to Lush, and over a series of presentations gave Lush the insights to see the problem and the solution.

It has been stated that Lush once said, "Hazel was the first student who wrote a thesis I couldn't altogether understand".

Whether true or not, it made a good story that was often repeated.

As stated by Hazel

"The computational work on selection index was carried out on a 'flip-flop' manual calculator and the one motorized Monroe calculator which were available to graduate students with priority, amounting to fairly full use from 10 pm to 7 am".

This work of Hazel (and Lush), even after more than 70 years of its development continues to be pertinent and vital, and is considered a classical work in Animal Breeding.

Hazel (along with Lush) also examined theoretical efficiencies of the three methods of selection for the genetic improvement of several traits. Tandem selection was found to be least efficient of the three methods and selection index to be more efficient than the independent culling method because it allows the individuals, which are superior in some traits to be selected even though they may be slightly lacking in one or more of the other traits. The relative efficiency of selection index over independent culling increases with the increase in the number of traits to be improved simultaneously. In actual practice of animal breeding, selection index may sometimes be supplemented by some independent culling method.

Theory of Selection Index

The total economic (monetary) returns from farm animal enterprise depend on several traits but each trait may not have equal economic importance. The **economic value** of a trait (designated as 'a') represents the contribution of that trait towards determining economic returns (profit as a monetary value). The economic value assigned to a trait represents the increment (change) in profit due to one unit increase in performance for that trait, independent of the improvement in other trait(s). For example, if an improvement of 1 kg in milk yield fetches Rs. 40 while an increase of 1% in fat content produces a price rise of Rs. 5, then the relative economic values of the two traits can be expressed by the ratio of '1 : 8' for fat and milk yield respectively.

Aggregate breeding value (aggregate genetic merit) : The concept of breeding value for a single trait (A) can be extended to have an aggregate / total breeding value involving several traits. The aggregate breeding value is also referred to as '*net genetic merit*' or '*aggregate genetic merit*' or 'total *genotypic-economic*' score and is designated as 'H'. The aggregate breeding value is sum of breeding values of different traits weighted by their respective economic values. The aggregate breeding value is represented by the following equation:

Aggregate breeding value: $H = a_1BV_1 + a_2BV_2 + \ldots\ldots\ldots + a_nBV_n$

Where: H refers to aggregate breeding value of an individual

BV_i refers to breeding value of that animal for a specific trait ('i'th trait)

a_i refers to economic value assigned to a particular trait. ('i'th trait)

Since we cannot determine the true breeding values of different traits, 'H' cannot be used as the criterion of selection. What we know about different animals are their phenotypic values (performance) for various traits. The 'total score' or 'selection index value' is computed as weighted performance of different traits:

$$I = b_1X_1 + b_2X_2 + \ldots\ldots\ldots\ldots\ldots\ldots + b_nX_n$$

The accuracy of selection index is measured as the correlation between the aggregate breeding value (H) and selection index value (I) and is designated as R_{HI} or R_{IH}.

Selection index maximizes the correlation between the net genetic merit (H) and selection index value (I) so that the selected individuals are superior in terms of their net genetic merit. In this way, the net genetic merit of the herd or flock is improved from one generation to the next. For this reason, selection index is also referred to as **'economic selection index'**.

The computation of selection index value requires knowing the performance records (X_i) of the animals for the traits included in selection index and the weighting coefficients (b_i). The weighting coefficients assigned to different traits depend on:

- relative economic value of each trait
- heritabilities of the traits (or alternately additive genetic and phenotypic variances)
- genetic and phenotypic correlations among traits (or alternately, genetic and phenotypic covariances among traits).

Once performance records of animals and the economic values for different traits are known, the calculation of selection index value requires computation of the weighting coefficients ('b' values). The calculations involved in computation of the 'b' values for use in selection index are shown in the next chapter.

Simpler Forms of Selection Index: In some situations, the complete information required for calculation of 'b' values may not be available. In such a situation, some simpler forms of selection index, given below, can be used.

- *Uncorrelated traits with equal heritabilities* When the traits considered for selection are independent of each other ($r_G = r_P = 0$), and different traits have more or less equal heritabilities or when this information is not available, then the best weights ('b' values) given to different traits are their relative economic values ($b_1 = a_1$, $b_2 = a_2$, ..., $b_n = a_n$) as follow:

  $$I = a_1X_1 + a_2X_2 + \text{........} + a_nX_n$$

  Where: a_1 .... a_n are relative economic values for different traits.

  X_1 .... X_n are performance records for different traits

- *Uncorrelated traits with equal economic values:* When the traits considered for selection are uncorrelated ($r_G = r_P = o$) and have equal economic importance, (i.e. $a_1 = a_2 = a_n$), then the optimum weights assigned to different traits are their heritabilities ($b_1 = h_1^2$; $b_2 = h_2^2$ etc.) as follow: :

  $$I = h_1^2X_1 + h_2^2X_2 + \text{........} + h_n^2X_n$$

- *Uncorrelated traits with unequal heritabilities and economic values:* When the traits are independent ($r_G=r_P=0$) but have different heritabilities and economic values, then the appropriate weightage given to each trait is product of h^2 and relative economic value of each trait (i.e. $b_1 = a_1h^2$; $b_2 = a_2h^2$ etc.):

  $$I = (a_1h_1^2)X_1 + (a_2h_2^2)X_2 + \text{........} + (a_nh_n^2)X_n$$

Obviously, a trait having very low h^2 or very low economic value contributes little to the index.

Under practical situations, these simpler forms of selection index are usually not applicable because different traits have unequal heritabilities and economic values, and are phenotypically and genetically correlated with each other. The actual computations involved for working out the weighting coefficients (or selection index) are given in next chapter.

Merits and Constraints: Theoretically, selection index is the ideal method for simultaneous genetic improvement of several traits. This is so as it combines different traits in an optimal manner to provide best estimate of aggregate breeding value (economic-genetic value) of the animal (H). It overcomes the demerits of independent culling levels as it balances the inferiority of an animal in one trait by its superiority in the other trait. Nonetheless, there are a few problems in the application of selection index in livestock:

- The chief problem with selection index lies in determining the economic values of different traits. These values are based on analysis of costs and returns (economic analysis), and may therefore vary from one market to the other. The economic values may also change overtime due to changes in market requirements.
- Sometimes, the selection index may select some animals that are so poor in one trait that their inclusion in the selected lot leads to serious problems. For example, a selection index involving egg number and egg weight may select a few birds that are dangerously poor in egg size such that eggs of their progeny may not be of saleable size.

In summary, selection index is the most efficient method of multi-trait improvement under most situations. However, there is no standard rule that one must always use it alone. In fact, in practice, breeders use selection index for the most important traits but use some independent cullings for some other traits as per their requirements.

Review Questions

1. Define or explain the following :
   (a) Economic value
   (b) Tandem selection
   (c) Independent culling level
   (d) Aggregate breeding value
2. Why selection has to be practiced for several traits? Cite two examples of importance of more than one trait in selection for each of the following species: poultry, pig, sheep, and cattle/buffaloes.
3. List the methods of selection that can be used for the improvement of more than one trait.
4. What are the merits and demerits of tandem selection?
5. Effectiveness of tandem selection for improvement of two traits depends on the direction and magnitude of genetic correlation between them. Justify.
6. What are the merits and demerits of independent culling level?
7. What factors determine the culling levels to be used for the improvement of a trait?
8. What is meant by selection index? Give the selection index equation.
9. What information is required for construction of selection index?
10. Which is the most effective method of selection for simultaneous improvement of several traits? Explain.

11. What factors determine the weighting coefficients to be assigned to different traits in a selection index?
12. What will the form of selection index equation when :
    (a) the traits are uncorrelated and have equal heritabilities.
    (b) the traits are uncorrelated but have equal economic values.
    (c) the traits are uncorrelated but have unequal heritabilities and economic values
13. What are the major constraints in application of selection index?
14. What is sire index? Why is it calculated?
15. List the different method of sire evaluation.
16. What is simple daughter average index of sire evaluation? What are its merits and demerits?
17. Explain the equi-parent (daughter-dam comparison) method of sire evaluation. What are its advantages?
18. Explain the herd-mate comparison method of sire evaluation. What are its advantages over other methods of sire evaluation?

●●●

10

Computation of Sire indices and Construction of Selection Index

Sire index which is an indicator of the breeding worth (value) of the sire is used to rank the sires for use in genetic improvement programs. Different methods of sire indexing have been proposed from time to time so as to minimize the variation between progeny groups due to other sources than due to the differences in sire's breeding values viz; genetic level of the dams, randomly distributed sources and systematic sources like age, year, season etc. There are essentially three methods that have been used for indexing of sires. In this chapter the computational steps involved in three types of sire indices have been outlined. Additionally this chapter also covers the construction of selection index, its use in selection for multiple traits and examines the effectiveness of selection index.

Computation of Sire Indices

10.1. Three buffalo bulls were each mated to four randomly chosen females (dams) from a herd in the same year. First lactation milk yield (kg) of dams and their daughters is given below:

Bull ID	Dam ID	Milk yield of dam	Milk yield of daughters	Bull ID	Dam ID	Milk yield of dam	Milk yield of daughters
	1	2105	2105		9	2060	2090
A	2	2130	2140	C	10	2080	2095
	3	2115	2125		11	2080	2100
	4	2110	2100		12	2000	2105
	5	2090	2110				
B	6	2100	2120				
	7	2094	2105				
	8	2084	2095				

Compute the sire indices by simple daughter average index (I_1) and equi-parent index (I_2), and rank the bulls in order of their genetic merit. Comment on the ranking of bulls by the two methods.

Solution : Simple daughter average index (I_1) of a sire is calculated as the average of its daughters:

$$I_1 = \text{Daughter average}$$

The equi-parent index (I_2) of a sire is calculated as:

$$I_2 = (2 \times \text{Daughters' average}) - \text{Dam's average}$$

For bull 'A', the calculations for I_1 and I_2 are shown below:

$$I_1 = \left[\frac{(2105+2140+2125+2100)}{4}\right] = 2117.5 \text{ kg.}$$

$$I_2 = \left[\frac{2\times(2105+2140+2125+2100)}{4}\right] - \left[\frac{(2105+2130+2115+2110)}{4}\right]$$

$$= 2120.0 \text{ kg.}$$

Sire indices for the three bulls by the two methods, along with their rank order, are tabulated below:

Bull	Sire index :		Rank on:	
ID	I_1	I_2	I_1	I_2
A	2117.5	2120	I	II
B	2107.5	2123	II	III
C	2097.5	2140	III	I

Comments: The two methods have ranked the bulls in different order. This is due to the reason that equi-parent index (I_2) takes into consideration the performance of dams mated to a sire whereas the first method (I_1) ignores the performance of dams. Consequently the genetic merit of sires by the 2nd method is not biased due to variable performance level of dams mated to different sires. For example, average performance of daughters of sire 'A' is highest (2117.5 Kg) while that of sire 'C' is lowest (2097.5 Kg). A comparison of the average performance of dams mated to these two sires indicates that sire 'A' was mated to superior dams (with average= 2115 Kg) whereas sire 'C' was mated to inferior dams (with average= 2055 Kg).

Inference: The equi-parent index, by adjusting for variable performance of dams, has yielded more reliable sire index than simple daughter average.

10.2. Given below are the months and year of calving , and first lactation milk yield records of 15 daughters of three dairy sires in a progeny testing program. Calculate the herd-mate average of the three bulls.

Bull ID	Daughter ID	Month & year of calving	Lactation milk yield (kg)	Bull ID	Daughter ID	Month & year of calving	Lactation milk yield (kg)	Bull ID	Daughter ID	Month & year of calving	Lactation milk yield (kg)
	1	4/2013	2500		6	5/2013	2300		11	2/2013	2900
	2	7/2013	2600		7	8/2013	2100		12	6/2013	3000
I	3	6/2013	2500	II	8	2/2013	2000	III	13	7/2013	3100
	4	3/2013	2700		9	9/2013	2200		14	8/2013	2700
	5	8/2013	2500		10	10/2013	2300		15	3/2013	2800

Solution : Herd-mate average of a daughter refers to the average performance of daughters of all other bulls which have calved in the same month or preceding or succeeding two months (± 2 months) as the daughter of the sire being evaluated. For example, herd mates of 1st daughter (ID-1) are daughters numbered 6, 8 (belonging to sire II) and 11, 12 and 15 (belonging to sire III). The average of these 5 daughters is 2600 Kg [(2300 + 2000 +2900 +3000 + 2800)/5]. Similarly, herd-mate averages of all the daughters are calculated.

Herd-mate average of a sire is calculated as the average performance of herd-mates of all the daughters of that sire. For example, herd-mate average of 1st sire is 2574.7 Kg [(2600+2566.7+2640+2500+2566.7)/4= 2574.7 kg].

The completed table for herd-mate average is shown below:

Sire ID	Daughter ID	Herd-mate average of daughter (kg)	Herd-mate average of sire (kg)	Sire ID	Daughter ID	Herd-mate average of daughter (kg)	Herd-mate average of sire (kg)	Sire ID	Daughter ID	Herd-mate average of daughter (kg)	Herd-mate average of sire (kg)
	1	2600.0			6	2742.8			11	2400.0	
	2	2566.6			7	2733.3			12	2416.6	
I	3	2640.0	2574.6	II	8	2725.0	2705.2	III	13	2366.6	2385
	4	2500.0			9	2725.0			14	2366.6	
	5	2566.6			10	2600.0			15	2375.0	

10.3. Given below are average milk yield (kg) records of daughters of six dairy sires along with average milk yield records of their herd mates. Herd average is 2000 kg and $h^2 = 0.25$.

Sire ID	No. of daughter(s)	Avg. milk yield of: Daughters	Herd mates
A	1	2800	2300
B	3	2710	2410
C	5	2600	2400
D	5	2700	2480
E	10	2550	2400
F	20	2490	2290

(a) Compute the sire indices using herd-mate comparison method and rank the sires in order of genetic merit

(b) Compute the sire indices assuming that herd-mate average is not available (therefore sire index is calculated on simple daughter average) and compare this rank with that of (a) and comment on the relative rankings.

Solution : The sire index by herd-mate comparison (I_3) is calculated as follow:

I_3 = herd avg. + b (daughters avg – herd-mate avg)

or

$$I_3 = \overline{H} + b\,(\overline{D} - \overline{H}_M)$$

Where:
$$b = \frac{0.5\, nh^2}{1 + (n-1)\,(0.25h^2)}$$

n = no. of daughters of the sire

$\bar{H}$ = herd average

$\bar{D}$ = daughter average

$\bar{H}_M$ = herd-mate average

Alternately, the sire index can be written as:

$$I_3 = \bar{H} + \frac{0.5\ nh^2}{1 + (n-1)(0.25h^2)} (\bar{D} - \bar{H}_M)$$

Computation for sire index:

Sire ID	Herd avg	No. of daughters	Avg milk yield of: daughter (s)	Avg milk yield of: herd mates	Weighting coefficient. (b)	Sire index value using: herd mate avg	Sire index value using: simple daughter avg	Rank of sires by: herd mate avg	Rank of sires by: simple daughter avg
A	2200	1	2800	2300	0.125	2262.5	2800	V	I
B	2200	3	2710	2410	0.333	2300	2710	IV	II
C	2200	5	2600	2400	0.500	2300	2600	IV	IV
D	2200	5	2700	2480	0.500	2310	2700	III	III
E	2200	10	2550	2400	0.800	2320	2550	II	V
F	2200	20	2490	2290	1.143	2428.6	2490	I	VI

Comments: The ranking of sires by the two methods is different : the herd-mate average index adjusts the performance of daughters of a sire for variable environmental conditions and therefore, provides a reliable measure of the genetic merit of a sire.

Multi Trait Selection

10.4. Given below are the performance records of five rams for fleece weight and staple length:

Ram ID	Fleece weight (lbs)	Staple length (cm)
I	5.0	2.9
II	5.1	2.8
III	3.8	2.7
IV	4.9	3.0
V	4.7	2.9

The breeder has set the following culling levels for the two traits:

Fleece weight > 5.0 lb ; Staple lengths > 2.7 cm.

(a) Which ram(s) meet(s) the performance standards for the two traits ?

(b) Examine the main problem associated with use of independent culling levels as outlined in the theory in the last chapter.

Solution:

(a) Ram number 2 (ID-II) fulfills the standards fixed for both the traits.

(b) Fleece weight of rams I and IV are close to ram II, and their staple lengths are higher than that of ram I. But the minimum standard (culling level) fixed for fleece weight has led to the exclusion of rams I and IV from selection. This is the major problem associated with independent culling levels *i.e.* some of the animals which are superior for one trait but do not fulfill the minimum standard for the other trait get rejected.

Inference : Culling levels should be carefully defined otherwise some of the really good animals may get rejected.

10.5. A poultry breeder has developed the following selection index (I) for body weight (X_1) and egg weight (X_2):

$$I = 0.9\ X_1 + 1.4\ X_2$$

(a) Compute the selection index values for the following five hens and rank them in descending order of genetic merit, and draw appropriate inferences about the selection index (total score) values (I) and the weighting coefficients (b).

Hen ID	Body weight (Kg)	Egg weight (g)
1	1.7	57
2	1.4	56
3	1.6	58
4	1.5	54
5	1.5	55

(b) Show how can the 'b' values and the index values (I) be re-scaled to make the computations easier or meaningful ?

Solution :

(a) **Calculation of selection index values :** Selection index (total score) value for 1st hen (ID-1) would be :

$$I_{1st\ hen} = 0.9\ X_1 + 1.4\ X_2 = (0.9 \times 1.7) + (1.4 \times 57) = 81.33$$

Index values for the five hens, calculated as per the above procedure, are tabulated below:

Hen ID	(b_1×body wt)	+	(b_2×egg wt)	=	Index value	Rank
1	(0.9 × 1.7)	+	(1.4 × 57)	=	81.33	II
2	(0.9 × 1.4)	+	(1.4 × 56)	=	79.66	III
3	(0.9 × 1.6)	+	(1.4 × 58)	=	82.64	I
4	(0.9 × 1.5)	+	(1.4 × 54)	=	76.95	V
5	(0.9 × 1.5)	+	(1.4 × 55)	=	78.35	IV

Inferences :

- The total score (selection index) values per se do not carry any meaning except for ranking of animals for use in selection.
- The weighing coefficient assigned to egg weight (b_2 = 1.4) is greater than that for body weight (b_1 = 0.9). For this reason, the hen with highest egg weight (ID No. 3) but with moderately high body weight has the highest index value (index value = 82.64).

(*b*) **Re-scaling of weighting coefficients:** Selection index values can also be computed by rescaling the weighting coefficients. This is done to make the calculations easy, and to make the weighting coefficients more interpretable or meaningful. The most commonly used re-scaling is to set the lowest '*b*' value equal to unity (b_1=1), and then to re-calculate the b_2. The index values are then obtained on the re-scaled scale.

Note: Though any of the '*b*' values (either b_1 or b_2) can be set equal to unity but normally the lowest '*b*' value is set equal to unity. For the present example, if 'b_1' is divided by 0.9., then the re-scaled 'b_1' becomes unity. The 'b_2' is also divided by 0.9 which yields the re-scaled 'b_2' as 1.56 (b_2 = 1.4 / 0.9 = 1.56). Thus the weighting coefficients, b_1 and b_2, on re-scaling become 1.0 and 1.56 for traits X1 and X_2. respectively .The selection index equation becomes:

$$I = X_1 + 1.56\ X_2$$

The re-scaled values indicate that the weighting coefficient assigned to second trait (X_2) is 1.56 times higher than that assigned to the first trait. The index values for the five hens on this scale, along with their rankings, are listed below:

Hen ID	Index value	Rank
1	90.62	II
2	88.76	III
3	92.08	I
4	85.74	V
5	87.3	IV

It can be seen that the rank order of hens using the rescaled index ($I = X_1 + 1.56\ X_2$) is same as on the original selection index equation ($I = 0.9X_1 + 1.4X_2$).

Inference: The total score values per se do not carry any biological meaning except for their use in ranking of animals.

Re-scaling of selection index values (I): The selection index values calculated using the original index equation (or the re-scaled equation) can also be re-scaled by addition/ subtraction or multiplication/division by a constant so that 'I' values fall in some desired range. The most common re-scaling is to subtract the mean of index value from the index value of each animal.

The mean of index values (un-scaled) is :

$$= \frac{(81.33 + 79.66 + 82.64 + 76.95 + 78.35)}{5} = 79.786$$

The re-scaled index values for different hens would be:

Hen ID	Index value	Rank
1	81.33 – 79.78 = 1.544	II
2	79.66 – 79.78 = – 0.126	III
3	82.64 – 79.78 = 2.854	I
4	76.95 – 79.78 = – 2.836	V
5	78.35 – 79.78 = – 1.436	IV

Note: The sum of index values, expressed in deviation terms, equals zero (1.544 – 0.126 + 2.854 – 2.836 – 1.436 = 0.0). The index values in deviation terms imply that an animal with positive index value is better than the flock average while the one with negative index value is below the flock average. But the ranking of hens using the re-scaled index values remains the same as using the un-scaled values.

10.6. A poultry breeder is to select hens for the simultaneous genetic improvement of egg number to 40 weeks of age (EN) and egg weight (EW). The following parameters are available:

Parameter :	Trait	
	Egg no.	Egg wt. (g)
h^2 :	0.15	0.50
σ^2_P :	144	16
Economic value :	3.0	1.0
r_G = – 0.15 and r_P = – 0.20		

(a) Construct the selection index

(b) Interpret the values of weighting coefficients ('b')

(c) Compute the selection index values (scores) for the following 20 hens of a flock, and rank them in order of their genetic merit :

Hen ID :	1	2	3	4	5	6	7	8	9	10	11	12	13	14	15	16	17	18	19	20
Egg no.:	81	91	90	86	94	87	90	83	91	84	89	93	87	85	87	92	91	93	88	87
Egg wt (g):	60	56	54	56	52	56	58	60	53	56	54	54	57	55	56	55	53	52	55	58

Flock mean for egg number is 88.5 and for egg weight is 55.5 g.

(d) Best four hens are to be retained for breeding using selection index of egg number and egg weight. List the four hens that will be selected by selection index. Work out the selection differential, response to selection (genetic gain) and the expected mean in the progeny generation for the two traits, and comment on the usefulness of selection index in the simultaneous improvement of the two negatively correlated traits.

(e) Compare the effectiveness of selection index method, as outlined above, with individual selection (tandem selection) by selecting the best four hens based on their: (i) egg number only, (ii) egg weight only .The effectiveness of the three procedures (viz. selection index, individual selection for egg number and individual selection for egg weight) is to be compared in terms of selection differential and expected genetic change.

Solution :

(a) **Construction of selection index equation:**

For the calculation of index weights (b_1 and b_2), (additive) genetic variance (σ^2_G), phenotypic variance ((σ^2_P), phenotypic covariance (Cov_P) and genetic covariance (Cov_G) are required. Values for these varitinces and covariances are not given in the problem but can be calculated from estimates of h^2 , r_G, r_P and σ^2_P (given in the problem) as follow:

$$\sigma^2_G \text{ for egg number} = h^2 \times \sigma^2_{P_1} = 0.15 \times 144 = 21.6 \text{ number}^2$$

$$\sigma^2_G \text{ for egg weight} = h^2 \times \sigma^2_{P_2} = 0.50 \times 16 = 8.0 \text{ g}^2$$

Since, $r_G = \frac{Cov_G}{\sqrt{\sigma^2_{G_1} \times \sigma^2_{G_2}}}$, Cov_G can be obtained by re-arrangement as follow:

$$Cov_G = r_G \times \sqrt{\sigma^2_{G_1} \times \sigma^2_{G_2}} = (-0.15) \times \sqrt{21.6 \times 8.0} = -1.972$$

Similarly, Cov_P can be obtained as follow :

$$Cov_P = r_P \times \sqrt{\sigma^2_{P_1} \times \sigma^2_{P_2}} = (-0.2) \times \sqrt{144 \times 16.0} = -9.6$$

The normal equations for obtaining b_1 and b_2 are:

$$(b_1 \times \sigma^2_{P_1}) + (b_2 \times Cov_P) = (a_1 \times \sigma^2_{G_1}) + (a_2 \times Cov_G)$$

$$(b_1 \times Cov_P) + (b_2 \times \sigma^2_{P_2}) = (a_1 \times Cov_G) + (a_2 \times \sigma^2_{G_2})$$

Putting the numerical values of various parameters (viz; σ^2_P ; σ^2_G ; a_1, a_2 ; Cov_G and Cov_P), we get the following equations:

$$(b_1 \times 144) + (b_2 \times -9.6) = (3.0 \times 21.6) + [1.0 \times (-1.972)] = 62.83$$

$$[b_1 \times (-9.6)] + (b_2 \times 16] = [3.0 \times (-1.972)] + (1.0 \times 8.0) = 2.084$$

After rearrangement the normal equations become:

$$144b_1 - 9.6b_2 = 62.83 \quad \text{.... (Eqn. no. 1)}$$

$$-9.6b_1 + 16b_2 = 2.084 \quad \text{.... (Eqn. no. 2)}$$

The normal equations are solved for b_1 and b_2 as follow:

Multiply eqn. no. 1 with 16, and eqn. no. 2 with 9.6, and then adding the two equations:

$$16 \times 144\ b_1 - 16 \times 9.6\ b_2 = 16 \times 62.83 \quad \text{.... (Eqn. no. 3)}$$

$$-9.6 \times 9.6\ b_1 + 9.6 \times 16\ b_2 = 9.6 \times 2.084 \quad \text{.... (Eqn. no. 4)}$$

We get:

$$2304\ b_1 - 153.6\ b_2 = 1005.28$$

$$-92.16\ b_1 + 153.6\ b_2 = 20.0064$$

Sum: $2211.84\ b_1 + 0.0 = 1025.2864$

$$\therefore\ b_1 = \frac{1025.2864}{2211.84} = 0.464$$

Putting the value of 'b_1' in equation no. 2 and solving for b_2 :

$$-9.6 \times 0.4635 + 16\ b_2 = 2.084$$

$$\therefore\ b_2 = \frac{2.084 + (9.6 \times 0.4635)}{16} = \frac{6.5336}{16} = 0.408$$

Selection index equation is:

$$I = 0.464\ X_1 + 0.408\ X_2$$

(*b*) **Interpretation of 'b' values:**

- The weighting coefficients (b_1 and b_2) for both the traits are positive which means that both the traits need to be increased for improving the economic profitability of the flock.
- The weighing coefficients for the two traits are nearly equal (b_1 = 0.464 and b_2 = 0.408) which means that both traits are equally important for improving profitability

(*c*) **Calculation of index values for the 20 hens:** Selection index value (I) for 1st hen is obtained as follow:

$$I_{1st\ hen} = 0.464\ X_1 + 0.408\ X_2.$$

The performance values for X_1 (egg number) and X_2 (egg weight) of the 1st hen are inserted in the equation of selection index for obtaining its index value:

$$I_{1st\ hen} = (0.464 \times 81) + (0.408 \times 60) = 62.1$$

Similarly, selection Index values for all the hens are obtained, which along with their rank order, are tabulated below:

Hen ID	Egg no.	b_1	Egg wt (g)	b_2	Selection index value (I)	Rank
1	81	0.464	60	0.408	62.1	18
2	91	0.464	56	0.408	65.1	4
3	90	0.464	54	0.408	63.8	10
4	86	0.464	56	0.408	62.8	17
5	94	0.464	52	0.408	64.8	5
6	87	0.464	56	0.408	63.2	15
7	90	0.464	58	0.408	65.4	1
8	83	0.464	60	0.408	63.0	16
9	92	0.464	53	0.408	64.3	7
10	84	0.464	56	0.408	61.8	20
11	89	0.464	54	0.408	63.3	12
12	93	0.464	54	0.408	65.2	2
13	87	0.464	57	0.408	63.6	11
14	85	0.464	55	0.408	61.9	19
15	87	0.464	56	0.408	63.2	14
16	92	0.464	55	0.408	65.1	3
17	91	0.464	53	0.408	63.8	9
18	93	0.464	52	0.408	64.4	6
19	88	0.464	55	0.408	63.3	13
20	87	0.464	58	0.408	64.0	8

(*d*) **Selection of best 4 hens on the basis of selection index values:** The best four hens selected on the basis of selection index values are listed below along with the related information:

Hen ID	Selection Index value (I)	Rank	Egg number	Egg weight (g)
7	65.4	1	90	58
12	65.2	2	93	54
16	65.1	3	92	55
2	65.1	4	91	56
	Mean of selected hens:		91.5	55.75
	Flock mean:		88.5	55.5

Selection differential for:

egg number : 91.5 – 88.5 = 3.00 eggs

egg weight (g): 55.75 – 55.5 = 0.25 g

Expected genetic change (ΔG)[1] in :

egg number: S x h^2 = 3.00 x 0.15 = 0.45 eggs

egg weight: S x h^2 = 0.25 x 0.50 = 0.125 g.

Expected mean of next generation for:

egg number = mean of parental generation + Δ_G (EN)
= 88.5 + 0.45 = 88.95 eggs

egg weight = mean of parental generation + Δ_G (EW)
= 55.5 + 0.125 = 55.63 g.

Comments: Selection index is useful in identifying the animals which have favorable combination of the two traits such that the net genetic-economic worth of flock is improved (from parental to next generation).

(*e*) **Individual selection for one trait at a time (tandem selection)**

(i) **Selection for egg number only:** Based on their egg number, the four hens selected are: ID no. 5, 18, 12, and 16. Information about the selected hens for egg number (the selected trait) and egg weight (the unselected trait) along with mean of the selected hens is tabulated below:

Hen ID	Egg number	Egg wt (g)
5	94	52
18	93	52
12	93	54
16	92	55
Mean of selected hens:	93.0	53.25
Flock mean:	88.5	55.5

Selection differential for:

egg number: 93.0 – 88.5 = 4.5 eggs

egg weight (g): 53.25 – 55.5 = – 2.25 g.

Genetic change (ΔG) in :

egg number: S x h^2 = 4.5 x 0.15 = 0.675 eggs

egg weight: S x h^2 = (– 2.25) x 0.5 = – 1.125 g.

(ii) **Selection for egg weight only:** Based on their egg weight, the four hens selected are ID no. 1, 8, 7 and 20.

Hen ID	Egg wt.(g)	Egg number
1	60	81
8	60	83
7	58	90
20	58	87
Mean of selected hens	59.0	85.25
Flock mean	55.5	88.5

1. The expression for expected genetic gain, ΔG, when using selection index is not simply 'S × h^2'. But for sake of simplicity (for undergraduate level), the usual expression of 'S × h^2' has been used.

Selection differential for:

egg weight (g): 59.0 – 55.5 = 3.5 g.

egg number: 85.25 – 88.5 = – 3.25 eggs

Expected genetic change, *(ΔG) in:*

egg weight: $S \times h^2 = 3.5 \times 0.5 = 1.75$ g.

egg number: $S \times h^2 = (-3.25) \times 0.15 = -0.4875$ eggs

Comparison of effectiveness of the three selection procedures: Summary of the results of the three selection procedures viz. selection index, individual selection (tandem) for egg number and individual selection (tandem) for egg weight is given below:

Parameter:	Selection procedure :					
	I Selection Index		II Individual selection for egg number		III Individual selection for egg weight	
	Egg no.	Egg wt (g)	Egg no.	Egg wt (g)	Egg no.	Egg wt. (g)
S	3.00	0.25	4.5	– 2.25	– 3.25	3.5
Δ_G	0.45	0.125	0.675	– 1.125	– 0.4875	1.75

Comments :

- Selection index yields desirable selection differential for both the traits though the genetic and phenotypic correlations between egg number and egg weight are negative and undesirable (antagonistic).
- Due to the existence of negative genetic correlation between egg number and egg weight, direct selection for either of the two traits (selection procedures II and III) results in desirable change in one trait (i.e. the trait being directly selected) but results in undesirable change in the other trait (the correlated trait).
- Selection based on index of egg number and egg weight(selection index), which combines both traits in an optimal fashion, is better than selection for either of the two traits at a time (tandem selection)

Supplementary Problems

10.7. Given below are the months of calving along with the first lactation milk yield (kg) of nine daughters of three buffalo sires in a progeny test programme:

Sire ID	Daughter's : ID	Month of calving	Milk yield	Dam's milk yield
	1	4/12	2400	2500
1	2	3/12	2600	2600
	3	8/12	2400	2700
	4	5/12	2200	2300
2	5	2/12	2000	2000
	6	9/12	2100	2200
	7	2/12	2800	2400
3	8	3/12	2700	2500
	9	8/12	2600	2500

$h^2 = 0.2$, Herd average = 2400 Kg.

Calculate the sire indices for the three bulls by simple daughters average (SDA), equal parent index (EP) and herd mate comparison (HMC) methods.

10.8. Growth rate and feed intake up to 6 weeks of age in broiler chicken are economically important traits. A selection index is to be constructed for the simultaneous genetic improvement of both the traits from the information given below:

Parameter :	Growth rate (X_1)	Feed intake (X_2)
h^2 :	0.5	0.30
σ^2_P :	100	300
Economic value :	1.0	– 2.0
r_G = 0.60 and r_P = 0.50		

Calculate the index weights and write the selection index equation.

10.9. Daily gain in weight (DG) and feed conversion ratio (FCR) are to be improved in a flock of swine. Construct the selection index from the parameters given below:

Parameter:	DG	FCR
h^2 :	0.4	0.30
σ_P^2 :	1.44	6.25
Economic value:	1.0	4.0

Cov_P = 2.4, Cov_G = 0.7275

10.10. Weighting coefficients for fleece weight (FW) and staple length (SL) in a flock of sheep have been calculated to be 2.0 and 1.0, respectively. Fleece weight and staple length of 20 rams and 20 ewes of a flock alongwith their selection index values (calculated using the above listed weighting coefficients) are tabulated below:

Rams:				Ewes:			
ID	FW (kg)	SL (cm)	Index	ID	FW (kg)	SL (cm)	Index
1	5.1	2.8	13.0	1	4.9	3.0	12.8
2	3.8	2.8	10.4	2	5.1	2.5	12.7
3	5.0	3.0	13.0	3	4.0	2.8	10.8
4	4.7	3.0	12.4	4	4.3	2.7	11.3
5	5.7	2.6	14.0	5	4.6	2.6	11.8
6	4.7	2.9	12.3	6	3.9	2.6	10.4
7	4.8	3.4	13.0	7	5.1	2.9	13.1
8	4.2	3.0	11.4	8	3.8	2.7	10.3
9	4.5	3.1	12.1	9	4.6	2.5	11.7
10	5.0	2.9	12.9	10	3.9	2.8	10.6
11	4.6	3.3	12.5	11	4.4	2.6	11.4
12	4.7	2.9	12.3	12	4.7	2.5	11.9
13	5.6	3.5	14.7	13	4.2	2.7	11.1
14	5.1	3.2	13.4	14	3.7	2.4	9.8
15	5.3	3.0	13.6	15	5.0	2.6	12.6
16	5.4	2.7	13.5	16	4.5	3.1	12.1
17	4.5	3.0	12.0	17	4.5	2.7	11.7
18	6.1	2.7	14.9	18	4.4	3.0	11.8
19	5.2	2.9	13.3	19	6.6	3.2	16.4
20	4.7	3.0	12.4	20	4.8	2.7	12.3

The means for the two traits are:

Trait	Rams	Ewes
FW(kg) :	4.935	4.55
SL(cm) :	2.985	2.73

The best two rams and best 10 ewes are to be selected from this flock for the simultaneous genetic improvement of fleece weight and staple length. If two animals with same values for a trait are available, one of the two that has higher value for the second trait can be selected. Select the required number of breeding candidates based on the following procedures:

(a) individual selection for FW only

(b) individual selection for SL only

(c) selection index of FW and SL

Compute the selection differential for both the traits for the three selection procedures and comment on the effectiveness of the three procedures.

10.11. Following independent culling levels have been set for the two traits listed in problem 10.10

Rams: FW > 5.0 Kg; SL > 3.0 cm

Ewes: FW > 4.3 Kg; SL > 2.6 cm

List the rams and ewes which will be selected.

10.12. Economic returns in mutton-type sheep are dependent not only on body weight (BW) at marketing but also on the amount of fat (measured as ultrasonic fat depth, FD) and muscle depth (MD). For improving the net genetic worth of sheep, body weight and muscle depth need to be increased while fat depth needs to be decreased. A sheep breeder has developed the following selection index:

$$I = 0.10\ BW - 0.45\ FD + 0.30\ MD$$

Given below are the measurements recorded on 10 sheep for body weight (BW), ultrasonic fat depth (FD) and ultrasonic muscle depth (MD):

Sheep ID	Body wt. (kg)	Fat depth (mm)	Muscle depth (mm)	Sheep ID	Body wt. (kg)	Fat depth (mm)	Muscle depth (mm)
1	50	3	25	6	60	4	20
2	55	4	25	7	60	4	30
3	60	4	25	8	60	5	30
4	60	3	25	9	58	4	28
5	60	5	25	10	56	4	30

Fifty per cent of sheep need to be selected by the selection index procedure.

(a) List the animals which will be chosen by the selection index method, along with their means for three traits.

(b) Compute the selection differentials for the three traits.

Answers to Supplementary Problems

10.7.

Sire ID	SDA Index	EP Index	HLC index
1	2466.6	2333.3	2400.000
2	2100.0	2033.3	2272.324
3	2700.0	2933.3	2484.098

10.8. $b_1 = 0.13$, $b_2 = -0.5$

Selection index equation : $I = 0.13\ X_1 - 0.5X_2$

10.9. $b_1 = 0.630$, and $b_2 = 1.074$.

Selection index equation : $I = 0.630\ X_1 + 1.074\ X_2$

10.10. (*a*) *Selection on own performance for FW:*

Male selection: Two rams (ID no 18 and 5) would be selected. Mean of selected rams for FW = 5.9 kg and selection differential = 0.965 kg. Correlated selection differential in staple length : – 0.335 cm.

Female selection: Ten ewes selected on EW would be : ID no 1, 2, 5, 7, 9, 12, 15, 16, 19 and 20. Mean of selected ewes = 4.99 kg and selection differential = 0.44 Kg. Correlated selection differential in staple length = 0.03 cm.

(*b*) *Selection on own performance for SL:*

Male selection: Two rams (ID no. 7 and 13) would be selected. Mean of selected rams for SL = 3.45 cm and selection differential = 0.465 cm. Correlated selection differential in fleece weight = 0.265 kg.

Female selection: Ten ewes selected on staple length would be: ID no. 1, 3, 4, 7, 10, 16, 17, 18, 19 and 20) (It may be noted that SL of animals with ID no. 4, 8, 17 and 20 was same but ewe no. 8 had the lowest FW therefore, it was not selected). Mean of selected ewes for SL = 2.89 cm and selection differential = 0.16 cm. Correlated selection differential in fleece weight = 0.15 kg.

(*c*) *Selection on index:*

Two rams selected would be ID no. 13,18 , and ten ewes selected would be: ID no. 1, 2, 5, 7, 12, 15, 16, 18, 19 and 20. Means of selected parents and the selection differential are tabulated below:

	Rams		Ewes	
	FW (kg)	SL (cm)	FW (kg)	SL (cm)
Mean of selected:	5.85	3.10	4.97	2.81
Flock mean:	4.935	2.985	4.55	2.73
Selection differential:	0.915	0.115	0.42	0.08

Comparison of the three procedures:

Selection differentials by the three procedures for the two traits are tabulated below:

Procedure:	Selection differential in rams for :		Selection differential in ewes for :	
	FW (kg)	SL (cm)	FW (kg)	SL (cm)
Individual selection for FW	0.965	– 0.335	0.44	0.03
Individual selection for SL	0.265	0.465	0.15	0.16
Selection index of FW and SL	0.915	0.115	0.42	0.08

Comments : Selection based on index of FW and SL is more efficient than selection either on FW or on SL only as it yields high desirable selection differential for both the traits.

10.11. Two rams (ID number 13 and 14) and 7 ewes (ID number 1, 7, 16, 17, 18, 19 and 20) would be selected.

10.12. (a) The best 5 animals selected would be : ID no 8, 7, 10, 9 and 5 since their index values are higher than of others.

(b) The means for BW, FD and MD of the flock are 57.9 Kg; 4.0 mm and 26.3 mm respectively while those of the selected animals are 58.8 Kg ; 4.4mm and 28.6 mm respectively.

(c) Selection differential would be : 0.9 kg for BW; 0.4 mm for FD and 2.3 mm for MD.

●●●

SECTION - 4

MATING SYSTEMS

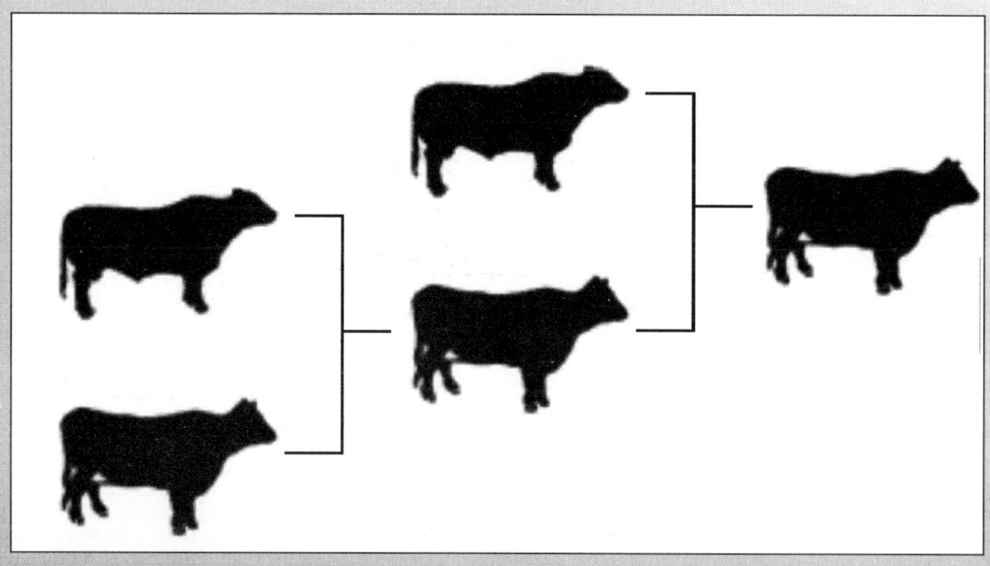

11

Inbreeding and Line Breeding

Genetic improvement of livestock involves two broad steps:

- *Choice of parents (selection)* : Choice of genetically superior animals to become parents of next generation, called selection, is the first and foremost step in genetic improvement.
- *Choice of mates (mating system)* : The second step in genetic improvement involves a system by which selected parents are mated to produce the progeny generation. The system or strategy for mating of the (selected) animals to produce the progeny generation constitutes a mating system. A mating system can be considered in individual or population context. In individual context, it means strategies for mating of specific animals within a population i.e. which sire is to be mated to which dam ; in population context it refers to strategies for mating (crossing) between different populations (e.g. which breeds/strains are to be crossed and in which combination).

A combination of selection and mating system constitutes a **breeding plan.** A comparison of selection and mating system is made in Table 11.1.

A mating system is designed to change the frequency of different genotypes such that the frequency of desired genotype (s) increases and those of undesired genotype (s) decreases. Changes in genotypic frequencies brought about by a mating systemm, however, are transient because the 'desirable' gene combinations created through it are broken down in the progeny generation. In order to capitalize 'gene combinations' (non-additive genetic effects), choice of mates has to be made each generation. In the preceding chapters, we dealt with some various bases and methods of selection of genetically superior animals. This and some subsequent chapters deal with the genetic and phenotypic consequences, and usefulness and limitations of the commonly used mating systems.

Table 11.1 Selection vs. mating system.

Selection	Mating system
1. Choice of genetically superior animals to become parents of progeny generation.	1. Choice of mates to produce progeny generation.
2. Fundamental genetic effect is a change in allele/gene frequency (i.e. increase in frequency of the desirable alleles).	2. Fundamental genetic effect is a change in genotypic frequencies (i.e. increase in frequency of desirable genotypes).
3. Utilizes additive component of genetic variance.	3. Utilizes non-additive component of genetic variance.
4. Change brought about by selection in performance of a population is relatively of permanent nature.	4. Change brought about by a mating system is of transient nature because gene combinations break down from one generation to next.

Classification of Mating Systems

An overview of classification of mating systems is given below:

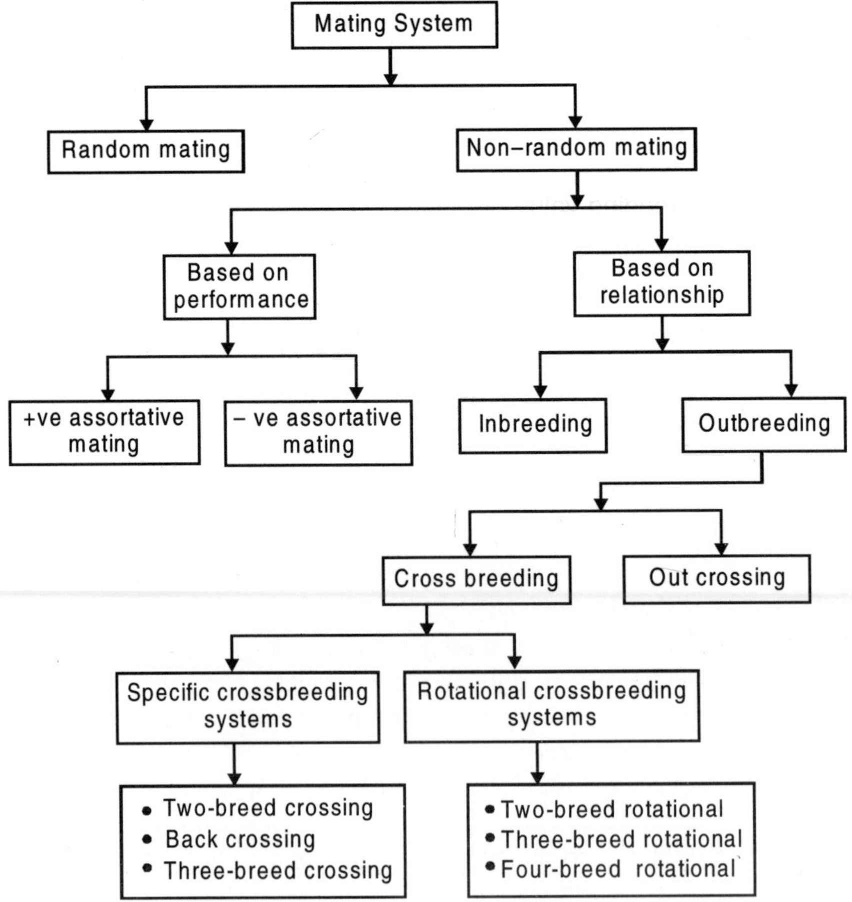

Fig. 11.1 Classification of mating systems.

Mating systems are broadly classified into two categories: random mating and non-random mating.

Random mating (Panmixia)

A system in which mates are chosen without any consideration to their genotype or phenotype is called random mating. Under such a system, an individual of one sex has an equal chance of mating with any individual of the opposite sex. Two important misconceptions about random mating are:

- Random mating does not imply 'unplanned mating' as its aim may sometimes be to keep the genetic constitution of a population stable over time (e.g. a genetic control population).
- Random mating by itself does not imply absence of selection; a population may be undergoing selection but still the selected parents may be intentionally mated at random (a usual practice in within-population selection).

Random mating nevertheless has certain advantages especially in populations undergoing selection:

- *Simple and inexpensive*: It is one of the easiest systems of mating within a population as it does not require parentage or animal identification; it also does not require knowing performance of mates. Once selection has been conducted within a population, little time is taken to make mating decisions. Its simplicity makes it the first system to be employed for mating of selected animals within a population.
- *Useful in genetic evaluation*: Random mating is a pre-requisite for accurate evaluation of breeding values of the animals. For example, breeding value of a sire can only be accurately determined if it is mated to a large number of dams taken at random from the population. But if the dams mated to a sire are not a random set of dams from the population, breeding value of the sire would be over- or under-estimated.

Random mating also serves as a benchmark for comparing the genetic consequences of other systems of mating. Ideally, random mating be carried out in pairs but in practice this is not possible in livestock because a smaller number of sires than dams are used for reproduction. Under such a situation, care should be taken to ensure that each sire is mated with approximately the same number of dams.

Non-random Mating and Its Classification

When the choice of mates is based on their phenotype or genotype, the system is referred to as non-random mating. Under such a system, the mates are related to each other more than the average relationship in the population. Non-random mating is further classified on the basis of phenotypic and genetic relationships:

Matings Based on Phenotypic Relationship: Assortative Mating[1]

Mating based on either similarity or dissimilarity is referred to as assortative mating. Assortative mating is difficult to be arranged as compared to random mating as it requires knowing performance of the mates. Also it can't be applied for sex-limited traits, and its application also becomes difficult when parents had been selected on multi-trait basis. Assortative mating can be either positive or negative:

(*a*) ***Positive assortative mating*** ('mating like to like'): Mating of phenotypically alike individuals (e.g. mating 'best' sires to 'best' dams ; 'worst' sires to 'worst' dams,and

[1]Assortative mating can be categorized as 'phenotypic' or 'genetic' assortative mating but here we restrict it to phenotypic assortative mating.

the 'average' sires to 'average' dams)[2] is termed positive assortative mating. Positive assortative mating (marriage) is common in some human societies where choice of marriage partners takes into account physical characteristics, professional attributes, religious practices and so on. For example, there is a tendency towards positive assortative mating (marriage) for skin color and height in humans. Two important effects of positive assortative mating are:

- *Increased variability*: Positive assortative mating causes more variation in the progeny than would be found among the progeny of a random mating population. Increased variability (or decreased uniformity) in the progeny can be 'desirable' as well as 'undesirable'. If progeny are to be used for commercial purposes, decreased uniformity is an undesirable attribute. On the other hand, if a population is undergoing selection, then the increased variability is useful as it increases the scope of applying high selection differential.
- *Higher chance of producing superior progeny:* Positive assortative mating increases the probability of producing some exceptionally superior offspring. Therefore, if the progeny are to be used for commercial purposes, mating of 'like' to 'like' on the average, is more likely to produce superior offspring. For example, a bull which has been progeny tested for milk production may be mated to high yielding cows so that its daughters are high yielders. Similarly, mating of a highly rated stallion for racing with highly rated mare is more likely to produce high performing offspring. In uniparous species, a single outstanding offspring, especially if it is a male, is likely to be beneficial. There is a general tendency among breeders of pet or companion animals to practice positive assertive mating for 'ideal breed-type' traits (e.g. color markings, conformation of body parts etc.) with the hope of producing greater number of animals of desired type for displaying at fairs/expositions.

(*b*) ***Negative assortative mating*** (mating between unlike): Mating between phenotypically dissimilar parents is referred to as negative assortative mating (e. g. mating of 'heavy' bulls to 'light' cows, and of 'light' bulls to 'heavy' cows). Such a system is likely to produce progeny that are intermediate to both the parents, and it is therefore also called **'corrective'** or **'complementary' mating**. This is so because the 'fault' of one parent, particularly for physical or visual traits, is likely to be corrected in its progeny if it is mated to a 'normal' mate.

Historically, one of the best examples of such a system is the mating of famous Sheepdog 'Roy' (who had an eye defect referred to as 'free eye') to 'Meg' (the bitch with 'strong eye') to produce 'Old Hemp'[3] (in 1893). This mating served as foundation of many strains of Border Collie dogs found throughout the world. Negative assortative mating decreases the variation in progeny compared to random mating.

Mating Based on Degree of Genetic/Pedigree Relationship

There are two mating systems which use degree of genetic relationship as the basis for choice of mates: inbreeding[4] and out breeding. In this chapter we will deal with inbreeding and its milder form, line breeding. Outbreeding will be dealt in a subsequent chapter.

---

[2]The terms 'best', 'worst' and 'average' refer to any visual, morphological or economic trait.

[3]Old Hemp (1893 –1901) is considered to be foundation sire of Border Collie breed. He was owned by an English man, and was used as a working dog to herd sheep as he worked more quietly than other sheepdogs of the time. He was used by other breeders and trainers and his descendants became successful international sheepdog champions.

[4]Inbreeding is positive genetic assortative mating while outbreeding is negative genetic assortative mating.

Inbreeding

Simply defined, inbreeding is the mating between relatives (i.e. mating between animals which are related to each other through common ancestry). However, all the members of a closed population are related to each other because they have had a common ancestor somewhere in the pedigree, inbreeding is, therefore, technically defined as *mating between individuals who are more closely related than the average relationship in the population.* Mating between relatives in humans is referred to as "consanguineous mating/marriage" (from Latin words meaning "*of the same blood*").

Common Types of Close Matings: Degree of inbreeding is a function of "how closely the mates are related to each other". The most common types of close matings are:

- *Selfing or* self-fertilization: Mating of an individual to itself is the most intense form of inbreeding, and occurs naturally in many species of plants.
- *Full-sib mating* (brother-sister mating): Mating between individuals having both parents common is the most intense form of inbreeding possible in farm animals.
- *Half-sib mating (half brother-sister mating)*: Mating between individuals having one common parent (either sire or dam).
- *Parent-offspring mating:* Mating of a sire to its daughters or of a dam to her sons.
- *Double first-cousin mating*: Mating between individuals having both sets of common grandparents (both maternal as well as paternal i.e. four common grandparents).
- *First-cousin mating*: First cousins are the individuals having only one set of common grandparents (i.e. two common grandparents), and the mating between such individuals is referred to as first-cousin mating.
- *Second-cousin mating* (or half-cousin mating): Mating between individuals having one common grandparent (either maternal or paternal).

Some of the common types of close matings are shown in Fig. 11.2.

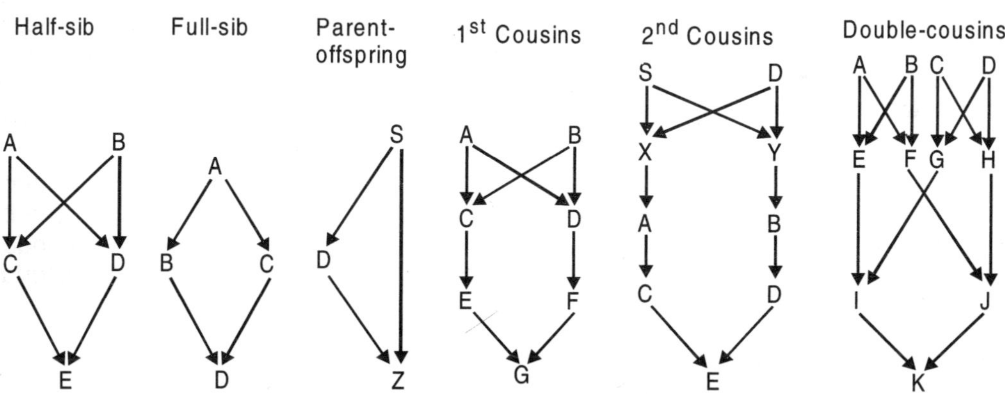

Fig. 11.2 Examples of close matings in farm animals.

Regular vs. Irregular Systems of Inbreeding: A regular system of inbreeding is the one in which mating between same types of relatives occur every generation (e.g. mating between full-sibs every generation). When matings do not occur between the same types of relatives every generation (i.e. matings in different generations occur between different types of relatives), it constitutes an irregular system of inbreeding.

Further inbreeding can be practiced intentionally for some specific purposes (e.g. inbreeding-hybridization) or it may occur unintentionally due to small size of the population.

Fundamental Genetic Consequence of Inbreeding: Increased Homozygosity

The primary genetic effect of a mating system is to bring about change in frequencies of different genotypes. Consequently, a population undergoing inbreeding departs from the Hardy-Weinberg proportions. A study of genetic effects of inbreeding is, therefore, fundamental to the general understanding of the genetic consequences of any mating system.

We examine here the fundamental genetic consequence of inbreeding by taking selfing as an inbreeding system. We begin with a population of individuals that are all heterozygous (Aa). If each individual reproduces by self-fertilization, half of its progeny will be heterozygous (Aa); one-fourth will be 'AA' and another one-fourth will be 'aa' (Fig 11.3). When the 1st generation also reproduces by self-fertilization, the homozygotes will produce only homozygote offspring, and the heterozygotes will produce offspring in proportion of 0.25 AA, 0.5 Aa and 0.25 aa.

Genetic consequences of repeated self-fertilization are shown in Fig 11.3. With repeated self-fertilization, the frequency of homozygote's increases by 50% of each generation and those of heterozygotes decreases by 50% each generation. Eventually, the population attains complete homozygosity. It can be seen from this example that inbreeding by itself does not change the allelic frequency (the frequency of each of the two alleles remains 0.5 every generation).

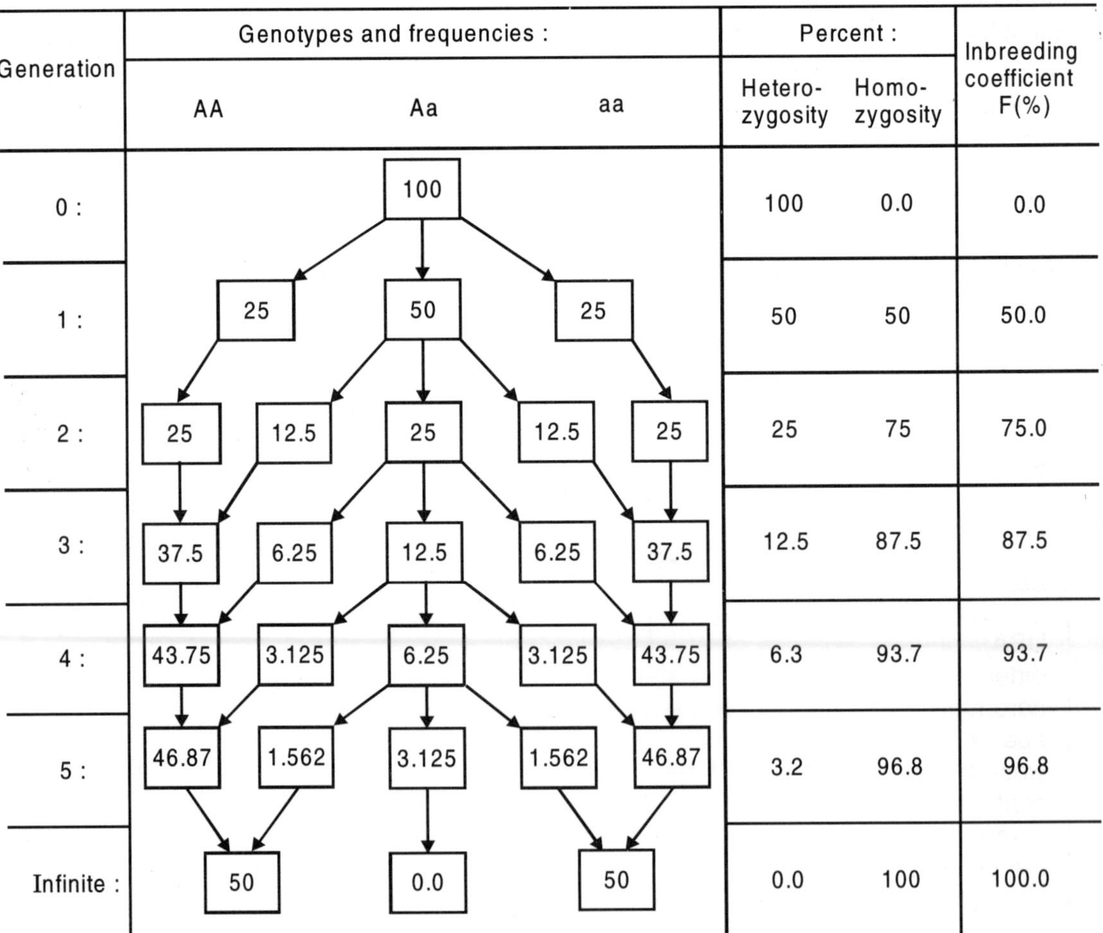

Generation	Genotypes and frequencies :					Percent :		Inbreeding coefficient F(%)
	AA		Aa		aa	Hetero-zygosity	Homo-zygosity	
0 :			100			100	0.0	0.0
1 :	25		50		25	50	50	50.0
2 :	25	12.5	25	12.5	25	25	75	75.0
3 :	37.5	6.25	12.5	6.25	37.5	12.5	87.5	87.5
4 :	43.75	3.125	6.25	3.125	43.75	6.3	93.7	93.7
5 :	46.87	1.562	3.125	1.562	46.87	3.2	96.8	96.8
Infinite :	50		0.0		50	0.0	100	100.0

Fig 11.3 Effect of inbreeding on degree of homozygosity, heterozygosity and inbreeding coefficient.

The general inferences from selfing (inbreeding) as shown above are :

- Inbreeding increases homozygosity which occurs at the cost of decrease in heterozygosity.
- Inbreeding by itself does not change allelic frequencies provided selection (either natural or artificial) is not operating against any of the genotypes.

It has, however, been observed that as inbreeding proceeds to uncover deleterious genes, natural selection tries to eliminate such genes which are mostly recessive. So some changes in gene frequencies do occur particularly when inbreeding accumulates to high levels.

The genetic consequences of inbreeding in this example have been demonstrated for a single locus but the same holds true for all other pairs of independently segregating genes. For example, If a population containing triple heterozygote's (AaBbCc) is selfed for a number of generations, the population will become homozygous for all the three loci. Since self-fertilization is not possible in farm animals, full-sib mating is the most intense form of inbreeding possible. Rate of inbreeding is therefore much slower in farm animals compared to plants. Nonetheless, the fundamental genetic effect of inbreeding is same for all the types of close matings. The magnitude of increase in homozygosity, however, differs for different types of close matings: closer the degree of relationship between mates higher the increase in homozygosity and *vice-versa*. The increased homozygosity resulting from inbreeding has some other genetic consequences which are discussed later in this chapter.

Measurement of Inbreeding: Inbreeding Coefficient

Genetic consequences of inbreeding in Mendelian terms were worked out immediately after the rediscovery of Mendelism in early 1900's. However, there was no universally accepted method for measuring the degree of inbreeding. In laboratory animals (e.g. mice and guinea pig), the amount of inbreeding was expressed in terms of the number of generations of full brother-sister mating; and in plants on the basis of number of generations of selfing. Wright (1921), an American population geneticist (Box 11.1), using Guinea pigs as a laboratory animal, developed a mathematical measure for determining the degree of inbreeding, called the **'inbreeding coefficient'** (designated as 'F' for fixation coefficient). The term **'fixation'** signifies that the individual has become homozygous (i.e. the allele has been fixed).

Box 11.1: Sewell Wright, inbreeding coefficient and Guinea Pig.

Sewell Green Wright (December 21, 1889 – March 3, 1988, USA), born in a family of scientists/academicians, was the oldest of the three gifted brothers—the others being an aeronautical engineer and a political scientist. From an early age he had a love and talent for mathematics. In school he was able to extract the cube root which was frustrating to his contemporaries. Due to this Wright had to avoid participation in many such activities in school for fear of sparking the ire of the other boys with his knowledge.

After graduating with a Master's degree in 1912, he accepted an opportunity to join Harvard University. There he worked with the pioneering mammalian geneticist Castle investigating the inheritance of coat color in mammals. His mathematical talent was appreciated by his fellow students. in particular by Harold Fish a fellow PhD student and roommate. Fish had read a recent article by a renowned biostastical geneticist (Raymond Pearl) on the effect of continued brother-sister mating on homozygosity but Fish was not convinced with Raymond's results. Therefore he started his own calculations/derivations. While doing these calculations, Fish had covered their room with pages of figures. Wright got interested in the figures/calculations, and based on these worked out a formula for calculating the genetic consequences of brother-sister mating for any number of generations. *It is thus said that Raymond Pearl was indirectly responsible for Wright's entry into the field of population genetics.*

On completing his PhD, he was employed by the Animal Husbandry Division of the U.S. Bureau of Animal Husbandry (1915 to 1925). His main project was to investigate the inbreeding that had occurred in the leading breeds of beef cattle. Concurrently he also performed experiments with thousands of guinea pigs belonging to 23 inbred lines produced by brother-sister mating. He developed a mathematical method of computing the inbreeding coefficient by analyzing the degree of relatedness in the pedigree by tracing the paths (hence called "path method"). This continues to be a standard method for measurement of inbreeding till date. Its validity has been verified by biological and molecular methods of measuring the inbreeding. Interestingly his parents were first cousins, an interesting fact in light of Wright's research on inbreeding.

After working with US Department of Agriculture for about 10 years, he returned to the work that he loved best, teaching, and joined the faculty of the University of Chicago. Upon mandatory retirement from the University of Chicago in 1955 (at the age of 65), Wright entered the University of Wisconsin, where he remained Leon J. Cole Professor of Genetics for 5 years. In addition to his genetic research, he was also interested in philosophy. On the philosophy of science, he expressed :

> *"It is the task of science, as a collective human undertaking, to describe from the external side, such statistical regularity as there is in a world in which every event has a unique aspect, and to indicate where possible the limits of such description..........".*

Wright was known to be shy and unassuming but intense in academic matters. James Crow of the University of Wisconsin's Genetics Department relates an incidence:

> *"I alongwith a fellow PhD student travelled from Texas to Chicago to ask Wright some questions on population genetic experiments. My fellow student was rather disappointed when Wright responded to each question that he was 'unable to answer' it. We left disillusioned, with my fellow student remarking that the fate of population genetics is not in safe hands. But within a few days, we were surprised to find that Wright had given thought to each question, and had written a 14 page response attending to every possible intricacy of each question".*

Dean of the University of Wisconsin's Medical School stated that Wright was "a remarkable man who lived his entire life for population genetics and related areas of research." He was actively working till his death at the age of 98 years.

An anecdote about Wright, disclaimed by Wright himself, describes a lecture during which Wright tucked an unruly guinea pig under his armpit, where he usually held a chalkboard eraser. At the conclusion of the lecture, Wright absent-mindedly, began to erase the blackboard using guinea pig.

Inbreeding coefficient is a measure of the heterozygosity that has been lost due to mating between relatives as compared to the base generation which is considered 100% heterozygous. For example, if a population were originally 100% heterozygous for a locus (i.e. all individuals were 'Aa'), selfing in such a population for one generation will result in the three genotypes, 'AA: Aa: aa', occurring in the ratio of '$\frac{1}{4}$: $\frac{1}{2}$: $\frac{1}{4}$'. The percentage heterozygosity in the progeny generation is 50% which means that there has been a loss of 50% of heterozygosity due to one generation of selfing; and hence 'F' equals 0.5 or 50%.

In animal breeding, we are more interested in calculating the inbreeding coefficient of a particular individual (e.g. sire); therefore we need a formal definition for inbreeding coefficient of an individual. Such a definition was provided by Malecot (a French Geneticist-cum-Mathematician) as: **Inbreeding coefficient** of an individual is the probability that two alleles at a locus in that individual are identical by descent. Alleles at a locus can be categorized into two types:

- *Identical by descent* – the alleles which are exact copies of a single allele (i.e. A or a) which was carried by a common ancestor. In molecular terms, they are exact copies of the same segment of DNA. The homozygosity resulting from alleles which are identical by descent is called autozygosity. The concept of identity of alleles by descent is diagrammatically shown below:

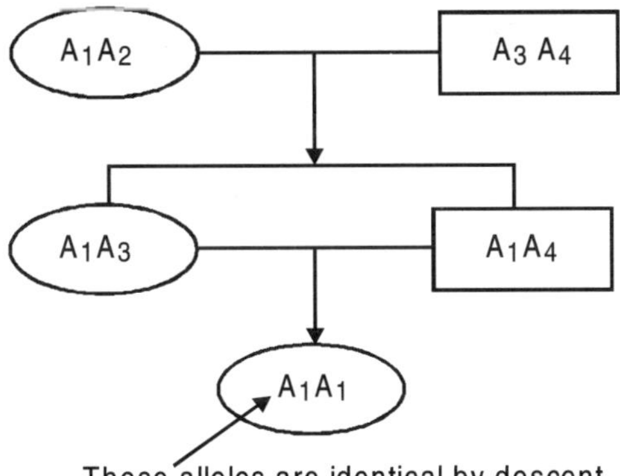

- *Identical by state or kind* – the alleles which are of the same kind (i.e. either A or a) and thus perform the same function but have not come from a common ancestor. The homozygosity resulting from alleles which are identical by state (function) is called allozygosity.

To illustrate the concept of inbreeding coefficient as the probability that two alleles at a locus in an individual are identical by descent, we consider the example of half-sib mating as shown below where individual 'X' results from the mating between half-sibs (D and E):

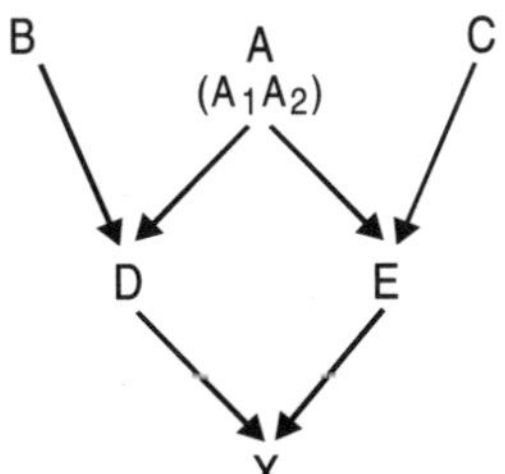

Since the parents of 'X' (D and E) are only related through their common parent 'A' (genotype 'A_1A_2'), therefore we only consider the transmission of alleles from A to X through D and E. Inbreeding coefficient of X (designated as F_x) is determined by computing the probability that individual 'X's genotype is either 'A_1A_1' or 'A_2A_2' (i.e. it two alleles are identical by descent). The probability that 'A_1' is transmitted from grandparent 'A' to parent 'D' is ½, and that it is then transmitted from 'D' to 'X' is further ½. Similarly the probability that 'A_1' is transmitted from grandparent 'A' to parent 'E' is ½, and that it is then transmitted from 'E' to 'X' is also ½.Thus the probability that 'X' has 'A_1A_1' genotype is the product of the probabilities of the four paths:

$$P\ (X = A_1A_1) = \left(\frac{1}{2}\right)^4 = \frac{1}{16}$$

Similarly the probability that 'X''s genotype is 'A_2A_2' : $= \left(\frac{1}{2}\right)^4 = \frac{1}{16}$

Therefore, $F_x = P\ (X = A_1A_1 \text{ or } A_2A_2) = \frac{1}{16} + \frac{1}{16} = \frac{1}{8} = 0.125$

Range of inbreeding coefficient: Inbreeding coefficient varies between zero and one (or 0.0 and 100%). In a completely random-mating population in which none of the animals is presumed to be related to each other through ancestry, 'F' equals zero. Inbreeding coefficient increases toward unity as proportion of heterozygosity is lost due to inbreeding. An inbreeding coefficient of unity means that the individual has become homozygous for all the loci.

Box 11.2: Gustave Malecot and definition of Inbreeding coefficient.

Malecot (1911-1998), a French mathematician and population geneticists, made most significant contributions to mathematical population genetics. Recognition of his work and achievements was however rather slow because of the language in which his work was published. It has been stated that if a fourth member to the "fathers "of population genetics is to be accepted (the others being Fisher, Wright and Haldane), it would surely be Malecot.

His professional career in population genetics began in 1939 on "the correlation between relatives". This work was expanded further, which led to one of most basic and fruitful concepts in population genetics: gene identity by descent (1941). This fundamental idea lies at the heart of the powerful genealogical approach to population genetics.

Introduction of inbreeding coefficient by Wright was extremely important and highly creative contribution to population genetics (defined as 'correlation between uniting gametes') but it involved too many numerical values thereby concealing the basic fact that inbreeding coefficient is a function of only the ancestry, and it is therefore the same for all the (autosomal) loci. Malecot's definition of inbreeding provides a more concise, lucid and rigorous theory of inbreeding, and hence it is a preferred approach for modern research on inbreeding.

Malecot, like many other population geneticists, had been active till the time of his death. Malecot's intellectual independence is reflected in the fact that he was the sole author of every one of his papers in population genetics.

Computation of Inbreeding Coefficient of an Individual: There are two broad methods for calculating the inbreeding coefficient of an individual: the 'path method' and the 'covariance table' method. Here, we demonstrate the computation of inbreeding coefficient by the path method (devised by Wright). The path method, which is based on counting the number of paths taken by the identical genes as they flow from ancestors to descendants, is illustrated below for the pedigree of the offspring 'X' resulting from mating between two half-sibs (S and D) :

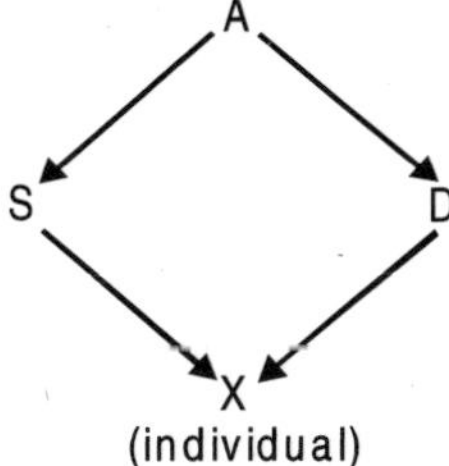

Inbreeding coefficient, Fx, is calculated by the following formula:

$$F_X = \frac{1}{2}\,\Sigma\left[\left(\frac{1}{2}\right)^n (1 + F_a)\right]$$

Where: Fx = inbreeding coefficient of individual 'X'

Σ = symbol to indicate sum of all the paths connecting parents of 'X' through common ancestor (A)

n = possible number of paths connecting parents of 'X' through common ancestor

Fa = inbreeding coefficient of common ancestor 'A'

The '½ 'which appears in the formula accounts for the fact that there is 50% probability of a copy of any particular allele being passed from a parent to an offspring (we have already demonstrated the probabilistic approach for calculating the inbreeding coefficient of an individual). The above formula of calculating F_X only simplifies the calculation but its logic is same as described earlier.

If the common ancestor, 'A', is not inbred or if its pedigree is not known, then its inbreeding coefficient is set or assumed to be zero (Fa = 0), then 'F_X' simplifies to:

$$F_X = \frac{1}{2}\Sigma\left(\frac{1}{2}\right)^n$$

In the above example, there is only one common ancestor ('A'), and the number of paths connecting 'S' and 'D' through 'A' are two (S ← A → D) and Fa = 0 ; F_X is calculated as follow:

$$F_X = \frac{1}{2}\left(\frac{1}{2}\right)^2 (1 + 0) = 0.125 \text{ or } 12.5\%$$

F_X of 0.125 means that the probability that two alleles at a locus in individual 'X' are identical by descent is 0.125. Alternatively, it means that individual 'X' is expected to have 12.5% more homozygous gene-pairs than the non-inbred individual (of the same population)

Relationship coefficient: Another concept closely related to inbreeding coefficient is the coefficient of relationship (R). **Relation coefficient** between two individuals is the probability that they both carry a particular allele due to common ancestry. Relationship coefficient between two non-inbred individuals, say 'S' and 'D', designated as R_{SD}, can also be calculated through the path method by the following formula:

$$R_{SD} = \Sigma\left[\left(\frac{1}{2}\right)^n (1 + Fa)\right]$$

When the common ancestor is not inbred, R_{SD}, becomes:

$$R_{SD} = \Sigma\left[\left(\frac{1}{2}\right)^n\right]$$

An inspection of the formulas of inbreeding and relationship coefficients indicates that inbreeding coefficient of an individual is one-half of the relationship coefficient between its parents (when the parents are not inbred):

$$F_X = \frac{1}{2} R_{SD}$$

Relationship coefficients between some common types of relatives along with the inbreeding coefficient of offspring resulting from their mating are given below:

Relatives	R	F
Unrelated :	0.0	0.0
Brother-sister :	0.5	0.25
Parent-offspring :	0.5	0.25
Double 1st cousins :	0.25	0.125
Half sibs :	0.25	0.125
Uncle-niece :	0.25	0.125
1st cousins :	0.125	0.0625
2nd cousins :	0.031	0.016

Accumulation of inbreeding under regular systems: Inbreeding continued over generations eventually leads to complete homozygosity when 'F' attains a value of one. Effect of continued inbreeding on the level of inbreeding coefficient for selfing, full-sib and half-sib mating is shown in Fig 11.4

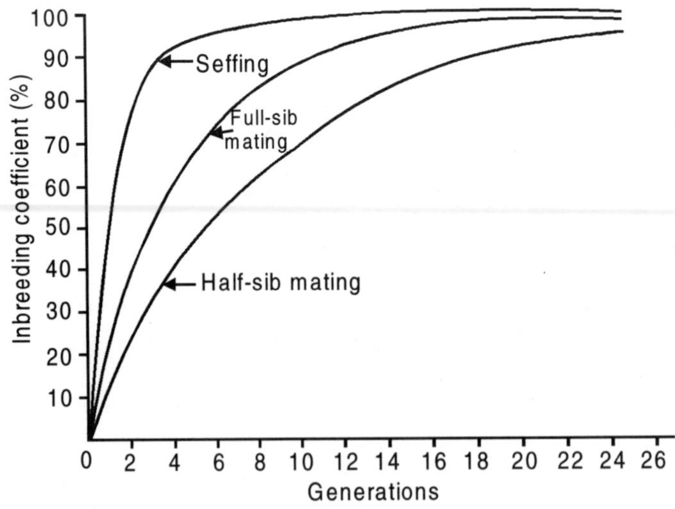

Fig. 11.4 Increase in inbreeding under some regular systems.

Selfing, which is the most intense form of inbreeding occurs naturally in some species of plants. Ten generations of selfing leads to complete homozygosity (F = 1.0). Selfing being not possible in farm animals, full-sib mating is the most intense form of inbreeding possible in farm animal species. It is apparent that it takes about 20 generations of full- sib mating to attain inbreeding coefficient of unity. In laboratory species like mice where inbred lines are produced for experimental purposes, it thus takes double the time than under selfing to designate a line as inbred.

Consequences of Increased Homozygosity (caused by inbreeding)

The fundamental genetic effect inbreeding is to increase homozygosity, and in turn, to decrease heterozygosity. The increased homozygosity resulting from inbreeding has a number of other genetic consequences:

- **Uncovering of recessive genes leading to appearance of hereditary defects**: A number of recessive genes, which in homozygous state are deleterious to the organism, are known to occur in different species of farm animals (Table 11.2). Such genes in homozygous state produce a number of developmental defects which lead to sub-vitality, semi-lethality or lethality. In a non-inbred population such genes remain hidden in heterozygous individuals. Inbreeding by increasing the frequency of homozygotes brings the deleterious genes to light. Since most of the deleterious recessive genes originate from mutations, they occur with very low frequencies. But inbreeding increases the frequency of homozygote recessives many-fold. The change in frequency of homozygote's due to inbreeding is a function of inbreeding coefficient, F, as shown below:

Population	Genotypic frequency :		
	f(AA)	*f(Aa)*	*f(aa)*
Non-inbred :	p^2	2pq	q^2
Inbred:	$(p^2 + pqF)$	$(2pq - 2pqF)$	$(q^2 + pqF)$

Illustrative example: Using the expression given above, we consider a numerical example where initial frequencies of the three genotypes are in the proportion of 0.25, 0.5 and 0.25 in a non inbred population. The changes in genotypic frequencies due to increasing level of inbreeding would be as follow:

Population	Genotypic frequency :		
	f (AA)	*f (Aa)*	*f (aa)*
Inbred:	$p^2 + pqF$	$2pq - 2pqF$	$q^2 + pqF$
F = 0.0	0.25	0.50	0.25
F = 0.5	0.375	0.25	0.375
F = 1.0	0.50	0.0	0.50

It is apparent that inbreeding decreases the frequency of heterozygotes as compared with their frequency in a non inbred population. And, as 'F' approaches unity, the population becomes completely homozygous. Most of the deleterious genes are recessive, and occur with very low frequencies. For example, if the frequency of a recessive allele in a base population were 0.02, the frequency of homozygote recessives would be 0.04 (i.e. 0.4 in 1000). The effect of increase in the degree of inbreeding on the frequency of homozygotes would be as follow:

Inbreeding coefficient (F) :	0.0	0.125	0.25	0.5
f (aa)= (q^2 + pqF) :	0.04% (0.4 in 1000)	0.29% (2.9 in 1000)	0.53% (5.3 in 1000)	1.02% (10.2 in 1000)

Its apparent, from the examples cited above, that as inbreeding coefficient increases, the frequency of homozygote recessives also increases. Such homozygous individuals manifest sub-vitality or lethality and thus constitute a **genetic load** on the population.

- ***Increased genetic resemblance and prepotency:*** **Prepotency** refers to the ability of an individual (parent) to pass on its characters to its offspring. Since inbreeding increases homozygosity, inbred animals produce only limited types of gametes. Therefore, if an individual has become homozygous for a certain number of dominant genes, its progeny is likely to resemble it irrespective of whether that individual is mated to individuals carrying dominant or recessive alleles. Similarly, all the offspring of that parent, because of homozygosity, tend to resemble each other more. This is another way of saying that inbred animals, particularly for the dominant gene/allele, are likely to be prepotent and will consequently be more uniform in performance.

  However the increase in homozygosity occurs independent of whether the genes are dominant or recessive: some individuals will become homozygous for the dominant allele, others for the recessive allele. Similarly, some will become homozygous for dominant alleles at two loci (AABB); others will become homozygous for recessive alleles at both the loci (aabb). Still others may become homozygous for several dominant genes; these individuals will be prepotent for a number of traits.

- ***Decreased genetic variability:*** Inbreeding by decreasing the proportion of heterozygotes reduces genetic variability within a population. Since inbred individuals have fewer heterozygous loci than non-inbreds, they will produce fewer types of zygotes leading to decreased genetic variability. Additive genetic variance in a line having undergone inbreeding would be:

$$\sigma^2_{A\ (\text{inbred})} = \sigma^2_{A\ (\text{Non-inbred})} - F\sigma^2_{A\ (\text{Non-inbred})}$$

As inbreeding coefficient ('F') approaches unity, the genetic variation within the line approaches zero. This means that all the individuals in a completely inbred line would have identical genotypes. Due to the loss of genotypic variability, highly inbred lines do not respond to selection as much as do the non-inbred lines.

Effect of Inbreeding on Environmental Variance: Decreased Homeostasis

Inbreeding within a population reduces genetic variability such that animals tend to resemble each other more genotypically than they do phenotypically. Inbred animals, however, become more sensitive to changes in environmental conditions which increases the environmental variance within the population. The term **'homeostasis'** is used to describe the ability of an organism to maintain its physiological processes within normal limits in spite of fluctuations in environmental conditions. The homozygosity caused by inbreeding reduces homeostatic ability. Inbred animals or lines thus become less 'buffered' against adverse environmental conditions, and need good 'environment' for their maintenance or survival.

Table 11.2 List of a some lethal/ semi lethal genes in livestock.

Autosomal incompletely/completely dominant:

- *Bull dog calves*, characterized by short legs, shortened head, protruding tongue

Autosomal recessive:

Pigs

- *Legless condition in Poland china piglets*

Cattle

- *Paralysis in Red Danish cattle*
- *Hydrocephalus condition in calves*; characterized by accumulation of fluid in ventricles of brain; affected calves are usually born alive at full term and are unable to stand, and therefore die within days.
- *Amputated calves*; born after a normal pregnancy; usually die at birth or within few days after birth.
- *Congenital dropsy in Swedish Friesian cattle* is characterized by large accumulation of fluid in body cavities and subcutaneous tissues so that such calves are extremely swollen; fetuses are aborted.

Horse

- Hyperkalemic periodic paralysis (HYPP): a muscular disorder in which uncontrolled muscle twitching occurs during the attack, causing intense weakness; defect disrupts sodium ion channel's ability to open or close; semi lethal, defect occurs in Quarter horses, Combined Immunodeficiency Disease (CID): a devastating condition seen in Arabian horses in which foals are born without a functioning immune system; at first appear normal but once the colostrum immunity wears off, they succumb to the first viral or bacterial infection.
- *Flexed forelegs in horse, abnormality of musculoskeletal system of newborn foals, semi lethal.*
- *Ataxia in horse*, neurological disorder, a loss of the sense of where their feet are placed and may appear "drunk" as the horse lacks perception of where its limbs are
- *Overo lethal white syndrome in Paint horse; lethal*

Chicken

- *Congenital loco,* a type of nervous disorders, occurs in newly hatched chicks and *is* characterized by the chicks being unable to stand immediately after hatching, the head is drawn back and the chicks fall over; when they get up the same procedure is repeated.
- *Shaker condition* ; semi-lethal
- *Wingless condition*;

Sex-linked dominant :

- *Streaked-hairlessness in Holstein-Friesian cattle*; hair completely lacking on vertical streaks over the hip joints, and sometimes extending up the sides.

Sex-linked recessive :

- *Hemophilia in dogs*
- *Bird tongue in dog*
- *Nakedness in poultry*, semi-lethal

Phenotypic Effects of Inbreeding: Inbreeding Depression

Undesirable phenotypic effects of mating between close relatives had been known since the beginning of civilization. In many human societies, there has been a general taboo against marriages between close relatives (called '*consanguineous marriage*'). The exact reasons which may have led to this social pattern are not known. It is thought that one of the reasons may have been the appearance of inherited defects in children born from consanguineous marriages. Breeders of domestic animals had also recognized the dangers of inbreeding on performance and health of progeny but these effects were not documented until late in 19th century. Darwin (1868) in his book on "variation in plants and animals under domestication" reported that: "close interbreeding carried for several generations reduces size, constitutional vigor and fertility, and increases the incidence of malformations." He further stated that "Nature thus tells us, in the most emphatic manner, that she abhors perpetual self-fertilization"

Inbreeding in animals : A quote on fanciful mixture of science and myth

Inbreeding (incest) has been avoided in many human societies, and there has been speculation about the same in animals. Aristotle quoted a story :

"A story goes that the king of Scythia had a highly-bred mare, and that all her foals were splendid; that wishing to mate the best of the young males with the mother, he had him brought to the stall for the purpose; that the young horse declined; that, after the mother's head had been concealed in a wrapper he, in ignorance, had inter-course; and that, when immediately afterwards the wrapper was removed and the head of the mare was rendered visible, the young horse ran away and hurled himself down a precipice"

At the beginning of 20th century, a large number of experimental studies on inbreeding in various species of livestock, notably in swine and poultry, were initiated. The results of these studies indicated that continued inbreeding led to a general decline in performance. The reduction in performance for different traits (e.g. growth, production, and most notably reproduction) due to inbreeding is referred to as **inbreeding depression**.

Box 11.3: Inbreeding in humans : Reasons, incidence and effects.

In modern human societies, marriages between close relatives **(consanguineous** marriages) are fairly rare; its incidence depends on ethnic and cultural traditions, and on geography.

- In Chinese societies, marriages between first cousins who bear the same family name (i.e. children of full brothers) are forbidden because they are considered as 'brothers and sisters'. But at the same time, children of full sisters or of a brother and sister can marry because they bear different family names. This despite the fact that both types of cousins bear same degree of genetic relationship (R= 0.125). Even marriage between paternal half-cousins is not allowed because they bear the same surname.
- In England, marriage between double first cousins is perfectly legal but that of uncle and niece is not only illegal but criminal as well. And, the marriage between half-brother and half-sister is criminal. Genetically, the three types of relationships are equally close, each having a relationship coefficient of 0.25.
- A study showed that only 0.08 % of Roman Catholic marriages recorded in 1959-60 in USA were between first cousins and only 0.11 % between second cousins. By comparison,

in Fouta-Djallon, Guinea (in 1955), 19.08% marriages were between first cousins and 6.22% between second cousins.

- In the past inbreeding has been seen to occur frequently in many royal families. Rather in some Royalties consanguineous marriages were not only accepted but even encouraged (e.g. Ancient Egyptian, Incan, Hawaiian, and many European royalties). Several factors may explain such a practice : in many cases it was followed to preserve royal blood (inheritance) lineage : in other cases it was intended to keep land, material possessions and resources within the family. Moreover, brother-sister royal marriages also allowed succession of the throne to both male and female sides of pedigree.
- There are also cases in which inbreeding has been a part of a culture which linked it to legends or myths. One such documented case was seen in the Incan culture of the 16th century. The Incan king married his full-sister. This was done to emulate the king's mythical ancestor, the Sun, who married his sister, the Moon. And this was thought to preserve the purity of the divine royal blood line.
- There are some groups that are associated with inbreeding because of religion, culture and geographic isolation. These are the small Anabaptist populations in North America which include the Amish, the Mennonites, and the Hutterites. These groups, which settled in North America in the 18th and 19th centuries, have registered increase in consanguineous marriages over time. By 1950, the level of consanguineous marriages in these ethnic groups reached as high as 85%. The reason for the high levels of inbreeding was not only religion but isolation of the group from mainstream. These factors of religion and small communal societies collectively led to limited choices when searching for possible marriage parterns.
- The earliest recorded study of inbreeding and its effects on human health was reported by Bemiss in 1858. Following this, Charles Darwin, through numerous, thorough and careful observations/experiments, was one of the first scientists to demonstrate the undesirable effects of inbreeding. *His interest in inbreeding is said to be due to the fact that he had married his first cousin, Emma Wedgwood.* They had ten children, but three died before the age of ten. Of the surviving children, three of the six who had long-term marriages did not have any children.

His book contains a large number of examples with the overall general conclusion that inbreeding is generally deleterious and cross-fertilization (crossing) is beneficial.

Factors Determining Magnitude of Inbreeding Depression

Genotypic mean of a random mating (non-inbred) population, M_{RM}, is given by the following expression (single locus):

$$M_{RM} = a\ (p - q) + 2pqd$$

Under inbreeding, the population mean, M_F, becomes:

$$M_F = a\ (p - q) + 2pqd - 2pqFd$$

Inbreeding depression, given by the difference between 'M_F' and 'M_{RM}', becomes:

$$-\ 2pqFd$$

The magnitude of inbreeding depression is thus dependent on:

- Allelic frequencies (p, q): inbreeding depression is highest when the allelic frequencies are at intermediate values (p = q = 0.5).
- Inbreeding coefficient (F): higher the intensity of inbreeding, greater will be the inbreeding depression.

- Level of dominance (d) :
  - (i) When 'd' = 0 (gene action is additive), there will be no inbreeding depression
  - (ii) When 'd' is more than zero (i.e. when genes show some degree of dominance), there will be some inbreeding depression:
    - When 'd' < 'a' (but more than zero)- partial dominance
    - When 'd' = 'a' (but more than zero)- complete dominance
    - When 'd' > 'a' - over dominance

Illustrative example: Consider a hypothetical example, given below, to demonstrate the changes in mean performance over successive generations of inbreeding under different types of dominance (gene action) using selfing as a system of inbreeding :

The following performance values (in hypothetical units) are assumed for different genotypes:

Additive gene action : aa = 160 ; AA = 200 ; Aa = 180

Complete dominance : aa = 140 ; AA = 180 ; Aa = 180

Partial dominance (of A over a) : aa = 140 ; AA = 200 ; Aa = 180

Over dominance : aa = 140 ; AA = 160 ; Aa = 180

The values for different genotypes have been assigned such that the mean of the population (group) at beginning is 180 units (with frequencies of AA, Aa and aa genotypes as 0.25, 0.50 and 0.25 respectively).

Generation	Genotypic frequency:			Types of gene action:			
	AA	Aa	Aa	Additive	Type of dominance:		
					Partial	Complete	Over
0	0	100	0	180	180	180	180
1	0.25	0.50	0.25	180	175	170	165
2	0.375	0.25	0.375	180	173	165	158
3	0.437	0.125	0.437	180	171.3	162.5	154
4	0.468	0.063	0.468	180	170.6	161.35	152

The values of performance as given under different types of dominance over four generations of inbreeding are diagrammatically shown below:

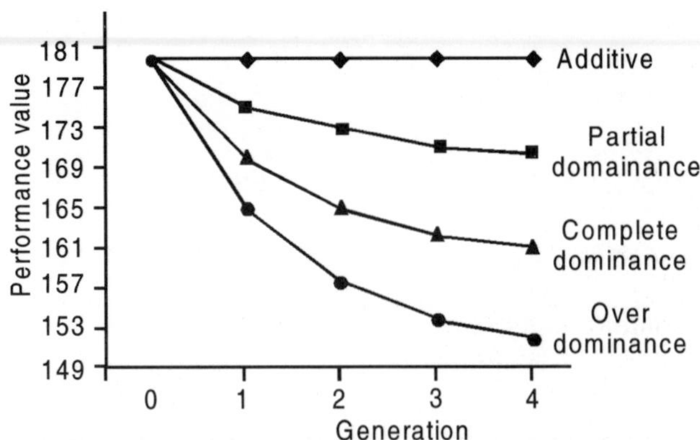

It is apparent from the expression for inbreeding depression (-2pqFd) as well as from the Fig above that:

- the traits which are primarily controlled by additive effect of genes do not show inbreeding depression.
- the traits which are governed by some degree of dominance (non-additive gene action) show inbreeding depression in order of over dominance (highest decline), complete dominance (intermediate decline) and partial dominance (lowest decline).

Effects of inbreeding have been experimentally studied on several traits across many farm animal species. A summary of these, based on several studies, is given in Tables 11.3 to 11.6.

Table 11.3 Effect of inbreeding in Dairy cattle (US & Canadian).

Change in yield of milk & milk constituents (Kg. per 1% increase in F):

- Decrease in 305- day milk yield (20 to 30 Kg.)
- Decrease in 305- day fat & protein yield (0.8 to 1.2 Kg.)

Change in reproduction and viability traits :

- Increase in calving interval, service period, services per conception
- Decrease in days open, conception rate, livability, productive life

% decline in mean performance per 1% increase in F :

- Milk yield traits decrease by 0.5%
- Reproductive viability traits decrease by more than 1.0%.
- Somatic cell score increases by 0.5%.

Table 11.4 Effect of inbreeding in swine, sheep and poultry.

Swine:

Change per 10% increase in F :

- No. of full formed pigs: – 0.4 pigs
- No. of pigs born alive: – 0.3 pigs
- No. of pigs weaned: – 0.2 pigs
- Survival rate to 21 days: – 2.5%
- Litter weaning weight: – 1.0 Kg.

Change as % of mean (per 1% increase in F):

- No. pigs born, weaned litter size, weight gain: – 0.5 to – .0%
- Back fat thickness: – 0.2%

Sheep:

- Decrease in greasy & fleece weight., staple length, weaning weight., post-weaning gain & lamb survival
- Decrease in gross life-time income per ewe

Poultry:

- Increased incidence of embryonic malformations
- Decrease in fertility, hatchability, viability and egg production; little or no effect on sexual maturity, body weight and egg weight

Table 11.5 Effect of Inbreeding in horse and dog.

Horse:	
Spanish Andalusian:	• Little or no effect on morphological traits viz., wither and Chest height; leg and body length; chest width; heart girth circumference
Arab mares:	• No effect on body conformation traits (wither height, chest circumference and cannon bone circumference)
Friesian stallions:	• Small negative effect semen traits viz.; ejaculate volume, sperm cell concentration, motile sperms and morphologically normal sperm
Friesian mares:	• Increased incidence of retained placenta (0.12% per 1% increase in F)
Norwegian trotter :	• Foaling rate decreased by 0.4% per 1% increase
	• Early abortions increased by 1.27% per 1% increase.
Dog:	
Sheep dog	• Increased incidence of hip dysplasia
Pointer dogs	• No effect on human aversion traits

Table 11.6 Effect of Inbreeding in Fish.

Zebra	
(Brachydanio rerio)	• Inbreeding coefficients of 12.5 and 25% reduced fertility, survival to 30 days, and length at 30 days and increased the incidence of crippled fry.
	• No effect on number of eggs per swan or hatchability
Guppy	
(Poecilia reticulata)	• Decreased survival (up to 120 day) and decreased salinity tolerance (measured as survival time in 35-ppt sea water) by 9.1% per 10% increase in F.
	• Little or no effect on body length (birth & 120 days)
	• Reduced male sexual activity
Rainbow Trout	One generation of full sib mating: (F=0.25)
(Salmo gairdneri)	• Decreased egg hatchability (1.7 to 4.4%)
	• Reduced% fry survival (16.1 to 22.2%)
	• Reduced weight at 150 days (9.8 to 12.0%)
Rainbow Trout	• Reduced body weight (0.2% per 1% increase in F)
(Onchorynchus mykiss)	

Based on a large number of experimental investigations conducted in different farm and laboratory animals, the following general conclusions have been drawn:

Nature of trait	Gene action	Inbreeding depression	h^2
Fitness traits :			
Fertility, survival, disease resistance, etc.	Mainly non-additive	High	Low
Production and growth traits:			
Milk, egg & wool production; growth rate; feed efficiency etc.	Both additive & non-additive	Moderate	Moderate
Morphological & quality traits:			
Body measurements & conformation; carcass traits; milk quality & constituents egg quality and constituents	Mainly additive	Low	High

It is evident that there is an inverse relation between the nature of trait vis-a-vis their h^2 and inbreeding depression: inbreeding depression is most pronounced for traits associated with reproduction, moderate for growth and production but relatively little for quality and morphological traits. The magnitude inbreeding depression is thus exactly the reverse of magnitude of heritability for these three general classes of traits. This is logical because heritability is a measure of that part of phenotypic variation which is due to additive effects while inbreeding depression is due to dominance and other types of non-additive genetic effects.

Since most economic traits exhibit low to moderate heritabilities, some degree of inbreeding depression invariably occurs for these traits. But the magnitude of inbreeding depression, varies depending on the previous selection history of the stock (as reflected in gene frequencies).

The highest sensitively of reproductive traits to inbreeding depression has also been well authenticated by experimental studies in rat, mice, guinea-pig, poultry and swine. Inbred animals in these studies displayed delayed puberty, lowered ovulation rate, increased embryo mortality, lowered libido and lowered vigor. Inbred lines (i.e. the lines developed through continued inbreeding over successive generations) also showed higher incidence of genetic abnormalities. Loss in reproductive performance of inbred lines due to continued inbreeding over generation in many cases was so severe that some of the inbred lines were wiped out (see Fig. 13.6 chapter 13). In fact a decrease in fertility, survivable, and an increase in embryonic/fetal mortality in a flock over successive generations could be an indication of accumulation of inbreeding. And, for these reasons, inbreeding is looked at unfavorably by breeders of most species of farm and companion animals.

Inbreeding in Closed Populations of Finite Size

A closed population is the one into which outside (i.e. unrelated) breeding animals are not introduced. Most closed populations of farm animals are of finite size. Such closed population of restricted size invariably undergoes some degree of inbreeding even though mating between close relatives is avoided.

Rate of increase in inbreeding coefficient per generation (ΔF) in a random mating closed population is a function of the number of sires (S) and dams (D) used for reproducing the population:

$$\Delta F = \frac{1}{8S} + \frac{1}{8D}$$

B-

Thus larger the number of parent (S and D) used for reproduction, lower the chance of mating between relatives, and in turn, lower the increase in inbreeding. The number of sires used in livestock breeding programs is much smaller than the number of dams; therefore, the relative contribution of dams to increase in inbreeding is very small. In other words, it is numbers of sire (the rarer sex) that influences the rate of inbreeding the most (Table 11.7). For example, increase in number of dams from 2 to 200 but keeping the same number of sires (two) does not make as large reduction in inbreeding as does the increase in number of sires but keeping the number of dams constant.

Table 11.7 Effect of number of sires and dams on inbreeding.

No. of sires (S)	No. of dams (D)	Contribution to ΔF (%) by:		Total ΔF (%) /generation
		Sires	Dams	
2	2	6.25	6.25	12.5
2	4	6.25	3.13	9.4
2	200	6.25	0.063	6.3
5	200	2.5	0.063	2.56
20	200	0.625	0.063	0.69
20	400	0.625	0.03	0.66

In some species like rat, guinea pig, poultry, pair-matings (i.e. one sire mated to one dam) can be made. The rate of inbreeding with different number of paired matings over successive generations is shown in Fig. 11.5.

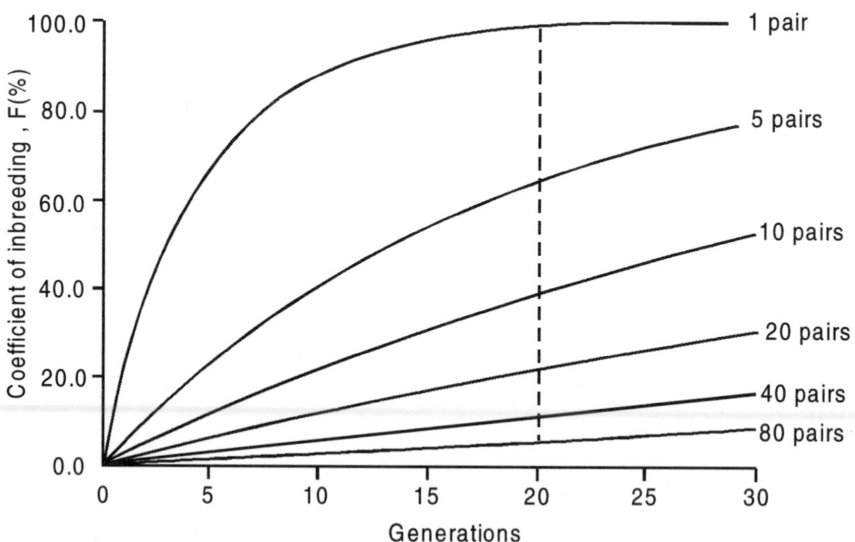

Fig. 11.5 Expected level of inbreeding as a result of finite population size (with pair mating)

Two observations are worth noting:

- When the number of parents used is small, inbreeding accumulates over generations even though mating between relatives is not intentionally practiced.
- Increase in 'F' over generations is inversely proportional to number of pairs of parents: more the number of pairs, lower the increase in inbreeding and *vice-versa*.

Level of Inbreeding in Livestock Breeds

A breed is a recognized group of interbreeding animals within a species which is isolated from other such groups. Animals of a breed are genetically related to each other more than they do to members of another breed. Some degree of inbreeding has inevitably occurred in every livestock breed for three reasons:

(a) Small number of founding animals and inheritance lineage: Members of a breed owe their lineage to a common genetic base because each breed has originated from a limited number of founder animals. If pedigree of animals of a breed at any time could be traced back far enough, it will be seen that a large number of animals have a number of common ancestors though the degree of relationship with the founders may be very small. This is despite the fact that deliberate matings between close relatives are normally avoided.

(b) Intentional inbreeding: Inbreeding in farm animals is sometimes classified as 'active' where animals are mated according to family relatedness, and 'passive' which results from use of a limited number of sires and dams. In some species of livestock, some outstanding ancestors are more extensively used within a breed through matings between 2nd and 3rd degree relatives (a milder form of inbreeding, called 'line breeding'). Such a system is quite common in breeding of horse and dog where the sires, who have won many "shows " or "races" and earned many titles, are used more extensively than their contemporaries. Over generations more and more animals of a breed/flock become related to each other, and the chance of mating between related animals increases. Over long period, it leads to accumulation of high levels of inbreeding. A few illustrative examples are :

- A renowned Holstein Friesian bull named *'Marshal'*, has contributed offspring virtually in every country, and the breeders in turn have attempted to concentrate his inheritance in their herds.
- A famous cat named *"Fan Tee Cee*" has been much sought after, and has appeared in Siamese pedigree many times (between 1960's and 1970's) ; sometimes several times in a single pedigree as the breeders were anxious to make their cat lines more 'typey'.

(c) Intensive selection: Most breeds of livestock have a long history of selection. Selection, particularly when based on accurate estimates of breeding value (e.g. using several types of relatives), picks up the animals which belong to a few families. This is because related animals have genes in common and hence their breeding values are more likely to be alike than those of unrelated animals. Higher the intensity of selection, more is the chance of selection of animals belonging to same families and hence more the chance of mating between relatives.

Some breeding and reproductive technologies have, and/or are further hastening the rate of inbreeding:

- Long-term and extensive use of a few outstanding sires through artificial insemination. For example the number of different inheritance lines (bloodlines) used by A.I. companies in Holstein breed of cattle has been greatly reduced. Some of the Holstein bulls used in AI have had 2, 50,000 milking daughters or 5,000 progeny-tested sons. Additionally 50% of the roughly 5,000 young Holstein bulls that were progeny tested each year globally were offspring of the 10 most popular sires.

- Embryo transfer technologies (e.g. multiple ovulation embryo transfer, MOET,) hold the potential of increasing the rate of inbreeding through the female path (as AI has done in males).
- Modern genetic evaluation methods (such as 'Best Linear Unbiased Predictor, BLUP', animal model) and the use of 'genetic markers' increase the possibility of selection of animals which belong to a few outstanding families. A faster rate of increase in inbreeding is anticipated if selection for low heritablity traits is pursued by giving more weightage to family performance (family and combined selection).

As a result of these contributing factors, considerable degree of inbreeding has accumulated in some breeds of livestock over a long period. For example, the history of Hereford breed (in USA) indicates that the breed was established by importing few animals from England in mid-19th century. By 1930, inbreeding in some of its strains accumulated to 8.1%, and by 1947 one strain attained inbreeding coefficient of 31%. Similarly a study conducted in major U.S. swine breeds showed that inbreeding has been consistently accumulating in all the breeds though at different rates (Fig 11.6).

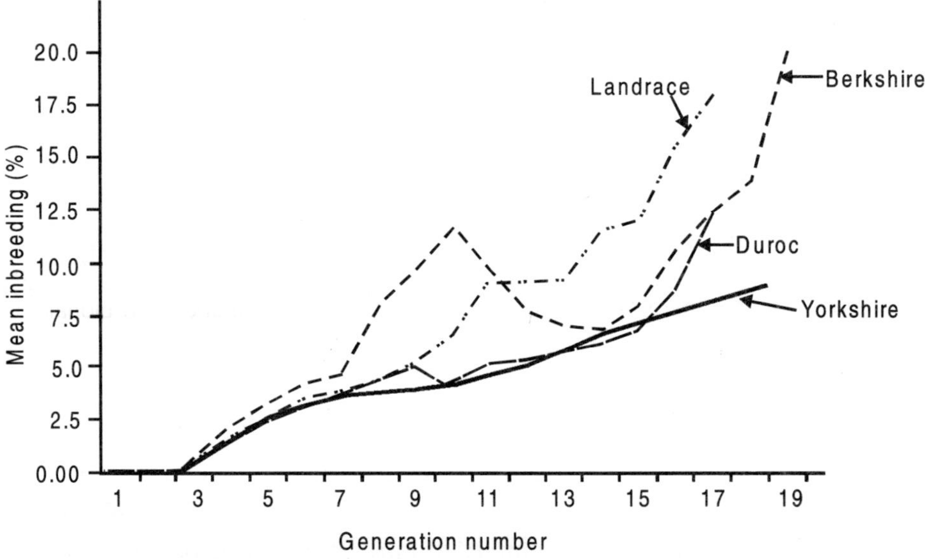

Fig 11.6 Inbreeding in some swine breeds in USA.

So has been the case in major dairy cattle breeds of USA in particular in Holstein which is most popular globally (Fig 11.7).

As is evident from Fig 11.7, in recent time, 2007 to 2012, inbreeding coefficients for Holsteins in North America and globally increased at the rate of 0.36% and 0.33% per year respectively. These levels are the highest in modern Holstein breeding history. This is due to severe reduction in number of different inheritance lines used by AI companies. Some of the Holstein bulls used in AI have contributed over 250,000 milking daughters or 5,000 progeny-tested sons. Additionally 50% of the roughly 5,000 young bulls that were progeny tested each year globally are the offspring of the 10 most popular sires.

Estimates of level of inbreeding (F) as well as the rate of increase in inbreeding (ΔF) in livestock breeds have been made (Table 11.8).While examining these values, it should be remembered that pedigree of many breeds may not be traceable too far back. Computed levels of inbreeding, therefore, are relative to some arbitrary starting point (initial reference year/ generation) from which onward the inbreeding coefficients are calculated. It is assumed that none of the animals in this initial population were inbred. Rate of increase in inbreeding per generation (ΔF) rather than the absolute inbreeding levels (F) is a better way of describing the consequences of inbreeding in a breed/population. The absolute values of inbreeding (F) cited in the Table are thus lower than actually prevalent in that breed.

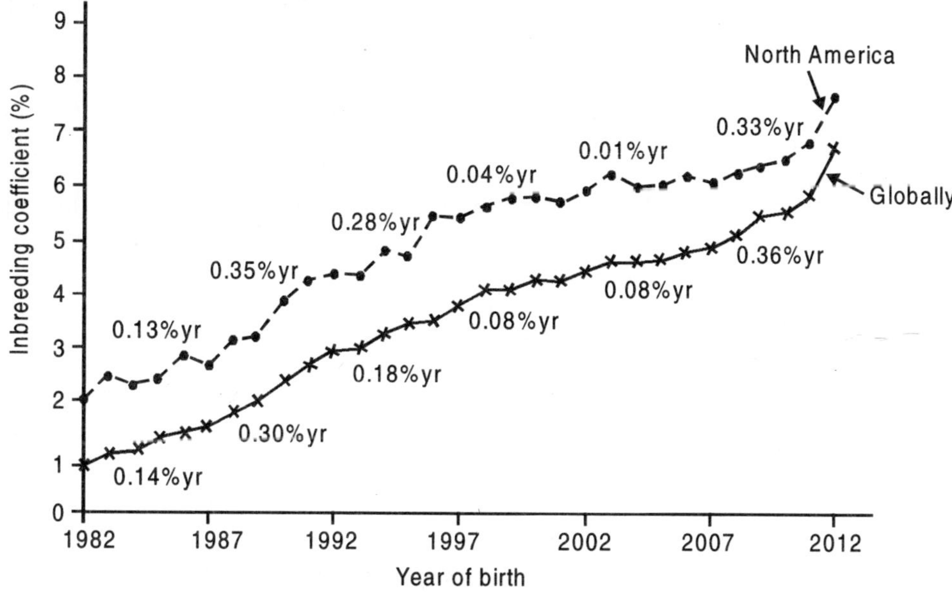

Fig 11.7 Inbreeding in Holstein Friesian breed in North America and globally (%F per year are listed in Fig.).

Table 11.8 Level of inbreeding in some farm animal species.

Species/Breed	Period	F (%)	ΔF (%)
Cattle :			
• Shorthorn (US)	1780 - 1920	24.9	1.54
• US Herefords	1860 - 1930	4.8	0.37
• US Herefords	1900 - 1930	2.5	–
	1930 - 1966	11.5	–
	1966 - 2002	10.0	–
• Holstein (US)	1880 - 1930	1.7	0.15
• Danish Holstein	1983 - 1992	–	0.74
	1993 - 2003	–	1.03
• Jersey	1977 - 1991	-	0.57
	1993 - 2003	-	0.94
• Red	1977 - 1998	-	0.32
	2001 - 2003	-	1.07
• Canadian Holstein	1950 - 1990	2.5	-
• Irish HF	1975 - 2004	1.49	-

Swine:			
• British Landrace	1965 - 1976	2.5	0.05
• Yorkshire	1978 - 1991	3.7	-
• Landrace	1982 - 1993	5.4	-
• Hampshire	1979 - 1991	3.0	
• Australian Landrace	1970 - 1980	7.5	
US-Dan bred			
• Landrace	1995 - 1999	3.7	0.74
• Yorkshire	1995 - 1999	1.9	0.38
• Duroc	1995 - 1999	3.5	0.70
• Hampshire	1995 - 1999	3.7	0.74
Horse:			
• Thoroughbreds	1930	8.8	1.64
• German trotter	1900 - 1970	3.1	0.41
• Halfinger	-	5.7	0.52
• Spanish Analusian	-	8.2	-
• Lipizzan	20 generations	1.7	-

Table 11.9 Level of inbreeding (%) in different breeds of dogs.

German Shepherd	Labrador Retrievers	Akita Inu	Boxer	English bulldog	Chow Chow	Rough Collie	Golden Retriever	Greyhound	German Shepherd	Labrador Retriever	Springer Spaniel
25.0	15.0	3.8	4.8	5.7	5.1	7.3	3.5	5.8	3.3	2.4	3.3

It is apparent from the values of inbreeding given above that every closed population/breed has experienced some degree of inbreeding. Further though accumulated level of inbreeding, ('F'), for the whole population may be small over a specific period (e.g. 2 to 4%) but individual animals of that breed or strain may have accumulated quite variable levels of inbreeding around average 'F' of the population. A few illustrative examples are given below:

Species & Breed		Cumulative F(%)	Range of F(%)
Cattle:	Holstein	for bulls: 3.0	0.0 - 38.0
		for cows: 2.0	0.0 - 31.0
Swine:	Yorkshire	3.7	0.0 - 50.0
	Landrace	5.4	0.0 - 46.0
	Hampshire	3.0	0.0 - 44.0
Horse:	Spanish	8.2	0.0 - 54.0
Dog (avg. of 9 breeds) :		5.0	0.0 - 45.0

The animals that have accumulated high level of inbreeding are thus more critical for causing inbreeding depression.

Solutions to Minimize Inbreeding

Given the detrimental effects of inbreeding, breeders attempt to reduce level of inbreeding in their herds / flocks through:

(*a*) *Increasing the number of sires and dams:* Increasing the number of parents used for reproduction, particularly the number of sires, reduces the level of inbreeding.

(*b*) *Introduction of replacement animals:* Level of inbreeding can be reduced by introduction of replacement animals from other herds or flocks. Effect of such an introduction, however, depends on the number of animals introduced, and the extent of their use in the herd. Two examples illustrate the effect of outside replacement on reduction of inbreeding :

- Inbreeding in two closed nucleus flocks of swine viz; Yorkshire and Hampshire attainted 'F' of 3.7 and 3.0% respectively over a period of 13 years (Fig 11.8). Introduction of replacements (parents) from outside flocks decreased these levels to 3.0 and 2.3% respectively in the subsequent two years.
- In 1990, Europe used 95% of sires from North America in its own cattle population but by 1995 it decreased their percentage to 50%. This change led to diversification of European gene pool and thus reducing inbreeding.

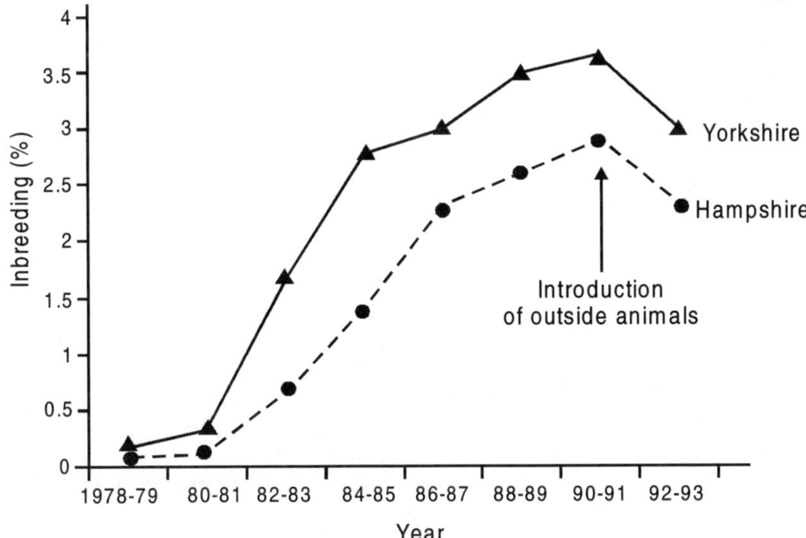

Fig. 11.8. Changes in inbreeding in two closed nucleus flocks of Swine over years.

While introducing animals from other herds/flocks, it is important that:

- Genetic merit of outside animals be reasonably high so that their introduction does not cause decline in performance of the flock / herd.
- Once the herd /flock has been closed following introduction of outside animals, inbreeding would again accumulate. Therefore, the breeder has to continue monitoring the level of 'F' of the herd as a whole as well as of the individual animals so that appropriate remedial measures are taken.

Selection programs should also be designed to achieve a balance between selection intensity and level of inbreeding as the two bear an antagonistic relationship.

Acceptable Level of Inbreeding: According to FAO guidelines, an increase of more than 1% in inbreeding per generation be considered the upper limit in order to maintain fitness of a flock/ breed. Some researchers put 0.5 % as the upper limit of inbreeding for continuation of the present rate of genetic improvement. A perusal of Table 11.8 to 11.9 shows that many breeds/flocks had lower rates of inbreeding but individual animals within the breed may have accumulated high level of F.

Potential Usefulness of Inbreeding

Inbreeding, in general, is looked upon unfavorably by animal breeders but it has some potential uses:

- **Detection and elimination of undesirable recessive genes:** A number of examples of genes which in homozygous form lead to abnormalities are known to occur in livestock (Table 11.2). Inbreeding can be used to identify carriers of recessive genes. Most of such genes are present in low frequencies in non-inbred (outbred) herds/flocks and are covered by their dominant alleles. But inbreeding, by causing increased homozygosity, allows these to be phenotypically expressed, and in turn makes their elimination possible.
- **Fixation of a desirable gene (trait):** Inbreeding can also be used by livestock breeders to *'fix' a desirable trait* (phenotype) within a population. But this can only be achieved by simultaneous culling of undesirable animals. Nonetheless the cost of carrying out inbreeding for this purpose may not be justifiable because of the problems associated with inbreeding.
- **Development of distinct families or inbred lines vis a vis increased genetic uniformity:** Increased homozygosity caused by inbreeding implies high genetic uniformity. Individuals of highly inbred lines (F=1.0) are equivalent to identical twins or clones; therefore, genotyping of a single animal in such a population is equivalent to genotyping the whole line. In biomedical research (e.g. immunology, diabetes, cancer studies etc.), genetic uniformity of experimental animals (e.g. mice, guinea pig) is of utmost importance so that the effect of drugs or other such treatments can be more effectively studied or delineated. Because of genetic similarity, fewer animals are required to achieve a given level of statistical confidence. Once the lines approach complete homozygosity, these stay constant over generations; the only way the inbred lines can change is as a result of new mutations which are relatively rare. Further each inbred line has unique characteristics some of which may be of value in particular research areas viz; models of specific human disease like cancer, auto-immune diseases, diabetes etc.

If we begin with an initial base population, inbreeding over successive generations would result in formation of a number of lines (i.e. sub-lines or families), each differing genetically from the other. Effect of continued inbreeding on sub division of a population into inbred lines is shown in Fig 11.9.

Continued inbreeding within these lines would tend to change the frequency of some of the genes found in the base population. For example, if a particular gene was present in only 1% of the animals in the base population, inbreeding and the development of distinct lines could result in the gene being present in nearly all animals in some lines, and in none or only a few of the animals in other lines.

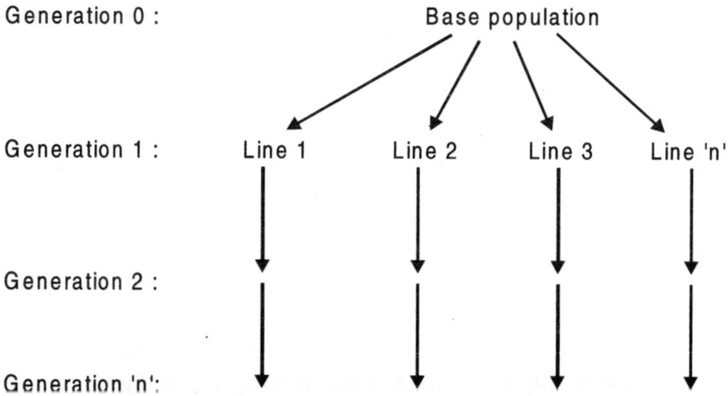

Fig. 11.9. Division of the base population into inbred lines

Inbred lines are most notably used in the development of hybrid animals; the procedure is called inbreeding-hybridization. In plant breeding, inbred lines have been developed and used for producing hybrids to make use of heterosis. In theory, the same procedure can be used in livestock, and has indeed been used in the early part of the 20th century, especially in poultry and swine.

***Use of inbreeding in the past*:** Robert Bakewell, the pioneer animal breeder of 19th century, used inbreeding for fixing desirable traits/phenotypes in his herd/flock which made the animals breed true-to-type. To some extent, inbreeding was also used in the formation of some of the new breeds.

During the first half of the 20th century, inbreeding was used for the development of highly homozygous lines (inbred lines), particularly in poultry and swine, to be used for hybridization. Eventually, the system of inbreeding followed by hybridization (called inbreeding–hybridization) was also replaced by other systems mainly due to the huge cost involved in it. The undesirable effects of inbreeding on performance, especially on reproductive traits, had been so common in all species of animals that inbreeding is no more used as a system for the production of commercial livestock.

Line Breeding

Since the undesirable effects of inbreeding on performance and health are well-documented, breeders try to avoid it as much as possible in their stocks. In farm animals, this is achieved by avoiding mating between close relatives e.g. avoidance of full-sib, half-sib and parent-offspring matings. Sometimes, however, breeders may like to accumulate inheritance/genes of some outstanding ancestor (s) in their herds/flocks. This is achieved by a system of mating referred to as line breeding.

Line breeding is a system which is designed to maintain a high degree of genetic relationship to an outstanding ancestor (or a line of ancestors) while keeping the level of inbreeding low. The outstanding ancestor whose inheritance is to be concentrated in animals of a given generation is usually a male (sire) rather than a female because:

- A male can produce very larger number of offspring particularly through semen cryopreservation and artificial insemination while a female can produce only a limited number of offspring in its lifetime.

- A male, through semen freezing and artificial insemination, can be used even if it is old or dead.
- A male (sire) can be progeny tested, therefore, its genetic worth (breeding value) is known more accurately.
- A male (sire) can be more effectively tested for the presence or absence of lethal/semi lethal genes.

Line breeding is a milder form of inbreeding by which the breeder attempts to take the benefit of accumulating the inheritance of an outstanding ancestor but at the same time trying to avoid the troubles arising from intense inbreeding.

Line breeding differs from inbreeding in two respects:

- *Objective*: The objective of a line breeding program is to maintain a high degree of relationship of individuals of a given generation with some outstanding ancestor (high genetic merit). On the other hand, inbreeding is practiced to achieve high degree of homozygosity, and to produce prepotent individuals.
- *Merit of mates*: Inbreeding involves mating between close relatives without any consideration to genetic merit of mates. In contrast to this, line breeding is practiced by considering performance of the ancestor into consideration.

Nonetheless, the fundamental genetic effect of line breeding is same as that of inbreeding (i.e. increased homozygosity). But the increase in homozygosity resulting from line breeding is much lower than with intense inbreeding. Consequently, line breeding does not expose undesirable recessive genes as extensively as does mating between close relatives. For this reason, line breeding is generally a safer program than intense inbreeding.

Several different types of matings can be arranged in a line breeding program. Fig 11.10 shows an example where a sire, designated as 'A', has been identified to be genetically superior. His progeny (designated as 'B', 'C', 'D') are related to him by one-half, and his grand progeny (E, F, G) are related to him by one-fourth. By making mating among the grand progeny (between F and G), genetic relationship of offspring 'H' to the outstanding ancestor 'A' has

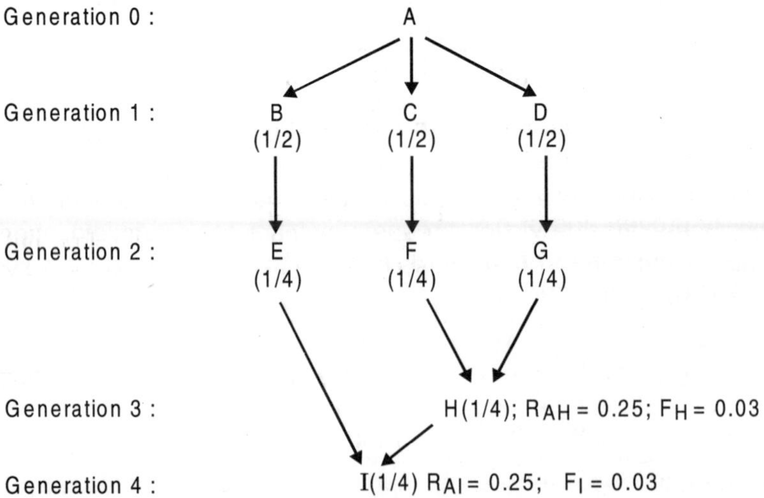

Fig 11.10 Example of a line breeding (figures in parentheses represent fraction of inheritance received from ancestor 'A').

become one-fourth even though animal 'H' is separated from 'A' by three generations. The relationship coefficient between 'A' and 'H' is 0.25 (R=0.25) but the inbreeding coefficient of animal 'H' (F_H) is only 0.03 (3%). So is the case of animal 'I'.

The greatest advantage of line breeding, as can be seen from the example illustrated in Fig. 11.10 is that line breeding helps in maintaining a high degree of genetic relationship to a valuable ancestor for many generations while keeping inbreeding at a low level. Many other types of mating can be arranged to increase the relationship of individuals of a generation with an outstanding ancestor but keeping the level of inbreeding low. Two of these are shown below:

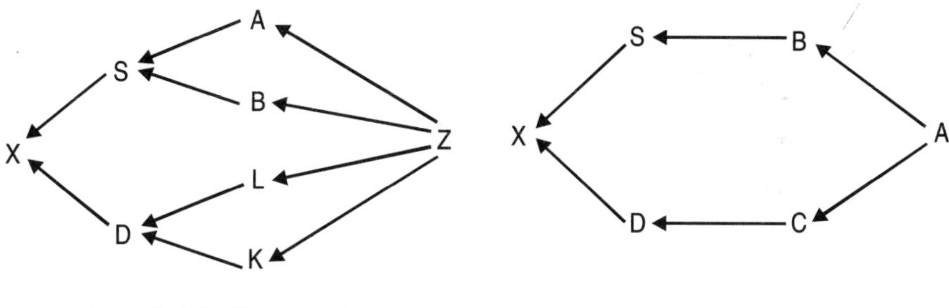

$F_X = 0.125$; $R_{XZ} = 0.48$ $F_X = 0.031$; $R_{XA} = 0.2465$

When to Line breed

A line breeding program is recommended when there are sound reasons for its use. Some of the situations where line breeding can be effectively employed are:

- **Availability of an outstanding sire or ancestor:** Line breeding is indicated when an individual of high genetic merit (breeding value) is available. Genetic merit of the individual to be used in a line breeding program should have been evaluated by some reliable criterion (e.g. progeny testing) on which high confidence can be placed.
- **Outstanding ancestor is dead:** Line breeding is most useful when the outstanding male (i.e. of proven genetic merit) is dead, and thus not available for use in the herd. Matings can be arranged in such a way that these increase the genetic relationship of the individuals of the herd/flock with the outstanding ancestor. Structure of a line breeding program can take many different forms but with the primary purpose of retaining high proportion of genes of the ancestor in its descendants.
- **When the herd is better than many herds:** Line breeding program is best suited for a closed herd which is better than many herds of the same breed. Since the use of sires from other herds into herd of high-merit can deteriorates its own performance, it would be reasonable to ensure that the sire to be used in a line breeding is genetically proven and of high genetic merit.
- **Inability to purchase an outstanding sire:** When an outstanding sire cannot be purchased by a breeder from other herds, it may be more appropriate to purchase some good male offspring of that sire to be used in the herd.

The increase in inbreeding under line breeding is low, therefore, it allows time to discard inferior individuals, and consequently helps in reducing the undesirable effects of inbreeding. *In fact line breeding is probably, the best known use of inbreeding.* Line breeding has been followed in most breeds of cattle; horse etc during early stages of their development. The modern race horses, for example, are all descendents of some outstanding Arabian stallions which were mated with local mares. The fast runners from F_1 were selected and inbred mildly. The F_2 animals served as base for the pedigree of almost all the modern breeds of race horses.

Inbreeding in Endangered Species and Zoo Animals

Many of the wild species of animals are becoming endangered[5] due to hunting, habitat destruction caused by urbanization, and related factors. Concerted efforts are now being made to conserve such species through captive breeding[6] and multiplication with the objective of releasing them into their wild habitat when conditions become conducive. Zoos, besides education, are serving an important role in conservation and multiplication of wild animals by keeping them under conditions that are as close as possible to their natural habitat with plenty of space, food and hygienic conditions.

Minimizing inbreeding and the loss of genetic variation are major concerns of captive breeding programs for endangered species. Even in natural populations whose sizes are becoming smaller, inbreeding is becoming a factor of serious concern. A number of studies and surveys including excellent review articles are available on degree of inbreeding, and the effect of inbreeding on reproduction and survival of wild animals. Since the topic does not fall within the domain of this publication, a few examples are cited below as illustrations:

- *1st study:* Conducted on several species gave the following results:

Species	Mortality (%)		'F'
	Non-inbred	Inbred	(%)
Zebra :	26	40	25
Dear :	25	100	25
Giraffe :	20	60	25
Oryx :	5	100	25
Doras gazelle.	28	60	27

- *2nd study (2003):* Study of a population of David's Deer maintained in Zoological Parks of Europe (from 1947-2000) showed that inbreeding coefficient of different individuals varied from 0.0 to 28.1% with average of 2.4% for the whole population of over two thousand animals. Each 2.8% increase in 'F' resulted in 24-40 days shorter length of life. Whole of the current population of Deer was found to have been derived from 18 founder animals.
- *3rd study (2005)*: Study of a small, naturally restored wild population of grey wolves in Scandinavia (founded in 1983 from three individuals) showed that 'F' varied between 0.0 and 41%. Each 10% increase in 'F' reduced the number of pups (in winter) per litter by 1.2. The Scandinavian wolf population appears to have a gloomy future unless it can be purged of its genetic load.
- *4th study (2007):* A survey of 88 species of mammals, birds, reptiles and amphibians in 119 Zoo populations (using meta analysis) showed that inbreeding occurred in all the populations in varying degree. Inbreeding reduced the neonatal survival (up to 7 days) in all species but the severity of inbreeding depression varied in different species/ populations.

[5]An endangered species is the one which has been categorized so by the International Union for Conservation of Nature (IUCN) because it faces a very high risk of extinction in the near future. Endangered is the second most severe conservation status for wild populations in the IUCN's scheme after 'critically endangered.

[6]Captive breeding is the process of breeding rare or endangered species in human controlled environments with restricted settings, such as wildlife preserves, zoos and other conservation facilities. It is meant to save species from extinction and so stabilize the population of species that it will not disappear.

General Inference: Inbreeding depression is a common phenomenon in zoo populations. Serious concerns which have been raised in certain quarters about the future of zoo and wild populations are justified. Inbreeding depression has often been found more severe under stressful conditions. It is therefore important that zoo conditions be made more comfortable for neonatal survival. Regular exchange of animals or semen between different zoos offers one way of overcoming the accumulation of inbreeding.

Review Questions

1. Explain/define the following:
   - I. Inbreeding
   - II. Inbreeding coefficient
   - III. Coefficient of relationship
   - IV. Common ancestor
   - V. Prepotency
   - VI. Inbreeding depression
   - VII. Line breeding
   - VIII. Negative assortative mating
   - IX. Positive assortative mating
   - X. Mating system
   - XI. Breeding plan
   - XII. Inbred line
2. Differentiate between the following with suitable examples wherever possible:
   (a) Selection and mating system
   (b) Positive and negative assortative mating
   (c) Regular and Irregular systems of Inbreeding
   (d) Alleles identical by descent (autozygosity) and alleles identical by state (allozygosity)
3. What are the two broad means of bringing about genetic improvement in livestock?
4. Explain the meaning of a mating system in 'individual' and 'population' context.
5. If animals within a population are randomly mated, does that mean the absence of selection too? Explain.
6. Does a mating system bring about a change in gene or genotypic frequency? Explain
7. Which type of the gene action is primarily exploited by a mating system?
8. Classify the mating systems into different types.
9. What is random mating? What are the benefits or advantages of random mating as a mating system?
10. Describe the effect on genetic and phenotypic variation of the following mating strategies: (a) positive assortative mating (b) negative assortative mating.
11. Why is negative assortative mating also the called compensatory mating ?
12. Which two systems of mating utilize genetic relation between mates as the basis for mating?
13. What are the common types of close matings? Show these diagrammatically. Which of these cannot be used in farm animals?
14. Which is the most common type of close mating generally possible in: (a) poultry (b) cattle.
15. What is the primary genetic consequence of Inbreeding? Explain
16. How does inbreeding affect the genotypic frequency? Show diagrammatically using selfing as an inbreeding system.

17. If inbreeding is carried out successively over generations, what will its ultimate effect on genotypic and gene frequencies?
18. How does inbreeding affects : (a) gene frequency (b) degree of heterozygosity?
19. Who developed the mathematical procedure for measurement of inbreeding?
20. Inbreeding coefficient is a measure of the heterozygosity that is lost due mating between relatives. Explain
21. Why is inbreeding coefficient designated by 'F'.
22. What is the range of inbreeding coefficient? Explain.
23. Which of the following is true: (a) F = ½ R (b) R = ½ F (c) F =R.
24. Give the numerical values of relationship coefficient for : (a) full-brother and -sister (b) half-sister (c) first cousins (d) mother and son (e) uncle and niece (f) husband and wife.
25. Which of the following is usually true: (a) R is more than F (b) F is more than R (c) R equals F.
26. What are the consequences of increased homozygosity caused by inbreeding ?
27. How does inbreeding cause appearance of hereditary defects? Show with an illustration.
28. If an inbred line attains inbreeding coefficient of unity, what will be the status of genetic variability within that line?
29. Inbreeding coefficient measures the degree of inbreeding relative to its level in the base population. Explain.
30. Why highly inbred lines do not respond to selection as much as do the non-inbred lines?
31. Why are inbred lines valuable for biological/veterinary experiments designed to compare genetic and environmental variation?
32. When a line attains an inbreeding coefficient of unity, would the parents produce completely homozygous or heterozygous progeny? Explain.
33. A particular type of dwarfism in cattle is caused by a recessive gene. In a herd of cattle, a dwarf calf appeared. The herd was of high merit, and was therefore closed to outside inheritance. Whether the frequency of the dwarf phenotype would increase or decrease over successive generations. Explain.
34. How does inbreeding affects the : (a) genetic variance (b) environmental variance ?
35. On what factors does the magnitude of inbreeding depression depend?
36. Higher the magnitude of inbreeding, greater will be the inbreeding depression. Explain
37. If a trait is exclusively determined by additive gene action, will it exhibit inbreeding depression? Explain
38. Rank the following traits with respect to inbreeding depression (in descending order): (a) a trait governed by overdominance (b) a trait governed by partial dominance (c) a trait governed by complete dominance.
39. Which of the following is true: (a) higher the h^2 greater the inbreeding depression (b) higher the h^2 lower the inbreeding depression (c) no relation between h^2 and inbreeding depression.
40. Traits with low heritability show the highest inbreeding depression. Explain.
41. Rank the following three category of traits in order of inbreeding depression (highest to lowest): (a) production and growth (b) fitness (c) morphological and quality.

42. If inbreeding is practiced over generations, what will happen to performance of a trait which has complete heritability (h^2 = 1.0) ?

43. Larger the number of parents (S and D) used for reproduction, lower the increase in inbreeding - Justify.

44. Which of the two sexes (sire or dam) is more critical to increase in rate of inbreeding within a herd/flock ?

45. Why can the accumulation of inbreeding not be avoided in small-sized closed herds?

46. Which breeding and reproductive technologies have contributed to accumulation of inbreeding in cattle? Explain

47. How can the rate of increase in Inbreeding in a herd/flock be reduced?

48. Introduction of replacement animals from outside into a herd is recommended as a measure to reduce the accumulation of inbreeding. What precautions need to be taken while making such an introduction?

49. What are the potential uses of inbreeding in livestock? Explain.

50. Can inbreeding be used to eliminate dominant genes from a population? Explain

51. How can inbreeding be used for gene fixation?

52. How can inbreeding be used for elimination of recessives genes from a population?

53. How can inbreeding be used for development of distinct families for increased genetic uniformity?

54. How does inbreeding increases prepotency?

55. Why do inbred animals tend to breed true to type?

56. Can detrimental genes be fixed in an inbred line?

57. How does inbreeding influence genetic variation within a population?

58. What is line breeding?

59. The outstanding ancestor used in a line breeding program is usually a sire. Explain

60. How does line breeding differ from intensive inbreeding?

61. Livestock breeders look favorably to line breeding but avoid inbreeding- Justify.

62. Show a mating depicting line breeding.

63. What are the situations where line breeding is used in livestock breeding ?

64. Which is the most practical use of inbreeding in cattle and buffaloes?

12

Measurement of Inbreeding and Relationship Coefficients

Inbreeding coefficient (F) of an individual is computed from its pedigree by analyzing the degree of relatedness between its parents. The parents are considered genetically related to each other if they have a common ancestor in their pedigree. *A* **common ancestor**, *is the one that appears on both the sire and dam sides of pedigree of the individual (whose 'F' is to be calculated).* In other words, a common ancestor is the one who has transmitted his/her genes to both the parents (sire and dam) of the individual.

The computation of 'F' involves counting the number of individuals or paths connecting the parents of the individual through common ancestor(s).

The formula of inbreeding coefficient is:

$$F_X = \frac{1}{2}\sum\left[\left(\frac{1}{2}\right)^n (1+F_{CA})\right]$$

Where:

F_x = inbreeding coefficient of individual 'X' (i.e. the individual whose 'F' is to be calculated).

Σ = represents the sum of all the paths connecting the parents (sire and dam) of 'X' through common ancestor(s) *i.e.* we add together '$[(½)^n(1+F_{CA})]$' for each path passing through the common ancestor.

n = number of paths connecting the parents of individual 'X' through each common ancestor.

F_{CA} = inbreeding coefficient of the common ancestor.

If F_{CA}= 0 (*i.e.* the common ancestor is not inbred), then the formula of inbreeding coefficient reduces to:

$$F_X = \frac{1}{2}\Sigma\left(\frac{1}{2}\right)^n$$

Steps and Rules in Calculation of Inbreeding Coefficient

I. Preparation of path (arrow) diagram from the pedigree

A pedigree is a systematized list of ancestors of an animal. In livestock, it is a common practice to assign some names to different individuals in the pedigree. The following rules are important in generating a path or arrow diagram:

(*i*) Replace the names of all members of the pedigree by suitable alphabets or numbers ensuring that each animal's name or number appears only once in the diagram.

(*ii*) Draw arrows from parents (sire and dam) to the individual (X) with the arrows pointing towards the individual (the direction of arrow signifies the direction of flow of genes).

(*iii*) Locate the common ancestor(s), and draw arrows from common ancestor(s) to the parents of the individual. Remember, in some pedigrees, there may be more than one path connecting the sire and dam through each common ancestor.

II. Tracing and counting the number of paths joining the parents

The following rules are important for tracing the paths joining the parents of the individual through common ancestor(s):

(*i*) Each path should connect the two parents (sire and dam) either directly or through the common ancestor (i.e. each path starts from one parent and ends at the second parent).

(*ii*) Each path moves upward from the first parent, reaches the common ancestor and then comes downward to the second parent. *Remember, once the path starts its downward journey (from the common ancestor), it can't go upward again because the flow of genes occurs from parent to offspring and not the other way.*

(*iii*) No individual should appear twice in any one path connecting the two parents.

(*iv*) Count all the arrows (individuals) connecting the parents through each common ancestor.

(*v*) In some pedigrees, one parent of the individual may also become its own common ancestor (this happens while calculating the inbreeding coefficient of an offspring resulting from 'sire x daughter' mating).

III. Calculating the inbreeding coefficient of common ancestor(s)

(*i*) When a common ancestor is inbred, its inbreeding coefficient is calculated by tracing the paths in its pedigree as outlined above (in step II).

(*ii*) In some pedigrees, there are certain intermediate ancestor(s) who themselves are inbred but are not common ancestors. In such cases, the inbreeding coefficient of intermediate inbred ancestor is not considered (since it is not a common ancestor).

(*iii*) Inbreeding coefficients of the parents are not considered unless one of the two parents is also the common ancestor.

The following exponential values of $(\frac{1}{2})^n$ are useful in calculations of inbreeding and relationship coefficients:

Table 12.1. Exponential values of (½)^n.

$(½)^2 = 0.25000$	$(½)^7 = 0.00781$
$(½)^3 = 0.12500$	$(½)^8 = 0.00391$
$(½)^4 = 0.06250$	$(½)^9 = 0.00195$
$(½)^5 = 0.03125$	$(½)^{10} = 0.00098$
$(½)^6 = 0.01563$	$(½)^{11} = 0.00049$

Preparation of Path or Arrow Diagram

12.1. From the pedigree of an outstanding bull named 'Prince' given below, prepare a path diagram.

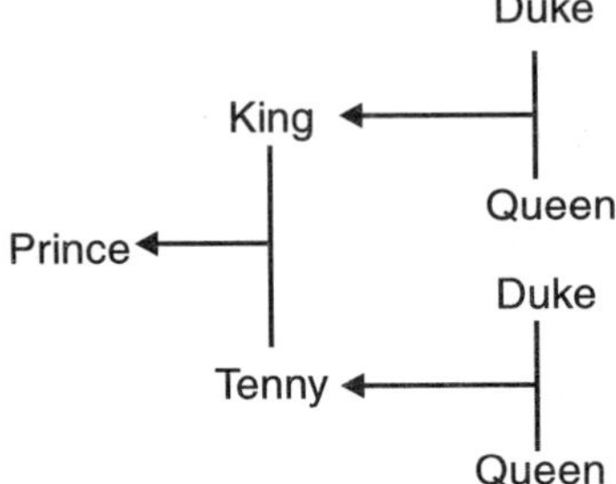

Solution : The path diagram for the above pedigree, after replacing the names of Prince, King, Tenny, Duke, Cooky and Queen respectively with P, K, T, D, C and Q and after eliminating the Individuals (*i.e.* C and D) who are not common aucesfors become :

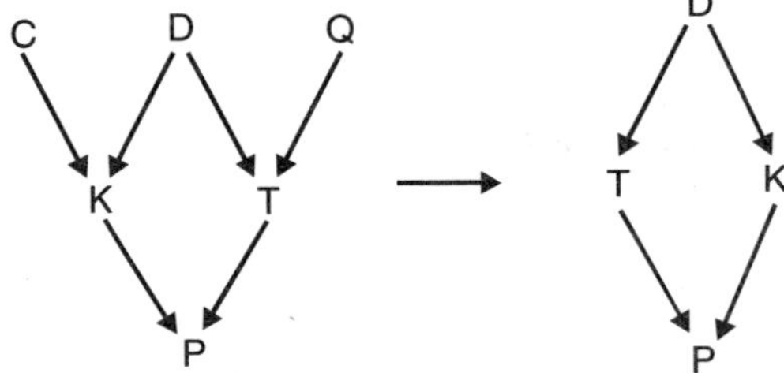

12.2. From the pedigrees of two puppies, named Tom and Rocky : (a) generate the path diagrams for calculation of their inbreeding coefficients (b) what types of mating (*i.e.* full or half sib etc.) is involved in producing Tom and Rocky?

(*i*) Pedigree of Tom **(*ii*) Pedigree of Rocky**

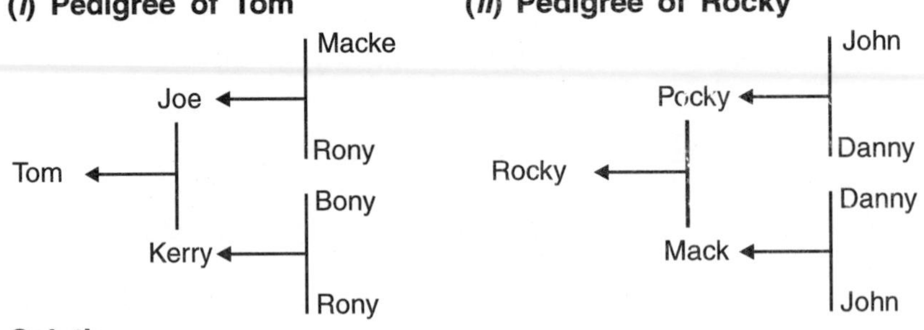

Solution:

(a) Path diagrams:

(i) Path diagram for the pedigree of Tom depicted above in (i) above is shown below: the left side shows all the animals of the pedigree; and the one on right side excludes the animals who do not contribute to the parents of 'Tom' viz. 'M' and 'T'.

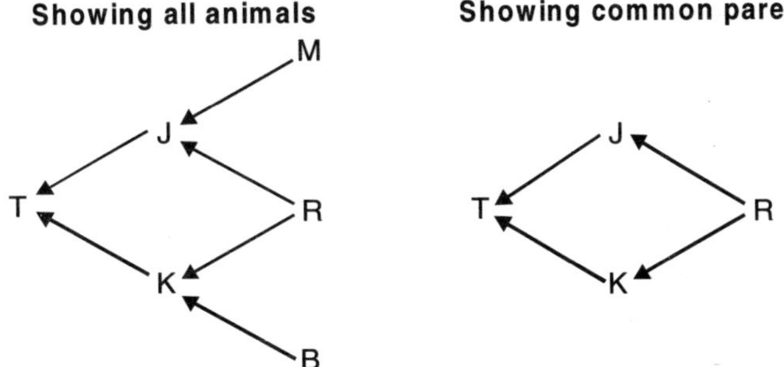

(ii) Following the same rule, the path diagram of Rocky generated from pedigree given in (ii) becomes:

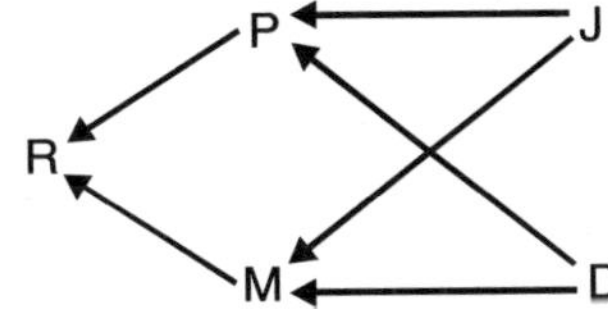

(b) Types of mating:

(i) The parents of Tom, as shown in (i), have one common ancestor, named 'Rony', therefore, individuals 'J' and 'K' are half-sibs. The mating of 'J' and 'K' to produce 'T' is a half-sib mating.

(ii) The parents of Rocky (second pedigree) come from same parents; therefore, individuals 'P' and 'M' are full sibs. The mating of 'P' and 'M' to produce 'R' is a full-sib mating.

12.3. Prepare an arrow/path diagram for the following pedigree in which "X' is the individual whose inbreeding coefficient is to be calculated.

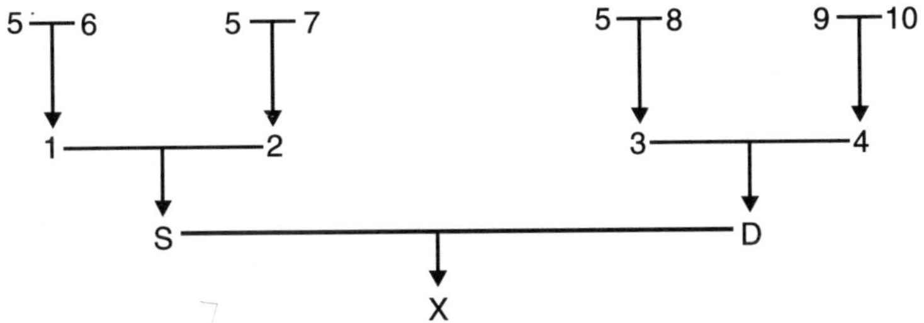

Solution : The parents of individual 'X' are 'S' and 'D'. For identifying the common ancestor(s), therefore, we have to examine the sire and dam sides of pedigree of 'X'. Such an examination shows that only animal numbered '5' appears on both sides of the pedigree. Therefore in making the path diagram, the ancestors 4, 6, 7, 8, 9 and 10 are eliminated. The path diagram is shown on right hand side :

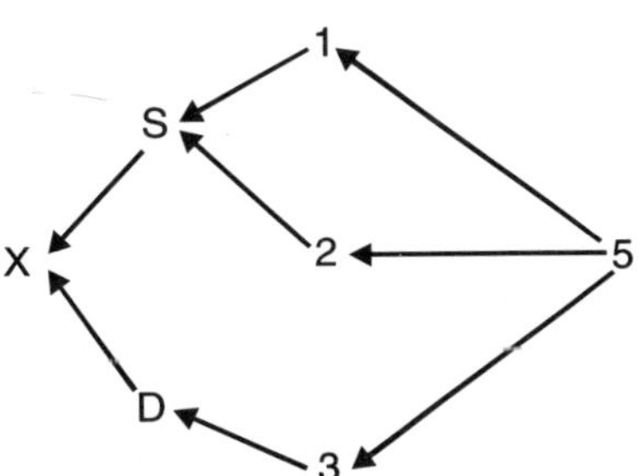

Calculation of Inbreeding Coefficient (F)

Regular Systems

12.4. Calculate the inbreeding coefficient of individual 'X' or 'Z' from the pedigrees of various types of close matings shown below:

Solution :

(a) Half-sib mating:

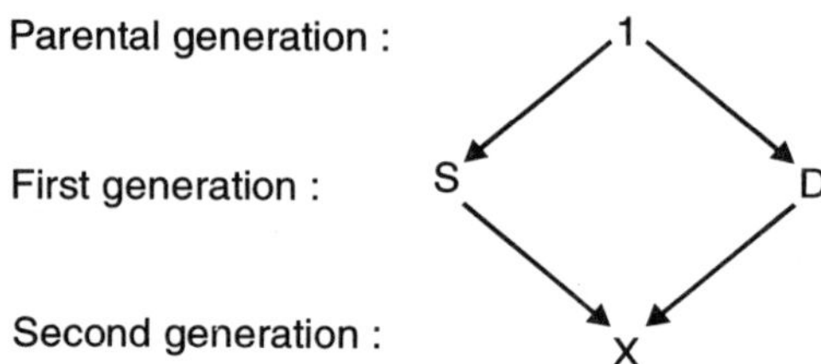

Common ancestor of X is '1' which is not inbred. There is only one path through the common ancestor:

CA	Path no.	Path chain	n	$(1/2)^n (1+F_{CA})$	Contribution
1	1	X←S←[1]→D→X	2	$(1/2)^2(1+0)$	0.25
				Summation =	0.25

$$F_X = \frac{1}{2}\sum(\tfrac{1}{2})^n (1+F_{CA}) = \frac{1}{2}(0.25) = 0.125$$

(b) Full-sib mating:

Parental generation : 1 2

First generation : S D

Second generation : X

The common ancestors of 'X' are '1' and '2' which are assumed to be non-inbred because nothing is known about their pedigree. The paths through common ancestors are:

CA	Path no.	Path chain	n	$(1/2)^n (1+F_{CA})$	Contribution
1	1	X←S←[1]→D→X	2	$(1/2)^2(1+0)$	0.25
2	2	X←S←[2]→D→X	2	$(1/2)^2(1+0)$	0.25
				Summation =	0.50

$$F_X = \frac{1}{2}\sum(\tfrac{1}{2})^n (1+F_{CA}) = \frac{1}{2}(0.5) = 0.25$$

(c) Parent-offspring mating:

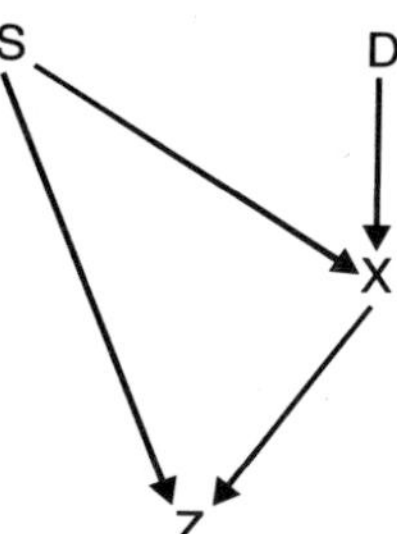

Common ancestor of 'Z' is 'S' which is not inbred. There is only one path through common ancestor:

CA	Path no.	Path chain	n	$(1/2)^n (1+F_{CA})$	Contribution
S	1	Z←[S]→X→Z	1	$(1/2)^1(1+0)$	0.50
				Summation =	0.50

(d) Selfing:

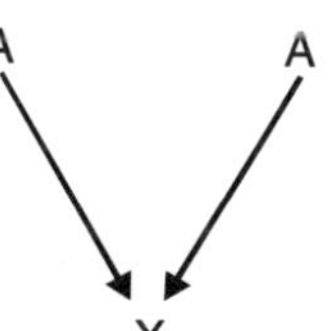

Since 'X' has been produced by selfing of 'A', the common ancestor of 'X' is 'A'. Hence the number of paths connecting the parents of 'X' is zero.

CA	Path no.	Path chain	n	$(1/2)^n (1+F_{CA})$	Contribution
—	1	X←[A]→X	0	$(1/2)^0(1+0)$	1.0
				Summation =	1.0

$$F_X = \frac{1}{2}\sum\left(\frac{1}{2}\right)^n (1+F_{CA}) = \frac{1}{2}(1.0) = 0.50$$

(e) Single first cousin mating:

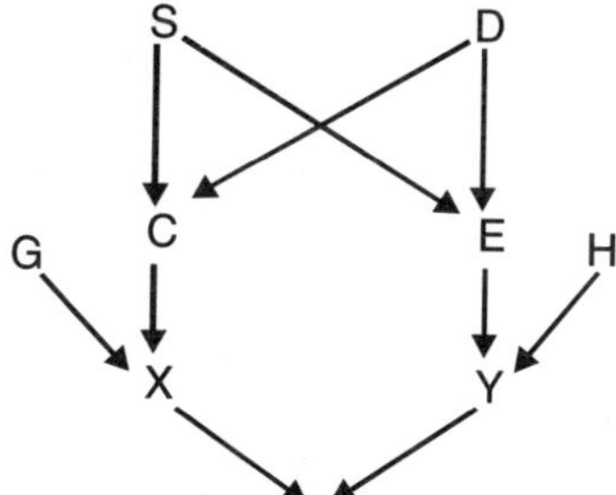

The common ancestors of Z are 'S' and 'D' which are not inbred. The paths through common ancestors are:

CA	Path no.	Path chain	n	$(1/2)^n(1+F_{CA})$	Contribution
S	1	Z←X←C←[S]→E→Y→Z	4	$(1/2)^4(1+0)$	0.0625
D	2	Z←X←C←[D]→E→Y→Z	4	$(1/2)^4(1+0)$	0.0625
				Summation =	0.125

$$F_Z = \sum (1/2)^n (1+F_{CA}) = \frac{1}{2}(0.125) = 0.0625$$

(f) Half cousin mating:

Parental generation : D_1 S D_2

First generation : G E F H

Second generation : X Y

Third generation : Z

The common ancestor of 'Z' is 'S' which is not inbred. The paths are:

CA	Path no.	Path chain	n	$(1/2)^n(1+F_{CA})$	Contribution
S	1	Z←X←E←[S]→F→Y→Z	4	$(1/2)^4(1+0)$	0.0625
				Summation =	0.0625

$$F_Z = \frac{1}{2}\sum\left(\frac{1}{2}\right)^n (1+F_{CA}) = (1/2)\ (0.0625) = 0.03125$$

Irregular Systems

12.5. Calculate the inbreeding coefficient of 'A' from the following pedigree:

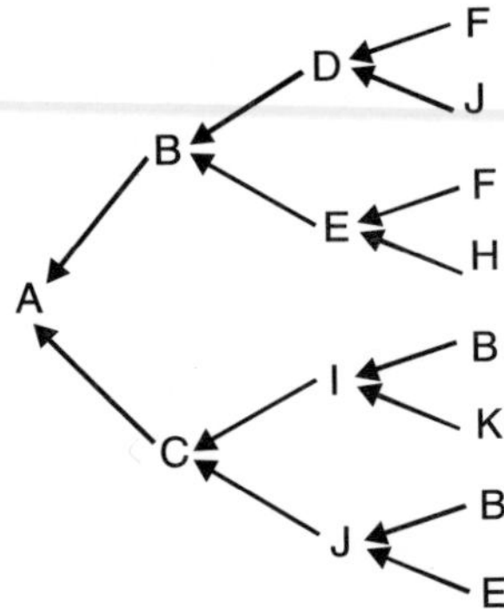

Solution : The path diagram by excluding the ancestors which do not contribute genes to 'A' through both sides of pedigree becomes:

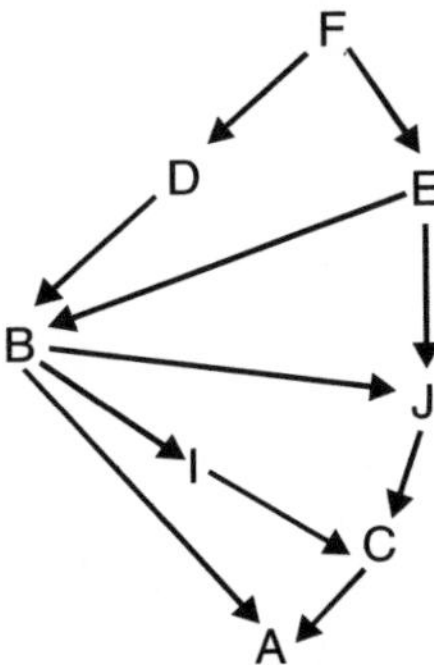

Calculations for inbreeding coefficient of 'A' are shown below in steps:

- *Identifying parents and common ancestors*: The parents of 'A' (whose 'F' is to be calculated) are 'B' and 'C'.
  Common ancestors of 'A' (i.e. the animals that appear on both sides of pedigree of 'A') are 'B', 'E' and 'F'.
  Note: Animal 'B' is parent as well as common ancestor of 'A'.
- *Inbreeding coefficient of common ancestor*: Of the three common ancestors, only 'B' is inbred (because it's parents, 'D and 'E', are related to each other through 'F').

The path for calculating inbreeding coefficient of 'B' is:

B←D←[F]→E→B, hence, $n = 2$

$$\therefore \quad F_B = (1/2)\ [(1/2)^2(1+0)] = 0.125$$

The paths through common ancestor of 'A' are:

CA	Path no.	Path chain	*n*	$(1/2)^n(1+F_{CA})$	Contribution
B	1	A←[B]←J→C→A	2	$(1/2)^2(1+0.125)$	0.28125
	2	A←[B]→I→C→A	2	$(1/2)^2(1+0.125)$	0.28125
E	3	A←B←[E]→J→C→A	3	$(1/2)^3(1+0)$	0.125
F	4	A←B←D←[F]→E→J→C→A	5	$(1/2)^5(1+0)$	0.03125
					Summation = 0.71875

$$F_A = \frac{1}{2}\sum\left[\left(\frac{1}{2}\right)^n(1+F_{CA})\right] = (1/2)\ (0.71875) = 0.359375$$

12.6. Calculate the inbreeding coefficient of 'I' from the pedigree given below. Given, animal 'A' is inbred with inbreeding coefficient of 0.01.

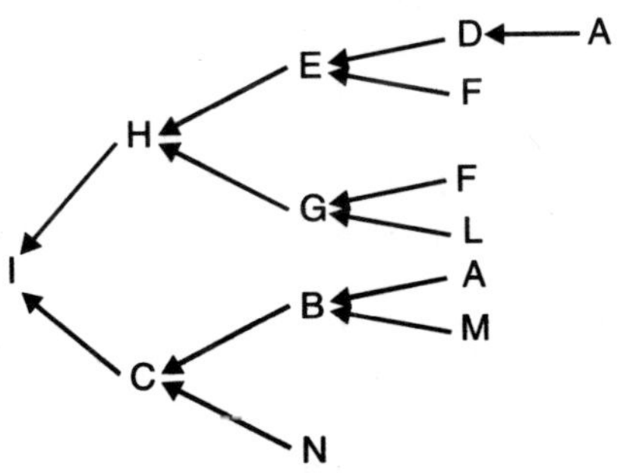

Solution : The path diagram is:

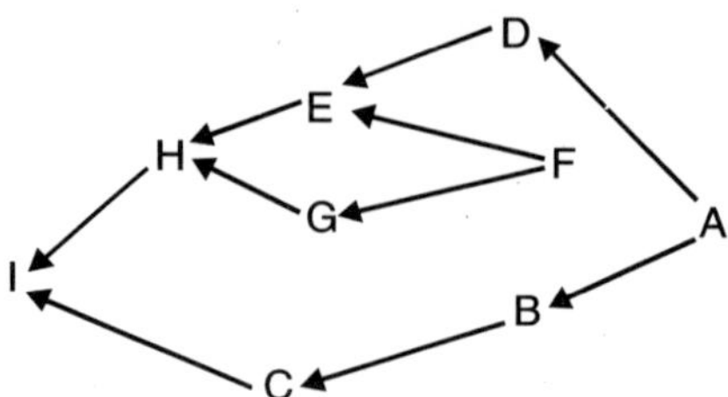

The parents of 'I' are 'H' and 'C'. The only ancestor that appears in the pedigree of both the parents ('H' and 'C') is 'A'. There is only one path connecting 'H' and 'C' through 'A' (the common ancestor):

CA	Path no.	Path chain	*n*	$(1/2)^n (1+F_{CA})$	Contribution
A	1	I←H←E←D←[A]→B→C→I	5	$(1/2)^5(1+0.01)$	0.03156
					Summation = 0.03156

$$\therefore \quad F_I = (1/2)\,(0.03156) = 0.0157$$

12.7. From the pedigree given below, calculate the inbreeding coefficient of 'X'

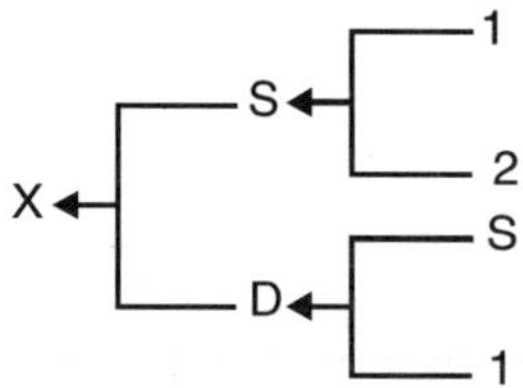

Solution : The path diagram by eliminating the animals who do not contribute genes to both the parents (i.e. 'S' and 'D') is:

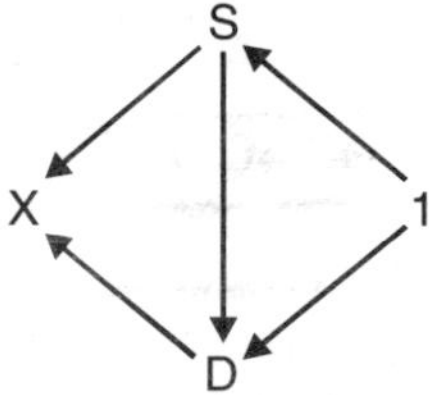

The stepwise calculations are shown below:

- *Identifying common ancestor(s) of 'X'*: Animal '1' and 'S' are common ancestors of 'X'.

  Note: Here 'S' is not only the parent but also the common ancestor (which illustrates rule V of step II of the procedure for calculation of 'F').

- *Tracing and counting the possible paths*:

CA	Path no.	Path chain	*n*	$(1/2)^n (1+F_{CA})$	Contribution
1	1	X←S←[1]→D→X	2	$(½)^2(1+0)$	0.25
S	2	X←[S]→D→X	1	$(½)^1(1+0)$	0.50
					Summation = 0.75

$$F_X = \frac{1}{2}\sum\left[\left(\frac{1}{2}\right)^n (1+F_{CA})\right] = ½(0.75) = 0.375$$

12.8. From the path diagram given below calculate: (a) F_X (b) F_S

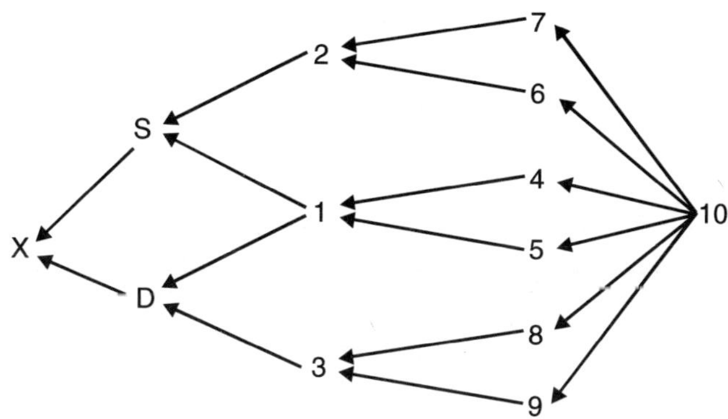

Solution :

(a) Calculation of F_X:

- *Identification of common ancestors of 'X'*: Common ancestors of 'X' are '1' and '10' as they appear on maternal (D) as well as on paternal (S) sides of pedigree of 'X'.
- *Calculation of the inbreeding coefficient of common ancestor(s)*:

  Of the two common ancestors, only '1' is inbred because its parents (4 and 5) are genetically related to each other through a common ancestor (10)

Calculation of inbreeding coefficient of '1' (i.e. the inbred common ancestor) is shown below:

CA	Path no.	Path chain	*n*	$(1/2)^n (1+F_{CA})$	Contribution
10	1	1←4←[10]→5→1	2	$(½)^2(1+0)$	0.25
				Summation =	0.25

$$F_1 = \frac{1}{2}\sum\left[\left(\frac{1}{2}\right)^n (1+F_{CA})\right] = ½(0.25) = 0.125$$

Calculation of inbreeding coefficient of 'X': Common ancestors of 'X' are '1' and '10'. The possible numbers of paths connecting parents of 'X' through common ancestors are:

CA	Path no.	Path chain	*n*	$(1/2)^n (1+F_{CA})$	Contribution
1	1	X←S←[1]→D→X	2	$(½)^2(1+0.125)$	0.28125
10	2	X←S←2←7←[10]→4→1→D→X	6	$(½)^6(1+0)$	0.015625
	3	X←S←2←7←[10]→5→1→D→X	6	$(½)^6(1+0)$	0.015625
	4	X←S←2←7←[10]→8→3→D→X	6	$(½)^6(1+0)$	0.015625
	5	X←S←2←7←[10]→9→3→D→X	6	$(½)^6(1+0)$	0.015625
	6	X←S←2←6←[10]→9→3→D→X	6	$(½)^6(1+0)$	0.015625

Contd....

CA	Path no.	Path chain	n	$(1/2)^n(1+F_{CA})$	Contribution
	7	X ←S ←2←6← [10] →5→1→D →X	6	$(½)^6(1+0)$	0.015625
	8	X ←S ←2←6← [10] →8→3→D →X	6	$(½)^6(1+0)$	0.015625
	9	X ←S ←2←6← [10] →4→1→D →X	6	$(½)^6(1+0)$	0.015625
	10	X ←S ←1←4← [10] →8→3→D →X	6	$(½)^6(1+0)$	0.015625
	11	X ←S ←1←4← [10] →9→3→D →X	6	$(½)^6(1+0)$	0.015625
	12	X ←S ←1←5← [10] →8→3→D →X	6	$(½)^6(1+0)$	0.015625
	13	X ←S ←1←5← [10] →9→3→D →X	6	$(½)^6(1+0)$	0.015625
					Summation = 0.46875

$$F_X = \frac{1}{2}\sum\left[\left(\frac{1}{2}\right)^n(1+F_{CA})\right] = ½(0.46875) = 0.234375$$

Note: *In this pedigree there are two intermediate ancestors viz. 2 and 3 who are inbred but they do not qualify as common ancestors of 'X'. Inbreeding coefficients of such intermediate inbred ancestors are, therefore, not required for calculations* (rule II of step III).

- ***Inference***: Inbreeding coefficient of an animal that appears in the pedigree as an intermediate common ancestor is not required for calculations.

(b) Calculation of F_S:

Common ancestor of 'S' is '10' who is not inbred. The possible numbers of paths connecting the parents of 'S' through common ancestor are:

CA	Path no.	Path chain	n	$(1/2)^n(1+F_{CA})$	Contribution
10	1	S ←2←7← [10] →4→1→S	4	$(½)^4(1+0)$	0.0625
	2	S ←2←7← [10] →5→1→S	4	$(½)^4(1+0)$	0.0625
	3	S ←2←6← [10] →4→1→S	4	$(½)^4(1+0)$	0.0625
	4	S ←2←6← [10] →5→1→S	4	$(½)^4(1+0)$	0.0625
					Summation = 0.250

$$F_S = \frac{1}{2}\sum\left[\left(\frac{1}{2}\right)^n(1+F_{CA})\right] = ½\ (0.25) = 0.125$$

12.9. Calculate the inbreeding coefficient of 'A' from the path diagram given below:

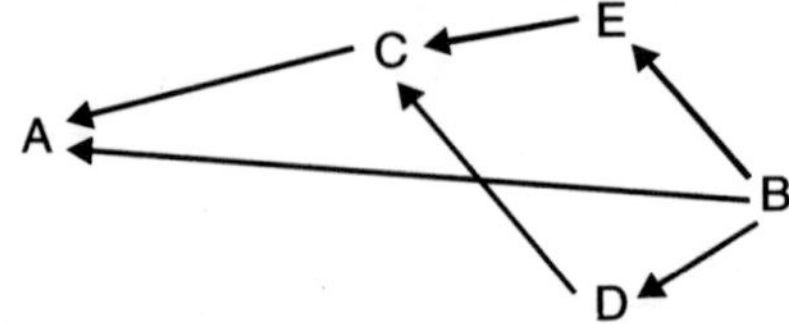

Solution :

Common ancestor of 'A' is 'B' who is not inbred. The paths are:

CA	Path no.	Path chain	n	$(1/2)^n(1+F_{CA})$	Contribution
B	1	A←C←E←[B]→A	2	(½)²(1+0)	0.25
	2	A←C←D←[B]→A	2	(½)²(1+0)	0.25
				Summation =	0.50

$$F_A = \frac{1}{2}\sum\left[\left(\frac{1}{2}\right)^n (1+F_{CA})\right] = ½\ (0.50) = 0.25.$$

12.10. Compute the inbreeding coefficient resulting from two successive generations of:

(a) Full-sib mating (b) Half-sib mating

(c) Sire-daughter mating (d) Selfing

Solution :

(a) Two generations of full-sib mating:

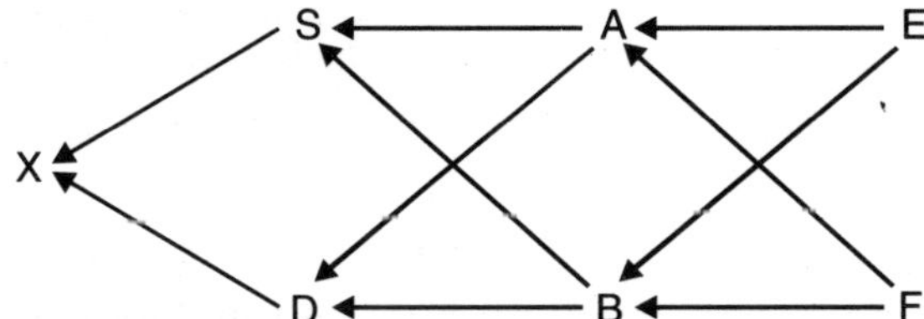

- *First generation of full-sib mating*: Mating of 'A' and 'B' to produce 'S' and 'D': The paths are:

CA	Path no.	Path chain	n	$(1/2)^n(1+F_{CA})$	Contribution
E	1	S←A←[E]→B→S	2	(1/2)²(1+0)	0.25
F	2	S←A←[F]→B→S	2	(1/2)²(1+0)	0.25
				Summation =	0.50

Inbreeding coefficient ($F_S = F_D$) = (1/2) x 0. 50 = 0.25

- *Second generation of full sib mating*: Mating of 'S' and 'D' to produce 'X'. The paths are:

CA	Path no.	Path chain	n	$(1/2)^n(1+F_{CA})$	Contribution
A	1	X←S←[A]→D→X	2	(1/2)²(1+0)	0.25
B	2	X←S←[B]→D→X	2	(1/2)²(1+0)	0.25
E	3	X←S←A←[E]→B→D→X	4	(1/2)⁴(1+0)	0.0625
E	4	X←D←A←[E]→B→S→X	4	(1/2)⁴(1+0)	0.0625
F	5	X←S←A←[F]→B→D→X	4	(1/2)⁴(1+0)	0.0625
F	6	X←D←A←[F]→B→S→X	4	(1/2)⁴(1+0)	0.0625
				Summation =	0.75

Inbreeding coefficient (F_X=F_Y) = (1/2) x 0.750 = 0.375

(b) Two generations of half-sib mating:

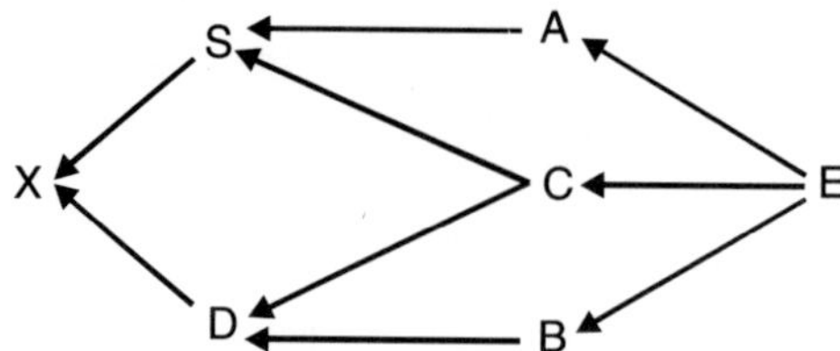

- *First generation of half sib mating*: Mating among 'A', 'C' and 'B' to produce 'S' and 'D'.

  Paths for calculation of inbreeding coefficient of 'S' are:

CA	Path no.	Path chain	n	$(1/2)^n(1+F_{CA})$	Contribution
E	1	S←A← [E] →C→S	2	$(1/2)^2(1+0)$	0.25
				Summation =	0.25

  Inbreeding coefficient ($F_S = F_D$) = (1/2) x 0.25 = 0.125

- *Second generation of half sib mating*: Mating of 'S' and 'D' to produce 'X' Paths for calculation are:

CA	Path no.	Path chain	n	$(1/2)^n(1+F_{CA})$	Contribution
E	1	X←S←A← [E] →B→D→X	4	$(1/2)^4(1+0)$	0.0625
	2	X←S←A← [E] →C→D→X	4	$(1/2)^4(1+0)$	0.0625
	3	X←S←C← [E] →B→D→X	4	$(1/2)^4(1+0)$	0.0625
C	4	X←S← [C] →D→X	2	$(1/2)^2(1+0)$	0.25
				Summation =	0.4375

Inbreeding coefficient (F_X) = (1/2) x 0.4375 = 0.218

(c) Two generations of sire-daughter mating:

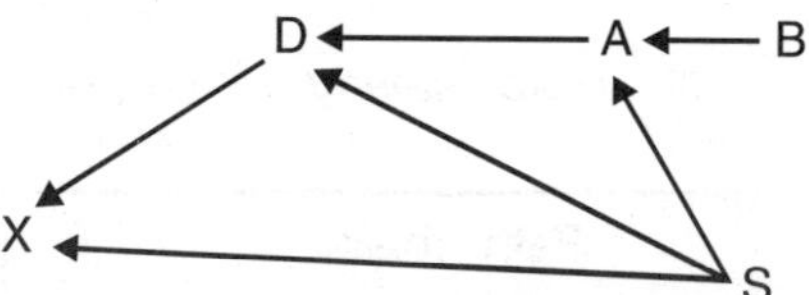

- *First generation of sire-daughter mating*: Mating of 'S' and 'A' to produce 'D'.

  Paths for calculation are:

CA	Path no.	Path chain	n	$(1/2)^n(1+F_{CA})$	Contribution
S	1	D←A← [S] →D	1	$(1/2)^1(1+0)$	0.50
				Summation =	0.50

  Inbreeding coefficient (F_D) = (1/2) x 0.50 = 0.25

- *Second generation of sire-daughter mating*: Mating of 'S' and 'D' to produce 'X'.

  Paths for calculation are:

CA	Path no.	Path chain	n	$(1/2)^n(1+F_{CA})$	Contribution
S	1	X←D←A←[S]→X	2	$(1/2)^2(1+0)$	0.25
	2	X←D←[S]→X	1	$(1/2)^1(1+0)$	0.50
				Summation =	0.75

Inbreeding coefficient (F_X) = (1/2) x 0.75 = 0.375

(d) Two generations of selfing:

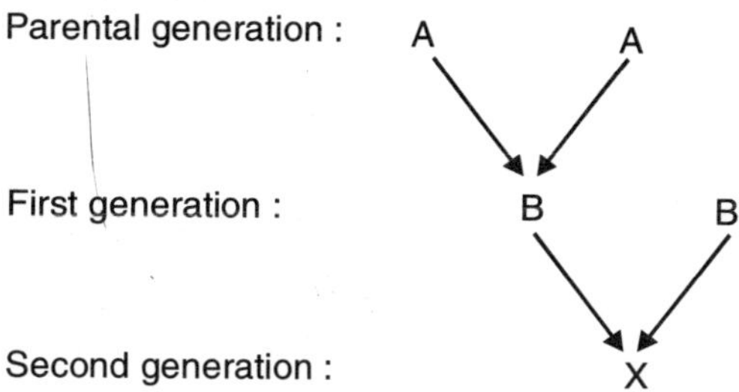

- *First generation of selfing*: Mating of 'A' with 'A' to produce 'B':

CA	Path no.	Path chain	n	$(1/2)^n(1+F_{CA})$	Contribution
A	1	B←[A]→B	0	$(1/2)^0(1+0)$	1.0
				Summation =	1.0

F_B = (1/2) (1.0) = 0.50

- *Second generation of selfing*: Mating of 'B' with 'B' to produce 'X':

CA	Path no.	Path chain	n	$(1/2)^n(1+F_{CA})$	Contribution
B	1	X←[B]→X	0	$(1/2)^0(1+0.5)$	1.5
				Summation =	1.5

Inbreeding coefficient (F_X) = (1/2) (1.50) = 0.75

Calculation of Relationship Coefficient (R)

The basis for measuring the relationship (genetic) between two animals is determining the closeness of their relationship through common ancestor(s). Calculation of relationship coefficient also involves tracing and counting the number of paths (generations) connecting the concerned individuals through common ancestor(s), and is thus similar to that used for inbreeding coefficient.

The formula for relationship coefficient between two individuals, 'X' and 'Y', designated as 'R_{XY}' is:

$$R_{XY} = \frac{\sum\left\{\left(\frac{1}{2}\right)^n(1+F_{CA})\right\}}{\sqrt{(1+F_X)(1+F_Y)}}$$

Where:

R_{XY} = Relationship coefficient between individuals 'X' and 'Y'

Σ = Addition for all the possible paths connecting 'X' and 'Y' (*i.e.*, if there are more than one paths connecting individuals 'X' and 'Y', the number of paths are computed separately and their coefficients are added together).

n = Number of arrows/paths connecting 'X' with 'Y'

F_X = Inbreeding coefficient of 'X'.

F_Y = Inbreeding coefficient of 'Y'.

F_{CA} = Inbreeding coefficient of the common ancestor.

The numerator of the expression for relationship coefficient corresponds to the fraction of inheritance shared by the two individuals or received by an individual from specific ancestor. Fraction of inheritance received by an individual from some common outstanding ancestor is required for designing a line breeding program. It is worked out by the following expression:

$$\text{Fraction of inheritance} = \Sigma\left(\frac{1}{2}\right)^n (1+F_{CA})$$

It can be converted to percent by multiplying it by 100.

If 'X' and 'Y', and their common ancestors are not inbred, then the formula of relationship coefficient reduces to:

$$R_{XY} = (½)^n$$

(a) Direct Relationships

12.11. From the pedigree given below, calculate: (a) R_{AI} (b) fraction of inheritance received by 'I' from 'A'.

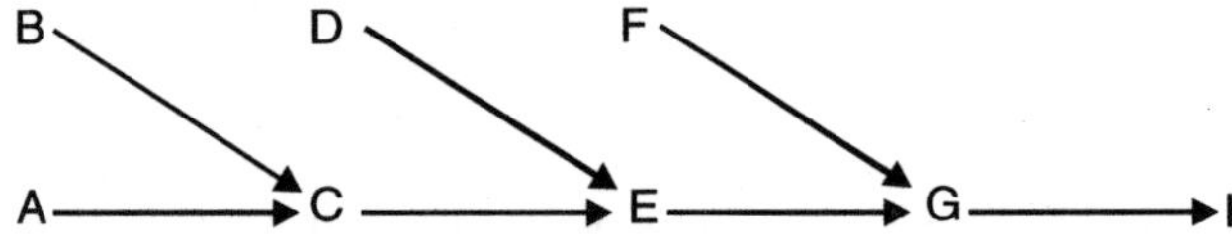

Solution :

(a) There is only one path connecting 'A' and 'I':

Path no.	Path chain	n	$(1/2)^n (1+F_{CA})$	Contribution
1	A→C→E→G→I	4	$(1/2)^4(1+0)$	0.0625
			Summation =	0.0625

Inbreeding coefficients of 'A' and '*I*' are zero as they are not inbred :

Relationship coefficient between 'A' and 'I' is:

$$R_{AI} = \frac{\Sigma\left(\frac{1}{2}\right)^n (1+F_{CA})}{\sqrt{(1+F_A)(1+F_I)}} = 0.0625$$

(b) Fraction of inheritance: $= \Sigma\left(\frac{1}{2}\right)^n (1+F_{CA})$

$= 0.0625$

Inference: When the individual and its common ancestor(s) are not inbred, the relationship coefficient equals the fraction of inheritance.

(b) Collateral Relationships

12.12. Compute the relationship coefficients for the following types of relatives: (a) Half sibs, (b) Full sibs, (c) Single first cousins, (d) Single second cousins, (e) Double first cousins, (f) Uncle and niece.

Solution :

(a) Half-sibs: (Individuals with one common parent)

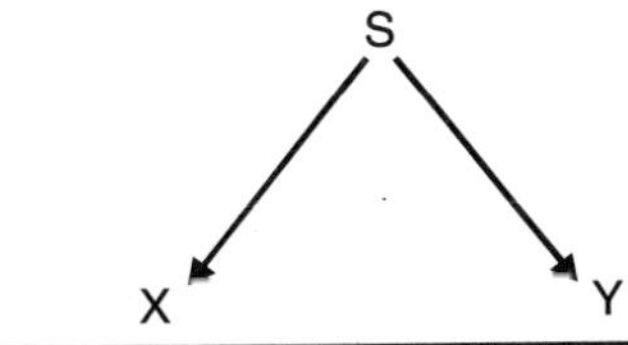

Path no.	Path chain	n	$(1/2)^n(1+F_{CA})$	Contribution
1	X← S →Y	2	$(1/2)^2(1+0)$	0.25

$$R_{XY} = \frac{\sum\left(\frac{1}{2}\right)^n(1+F_{CA})}{\sqrt{(1+F_X)(1+F_Y)}} = \frac{0.25}{\sqrt{(1+0)(1+0)}} = 0.25 \quad (\text{since } F_X = F_Y = 0)$$

(b) Full sibs: (Individuals with both parents common)

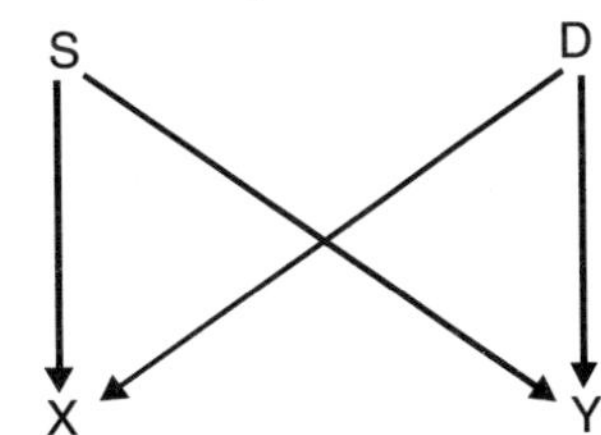

Path no.	Path chain	n	$(1/2)^n(1+F_{CA})$	Contribution
1	X← S →Y	2	$(1/2)^2(1+0)$	0.25
2	X← D →Y	2	$(1/2)^2(1+0)$	0.25
			Summation =	0.50

$$R_{XY} = \frac{\sum\left(\frac{1}{2}\right)^n(1+F_{CA})}{\sqrt{(1+F_X)(1+F_Y)}} = \frac{0.50}{\sqrt{(1+0)(1+0)}} = 0.5 \quad (\text{since } F_X = F_Y = 0)$$

(c) Single first cousins: (Individuals having two common grand-parents)

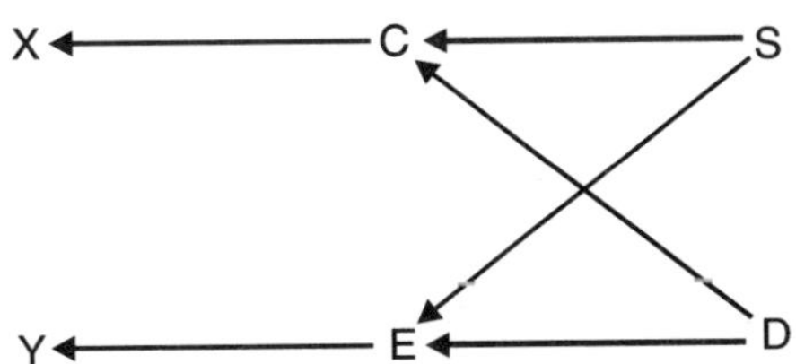

Path no.	Path chain	n	$(1/2)^n(1+F_{CA})$	Contribution
1	X←C←[S]→E→Y	4	$(1/2)^4(1+0)$	0.0625
2	X←C←[D]→E→Y	4	$(1/2)^4(1+0)$	0.0625
			Summation =	0.125

$$R_{XY} = \frac{\sum\left(\frac{1}{2}\right)^n(1+F_{CA})}{\sqrt{(1+F_X)(1+F_Y)}} = \frac{0.125}{\sqrt{(1+0)(1+0)}} = 0.125$$

(d) Single second cousins: (Individuals having two common great grand-parents)

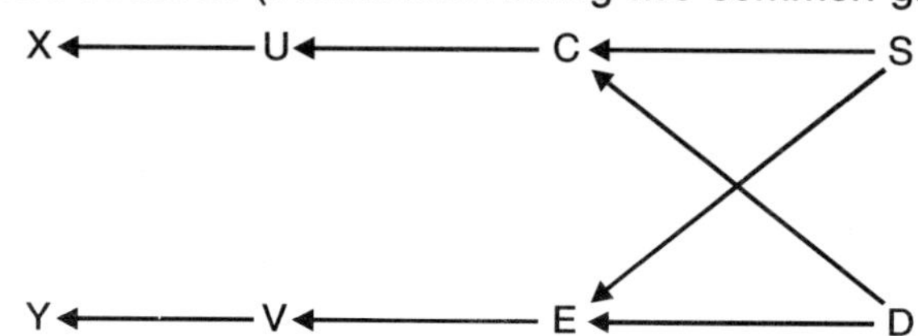

Path no.	Path chain	n	$(1/2)^n(1+F_{CA})$	Contribution
1	X←U←C←[D]→E→V→Y	6	$(1/2)^6(1+0)$	0.0156
2	X←U←C←[S]→E→V→Y	6	$(1/2)^6(1+0)$	0.0156
			Summation =	0.0312

$$R_{XY} = \frac{\sum\left(\frac{1}{2}\right)^n(1+F_{CA})}{\sqrt{(1+F_X)(1+F_Y)}} = \frac{0.0312}{\sqrt{(1+0)(1+0)}} = 0.0312$$

(e) Double first cousins: (Individuals with four common grand-parents)

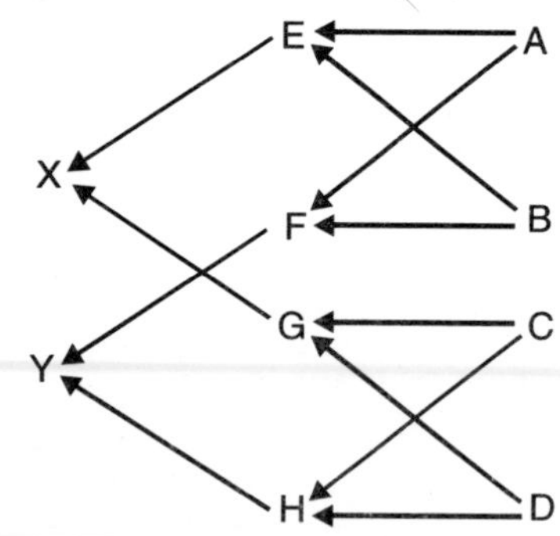

Path no.	Path chain	n	$(1/2)^n(1+F_{CA})$	Contribution
1	X←E←[A]→F→Y	4	$(1/2)^4(1+0)$	0.0625
2	X←E←[B]→F→Y	4	$(1/2)^4(1+0)$	0.0625
3	X←G←[C]→H→Y	4	$(1/2)^4(1+0)$	0.0625
4	X←G←[D]→H→Y	4	$(1/2)^4(1+0)$	0.0625
			Summation =	0.25

$$R_{XY} = \frac{\sum\left(\frac{1}{2}\right)^n (1+F_{CA})}{\sqrt{(1+F_X)(1+F_Y)}} = \frac{0.25}{\sqrt{(1+0)(1+0)}} = 0.25$$

(f) Uncle and niece: D is uncle & E is niece.

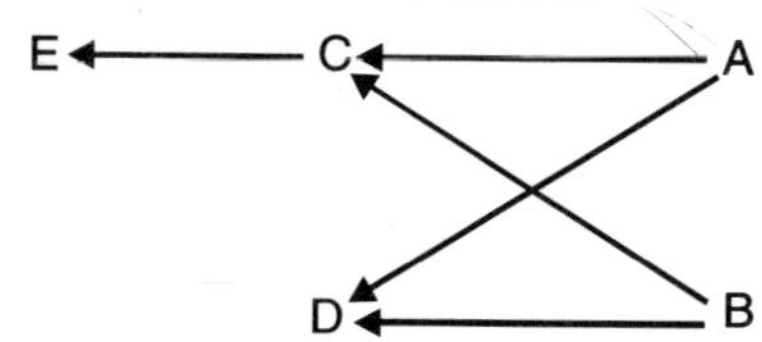

Path no.	Path chain	n	$(1/2)^n (1+F_{CA})$	Contribution
1	E←C←[A]←D	3	$(1/2)^3(1+0)$	0.125
2	E←C←[B]←D	3	$(1/2)^3(1+0)$	0.125
			Summation =	0.25

$$R_{DE} - \frac{\sum\left(\frac{1}{2}\right)^n (1+F_{CA})}{\sqrt{(1+F_D)(1+F_E)}} = \frac{0.25}{\sqrt{(1+0)(1+0)}} = 0.25$$

(c) Mixed Relationships

12.13. Compute: (a) R_{SD}, (b) R_{X1}, (c) percentage of inheritance received by 'X' from '1' for the pedigree given below.

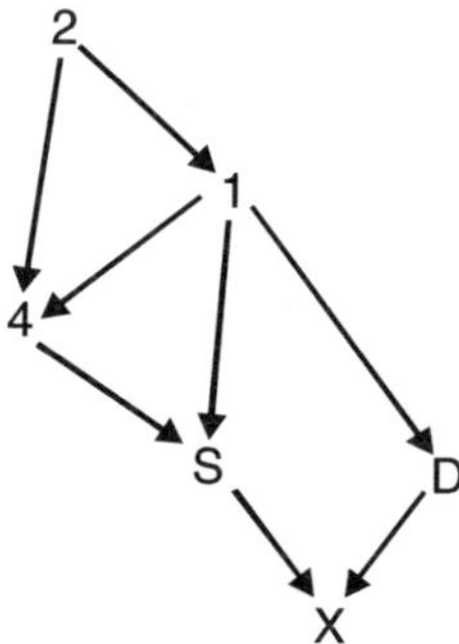

Solution :

(a) Calculation of R_{SD}: Paths connecting 'S' and 'D' are:

Path no.	Path chain	n	$(1/2)^n (1+F_{CA})$	Contribution
1	D←[1]→4→S	3	$(1/2)^3(1+0)$	0.125
2	S←[1]→D	2	$(1/2)^2(1+0)$	0.250
3	S←4←[2]→1→D	4	$(1/2)^4(1+0)$	0.0625
			Summation =	0.6875

For the calculation of R_{SD}, the inbreeding coefficient of 'S' and 'D' are required: There is a single path connecting parents of 'S':

'S←[1]→4→S'

$F_S = (1/2)ó\,(1/2)^1 = 0.25$; $F_D = 0.0$

$$R_{SD} = \frac{\sum\left(\frac{1}{2}\right)^n (1+F_{CA})}{\sqrt{(1+F_S)(1+F_D)}} = \frac{0.4375}{\sqrt{(1+0.25)(1+0)}} = 0.391$$

(b) Calculation of R_{X1}: Paths connecting 'X' and '1' are:

Path no.	Path chain	n	$(1/2)^n(1+F_{CA})$	Contribution
1	X←S←1	2	$(1/2)^2(1+0)$	0.250
2	X←S←4←1	3	$(1/2)^3(1+0)$	0.125
3	X←D←1	2	$(1/2)^2(1+0)$	0.250
4	X←S←4←2←1	4	$(1/2)^4(1+0)$	0.0625
			Summation =	0.6875

Individual 'X' is inbred and its inbreeding coefficient is calculated below:

Path no.	Path chain	n	$(1/2)^n(1+F_{CA})$	Contribution
1	X←S←[1]→D→X	2	$(1/2)^2(1+0)$	0.25
2	X←S←4←[1]→D→X	3	$(1/2)^3(1+0)$	0.125
3	X←S←4←[2]→1→D→X	4	$(1/2)^4(1+0)$	0.0625
			Summation =	0.4375

$$F_X = \frac{1}{2}\sum\left(\frac{1}{2}\right)^n (1+F_{CA}) = (1/2)\ (0.4375) \quad = 0.21825 \text{ and } F_1 = 0$$

$$R_{X1} = \frac{\sum\left[\left(\frac{1}{2}\right)^n (1+F_{CA})\right]}{\sqrt{(1+F_X)(1+F_i)}} = \frac{0.6875}{\sqrt{(1+0.21825)(1+0)}} = 0.62275$$

(c) The numerator of R_{X1} is the percentage of inheritance received by 'X' from '1':
Percentage of inheritance = 100 × 0.735 = 73.5

12.14. From the pedigree diagram given below, compute:

(a) F_B, F_A, F_z (b) R_{AF} (c) R_{AE} (d) R_{AB} (e) R_{JF} (f) R_{AJ} (g) R_{CD}

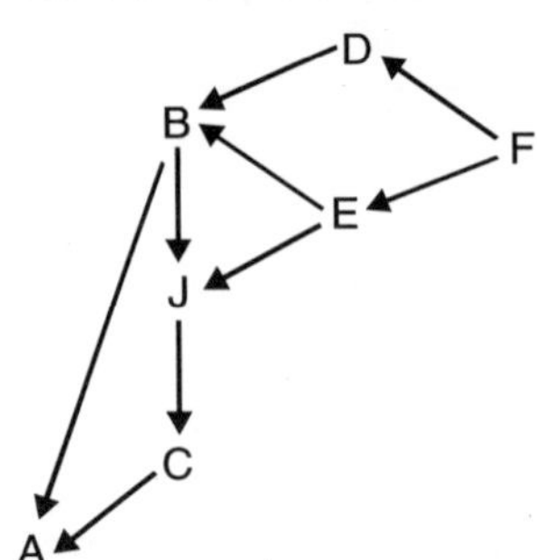

Solution :

(a) Calculation of F_B , F_A and F_J:

F_B: Common ancestor of 'B' is 'F' , and the paths are:

CA	Path no.	Path chain	n	$(1/2)^n (1+F_{CA})$	Contribution
F	1	B←D← [F] →E→B	2	$(1/2)^2(1+0)$	0.25
				Summation =	0.25

$$F_B = \frac{1}{2}\sum\left[\left(\frac{1}{2}\right)^n (1+F_{CA})\right] = (1/2)\ (0.250) = 0.125$$

F_A: Common ancestors of 'A' are 'B', 'E' and 'F'. Of these, only 'B' is inbred whose common ancestor is 'F'. Inbreeding coefficient of 'B' is 0.125 (as calculated above). Paths for calculation of F_A are:

CA	Path no.	Path chain	n	$(1/2)^n (1+F_{CA})$	Contribution
B	1	A← [B] →J→C→A	2	$(1/2)^2(1+0.125)$	0.281
E	2	A←B← [E] →J→C→A	3	$(1/2)^3(1+0)$	0.125
F	3	A←B←D← [F] →E→J→C→A	5	$(1/2)^5(1+0)$	0.03125
				Summation =	0.4375

$$F_A = \frac{1}{2}\sum\left(\frac{1}{2}\right)^n (1+F_{CA}) = (1/2)\ (0.4375) = 0.21875$$

F_J: Paths are:

CA	Path no.	Path chain	n	$(1/2)^n (1+F_{CA})$	Contribution
E	1	J←B← [E] →J	1	$(1/2)^1(1+0)$	0.50
F	2	J←B←D← [F] →E→J	3	$(1/2)^3(1+0)$	0.125
				Summation =	0.625

$$F_J = \frac{1}{2}\sum\left(\frac{1}{2}\right)^n (1+F_{CA}) = (1/2)\ (0.625) = 0.3125$$

(b) R_{AF}: Inbreeding coefficients of 'A', 'B' and 'J' calculated above (F_A= 0.21875, F_B= 0.125 and F_J= 0.3125) are used below for computing the relationship coefficients. The paths connecting 'A' and 'F' are:

Path no.	Path chain	n	$(1/2)^n (1+F_{CA})$	Contribution
1	F→D→B→A	3	$(1/2)^3(1+0)$	0.125
2	F→E→B→A	3	$(1/2)^3(1+0)$	0.125
3	F→E→B→J→C→A	5	$(1/2)^5(1+0)$	0.03125
4	F→D→B→J→C→A	5	$(1/2)^5(1+0)$	0.03125
5	F→E→J→C→A	4	$(1/2)^4(1+0)$	0.0625
			Summation =	0.375

$$R_{AF} = \frac{\sum\left(\frac{1}{2}\right)^n (1+F_{CA})}{\sqrt{(1+F_A)(1+F_F)}} = \frac{0.375}{\sqrt{(1+0.21875)(1+0)}} = 0.339$$

(c) R_{AE}:

The paths connecting 'A' and 'E' are:

Path no.	Path chain	n	$(1/2)^n (1+F_{CA})$	Contribution
1	E→B→A	2	$(1/2)^2(1+0)$	0.25
2	E→B→J→C→A	4	$(1/2)^4(1+0)$	0.0625
3	E←F→D→B→A	4	$(1/2)^4(1+0)$	0.0625
4	E←F→D→B→J→C→A	6	$(1/2)^6(1+0)$	0.015625
5	E→J→C→A	3	$(1/2)^3(1+0)$	0.125
			Summation =	0.5156

$$R_{AE} = \frac{\sum\left(\frac{1}{2}\right)^n (1+F_{CA})}{\sqrt{(1+F_A)(1+F_E)}} = \frac{0.5156}{\sqrt{(1+0.21875)(1+0)}} = 0.467$$

(d) R_{AB}: 'B' is inbred, F_B= 0.125(calculated above).

The paths connecting 'A' and 'B' are:

Path no.	Path chain	n	$(1/2)^n (1+F_{CA})$	Contribution
1	B→A	1	$(1/2)^1(1+0.125)$	0.5625
2	B→J→C→A	3	$(1/2)^3(1+0.125)$	0.140625
3	A←C←J←E→B	4	$(1/2)^4(1+0)$	0.0625
4	A←C←J←E←F→D→B	6	$(1/2)^6(1+0)$	0.015625
			Summation =	0.78125

$$R_{AB} = \frac{\sum\left(\frac{1}{2}\right)^n (1+F_{CA})}{\sqrt{(1+F_A)(1+F_B)}} = \frac{0.78125}{\sqrt{(1+0.21875)(1+0.125)}} = 0.667$$

(e) R_{JF}:

The paths connecting 'J' and 'F' are:

Path no.	Path chain	n	$(1/2)^n (1+F_{CA})$	Contribution
1	F→D→B→J	3	$(1/2)^3(1+0)$	0.125
2	F→E→B→J	3	$(1/2)^3(1+0)$	0.125
3	F→E→J	2	$(1/2)^2(1+0)$	0.25
			Summation =	0.50

$$R_{JF} = \frac{\sum\left(\frac{1}{2}\right)^n (1+F_{CA})}{\sqrt{(1+F_J)(1+F_F)}} = \frac{0.50}{\sqrt{(1+0.3125)(1+0)}} = 0.436$$

(f) R_{AJ}:

The paths connecting 'A' and 'J' are:

Path no.	Path chain	n	$(1/2)^n(1+F_{CA})$	Contribution
1	J→C→A	2	$(1/2)^2(1+0)$	0.25
2	J←B→A	2	$(1/2)^2(1+0.125)$	0.28125
3	A←B←D←F→E→J	5	$(1/2)^5(1+0)$	0.03125
4	A←B←E→J	3	$(1/2)^3(1+0)$	0.125
			Summation =	0.6875

$$R_{AJ} = \frac{\sum\left(\frac{1}{2}\right)^n(1+F_{CA})}{\sqrt{(1+F_A)(1+F_J)}}$$

$$= \frac{0.6875}{\sqrt{(1+0.21875)(1+0.3125)}} = 0.543583$$

(g) R_{CD}:

The paths connecting 'C' and 'D' are:

Path no.	Path chain	n	$(1/2)^n(1+F_{CA})$	Contribution
1	D→B→J→C	3	$(1/2)^3(1+0)$	0.125
2	D←F→E→J→C	4	$(1/2)^4(1+0)$	0.0625
3	D←F→E→B→J→C	5	$(1/2)^5(1+0)$	0.03125
			Summation =	0.21875

$$R_{CD} = \frac{\sum\left(\frac{1}{2}\right)^n(1+F_{CA})}{\sqrt{(1+F_C)(1+F_D)}} = \frac{0.21875}{\sqrt{(1+0)(1+0)}} = 0.21875$$

12.15. From the pedigree given below, compute: (a) Fx (b) R_{X5} (c) Percentage of inheritance received by 'X' from '5'.

Solution :

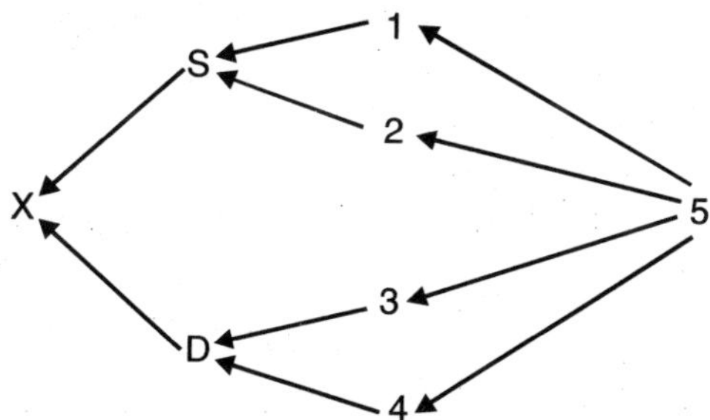

(a) Fx : Common ancestor of 'X' is '5' who is non-inbred.

The paths through the common ancestor '5' are :

Path no.	Path chain	n	$(1/2)^n(1+F_{CA})$	Contribution
1	X←S←1←[5]→4→D→X	4	$(1/2)^4(1+0)$	0.0625
2	X←S←1←[5]→3→D→X	4	$(1/2)^4(1+0)$	0.0625
3	X←S←2←[5]→3→D→X	4	$(1/2)^4(1+0)$	0.0625
4	X←S←2←[5]→4→D→X	4	$(1/2)^4(1+0)$	0.0625
			Summation =	0.250

$$F_X = \frac{1}{2}\sum\left(\frac{1}{2}\right)^n(1+F_{CA}) = 1/2\ (0.250) = 0.125$$

(b) R_{X5}: For calculation of coefficient of relationship, we need F_x and F_5.

F_X has been calculated above in (a) to be 0.125

Ancestor '5' is not inbred ($F_5 = 0.0$).

The paths connecting 'X' and '5' for computation of 'R_{X5}' are:

Path no.	Path chain	n	$(1/2)^n(1+F_{CA})$	Contribution
1	5→1→S→X	3	$(1/2)^3(1+0)$	0.125
2	5→2→S→X	3	$(1/2)^3(1+0)$	0.125
3	5→3→D→X	3	$(1/2)^3(1+0)$	0.125
4	5→4→D→X	3	$(1/2)^3(1+0)$	0.125
			Summation =	0.50

$$R_{X5} = \frac{\sum\left(\frac{1}{2}\right)^n(1+F_{CA})}{\sqrt{(1+F_X)(1+F_5)}} = \frac{0.50}{\sqrt{(1+0.125)(1+0)}} = 0.47$$

(c) Percentage of inheritance received by 'X' from '5': The numerator of the relationship coefficient denotes the percentage of inheritance, which equals 0.5 or 50%.

12.16. From the pedigree given below, calculate:

(a) F_X, F_S, F_D (b) R_{SD}, R_{X9} .

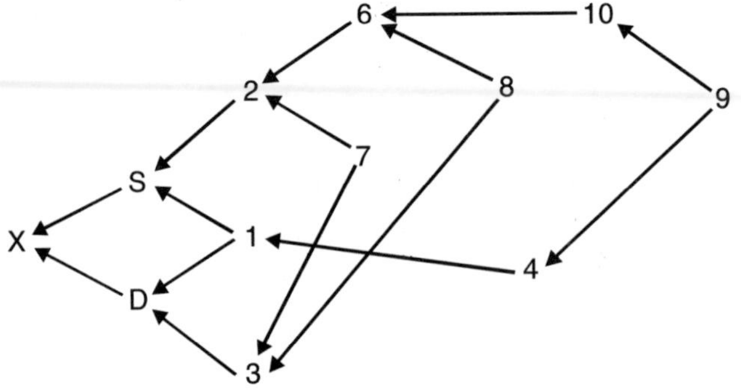

Solution :

(a) F_X: Common ancestors of 'X' are 1, 7, 8 and 9 and none of them is inbred. The paths are:

CA	Path no.	Path chain	n	$(1/2)^n (1+F_{CA})$	Contribution
1	1	X←S←[1]→D→X	2	$(½)^2(1+0)$	0.25
7	2	X←S←2←[7]→3→D→X	4	$(½)^4(1+0)$	0.0625
8	3	X←S←2←6←[8]→3→D→X	5	$(½)^5(1+0)$	0.03125
9	4	X←S←2←6←10←[9]→4→1→D→X	7	$(½)^7(1+0)$	0.0078125
				Summation =	0.35155

$F_X = ½\ \Sigma\ (½)^n\ (1+F_{CA}) = ½(0.35155) = 0.175775$

F_S: Common ancestor of 'S' is 9 who is not inbred. There is a single path:

CA	Path no.	Path chain	n	$(1/2)^n (1+F_{CA})$	Contribution
9	1	S←2←6←10←[9]→4→1→S	5	$(½)^5(1+0)$	0.03125
				Summation =	0.03125

$F_S = ½\ \Sigma(½)^n\ (1+F_{CA}) = ½\ (0.03125) = 0.015625$;

$F_D = 0$, as it has no common ancestor

(b) R_{SD}: Paths connecting 'S' and 'D' are:

Path no.	Path chain	n	$(1/2)^n (1+F_{CA})$	Contribution
1	S←1→D	2	$(½)^2(1+0)$	0.25
2	S←2←7→3→D	4	$(½)^4(1+0)$	0.0625
3	S←2←6←8→3→D	5	$(½)^5(1+0)$	0.03125
4	S←2←6←10←9→4→1→D	7	$(½)^7(1+0)$	0.00781
			Summation =	0.35156

$$R_{SD} = \frac{\sum\left(\frac{1}{2}\right)^n (1+F_{CA})}{\sqrt{(1+F_S)(1+F_D)}} = \frac{0.35156}{\sqrt{(1+0.015625)(1+0)}} = 0.3488$$

R_{X9} : Paths connecting 'X' and '9' are:

Path no.	Path chain	n	$(1/2)^n (1+F_{CA})$	Contribution
1	X←S←2←6←10←9	5	$(½)^5(1+0)$	0.03125
2	X←D←1←4←9	4	$(½)^4(1+0)$	0.0625
3	X←S←1←4←9	4	$(½)^4(1+0)$	0.0625
			Summation =	0.15625

$$R_{X9} = \frac{\sum\left(\frac{1}{2}\right)^n (1+F_{CA})}{\sqrt{(1+F_X)(1+F_9)}} = \frac{0.15625}{\sqrt{(1+0.175775)(1+0)}} = 0.1441$$

12.17. From the pedigree given below, calculate : (a) F_X (b) R_{SD} (c) Percentage of inheritance received by 'X' from 'S'.

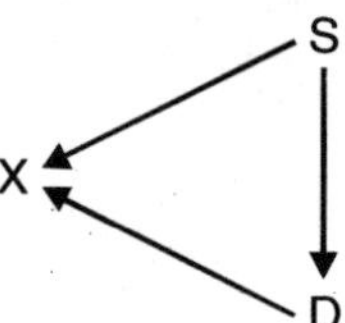

Solution :

(a) F_X: The common ancestor of 'X' is 'S', who is not inbred

CA	Path no.	Path chain	n	$(1/2)^n (1+F_{CA})$	Contribution
S	1	X←S→D→X	1	$(½)^1(1+0)$	0.5
				Summation =	0.5

$F_X = ½(0.5) = 0.25$

(b) R_{SD}: There is a single path connecting 'S' and 'D':

Path no.	Path chain	n	$(1/2)^n (1+F_{CA})$	Contribution
1	S→D	1	$(½)^1(1+0)$	0.5
			Summation =	0.5

$$R_{SD} = \frac{\sum\left(\frac{1}{2}\right)^n (1+F_{CA})}{\sqrt{(1+F_S)(1+F_D)}} = \frac{0.5}{\sqrt{(1+0)(1+0)}} = 0.5$$

12.18. From the pedigree given below, calculate : (a) F_X, F_S, F_D (b) R_{SD}, R_{X5} (c) Percentage of inheritance received by 'X' from '5'.

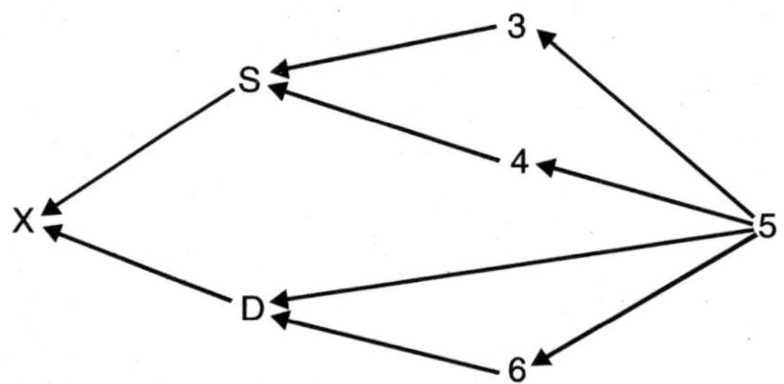

Solution :

(a) F_X : Common ancestor of 'X' is '5' who is not inbred. The paths are:

CA	Path no.	Path chain	n	$(1/2)^n (1+F_{CA})$	Contribution
5	1	X←S←3←[5]→D→X	3	$(½)^3(1+0)$	0.125
	2	X←S←3←[5]→6→D→X	4	$(½)^4(1+0)$	0.0625
	3	X←S←4←[5]→D→X	3	$(½)^3(1+0)$	0.125
	4	X←S←4←[5]→6→D→X	4	$(½)^4(1+0)$	0.0625
				Summation =	0.375

$F_X = ½ (0.375) = 0.1875$

F_S: Common ancestor of 'S' is '5' who is not inbred. The paths are:

CA	Path no.	Path chain	n	$(1/2)^n (1+F_{CA})$	Contribution
5	1	S←3←[5]→4→S	2	$(½)^2(1+0)$	0.25
				Summation =	0.25

$F_S = ½ (0.25) = 0.125$

F_D : Common ancestor of 'D' is '5' who is not inbred. The paths are:

CA	Path no.	Path chain	n	$(1/2)^n (1+F_{CA})$	Contribution
5	1	D←[5]→6→D	1	$(½)^1(1+0)$	0.5
				Summation =	0.5

$F_D = ½ (0.5) = 0.25$

(b) R_{SD}: The paths connecting 'S' and 'D' are:

CA	Path no.	Path chain	n	$(1/2)^n (1+F_{CA})$	Contribution
5	1	S←3←[5]→D	3	$(½)^3(1+0)$	0.125
	2	S←4←[5]→D	3	$(½)^3(1+0)$	0.125
	3	S←3←[5]→6→D	4	$(½)^4(1+0)$	0.0625
	4	S←4←[5]→6→D	4	$(½)^4(1+0)$	0.0625
				Summation =	0.375

$$R_{SD} = \frac{\sum\left(\frac{1}{2}\right)^n (1+F_{CA})}{\sqrt{(1+F_S)(1+F_D)}} = \frac{0.375}{\sqrt{(1+0.125)(1+0.25)}} = 0.316$$

R_{X5}: The paths connecting 'X' and '5' are:

Path no.	Path chain	n	$(1/2)^n (1+F_{CA})$	Contribution
1	X←S←3←5	3	$(½)^3(1+0)$	0.125
2	X←S←4←5	3	$(½)^3(1+0)$	0.125
3	X←D←5	2	$(½)^2(1+0)$	0.25
4	X←D←6←5	3	$(½)^3(1+0)$	0.125
			Summation =	0.625

$$R_{X5} = \frac{\sum(\frac{1}{2})^n (1+F_{CA})}{\sqrt{(1+F_X)(1+F_5)}} = \frac{0.625}{\sqrt{(1+0.1875)(1+0.0)}} = 0.5735$$

(c) Percentage of Inheritance received by 'X' from '5'

$= \Sigma[(½)^n(1+F_A)] = 0.625$ or 62.5%

12.19. From the pedigree given below, calculate: (a) F_X , F_S , F_D (b) R_{SD}, R_{XS}

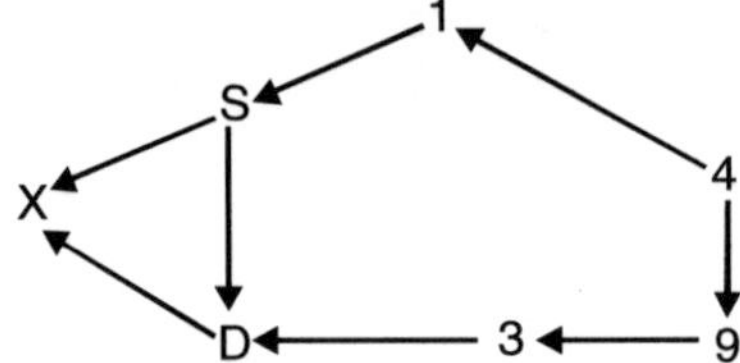

Solution :

(a) F_X: Common ancestors of 'X' are 'S' and '4':

CA	Path no.	Path chain	n	$(1/2)^n(1+F_{CA})$	Contribution
S	1	X←[S]→D→X	1	(½)¹(1+0)	0.5
4	2	X←S←1←[4]→9→3→D→X	5	(½)⁵(1+0)	0.03125
					Summation = 0.53125

F_X = ½ (0.53125) = 0.2656 ; F_S = 0 as it is not inbred

F_D: Common ancestor of 'D' is '4' who is not inbred. The paths are :

CA	Path no.	Path chain	n	$(1/2)^n(1+F_{CA})$	Contribution
4	1	D←S←1←[4]→9→3→D	4	(½)⁴(1+0)	0.0625
					Summation = 0.0625

F_D = ½ (0.0625) = 0.03125

(b) R_{SD}: The paths connecting 'S and 'D are:

Path no.	Path chain	n	$(1/2)^n(1+F_{CA})$	Contribution
1	S→D	1	(½)¹(1+0)	0.50
2	S←1←4→9→3→D	5	(½)⁵(1+0)	0.03125
				Summation = 0.53125

$$R_{SD} = \frac{\sum(\frac{1}{2})^n(1+F_{CA})}{\sqrt{(1+F_S)(1+F_D)}} = \frac{0.53125}{\sqrt{(1+0.0)(1+0.03125)}} = 0.5231$$

R_{XS}: The paths connecting 'X and 'S' are:

Path no.	Path chain	n	$(1/2)^n(1+F_{CA})$	Contribution
1	X←S	1	(½)¹(1+0)	0.5
2	X←D←S	2	(½)²(1+0)	0.25
3	X←D←3←9←4→1→S	6	(½)⁶(1+0)	0.0156
				Summation = 0.7656

$$R_{XS} = \frac{\sum(\frac{1}{2})^n(1+F_{CA})}{\sqrt{(1+F_X)(1+F_S)}} = \frac{0.7656}{\sqrt{(1+0.2656)(1+0.0)}} = 0.6805$$

12.20. From the pedigree given below, calculate:

(a) F_X, F_S, F_D (b) R_{SD}, R_{X4}

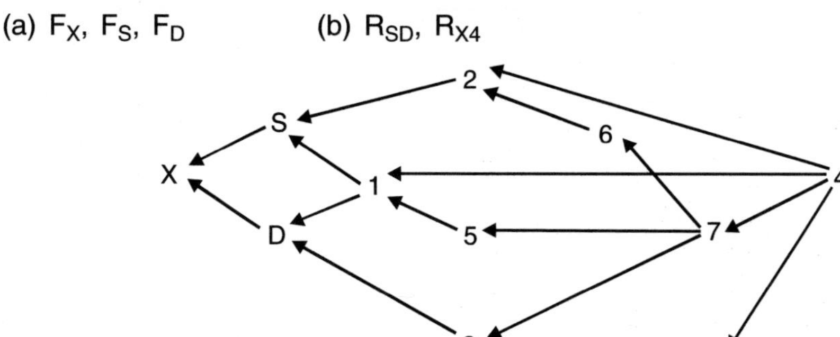

Solution :

(a) F_X: Common ancestors of 'X' are 1, 4, 7. Of these only '1' is inbred

F_1: Common ancestor of '1' is '4' who is not inbred

CA	Path no.	Path chain	n	$(1/2)^n(1+F_{CA})$	Contribution
4	1	1←[4]→7→5→1	2	(½)²(1+0)	0. 25
				Summation =	0. 25

F_1 = ½ (0. 25) = 0. 125

Paths for calculation of F_X are:

CA	Path no.	Path chain	n	$(1/2)^n(1+F_{CA})$	Contribution
1	1	X←S←[1]→D→X	2	(½)²(1+0.125)	0.28125
4	2	X←S←2←[4]→1→D→X	4	(½)⁴(1+0)	0.0625
4	3	X←S←2←[4]→7→5→1→D→X	6	(½)⁶(1+0)	0.015625
4	4	X←S←2←[4]→7→3→D→X	5	(½)⁵(1+0)	0.03125
4	5	X←S←2←[4]→8→3→D→X	5	(½)⁵(1+0)	0.03125
4	6	X←S←1←[4]→7→3→D→X	5	(½)⁵(1+0)	0.03125
4	7	X←S←1←[4]→8→3→D→X	5	(½)⁵(1+0)	0.03125
4	8	X←S←2←6←7←[4]→8→3→D→X	7	(½)⁷(1+0)	0.0078125
4	9	X←S←1←5←7←[4]→8→3→D→X	7	(½)⁷(1+0)	0.0078125
4	10	X←S←2←6←7←[4]→1→D→X	6	(½)⁶(1+0)	0.015625
7	11	X←S←2←6←[7]→3→D→X	5	(½)⁵(1+0)	0.03125
7	12	X←S←2←6←[7]→5→1→D→X	6	(½)⁶(1+0)	0.015625
7	13	X←S←1←5←[7]→3→D→X	5	(½)⁵(1+0)	0.03125
				Summation =	0.59375

F_X = ½ (0.59375) = 0.296875

F_S: Common ancestors are '4' and '7' and they are not inbred.

The paths are:

CA	Path no.	Path chain	n	$(1/2)^n(1+F_{CA})$	Contribution
4	1	S←2←[4]→7→5→1→S	4	$(½)^4(1+0)$	0.0625
4	2	S←2←[4]→1→S	2	$(½)^2(1+0)$	0.25
4	3	S←1←[4]→7→6→2→S	4	$(½)^4(1+0)$	0.0625
7	4	S←2←6←[7]→5→1→S	4	$(½)^4(1+0)$	0.0625
				Summation =	0.4375

F_S = ½ (0.4375) = 0.2187

F_D: Common ancestors of 'D' are '4' and '7', and they are not inbred. The paths are:

CA	Path no.	Path chain	n	$(1/2)^n(1+F_{CA})$	Contribution
4	1	D←1←[4]→7→3→D	3	$(½)^3(1+0)$	0.125
4	2	D←1←[4]→8→3→D	3	$(½)^3(1+0)$	0.125
4	3	D←1←5←7←[4]→8→3→D	5	$(½)^5(1+0)$	0.03125
7	4	D←1←5←[7]→3→D	3	$(½)^3(1+0)$	0.125
				Summation =	0.40625

F_D = ½ (0.40625) = 0.203125

(b) R_{SD} : The paths connecting 'S' and 'D' are the same as shown for calculation of F_X above in (a):

$$R_{SD} = \frac{\sum\left(\frac{1}{2}\right)^n(1+F_{CA})}{\sqrt{(1+F_S)(1+F_D)}} = \frac{0.59375}{\sqrt{(1+0.2187)(1+0.203125)}} = 0.490343$$

R_{X4}: The paths connecting 'X' and '4' are:

Path no.	Path chain	n	$(1/2)^n(1+F_{CA})$	Contribution
1	X←S←2←4	3	$(½)^3(1+0)$	0.125
2	X←S←1←4	3	$(½)^3(1+0)$	0.125
3	X←S←2←6←7←4	5	$(½)^5(1+0)$	0.03125
4	X←S←1←5←7←4	5	$(½)^5(1+0)$	0.03125
5	X←D←1←4	3	$(½)^3(1+0)$	0.125
6	X←D←1←5←7←4	5	$(½)^5(1+0)$	0.03125
7	X←D←3←8←4	4	$(½)^4(1+0)$	0.0625
8	X←D←3←7←4	4	$(½)^4(1+0)$	0.0625
			Summation =	0.59375

$$R_{X4} = \frac{\sum\left(\frac{1}{2}\right)^n(1+F_{CA})}{\sqrt{(1+F_X)(1+F_4)}} = \frac{0.59375}{\sqrt{(1+0.2968)(1+0.0)}} = 0.52138$$

12.21. From the pedigree given below, calculate (a) F_E, F_A (b) R_{AJ}

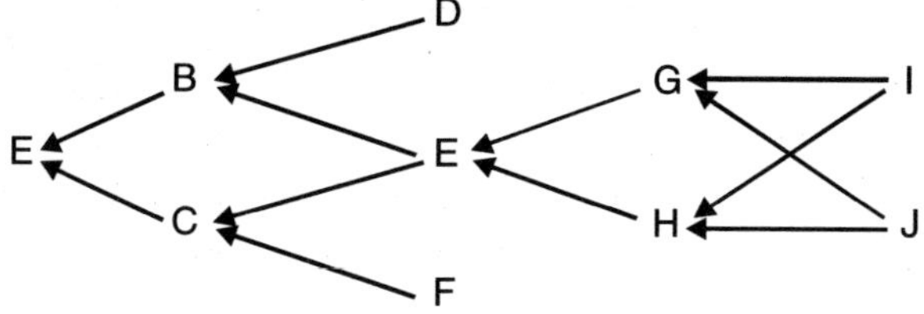

Solution :

(a) F_E: Common ancestors of 'E' are 'I' and 'J' who are not inbred. The paths are:

CA	Path no.	Path chain	n	$(1/2)^n(1+F_{CA})$	Contribution
I	1	E←G←I→H→E	2	$(½)^2(1+0)$	0.25
J	2	E←G←J→H→E	2	$(½)^2(1+0)$	0.25
				Summation =	0.50

$$F_E = ½\ (0.50) = 0.250$$

F_A: Common ancestor of 'A' is 'E' who is inbred with F_E = 0.250. The paths are:

CA	Path no.	Path chain	n	$(1/2)^n(1+F_{CA})$	Contribution
E	1	A←B←E→C→A	2	$(½)^2(1+0.250)$	0.3125

$$F_A = ½\ \Sigma\ (½)^n\ (1 + F_{CA}) = ½\ (0.3125) = 0.15625$$

(b) R_{AJ}: The paths connecting 'A' and 'J' are:

Path no.	Path chain	n	$(1/2)^n(1+F_{CA})$	Contribution
1	A←B←E←G←J	4	$(½)^4(1+0)$	0.0625
2	A←B←E←H←J	4	$(½)^4(1+0)$	0.0625
3	A←C←E←G←J	4	$(½)^4(1+0)$	0.0625
4	A←C←E←H←J	4	$(½)^4(1+0)$	0.0625
			Summation =	0.25

(F_J = 0 and F_A = 0.15625)

$$R_{AJ} = \frac{\sum\left(\frac{1}{2}\right)^n (1+F_{CA})}{\sqrt{(1+F_A)(1+F_J)}} = \frac{0.25}{\sqrt{(1+0.15625)(1+0.0)}} = 0.2325$$

12.22. From the pedigree given below, calculate (a) F_J , F_B, F_C,F_A (b) R_{AJ} and R_{AM}

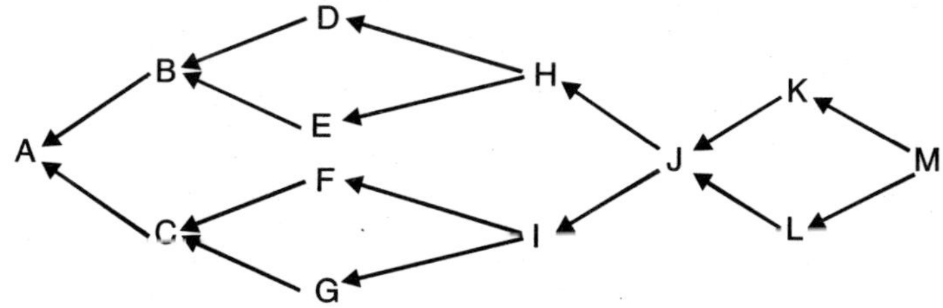

Solution :

(a) F_J: Common ancestor is 'M' who is not inbred:

CA	Path no.	Path chain	n	$(1/2)^n(1+F_{CA})$	Contribution
M	1	J←K← [M] →L→J	2	$(½)^2(1+0)$	0.25
				Summation =	0.25

$F_J = ½(0.25) = 0.125$

F_B: Common ancestor is 'H' who is not inbred:

CA	Path no.	Path chain	n	$(1/2)^n(1+F_{CA})$	Contribution
H	1	B←D← [H] →E→B	2	$(½)^2(1+0)$	0.25
				Summation =	0.25

$F_B = ½(0.25) = 0.125$

F_C: Common ancestor is 'I' who is not inbred:

CA	Path no.	Path chain	n	$(1/2)^n(1+F_{CA})$	Contribution
I	1	C←F← [I] →G→C	2	$(½)^2(1+0)$	0.25
				Summation =	0.25

$F_C = ½(0.25) = 0.125$

F_A: Common ancestor is 'J' who is inbred (F_J=0.125, as shown above):

CA	Path no.	Path chain	n	$(1/2)^n(1+F_{CA})$	Contribution
J	1	A←B←D←H← [J] →I→F→C→A	6	$(½)^6(1+0.125)$	0.017578
	2	A←B←D←H← [J] →I→G→C→A	6	$(½)^6(1+0.125)$	0.017578
	3	A←B←E←H← [J] →I→F→C→A	6	$(½)^6(1+0.125)$	0.017578
	4	A←B←E←H← [J] →I→G→C→A	6	$(½)^6(1+0.125)$	0.017578
				Summation =	0.070312

$F_A = ½\ (0.070312) = 0.03515625$

(b) R_{AJ}: The paths connecting 'A' and 'J' are:

Path no.	Path chain	n	$(1/2)^n(1+F_{CA})$	Contribution
1	A←B←D←H←J	4	$(½)^4(1+0.125)$	0.0703125
2	A←B←E←H←J	4	$(½)^4(1+0.125)$	0.0703125
3	A←C←F←I←J	4	$(½)^4(1+0.125)$	0.0703125
4	A←C←G←I←J	4	$(½)^4(1+0.125)$	0.0703125
			Summation =	0.28125

$$R_{AJ} = \frac{\sum\left(\frac{1}{2}\right)^n(1+F_{CA})}{\sqrt{(1+F_A)(1+F_J)}} = \frac{0.28125}{\sqrt{(1+0.035)(1+0.125)}} = 0.2606\ (F_J=0.125;\ F_A=0.0352)$$

R_{AM}: The paths connecting 'A' and 'M' are:

Path no.	Path chain	n	$(1/2)^n(1+F_{CA})$	Contribution
1	A←B←D←H←J←K←M	6	$(½)^6(1+0)$	0.015625
2	A←B←E←H←J←K←M	6	$(½)^6(1+0)$	0.015625
3	A←C←F←I←J←K←M	6	$(½)^6(1+0)$	0.015625
4	A←C←G←I←J←K←M	6	$(½)^6(1+0)$	0.015625
5	A←B←D←H←J←L←M	6	$(½)^6(1+0)$	0.015625
6	A←B←E←H←J←L←M	6	$(½)^6(1+0)$	0.015625
7	A←C←F←I←J←L←M	6	$(½)^6(1+0)$	0.015625
8	A←C←G←I←J←L←M	6	$(½)^6(1+0)$	0.015625
			Summation = 0.125	

$$R_{AM} = \frac{\sum\left(\frac{1}{2}\right)^n(1+F_{CA})}{\sqrt{(1+F_A)(1+F_M)}} = \frac{0.125}{\sqrt{(1+0.035)(1+0.0)}} = 0.122856\ (F_M = 0;\ F_A = 0.0352)$$

Supplementary Problems

12.23. Calculate (a) R_{X1}; (b) Fraction of inheritance received by 'X' from '1'.

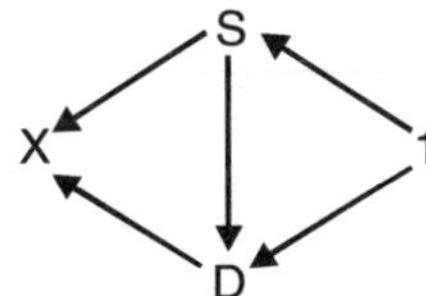

12.24. Calculate (a) R_{SD}, (b) R_{X1}; (c) R_{X10}; given, $F_X = 0.23443$, $F_S = F_D = F_1 = 0.125$.

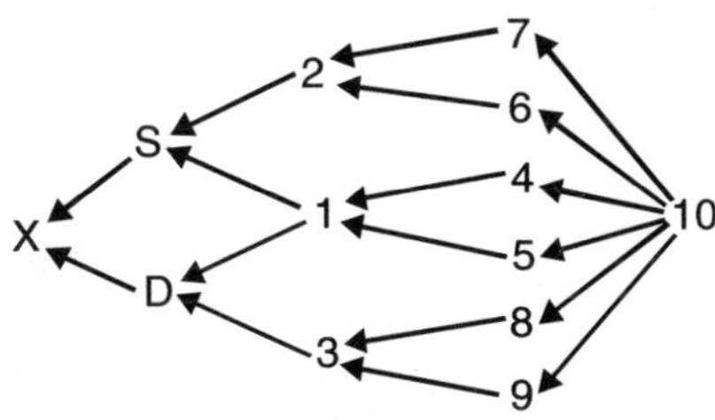

12.25. Calculate: (a). F_X, F_S, F_D (b). R_{SD}

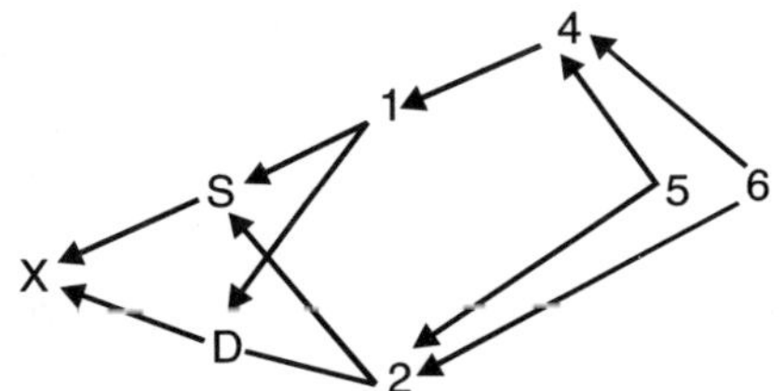

12.26. Calculate (a) F_B (b) F_A

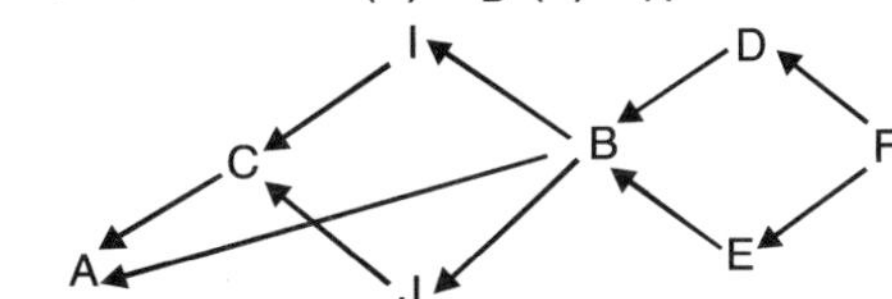

12.27. Calculate (a) F_X, (b) R_{CD}

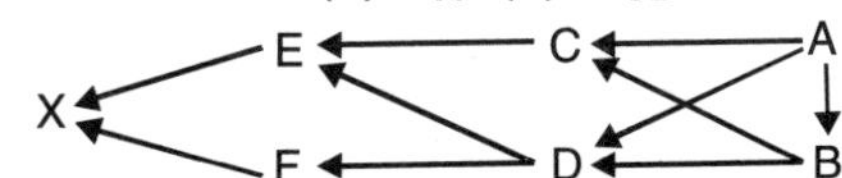

12.28. Calculate inbreeding coefficient resulting from three consecutive generations of half sib mating:

12.29. (a) Prepare A path diagram; (b) Calculate F_X, R_{X5} and calculate the percentage of inheritance received by 'X' from '5'.

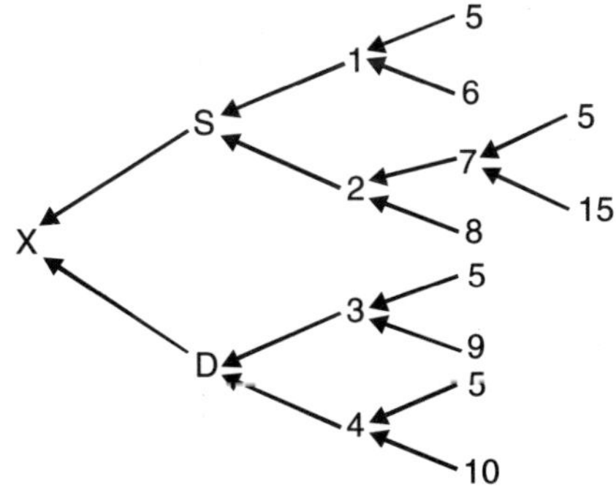

12.30. (a) Draw the path diagram; (b) calculate F_X; R_{X5} and the percentage of inheritance received by 'X' from '5

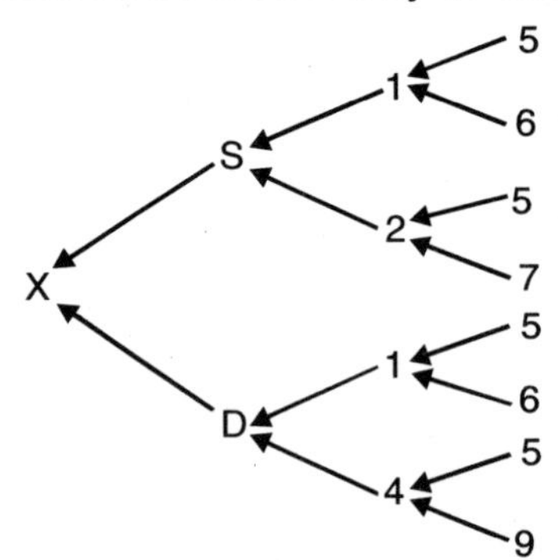

12.31. Calculate (a) F_X; (b) R_{X5}, R_{SD}

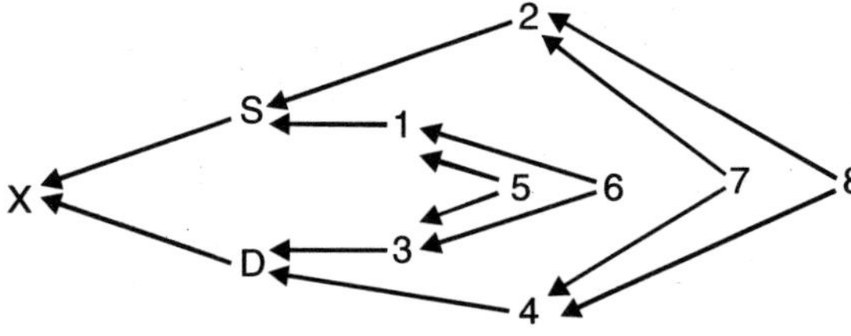

12.32. Compute (a) F_X, (b) R_{X5}; (c) percentage of inheritance received by 'X' from '5'.

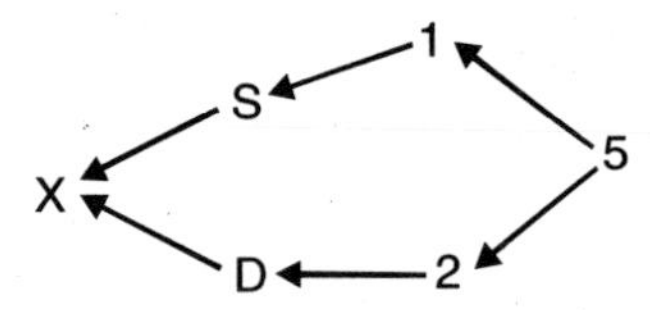

12.33. Compute (a). F_X, (b) R_{X5};

(c) percentage of inheritance received by 'X' from '5'.

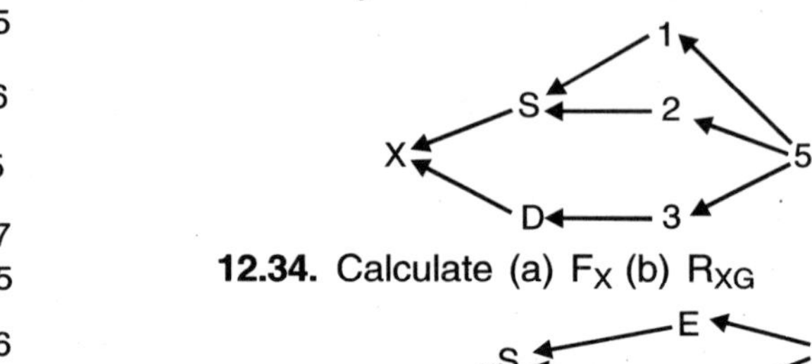

12.34. Calculate (a) F_X (b) R_{XG}

12.35. Calculate F_X ; R_{X1}

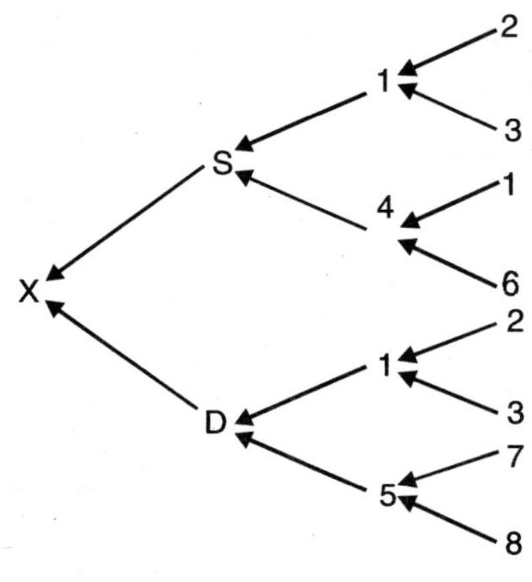

Answers to Supplementary Problems

12.23. (a) $F_X = 0.375$, $F_S = 0$, $F_D = 0.25$, $R_{X1} = 0.5330$

(b) Percentage of inheritance received by 'X' from 'I' = 62.5

12.24. (a) $R_{SD} = 0.4166$ (b) $R_{X1} = 0.5834$ (c) $R_{X10} = 0.45$

12.25. (a) $F_X = 0.3125$, $F_S = 0.125$, $F_D = 0.125$ (b) $R_{SD} = 0.55$

12.26. (a) $F_B = 0.125$ (b) $F_A = 0.28125$

12.27. (a) $F_X = 0.25$ (b) $R_{CD} = 0.60$

12.28. (a) The path diagram for three generations of half-sib mating is:

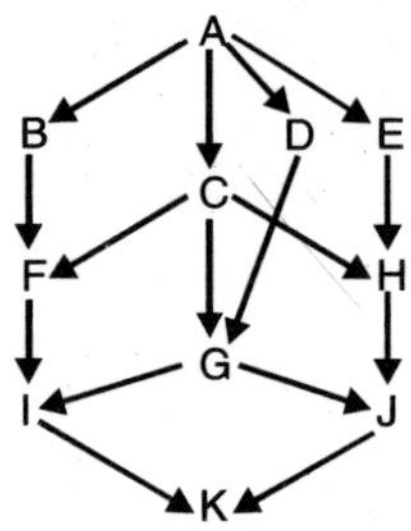

(b) $F_K = 0.3046874$

12.29. (a) The path diagram is:

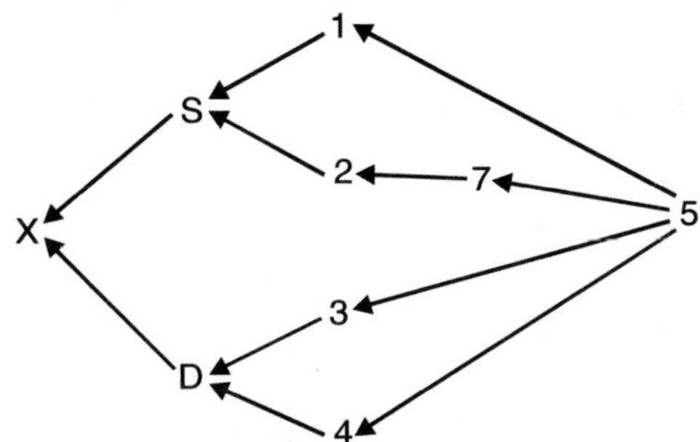

(b) $F_X = 0.125$ (b) $R_{X5} = 0.412479$

(c) Percentage of inheritance received by 'X' from '5' = 43.75

12.30. $F_X = 0.219$, $R_{X5} = 0.45$, percentage of inheritance from 5 to 'X' is 50%.

12.31. (a) $F_X = 0.125$ (b) $R_{X5} = 0.2358$ (c) $R_{SD} = 0.25$

12.32. $F_X = 0.03125$, $R_{X5} = 0.246$, percentage of inheritance got by 'X' from '5' = 25

12.33. (a) $F_X = 0.0625$ (b) $R_{X5} = 0.36386$ (c) percentage of inheritance received = 37.5

12.34. (a) $F_X = 0.28125$ (b) $R_{XG} = 0.331$

12.35. $FX = 0.1875$, $R_{X1} = 0.5735$

13

Outbreeding and Heterosis

Mating between animals which are less closely related than the average relationship of the population is called **outbreeding**. In principle, all the animals of a given population (breed) are related to each other, though remotely, therefore, for practical purposes, a mating in which mates are not related to each other for at least four to six generation is considered an outbred mating.

Outbreeding is further classified into as outcrossing and crossbreeding: Mating between unrelated animals within the same breed is called **outcrossing** while mating between animals of different breeds is called **crossbreeding.**

Under broad categorization, however, outbreeding includes mating of unrelated animals within the same population (breed / strain etc); crossing of breeds; crossing of inbred lines; crossing of strains; and even crossing of different species. This chapter deals with the genetic and phenotypic consequences of outbreeding, the genetic basis of heterosis including its measurement, types of heterosis, and the various systems used for the utilization of heterosis.

Terminology

Species: Animal kingdom, according to Zoological systemic, is divided into Phylum, Class, Order, Family, Genus and Species. A **species** is a group of interbreeding animals that is reproductively isolated from other such groups.

One species differs from the other for one or more distinct morphological and or physiological characters; the most distinguishing feature being reproductive discontinuity between species (i.e. two species do not interbreed, or do not produce fertile progeny). Reproductive discontinuity can have different causes, and may also vary in degree of its manifestation as the examples cited below illustrate:

- Horse and ass, through different species, can be mated and produce fully viable but sterile offspring (mule). A few rare cases of fertile females (mules) have also been reported. There is thus a clear cut demarcation between these species.

- The Tibetan yak *(Bos grunniens)* and domestic cattle *(Bos taurus/indicus)*, though different species, can be mated and produce offspring; the females are fertile but males are sterile.
- Zebu cattle *(Bos indicus)* and European cattle *(Bos taurus)*, though classified as different species, do not have reproductive discontinuity; the offspring resulting from crosses between them are fully fertile. But these two species of the family *Bovidae* have some distinct morphological features (e.g. presence of hump and dewlap in Zebu). In view of reproductive continuity between them, it is being increasing suggested that these may be considered as subspecies of the same species. This underlines the fact that the dividing line between species is not always distinct.

Box 13.1: Species concept.

The term species refers to fundamental taxonomic unit. Three concepts of species have been proposed: biological, morphological and evolutionary.

- *Biological:* According to this concept, a species is a group of individuals who interbreed, or have the potential to interbreed in nature. According to this definition, a new species evolves when individuals of a population become distinctly different from others, and are no more able to interbreed with individuals of other populations.
- *Morphological:* The morphological concept defines a species as a group of individuals with common shared morphological features. If a group of organisms thus appear distinct from other groups, it constitutes a separate species. This concept has limited implication as one might consider dozens of species of dog by virtue of morphological distinctions (e.g. coat color, its pattern, body shape, size etc)
- *Evolutionary:* The evolutionary concept of species, also known as phylogenetic species concept, categorizes organisms on the basis of their shared evolutionary history. By this definition, two species might actually be able to interbreed but due to their different histories, they are considered different species. The greatest limitation of this concept is that it may be difficult to trace the complete history of a species.

Most workers apply the biological species concept when it is practical; otherwise they use either one of the other concepts.

Breed: A species is further differentiated by biologists/geneticists, into breeds (in animals), varieties (in plants) and races (in humans). However in Linnaean taxonomic classification, grouping of a species into breeds does not exist. A breed of livestock has been defined in several different ways but the definitions provided by FAO are widely accepted, and read as follow:

- Breed is a sub-specific group of domestic livestock with definable and identifiable external characteristics that enable it to be separated by visual appraisal from other similarly, defined groups within the same species.
- Breed is a group for which geographical and or cultural separation from phenotypically similar groups has led to its acceptance as a separate identity.

Some important features of a livestock breed, as pointed out in above definitions, are important:

- Animals of a breed differ from those of other breeds for genetically determined/ identifiable characters. These definable attributes enable one breed to be distinguished or separated from the other by visual appraisal. For example, Nili Ravi breed of Buffalo is distinguished from other breeds by the presence of distinct white markings on extremities. In most cases, the distinguishing traits of a breed are qualitative in nature such morphological attributes (e.g. hair type, coat color, shape of horns etc.); in other cases, the traits may be quantitative such as body size and production attributes. Animals of a breed are expected to breed true to type for one or more of the distinguishing traits of that breed.

- There are cases in which animals are considered to belong to a breed by virtue of their geographical location or due to the cultural belief associated with them though their external features may not be clearly demarcated from another well-defined breed.
- With regard to quantitative traits (growth and production), considerable overlapping between different breeds may exist even though the average performance of different breeds may be quite different.

Concept of a breed in applied livestock breeding is thus conventional rather than biological or genetic. Genetically speaking, there are no pure breeds of farm animals. Division of species into breeds has been justified as different breeds have specialized for different purposes under the geographical conditions of an area.

Purebreds and Crossbreds: Animals resulting from mating between different breeds are referred to as **crossbreds** while those resulting from mating within the same breed are referred to as **purebreds** or **straightbreds**. In some species of livestock, the term 'purebred' is used specifically to refer to animals which are registered, or are eligible for registration in the herd book of a breed.

Strain: In some species of domestic animals, a breed may be further differentiated into strains. From a breeding standpoint, a **strain** is an interbreeding group of animals within a breed that is isolated from other strains due to differences in selection and breeding objectives. For example, many of strains of black and white Friesian cattle (or Holstein Friesian) found around the world have originated from the Dutch Friesian (black pied animals). These strains differ from one another because they have undergone selection for different breeding objectives (i.e. a set of traits). The concept of strains is more prevalent in poultry than in other species of livestock.

Line: A line refers to a group of animals within a breed/strain who as a result of close mating (due to inbreeding and or selection) has become more closely related to each other than to other such groups within the strain/breed. When a line attains an inbreeding coefficient of at least 0.375, (corresponding to two generations of full-sib mating), such a line in livestock species is then designated as an **inbred line.**

A line is further made up of families. A family is a group of related animals like full sibs or half-sibs. All the animals within a family are equally related to each other. In old literature, the term family was used to refer to cow families, boar families etc, to depict a line of descent to some specific outstanding ancestor.

Genetic Effects of Outbreeding

Outbreeding is just the opposite of inbreeding; therefore, its genetic effects are also exactly the opposite of inbreeding: inbreeding increases homozygosity while outbreeding Increases heterozygosity.

If two populations (breeds, strains / lines) are assumed to possess different alleles of the same locus, for example, one population is homozygous for the dominant allele (AA) while the second is homozygous for recessive allele (aa), then all the offspring resulting from crossing of these populations would be heterozygous for that locus (100% heterozygosity). The resultant heterozygosity applies to all the loci for which the parental populations were homozygous. The following single-, two-, and three- loci hypothetical examples illustrate the concept of increased heterozygosity resulting from outbreeding:

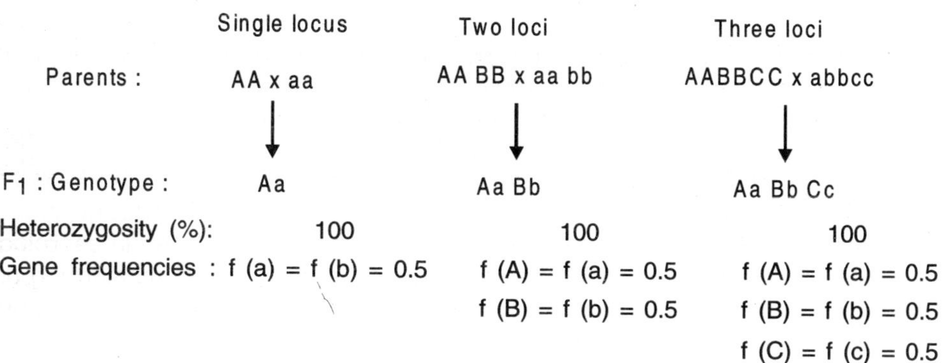

Heterozygosity (%): 100 100 100

Gene frequencies : f (a) = f (b) = 0.5   f (A) = f (a) = 0.5   f (A) = f (a) = 0.5

f (B) = f (b) = 0.5   f (B) = f (b) = 0.5

f (C) = f (c) = 0.5

The hypothetical examples considered above are extreme cases where each population (breed / strain) has been assumed to be completely homozygous for one allele/locus while the other is homozygous for the alternative allele or other locus. Realistic situation may not be so when a large number of loci are simultaneously considered. Consider a more realistic example of six unlinked loci where each of the two parental populations is homozygous for four loci and heterozygous for two loci:

Parental population: AA Bb cc DD Ee ff × Aa bb CC dd EE Ff

The offspring resulting from the above mating, on the average, are expected to be heterozygous for four loci though some (1/16) would be heterozygous for all the six loci while some others (1/16) would be heterozygous for two loci only (the students should check the outcome of mating by making a checkerboard).

Thus the primary genetic effect of mating between genetically unrelated animals (outbreeding), on the average, is to increase heterozygosity in the F_1 generation. Outbreeding, however, does not change the allelic frequencies. Other genetic consequences of outbreeding, which arise from increased heterozygosity, are:

- **Masking of deleterious recessive alleles**: Inbreeding by increasing homozygosity increases the likelihood of deleterious recessive alleles becoming homozygous, and hence it increases the chance of their expression. On the other hand, outbreeding by increasing heterozygosity tends to mask the expression of the deleterious recessive alleles. This is the reason that outbred animals exhibit lesser genetic defects compared to purebreds. One of the obvious examples of masking of undesirable recessives is seen in dogs, where mongrels/mutts (i.e. the dogs resulting from crossing between different established breeds, called mixed breed dogs) exhibit fewer genetic abnormalities than purebred dogs (i.e. the one's belonging to recognized breeds). Pigeon keepers also claim that city pigeons who are rather hardy. But outbreeding by itself does not eliminate the deleterious recessives; it only blocks their expression.

- **Outbreed animals are less likely to breed true**: Outbred animals possess different alleles; therefore, they can produce many different types of gametes. For example, an individual of 'Aa Bb Cc Dd' genotype can produce 16 kinds of gametes while a completely homozygous individual for these four loci (either 'AA BB CC DD' or 'aa bb cc dd') can only produce one type of gametes. Outbred animals are thus less likely to breed true to their type as compared to inbred animals.

Phenotypic Effects of Outbreeding: Hybrid Vigor

The most pronounced phenotypic effect of outbreeding for polygenic traits is the improved performance of crossbreds/outbreds relative to purebreds (i.e. their parental populations/ breeds/ strains etc). The improved performance of progeny (crossbred) resulting from crossing of different parental populations (breeds/strains, lines etc.) over the performance of parental populations (purebreds) is referred to as **heterosis** (this is just the opposite of inbreeding depression).

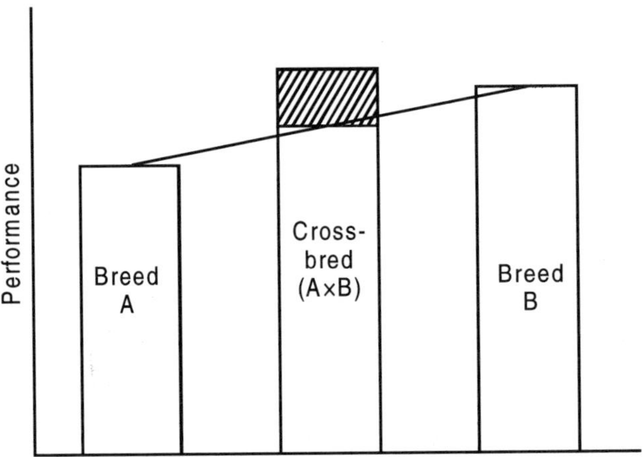

Fig 13.1 Illustration of heterosis expressed as superiority of offspring over average of parental breeds.

The phenomenon of improved performance of hybrids/crossbreds resulting from breed, strain, line or species crosses had been known to biologists and breeders, much before the actual term 'heterosis' was coined. Darwin in his book on natural selection (1876) reported that: "Hybrids show a much better adaptation to the environment and also show higher fertility, viability and size compared to inbreds". The formal term 'heterosis' was coined by Shull, an American plant geneticist (Box 13.2), to refer to increased size, yield and vigor of plants resulting from crossing of inbred varieties of maize. The words 'hybrid vigor' and 'heterosis' are used interchangeably.

Technically heterosis refers to the additional/extra performance, if any, shown by the first generation offspring (crossbred) above the mean performance of their parental populations (breeds). Sometimes performance of F_1 progeny may exceed the performance of best parental population but there are cases when it may be poorer than average of parental populations (this has been termed 'negative heterosis' but this expression does not have a general acceptance).The populations being crossed may be different breeds, strains or lines of the same breed or even different species. Further the increased performance may refer to improvement in reproduction, production, growth and in other polygenic traits.

Box 13.2: Heterosis and George Shull (1874-1954).

George Harrison Shull (1874-1954), an American plant breeder and geneticist, has been one of the pioneers in elaboration of concept of heterosis, and its application in plant breeding. After obtaining his MS and PhD from University of Chicago, he joined the eminent Geneticist Davenport's group at Cold Spring Harbour in 1904 when Davenport moved to the position of Director. In 1915, he moved to Princeton University where he remained till his retirement.

While working on development of inbred lines, Shull (1908) reported that inbred lines of maize showed general deterioration in yield and vigor, and the hybrids resulting from their crossing not only immediately recovered from inbreeding depression but in many cases their yield exceeded that of the populations (varieties) from which the inbred lines were derived. The hybrids also had high uniformity, a desirable attribute for farmers for harvesting. In a following paper he outlined the procedures for production of hybrids.

The word 'heterosis' was introduced by him in 1912 as a shorthand to replace the more cumbersome, long and complicated expressions which had till then been used for the improved yield of hybrids (e.g. stimulation of heterozygosis).

Shull's idea of crossing of inbred lines for production of hybrids spread rapidly in the maize producing regions of USA which was immediately adopted by private companies. This ushered in a new era in maize production. Before the introduction of hybrids the yield of maize had stabilized at around 30 bushels per acre. But introduction of hybrids sharply increased maize yield, and by 2000 its increase over pre-hybrid era was about five-fold. A part of this increase has resulted from better agronomic practices including use of fertilizers but it has been estimated that 60% or more of the increased yield was attributed to hybrids including their improvement (genetic increase).Not only the success of hybrid maize been replicated virtually in all the major maize producing countries but the word 'hybrid' has become associated with improved plant/animal varieties/breeds/strains.

Shull was very particular about the use of the term heterosis. According to him the word 'heterosis' was to be used for 'increase of size, yield, vigor etc' If no such improvement resulted from crossing, there is no heterosis. A decrease in these and related productivity attributes should not be treated as 'negative heterosis'. He was very vocal against the use of negative heterosis, and stated if such a term is to be used for decline in productivity etc then term 'negative respiration' would also have to be justified for photosynthesis.

Genetic Basis of Heterosis

The commercial importance of crossbred animals or hybrid plants has stimulated considerable research into understanding the genetic basis of heterosis. Genetic explanation of heterosis is based on the heterozygosity involving genes with non-additive effects. The non-additive gene action involves interaction between alleles at the same locus (dominance) and interaction between alleles of different loci (epistasis). Three hypotheses have been proposed to explain heterosis:

Favorable dominance hypothesis (Davenport, 1908, and later by Lerner)**:** Development of different populations (breeds / strains / lines) is the result of selection that had been going on over a long period of time. Some degree of inbreeding during this period has invariably occurred in such populations. Both selection and inbreeding must have led to homozygosity at a number of loci in different lines/ breeds: some lines/breeds may have become homozygous for 'favorable' dominant genes while others may have attained homozygosity for 'unfavorable' recessive genes.

When two breeds, which are presumably homozygous for different genes/loci, are crossed the resulting offspring (F_1) would be heterozygous at several loci. The 'favorable' dominant genes possessed by the heterozygous individuals mask the expression of 'unfavorable' recessive genes. This is referred to as the 'favorable dominance hypothesis' and implies that F_1 generation (crossbred) is likely to contain a higher proportion of 'favorable' dominant genes than either of parental populations. The hypothesis is illustrated below for a three-locus model by assuming that each dominant genotype contributes two units to phenotype while a recessive genotype contributes one unit:

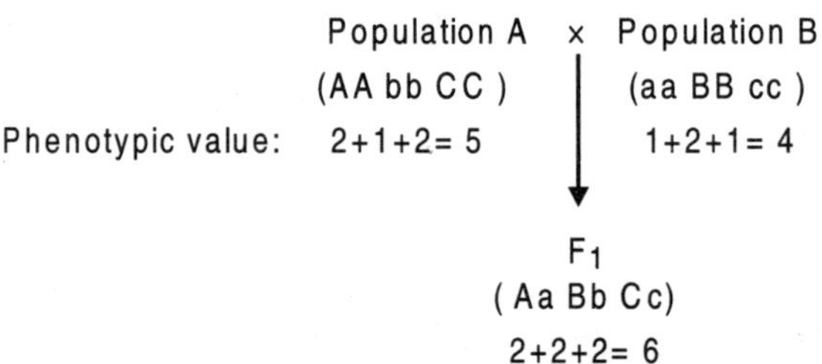

Thus if we assume that each dominant allele has a more favorable effect than its alternative recessive allele, the F_1 would be better than the average of parental groups by virtue of possessing favorable (dominant) alleles at the three loci. This explanation of heterosis assumes it to be largely the result of 'covering-up' effects of the 'undesirable' (recessive) by the dominant alleles. Heterozygosity of the F_1 is only incidental and not the cause of heterosis.

If dominance were the only cause of heterosis, then it should be possible (by crossing and subsequent selection) to accumulate all the favorable (dominant) alleles in some populations, and all the recessive alleles in some other populations (lines/breeds etc). Then the population containing dominant homozygotes would have performance equal to those of F_1 crossbreds (heterozygotes) but such a population would breed true to type. This would thus make outbreeding / crossbreeding unnecessary. However, it has not been possible to produce such populations (breeds/lines). Dominance is, therefore, only a partial explanation of heterosis.

Overdominance hypothesis (Shull and East, 1908, and later by Mather and Jinks): Overdominance is the interaction between alleles of the same locus such that the heterozygotes are better than either of the two homozygotes (Fig.13.2). According to this hypothesis, heterosis occurs because the F_1 progeny are heterozygous for the genes displaying overdominance. According to this hypothesis, heterozygosity per se is the cause of heterosis.

To illustrate this phenomenon, consider a trait controlled by three loci where a homogygote recessive, a heterozygote and a homozygote dominant contribute one, two, and one- and -a half units to phenotypic value as shown below:

	Population A	×	Population B
	(AA bb CC)		(aa BB cc)
Phenotypic value:	1½ +1+2= 4½	↓	1+1 ½ +1= 3½
		F_1	
		(Aa Bb Cc)	
		2+2+2=6	

When overdominance is operative, the performance of F_1 would be better than the performance of the best parental population, and the magnitude of heterosis will be high. If overdominance were the cause of heterosis, it will not be possible to produce breeds, strains or lines which could give high performance in pure forms. Therefore, maximum productivity can only be obtained by making crosses every generation. It is, however, not known how much overdominance is exhibited by the genes that control various economic traits of livestock. Although some examples of overdominant gene pairs are known, the phenomenon of overdominance is not as frequently encountered as is the case with partial to complete dominance.

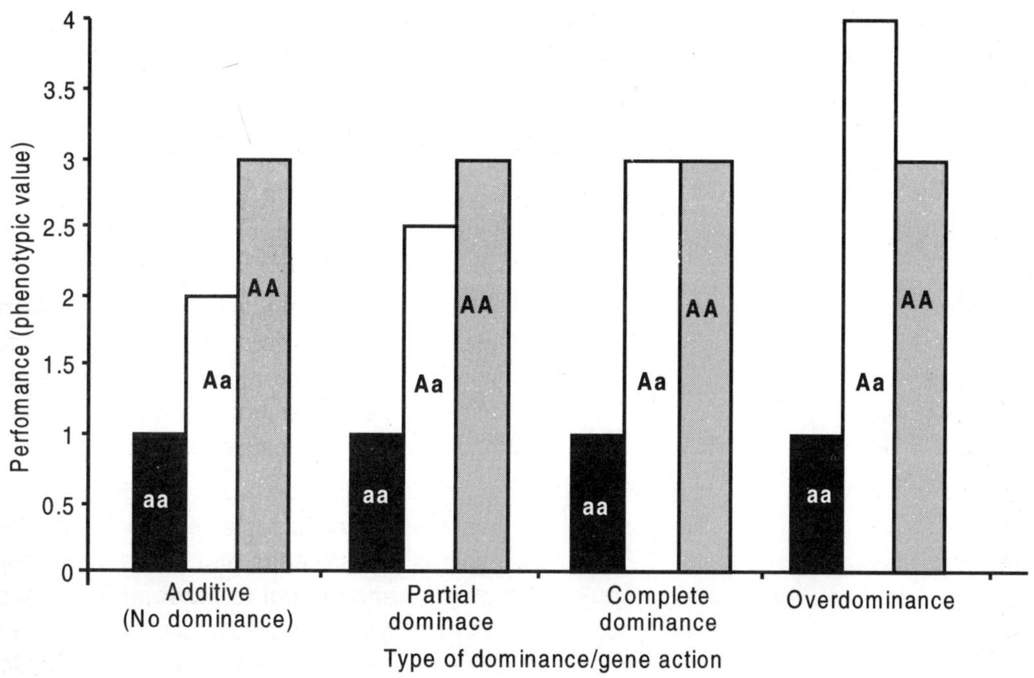

Fig 13.2 Illustration of type of gene action at a single locus.

Epistatic hypothesis: Interaction between non-allelic genes, referred to as epistasis, could be another explanation of heterosis. For example, consider the crossing of 'AAbb' and 'aaBB' to produce 'AaBb' offspring. The 'Aa Bb' combination of genes may be such that the presence of both 'A' and 'B' genes gives a better performance than would be expected on the basis of average performance of 'AAbb' and 'aaBB'. Since most of the economically important traits of farm animals are governed by several loci, interaction between genes at different loci could be expected to influence their expression. Epistasis is therefore believed to play some part in exerting the heterotic effect. A genotype heterozygous at many loci is likely to produce new epistatic combinations by virtue of it carrying a greater variety of genes.

It has not been possible to experimentally determine which of these three kinds of non-additive effects are responsible for the heterosis exhibited by different traits of farm animals. Probably various levels of dominance including overdominance, and many types of non-allelic interactions (epistasis) together account for the heterosis observed for various traits of farm animal species. The relative contribution of different types of non-additive genetic effects to heterosis may vary from one trait to another. Biochemical reactions occurring within the living organism are initiated by the action of genes through production of enzymes; therefore, heterosis may be physiologically expressed due to highly efficient metabolic system resulting from heterozygosity or the combination of different genes/loci.

Factors Determining Magnitude of Heterosis

The magnitude of heterosis, (H), measured as difference between performance of F_1 ($\overline{X}_{F1}$) and average performance of parental groups ($\overline{X}_P$), is given by the following expression[1]

$$H = \overline{X}_{F_1} - \overline{X}_P = dy^2$$

Where:

'd' is a measure of degree of dominance

'y^2' is square of difference in gene frequency between the parental populations.

The value of 'd' depends on the type of gene action while 'Y' is a measure of the genetic diversity between the parental populations.The magnitude of heterosis observed for a trait thus depends on two factors:

- **Type of gene action:** If the gene action governing a trait is purely additive, (d = o) no heterosis will be exhibited. That is, the degree of heterosis is dependent on the existence of some degree of non-additive gene action:

		Level of dominance	Heterosis
$d = 0$	:	No dominance-(additive gene action)	Nil
$d > 0 \leq a$	:	Partial dominance	Low
$d > a$	:	Over dominance	Highest

- **Genetic diversity between populations/ breeds:** The difference in gene frequency between the breeds being crossed (y) is a measure of the genetic diversity between them. Larger the value of 'y' larger will be the heterosis. This leads to the generalization that wider the cross, the greater will be the magnitude of heterosis. Breeds that have similar origins and that have been subjected to similar types of selection during their development will be expected to be genetically much more alike than the breeds having different origin and having different selection history (purpose). Higher heterosis would be exhibited when exotic breeds (*Bos taurus*) are crossed with indigenous breeds (*Bos indicus*) than the crossing between different breeds of indigenous cattle. Similarly, higher heterosis is expected from breed crossing than from strain crossing.

Therefore, the two basic genetic requirements for a trait to exhibit heterosis are:

– the trait must be influenced by some degree of dominance (d # 0)

– the populations being crossed must be genetically divergent from one another (y # 0)

[1]This expression pertains to a single locus model, and does not include the contribution of epistasis as cause of heterosis.

If either of the two conditions is not fulfilled, the trait will not manifest heterosis. In such case, the performance of F_1 (crossbred) would simply be the average of two purebred groups involved in the cross.

A logical question in this context is "how to maximize heterosis". Theoretically it can be achieved either by increasing 'd' or 'y' or both. However, 'd' ; can't be changed within the particular populations being crossed. But 'y' can be made high by choosing parental populations which are genetically divergent from each other. Theoretically, if one allele is fixed in one parental population, and the other allele in the second population, heterosis will be maximized. One such system involves the development of inbreed lines; and their crossing (inbreeding-hybridization).

Fate of Heterosis on Inter-se-mating

Consider crossing of two parental populations, P_1 and P_2, to produce the F_1 crossbreeds and the subsequent inter-se-mating F_1 to produce the F_2 generation. It is assumed that the parental populations are completely homozygous: one for the dominant and other for the recessive allele:

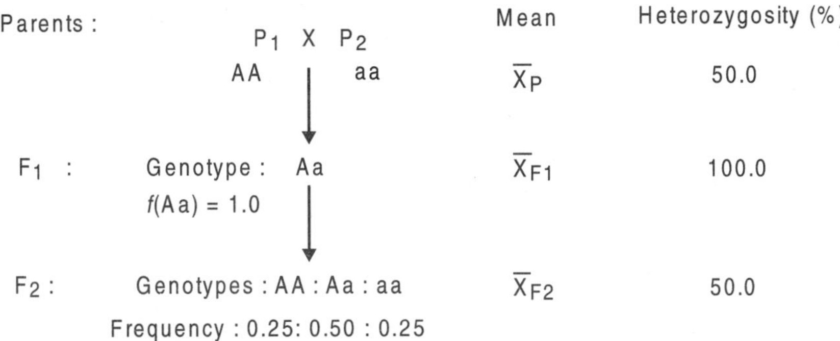

Average heterozygosity in the parental populations is 50% which serves as the baseline for comparison with heterozygosity in subsequent generations. The F_1 exhibit 100% heterozygosity representing an increase of 50% over parental populations. This increased heterozygosity is reflected as increased performance of F_1 over average of the parental populations (referred to as heterosis, designated as H or H_{F1}). Inter-se-mating among F_1 to produce F_2 leads to 50% loss in heterozygosity:

$$H_{F_2} = \frac{1}{2} H_{F_1}$$

Thus the heterosis exhibited by F_2 is half of that present in F_1 generation. The loss of heterosis resulting from inter-mating of F_1 (crossbreds) indicates that improved performance of crossbreds over the parental breeds (i.e. heterosis) is transient; hence it is not possible to fix or perpetuate heterosis. This is the reason that farmers have to purchase hybrid seeds, chicks from breeders every year. Consequently, to take full benefit of heterosis, crosses between breeds/strains have to be made fresh each generation.

The heterosis which remains in F_2 generation (resulting from inter-mating of F_1) is referred to as **'retained heterosis'** (H_{F_2}). It can be seen that if F_2 are mated among them, the level of heterozygosity in F_3 would remain the same as was in the F_2 generation (the students can check this by making a checkerboard of the mating of F_2 among themselves (i.e. $F_2 \times F_2$).

The conclusion is that though heterosis is halved between F_1 and F_2 generations but it remains constant in subsequent generations (i.e. from F_2 onwards).

Types of Heterosis

When parents from two different breeds, strains or lines are crossed, the resulting offspring (F_1) exhibit variable amount of heterosis depending on the magnitude of genetic diversity between the parental populations and the magnitude of non-additive gene action. The first traits to manifest heterosis would be pre-natal and neonatal survival, and early growth. Once the F_1 animals attain sexual maturity, heterosis may be evident for another set of traits associated with production and reproduction including 'maternal / mothering' ability. Consequently, some of the benefits of heterosis are obtained directly by the crossbred animals them (F_1), and some by the offspring by virtue of dam's heterosis (Fig 13.3).

Heterosis is categorized into three types on the basis of nature of the trait exhibiting it:

- **Individual heterosis**: Increased performance of offspring by virtue of them being crossbred is referred to as individual heterosis. In other words, the trait exhibiting heterosis is measured on an individual, and its benefits are also obtained by the animal itself.
- **Maternal heterosis:** When the dams are crossbred, the benefits occurring from their heterotic effects (for maternal attributes) are obtained by their offspring, the heterosis is referred to as maternal heterosis. That is, the benefit of heterosis, which is manifested by crossbred dams, accrues to their offspring though offspring themselves are not crossbreds. Maternal heterosis occurs for pre- and post-natal survival, early growth traits (e.g. birth weight, weaning weight etc. which are influenced by the mothering ability of the dam).

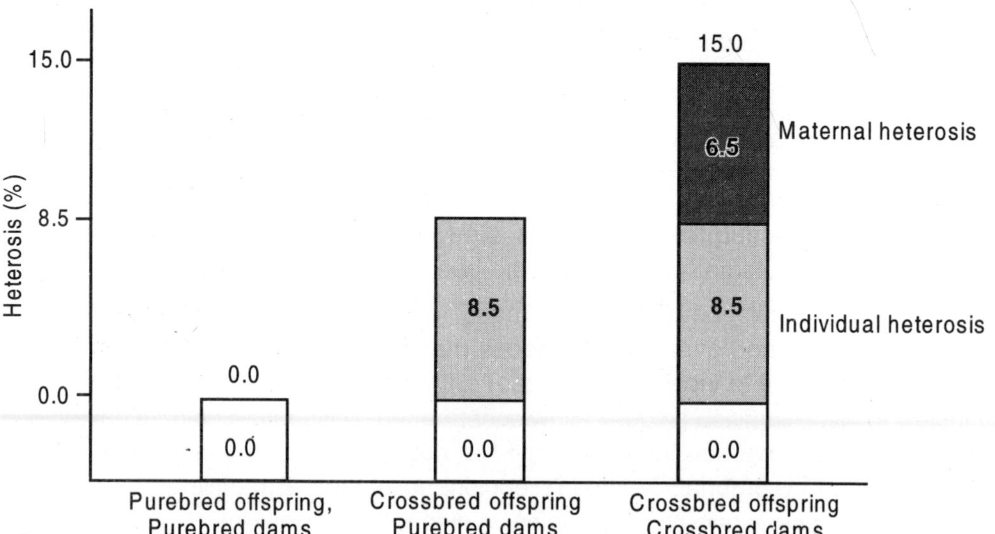

Fig 13.3 Illustration of cumulative advantage of individual and maternal heterosis

Some specific examples of maternal heterosis are :

- o Crossbred sows exhibit better mothering ability which is reflected in higher survival and growth of their offspring (piglets).
- o Heterosis for milk production may be reflected as maternal heterosis for pre-weaning growth rate.

  - o Heterosis of crossbred hens for egg size influences embryonic and post-embryonic growth of chicks.
- **Paternal heterosis:** Heterosis exhibited by crossbred males for paternal traits (e.g. libido, semen traits etc) is referred to as paternal heterosis.

Of the three types, individual heterosis is most important and of practical significance as the animals used for commercial production are mostly crossbreds. Maternal heterosis is of use in litter-bearing and meat-producing animals (e.g. swine, beef cattle and sheep) where early survival and growth of offspring are dependent on the mothering ability of dam. Paternal heterosis is economically the least important since sires influence performance of their offspring through genes only. The cumulative benefits of heterosis primarily result from its individual and maternal components (Fig 13.3). A good crossbreeding system should therefore attempt to take advantage of both the individual and maternal components of heterosis.

Typical Values of Heterosis

A large number of well-designed research studies have been conducted to estimate heterosis for a number of economically important traits. Typical values of heterosis range up to 10%.The average estimates of heterosis based on several studies are presented in Table 13.1.

Table 13.1 Typical values of heterosis in some livestock species.

Species	Trait	Heterosis (%)	
		Individual	Maternal
Beef cattle :	Birth weight	3.0	2.0
	Weaning weight	5.0	8.0
	Survival to weaning	4.0	2.0
	Conception rate	6.0	–
	Age at puberty	5.0	–
Dairy cattle :	Milk yield	5.0	–
	Fat yield	1.0	–
	Protein yield	4.0	–
	Services per conception	10.0	–
	Service period	10.0	–
	Calf survival (%)	15.0	–
Swine :	Little size	2.0	5.0
	Birth weight	3.0	2.0
	Weight at 21 day	3.0	7.0
	Weaning weight	4.0	8.0
	Feed conversion	2.0	–
	Carcass length	1.0	–
	Back fat thickness	2.0	3.0
	Loin eye area	1.5	1.5
Sheep :	Birth weight	3.0	5.0
	Weaning weight	5.0	7.0
	Fertility	6.0	–
	Prolificacy	2.5	8.0
	Lamb survival rate	3.0	5.0

Poultry :	Body weight (5-7 week)	3.0	–
	Feed conversion	8.0	–
	Age at sexual maturity	8.0	–
	Egg production	8.0	–
	Egg weight and quality	3.0	–
	Fertility and hatchability	10.0	–
	Livability	8.0	–

Not all the traits exhibit the same degree of heterosis. Traits can be grouped into three broad classes with respect to magnitude of heterosis. In general the traits associated with reproduction and viability show high levels of heterosis, growth and production traits exhibit moderate levels of heterosis, and quality traits show lowest heterosis.

There may be some exceptions to these generalizations (mainly because of degree of genetic diversity between the breeds/strains involved in crossing) but this is a general pattern. It should be pointed out that this pattern of heterosis for different categories of traits is exactly the reverse of the general levels of heritability for these classes of traits. For example, carcass/quality traits show low level of heterosis but have high h^2. Also, the traits that express higher heterosis also show high inbreeding depression. This is logical as h^2 depends on the magnitude of additive gene action while heterosis depends on the magnitude of non-additive gene action. A summary of these observations is given below:

Category of trait	Level of heterosis	h^2	Inbreeding depression
Reproduction & viability :	Highest	Very low	Highest
Production including growth :	Moderate	Low	Moderate
Composition & quality :	Low or very low	High	Low or very low

Measurement of Heterosis

Heterosis refers to improved performance of crossbred (F_1) over their parental populations. The computation of heterosis, therefore, involves the comparison of performance of crossbreed and purebred animals under similar conditions of environment. Further, though heterosis is expressed by individual crossbred animals but an accurate estimate of heterosis cannot be made from a single or a few matings. This is so because non-genetic factors may cause considerable variation among progeny of a single or few matings. Therefore to obtain reliable estimates of heterosis, average performance must be measured on large number of crossbred and purebred animals reared under similar types of environmental / managemental conditions. The animals being used for crossing should be representative of the breeds involved in the crossing so that their average performance equals the breed (s) average. When purebreds and crossbreds originate from different parental sources or do not perform under similar managemental and environmental conditions, estimate of heterosis will not be dependable. Heterosis can be expressed in the units in which the trait has been measured or as percentage increase over the average the parental populations.

Two ways of expression of heterosis (Fig 13.4):

- **Better-parent heterosis:** The amount by which crossbreds exceed performance of the best parental population:

  Heterosis = Crossbred average – Best parental population average

$$\text{Heterosis}(\%) = \frac{(\text{Crossbred average} - \text{Best parental population avg.})}{\text{Best parental population avg.}} \times 100$$

- **Mid-parent heterosis:** The amount by which crossbreds exceed average of parental populations (mid-parent value):

  Heterosis = Crossbred average – Average of parental populations

$$\text{Heterosis}(\%) = \frac{(\text{Crossbred average} - \text{Average of parental populations})}{\text{Average of parental populations}} \times 100$$

Heterosis can be expressed in either of these two ways. The second measure of heterosis (i.e. superiority of crossbreds over average of parental populations) is genetically more reasonable. As indicated earlier, one of the features of additive gene action is that average (mean) of F_1 lies exactly mid-way between average of parental populations. Therefore, the mean of F_1 for a trait which is exclusively governed by additive gene action, should be same as the mean of the parental populations. On the other hand, when the gene action is non-additive, the mean performance of F_1 would be either more or less than the mean of parental groups.

Comparison of mean of F_1 with mean of parental populations under similar environment indicates the importance of non-additive gene action. If the performance of F_1 exceeds the performance of the best parental population, it is indicative of the overdominant type of non-additive gene action. Since Heterosis for a trait is the collective result of the three types of non-additive genetic effects, therefore, the second measure of heterosis is more suitable.

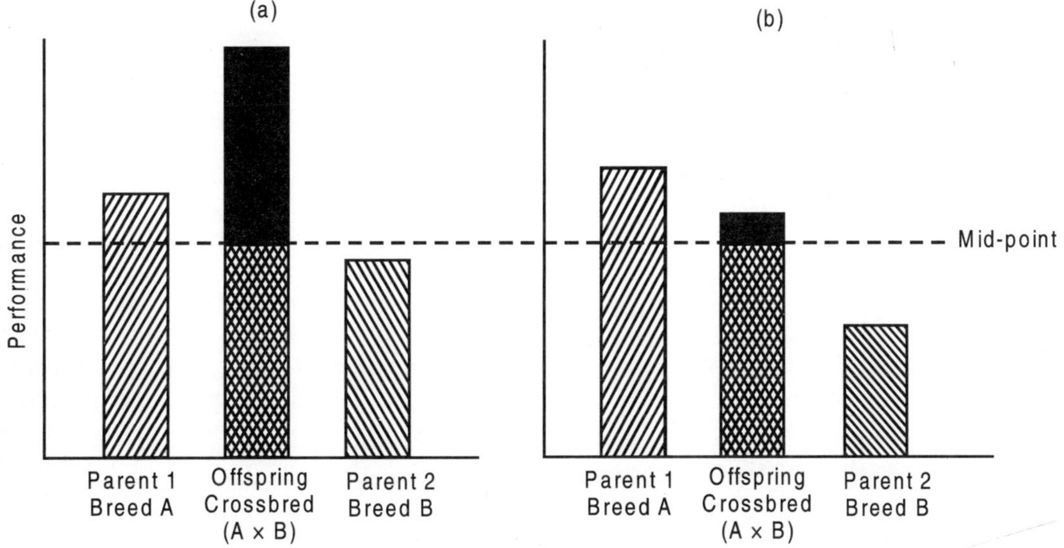

Fig 13.4 Measurement of heterosis as : (a) superiority over best parental breed (b) superiority over average of both parental breeds.

Reciprocal Cross Differences

The F_1 offspring can result from two types of crosses: crossing of males of 1st population with females of 2nd population: $P_1♂ \times P_2♀$, or crossing of males of 2nd population with females of 1st population: $P_2♂ \times P_1♀$. These two types of crosses, in which the breed (strain) of male and female parents is reversed, are referred to as **reciprocal crosses**.

Symbolically, these two crosses are shown below:

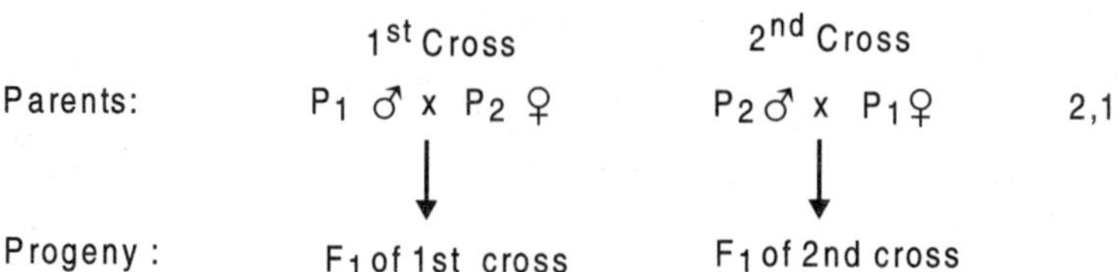

The difference in average performance of reciprocally produced crossbreds (F_1), referred to as **'reciprocal effects'**, can result from sex-linked and or maternal effects:

- **Sex-linked effects**: The F_1 (crossbred) offspring receive equal proportion of their autosomal genes from each of the parental populations (breeds). But it is not so for sex-linked genes because one of the sex chromosome does not carry any active genes (e.g. Y chromosome in mammals and the Z chromosome in poultry). Performance of reciprocally produced F_1 offspring may differ from each other if the trait is controlled by genes located on the sex-chromosome. In mammals ((XY system) if sex-linked effects are important, only the male offspring resulting from reciprocal crossing will manifest the difference. Sex-linked effects do not contribute to the difference in performance of the F_1 females of the reciprocal crosses because they receive one of their X chromosome from the sire and the other X from the dam. In poultry, the situation is reversed.
- **Maternal effects**: Maternal effects refer to the influence of mother's genotype on the phenotype (performance) of her offspring through the environment provided by her (apart from the direct influence of her genes):
    - Firstly, the dam contributes most of its cytoplasm to offspring which influences embryonic and fetal growth.
    - Secondly, the performance of offspring, especially for traits manifested at birth or early in life (neo-natal and post-natal growth), may be influenced by the environment provided by dam in terms of her mothering ability (e.g. milk production). Birth weight in cattle, for example, is influenced by body size of dam. Crossing of Holstein Frisian (heavy breed) with a Jersey (light breed) illustrates the effect of maternal effects, as shown below:

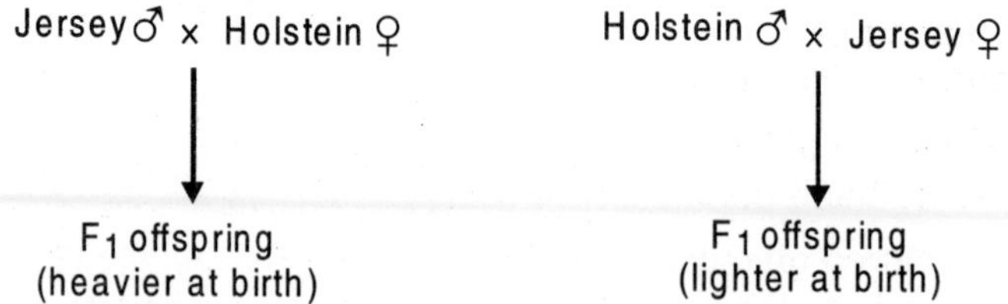

Similarly, early growth and weaning weight in beef cattle and swine are influenced by the amount of milk produced and other mothering ability traits of dam. Therefore, the crossbreds raised by a high milk-producing maternal breed are expected to be heavier at weaning.

Maternal effects, in general, are thought to be the major cause of differences between the reciprocals crosses. This is so because fewer genes are located on X chromosome, and moreover, these are manifested in one sex. Maternal effects are most pronounced for early growth in all species of mammals. In poultry maternal effects for early growth and viability are mediated through egg size.

Systems of Utilization of Heterosis

Heterosis results from increased heterozygosity when genetically divergent groups like breeds, strains, lines etc are crossed. The procedures which aim at utilization of heterosis, therefore, involve crossing of genetically distinct groups to produce progeny generation carrying high degree of heterozygosity. The parental groups may be inbred lines or strains or breeds or even species.

Crossing of Inbred Lines: Inbreeding-Hybridization

Heterosis has been commercially most extensively exploited in maize breeding where the seeds meant for commercial production are hybrid resulting from crossing of inbred lines. Development of hybrid maize made remarkable improvement in productivity and transformed the corn production industry. For example, in USA maize (corn) yield before extensive use of hybrids (by early 1930) was around 1250 kg/hectare. The yield quadrupled to 4850 kg/hectare following adoption of hybrids by early 1970's. More than 60% of yield increase was attributed to hybrid genotype, the remainder to improved production (agronomic) practices.

Following the success of inbreeding and hybridization in maize, development of inbred lines in livestock species, particularly of poultry and swine, for the production of hybrids was initiated. Procedure of development of inbred lines and crossing between them to utilize hybrid vigor is referred to as **inbreeding-hybridization.** Between 1930 and 1950's, inbred lines of chicken and swine were developed, and crossed for production of commercial progeny. In earlier literature, two terms viz; in-crossbreeding and incrossing have been used for crossing of inbred lines, particularly of poultry and swine: crossing of inbred lines which belong to different breeds is called **in-crossbreeding** while the crossing of inbred lines of same breed is referred to as **incrossing.**

Development and maintenance of inbred lines: Development of inbred lines in animals, as compared to plants, is expensive and time consuming. Plants can be selfed, produce more progeny (seeds) per individual, involve lesser cost in growing, have shorter generation interval, and can be grown at one time every year or even twice in a year (by growing in different locations). But such conditions do not prevail in farm animals, therefore, its cumbersome and very costly proposition.

Despite the difficulties associated with formation of inbred lines in farm animals, serious efforts were made in this direction particularly in chickens and swine where reproductive rate is high, and the cost of rearing of animals is relatively lesser than in other livestock species. Some important steps and consideration in this procedure are:

- Program of development of inbred lines is initiated with a large number of lines derived from either the same base population or from different populations.
- The lines are then inbred generation after generation through some regular system of inbreeding (e.g. full-sib, half-sib matings).
- A number of lines survive only for a short time either due to appearance of lethal genes or due to severe decline in their reproductive capacity (Fig 13.5). The extent of appearance of defects and the loss in reproduction depends on the types of genes possessed by the base (original) population from which the inbred lines were developed.
- Some of the lines that may survive initial cycles of inbreeding may have poor performance. The lines showing greater decline in reproduction and lower performance get eliminated or are discarded during early stages.

- In general, those lines which perform well and posses' normal vigor are further inbred to attain still higher levels of homozygosity. Since the performance of inbred lines can not be predicted beforehand, therefore, the only certain way to know what will happen on inbreeding is to do "inbreeding and observe the results".

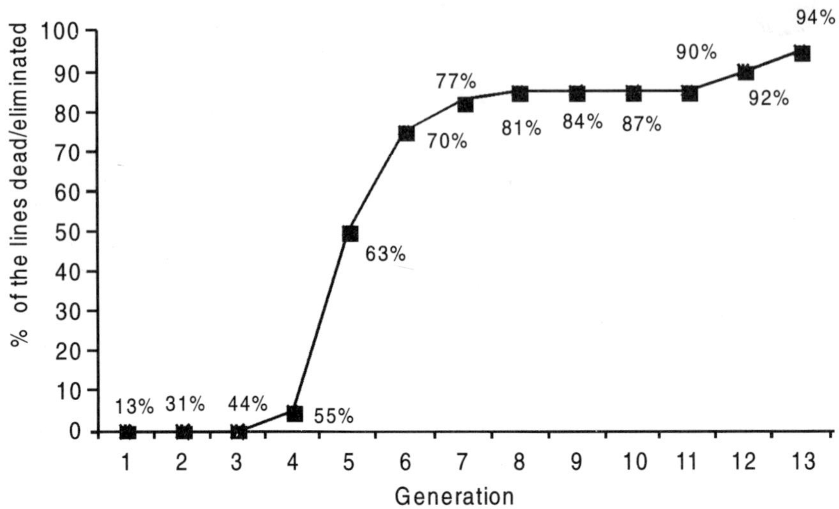

Fig 13.5 Results of one study on development of inbred lines in mice.The program was started with 20 lines which were inbred each generation through cousin matings. The values alongside dots are the inbreeding coefficients (%) of that generation.

Precautions in development of inbred lines: A few precautions to be followed during the development and maintenance of inbred lines are:

- Lines should be started from the parents having good performance and with no obvious genetic defects.
- High rate of inbreeding is recommended (e.g. full-sib or parent-offspring mating; ΔF=25% in 1st generation) during early generations as it helps in early uncovering of detrimental or lethal recessive genes.
- The lines which show higher incidence of defects as well as poor reproductive and production performance are discarded so that all the resources and efforts are diverted towards other lines.
- In subsequent generations, the rate of inbreeding is slowed down which helps in partially ameliorating some of the negative effects of inbreeding, and also in further propagation of superior lines.
- If possible, mild selection may be practiced within the inbred lines so as to improve their performance.

Diallel Crossing and Combining Ability

The primary objective of developing the inbred lines is to use them for crossing to obtain hybrids (i.e. to utilize heterosis). The inbred lines which survive successive generations of inbreeding and show good performance are then crossed, if possible, in all the possible combinations. The lines showing best performance in crosses (i.e. the ones that "nick well") are selected, propagated, and used for production of hybrids which are marketed. In general, the inbred lines which produce the most superior progeny are the ones that are homozygous for contrasting pair of alleles such that their progeny are heterozygous for many gene pairs.

The hybrids resulting from crossing of unrelated inbred lines would show some heterosis but which combination of inbred lines gives the best performance cannot be predicted beforehand. The value of any inbred line ultimately depends on its ability to produce superior hybrids in combination with other inbred lines Evaluation of performance of inbred lines can be systematically studied by crossing them in all the possible combinations including the reciprocal crosses, which is referred to as **diallel crossing (mating).** The term 'diallel'[2] was first proposed by Danish Geneticist, Schmidt. Diallel crossing makes it possible to systematically study which combination of lines gives the best cross performance. A diallel cross of 'n' lines results in 'n × n' possible combinations:

- 'n' are the number of lines (inbred lines)
- 'n (n–1)' are the number of crosses, out of which 'n (n-1)/2' will be one type of combinations, and the remaining 'n (n-1)/2' will be the reciprocal combinations.

A diallel cross involving four lines (designated as A, B, C and D), where the average performance of the progeny of pure- and line- crosses is designated with a 'bar' on 'X' ($\overline{X}$), is shown below:

Males from line :	Females from line:				Mean for row (male line)
	A	B	C	D	
A	$\overline{X}_{AA}$	$\overline{X}_{AB}$	$\overline{X}_{AC}$	$\overline{X}_{AD}$	$\overline{X}_{A\bullet}$
B	$\overline{X}_{BA}$	$\overline{X}_{BB}$	$\overline{X}_{BC}$	$\overline{X}_{BD}$	$\overline{X}_{B\bullet}$
C	$\overline{X}_{CA}$	$\overline{X}_{CB}$	$\overline{X}_{CC}$	$\overline{X}_{CD}$	$\overline{X}_{C\bullet}$
D	$\overline{X}_{DA}$	$\overline{X}_{DB}$	$\overline{X}_{DC}$	$\overline{X}_{DD}$	$\overline{X}_{D\bullet}$
Mean for column (female line)	$\overline{X}_{\bullet A}$	$\overline{X}_{\bullet B}$	$\overline{X}_{\bullet C}$	$\overline{X}_{\bullet D}$	$\overline{X}_{Overall}$

The first subscript following $\overline{X}$ refers to the male (dam) line, and the second to female (dam) line.The following points may be noted from the diallel set-up shown above:

(*i*) The diagonal elements of the table (viz; $\overline{X}_{AA}$, $\overline{X}_{BB}$, $\overline{X}_{CC}$, $\overline{X}_{DD}$) represent average performance of pure lines (inbred lines).

(*ii*) The off-diagonal elements refer to crosses between the four lines:

- Above diagonal elements represent average performance of different crosses with the first letter following '$\overline{X}$' referring to line of male parent (sire line) and the second letter referring to line of the female parent (dam line). For example, $\overline{X}_{AB}$ refers to performance of cross generated by crossing 'A' line males with 'B' line females

[2]The term 'diallel', derived from Greek word, was originally used by Schmidt (1919) to mean successive mating of two sires to the same dam in order to calculate their breeding values more accurately. This term has now come to designate crosses of inbred lines (or even strains/breeds) in all the possible combinations.

- Below diagonal elements represent average performance of reciprocal crosses of the crosses shown above diagonal.For example, $\overline{X}_{AB}$ refers to performance of cross generated by crossing 'A' line males with 'B' line females whereas $\overline{X}_{BA}$ refers to performance of reciprocal cross i.e. 'B' line males with 'A' line females. The difference between the reciprocal crosses is referred to as **reciprocal effect.**

  Reciprocal effect for cross between lines A and B = $\overline{X}_{AB} - \overline{X}_{BA}$

(*iii*) The marginal means refer to average performance of a line as either male line (last column of each row) or as female line (last row of each column). For example, $\overline{X}_{A\bullet}$, refers to average performance of line 'A' when used as male (sire) line and $\overline{X}_{\bullet A}$ refers to average performance of line 'A' when used as female (dam) line.

The diallel system of mating allows us to calculate the average performance of a line in crosses with all other lines. This is known as **general combining ability** (GCA) of the line, and is defined as the average performance of a line in crosses with all the other lines.GCA is useful for deciding as to which lines to be used in a general crossing program. It is calculated by taking the average of two means when that line is used as male and female line. For example, general combining ability of line 'A', $GCA_{(A)}$, is calculated as:

$$GCA_{(A)} = \frac{1}{2}\left(\overline{X}_{A\bullet} + \overline{X}_{\bullet A}\right)$$

Performance of a particular cross is referred to as **specific combining ability** of the cross (SCA). SCA helps us in deciding as to which is the best cross for producing the cross-line progeny. For example, SCA of lines 'A' and 'B', designated as SCA (AB), is calculated as follow[3] :

$$SCA_{(AB)} = \frac{1}{2}\left(\overline{X}_{AB} + \overline{X}_{BA}\right)$$

The concept of combining ability including the two terms, GCA and SCA was developed by Sprague and Tatum (1942) in connection with studying and comparing the performance of inbred lines of plants in hybrid combinations.

The mean genotypic value of a particular cross, $G_{(AB)}$, is composed of:

$$G_{(AB)} = GCA_{(A)} + GCA_{(B)} + SCA_{(AB)}$$

Where: $G_{(AB)}$: genotypic value of cross 'AB'

$GCA_{(A)}$: general combining ability of line 'A'

$GCA_{(B)}$: general combining ability of line 'B'

$SCA_{(AB)}$: specific combining ability of cross 'AB'

Genetically, GCA is caused by additive gene effects while SCA is caused by dominance and epistatic gene effects (non-additive). A comparison of the variances caused by GCA and SCA for a trait indicates the relative influence of additive and non-additive genetic variances, and thus helps in choice of the selection procedure to be followed for the improvement of cross performance.

[3]This is a simplified version of calculation of GCA and SCA, and is given for purpose of illustration only (technically GCA and SCA are expressed as deviations from mean of all crosses, and by different models)

Box 13.3: Concept of combining ability : George Sprague (1902-1998).

George Frederick Sprague (1902-1998), obtained his BS and MS from University of Nebraska (USA), and joined Cornell University where he completed his PhD (1930) in plant breeding and genetics under renowned corn (maize) geneticist Emerson. Sprague worked for the United State's Department of Agriculture (USDA), first at University of Missouri (1934-1939), and later at Iowa state University (1939-58). His period at Iowa was most fruitful where he, besides doing research on Corn with USDA, became a collaborator in teaching and student-guidance in plant breeding and genetics. His course on corn breeding and genetics was well received by students from across the university. His interactions with geneticists working in various field, in particular with Lush, an animal breeder and Kempthorne, a statistical geneticist, led to interrelating basic studies in genetics and statistics with applied breeding. *He served on the PhD committee of renowned animal breeder* ***Henderson.*** Henderson also used the idea of combining ability in swine for his dissertation.

His work on genetic improvement of corn has been a landmark. It was in this area where he advocated the concept of combining ability for which he is not only known in plant breeding but in animal breeding as well. Sprague and Tatum introduced and defined the terms "general and specific" combining ability.

Sprague (1902-98)

Combining ability of an inbred line was a phrase used to define the potency of an inbred line in hybrids. Sprague and Tatum partitioned combining ability among diallel crosses of inbred lines into (i) average performance of a line in crosses with all other lines (general combining ability), and (ii) performance of the cross of specific pairs of lines relative to their average performance (specific combining ability). The concepts of general and specific combining abilities stimulated theoretical research for types of genetic effects important for each and a method to evaluate inbred lines in hybrids.

His paper on GCA and SCA, where he coined the two terms, continues to be a classic citation, and for which he is more popular. Recalling about this paper, Sprague (1986) stated (slightly redrafted) :

" —It has become obvious to corn breeders that inability to achieve further genetic improvement was not due to lack of genetic variability; rather it indicated lack of our knowledge of types of gene action involved in combining ability. We postulated that the data on inbred lines and their testing in hybrid combinations could provide useful estimates of different types of gene action involved in heterosis for corn yield. We realized that hybrid component could be further partitioned into: variation between lines and variation within lines. Such a breakdown was useful but it did not serve our objective which was to obtain 'estimates' which could be indicative of usefulness in selecting the parental combinations. Hence a new definition of terms was necessary to characterize the estimates obtained. We chose the terms 'GCA' and ; "SCA" The frequency of citation of this paper stems from the fact that it was one of the first to utilize the concept in plant breeding.The concept of GCA and SCA has later proven useful in many subsequent studies"

Sprague after formal retirement from USDA accepted appointment at University of Illinois where he continued till 1994, but was active in scientific work till his death.

Strain and Breed Crossing

Inbreeding and hybridization is not only time consuming but costly as well. Therefore such a system was not found to be economically viable for exploitation of heterosis in farm animals, and was replaced by crossing between strains as strain crossing does not involve the major effort of developing and maintenance of inbred lines.

From breeding standpoint, a strain is an interbreeding group of animals within a breed that is isolated from other such groups due to its history of selection (breeding objectives). Concept of strain is very common in poultry breeding where a **strain** it is defined as a group of birds within a breed which have been kept by a breeder as a closed population and has been selected for some trait (s) for at least five generations. Because of the consistent selection practiced within a strain, and its closed nature, a strain is also expected to be homozygous for some of the loci, and hence it can be considered as mildly inbred. Strain crossing is common practice for the production of strain-cross birds because of its economic benefits over inbreeding and hybridization. This topic shall be discussed in more detail in the chapter on genetic improvement of poultry.

Another method used for the exploitation of heterosis is crossing between breeds. Breed crossing is common in cattle, swine and sheep. There are number of systems of making crosses between breeds and are covered in detail in the next chapter. The concept of diallel crossing and of GCA and SCA are equally applicable for strain crossing or even breed crossing.

Species Hybridization

The widest possible type of outbreeding is crossing between species, referred to as **species hybridization**. The best example of species hybridization is crossing of male ass (Jack of *Equus asinus*) and female horse (mare of *Equus caballus*) to produce mule. Mule has been valued for centuries for it's: ability to work under harsh conditions particularly hot climate and to withstand abuse because of it's high endurance

Mules are sterile due to abnormal separations of chromosomes during cell division due to which gametes are either not formed or are not capable of fertilization, and due to unequal number of chromosomes. The reciprocal cross between these two species (i.e. male horse and female ass) produces 'hinny' which has more resemblence to horse.

Other examples of species hybridization which are only of Zoological interest are:

- Ass ♂ × Zebra ♀ : Offspring called 'Asbra'
- Zebra ♂ × Horse ♀ : Offspring called 'Zebroid' ; popular in tropics, resistant to diseases and stress, docile.
- Cattle ♂ × Yak ♀ : Offspring called "Pie niu'; popular in Upper Himalayas; excellent hardiness.
- Cattle × Buffalo : Offspring called 'Cattalo' or 'Beefalo'.
- Goat × Sheep : Offspring called 'Geep'.

Selection for Combining Ability

In order to improve combining ability of lines/strains, two selection (breeding) schemes, referred to as 'recurrent selection' (RS) and 'reciprocal recurrent selection' (RRS) have been proposed. The lines used in these two schemes may not be inbred at all or may be mildly inbred (like strains/breeds), and hence they have considerable genetic variability.

Recurrent Selection (RS)

The procedure of recurrent selection was proposed by Hull (1945) for improvement of specific combining ability of corn (maize) lines. The procedure involves two populations: one is a segregating population (also called the 'line to be tested') and the other is a highly inbred line, called **'tester line'**. The inbred or tester line is reproduced each generation without any selection while the segregating population, which could be a line, strain or breed, is selected on the basis of performance of its test-cross progeny (Fig. 13.6)

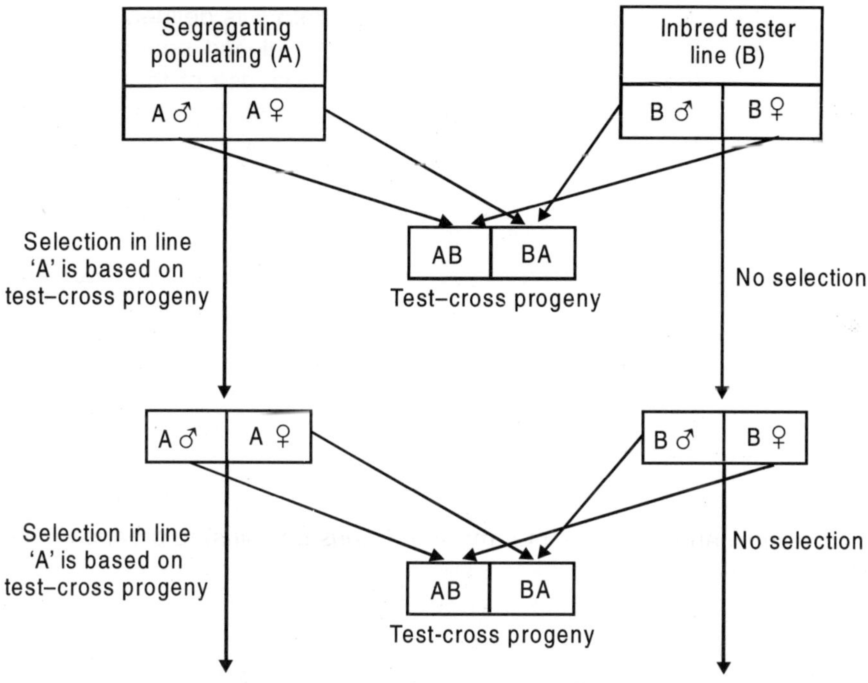

Fig 13.6 Recurrent selection scheme

Operational steps in RS : Recurrent selection involves following steps :

- Males and females of the segregating population are selected on the basis of average performance of their test-cross progeny.
- The selected males and females are mated among themselves to produce next generation of the segregating population.
- The males and females of the inbred/tester line are mated among themselves (without any selection) to produce next generation of this line.

This type of selection cycle is repeated generation after generation. Since selection in segregating population is based on its ability to give best performance in cross with the tester line, selection is directed for the gene combination which give high performance in the cross.

The greatest limitation of this system is the difficulty and cost of maintaining the highly inbred line (tester line). Being highly inbred, its performance is poor but it has to be maintained without serving any economic purpose. It can be noted that this procedure of selection is just like progeny testing (except that selection is based on cross-bred performance, test cross progeny); therefore it prolongs the generation interval.

Reciprocal Recurrent Selection (RRS)

This system of selection (breeding), proposed by Comstock and co-workers (1949) for plants, aims at overcoming the first short coming of recurrent selection i.e. RRS it does use an inbred tester line. In this system both the lines/ populations are segregating, and therefore both are subjected to selection based on their cross-line progeny (Fig 13.7).

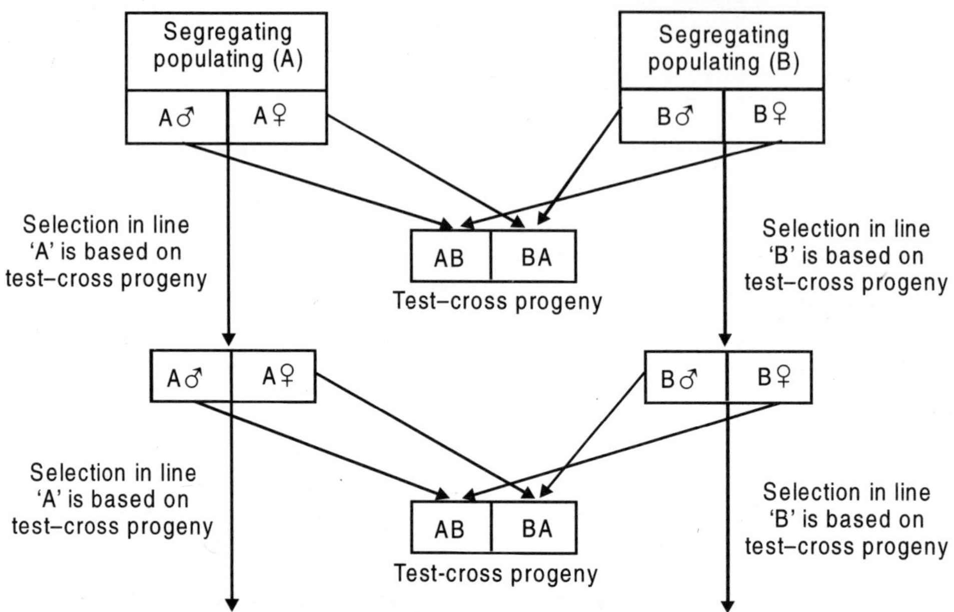

Fig 13.7 Reciprocal recurrent selection scheme

Operational steps in RRS: A reciprocal recurrent selection program involves to following steps :

- Males and females of the two populations, designated 'A' and 'B', are mated to produce test-cross progeny, both 'AB' and 'BA'.
- Males and females of each of the two populations are selected on the basis of their test-cross progeny
- The selected males of each of the two populations are then mated within their own population to produce the next generation.

This cycle is repeated generation after generation. It may be noted that the test-cross progeny are raised only for testing the combining ability of the two populations, and are discarded once the parents have been selected.

The populations which are used for increasing the combining ability are the ones that have already demonstrated their ability to combine well from previous testing. Since selection in both the populations is for their ability to combine well, it results in accumulation of complementary genes in both the populations. The greatest advantage of this system over recurrent selection is that both the populations are being simultaneously improved for combining ability. Selection based on performance of test-cross (crossbred) progeny leads to improvement in both the 'GCA' and 'SCA', and hence utilizes both the additive and non-additive effect of genes. Reciprocal recurrent selection, like RS, doubles the generation interval, and is thus used in highly prolific species like poultry and swine.

Review Questions

1. Define/ Explain the following terms:

   I. Outbreeding
   II. Outcrossing
   III. Species
   IV. Breed
   V. Purebred/straightbred and crossbred
   VI. Strain
   VII. Line
   VIII. Heterosis
   IX. Individual heterosis
   X. Maternal heterosis
   XI. Paternal heterosis
   XII. Retained heterosis
   XIII. Inbreeding-hybridization
   XIV. Diallel crossing/mating
   XV. Species hybridization

2. Differentiate between the following with suitable example(s):

   I. General combining ability and specific combining ability
   II. Outbreeding and outcrossing
   III. Recurrent and reciprocal recurrent selection

3. Is it justified to consider Zebu cattle and European cattle as distinct species? Explain
4. What are the important features of a livestock breed?
5. What is the primary genetic consequence of outbreeding/crossing?
6. What are two genetic consequences of outbreeding which result from increased heterozygosity?
7. How does outbreeding cause expression of deleterious recessive alleles to be 'masked' or 'covered-up' in heterozygotes?
8. Outbred animals do not breed true to their type. Explain.
9. What is the main phenotypic effect of outbreeding?
10. Give examples where heterosis has been used to greatest advantage in farm animals.
11. Explain the possible genetic bases of heterosis or What types of gene expression are responsible for heterosis ? or Outline the theories explaining heterosis.
12. What appears to be the physiological basis of heterosis?
13. What is meant by 'favorable dominant gene' hypothesis of heterosis? What is the objection to this as the only cause of heterosis ?
14. What factors determine the degree of heterosis?
15. Why do crosses of closely related populations show less hybrid vigor than crosses of less related populations?
16. How much heterosis, as a general rule, is expressed for: lowly heritable traits; highly heritable traits? Why?
17. Why more heterosis is exhibited by fitness traits?
18. What general category of traits of animals appear to be affected most by heterosis?
19. Give two examples of maternal heterosis.
20. What are the two causes of differences in performance between reciprocal crosses? Explain.

21. Which is the most likely cause of reciprocal cross differences in livestock species?
22. What are the two ways of expression or measurement of heterosis?
23. Why does the proportion of heterozygosity decrease in F_2 generation as compared to F_1?
24. Which of the following is true : (a) F_2 heterosis is more than F_1 heterosis (b) F_2 heterosis is less than F_1 heterosis (c) F_2 heterosis equals F_1 heterosis.
25. What procedure should be followed for utilization of heterosis to the greatest possible extent in farm animals?
26. Explain the procedure of inbreeding and hybridization.
27. Explain the procedure of development and maintenance of inbred lines for crossing.
28. Why is it more difficult to produce inbred lines in farm animals than in plants?
29. Explain the procedure of diallel crossing using appropriate diagram and equations/ expressions.
30. Give the expression for genotypic value of a cross in terms of general and specific combining abilities.
31. Why has mule been valued for centuries?
32. Why are species hybrids usually sterile?
33. What are procedures used for improving combining ability of lines, strains or breeds?
34. Explain the procedure of reciprocal recurrent selection.
35. What is the greatest limitation of recurrent selection?
36. What is the limitation of reciprocal recurrent selection?

●●●

14

Estimation of Heterosis

The theory of heterosis including its genetic basis has been outlined in the last chapter. In this chapter the computation of compute heterosis in two- and three-breed crosses, prediction of the average performance of crossbreds (F_1 and F_2) resulting from two- and three-breed crossing has been outlined.

Heterosis in Two Breed Crosses

14.1. Consider a hypothetical example of two breeds 'A' and 'B'. Suppose each of these is homozygous for contrasting alleles of a single locus (A-locus) : individuals of breed 'A' are all 'AA' while those of breed 'B' are all 'aa'. Assume allele 'A' is completely dominant over 'a'; and let the genotypic values for the three genotypes be :

AA = 100 units, Aa = 100 units, aa = 90 units.

Calculate the absolute and percentage heterosis exhibited by (F_1) crossbreds over the average of the two parental breeds.

Solution :

	Breed :		Crossbred
	A	B	(A x B)
Genotypes :	AA	aa	Aa
Values :	100	90	100

Average of purebreds $= \frac{(100+90)}{2} = 95$ units

Average of crossbreds = 100 units

Heterosis (absolute) = Crossbred average – Purebred average

= 100 – 95 = 5 units

$$\text{Heterosis (\%)} = \frac{(\text{Crossbred average} - \text{Purebred average})}{\text{Purebred average}} \times 100$$

$$= \frac{(100-95)}{95} \times 100 = 5.26$$

14.2. Average lactation milk yield of two breeds of cattle viz ; Holstein-Friesian and Sahiwal, and their cross (HF x Sahiwal) was 5500, 2500 and 4450 kg, respectively.

(*a*) Calculate the heterosis (%) exhibited by the crossbreds by both the methods viz; (*i*) best-parent and (*ii*) mid-parent methods

(*b*) Is the heterosis exhibited by the cross (HF × Sahiwal) desirable? Comment.

Solution :

(*a*) Average of parental breeds : $\overline{X} = \frac{5500+2500}{2} = 4000$ kg.

(*i*) Heterosis (%) with respect to performance of best parent :

$$= \frac{[\text{Crossbred average} - \text{Average of best parental bread}]}{\text{Average of best parental breed}} \times 100$$

$$= \frac{(4450-5500)}{5500} \times 100 = -19.09$$

(*ii*) Heterosis (%) with respect to average of the two parental breeds :

$$= \frac{[\text{Crossbred average} - \text{Average of parental breeds}]}{\text{Average of parental breeds}} \times 100$$

$$= \left[\frac{(4450-4000)}{4000}\right] \times 100 = \frac{(450 \times 100)}{4000} = 11.25$$

(*b*) Average performance of crossbreds does not exceed the performance of HF breed (*i.e.* the best parental breed) therefore, heterosis as expressed with respect to performance of best parent, is not desirable. However, with respect to average of the two parental breeds, heterosis is desirable as the crossbreds have higher milk yield than average of parental breeds.

14.3. In a broiler chicken crossing program, average six-week body weight of two strains was 1800 and 1600g. The strain-cross offspring exhibited 7% heterosis over the mid-parental value. Compute the average performance of F_1 offspring.

Solution :

Average of parental strains : $\overline{X} = \frac{(1800+1600)}{2} = 1700$ g

The average of strain-cross offspring, $\overline{X}_{F1}$, can be calculated by rearranging the equation of heterosis (%) :

$$\text{Heterosis (\%)} = \frac{\overline{X}_{F1} - \overline{X}}{\overline{X}} \times 100$$

Rearranging the equation and putting the values of $\overline{X}$ and % heterosis :

$$\overline{X}_{F1} = \left(\frac{\text{Heterosis \%} \times \overline{X}}{100}\right) + \overline{X} = \left(\frac{7 \times 1700}{100}\right) + 1700 = 1819 \text{ g}$$

14.4. Average age at first calving of Holstein-Friesian and Sahiwal breeds and their cross was 800, 1200 and 950 days, respectively. (*a*) Calculate the percentage heterosis with respect to mid-parental value. (*b*) Is the direction of heterosis favourable? (*c*) Predict the average age at first calving of the F_2 produced from inter-se-mating of the F_1 (crossbreds).

Solution :

(*a*) Average of parental breeds : $\overline{X} = \dfrac{800+1200}{2} = 1000$ days

Heterosis (%) with respect to mid-parent value :

$$= \frac{(\text{Crossbred average} - \text{Average of parental breeds})}{\text{Average of parental breeds}} \times 100$$

$$= \frac{(950-1000)}{1000} \times 100 = -5.0$$

(*b*) The negative heterosis exhibited for age at first calving is desirable.

(*c*) Heterosis (absolute) in F_1 = 950–1000 = –50days

Heterosis (absolute) in $F_2 = \dfrac{1}{2}$ (Heterosis in F_1) $= \dfrac{(-50)}{2}$ = 25 days

Average age at first calving in $F_2 = \overline{X}_{F1} + \dfrac{1}{2}$ (F_1 heterosis) = 950 + (–25) = 925 days

14.5. In a swine crossbreeding program involving Chester (C) and Yorkshire (Y) breeds, average 70-day body weight of different groups was as follows:

Genetic group		Body wt. (kg)
Purebreds :		
	Chester (C) :	17.0
	Yorkshire (Y) :	18.0
F1 (Crossbreds) :		
	C♂ x Y♀:	19.0
	Y♂ x C♀:	17.0

(a) Calculate the % heterosis (using mid-parent value) resulting from crossing of : (*i*) Chester males with Yorkshire females and (*ii*) Yorkshire males with Chester females.

(b) Calculate the overall heterosis (%) (using mid-parent value) resulting from two-breed crossing including reciprocal crosses

(c) Are the reciprocal effects present ? Comment.

(d) What would be the average 70-day body weight of the F_2 resulting from interse mating of all the F_1 (*i.e.* F_1 resulting from 'C × Y' and 'Y × C')

Solution :

(a) Average of parental breeds :

$$\overline{X} = \frac{(17+18)}{2} = 17.5 \text{ kg}$$

Average of the two crossbred groups: $= \dfrac{\text{Avg. of } (C\square \times Y\square) + \text{Avg. of } (Y\square + C\square)}{2}$

$$= \frac{19+17}{2} = 18.0 \text{ kg}$$

(*i*) Heterosis (%) from mating of Chester males with Yorkshire females (*w.r.t.* mid-parent value) :

$$= \frac{\overline{X}_{F1} - \overline{X}}{\overline{X}} \times 100 = \frac{(19-17.5)\times 100}{17.5} = 8.57$$

(*ii*) Heterosis (%) from mating of Yorkshire males with Chester females (*w.r.t.* mid-parent value) :

$$= \frac{\overline{X}_{F1} - \overline{X}}{\overline{X}} \times 100 = \frac{(17-17.5)\times 100}{17.5} = -2.86$$

(b) Overall heterosis (%) resulting from all the crosses between the two breeds :

$$= \frac{(\text{Average of two crossbred groups} - \text{Average of purebred groups})}{\text{Average of purebred groups}} \times 100$$

$$= \frac{(18.0-17.5)}{17.5} \times 100 = 2.86$$

(c) Reciprocal effects are present because average performance of the two reciprocal crosses differs from one another. The cross 'C (♂) x Y(♀)' has higher body weight than its reciprocal cross 'Y(♀) x C(♂)'. The differences in reciprocal crosses can be caused by maternal and sex-linked effects. The crossbreds resulting from C(♂) x Y(♀) are recommended for commercial purposes.

(d) Absolute heterosis in F_1

= Crossbred average – Purebred average = 18.0 – 17.5 = 0.5 kg

Average body weight of F_2 resulting from inter-se mating of F_1 :

$$\overline{X}_{F2} = \overline{X}_{F1} + \frac{1}{2} (F_1 \text{ heterosis}) = 18 + ½(0.5) = 18.25 \text{ kg.}$$

14.6. Chester (C), Duroc (D) and Yorkshire (Y) breeds of swine were crossed in all possible pair-wise combinations (diallel crosses). The traits recorded were body weight at 154 days (BW), the number of pigs per litter at 154 days (N) and total weight per litter (TW). The total weight per litter is calculated as the product of the body weight and litter size (*i.e.* BW x N), and is a measure of overall efficiency of pig production. The averages for the three traits for various purebred and crossbred combinations are tabulated below :

	Purebreds :			Crossbreds :					
Trait	C	D	Y	C♂xD♀	D♂xC♀	C♂xY♀	Y♂xC♀	D♂xY♀	Y♂xD♀
BW	84	94	90	98	96	96	89	92	94
N	6.4	6.1	7.7	7.2	7.1	8.0	8.2	7.1	6.9
TW	537.6	573.4	693	705.6	681.6	768	729.8	653.2	648.6

(*a*) Calculate the heterosis for total weight per litter (TW) in absolute and percent terms for each cross by both the methods viz; best parent and mid-parent.

(*b*) Calculate the overall heterosis resulting from all the possible crosses for the three traits using mid-parent value.

(*c*) What type of non-additive gene action could be responsible for the heterosis obtained from all the crosses for each of the three traits assuming epistasis to be absent?

(*d*) Are reciprocal effects present ? If so, what could be their causes ?

(*e*) Which one cross would you recommend for commercial use if improvement is desired in : (*i*) body weight (*ii*) litter size (*iii*) efficiency of pig production (*i.e.* total litter weight) ?

Solution :

(*a*) Absolute heterosis using best parent = $\overline{X}_{F1}$ – Best parental breed average

Absolute heterosis using mid-parent value = $(\overline{X}_{F1} - \overline{X})$

Heterosis (%) using best parent = $\dfrac{\overline{X}_{F1} - \text{Best parental breed average}}{\text{Best parental breed average}} \times 100$

Heterosis (%) using mid-parent value = $\dfrac{\overline{X}_{F1} - \overline{X}}{\overline{X}} \times 100$

Using the expressions listed above, heterosis values alongwith the calculations involved for total weight (TW) are shown below :

	Purebreds :					
	Parental breeds:		Parental breeds:		Parental breeds:	
	C	D	C	Y	D	Y
Breed average :	537.6	573.4	537.6	693.0	573.4	693.0
Average of parental breeds :	$\dfrac{537.6+573.4}{2}$ = 555.5		615.3		633.2	
	Crossbred Combinations :					
	C x D	D x C	C x Y	Y x C	D x Y	Y x D
Average of F_1 (crossbreed) :	705.6	681.6	729.8	768	724.2	748.8
Average of the two reciprocal crosses :	$\dfrac{705.6+681.6}{2}$ = 693.6		748.9		736.5	
Absolute heterosis (w.r.t. best parent) :	693.6 – 573.4 = 120.2		55.9		43.5	
% heterosis (w.r.t. best parent) :	$\dfrac{120.2\times100}{573.4}$ = 20.96		8.07		6.28	
Absolute heterosis (w.r.t. mid-parent) :	693.6 – 555.5 = 138.1		133.6		103.3	
% heterosis (w.r.t. mid-parent) :	$\dfrac{138.1\times100}{555.5}$ =24.86		21.71		16.31	

(*b*) Computation of overall heterosis

Overall parental breed average :

$$\overline{X} = \frac{(\text{Chester average} + \text{Duroc average} + \text{Yorkshire average})}{3}$$

Parental breeds average for body weight : $\overline{X} = \frac{(84+94+90)}{3} = 89.3$ Kg

Average of F_1 (all crosses) :

$$\overline{X} = \frac{(C\times D \text{ avg.})+(D\times C \text{ avg.})+(C\times Y \text{ avg.})+(Y\times C \text{ avg.})+(D\times Y \text{ avg.})+(Y\times D \text{ avg.})}{6}$$

$$= \frac{(98+96+96+89+102+104)}{6} = 97.5 \text{ kg}$$

Overall heterosis for body weight :

Absolute heterosis $= \overline{X}_{F1} - \overline{X} = 97.5 - 89.3 = 8.2$ kg

Percent heterosis $= \frac{\overline{X}_{F1} - \overline{X} \times 100}{\overline{X}} = \frac{(97.5-89.3)\times 100}{89.3} = 9.18$

Mean values and heterosis for the three traits, calculated as per the procedure shown above, are tabulated below :

Trait	Best parent average	Average of parental breeds	Average of all F_1	Heterosis (absolute)	Heterosis (%)
BW(kg)	94	89.33	97.50	8.2	9.14
N	7.7	6.73	7.47	0.73	10.89
TW (kg)	69.3	601.33	726.33	125.0	20.79

(*c*) Inference about type of gene action :

- Crossbred average exceeded the average of the parental breeds for all the traits which meant that heterosis is exhibited by all the traits.
- Crossbred average is better than the best parent average for BW and TW, therefore, overdominance type of gene action is operative for these two traits.
- For litter-size, the average of crossbreds is better than the average of parental breeds only, therefore, dominance is operative for litter-size.

(*d*) Reciprocal effects are present since the average of the reciprocally crossed F1 differs from each other for the three traits. This indicates that maternal and sex-linked effects are important for the three traits.

(*e*) Use of crossbreds for commercial pig production :

- For commercial improvement of body weight, Yorkshire ♂ x Duroc ♀ cross is recommended,
- For commercial improvement in litter size, Yorkshire ♂ x Chester White ♀ cross is recommended.
- For commercial improvement of total litter weight, Chester ♂ x Yorkshire ♀ cross would be most suited.

Heterosis in Three Breed/Strain Crosses

14.7. Outline the procedure for calculation of heterosis from crossing of three breeds, A, B and C, as shown below :

Parental generation : A x B
↓
1st generation : AB x C
(Two-breed cross) (½ A + ½ B) ↓
2nd generation : (AB)C
(Three-breed cross) (¼A + ¼B + ½C)

Where :

'AB' refers to two-breed cross

'(AB)C' refers to three-breed cross

Solution : The heterosis shown by F_1 crossbred resulting from crossing of 'A' and 'B' is referred to as two-breed heterosis, and the one resulting from mating of F_1 (*i.e.* AB) with 'C' is called three-breed heterosis.

The parental breed average for three-breed cross, '(AB)C', would be :

$$= (\tfrac{1}{4}\bar{A} + \tfrac{1}{4}\bar{B} + \tfrac{1}{2}\bar{C})$$

where $\bar{A}$, $\bar{B}$ and $\bar{C}$ refer to average of A,B and C breeds respectively.

Heterosis (absolute) from three-breed crossing (H) :

$$= \text{Average of '(AB)C' crossbred} - \frac{(\text{Average of 'AB'} + \text{Average of 'C'})}{2}$$

Heterosis (%) from three breed crossing would be :

$$= \frac{\text{Average of '(AB)C' crossbred} - \frac{1}{2}(\text{Average of 'AB'} + \text{Average of 'C'})}{\frac{1}{2}(\text{Average of 'AB'} + \text{Average of 'C'})}$$

Heterotic advantage (%) due to use of 3rd breed (over two-breed cross) :

$$= \frac{\text{Average of '(AB)C' crossbred} - \text{Average of two breed crossbreds}}{\frac{1}{2}(\text{Average of two breed crossbreds} + \text{Average of 3rd breed})}$$

14.8. Average six-week body weight (g) for three breeds of chicken along with their of 2-way and 3-way crosses is given below :

Breed/Cross :	Genetic group designation	6-wk body wt. (g)
New Hampshire:	N	1500
Barred Plymouth Rock:	B	1530
Rhode Island Red:	R	1650
2-way cross :	(N♂ x B♀)	1590
3-way cross :	(N x B)♂ x R♀	1660

Calculate the :

(*a*) heterosis (%) in two-way and three-way crosses (with respect to average of parental breeds).

(*b*) heterotic (%) advantage due to use of 3rd breed (R)in the crossing.

Solution :

(*a*) Average of parental breeds, N and B = $\frac{(1500+1530)}{2}$ = 1515 g.

Average of three parental breeds, N, B and R = $\frac{(1500+1530+1650)}{3}$ = 1560 g.

Heterosis (%) from two-breed crossing (N×B) :

$$= \frac{(\text{Average of two way cross} - \text{Average of parental breeds})}{\text{Average of parental breeds}} \times 100$$

$$= \frac{1590-1515}{1515} \times 100 = 4.95$$

Heterosis (%) from three-breed crossing (NB×R) :

$$= \frac{(\text{Average of three way cross} - \text{Average of three parental breeds})}{\text{Average of three parental breeds}} \times 100$$

$$= \frac{1660-1560}{1560} \times 100 = 6.41$$

(*b*) Heterotic advantage (%) resulting from three-breed crossing :

$$= \frac{(\text{Average of three breed cross} - \text{Average of two breed cross})}{\frac{1}{2}(\text{Average of two breed cross} + \text{Average of 3rd breed})}$$

$$= \frac{(1660-1590)}{\frac{1}{2}(1590+1650)} \times 100 = 4.32$$

14.9. Average body weight at 154 days of age for Chester (C), Duroc (D) and Yorkshire (Y) breeds of swine were 35, 40 and 37 Kg respectively. Assume heterosis for each of the two-breed cross to be 5% and for three-breed cross (CD × Y) to be 15%.

Calculate the :

(*a*) expected weight of the crossbreds resulting from : (*i*) Chester and Duroc (*ii*) Duroc and Yorkshire, and (*iii*) Chester and Yorkshire.

(*b*) expected weight of crossbreds resulting from three-bred crossing : 'CD X Y'

Solution :

(*a*) Averages of different pure breed combinations :

Average of C and D $= \frac{35+40}{2}$ =37.5 Kg

Average of D and Y $= \frac{40+37}{2}$ = 38.5 Kg

Average of C and Y $= \frac{35+37}{2}$ = 36 Kg

Average of C, D and Y $= \frac{35+40+37}{3}$ = 37.3 Kg

Heterosis (%) in 2-breed cross

$$= \frac{(\text{Average of two breed crossbreds} - \text{Average of parental breeds})}{\text{Average of parental breeds}} \times 100$$

Let $\overline{X}$ be the average of the two-breed crossbreds progeny.

Putting the values of the known quantities in the above equation of heterosis (%) in 2-way cross, and solving for '$\overline{X}$', gives the predicted average of crossbreds:

(*i*) For cross of Chester and Duroc :

$$5 = \frac{(\overline{X} - 37.5) \times 100}{37.5}$$

$$\overline{X} = 39.38 \text{ kg.}$$

(*ii*) For cross of Duroc and Yorkshire

$$5 = \frac{(\overline{X} - 38.5) \times 100}{38.5}$$

$$\overline{X} = 40.42 \text{ kg}$$

(*iii*) For cross of Chester and Yorkshire:

$$5 = \frac{(\overline{X} - 36.0) \times 100}{36.0}$$

$$\overline{X} = 37.8 \text{ kg}$$

(*c*) Heterosis (%) in 3-breed cross ('CD X Y') :

$$= \frac{(\text{Average of three breed cross} - \text{Average of three parental breeds})}{\text{Average of three parental breeds}} \times 100$$

$$15 = \frac{(\overline{X} - 37.3) \times 100}{37.3}$$

$$\overline{X} = 43.1 \text{ kg}$$

∴ Average weight of the three breed cross is expected to be 43.1 Kg.

Supplementary Problems

14.10. Six-week body weight of two breeds of broiler chicken and their reciprocally produced crossbred offspring are given below :

Groups :		Body weight (g)
Purebred	Breed A	1650
	Breed B	1780
Crossbred	A x B	2040
	B x A	2090

Calculate the % heterosis on the basis of mid-parent value.

14.11. Average lamb weaning weight (kg) of Hampshire (H), Suffolk (S) and Dorset (D) breeds of sheep, and their all the possible two-way combinations is given below.

Breed/Cross Breed of : Sire	Dam	Weaning weight (Kg)	Breed/Cross Breed of : Sire	Dam	Weaning weight (kg)
H	H	30.0	D	S	34.0
D	D	26.0	H	S	32.0
S	S	33.0	H	D	29.0
S	H	31.0	S	D	31.0
D	H	30.0	–	–	–

Calculate the heterosis (%) for all the crosses on the basis of mid-parent average.

14.12. Two strains of White Leghorn chicken (PL1 and PL2) maintained at a poultry breeding farm of GADVASU, Ludhiana, Punjab (India) were reciprocally crossed. Average egg production and egg weight of different groups are given below.

Pure or Strain Cross	Trait : No. of eggs (annual)	Egg weight (g)
PL1	280	58
PL2	290	57
PL1 × PL2	305	58

Estimate the % heterosis on the basis of average of the two parental strains.

14.13. Two inbred lines of chicken, A and B, were crossed in pair-wise combinations. Average adult body weight (BW), egg number (EN) and egg weight (EW) of the two inbred lines, and of the F1 resulting from their crossing are shown below :

	Line A	Line B	F_1
BW(g) :	1800	1600	1900
EN :	220	260	300
EW(g) :	58	59	59

Which of the three traits exhibits the highest, and which the lowest heterosis?

Answers to Supplementary Problems

14.10 Twenty percent.

14.11 5.08%.

14.12 7.01% for egg number and 0.87% for egg weight.

14.13 11.7% for BW, 25.0% for EN and 0.85% for EW.

15

Systems of Crossbreeding

Crossbreeding is traditionally defined as mating between animals belonging to different established breeds. In some cases, the differences between breeds are not clearly demarked, and the breeds are further differentiated into strains/lines. The term crossbreeding, has, therefore, come to cover mating between animals of not only different breeds but also of genetically distinct groups such as strains, lines etc. The term crossing is also used to refer to all such types of matings.

Uses of Crossbreeding

Crossbreeding can be used for commercial as well as specialized purposes.

Crossbreeding for Commercial Production

1. **Utilization of heterosis:** One of most important practical use of crossbreeding is to take advantage of heterosis. Crossbred animals generally have better performance than average of their parental breeds, and in some situations may even exceed the performance of the best parental breed. Two features of heterosis make it especially suitable for commercial livestock production:
   - *Higher heterosis under unfavorable conditions*: The magnitude of heterosis is, theoretically, a function of genetic diversity between the breeds/lines being crossed and the magnitude of non-additive gene action. However, extensive experimental evidence from plants, laboratory animals, and to some extent from livestock species, has shown that heterosis is more pronounced under unfavorable conditions. This is due, most likely, to the fact that crossbreds (being heterozygous for most of the loci) are genetically better buffered to cope with unfavorable environmental conditions. Lerner's theory of **'genetic homeostasis'** states that the *"heterozygotes have a greater physiological ability to remain within the normal paths of development than*

the purebreds". Since commercial animals have to perform under varied agro-climatic conditions, the crossbreds would be better able to withstand the unfavorable environmental challenges, and still perform better.

- *Higher heterosis for efficiency of livestock production*: Performance of crossbreds for any single trait may sometimes be better than purebreds only a little. But the cumulative benefits obtained from small improvement in each of the economic trait may accumulate to large improvement in overall economic efficiency that is dependent on a number of traits.

In view of these features of heterosis, crossbreeding (crossing) is widely used for roduction of commercial livestock and poultry. The hybrid maize which is the result of crossing of inbred lines is the most successful example of utilization of heterosis. Introduction of hybrid maize in USA led to a quantum jump in productivity. In animals, the phenomenon of heterosis has been most successfully utilized in chickens, followed by swine, beef cattle and sheep.

2. **Breed complementarity:** The second major practical advantage of systematic crossbreeding is its ability to utilize complementarity of different breeds. One breed may have good performance for some traits and poor performance for others; the other breed may have strengths for some other traits and deficiencies for others (i.e. no one breed excels in all the traits of economic importance). Crossbreeding by combining desirable traits of two or more breeds may significantly contribute to increase in overall productivity. The concept of combining desirable traits of different breeds in such a manner as to maximize the genetic merit of crossbred (offspring) for total productivity is called '*breed complementarity*'.

Breed complementarity is formally defined as the ability of two or more breeds to produce offspring that have optimum level of performance for several traits. It is thus the additional advantage offered by crossbreds over and above heterosis. The breeds to be used for crossbreeding should therefore be the ones that partially cancel-out each other's deficiencies. Crossing any two breeds without consideration to their complementarities would not yield the desired benefits. A few illustrative examples of breed complementarity are:

- The Brahmans (the indigenous humped cattle of tropics) are resistant to heat and tropical disease (tick and tick-born diseases) but are poor in growth and carcass qualities. The British and US beef breeds are superior in growth and carcass quality. The crossbreds resulting from Brahmans and US and British breeds, in addition to heterotic benefit, complement the desirable traits of both the types.
- Similarly in chickens, crossing of Cornish breed, which is excellent in growth, with White Plymouth Rock, which is good in egg production but mediocre in growth, produces chicks of desirable type for commercial meat production.
- Sheep producers in western states of USA have capitalized on the vigor and hardiness of Rambouillet breed by crossing it with Suffolk, which is good in growth and carcass quality.

Frequently, performance of reciprocally produced crossbreds may differ from each other mainly due to maternal effects. The breed that is outstanding for maternal traits (e.g. fertility, litter-size, milk production and other mothering attributes) is used as female parent and is referred to as **dam line/breed**. The breed that is outstanding for market/commercial traits (e.g. growth, production, carcass attributes, feed efficiency etc.) is used as male parent and is referred to as **sire line/breed**. Currently abundant information is available on the basic competencies of many well-established breeds of swine, beef cattle and sheep. To accomplish complementarity, the breeds to be used in crossbreeding should be chosen in such a fashion that the crossbred offspring combine their desirable attributes in an optimum fashion.

In summary, crossbreeding utilizes the genetic variability in two ways:

- uses non-additive genetic effects (through hybrid vigor)
- uses additive genetic effects of the breeds being crossed (through complementarity)

Specialized Uses of Crossbreeding

1. **Formation of new breed:** Crossbreeding has been used to provide a broad genetic base for development of new breeds. The first step in development of a new breed involves crossing of two or more breeds. When two breeds are crossed, the F_1 are relatively uniform as half of each animal's genes come from each parental breed. However, when the F_1 are crossed among them (i.e. F_1 males mated to F_1 females), the resulting offspring (F_2), show wide variability in both appearance and performance because of segregation and recombination. This wide variability observed in F_2 provides opportunity for selection to operate. Selection for a number of generations for desirable traits, combined with some form of line breeding, may eventually lead to the formation of a recognized breed. Such a new breed would incorporate a number of desirable attributes that were not found in the original breeds utilized in crossing.

   Crossbreeding thus serves as an intermediate step in the synthesis of a new breed or strain. A large number of our present-day breeds have been formed by crossing of different breeds, and subsequently combining their desirable traits into a single population through selection. The topic of breed formation is discussed in chapter 16.

2. **Gene introgression (introduction of a specific allele):** Crossing can be used to introduce a specific allele into an already established breed/strain which otherwise lacks that allele. This process of introduction of a major gene into an existing population (breed/strain) which involves repeated backcrossing is referred to as **gene introgression** or simply as **introgression**. Examples of gene introgression include:

   - Introduction of 'polledness' allele into naturally horned breeds of cattle.
   - Introgression of 'Booroola' allele (that improves fecundity), originally found in Merino breed, into other breeds of sheep.
   - Similarly, the 'naked-neck' allele, (that confers heat tolerance), originally found in some wild stocks of chickens, has been introgressed into high producing strains of chickens.

   In addition to the uses listed above, crossing can also be used for infusion of genetic variability into existing breeds/strains. A breed that is numerically small is likely to have lesser genetic variability. Inbreeding, which inevitably occurs in small populations, further leads to uncovering of genetic defects. Introduction of genetically outstanding animals from other breeds into an existing breed infuses new variability which can then be capitalized through selection.

General Considerations in Crossbreeding

In planning a sound crossbreeding program, a number of points need to be taken into consideration. If ignored, these may become major constraints of the program.

- **Merit of component breeds**: While planning a crossbreeding program, attention should to be paid to average genetic merit of the breeds as well as to the pedigree of the animals to be actually used in crossing. Crossing any two or more breeds at random without consideration to their merit will not produce the benefits which a well-organized and planned crossbreeding system can provide. Many crossbreeding programs have failed as a result of unplanned choice of breeds.
- **Efficient use of purebred and crossbred animals**: A suitable use of purebred and crossbred animals, which are produced in a crossbreeding program, but are not the main product of program, needs to be made. For example, crossbred dairy bulls and cocks of egg-type chickens, do not serve any direct economic purpose. Also, the purebred females that are required for producing crossbreds have lower performance; and fetch lesser returns. In planning a crossbreeding program, the production of such 'undesired' animals must be kept to the barest minimum.
- **Purchase and maintenance of purebreds stocks**: All forms of crossbreeding depend on a ready availability of purebred stocks which are to be used for the production of crossbred animals. The purebreds stocks can be produced and maintained by a single breeder or can be purchased from some other breeders (sources) at frequent intervals. Purchase of replacement purebreds at regular intervals may entail all the associated health and production problems.

Systems of Crossbreeding

Crossbreeding, in one or the other form, is widely practiced for the production of farm animals to be used for commercial purposes. The actual system of crossing may vary from one species to the other due to differences in reproductive capacity and economic considerations. However, certain kinds of crosses are applicable in most species of livestock.

Genetic considerations: The two main genetic considerations in the choice of a crossbreeding system are:

- to generate and maintain a certain magnitude of heterosis (both individual and maternal).
- the need and extent of breed complementarity.

The main goal of a well-designed and systematic crossbreeding system is thus to simultaneously take advantage of heterosis and breed complementarily optimally. The crossbreeding system to be employed should be simple so that it is easy to be implemented. Complex systems cannot stay operational for a long due to the problems associated with management of various categories of animals produced. In addition to these twin objectives, the choice of a crossbreeding system should also take into consideration economics of implementing the program.

Crossbreeding systems fall into two broad categories: specific crossbreeding systems and rotational crossbreeding systems. Each has its advantages and disadvantages in terms of amount of heterosis accrued, the potential for breed complementarity, the source of replacement females, and the ease of management of the system.

Specific Crossbreeding Systems

Specific crossbreeding systems use a specific pattern of consistently mating a particular breed of sire to a particular breed or breed-cross of dam. The resulting crossbreeds offspring are meant for commercial purposes only, and do not enter back into the crossbreeding system.

Two-breed Crossing or Two-breed Specific Crossing

The simplest form of specific system of crossbreeding involves a two-breed cross. The purebred/straightbred animals of one breed, 'A', are mated to animals of a second breed, 'B', to produce crossbred offspring (referred to as 1st cross or F_1 crossbred progeny) as shown below:

The F_1 offspring are marketed for commercial production while the parental breeds are maintained in pure form for producing commercial (crossbred) animals on a regular basis. In the species, that are reared for meat (swine, beef cattle, broiler chicken), and wool (sheep), both the male and female offspring are commercially utilized. In dairy-cattle and egg-type chickens, however,only the females are commercially used. In such a two-breed crossing system, the crossbred progeny (F_1) never enter the breeding herd (as replacement) and so the cross is called a **'terminal cross'** with the sires from breed 'A' referred to as **'terminal cross sires'**.

The expected average genetic merit (performance) of the crossbred offspring ($\overline{X}_{F_1}$) resulting from crossing 'A♂ × B♀' depends on the average additive genetic value (AGV) of the two parental breeds plus contributions of individual heterosis (H_I) and maternal genetic effect of dam's breed (M_B), as shown below:

$$\overline{X}_{F_1} = \left(\frac{AGV_A + AGV_B}{2}\right) + \left[(H_{I(AB)}) + (M_{(B)})\right]$$

Where: AGV_A and AGV_B = average genetic merit of breeds 'A' and 'B' respectively.

$H_{I\,(AB)}$ = individual heterosis (i.e. of crossbred offspring,'AB')

M_B = maternal effect of breed 'B'

Since each of the two parental breeds contributes 50% of genes to crossbred offspring (AB), it is important that the breeds used in crossing are genetically superior. The crossbred offspring (F_1) receive their intra-uterine and early nutrition from the mother, therefore, the breed to be used as dam breed (B) should be good in maternal traits to utilize its maternal attributes (M_B). The breed to be used as male parent is chosen to complement the traits of maternal breed (complementarity). In swine, for example, boars from a breed known for fast growth, good feed efficiency and desirable carcass traits are mated to sows from a breed having good fertility, large litter-size, high milk production etc for utilizing breed complementarity.

Two requirements for the implementation of a two-breed crossbreeding system are:

- a portion of breed 'B' females has to be mated to males of the same breed to produce straight-/pure-bred replacement females.
- a regular source of males of breed 'A' should be available.

Merits: The two main merits of a specific two-breed crossing are:

- Two-breed crossing is simplest of all the crossbreeding systems, and is the first step involved in more sophisticated systems of crossbreeding.
- It takes full advantage of individual heterosis.

Demerits: Drawbacks of two-breed crossing include:

- *Inability to use F_1 offspring for replacement/multiplication:* The most serious drawback of a specific two-breed crossing system is the inability of F_1 crossbreeds to be used as replacement parents since 50% of heterosis would be lost if F_1 are inter-se-mated.
- *Maintenance or purchase of pure breeds*: A two-breed crossing requires maintenance of a relatively large number of animals of the parental breeds (in pure form) for regular production of crossbreds. And, eventually when the purebreds grow old, they have to be replaced by purebreds of the same breed. Therefore, a portion of breed 'B' females has to be mated to males of the same breed to produce replacement females. This requirement may severely reduce the proportion of females actually available for crossbreeding. For this reason, this system is more applicable in highly prolific species where fewer females can produce the required number of purebred replacements. Because of the requirement of large-sized purebred population of the dam breed, two-breed crossing is relatively difficult to be applied in species with low reproductive rate. Further, a source of breed 'A' males needs to be available from which purebred males can be regularly purchased.

Two-breed crossing is, therefore, difficult to be executed and is not economically viable. It is beneficial where only a single maternal/dam breed (for example breed 'B' in the illustration above) is well-adapted to the agro-climatic and economic conditions of the region.

Backcrossing

A two-breed system of crossing, in which crossbreds have to be regularly produced from pure breeds is difficult to be implemented, and is economically not very sound. The breeder has, therefore, to decide as what to do after producing F_1 crossbreds. There are a number of alternatives. Backcrossing is one such system that can overcome some of the drawbacks of two-breed crossing. Backcrossing involves mating of F_1 crossbreds, designated as 'AB', to one of the two parental breeds. Symbolically, backcrossing is represented below:

Parents :	A♂ x (AB)♀	or	B♀x (AB)♀
	↓		↓
Offspring :	A(AB)		B(AB)
(Proportion of genes)	$\left(\frac{3}{4}A + \frac{1}{4}B\right)$		$\left(\frac{3}{4}B + \frac{1}{4}A\right)$

Expected average performance (merit) of backcross offspring can be expressed as:

$$\overline{X}_{A(AB)} = \left[\frac{3}{4}AGV_A + \frac{1}{4}AGV_B\right] + \left[\frac{1}{2}H_I + H_{M(AB)}\right]$$

Where: AGV_A and AGV_B refer to average genetic merit of breeds 'A' & 'B' respectively.

H_I refers to individual heterosis exhibited by offspring.

$H_{M\,(AB)}$ refers to maternal heterosis manifested by 'AB' crossbreds.

The backcross offspring, which are produced from mating of F_1 crossbreds back to one of the parental breeds, manifest only half of the individual heterosis relative to that shown by F_1 individuals. This is because the backcross progeny, on the average, are 50% less heterozygous than the F_1 crossbreds. With backcrossing, however, one of the two parents is crossbred (in some species such crossbred animals are referred to as halfbreds), and hence shows 100% parental heterosis. Since maternal heterosis is generally more important than paternal heterosis, the crossbred parent used in a backcrossing is usually the female. That is, the crossbred females, 'AB', are mated (back crossed) to males of one of parental breeds (either 'A' or 'B'), and hence this system utilizes full maternal heterosis ($H_{M\ (AB)}$).

The greatest advantage of a two-breed backcrossing is that both the dams as well as offspring are crossbreds and hence it utilizes both maternal (milk production, reproduction, survival of young ones and other mothering ability traits) and individual heterosis (though only half of it). It is especially advantageous in species in which maternal heterosis is important (swine, beef cattle etc.). Backcrossing is often used when a particular breed is well suited to production environment of a country or region such as indigenous breeds of tropical countries (e.g. Indian breeds of cattle).

Three-breed Crossing

One-way of maintaining heterosis (individual heterosis) after producing a two-breed cross is to mate the F_1 crossbreds to a third breed to produce specific three-breed crossbreds. A three-breed cross consists of mating of two-breed crossbreds (AB) to a third breed, C. Generally, the female parent used in a three-breed crossing is crossbred (AB) to utilize maternal heterosis. A three-breed cross is symbolically illustrated below:

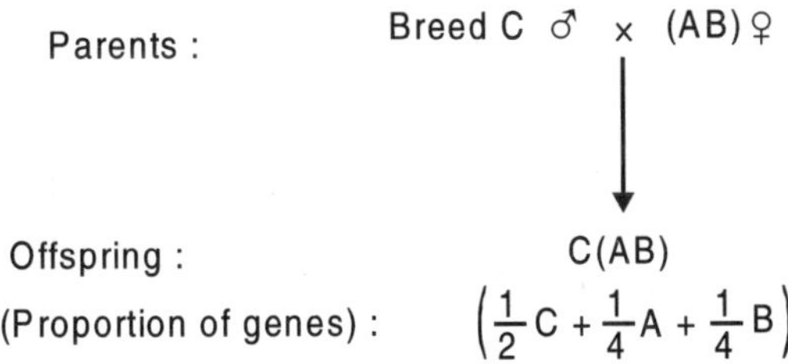

The three-breed crossbred animals are exclusively used for commercial purposes. This system of three-breed crossing, in which females of a particular two-breed cross are always used as female parent, and are mated to males of a third breed, is also referred to *a* **three-breed terminal sire-crossing system**. The expected average genetic merit of offspring of three-breed cross can be expressed as:

$$\overline{X}_{C(AB)} = \left[\frac{1}{2}AGV_C + \frac{1}{4}AGV_A + \frac{1}{4}AGV_B\right] + [H_I + H_M] + M$$

Where : AGV_A, AGV_B and AGV_C refer to average genetic merit of breeds 'A', 'B' and 'C' respectively

H_I refers to individual heterosis manifested by resulting offspring

H_M refers to maternal heterosis (shown by crossbreed dams, 'AB')

M refers to maternal genetic effects contributed by breeds 'A' and 'B'

Merits: A three-breed specific crossing offers two benefits:

- *Better utilization of heterosis*: A three-breed crossing exploits heterosis better than either backcrossing or two-breed crossing because it fully utilizes both the individual and maternal heterosis. The dams are 100% heterozygous, therefore, full heterosis for reproductive and mothering ability traits is exploited. Since the progeny are expected to be 100% heterozygous, (being 'AC' at one-half of their loci and 'BC' at other half of their loci), it enables full-utilization of individual heterosis. Although three-breed cross progeny maintain full level of individual heterosis compared to that shown by F_1 crossbreds but the absolute amount of individual heterosis will depend on the specific breeds involved; some combination of breeds will show higher heterosis than other combinations.

- *Three-breed complementarity*: In addition to better utilization of heterosis, it can also use complementarity of the three breeds as it combines genes from three different breeds (e.g. 50% from 'C' and 25% from each 'A' and 'B'.)

From a strict genetic consideration, three-breed cross is optimum as it combines the potential benefits from individual and maternal heterosis, and from breed complementation better than any other system of crossbreeding. The choice of breeds to be used as male and female parents is important. Breeds 'A' and 'B', in the example cited above (i.e. the ones whose crossbreds are to be used as female parent), could be the ones which are superior in maternal traits. The sire breed 'C' should be chosen for genetic merit in individual (commercial) traits (e.g. growth rate, feed efficiency and carcass traits) as sire influences offspring through genes only.

Three-breed crossing has been widely used throughout the world in many species of livestock (e.g. beef cattle, pig and sheep). The main requirement for a three-breed crossing is availability of a suitable third breed to be used as male parent.

Drawbacks: The major potential drawback of a three-breed crossing relates to management and economics of running such a program. If all the females to be used as replacements are being produced within the herd, then such a three-breed crossing would require maintenance of three types of groups viz; 'AB' crossbreeds, and 'A' and 'B' straightbreds. Such a diverse group of animals would have their own management requirements (feed, fodder, etc.). Alternatively, one needs to have a dependable source for regular supply of replacement females.

Rotational Crossbreeding Systems

In all the systems of specific crossing, as discussed above, the parents of commercial progeny are either purebreds or crossbreds resulting from purebred parents. Therefore, the breeder desiring to produce only the final commercial animals will have to obtain all the replacement animals (both male and female parents) from another source. This, on many instances, may not be desirable, particularly from disease or general management standpoint. Rotational crossbreeding systems aim to overcome some of the difficulties associated with specific crossing.

Rotational crossbreeding systems involve a specific cyclic-pattern of rotating the use of sire breeds on crossbred females (resulting from a preceding cross). It can involve two, three or even four breeds.

Two-breed Rotational Crossing: Crisscrossing

The simplest example of a rotational crossing system is a two-breed rotatio,n (also called 'crisscrossing'. A two-breed rotational crossing is an extension of specific two-breed crossing. In this system, the F_1 crossbred females ('AB') produced, are mated back to males of one of the parental breeds. In the next generation, the crossbred females, 'B' (AB)', resulting from backcrossing are mated to sires of breed 'B'. Thus, the sires from breeds 'A' and 'B' are mated alternately on successive generations of crossbred females. The system, along with the proportion of genes (inheritance) including average heterozygosity, is shown in Table 15.1.

Table 15.1 Expected inheritance (%) from each of the two breeds in offspring and average heterozygosity in successive generations of two-breed rotational crossing.

Generation*	Breed (constitution) of Sire	Dam		Progeny constitution	% of inheritance in offspring from breed: A	B	Average heterozygosity (%)
1	A ×	B	⟶	$(\frac{1}{2}A, \frac{1}{2}B)$	50.0	50.0	100.0
2	A ×	Cross $(\frac{1}{2}A, \frac{1}{2}B)$	⟶	$(\frac{3}{4}A, \frac{1}{4}B)$	75.0	25.0	50.0
3	B ×	Cross $(\frac{3}{4}A, \frac{1}{4}B)$	⟶	$(\frac{3}{8}A, \frac{5}{8}B)$	37.5	62.5	75.0
4	A ×	Cross $(\frac{3}{8}A, \frac{5}{8}B)$	⟶	$(\frac{11}{16}A, \frac{5}{16}B)$	68.8	31.2	63.0
5	B ×	Cross $(\frac{11}{16}A, \frac{5}{16}B)$	⟶	$(\frac{11}{32}A, \frac{21}{32}B)$	34.4	65.6	69.0
6	A ×	Cross $(\frac{11}{32}A, \frac{21}{32}B)$	⟶	$(\frac{43}{64}A, \frac{21}{64}B)$	67.2	32.8	66.0
7	B ×	Cross $(\frac{43}{64}A, \frac{21}{64}B)$	⟶	$(\frac{43}{128}A, \frac{85}{128}B)$	33.6	66.4	67.0
'n'	B ×	Cross $(\frac{43}{128}A, \frac{85}{128}B)$	⟶	$(\frac{1}{3}A, \frac{2}{3}B)$	67.0	33.0	67.0
'n+1'	A ×	Cross $(\frac{1}{3}A, \frac{2}{3}B)$	⟶	$(\frac{2}{3}A, \frac{1}{3}B)$	33.0	67.0	67.0

*Generation number refers to generation of progeny.For example generation one has been produced by mating of sire breed 'A' and dam breed 'B' of foundation 'base' generation.

Following 6-7 cycles of rotational crossing, inheritance percentage from the two breeds stabilizes. At this point, the offspring carry two-third (67%) of the genes from the sire breed last used; and one-third (33%) from the other breed.

The F_1 crossbred, being 100% heterozygous, express maximum individual heterosis but backcrossing of F_1 leads to 50% loss in individual heterosis. In subsequent cycles of rotational crossing, however, about 67% of the heterosis seen in F_1 generation is maintained. Thus, there is a loss of about one-third of individual heterosis relative to a specific two-breed cross.

Merits: Two-breed rotational crossing has some advantages over specific two-breed crossing:

- The dams are also crossbred; therefore, maternal heterosis is also utilized.
- The crossbred females (AB) can be used for breeding (as replacements) every generation, therefore, only the purebred males need to be purchased.

Demerits: The main disadvantages of the system are:

- Arranging matings for this system can be somewhat cumbersome.
- Breed complementarity is not obtained.
- Since the sire breeds alternate between generations, the use of specialized sire and dam breeds is not possible.

Three-breed Rotational Crossing

A three-breed rotational system is very similar to a two-breed system but with an additional breed included in the crossing. A three-breed rotational crossing involves mating of sires from each of the three breeds (say A, B and C) on crossbred females in succession as shown below:

- Base generation: 'A' and 'B' are mated to produce 1st generation (F_1) crossbred.
- First generation: F_1 crossbred females ('AB') are mated to males of 3rd breed ('C').
- 2nd generation : Three-breed crossbred females are mated to males of 1st breed, (A).

The sires from all the three breeds are then successively mated to crossbreed females. Since this system involves use of purebred sires from three different breeds, it is sometimes called a **three-breed sire-rotational crossing.** Three-breed rotational crossing is illustrated in Table 15.2.

Table 15.2 Expected percentage of genes from each breed in progeny and average heterozygosity in successive generations of three-breed rotational crossing.

Generation*	Sire breed		Dam breed /constitution		% of genes in offspring from breed: A	B	C	Average heterozygosity (%)
1	A	×	B	⟶	50	50	0	100
2	C	×	Cross	⟶	25	25	50	100
3	A	×	Cross	⟶	62	13	25	75
4	B	×	Cross	⟶	31	56	13	88
5	C	×	Cross	⟶	16	28	56	87
6	A	×	Cross	⟶	58	14	28	84
7	B	×	Cross	⟶	29	57	14	86
8	C	×	Cross	⟶	14	29	57	86
9	A	×	Cross	⟶	57	14	29	86
10	B	×	Cross	⟶	29	57	14	86

*Generation number refers to the generation of progeny.

After 7-8 cycles of rotational crossing, the percentage of inheritance from three breeds stabilizes at 57, 29 and 14. That is, approximately 57% inheritance comes from the last sire breed used, 29% from the previous sire breed (breed of maternal grandsire) and 14% from the third breed (the breed of maternal great grandsire).The genetic contribution of the respective breeds will, however, change in each generation as per pattern of rotational crossing.

Merits:

Highest amount of heterosis is observed in 2nd generation as both the individual and maternal heterosis is manifested. In subsequent generations of rotational crossing, however, some loss of heterosis, as compared to the original three-breed cross, occurs. Theoretically, the retained heterozygosity in each successive generation gives about 6/7th of the heterosis of F_I crossbred generation. However, once a rotational crossing scheme stabilizes, all the dams and offspring are crossbred and therefore benefit from both types of heterosis.

Demerits:

The main genetic disadvantage of rotational crossing (two or three breed), as compared to specific crossing, is that it does not utilize breed complementarity effectively as the breeds involved in crossing cannot be restricted to a single-purpose use. For example, in a three breed specific crossing, two breeds can be mated to sires of the third breed chosen for market (commercial) traits. In rotational crossing, each breed is eventually used both as male parent and as an ancestor of the female parent. None of the three breeds can be used for a single-purpose. Therefore, rotational crossing is beneficial only where the breeds available for crossing show little or no complementarity but exhibit heterosis.

In many classes of farm mammals (e.g. cattle and sheep) the generations overlap (i.e. at a given time offspring, parent and grandparents are available in the herd). In a three-breed rotational crossing, the herd will eventually come to contain animals pertaining to several generations containing variable proportion of inheritance and variable age groups. This can lead to difficulties in trait recording, arranging matings, and in managing and feeding of animals of different genetic groups. In order to have progeny of similar morphological attributes (color, body size etc), the breeds used in rotational crossing systems should be of similar biological types (size, color etc) as progeny of uniform color, size etc are easy to be marketed.

Four-breed Rotational Crossing

Similar to a two- or three-breed rotation, the commercial producer can use a four-breed rotational crossing. Four-breed rotation is, however, not used in practice because of two reasons:

- It is often difficult to find four breeds with good genetic merit.
- Four-breed crossing is much more difficult to be managed in terms of matings and the management of variable types of animals.

Rotational crossing is widely practiced in pig breeding industry of North America and Europe. Beef cattle breeding industries of several countries also use it. For example, traditional beef breeds of Britain viz; Aberdeen, Angus, Hereford and Shorthorn are used in rotational crossing system.

A summary of the percentage of heterosis expected in the progeny resulting from some of the commonly used crossbreeding systems is given in Table 15.3.

Table 15.3 Percentage of heterosis (measured as average heterozygosity expected under common types of cross breeding systems).

Types of cross	Percentage of heterosis :		
	Individual	Maternal	Paternal
Straight/purebred:	0.0	0.0	0.0
Two-breed cross:			
A♂ × B♀	100.0	0.0	0.0
Back cross :			
A♂ × AB♀	50.0	100.0	0.0
Three-breed cross :			
C♂ × AB♀	100.0	100.0	0.0
Four-breed cross:			
AB♂ × CD♀	100.0	100.0	100.0
Rotational cross :			
2 sire-breed	66.7	66.7	0.0
3 sire-breed	85.7	85.7	0.0

Grading-up (Repeated Backcrossing)

The term 'grading-up' was originally used to refer to mating of sires of a breed to 'nondescript', 'scrub' or 'native' females; and to their female offspring generation after generation. However the term grading up is now used to cover mating of sires of any given population (superior merit) to the females of another population (low merit), and to their female offspring generation after generation. Grading-up is accomplished by repeated backcrossing. The populations involved in grading-up may belong to different strains/lines of the same breed or they may be two different breeds. Grading-up does not aim to take advantage of either heterosis or breed complementation but is used with two objectives:

- **Complete replacement of an existing breed/population:** Grading-up is suited when a particularly 'low performing ' population, usually a native stock, is required to conform to another breed which is of high merit. A grading-up scheme involving mating of 'native' females with purebred sires of a superior breed as shown below (with percentage of inheritance in progeny obtained from 'pure' breed shown in the parentheses):

Generation	
Base :	Purebred sires × Native females
	↘
1st :	Purebred sires × 1st generation females (50% pure)
	↘
2nd :	Purebred sires × 2nd generation females (75% pure)
	↘
3rd :	Purebred sires × 3rd generation females (87.5% pure)
	↘
4th :	Purebred sires × 4th generation females (93.8% pure)
	↘
5th :	Purebred sires × 5th generation females (96.9% pure)

After about four generations, percentage of inheritance (genes) from the 'non-descript' population decreases to 6.25 with the remaining 93.75% coming from purebred population. A population containing 93.75% of genes from the pure breed is accepted as purebred by most breed societies for registration purposes. The ultimate consequence of grading up would be that a 'non-descript' or 'native' breed becomes like an improved pure breed.

The greatest improvement through grading-up is achieved with the first cross, with less and less improvement being made in later generations. A part of the improvement achieved in first cross comes from heterosis. In later generations, the amount of heterosis decreases each generation, and the main improvement comes from the additive effect of desirable genes.

It should be noted that after two generations of backcrossing, the average proportion of exotic inheritance in 'native' population would be ¾ but due to segregation, there might be some animals, which carry very high proportion (100%) of 'exotic' inheritance. In view of this, it has been suggested that grading up may be terminated after the 2nd backcross; and selection for traits of interest be started. Selection would maintain optimum proportion of exotic inheritance required for good performance under the agro-climatic conditions of the 'native' region/country. It will also retain the genes of native stock that are useful for their adaptation to local environment.

- **Introduction of a specific allele (s):** Sometimes a specific or unique allele from one population may be required to be incorporated into an existing or native population, which is otherwise superior but for the absence of that specific allele. Introgression of the specific allele would involve an initial cross between the two populations, followed by selection for the desired allele and repeated backcrossing to the native population. An example of introduction of polled allele (P) into an local 'horned' population (pp) of cattle which is genetically superior for other economic traits is shown below:

Generation	Mating type :	% of genes in offspring from 'native'
Base	Polled (PP) × Horned (pp) (native)	–
1st	Polled offspring (Pp) × Horned (native)	50.0
2nd	Polled offspring × Horned (native)	75.0
3rd	Polled offspring × Horned (native)	87.5
4th	Polled offspring × Horned (native)	93.5
5th	Polled offspring	96.3

At the onset, the 'native' population (horned) is mated to homozygous polled (PP) animals of the 'donor' breed (carrying 'P' allele). The F_1 progeny will be heterozygous polled (Pp) but would also have 50% of their genes from each of the two parental populations. The F_1 offspring are backcrossed to the existing 'horned' population. Half of their resulting offspring would be 'polled'. The polled are selected, and are again backcrossed to horned population. By about four generations of backcrossing, coupled with selection each generation, the new population would carry about 7/8 of the genes from the original population of high merit but will also carry the polled allele. Since the polled allele is dominant, the polled animals will be heterozygous, and will not breed true to type. In subsequent generations, carriers would have to be identified by making testcrosses, and culled.

If the allele to be introduced is recessive, then the formation of new population is easier because all the animals exhibiting the trait will breed true to type. Of the various examples of gene introgression, an illustrative is the incorporation of naked-neck dominant allele found in Desi/wild stocks of chicken. This gene has been incorporated into many high performing strains through repeated backcrossing.

Use of Grading-up in the Past for Breed Improvement

Grading-up was widely used, especially during 1960's through purchase of bulls or semen of established breeds to improve native populations of many countries. One of the earliest documented uses was in the Longhorn breed (cattle) in USA. Prior to 1900's, most of the local range cattle in USA were descendents of the Longhorn breed. However, many farmers considered animals of Hereford breed better. Due to economic considerations, farmers could not dispose off their Longhorns for replacing it with Herefords. Grading up was, therefore, used to incorporate the inheritance of Hereford in Longhorn. Indigenous dairy cattle breeds of many countries have been upgraded with Holstein-Friesian. In India, attempts on grading-up of hill cattle of Himachal Pradesh with Jersey were made during 1960's.

Grading-up is the quickest and most effective method of improving low-performing native breeds. Its success, however, depends on the choice of a suitable breed. In fact, the main problem with grading up relates to the adaptation of the 'new' graded animals to the local environment viz; agro-climatic, management and economic conditions. An example of such a scenario is the crossing of indigenous cattle with exotic (temperate) breeds. The first generation crossbreed resulting from crossing of Indian cattle breeds with temperate (exotic) breeds can adapt to local conditions. As the level of exotic inheritance increases, the animals are unable to withstand the climatic and management conditions prevalent in the tropical region.

Outcrossing: Mating of unrelated animals within the same pure breed is called outcrossing. For practical purposes, a mating is considered as outcross if mates are not related to each other for the last four to six generations. Outcrossing has been used by American livestock breeders for production of seed stocks (parent stocks). Since many breed associations do not allow introduction of genes from other breeds, outcrossing through the use of best-available but unrelated sires from the same breed is a common practice. It is practiced by the breeders with two main objectives:

- to offset the undesirable effects of accumulated inbreeding in small-sized herds.
- to introduce inheritance of some outstanding bulls ('show-ring', 'prize-winning' types), in their own herd which primarily serves the purpose that a 'well-known' name appears in the pedigree.

Review Questions

1. Define / explain the following:

I. Breed complementarity
II. Dam breed (line)
III. Sire breed (line)
IV. Gene introgression
V. Grading–up
VI. Outbreeding
VII. Outcrossing
VIII. Genetic homeostasis
IX. Backcrossing
X. Terminal cross
XI. Criss crossing

2. Write a short note on the following:

(a) Two breed crossing
(b) Three breed crossing
(c) Backcrossing
(d) Two-breed rotational crossing
(e) Three-breed rotational crossing
(f) Grading up

3. What are the two broad uses of crossbreeding for commercial livestock production? Explain

4. What are the specialized uses of crossbreeding? Explain
5. What are the general considerations to be taken into account while designing a crossbreeding program? Explain
6. What are the two genetic considerations involved in the choice of a crossbreeding system?
7. Explain a two-breed crossing.
8. What factors determine the average merit of crossbreed offspring resulting from a two-breed crossing?
9. Discuss the merits and demerits of two-breed crossing.
10. Explain a backcrossing system involving two breeds.
11. What factors determine the merit of backcross offspring resulting from crossing of 'AB' females with 'A' males ?
12. What is the greatest advantage of a two- breed backcrossing?
13. The crossbred parent used in backcrossing is usually the female. Explain
14. Explain a three-breed crossing (diagrammatically) showing the proportion of the genes (inheritance) carried by offspring from each of the three breeds, the type and magnitude of heterosis, and the maternal effects, if any.
15. What factors determine the merit of offspring resulting from a three-breed crossing?
16. Discuss the merits and drawbacks of a specific three-breed (specific) crossing program.
17. In a three-breed (specific) crossing, 'C♂ × AB♀, what should be the merits of the three breeds in terms of maternal, paternal and commercial traits?
18. What is the main advantage of a rotational crossing system over a specific breed crossing?
19. Explain the problems associated with a three-breed rotational crossing in cattle where generations overlap.
20. Explain a two breed rotational crossing program showing the percentage of inheritance received from the two breeds up to 4 cycles of rotation.
21. What are the major advantages and drawbacks of a two breed rotational crossing?
22. Explain a three-breed rotational crossing showing the proportion of genes in offspring received from each of the three breeds up to 4-5 generations.
23. What are the merits and demerits of a three-breed rotational crossing?
24. Why is four-breed rotational crossing not a common practice?
25. How can grading up be used for complete replacement of an existing breed?
26. Show with a suitable example introduction of specific allele into a native population by repeated backcrossing.
27. Grading up when used for breed replacement in species like cattle is recommended to be terminated after second or third backcross. Explain
28. Give few examples where grading up has been used in breed improvement programs.
29. Give two illustrative examples of breed complementarity.
30. Give two examples of gene introgression.
31. Which crossbreeding system (two-breed or three-breed crossing) would give the highest heterosis? Explain.

32. Why is it important to use high merit breeds in a crossbreeding program?

33. Why is crossbreeding practiced mostly by commercial producers but not by seed-stock (parent-stock) breeders?

34. If a trait exhibits considerable maternal heterosis but little individual heterosis, which is likely to perform better: a two-breed F_1 or a two-breed F_2. Explain.

35. Crossbreeding serves an intermediate step in the formation of a new breed. Explain.

36. Why is higher heterosis observed under unfavorable conditions?

37. Which of the following needs to be good for maternal traits: a sire line or a dam line?

38. Differentiate between specific and rotational crossbreeding systems.

39. Holstein-Friesian (HF) males are mated to Sahiwal (S) females to produce F_1 progeny. What will be the proportion of genes carried by F_1 from each of the two parental breeds?

40. The F_1 female progeny produced by mating of Holstein-Friesian (HF) bulls with Sahiwal (S) cows are mated back to HF bulls, what will be the proportion of genes carried by resulting progeny from each of the two breeds, the magnitude and type of heterosis and the maternal effects?

41. List the percentage of heterosis shown by different systems of crossbreeding as per following table:

Types of cross		Percentage of heterosis :		
		Individual	Maternal	Paternal
Two-breed cross:	A♂ × B♀	....	....	....
Back cross :	A♂ × B♀	....	....	....
Three-breed cross :	C♂ × AB♀	....	....	....
Four-breed cross:	AB♂ × CD♀	....	....	....
Rotational cross :	2 sire breed	....	....	....
	3 sire breed	....	....	....

42. A three breed crossing program involving Chester (C), Duroc (D) and Hampshire (H) breeds of swine is shown diagrammatically below:

Parents: Breed C♂ × (DH)♀

↓

Offspring: C(DH)

What will the average genetic merit of progeny, $\overline{X}C_{(AB)}$, in terms of AGV_C, AGV_B, $AGV_{C'}$ individual and maternal heterosis?

●●●

SECTION - 5

GERMPLASM RESOURCES

16

Breed Development, Structure and Comparison

Composites (Synthetics)

The use of crossbreeding as a means of improving productivity of commercial livestock is widely accepted. The regular systems of crossbreeding (e.g. rotational crossing) are often too cumbersome for small-scale breeders. An alternative to regular crossing is to make one or a few crosses between two or more breeds (strains) in order to produce a single population carrying genes from each of the breeds involved in crossing. A single population that is a mixture of many different populations (like breeds, strains) is called a composite or synthetic. A **composite or synthetic** is formally defined as a self-contained population, which is made up of two or more component breeds designed to retain heterosis in future generations without crossbreeding, and maintained as a purebred.

The concept of synthetics (synthetic varieties) was first suggested by Hayes and Garber (1919) in plants and defined in context to plants by Lonnquist (1961) as open-pollinated varieties derived from intercrossing of selfed plants or lines known to possess high general combining ability. Subsequently the concept found wide spread application in poultry and livestock.

Benefits of composites: Composites are mainly produced to accomplish two broad objectives:

- *Retaining benefits of crossbreeding:* Composites are developed to retain the twin benefits of crossbreeding viz; breed complementarity and heterosis but without actually performing crossbreeding. Composites provide their own male and female breeders (replacements) for propagation (thus self-contained), and are, therefore, for practical purposes like pure breeds.

- *Operational simplicity*: Traditional crossbreeding systems are complex because the breeder has to maintain a number of different breeds. There are management problems associated with maintenance of different types of animals, keeping identification of animals, and the difficulties in deciding which breed to use in producing the next generation.

Composites by combining breed complementarity and heterosis into one 'package' eliminate most of the problems associated with conventional systems of crossbreeding. Composites are especially useful for small-scale producers because they cannot afford to purchase animals of different breeds that are needed for a complex crossbreeding system.

Retained heterosis in composites: Composites do not have as much heterosis as do the F_1 crossbreds because inter-se-mating among crossbreds leads to some loss of heterozygosity (heterosis). But once a composite is established and closed to outside inheritance, and random mating occurs within the herd, the resulting heterosis would remain constant over generations provided inbreeding does not occur in such populations.

Table 16.1 presents the level of heterozygosity relative to F_1 that is retained after equilibrium has been reached for rotational systems and composites. Retention of initial heterozygosity in F_2 generation in a composite resulting from 'n' breeds is given by the following expression: retained heterosis = '[1-1/n]' of F_1 heterosis.This formula assumes equal contributions by each breed used to create the composite. For example, a four-breed composite (1/4th of each breed) would potentially have 75% heterozygosity relative to F_1. Retention of heterosis in composite is also influenced by inbreeding. If no inbreeding occurs, heterosis is retained for several generations.

Table 16.1 Heterozygosity in different mating types (crosses and composites).

Mating type	Heterozygosity relative to F_1 (%)
Pure breeds :	0.0
Crosses:	
Two-breed rotation	66.7
Three-breed rotation	85.7
Four-breed rotation	93.3
Two-breed composite:	
F_3 : $\frac{1}{2}$ A, $\frac{1}{2}$ B	50.0
F_3 : $\frac{5}{8}$A, $\frac{3}{8}$B	46.9
F_3 : $\frac{3}{4}$A, $\frac{1}{4}$B	37.5
Three-breed composite:	
F_3 : $\frac{1}{2}$ A, $\frac{1}{4}$ B, $\frac{1}{4}$ C	62.5
F_3 : $\frac{3}{8}$A, $\frac{3}{8}$B, $\frac{1}{4}$ C	65.6
Four-breed composite:	
F_3 : $\frac{1}{4}$ A, $\frac{1}{4}$ B, $\frac{1}{4}$ C, $\frac{1}{4}$ D	75.0
F_3 : $\frac{3}{8}$A, $\frac{3}{8}$B, $\frac{1}{8}$C, $\frac{1}{8}$D	68.8
F_3 : $\frac{1}{2}$A, $\frac{1}{4}$B, $\frac{1}{8}$C, $\frac{1}{8}$D	65.6

Mating type	Heterozygosity relative to F_1 (%)
Five-breed composite:	
F_3 : $\frac{1}{4}$A, $\frac{1}{4}$B, $\frac{1}{4}$C, $\frac{1}{8}$D, $\frac{1}{8}$E	78.1
F_3 : $\frac{1}{2}$A, $\frac{1}{8}$B, $\frac{1}{8}$C, $\frac{1}{8}$D, $\frac{1}{8}$E	68.8
Six-breed composite:	
F_3 : $\frac{1}{4}$A, $\frac{1}{4}$B, $\frac{1}{8}$C, $\frac{1}{8}$D, $\frac{1}{8}$E, $\frac{1}{8}$F	81.3
Seven-breed composite:	
F_3 : $\frac{3}{6}$A, $\frac{3}{16}$B, $\frac{1}{8}$C, $\frac{1}{8}$D, $\frac{1}{8}$E, $\frac{1}{8}$F, $\frac{1}{8}$G	85.2
Eight-breed composite:	
F_3 : $\frac{1}{8}$A, $\frac{1}{8}$B, $\frac{1}{8}$C, $\frac{1}{8}$D, $\frac{1}{8}$E, $\frac{1}{8}$F, $\frac{1}{8}$G, $\frac{1}{8}$H	87.5

*Heterozygosity = '$1 - 1/n$' where 'n' is number of breeds contributing equally to composite.

The simplest synthetic is the one derived from two breeds. The F_1 produced from crossing of two parental breeds are mated inter-se to produce F_2 generation. Inter-se mating is repeated generation after generation. The synthetic carries 50% heterozygosity in F_2 and in subsequent generations. The percentage of genes from each of the two parental breeds in the synthetic would be 50%, and would remain at the same level in subsequent generations. But individual animals will vary in the proportion of the genes that they carry from each of the two breeds.

Selection, if any, carried out in the synthetic may increase or decrease the average proportion of genes of the two breeds. Since magnitude of heterosis retained in composites is also influenced by inbreeding, the breeders of composites have to maintain large-sized herds with adequate representation of each breed to minimize inbreeding.

Disadvantages/Problems in composites: Though development of composite populations overcomes the problems associated with regular crossbreeding but composites do have few disadvantages and problems.

- The main disadvantage is the unavailability of extensive performance records of animals of composite population. Consequently, breeding values of composite animals are difficult to be computed.
- In the long term, a large fraction of heterosis will not be maintained in the composites. Therefore, some of the early excitement among breeders of composites due to their high performance resulting from heterosis and breed complementarity may wane when composites expand in numbers (size). Genetic segregation during this phase leads to considerable variation in performance and type traits.

Synthetic varieties are very popular in open-pollinated crops. Among livestock species, composites are quite popular in beef cattle particularly in North America where some composite breeds have their own breed associations with a governing body as well as herd book registration. Development of new strains in poultry in many cases encompasses synthetic/composite base populations.

Box 16.1: Composite vs. synthetic.

The terms composites and synthetics, which are used to signify new self-contained populations, are many a times used interchangeably. However some breeders differentiate the two terms:

Composite is used to refer to a population which carries certain percentage of genes (inheritance) from a fixed number of breeds while synthetic is a population in which breeds can be added at any time and with no fixed percentage of genes from each breed

Development of New Breeds

Breed Concept and Definition

Livestock species are sub-divided into distinct genetic groups called breeds. The definition of what constitutes a breed is complex and variable because grouping of a species into breeds does not exist in Linnaean taxonomic classification. Also, a universally acceptable definition of breed is difficult to be found in literature as is evident from some of the definitions given by different workers (Box 16.2). The definitions provided by Food and Agriculture Organization (FAO), which are suited for many purposes, read as follow:

- A sub-specific group of domestic livestock with definable and identifiable external characteristics that enable it to be separated by visual appraisal from other similarly defined groups within the same species.
- An interbreeding group of animals which are isolated, geographically or reproductively, from phenotypically similar groups, and the separation has led to the acceptance of its separate identity.

The total number of recognized livestock breeds worldwide is more than one thousand although some of these may be only local or national names of the same breed. The breed-website of Oklahoma State University, USA (www.okstate.edu/breed), describes a number of livestock breeds.

Box 16.2: What is a livestock breed ?

The earliest observation made about the definition of breed by Lloyd-Jones of Iowa State University, 1915, and later reiterated and cited by Lush indicates the vastness of breed concept as well as to why a good definition of breed is elusive:

"Breed is a group of domestic animals, termed such by common consent of the breeders, a term which arose among breeders of livestock, created one might say, for their own use, and no one is warranted in assigning to this word a scientific definition and in calling the breeders wrong when they deviate from the formulated definition. It is their word and the breeders common usage is what we must accept as the correct definition"

Subsequently a number of informal definitions of livestock breed have been put forward; a few of these are listed below:

- A homogeneous sub-specific group of domestic livestock with definable and identifiable characteristics that enable it to be separated by visual appraisal from other similarly defined groups within the same species. (Turton, 1974).
- Genetic entity developed over a long period by either following certain mating scheme or by selection characteristics. (KrauBlich,1998)

- A sub-group of a species possessing certain recognizable characteristics and maintained as a closed breeding population in a single geographical area (Carter and Cox, 1982).
- A group of animals related by descent and has similar characteristics which when mated together produce progeny of the same type (Alderson, 1985).
- A uniform population which when interbred reproduces itself (Sponenberg, 1991).
- Animals of same species, which are clubbed together due to their common origin and certain physical and physiological attributes (Brem et al. 1989).
- A group of animals of the same species, which have the same origin, share certain physical, physiological characteristics and are of commercial value (Mariante, 1982).
- A population of livestock and poultry in a given location with certain well-defined physical conformation, distinct local names and distinguished from other breeds in its vicinity (Acharya, 1990).

How Breeds Evolved ?

Many of the older breeds of livestock have mainly originated from physical isolation (mountains, seas, rivers etc.). Subsequently the forces of natural and artificial selection coupled with genetic drift made these isolated populations to appear distinct. Other breeds have evolved from crossing among already established breeds followed by selection for type traits. Social differences among breed users, fashion, fads etc have all helped to keep breeds separated from one another. Dog is possibly the best example of a species which has been sub-divided into a very large number of distinct breeds.

The critical element in formation and development of new breeds has been the isolation, which in earlier times was provided by geographical barriers. Breeds developed in Channel Island are some examples. After introduction of herd books, isolation was implemented by pedigree barriers as animals whose parents were not recorded in herd books were not accepted for registration.

Differentiation of Breed into Strains

A breed may be further differentiated into sub-populations referred to as strains. The strains result from the variable selection exerted within sub groups of a breed. One of the illustrative examples of a breed's differentiation is that of Holstein Friesian into strains. The Holstein of USA is a bigger in size, produces large quantity of milk but with lesser butterfat and protein contents than Friesian strains previously found in Europe which were developed into smaller dual purpose animals (in middle of 19th century). The Friesian strain of New Zealand is also unique, as it has been selected for increased milk production under pasture-based production systems. Examples of strains of breeds in Africa include the Boran/Borana strains in Kenya and Ethiopia and the strains of Djallonké sheep in several countries in West and Central Africa. Strains are most popular in poultry because it is much easier to conduct intensive selection due to high reproductive rate.

Lately there has been a general move to recognize strains found in different countries as distinct breeds. This is principally in response to the **'Convention on Biological Diversity'**, a global effort which emphasizes national ownership of genetic resources.

Breed Associations vis-à-vis Breed Development

The concept of livestock breed, and its development mainly originated and flourished in Great Britain. Breeders employed selection to produce herds which conformed to their personal liking (e.g. colour, form and performance). In order to protect and preserve the identity of their herds, breed societies/ organizations/associations which maintained the *'herd books'* were also formed. The breed associations which originated in Great Britain during the early part of 19th century, and spread to other developed countries, were formed to:

- Foster co-operation in breed promotional work
- Promote widespread acceptance of the view that 'purebred/registered' livestock were superior to non-pedigreed animals.

Breed societies/associations proved successful in promoting the idea of 'purebred supremacy', and in turn, it led to enactment of legislation to encourage and extend the use of purebred stocks. In this process of 'pure breeding', however, breeders overlooked requirements of commercial producers because they emphasized the qualities which could increase the prestige of their herd/flock in the eyes of other purebred breeders ('eye ball traits'). By 1930's, however, it became clear that philosophy of purebred industry did not match with the consumer demands and the production methods. Consequently, attempts at development of new breeds derived from crossbred foundation began. Since that time, many breeds of pigs, sheep and cattle have been developed in many countries.

Justification for Development of a New Breed

The real need for formation of a breed arises when demands of commercial livestock industry in a given geographic area or production system are not served by the existing breeds or by their crosses. For example, crossing of exotic *(Bos taurus)* with indigenous *(Bos indicus)* cattle in India formed the F_1 crossbred population. Since the F_1 crossbreds cannot perpetuate their own type, the logical course was development of a new breed to fulfill market demands under the agro climatic and input conditions of the country. Generally, a new breed is developed because existing breeds or their crosses do not to possess the optimal combination of traits for production requirements of a region/country. However, for a new breed to succeed, it should be able to meet the economic requirements of the market, and should be well adapted to the environmental and input conditions of the area. Breeds with low commercial or economic value gradually get eroded. The following examples illustrate how the requirements dictated the development of a new breed:

- Santa Gertrudis breed of beef cattle was developed in South Texas, (USA) to fulfill the requirement for a breed which was adaptable to hot environmental conditions of the area while keeping it as productive as British breeds for growth and carcass traits.
- Sheep producers of Western states of USA needed a sheep which was capable of high lamb production, acceptable production of medium grade wool but retaining the flocking instinct of the traditional Rambouillet breed. This led to development of Columbia and Targhee breeds of sheep (by the United States Department of Agriculture.)

Steps in Breed Development

The exact history of development of many breeds of the 18th or 19th century is not known. But most of the breeds during early part of 20th century seem to have originated from crossbred foundations. In fact, crossbreeding is now considered an intermediate and logical step in development of a new breed. Though there is no standard procedure for the creation of a new breed but it definitely involves crossing between a number of existing breeds, inter-se-mating followed by selection for the desired traits. The breeds which are to be included in initial crossing including the percentage of their genes in the new breed are the critical steps in breed development. The broad steps are:

1. **Choosing the foundation breeds**: Success of a new breed is largely dependent on the choice of breeds which will contribute their inheritance to the new breed. Not only the breeds but also the foundation animals chosen from those breeds should be the ones that are a representative sample of the concerned breed. In making choice of breeds, the following points need consideration:
   - The production environment (i.e. geographic area, managemental conditions and production systems), under which the new breed is to perform needs to be clearly defined. Local market demands, traditions, cultural beliefs of people etc. may also influence the choice of breed.
   - Information on the complementarily of breeds, if available, can also be useful in choosing the breeds to be included in breed formation. In developed countries, extensive breed summaries are available which are useful in choosing the breeds. There are, however, some traits like temperament for which objective information may not be readily available.
   - The sires and dams of each component breed must be chosen with care on the basis of their genetic potential for traits included in breeding goals of the new breed.
   - In addition to these, the financial and physical resources available at breeders' disposal are important considerations in optimizing the use of different breeds.
2. **Crossing of breeds***:* Once the breeds are chosen, crosses between component breeds are made to create all the possible types of genotypic combinations i.e. creating a heterogeneous population. For instance, if four breeds are involved (A, B,C,D), two different types of crossbred groups could be produced in the first generation (AB and CD). In the second generation, these two crossbred types could be mated to each other to produce offspring which on average, have one-fourth of the genes from each of the four original breeds (¼ A+¼ B+¼ C+¼ D). In other situations, some other types of crosses, in which contribution of different breeds is not equal, may be made. This crossbred material constitutes the base population which may then be closed to outside inheritance.

   Random mating within the crossbred base population would lead to wide variation in appearance and performance of individual animals (due to segregation and recombination). There will also be variation in the proportion of genes which individual animals receive from the original breeds. For any particular trait, some offspring will resemble one of the original parental breeds more while others will resemble other parental breed more. However, most of the offspring will fall somewhere in the middle. This wide variability provides the opportunity for applying intensive selection. If initial crossing is not carried on a large scale, problems associated with inbreeding may be encountered. This is particularly so when the number of animals coming from different breeds is too small or when they come from only single herds. Inbreeding might result in loss of genetic variability or in fixation of undesirable genes in the new breed. Such 'bottlenecks' at beginning have led to failures in formation and subsequent success of many breeds. It may also be noted that some loss of heterosis would also occur during interbreeding stage.
3. **Selection**: Once a sufficiently heterogeneous population has been created through initial crossings, a pedigree or herd book is created for keeping pedigree identification and maintenance of performance records. Selection is then conducted for productivity under the production environment of the area or region. In many cases, selection emphasis may also have be exerted to 'fix' some distinguishing characteristics like

color, conformation etc. Such attributes may not be relevant to overall productivity but they may serve as trademarks.

A schematic illustration of formation of a breed involving four component breeds, A, B, C and D, is shown in Fig 16.1.

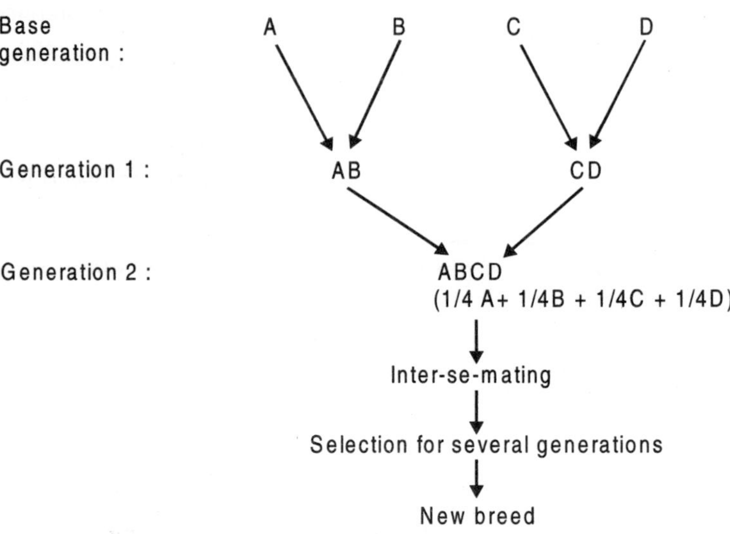

Fig. 16.1 Formation of a new breed from four breeds with each breed contributing equal inheritance.

Soon after initial crossing and selection, pedigree barrier, like herd book, would have to be enforced to prevent dilution of genetic improvement made within the new breed.

Breeds Developed in 19th and 20th Centuries

A number of breeds of cattle and sheep and other species of linestock and poultry have been formed during past two centuries though it is debatable whether these qualify to be labeled as breeds or they are only composites. A list of a few such breeds of cattle and sheep is given in Table 16.2.

Table 16.2 List of some breeds of cattle and sheep.

Species/breed	Breeds contributing inheritance:	Region
Cattle:		
Santa Gertrudis :	$\frac{5}{8}$ Shorthorn; $\frac{3}{4}$ Brahman	Texas, USA
Brangus Cattle :	Angus, Brahman	USA
Charbray :	Charolais, Brahman	USA
Luing cattle :	Shorthorn, Highland (crossing & interbreeding)	Scotland
Australian Milking : Zebu (AMZ)	Jersey, Zebu (Sahiwal and Red Sindhi)	Australia
Jamaica Hope :	Jersey, Sahiwal, Holstein Friesian	Jamaica
Siboney de Cuba :	Holstein (5/8), Zebu (3/8)	Cuba
Beef masters :	Brahman, Shorthorn, Hereford	USA

Species/breed	Breeds contributing inheritance:	Region
Sheep:		
Columbia :	Rambouillet, Longwool	USA
Panama :	Rambouillet, Lincoln	
Coopworth (dual-purpose) :	Border Leicester, Romney (crossing, interbreeding, selection)	New Zealand
Meat line :	Suffolk, Dorset Down, Ile de France, Berrichon du Cher and Charollais	England

Brief information on some of the well-known breeds of cattle developed in 20th century is given below:

1. **Tropical Milk Breed 'Siboney de Cuba':** Siboney de Cuba breed was originally created in 1970 to improve milk yield of the hardy but poor yielding Zebu (of Cuba) by incorporating Holstein Friesian inheritance. The mating system used was as follow:

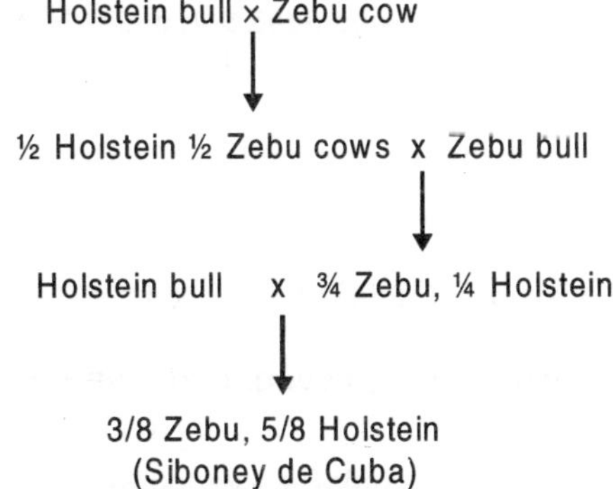

2. **Australian Milking Zebu (AMZ):** Australia originally did not have any indigenous breeds of cattle. *Bos taurus* (temperate-type) cattle were introduced by European settlers into Australia in late 18th century. The *Bos taurus* animals, however, did not perform well in the tropical areas of Australia, therefore, cattle of *Bos indicus* type (Sahiwal and Red Sindhi) were imported from Pakistan in 1950's as gift (ten males and eight females). A crossbreeding program involving Sahiwal and Red Sindhi as 'indicus type' and with Jersey as 'taurus type' was initiated with the aim of developing a tropical dairy breed of mixed 'taurus-indicus' type.

   After initial generations of crossing and inter-se- mating among crosses, the population was subjected to intensive selection for milk yield and milk-let down. During this stage, Red Sindhi inheritance got largely eliminated. Later a progeny-testing program of selection of bulls was also initiated. Screening tests for heat tolerance and tick resistance were applied on young bulls. The new breed named **'Australian Milking Zebu'** (AMZ) had been well accepted by farmers because it was found to be at par with Jersey for milk production while being tolerant to tropical climate. The AMZ had about 60% *Bos taurus* inheritance.

   A few years later, another breed named **'Australian Friesian Sahiwal'** (AFS) was also developed from crossing of Friesian ('taurus-type') and Sahiwal ('indicus-type').The breeding and selection procedure used for formation of AFS was similar to that used for AMZ except that heat tolerance tests were excluded. The major emphasis was on milk let-down and tick resistance.

3. **Jamaica Hope Breed:** In Jamaica, a comprehensive crossbreeding program was initiated in 1910 at the Hope Farm to develop a breed of dairy cattle that was well adapted to climate of the island. The program included *Bos taurus* (Holstein Friesian and Jersey) and *Bos indicus* (Sahiwal) breeds. In early 1950's the breed was closed to outside inheritance. The period from 1952-64, involved considerable attention to breed development when nucleus herd sires were selected on the basis of dam and progeny performance. The population was then stabilized. The breed developed from this combination, called **'Jamaica Hope',** contains 80% inheritance from Jersey, 5% from Holstein Friesian, and 15% from Zebu (mainly Sahiwal). Jamaican Hope has been claimed to be well adapted to local conditions but at the same time to be superior to HF in overall productivity under the Jamaican conditions. This breed, which is supported by a breed society, is widely distributed. Some animals of the breed had been exported to countries within the Caribbean area. There are a number of other examples of the new breeds developed in several farm animal species.

Problems and Obstacles in Breed Development

Development of a new breed of livestock is not as easy as may seem.Some of the common problems in such a program include:

- **Time and capital required:** Development of a new breed, especially in species with long generation interval (e.g. cattle) takes a long time. It involves choice of the appropriate breeds, maintaining animals of different breeds, systematic crossing, and selection for a number of generations. Usually very large number of animals are required to provide a broad genetic base for effective selection. For example, formation of Santa Gertrudis breed is said to have been based on a herd of 25,000 cattle.
- **Change in production environment and market demands:** Breed development involves defining the breeding objectives (desired level of performance for the traits contributing to market requirements) for the geographic and economic conditions in which the breed is to perform. It is, however, possible that by the time a breed has been developed, the market demands, managemental, nutritional and health practices may change considerably making the new breed to be of little practical utility. There may be difficulty in acceptability of the breed by producers of the country/region

Development of a new breed is therefore is not only expensive and long-term proposition but also requires courage, patience and foresight.There have been instances of development of new breeds of cattle which were later on suspended for one or the other reasons:

- In Tanzania, development of a new breed was started from the cross of European dairy cattle *(Bos taurus*) and African Zebus in 1946, and the breed was officially baptized 'Taurindicus'. By 1966, however, the breed got disbanded.
- In India, breeds named 'Jersind'; 'Brown Sind' and 'Karan Swiss' were developed from crossbred foundations in 1980-1990's. However, these could not be accepted as breeds. Similarly, in 1856, a breed named 'Taylor' was developed in Patna (India) by the then Commissioner Taylor by crossing of Shorthorn and Channel Island bulls with indigenous cows. The breed does not exist anymore.
- In Tropical South America, names of some breeds like 'OCampo', 'Jerdi' and 'Suisbu' have been heard but particulars of these breeds are no more known.

In fact, there are only limited examples of successful development of new breeds of dairy cattle. In some cases, new breeds were developed simply as a fad or gimmicky to get popularity or political mileage. Obviously, it is easier to name a breed than to maintain it and make it economically acceptable. The new breed has to be clearly superior to the existing ones in order to justify the extra costs involved in its development.

General remarks: Development of new breeds is not a cure for the genetic problems faced by livestock industry of a country or region. Rather it is a process to supplement other breeding strategies for genetic enhancement of productivity.

Box 16.3: Composite versus a new breed.

Composite is a self-contained population, and so is a breed (or 'pure breed). The difference between a composite and a pure-breed is rather thin. A composite is a breed that has been produced from some existing breeds, and retains a certain amount of heterosis and breed complementarity. This is so because no selection following its formation has taken place in the composite. However, when selection is the composite takes place for some generations and consequently some unique traits are fixed in the new population, it becomes a 'pure breed'.

Breed and Breeding Industry Structure

Structure of a breed and the breeding industry controls the way by which genetic improvement accomplished in the breeds is transmitted to end users (farmers). Traditionally most breeds of farm animals have hierarchical structure of two or more distinguishable levels, called **breeding pyramid**.

Traditional Breeding Pyramid: Closed Nucleus Breeding Scheme

The usual structure of a breed and breeding industry is like a pyramid, called breeding pyramid, which usually consists of sub-groups called tiers. In most situations, the breeding pyramid has three tires viz; nucleus, multiplier and commercial (Figure 16.2).This structure enables the genetic improvement achieved at the top of pyramid to be transmitted on to lower strata in the hierarchy, largely through sale of males.

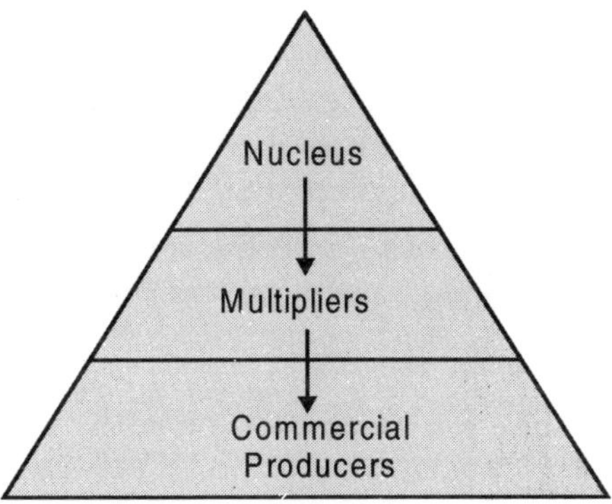

Fig. 16.2 Traditional breed structure (breeding pyramid): Closed nucleus breeding system

The area of the pyramid covered by each tier reflects the relative number of animals in that tier. The salient features of the three tiers are:

- ***Nucleus tier***: Nucleus tier consists of elite breeders ('seed stock'/'stud') who are self-contained as they produce their own male and female replacements. The elite breeders are the main center of genetic improvement and their role is supply of superior breeding stock to the next tier - the multipliers. In some cases, nucleus breeders may form a group and interchange sires and dams between them. For this category of breeders (nucleus tier), breeding values (genetic merit) of animals rather than their performance are of primary importance because their main objective is to bring about genetic improvement which is then to be transmitted to the next strata.

Usually elite breeders practice selection (coupled with some degree of mild inbreeding depending on the species concerned) for genetic improvement of their stocks. They may use crossing only to introduce new ('superior) genes into their closed stocks. Introduction of inheritance from other stocks however needs to be made with great care since such an introduction may reduce performance of their stocks, if the introduced animals are of low genetic merit.

- ***Multiplier tier:*** The next stratum of the traditional breeding pyramid consists of multipliers who purchase their replacements from nucleus herds. The multiplier groups which could be of one or more tiers, are much larger than the elite group. Multipliers may produce 'purebred/straightbred' or 'crossbred' stocks depending on the species. Their main aim is to procure improved stock from elite breeders and to produce or multiply the breeding stock in large numbers to meet demands of commercial tiers below them.
- ***Commercial tier:*** It consists of commercial producers who are primarily involved in production of milk, meat, eggs or fiber. They do not have any role in selling stock for further breeding.

In this traditional hierarchal breeding structure, the genetic improvement made by elite breeders (nucleus tier) flows downward i.e. it is disseminated from nucleus tier through multiplier to the commercial herds. This breeding scheme is based on the assumption that elite breeders are making genetic progress which is then being constantly transmitted to the next tier. Such a breeding scheme in which there is only one-way flow of genes (i.e. from top to down) is called **closed nucleus breeding scheme (CNBS)**.

Demerits of CNBS: There are two main disadvantages or problems with this scheme:

- There is transmission of genetic improvement from top to downward only, therefore, the progress in lower strata is solely dependent on the genetic superiority of nucleus herd/ flock. In species like cattle, the elite/nucleus units are usually too small to make genetic improvement, therefore, little genetic progress may be accomplished in lower tiers. Since there is no flow of genes into nucleus tier, rate of inbreeding in this tier also increases which also reduces genetic progress.
- There is a long time lag, called **'improvement lag'**, involved in transmission of genetic improvement from one strata to the next (usually 2-3 generations). For example, in pigs, the normal time lag from nucleus to commercial herd could be as long as 7-8 years. In cattle, it will be many fold more. Thus it takes a long time for genetic progress achieved in one tier to be transmitted to the next tier (long improvement lag).The size of improvement lag is affected by the age structure in the lower tier and the source and genetic merit of sires and dams in the lower tiers.The improvement lag can be reduced to a certain extent by:

  (*i*) keeping sires and dams in lower tiers for shorter periods before replacing them with younger stock

  (*ii*) transferring ♀ and ♂ between tiers.

  (*iii*) transferring some nucleus parents directly to commercial tier.

Closed Nucleus Breeding Scheme in Poultry

Closed nucleus breeding scheme is most common used in poultry followed by pigs because commercial breeding industries in these species are well developed and organized. Short generation interval and high prolificacy of the species also favor its application. Salient features of this structure in poultry and pigs are:

- The nucleus flocks are kept under strict quarantine to reduce chance of introduction of diseases from other breeders. The elite and multiplier tiers in these species are in hands of a few multinational breeding companies who control most of the business throughout the world.
- The elite purebred or synthetic strains (lines) maintained by the companies, which are several in numbers, are most intensively selected.
- The breeding stock from these lines moves down to company's own multiplier flocks or to the flocks contracted by them for multiplication purposes (a franchise system). The males and females from multiplier flocks produce the commercial stocks which are always crossbred (strain or line-crosses).

On the other hand, the elite tiers in dairy cattle, sheep and beef cattle are owned by private breeders rather than by large-scale breeding companies, therefore, they are not able to make large genetic progress. The closed nucleus breeding scheme in these species is therefore, not very successful in efficient transmission of genetic improvement to lower tiers.

Open Nucleus Breeding Scheme

If the nucleus or elite herds are opened to genetically superior animals from lower tiers of breeding pyramid, a two-way flow of genes becomes possible. Such a nucleus breeding scheme in which flow of genes is bidirectional i.e. from upper to lower, and *vice versa*, is called **open nucleus breeding scheme (ONBS).** Structure of a typical ONBS is shown in Fig 16.3.

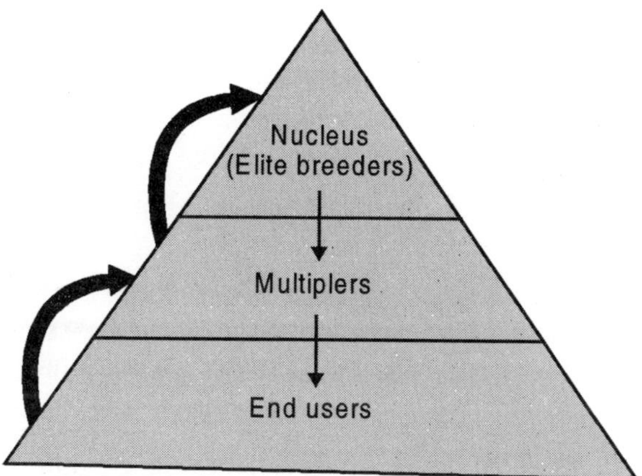

Fig. 16.3 Open nucleus breeding scheme

The ONBS is used to overcome one of the most serious limitations of CNBS which is lack of transfer of genetically superior animals to upper tiers. ONBS allows this movement from lower to upper strata so that sufficient genetic variation becomes available in upper tier for making further genetic improvement. This is based on the premise that, in general, there is sufficient variation within the lower tiers. Some of the animals born in lower tiers may have better performance than those in a higher tier. If higher performance of lower tier animals is due to their superior genetic merit (breeding value), it would be of great use to transfer such superior animals into a higher tier. And if they are sufficiently superior, these could be transferred into the nucleus tier. This opening of the nucleus tier to genetically superior animals of lower tiers gives rise to a two-way flow of genes.

Advantages of ONBS: There are two main benefits of ONBS:

- Annual rate of genetic improvement is increased due to access to greater genetic variability and superior animals.
- Rate of inbreeding in the nucleus tier is reduced because of introduction of animals from outside.

ONBS can take many different forms and structures. One of its popular form is a **'group breeding'** or **'cooperative breeding' scheme**.

Group Breeding Scheme

Breeders of only a few farm species, and these too in countries where livestock breeding has taken industrial shape operate on large scale. For example average size of dairy cattle herds, though varies from country to country, but in most locations it is less than 50 cows. Pedigreed beef herds in developed countries of Europe are less than 20 though they are much larger under ranch in North America. Under such a scenario which is prevalent in most of the developing nations, large scale organized breeding is not feasible. Developing countries have a number of other distinct disadvantages for setting up successful breeding programs. Infrastructure needed for performance testing is normally lacking because herd sizes are normally small and there is considerable variability between farms, farming systems and seasons. Further reproductive efficiency in such situations is low, due mainly to poor nutrition, especially in cattle. And, grazing excludes implementation of systematic breeding programs. The group breeding programs attempts to overcome these difficulties.

Group breeding involves a group of breeders who agree to cooperate in the formation and subsequent running of an open nucleus/elite herd/flock in return for a regular supply of breeding stock males from nucleus for use in their own herds. Since it involves participation of a number of breeders, it is also called as **cooperative breeding scheme (system)**.

The important element of this scheme is for a group of breeders to agree to pool their high merit animals into a nucleus flock. This scheme is based on the premise that each herd contains is a small number of genetically very superior animals. The important steps are:

- Genetically superior animals from different herds are brought together to form a nucleus. The average genetic merit of nucleus would be far greater than of any of the contributing herds.
- Once a nucleus is established, a proper system of recording of performance and selection is implemented.
- The best males are kept for breeding in the nucleus while other selected males are given to the base herds for breeding. Therefore, the genetic improvement is quickly disseminated throughout the group.

The nucleus can remain open to animals from the base herds such that the best females from these are admitted periodically and compared with those in the nucleus. Usually, only females are transferred from the base to the nucleus since male selection may not be practicable in base flocks due to managerial reasons. A group breeding scheme is illustrated in Figure 16.4.

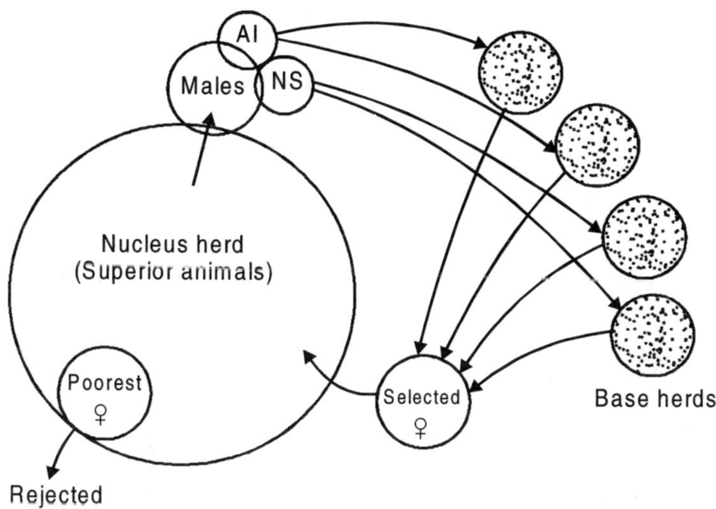

Fig. 16.4 Group breeding scheme.

The main advantage of such a group breeding nucleus scheme is that the genetic superiority of sires coming into base herds from nucleus is far greater than what is achievable in each of the base herds. It is particularly attractive in situations where within-herd selection programs are ineffective due to small population size or inadequate technical skill.

The nucleus breeding scheme shifts the responsibility of operating the breeding program from farmers to nucleus herd. Therefore, the scheme is useful where herd or flock sizes are small. In developing nations, organization of such schemes is mediated by Government agencies because cooperative programs among farmers may not always be practicable.

Sire Referencing Scheme

Sire referencing is a type of group breeding scheme in which different herds or flocks are linked together genetically by sharing a few sires in common without forming a central nucleus (like in nucleus breeding which is expensive to maintain). Once these links are established between herds, the genetic merit (breeding value) of animals of different herds/ flocks can be directly and accurately compared across flocks. This is one of the best strategies to select and use those 'reference sires' across herds to facilitate quick rate of genetic gain while minimizing inbreeding. Such a scheme has been highly effective in the UK sheep industry where flock sizes were small. A sire referencing scheme is shown in Fig. 16.5. Operationally the scheme involves the following :

- Selection of a group of sires (called 'panel of reference sires') which are to be used across herds of participating farmers (by either AI or natural service).
- Use of a few reference sires on a certain proportion of females of each of the member herds.
- Performance recording (on-farm) of offspring born in each herd for the traits of to be improved which have been agreed upon by members precipitating in the scheme.
- Evaluating the breeding values of animals (across flocks/years etc), and to select the future reference sires.

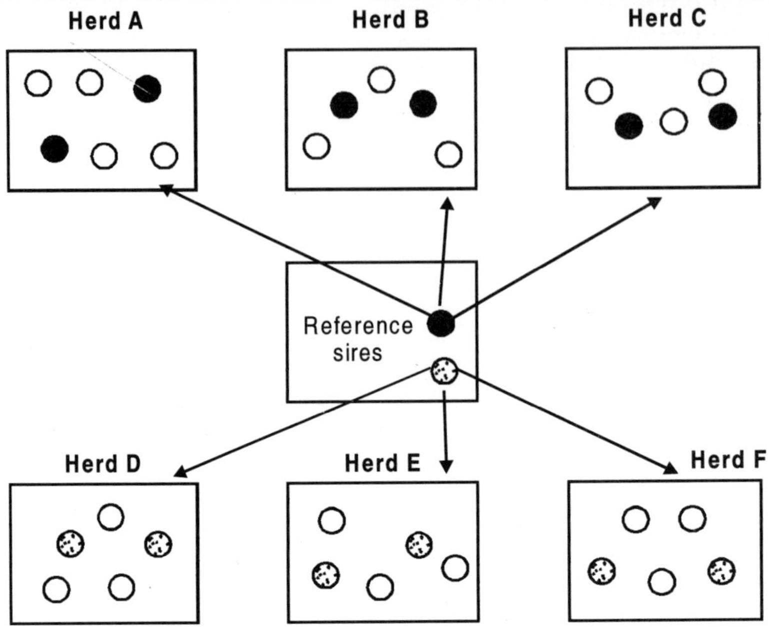

● : Represent progeny of 1st reference sire ◌ : Represent progeny of 2nd reference sire

○: Represent progeny of other sires (non-reference)

Fig. 16.5 A sire referencing scheme in which there is sharing of some sires across herds to develop genetic links between herds. Note: Each herd has progeny from two reference sires (shared sires, designated as '●' and '◌') as well as from other sires, designated as '○'.

Breed Comparison : Genotype-Environment Interaction

A number of breeds, or other such genetic groups have been developed in all species of farm animals. Choice of a breed/strain from among the several available ones for a production system, area, region, country etc. would require objective information on the performance of such breeds under the environment in which they are likely to perform. While making such comparisions, breeders often encounter the issue of genotype-environment interaction.

The basic quantitative genetics model assumes that performance of an individual or a group like breed, strain etc is the sum of its genotypic value (G) and environment which it experiences (E): P = G + E. The validity of this expression/model rests on two simplifying assumptions:

- First, there is no association between the genotype and the environmental conditions experienced by it. That is, genetically superior animals are not provided favorable environmental conditions and *vice versa*.
- Secondly, the genotype and environment act in an additive manner to produce the phenotype. In other words, the relative difference between performances of two genotypes remains same from one environment to another.

  However, when the magnitude of difference in performance between two genotypes changes from one environment to another, genotype-environment interaction, designated '*G* × *E*', is said to exist.

Genotype-environment interaction is alternately defined as the differential response of different genotypes to different environments. The 'G x E' may occur because one environment may permit better expression of one genotype while the other environment may not permit its expression to the same extent. The concept of genotype-environment interaction is applicable when two or more than two genotypes are compared in two or more than two environments.

Terms genotype and environment defined: The terms genotype and environment have so far been defined in context of an individual. With reference to 'genotype-environment' interaction, these terms need to be expanded. In this context, **genotype** may not only refer to the genetic make-up of an individual but it can refer to average genetic make-up of a family (e.g. sire or dam families), average genetic make-up of a strain or breed or inbred line. Therefore, the term genotype can be envisaged at individual, family (sire or dam), strain or breed levels.

Similarly, the term **'environment'** may refer to any single or combination of non-genetic factors which could be physical/external viz; locations (regions/countries), years, seasons, nutritional/feeding regimes, management/production systems etc. or these could be internal (biological, chemical etc) non-genetic attributes viz; sex, age, physiological status of the animal. Further environment could be classified as macro or micro. **Macro environment** includes clearly defined environmental factors such as farms, climate, season, level of nutrition, management type etc. **Micro environment** includes factors, which are difficult to be identified like social hierarchy within a flock.

Accordingly, we may have many types of genotype environment interactions viz; 'genotype-location', 'genotype-year', 'genotype-ration', 'genotype-management system', 'genotype-sex', 'genotype-age', 'genotype-production system', 'genotype-herd interactions'. There can be three- or higher-order interactions like 'genotype-location-production system', genotype-location-ration' interactions and so on. Our discussion here is however restricted to two-way interactions.

Classification of Genotype - Environment Interaction

Genotype-environment interactions can be classified in a number of ways but for our purpose, the following two broad types of genotype-environment interactions are important:

- **Change in rank order:** 'G×E' may involve a change in rank of genotypes in different environments. This is a severe form of interaction because one genotype gives higher performance in one environment but the second genotype outperforms in another environment (Fig 16.6 c).
- **Change in relative difference:** 'G×E' may involve only a change in magnitude of difference between genotypes in different environments. This is a milder form of interaction because one genotype is consistently superior over the other genotype across the entire test environments (Fig 16.6b).

These two types of genotype-environment interactions, along with the absence of interaction(*a*) are presented in Figure 16.6.

One illustrative example of existence "G×E' is frequently seen when tropical dairy breeds *(Bos indicus*) and breeds of temperate region *(Bos taurus*) are tested in tropical as well as in temperate climatic conditions. It is well known that indigenous Zebu cattle perform more satisfactorily than exotic cattle in tropics and semi-arid regions than they do in temperate climate. Similarly, it is fairly well established that small-sized beef and sheep breeds give relatively better performance in 'extensive grazing systems' while the heavy breeds tend to perform best in 'intensive systems' (with high level of concentrate feeding).

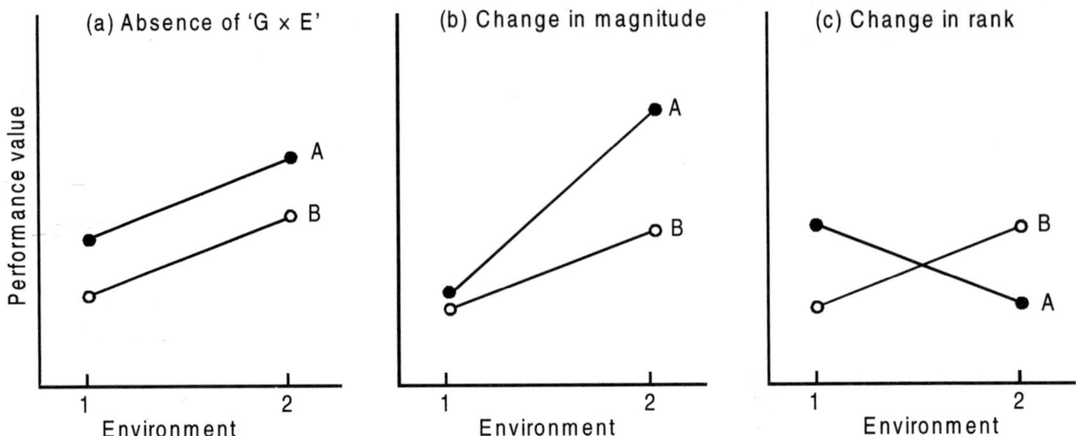

Fig. 16.6 Illustration of concept of genotype-environment interaction for two genotypes (A and B) tested in two environments (1 and 2)

The extent of the genotype- environment interactions, as a general rule, depends on:

- Extent of differences between genotypes
- Extent of differences between environments

If the genotypes differ too widely from each other (like indigenous vs. exotic cattle) or the environments differ too widely (e.g. extreme differences in climatic conditions like temperature, humidity, photoperiod, feeding, disease exposure etc), the likelihood of 'G×E' interaction is more. Further, some traits exhibit far greater 'G×E' than others. In general, the traits which are lowly heritable show greater 'G×E' than highly heritable traits. A number of studies on genotype environment interactions have been made across species particularly in context of genotypes developed in one country or continent when tested in another country, region etc.

Importance of Genotype-Environment Interaction in Breeding

The topic of genotype-environment interaction is important as the breeds, strains or crosses developed by breeders have to perform under a variety of environmental conditions than the one under which they were developed. The commercial producer has to choose genotype (breed, strain, or cross etc) that is best suited for his production conditions. Distribution of germplasm across the globe in the form of semen and embryos has made it crucial to decide as to which genotype to import into a country for enhancement of productivity under its production environment. Inappropriate choice of a genotype for a country may have disastrous consequences, as has been the experience of introduction of temperate dairy breeds in low-input production environment of tropics.

Study of existence and magnitude of genotype-environment interactions is important in animal breeding in two broad contexts:

1. **Type of environment vis a vis selection program:** In planning a selection program, the breeder has to decide about the type of environment in which the improved strain/ breed has to perform. This is an important decision which becomes more important when the range of the environments in which the breeds/strains are to commercially perform increases. Further it has become more critical as breeding industries are increasingly becoming global. When genotype-environment interactions exist, two options are available to be breeder for development of the strains/breeds or their crosses:

- To develop the breeds/strains that can perform satisfactorily under a wide range of climatic, nutritional, management and disease conditions. Obviously, this would involve testing and selection under all the relevant types of environments which is usually impracticable for most species of livestock.
- To develop breeds/strains which are specifically suited for a particular type of environmental conditions viz; low-input vs. high-input, tropic vs. temperate regions etc.

2. **Extent of testing of a new strain or cross** (as in poultry and swine): During developmental stages, the strains/varieties including their crosses that are to be used commercially, are tested under the range of environments in which they are likely to perform. The extent of their testing depends on the magnitude of genotype-environment interaction. If high 'G×E' exists, a strain/variety has to be more intensively tested under varied environmental conditions than if the 'G×E' were low.

Extent of Genotype-Environment Interaction

Lot of investigations have been made on comparing the performance of different genotypes (strains/breeds/ crosses) in different environments (countries, production systems, feeding regimes etc) to identify the most suited genotype for an environment. In general, it has been seen that within the usual range of environments existing in many commercial large-scale intensive production enterprises, the extent of 'G×E' is small. But under extensive or low-input production systems, 'G×E' can be quite large. For example in poultry, the same strains or their crosses are used commercially throughout the world by the well developed poultry breeding industry. Though most of these strains have been developed in North America and Europe but they are commercially used in developing tropical countries. The reason being that most commercial poultry flocks in large scale operations are reared such that most of environmental factors like housing, feeding, etc. are well controlled. Therefore the 'G×E' are much lesser in commercial poultry than in other livestock species like cattle where environmental conditions are less modifiable.

Genotype-environment interactions can thus be minimized either by using the genotypes that are most suited for a specific production system or by modifying the production system to suit a genotype. The intensive poultry and pig industries have adopted the latter approach (i.e. to modify the environment). In dairy cattle however breeds or crosses are chosen to suit the environmental conditions of a country or region.

Genotype-environment interactions can be studied through a number of methods like ANOVA, rank correlation coefficient between performance in two environments, genetic correlation between performance of related animals in two environments.

Germplasm evaluation programs in different species are conducted to make fair comparisons between breeds, strains or crosses to ascertain the comparative performance of various breeds and various crossbreds under well-defined environmental conditions in various regions, countries.Such information is useful for making choice of a breed, strain or cross for a particular environment. Such organized performance evaluation tests include central test stations in pigs, random sample tests in poultry, and international comparisons in dairy cattle like INTERBULL.

Review Questions

1. Define/ explain the following terms:
   (a) Composite
   (b) Breed as per FAO
   (c) Genotype-environment interaction
2. Write short note on:
   (a) Steps involved in the formation of a new breed.
   (b) Closed nucleus breeding scheme
   (c) Open nucleus breeding scheme
   (d) Group breeding scheme
   (e) Importance of genotype-environment interaction in breeding
3. What are the two broad objectives achieved by formation of composites?
4. How much of F_1 heterozygosity is retained in a two-breed composite once it attains equilibrium?
5. What factors (genetic) can influence the retained heterosis in a composite?
6. What are some of the problems or disadvantages of a composite?
7. In which livestock species composites are most popular?
8. How did livestock breeds evolve or originate?
9. Breeds are usually differentiated into strains. Explain with a suitable example.
10. Explain the role of breed associations/societies in breed development
11. What are the justifications for development of a new breed?
12. What are the major problems and obstacles in breed development? Explain
13. List a few examples of the new breeds formed in developing countries that could not succeed.
14. List at least two examples of successful formation of new breeds of cattle. Explain the formation of one of such breed.
15. What are some of the disadvantages of a closed nucleus breeding scheme?
16. Explain a closed nucleus breeding scheme in poultry.
17. What is a sire referencing scheme?
18. What two factors determine the extent of the genotype- environment interactions?
19. What are two broad categories of GE? Explain with a diagram.
20. Why are genotype- environment interactions less important in poultry than in cattle?

17

Germplasm Conservation and Designing of Breeding Program

Animal Genetic Resources

Germplasm[1] resources provide the basic material required to initiate a breeding program. Before launching a breeding program, the breeder has to evaluate the available germplasm resources, and then to make a choice as to which is to be used in genetic improvement program. Once sufficient genetic progress has been made, the promising genotypes (i.e. breeds, strains, lines, or their crosses) that are adapted to the region, country, or production system are released for commercial utilization. In an alternative scenario, breeders may like to generate new variability from the available germplasm, and then subject the newly created base population to genetic improvement. Therefore, germplasm resources of a species constitute the foundation material without which a breeding program is impossible to be undertaken.

This section deals with the status of animal diversity in terms of breeds, the causes of erosion of germplasm resources and the methods of its conservation.

Terminology in Germplasm Conservation

- **Animal genetic resources (AnGR)** or farm animal genetic resources include all the animal species, breeds and strains and their wild relatives that are of economic, scientific and cultural interest to humankind in terms of food and agricultural production for the present or in the future.
- **Biodiversity** or biological diversity refers to the range of differences observed within the living world in terms of the number and the variety of living organisms.

[1]The term germplasm is used to refer to all the genetic resources of a given species which are maintained for the purpose of animal breeding, preservation, and other research uses. These include not only the live animals of breeds, strains, lines etc, semen, embryos, DNA/gene banks.

- **Domestic animal diversity (DAD)** is the sum total of all the differences observed between species, between breeds within species, and within each breed, all of which are available for the sustainable food and agriculture production (defined by Food and Agriculture Organization, FAO).
- **Germplasm** is an alternative term used for genetic resources and includes animal genetic resources not only in the form of live animals but also in the form of semen, oocytes, embryos etc.

Vastness of Genetic Resources

Diversity is the rule of nature as is evident from the existence of wide types of fauna and flora. The vast genetic resources of domestic animals that are prevalent today are result of thousands of years of selection (natural and artificial) under varied environmental conditions as diverse as arctic and tropical; maritime and mountain; humid and arid/semi-desert ecological-zones.

There are more than 40 species of animals that have been domesticated or semi-domesticated during the past 10 to 12 thousand years. These animal species contribute directly through food and fiber, and indirectly through draft power, manure, transport, store of wealth etc. However, only six mammalian (buffalo, cattle, goat, sheep, pig and horse) and four avian (chicken, turkey, ducks and geese) species are used extensively throughout the world for commercial production. The distribution of world's mammalian and avian breeds is given in Fig 17.1. Cattle contribute 22 percent of the world's total number of recorded mammalian livestock breeds while buffalo accounts for 12%.In avian species, chicken contribute 63% followed by ducks.

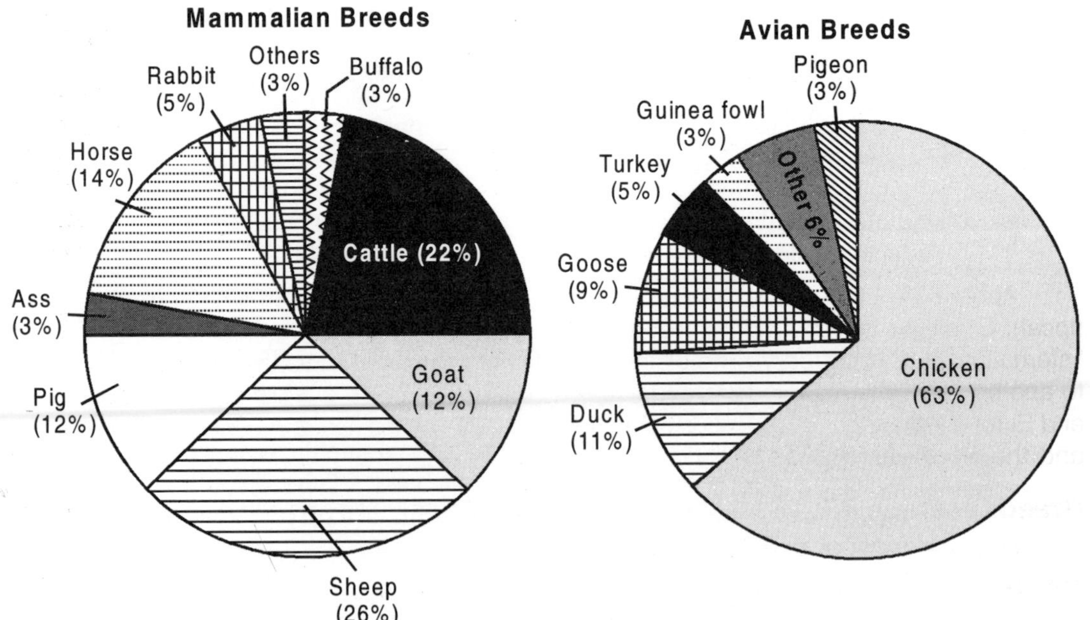

Fig. 17.1 Distribution of world's mammalian and avian breeds

Many other domesticated species such as camels, donkeys, and elephants, various poultry species, reindeer, rabbits, etc. are important in some cultures and regions of the world.

Since the first domestication, several thousand breeds of different species of animals have evolved or developed to serve various purposes. These AnGR are vital to economic development of majority of the countries in world. In developing countries, they additionally play an important role in the subsistence of several communities and the sustainability of livestock systems.

The Global Data Book of FAO of the United Nations Organization, based in Rome, lists a total of 7616 breeds of livestock and poultry (Table 17.1). These breeds are categorized as either 'local' or 'transboundary':

- **Local breeds** - the breeds that occur in only one country.
- **Transboundary breeds** - the breeds that occur in more than one country.

  The 'transboundary breed' category is further subdivided into:

  (i) **Regional transboundary** - the breeds which occur in more than one country within a single region.

  (ii) **International transboundary** - the breeds which occur in more than one region.

Table 17.1 Global status of biodiversity of major livestock and poultry species (FAO 2007).

Species	Number of breeds* :			
	Local	Transboundary		Total
		Regional	International	
Buffalo	122	10	05	137
Cattle	897	93	112	1102
Goat	512	47	40	599
Sheep	995	134	100	1229
Pig	541	25	33	599
Chicken	1077	55	101	1233
All species	6536	523	557	7616

*Excluding extinct breeds except all species figures.

About 87% breeds of all the major livestock and poultry species are specific to a country (local). Of all the transboundary breeds, about 48% are 'regional transboundary' and 52% are 'international transboundary'. The distribution of different breed categories varies from one region to another. For example, local breeds' make-up a major share of all the breeds in Africa, Asia, and Europe including the Caucasus compared to Latin America, North America and the Caribbean, and the Southeast Pacific.

Breed Diversity is More Important than Species Diversity

Genetic diversity within a species of livestock and poultry is reflected in the range of breeds or types as well as the variation present within each breed or type (e.g. beef vs. milk breeds of cattle, layer vs. broiler chickens; different strains of a breed like Holstein Friesian, strains etc). It has been estimated that about 50% of all the livestock genetic diversity is accounted for by differences between breeds, and the remaining 50% by differences between species. Since there is no direct and quantitative measure of diversity, the number of breeds within a species is taken as the best indicator of livestock diversity.

From conservation and utilization standpoint also, the breeds and possibly the strains within a breed are generally more important than species. This is so because it is the differentiation of species into breeds which has allowed the existence of livestock production throughout the world. Moreover, livestock species like cattle, buffalo, sheep, goat, pig, poultry etc are unlikely to become extinct but certain breeds within each of these species are likely to or are becoming extinct. For example, the buffalo breeds (the swamp type) which were used for draft in Thailand and other South-East Asian countries are threatened with extinction due to mechanization of paddy cultivation. For all these reasons, breeds are of greater importance than species for the conservation of livestock diversity.

Global Breed Risk Status

In order to accurately evaluate the status of a breed, it is necessary to have its census numbers including the number of breeding males and females, and the change in its population size (whether stable, decreasing or increasing). From conservation standpoint, the FAO has established a detailed scheme for classification of breeds into a number of categories.

- **Extinct**: A breed is categorized as extinct when there are no breeding males or breeding females remaining. Nevertheless, genetic material might have been cryopreserved which would allow recreation of the breed. Extinction is absolute when there are no breeding males (semen), breeding females (oocytes) nor embryos remaining. In reality, extinction may be realized well before the loss of the last animal or genetic material.
- **Critical**: A breed is categorized as critical if the total number of breeding females is less than or equal to 100 or the total number of breeding males is less than or equal to five; or the overall population size is less than or equal to 120 and decreasing and the percentage of females being bred to males of the same breed is below 80 percent, and it is not classified as extinct.
- **Critical-maintained**: These are those critical populations for which active conservation programs are in place or populations are maintained by commercial companies or research institutions.
- **Endangered**: A breed is categorized as endangered if the total number of breeding females is greater than 100 and less than or equal to 1 000 or the total number of breeding males is less than or equal to 20 and greater than five; or the overall population size is greater than 80 and less than 100 and increasing and the percentage of females being bred to males of the same breed is above 80 percent; or the overall population size is greater than 1 000 and less than or equal to 1 200 and decreasing and the percentage of females being bred to males of the same breed is below 80 percent, and it is not assigned to any of above categories.
- **Endangered-maintained**: These are those endangered populations for which active conservation programs are in place or populations are maintained by commercial companies or research institutions.
- **Breed at risk**: A breed that has been classified as either critical, critical-maintained, endangered, or endangered-maintained.

Global risk status of mammalian and avian breeds (livestock) indicates that of all the breeds of livestock and chicken combined, 9% breeds have become extinct and another 20% are at risk. And, of all the breeds across species for which the census figures are available, about 36% are at risk of being lost; the figure for chicken is 57% and for mammals is 26%. Higher risk for chicken could be due to industrialization of poultry breeding industry where only a few breeds are major contributors to egg and meat production. Global risk status of mammalian and avian breeds (livestock) is presented in Fig 17.2.

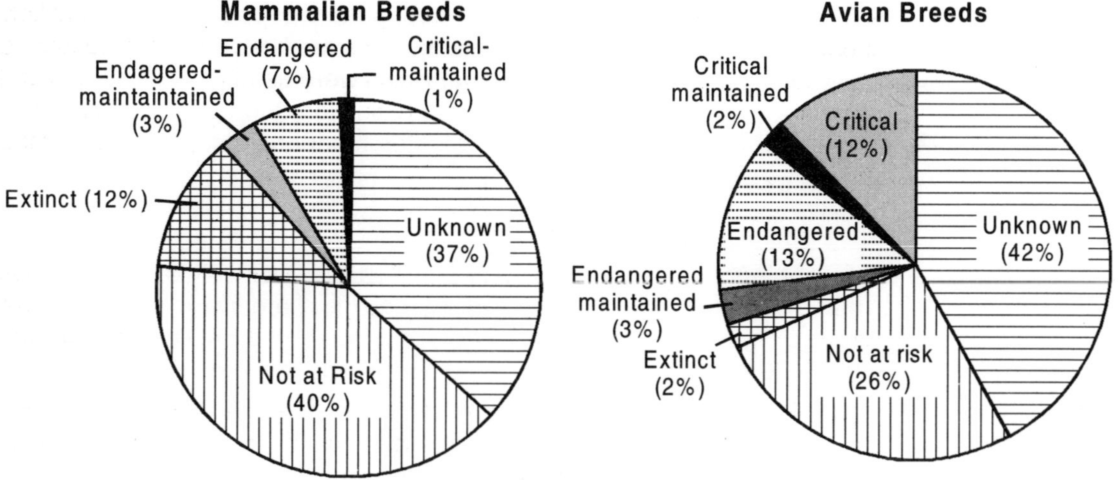

Fig. 17.2 Global risk status of mammalian and avian breeds (livestock).

Data for risk status of individual species of livestock and poultry on world basis as well as on continental basis are available in FAO documents. The salient features from these data are:

- There are large number of breeds for which risk status data are not available which imposes a serious constraint on breed conservation efforts.
- Cattle are the mammalian species with the highest number of breeds at risk. Also, in cattle highest number of breeds (209) are reported as extinct.
- Large numbers of extinct pig, sheep and horse breeds are also reported but for goat the number is small.
- Among avian species, chickens have by far the highest number of breeds at risk on a world scale which is partly related to high absolute number of chicken breeds in the world. Nonetheless the proportion of breeds at risk is also high in chickens because of industrialization of poultry industry in which only a limited number of breeds are commercially used.

However, these figures of risk status do not reflect a complete picture of breed extinctions because it is likely that several breeds may have been lost before being documented. A considerable number of breeds have been lost, and a more are likely to become extinct if the risk factors are not eliminated. In developing countries, factors like disease epidemic, drought etc may also affect the risk status even though the number of animals in a breed may be in thousands. For developing countries, the FAO recommends that a breed be considered for conservation when the number of breeding females drops to about 5,000 within the total population size of about 10,000.

Trends in risk status of breeds: The trends in risk status of breeds over time are assessed by comparing the current risk status of breeds with their risk status in the past. Such a comparison for local breeds (of all the countries) between 1999 and 2006 (as per FAO) across mammalian livestock and avian species gave following trends:

- Sixty breeds which were classified 'at risk' in 1999 were classified as 'not at risk' in 2006 indicating thereby the positive efforts in conservation.

- Fifty-nine breeds during this period moved into the 'at-risk' category.
- Sixty-two breeds got extinct during this period which amounted to a loss of about one breed per month.

These figures, coupled with risk-trends from various regions, show that there are valid reasons for conservation of livestock and poultry breeds.

Causes of Loss of Animal Genetic Resources

- **Replacement of indigenous breeds with crossbreds:** Traditionally, livestock in many developing countries have been reared under low-input conditions (in terms of feed resources, disease prevention etc.). The indigenous breeds due to the genes carried by them for disease, heat and stress tolerance etc are well adapted to low-input conditions and extensive or semi-intensive system of production; however, they are genetically poor producers of milk, meat, eggs etc. The demand of increasing human population for livestock products cannot be met by low producing breeds, as these are economically inefficient. Therefore, the low-producing indigenous breeds are being replaced by limited number high-output breeds through indiscriminate up-gradation and crossbreeding programs. These high producing (upgraded and crossbreed) populations require high inputs (concentrate feeds, good housing conditions, higher hygiene etc) for full expression of their genetic potential.
- **Intensification of agriculture:** Loss of grazing land caused by intensification of agriculture has increased the depletion of low producing breeds suited for extensive production system. The globalization of livestock market due to cheap and rapid transport systems has led to increased consumption of animal products. Most of the increasing milk and meat demand is being met by a few species and breeds reared under intensive production systems.
- **Shift of traditional livestock communities to other activities:** Many of the diverse indigenous breeds have traditionally been kept by small ethnic communities. Young people from these communities are migrating to cities for employment which is not only leading to erosion of indigenous breeds but also to loss of traditional knowledge acquired by these communities over a long period of time for rearing of livestock.

There is no doubt that animal genetic resources exist in relation to the ecological systems in which they live or perform. Since the ecological and production systems are undergoing constant changes, it is expected that over long periods, some new breeds/populations of livestock will emerge while others may be lost. However, the speed with which breeds are eroding is enhanced by human activities.

Why Conserve Genetic Diversity?: Rationale for Conservation

The word **'conservation'** (of farm animal genetic resources) includes all the human activities (i.e. plans, policies etc.) undertaken to ensure that the diversity of farm animal genetic resources is maintained to meet the needs for food production for the present and the future (FAO, 2007).

Two views on conservation : There are two views on the need to conserve livestock genetic diversity particularly of low- producing indigenous breeds:

- A small fraction feels that there is no need to be concerned about conservation because the market forces will automatically bring about conservation of the breeds that are suited for a region/country. High costs of organized conservation and the absence of any assured returns from indigenous breeds are cited as valid reasons against conservation.

- A majority, on the other hand, feels that animal genetic resources are part of earth's natural heritage therefore the unique genetic material available in the form of diverse breeds needs to be conserved for the future.

Reasons for conservation : The two main reasons for conservation of animal genetic resources are:

- **Economic - genetic insurance for future needs:** The chief and the strongest basis for conservation of animal genetic resources is practical: '*to meet the unknown future needs'*. Various breeds of livestock during their evolution and development, particularly in tropical countries, which have a wider range of environmental conditions, have been exposed to diseases, parasites, and nutritional and climatic stresses. This has allowed natural selection, coupled with reproductive and geographic isolation, to conserve the genes useful for adaptation to diverse environmental conditions. *Each breed may thus be considered a potential repository of unique genes and gene combinations.* In view of unpredictability of future needs, the genetically diverse breeds offer a great range of option for meeting future challenges viz; climatic changes, emerging diseases, new knowledge of human nutritional requirements, fluctuating market conditions or changing social needs.

- **Scientific and socio-cultural utility:** Diversity is the base on which biological research, particularly genetic research, depends. Studies on genetic diversity provide insight into the developmental, physiological and biochemical functions; gene structure and control; and the pattern of evolution and speciation, domestication. Some of the stocks carrying rare mutants like miniature pigs, dwarf and naked-neck chicken etc have been used as model in such medical areas like obesity, diabetes etc. Special genetic stocks (breeds, strains, lines) of livestock are essential material for research in immunology, disease resistance etc. Such scientific studies will beneficial to both animal and human health.

  Genetic diversity also has social and cultural value as some animal species or breeds are used for recreation, exhibitions. Genetically diverse stocks of livestock provide visual evidence of a nation's heritage just like historical buildings, monuments and geographic sites do. In Hungary and France, for example, traditional indigenous breeds are maintained in national parks for tourist attractions.

Obviously all the breeds of livestock and poultry that exist today cannot be conserved., therefore, we have to prioritize which breeds to conserve. Since some breeds occur in more than one country, there is no rationale for their being conserved in each country separately. A close cooperation between countries is thus required not only in conservation but for documentation on breeds as well.

Methods of Conservation

There are two methods for the conservation of animal genetic resources: 'in-situ' and 'ex-situ' conservation.

1. **Maintaining Living Herds or Flocks (In-situ Conservation):** Maintenance of live populations (breeds) in their natural environment or in an environment that is quite close to their adaptive environment is one of the best practical ways of conservation.

- **Advantages:** The major advantage of this method is that the breeds continue to be used for economic and scientific purposes, and at the same time are improved further through selection.
- **Disadvantages:** Small size of herds or flocks used for conservation leads to genetic drift and inbreeding. Drift causes gene fixation while inbreeding causes decline in reproductive and productive traits. Since inbred animals are more sensitive to environmental fluctuations, small-sized live populations may be at higher risk to diseases.

  The major problem associated with maintaining living populations for conservation is its high cost. In cases where some breeds are genetically close to each other, an alternative to individual breed conservation is to cross such breeds, and then to maintain their gene pool as a single interbreeding population. Although this method is suitable for conserving the unique genes of the breeds involved but the identity of different breeds in the new population is lost. Further, the gene combinations associated with specific breeds are also broken-up and lost.

2. **Cryopreservation: Ex-situ Conservation:** Ex-situ conservation involves conservation of genetic resources in a situation which different from their normal habitat. Cryopreservation of germplasm in the form of semen, oocytes and embryos (at ultra low temperature typically between –150° and –196°C) is a cost effective method of ex-situ conservation of genetic material for an indefinite period. Storage of somatic cells and DNA also fall under this category. From the cryopreserved germplasm, living populations of animals may be reestablished when needed.

   Advantages:

   (i) Cryopreserved genetic material is free from unintentional genetic change, and is also at a lower risk from disease outbreak and other natural disasters than the live animals.

   (ii) Cryopreservation can also complement the efforts to conserve breeds in their natural environment, and serves as a safeguard when population size becomes dangerously low.

   Disadvantages:

   (i) High initial investment though it becomes inexpensive in the end.

   (ii) Risk of technological failures but these can be minimized by splitting the genetic material, and storing it at different places.

   (iii) Cryogenic technologies required for preservation are not equally successful in all species, and may not be available at all the places.

Animal Genetic Diversity in India

Status: India has rich biodiversity including diversity in crop plants, wild plants, livestock, and microbial, aquatic and belowground diversity. Two out of the 25 priority global hot spots of biodiversity (identified by the Conservation International), namely the *Indo-Burma* and *Western Ghats/ Sri Lanka* exist here. The animal genetic resources are represented by a broad spectrum of native breeds of buffaloes, cattle, goat, sheep, swine, equines, camel and poultry. Presently, large number breeds of livestock and poultry of India (about 144)[2] are registered with National Bureau of Animal Genetic Resources (NBAGR). It includes 39 breeds of cattle, 13 of buffalo,

[2]The information on number of breeds with NBAGR keeps on changing as more breeds get included.

39 of sheep, 24 of goat, 6 of horse and ponies, 8 of camel, 3 of pig, 1 of donkey and 16 of poultry in addition to many more which have not yet been characterized and accredited so far. Additionally there are populations/breeds of other species like mules, yaks, mithuns, ducks, quails etc that are yet to be classified in to well-descript breeds.

India, by virtue of having a greater range of environments than the temperate zones, is a repository of a large segment of domestic animal biodiversity. This biodiversity is the product of thousands of years of natural and artificial selection under limited resources and harsh conditions. Heat tolerance of Indigenous (zebu) cattle, for example is well documented. However, this diversity is fast eroding as indicated by the population trends of different breeds/populations. In fact, the **World Watch List** for genetic resources has identified some of Indian breeds under the highest priority group for conservation by management: Sahiwal breed of cattle; Murrah and Nili Ravi breeds of buffalo; Jamunapari breed of goats. The indigenous breeds/populations showing declining trends are listed in Table 17.2.

Table 17.2 List of indigenous breeds showing declining trend in census numbers.

Species	Breeds showing declining trends	Percentage of total breeds
Cattle :	Vechur, Punganur, Krishna Valley, Bargur,Ponwar, Binjharpuri, Red Sindhi, Sahiwal, Tharparkar, Amritmahal.	27
Buffalo :	Nili-Ravi, Bhadawari, Toda, Chilika.	31
Sheep :	Ganjam, Bhakarwal, Gurez, Karnah, Niligiri, Poonchi, Pugal, Magra.	21
Goat :	Jamunapari, Beetal, Surti, Changthangi, Chegu, Attapady Black.	26
Camel :	Marwari, Mewati, Kutchi, Double humped camel.	44
Horse :	Bhutia, Manipuri, Zanaskari, Spiti, Marwari, Kathiawari.	100
Chicken :	All indigenous breeds of poultry except Aseel and Kadaknath.	87

Source: National Plan Action for management of AnGR, NBAGR, Karnal, India

Conservation Programs in India

The Indian Council of Agricultural Research (ICAR), the country's premier agriculture research organization, had established the **National Bureau of Animal Genetic Resources (NBAGR)** at Karnal (Haryana) in 1984 with the objectives of undertaking activities on identification, evaluation, characterization, conservation and utilization of livestock and poultry genetic resources in the country. Besides this, there are a number of species-specific National Research Institutes under the aegis of ICAR which are also involved in conservation activities. Some of the important activities/achievements of the NBAGR and other institutes in this direction include:

Identification and characterization: The Bureau has been conducting systematic surveys with active collaboration of State Animal Husbandry Departments, State Agriculture/Animal and Veterinary Universities, ICAR institutes and NGO's.Through such surveys:

- About 90% of total indigenous farm-animal breeds have been characterized both phenotypically (morphometric and production) and genetically (DNA based markers).
- More than 60% of the breeds have been documented in the form of bulletins and breed monographs. The information system on animal genetic resources of India has been developed and is being updated regularly.

Conservation: In-situ model of conservation, by providing technical inputs and incentives to the farmers, breeders in the breeding tract of the respective breeds, has been adopted under a Net-Work Project in which State Agricultural/Animal and Veterinary Universities, State Animal Husbandry Departments, ICAR Institutes and NGOs are collaborators. Eight in-situ conservation schemes involving Tharparkar cattle, Toda buffalo, Nilgiri, Magra and Kilakarsal sheep, Beetal and Surti goat and Spiti horse have been established.

Similarly, ex-situ programs are also in operation at NBAGR. About 50,000 cryopreserved semen doses representing important and endangered breeds are being maintained in the **'Animal Gene Bank'** of the Bureau. The breeds are:

- Cattle: Punganur, Amrit Mahal, Hariana, Gir, Kangayam, Dangi, Sahiwal, Tharparker and Red Sindhi.
- Buffalo: Bhadawari, Tarai, Surti, Murrah, Pandharpuri and Assameese populations.
- Sheep and Goat: Garole of sheep, Chegu and Black Bengal of goat.
- Camel: Jaisalmeri.

Additionally, **somatic cell lines** from different breeds of cattle (Sahiwal, Tharparkar), buffalo (Bhadawari and Murrah), sheep (Pugal and Allure), goats (Gaddi and Parbatsiri) and camel (Jaisalmeri) are maintained by the bureau. The bureau also has an **Animal Genomic Resource Bank** which has collection of genomic DNA from about 130 breeds of livestock and poultry.

Utilization: 'Utilization of AnGR' has been defined by FAO as 'the use and development of animal genetic resources for the production of food and agriculture'. The use and genetic improvement of AnGRs that already possess high levels of fitness to the environment concerned will ensure that well-adapted local breeds remain a functional part of production systems. This is accomplished through identifying the elite animals of breeds while conducting the survey for their characterization. The list of elite animals is supplied to State Animal Husbandry departments for making use of these elite animals in genetic improvement programs.

Accreditation: A mechanism for registration of livestock and poultry breeds of India has also been initiated with the objective of providing identity and protection to the valuable indigenous animal genetic resources. Accession numbers have been given to each of the known breeds of various species. Guidelines and application forms for registration of new breeds are being distributed to potential stakeholders for registration of new populations as breeds. There are many domestic animal populations having distinct characteristics and geographical isolated distribution, which have not yet been recognized as breeds. These are targeted to be registered as breeds for their protection and making benefits of **'Intellectual Property Rights'**.

In addition to the National Bureau and Central Institutes, there is a large extension network in the country in the form of Krishi Vigyan Kendras for dissemination of information generated by different research organizations on conservation. There are also a number of farms with NGOs, Muths, Pinjrapoles and Gaushalas, which maintain indigenous animals and are thus potential 'in-situ' and 'ex-situ' conservation centres.

ICAR Network and All India Coordinated Research Projects (AICRP) for genetic improvement of native farm animal breeds are in operation both under organized farms and under field conditions. Breeds covered under these projects are

- Buffalo : Murrah, Nili Ravi, Jaffarabadi, Pandharpuri, Surti, Bhadawari, and swamp buffalo.

- Cattle : Gir, Sahiwal, Hariana and Ongole.
- Goat : Marwari, Sirohi, Surti, Jamunapari, Barbari, Ganjam, Black Bengal, Osmanabadi, Malabari Chegu and Changthangi.
- Sheep : Gaddi, Nali, Chokla, Patanwadi, Nilgiri, Bonpala, Muzzafarnagri, Malpura, Sonadi, Deccani, Mandya and Nellore.
- Pig : Ghoongroo, Desi.

Designing of Breeding Program

Development of an optimal breeding program, in addition to genetic considerations of selection and mating, also needs to take account of a number of other factors. A single breeding plan cannot be applicable to all the species of livestock as well as to all the situations (geographical, economic etc) for a species. Following is, therefore, a general account of the different steps involved in developing a breeding program.

Box 17.1: Definitions and classification of production environment and systems.

Production environment includes biological, climatic, economic, social, cultural and political factors which combine to determine the performance potential of livestock enterprise. Two types of production environment are commonly recognized:

- **High-input production environment:** A production environment in which all the inputs required for animal production can be managed to ensure high levels of survival, reproduction and output. The risks associated with production in such an environment are constrained primarily by managerial decisions.
- **Low-input production environment:** A production environment in which one or more inputs required for animal production impose severe pressure on livestock resulting in low survival, reproduction or production. The risks associated with production in such an environment are exposed to major influences which may go beyond human management capacity

Production system : Livestock are generally produced and managed in two distinct ways:

- **Extensive production system**: A production and management system under which domestic livestock have traditionally been raised; generally under range conditions, (i.e. the animals obtain forage and water on open pasture/ grazing land with supplements such as hay). Nutritional supplementation and veterinary care are minimal. Animals are subjected to a variety of environmental and nutritional stresses and thus become adapted to their local environments. Small extensive systems are primarily found in developing countries.
- **Intensive production system:** The production and management system for domestic livestock that involve high levels of inputs, such as supplemental feed, veterinary care, good housing (shielded from the extremes of natural environment and protected from most disease)and thus require greater management effort. Intensive systems are predominantly found in developed/industrialized countries.

General Considerations in Breeding Program

The key elements involved in general considerations include agriculture or /animal husbandry developmental policy, environmental and market conditions, infrastructure including financial resources etc.

1. **Development policy:** A livestock-breeding program has to fit into the long-term agricultural/animal husbandry policy of the country/region. In general, the agricultural policies in developing countries aim at increasing food production and improved utilization of the available resources. The goals of the breeding program must match with the long-term developmental policy of the country or a region within the country.
2. **Production environment and systems including market conditions:** Development of a breeding program must take into consideration the production environment, production system (Box: 17.4) as well as the market conditions of a country/region. Intensive production system (with good management, feed and health-care facilities) offers greater opportunities for genetic improvement than the extensive system. For a breeding program to be sustainable, it has to be market-oriented both in terms of immediate as well as long-term demands. This is so because a considerable time lag exists between the start of a breeding program and realization of benefits resulting from genetic improvement. Further, a breeding program should be flexible enough so that it can be modified as per needs.
3. **Infrastructure:** The infrastructure required for implementation of a breeding program includes a broad range of essential inputs viz; trained staff, facilitates for housing, breeding and dissemination of germplasm to end-users (farmers), methods for recording, storage and handling of data (performance as well as pedigree) and finances etc.

   Lack of required infrastructure in developing countries is one of the serious constraints in developing sound breeding programs. In most developing countries, the initial investments in breeding programs are made by Government organizations because the benefits accruing from improving food production are national issues. Though the basic investments and organizational set-up are made by Government but the involvement of farmers in improvement program ensures that their needs are taken into consideration while setting the breeding goals. The farmers in turn can provide support for overall execution of the program (e.g. field progeny testing).

Genetic Considerations in Breeding Program

The broad genetic framework for development of breeding program is discussed below.

1. **Choice of foundation stock(s):** The stock(s) to be improved must have good level of performance for the traits of importance. At the same time, the stock(s) must have sufficient genetic variability so that it/they can respond to selection effectively. In making choice of the foundation stock(s), it is also important to define characterize the environment and production system under which improved stocks are to be reared. Further, it has to be evaluated whether some components of the production environment (e.g. feed, housing etc) can be modified to fully exploit the genetic potential of a particular breed/population.

   In most developing countries, there is usually a stratification of livestock breeds by climatic zones. The breed (s) that is (are) most suited to a specific climatic zone would be the first to be included in the breeding program. Generally, light breeds are more suited for harsher and low-input conditions. But if some components of the production environment or system can be modified, medium-sized breeds can be chosen as these can be accommodated in production environment/system.

The greatest problem in making choice the indigenous breeds in developing countries is the lack of their systematic characterization in terms of population structure, trait descriptions etc.

2. **Defining breeding goals**: Breeding goals or objectives refer to a combination of different traits, taking into consideration their relative economic importance, which are chosen to be included for genetic improvement in a given production environment. The general breeding goals for all farm species are a balanced performance profile which is suitable for efficient production (meat, milk, eggs, wool) under prevailing or future conditions of the region or country.

   Breeding goals (objectives) can be set at different levels: national or regional/local depending upon the species and the target end users:

   - Breeding goals must reflect the real needs of the area/farmers/end users. Abundant examples exist where breeding programs have failed due to ill-defined mixture of traits included in breeding goals.
   - Both the short-term and long-term goals are important; short term goals benefit farmers immediately and thus help them to become active participants. Inclusion of production traits in breeding goals is relatively easy because they directly influence economic returns. However, it is usually not so for fitness and adaptation traits because of the difficulty in assigning economic values to them.

3. **Trait measurement:** Accurate recording of performance of animals is of utmost importance for management as well as genetic-evaluation purposes. The ease, cost and accuracy of performance recording depends on the species and nature of the traits included in the breeding goal. For example, in species like poultry which can be kept under strict controlled conditions, it is much easier to record data than in species like cattle and buffaloes where recording may have to be made at farmer's level spread over a large area. Recording of performance for some traits (e.g. body weight, milk/egg/meat production etc) is quite easy whereas traits like conformation, meat quality or disease resistance may pose a real obstacle in recording; both the technical and cost of recording.

4. **Genetic parameter estimation:** Once the traits have been accurately recorded, it is reasonable to assess whether significant genetic improvement for these traits is possible. For this, the breeder has to estimate heritabilities so that choice of an appropriate basis of selection can be made. Also, the estimation of genetic correlations among different traits is important in designing the breeding program as several traits contribute to breeding goals.

   Genetic parameters (h^2 and r_G) can be more easily and efficiently estimated in prolific species (e.g. poultry swine) than in uniparous species (cattle, buffalo, horse etc) due to their short generation interval and high reproductive rate which make a large number of animals available for selection. Moreover, it is relatively easy to provide reasonably uniform conditions for breeding candidates in poultry and swine than in cattle and buffaloes. Once a breeding program becomes operational, periodic revision of genetic parameters should become a routine part of it.

5. **Which breeding tool to use (selection vs. crossing):** Choice of the breeding method (i.e. selection alone or crossing or a combination of the two) is one of the crucial genetic decisions to be made. It will vary depending upon the species, availability of local breeds (stocks), and the possibility of using exotic breeds/stocks.

   The major genetic consideration involved in choosing the breeding method is the nature and extent of genetic variability, and the types of genetic correlations between the trait (s) included in the breeding goals:

   - ***Genetic variability:*** If traits to be improved are sufficiently heritable, selection within the population will be effective. Breeding values of the animals are computed using different criteria (individual's own performance, relatives performance or a combination of the two) depending on the magnitude of heritability, availability of records of individual, relatives, and nature of the trait.

     When the heritability of a trait is very low, the extent of non-additive genetic variability needs to be assessed. For many lowly heritable traits (e.g. reproductive and viability), enough non-additive genetic variation has usually been found to capitalize heterosis. The utilization of heterosis involves testing of a number of stocks (breeds / strains/ lines etc) and suitable system of crossing to produce desirable genotypic combinations.

     For most traits of livestock and poultry, both additive and non-additive genetic variances have usually been found, therefore, selection and crossing may be simultaneously employed for their improvement.

   - ***Genetic antagonism***: If the traits to be improved have desirable genetic correlations, then selection becomes easier. However, when there are unfavorable genetic correlations between traits included in breeding goal, two courses of action are possible:

     (i) The first is to follow index selection wherein different traits are combined optimally to maximize net genetic-economic returns.

     (ii) The second option is to form two strains, lines or breeds (genetic groups), and to pursue selection in each group separately for different objectives and then to cross the two groups for producing commercial progeny (i.e. to develop separate sire and dam lines). In this case, different selection objectives are followed for the two groups. Operationally, it consists of choosing and evaluating the potential lines/strains for crossing, and at the same time to follow selection within lines.

The general frame work of a genetic improvement program is shown in Figure 17.3.

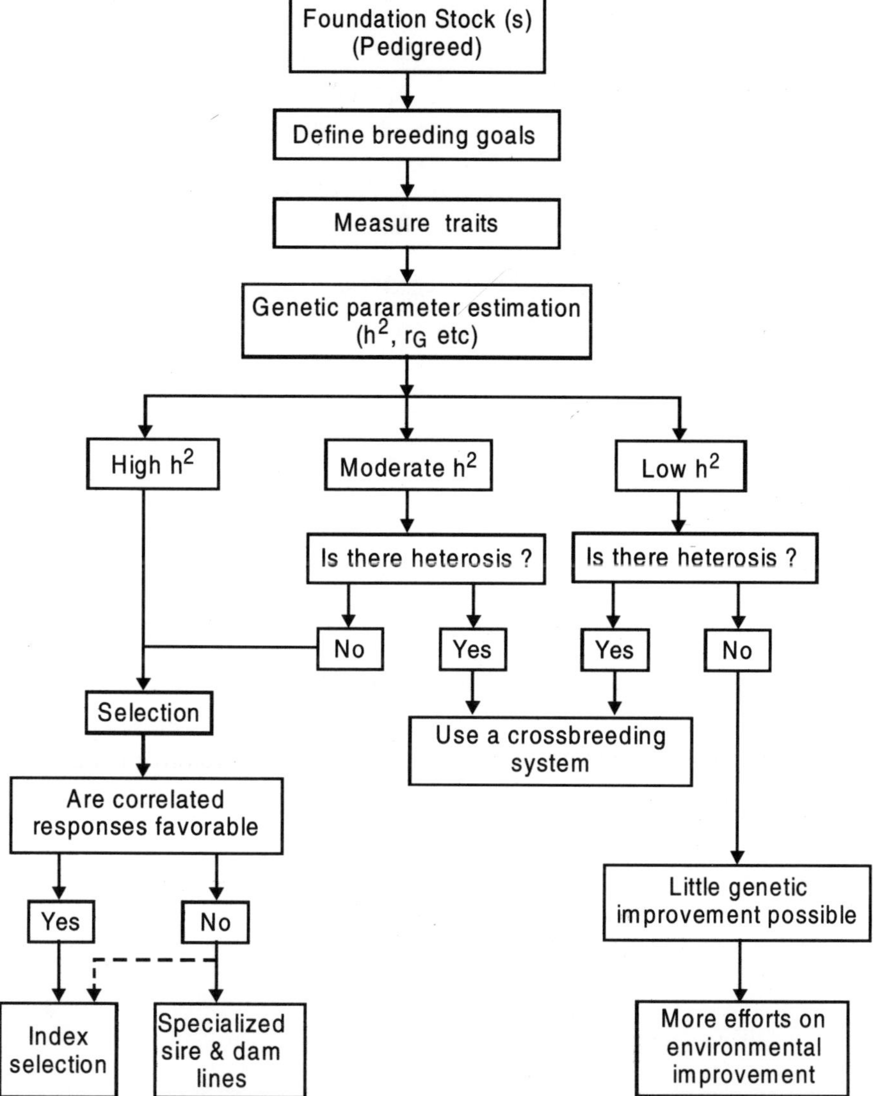

Fig. 17.3 Flow chart showing the steps involved in various decisions in a genetic improvement program.

6. **Dissemination of genetic improvement**: In most species of livestock, the breeding and production (commercial) populations are distinct (recall the breeding pyramid). Genetic improvement is made in breeding population (the top-tier) but the final aim of the breeding program is to improve productivity of the population which is used for commercial production. It has therefore to be ensured that the genetic progress is appropriately transmitted to the level of commercial producers.
7. **Monitoring and evaluation:** Once a breeding program has becomes operational, it is essential to regularly analyze its outcome in terms of amount of genetic improvement

achieved. The monitoring would indicate whether genetic improvement is being obtained in all the important traits included in breeding goals.

Regular monitoring would allow remedial measures to be taken to enhance the rate of genetic improvement. Evaluation of the breeding program should also consist of demonstrating the economic output in terms of total output of products in the entire population. In the long-term, the outputs may have to be related to inputs (i.e. efficiency). Demonstrating the success of a breeding program is also essential in obtaining financial support required for continuation of the genetic improvement program, especially when such programs are Government funded.

Review Questions

1. Define/ explain the following :
   (a) Animal genetic resources
   (b) Domestic animal diversity
   (c) Local breeds
   (d) Transboundary breeds
2. Write short note on:
   (a) Breed conservation programs in India
   (b) Genetic considerations in designing a breeding program.
3. Why breed diversity is more important than species diversity?
4. Why higher percentage of breeds of chicken than of other livestock species have been lost?
5. What are the main causes of loss of animal genetic resources? Explain.
6. What are the two divergent views on need to conserve livestock breeds?
7. What are two main reasons for conservation of animal genetic resources? Explain most important of the two.
8. Explain the two methods of conservation of animal genetic resources.

SECTION - 6

LIVESTOCK AND POULTRY BREEDING

18

Dairy Cattle and Buffalo Breeding

Cattle and buffaloes are the main stray of livestock production in India where they contribute to agriculture economy in a number of ways. However, their primary role is to country's milk production. Of the total milk produced in the country, cattle and buffaloes contribute 96%. There are a total of 290 million buffalo and cattle in the country; about 66 % of these are cattle and the remaining 44% are buffaloes. This chapter deals with the breeding of these two species as dairy animals in relation to Indian situation.

CATTLE BREEDING

Phylum: Chordata (back bone)
Class: Mammalia (suckle their young ones)
Order: Artiodactyla (hooved, even footed)
Suborder: Ruminantie
Family: *Bovine* (hollow-horned ruminants)
Subfamily: Bovinae
Genus: *Bos*
Species: *B.indicus* & *B taurus*

Domesticated cattle, which have descended from the aurochs/aurox, are classified into two major groups : Zebu (*Bos indicus*) and European (*Bos taurus*) cattle. These groups are considered to constitute two different species although they are believed to have descended from the same wild species(the ±). They have the same number of chromosomes (2n = 60)

Fig. 18.1 Zebu (left) and European (right) cattle

but differ in the morphology of Y chromosome. The Y chromosome in Zebu is small and acrocentric while it is metacentric in European. There are also substantial physical and physiological differences between the two groups. Nevertheless, they can readily be interbred, and produce fertile offspring of both sexes.

Zebu Cattle

Most of the cattle indigenous to the tropics belong to the Zebu species (*Bos indicus*). The external trait, which most clearly separates Zebu from European-type cattle, is the hump over the shoulders or the posterior part of the neck (Fig 18.1). Because of the presence of hump, the term 'humped cattle' is frequently used as a synonym to zebu cattle. The hump consists of muscle, connective tissue, and variable amounts of fat. The function of hump is not clearly understood though it has been suggested that it served as a store of fat for use during periods of feed scarcity. Other prominent characteristics of Zebu include a long face, upright horns, drooping ears, a dewlap, slender legs and thin hide. Size of Zebu animals varies over a wide range.

Fig. 18.2 Distribution of *Bos indicus* and *Bos taurus* cattle

Zebu are More Adapted to Tropical Environment

Zebu cattle, having been domesticated and subjected to natural selection under the tropical conditions (Fig 18.2) are well adapted to the tropical environment by virtue of mainly three attributes:

- **Heat tolerance:** Zebu have a high degree of heat tolerance derived partly from low heat production and partly from better capacity for heat dissipation. A high density of efficient sweat glands increases heat loss through evaporation and the short, sleek coat facilitates convection to the surrounding air. Small body size may by itself also contribute to heat tolerance, as small animals have a higher surface to body mass ratio.
- **Partial resistance to ticks:** They are less susceptible to ticks, and thus to the many tick-borne diseases occurring in tropical countries (e.g. Theileriosis, Babesiosis and Anaplasmosis). Zebu animals have the ability to repel ticks by movements of skin, but this is only part of the explanation of their higher resistance. It has been demonstrated that when animals are infested with tick larvae, fewer larvae develop into ticks in Zebu

than in European-type cattle. Zebu cattle are claimed to possess a certain degree of resistance to many other tropical diseases also.

- **Low nutritional requirements:** Zebu have low nutritional requirement, which is due to their small size, low metabolic rate, and possibly more efficient digestion at low feeding levels.

However, Zebu cattle have a number of unfavorable features as milch animals:

- **Low milk yield:** Milk production of most breeds of Zebu is poor. The cows usually need stimulus for milk let down (e.g. they are stimulated by calf sucking, and therefore adapt poorly to modern milking routines. Failure to milk let down when milked by hand or machine without the presence of calf usually leads to complete cessation of milk secretion and consequently to short or truncated lactations.
- **Late maturity:** Zebu animals are late maturing, both physiologically and sexually, and heat symptoms are weaker than in European cattle. Their calving intervals are also longer.

Classification of Zebu into Breeds

Zebu cattle are classified into a number of subgroups (breeds) according to external traits, such as coat color, size and shape of horns etc. Another classification of Zebu in India is made based on utility whereby they are broadly categorized into the three groups: milch (dairy), dual and draft type.

A large number of zebu breeds of milch-type, draft- type and dual-purpose cattle are recognized in India (29 are registered with National Bureau of Animal Genetic Resources Fig. 18.3) but only a few of them have gained any importance outside their home region (tract).

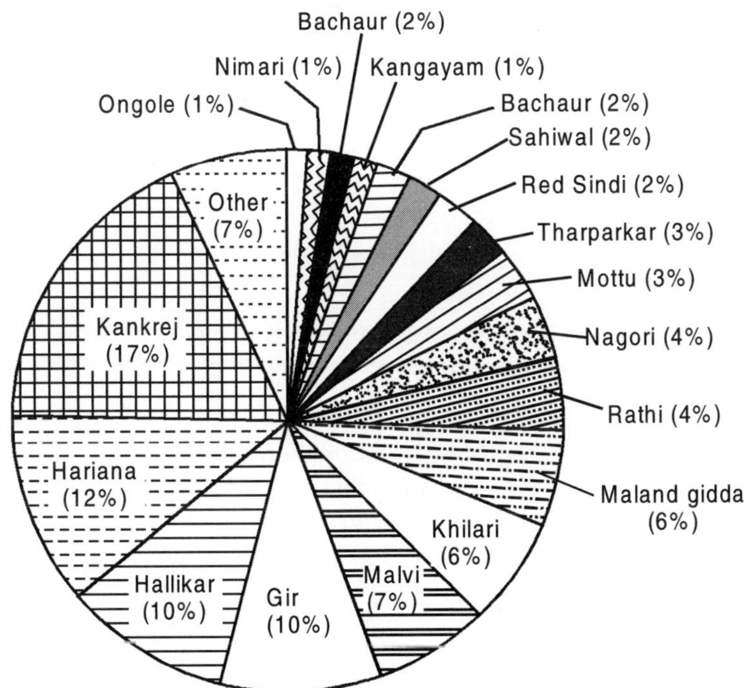

Fig. 18.3 Distribution (%) of total number of animals of recognized breeds of Indian cattle.

Of the various indigenous breeds, Sahiwal, Red Sindhi, Ongole, Gir, Tharparkar and Hariana have been used in organized or systematic crossbreeding programs. The characteristic features of some of the important milch and dual-purpose breeds that have been used in crossbreeding or up grading are discussed below. The important milch breeds (e.g. Sahiwal, Red Sindhi and Gir etc) generally calve between 40 and 50 months and produce 1400 to 2000 kg in lactation with calving interval of 400 to 470 days while the dual-purpose breeds (e.g. Hariana, Tharparkar, Kankrej, Rathi, Ongole etc.) calve later and have a longer calving interval.Some of the important breeds used for milk production are described below and values for their production traits given in Table 18.1.

Table 18.1 Average all lactation performance of some indigenous cattle breeds.

Breed	Age at first calving (months)	Weight at first calving (kg)	Service period (days)	Dry period (days)	Calving interval (days)	Lactation milk yield(kg)	Lactation length (days)	Peak yield (kg)
Sahiwal	42.2	343	148	156	403	2065 (1928-2632)	285 (270-300)	11.1 (10.1-12.1)
Red Sindhi	43.0	341	163	149	433	1839 (1488-2200)	285 (254-330)	11.0 (9.9-12.6)
Tharparkar	43.5	324	128	138	431	1749 (913-2147)	286 (240-377)	10.6 (6.5-11.4)
Gir	51.0	324	222	149	466	1367 (1200-2030)	361 (350-372)	—
Rathi	47.0	248	205	162	516	1573 (1063-2810)	336 (306-430)	8.7 (6.5-10.6)
Hariana	52.2	307	232	255	483	997 (692-1753)	272 (4.8-8.1)	7.3
Kankrej	54.3	—	195	182	487	1850 (423-2369)	304	—
Deoni	46.4	—	170	140	454	816 (284-868)	260 (150-293)	4.5 (2.3-5.8)
Malvi	49.2	—	208	140	507	916	260	—
Ongole	50.2	—	200	289	502	688 (475-1000)	231 (3.2-4.2)	3.5

Sahiwal

The breeding tract of this breed is Montgomery district in Pakistan (North West Punjab) which is now named Sahiwal (many farmers still call this breed Montgomery). This is the best dairy breed of the Indian subcontinent. The breed is a comparatively heavy with a symmetrical body and loose skin as compared to Red Sindhi with which it closely resembles. The main distinguishing feature between Sahiwal and Red Sindhi is the muzzle: Red Sindhi has dark colored muzzle while Sahiwal has lighter colored. The animals are usually long and fleshy and of heavier build. They are colored reddish dun or pale red, sometimes flashed with white patches. Muzzle and eyelashes are light in color. Few good herds of this breed are maintained in India; one such being at Bhaini sahib (in Ludhiana district of Punjab) by Vishav Namdhari Sangat. The milk yield in well-managed herds in is range of 2000 to 2600 kg. per lactation. The age at first calving ranges from three to four years and the calving interval from 400 to 580 days.

Sahiwal animals have been exported to Sri Lanka, Kenya, the West Indies and many other Latin American countries. A new breed called 'Jamaica Hope' has been evolved using Sahiwal- Jersey crossbreeds. The number of animals of this breed has been declining due to its crossbreeding with exotics, mainly Holstein and Jersey. Conservation programs for this breed have been launched by the Government of India.

Red Sindhi

The home tract of this breed is Karachi and Hyderabad districts of Pakistan (Sindh, Pakistan). Few herds of this breed exist in India. The animals of this breed are medium-sized well proportioned with heavy hump, dewlap and sheath. Horns are thick at the base, emerge laterally and curve upwards, and ears are medium sized and drooping. The udder is capacious. Body color is usually red, but shades varying from dark red to dim yellow are also seen. Milk production at institutional herds has been reported to be around 1800 kg per lactation. Red Sindhi animals have also been exported from its home tract to many other countries like Sri Lanka, Tanzania, Philippines, United States, Malaysia, Iraq, Burma and Indo-China.

Tharparkar

The original habitat of this breed is Tharparkar district in the province of Sindh, Pakistan. The animals of this breed are known as 'Thari'. The breed is also found in the adjoining tracts in Rajasthan State of India, particularly around Jodhpur and Jaisalmer where excellent milch specimens are found. They are strongly built, medium sized and well-proportioned body with short, straight and strong limbs. The fore and hindquarters are somewhat drooping. The coat color is white or grey and the switch is black. Animals of this breed are slightly larger than of Sahiwal and Red Sindhi breeds but have similar milk yields. The males are also good draught animals.

Gir

The original home of this breed is Gir forest and areas around it like Junagarh, Bhavnagar, Rajkot and Amreli districts of Gujarat. Though the breed is native to Gujarat (India) but is also found in Maharashtra and Rajasthan. The animals are well adapted to high temperature and high humidity environment. They possess well-proportioned body, prominent and broad forehead; curved horns turning back at the tip, long and pendulous ears; long whip like tail and prominent hip bones. Coat color varies from shades of red and white to almost black and white or entirely red. The Gir cows are good milkers; milk yield ranges from 1200 to 2000 kg per lactation. The age at first calving varies from 43 to 62 months and the intercalving period from 426 to 561 days. Gujarat is estimated to have only around 3,000 pure Gir cows left. This breed has now become a global one as it has been exported to other parts of the world. In Brazil where its large herds are found, it is known as Gyr. Brazil has also evolved a strain called 'Indubrasil', which is a cross between Gir and Kankrej. Gir animals are highly prized by Brazilian breeders who have also created a breed society. The breed has also been exported to USA, especially Texas, Florida and Louisiana states (the southern states characterized by high temperature conditions).

Rathi

The breed is named after a pastoral tribe called 'Raths' who lead a nomadic life. The home tract of this breed lies in the heart of Thar desert. Rathi breed is a mixture of Sahiwal, Red Sindhi, Tharparkar and Dhanni breeds with a preponderance of Sahiwal inheritance. The animals are medium-sized with a symmetrical body and a short and smooth body coat. Their udder is well developed with a prominent milk vein. The females are docile and good milkers.

Hariana

The native breeding tract of this breed covers large parts of Rohtak, Hissar, Karnal and Gurgaon districts of Haryana. It is a prominent dual-purpose breed of North India. Body is proportionate moderately long and compact in appearance. The horns are short, curving upwards and inwards; and ears are small and slightly pendulous. Dewlap is small. The udder is well developed. Color is white or light grey. Bullocks of Hariana are highly recognized for their vigor and persistency for draft. Hariana has been the local counterpart in many crossbreeding programs. Good specimens of cows yield up to 1500 kg of milk per lactation.

Kankrej

The home tract of this breed is Gujarat (Southeast of Rann of Kutch). It is heaviest of all the Indian breeds with powerful body, broad chest, straight back, well-developed hump, thick skin and moderately developed dewlap. These cows are average milkers. Bullocks are fast and powerful draft animals. This breed has been exported to a number of countries to be raised as purebred and for crossbreeding. Excellent herds of this breed are found in Brazil where it is known as 'Guzerat'. Many beef breeds in some Latin American countries and southern states of USA have some inheritance of Kankrej.

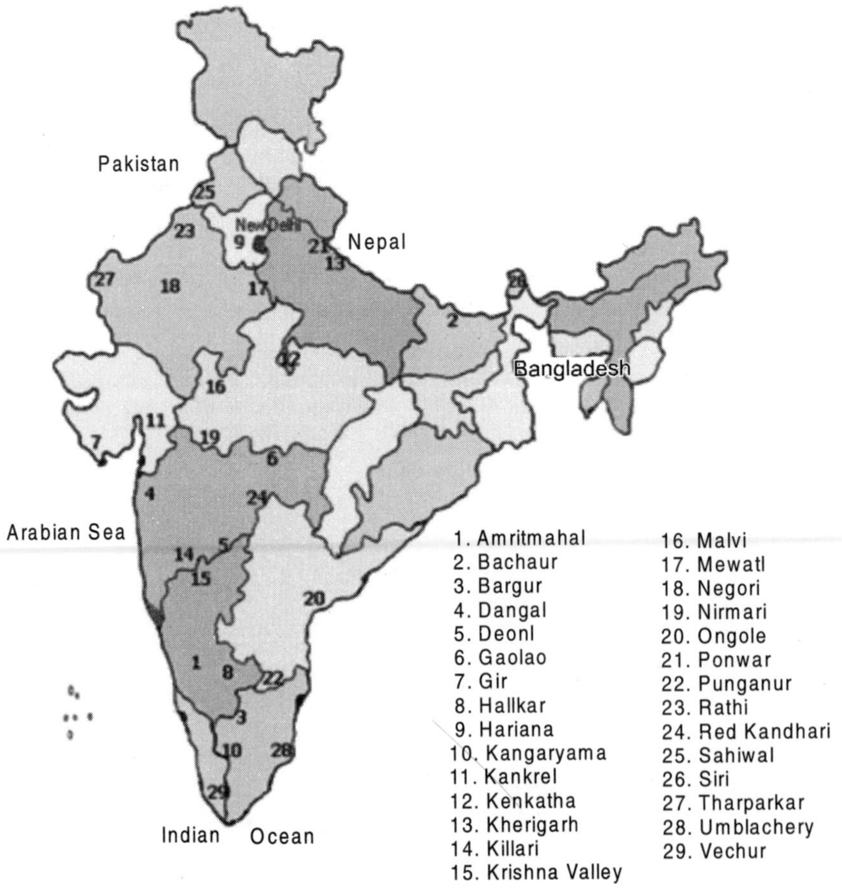

Fig. 18.4 Distribution of some indigenous breeds of cattle in the country

Ongole

This breed is found in some districts of Andhra Pradesh. Animals are large-sized with loosely knit frame and have long body, long ears and stumpy horns. Dewlap is large and fleshy. Color is white with dark grey markings on neck and head. Excellent animals of this breed have been exported to Brazil (known as 'Nellore' in Brazil). This breed has also been exported to other countries (e.g. Fiji, Indonesia, Malaysia and USA). The famous Santa Gertrudis breed developed in Texas State of USA carries its inheritance.

Deoni

These animals have medium sized, forehead, less pronounced but well developed dewlap and sheath. Horns take a characteristic outward and backward curve, and ears are pendulous but short without a notch near the tip. Color is black and white or red and white with irregular patches and spots.

European/Temperate (*Bos taurus*) Breeds of Cattle Used in India

Since the establishment of first herd books most European dairy cattle populations have evolved into clearly defined breeds. Selection for specialized dairy or dual purpose functions has been accompanied by an increase in uniformity of color and other external characteristics within each breed. In the recent past, the number of breeds of *Bos taurus* has declined considerably. More than half of the dairy cows in Europe and North America now belong to one breed type, the Holstein Friesian. Many temperate breeds have been used in crossbreeding programs in tropical countries. The important ones that have been used in India are described below:

Holstein Friesian (HF)

This breed, which originated in Holland, is the most renowned dairy breed of the world. Large-scale exports of this breed from 1860 onwards have led to its distribution in a number of European countries, and in North America. The breed has expanded so rapidly that its cows now account a very large fraction of all dairy cows in the world. In Europe, it is known as 'Friesian' or 'Black and White', and in North America as 'Holstein-Friesian or 'Holstein'. In USA over 90% of the cows in dairy operations are of Holstein. In fact, the inheritance of Holstein has been increasing in populations of most countries, and with increasing globalization, most dairy populations are getting 'Holsteinzed'. In UK's dairy cattle population, for example, the inheritance percentage of Holstein increased from about 20% in 1983 to over 75% in 1995. During the last century, significant differences in production traits between various national populations (strains) of Holstein Friesian have developed due to the selection practiced for different breeding objectives within each such group.

The prominent coat color of the breed is black and white but animals of red and white color also exist. Because of its prominent role in dairy, the Holstein Friesian has undergone intensive selection for dairy traits. This breed is the highest milk producer but its butter fat percentage is comparatively low. The average milk yield of different strains or national populations (in USA) of this breed is about 11000 kg. Age at first calving is between 24 to 26 months. The average fat percentage is 3.7. Average milk production for registered Holstein herds enrolled in the HF Breed Associations programs is higher: 12670 kg of milk, 469 kg of butterfat and 388 kg of protein per year. More than 22 million animals are registered in the Holstein Association's herd book in the USA. Holstein has been extensively used in crossbreeding programs of India.

Brown Swiss (BS)

This breed has its origin in Switzerland (the Alpine regions of several European countries). This breed is known as 'Braunvieh' in Switzerland, Austria and Southern Germany while in Italy it is known as 'Bruna Alpina' and in France as 'Brune des Alpes'. Internationally, the breed is much less numerous than Holstein Friesian. The milk yield of Brown Swiss is somewhat lower than of Holstein Friesian's but it's butterfat percentage is higher (about 4.0%) and it is slightly better in beef traits, and is therefore considered a triple-purpose breed (i.e. milk, meat and draft).US-Brown Swiss cows are large-framed and adaptable to relatively wide range of conditions.

Red Dane (RD)

This breed, which originated in Denmark, is medium to heavy in size with deep red color. Its current level of performance in Denmark is about 8600 kg milk, 4.2% fat. Calving interval is 13 months. Basically, it was intended to be a dual-purpose breed (milk and meat).This breed was used in earlier stages of crossbreeding in India. However, it has the limiting characteristic of pendulous udder that is very pronounced in high yielding and old cows and, therefore, its use in crossbreeding was discontinued.

Jersey

The breed, which originated on the Island of Jersey in the English Channel (UK), is the smallest of the dairy breeds. Based on its export, a number of populations (strains) of this breed now exist in New Zealand, Denmark, Great Britain, and USA. Coat of this breed is fawn colored with shades of brown to dark red. The characteristics of the breed are small body size (mature body weight equals 60% of Holstein Friesian); early sexual maturity; high butterfat (about 4.8%) but low milk yield (less than 9000 kg). They average 400 to 450 kg in weight and are capable of producing 16 times their body weight per lactation. Because of their small size, Jersey requires less feed and thus production costs are less. The breed is also noted for its excellent conformation, reproductive efficiency, early maturity and longevity. Jerseys have the highest fertility and are known for ease of calving. The breed adjusts to all climates, including severe heat and humidity, and adapts to every management system that has been designed for dairying. This breed has been used for crossbreeding in many countries including India. In Australia, it has been used as the basis for development of the 'Australian Milking Zebu' (AMZ), a synthetic 'Zebu × Taurus' dairy breed. Likewise, it has been a major contributor to the inheritance of 'Jamaica Hope' breed.

Ayrshire

A dairy breed of Scottish origin, it is now mainly represented by the Finnish population, though the breed is also found in Canada, USA, Britain and a number of other countries. It was also tried in crossbreeding programs in India during initial stages. The breed does very well under pasture conditions and, when pastures are poor, they need less grain to keep them in condition. Few other breeds can match the ability of Ayrshire to rustle and forage for themselves under adverse feeding or climatic conditions. They respond to good management and feeding practices. Other traits that make Ayrshires attractive to dairyman include the vigor of Ayrshire calves which are strong and easy to raise.

Economic Traits: Description, Measurement and Inheritance

Cattle in India are reared for milk, and in some regions they are used for draft though their role as draft animal has been increasingly declining due to mechanization. Genetic improvement programs of cattle are centered on improving their milk production. Therefore, their economic traits relate, directly or indirectly, to the attributes that contribute to milk production.

Reproductive traits are also important as they contribute to efficiency of milk production and therefore to economic returns. Growth traits are important only in relation to their contribution to milk and reproduction. Type and conformation traits are considered in breeding programs in dairy developed countries where linear scores for these have been developed. Milk production traits make a direct contribution to profitability therefore they receive the highest attention in genetic improvement programs throughout the world.

Milk Production Traits

Milk Yield: The total amount of milk produced, referred to as 'milk yield' or 'milk production', is the most important trait in a dairy cattle breeding program. It meets the three important criteria for inclusion in an effective breeding program viz;

(i) it is economically most important.

(ii) it is accurately measurable.

(iii) it is heritable enough to yield considerable genetic progress from selection.

The most important measure of milk yield/production (potential) of a cow is the amount of milk produced in a lactation (referred to as 'lactation milk yield' or simply as 'lactation yield') or in a standard lactation of 305 days (referred to as 305-day milk yield):

- **Lactation milk yield**: It is the total milk produced by a cow in a complete lactation (i.e. from calving until the cow dries off). Since cows with longer lactation length produce more milk, lactation yield *per se* does not allow a fair comparison of milk yield of different animals within the herd.
- **305-day milk yield**: It is the amount of milk produced in a standard lactation of 305 days. Conversion factors (as discussed in Chapter 2) are used to convert the actual milk yield of a cow 305-day lactation.

Heritability of milk yield is about 0.25, therefore, the trait can be improved by selection. Cows can be selected on the basis of their own milk yield while the sires can be preliminarily selected on their dam's yield, and subsequently progeny tested. Such selection would be effective in genetic improvement provided effective selection pressure can be exerted. Such selection schemes have been used in dairy developed countries.

Genetic improvement programs of dairy developed countries are well organized through involvement of a number of agencies with large sized herds. For example in USA, genetic evaluation program is coordinated through National Cooperative Dairy Herd Improvement Program (NCDHIP) and participation of breed associations, AI organizations, dairy record processing centres, milk testing labs and animal improvement program's laboratory of United States Department of Agriculture (USDA). Recording procedure involves supervisors (testers) who visit each participating herd once each month (monthly recording). Supervisors record milk yield of each cow of the herd, and collect samples of milk to be tested for milk composition alongwith cow identification and other information such as drying dates, breeding dates, freshening dates etc. This information is then fed into the computer, and these records are then used by the

US department of agriculture (USDA) to calculate breeding values of sires based on progeny performance. USDA publishes the performance values as well as breeding values for the cows and bulls from 1957 onward.

Because of selection, mainly selection of sires, milk yield has made tremendous genetic improvement during the last several decades (Fig 18.5). Further, the milk yield, despite the consistent selection applied for it for the last 50-60 years in developed countries, continues to respond to selection with no evidence to suggest that the trait is approaching any genetic limits. Heritability of milk yield remains as high as in beginning. In the US, the important dairy breeds like HF, Brown Swiss, Ayrshire and Jersey have been genetically improving at a rate of at least 100 kg per year. Sound genetic evaluation procedures coupled with high selection intensity made possible by AI and embryo transfer technology,and shortening of generation interval through use of younger cows and some virgin heifers as bull dams, the rate of genetic improvement has enhanced.

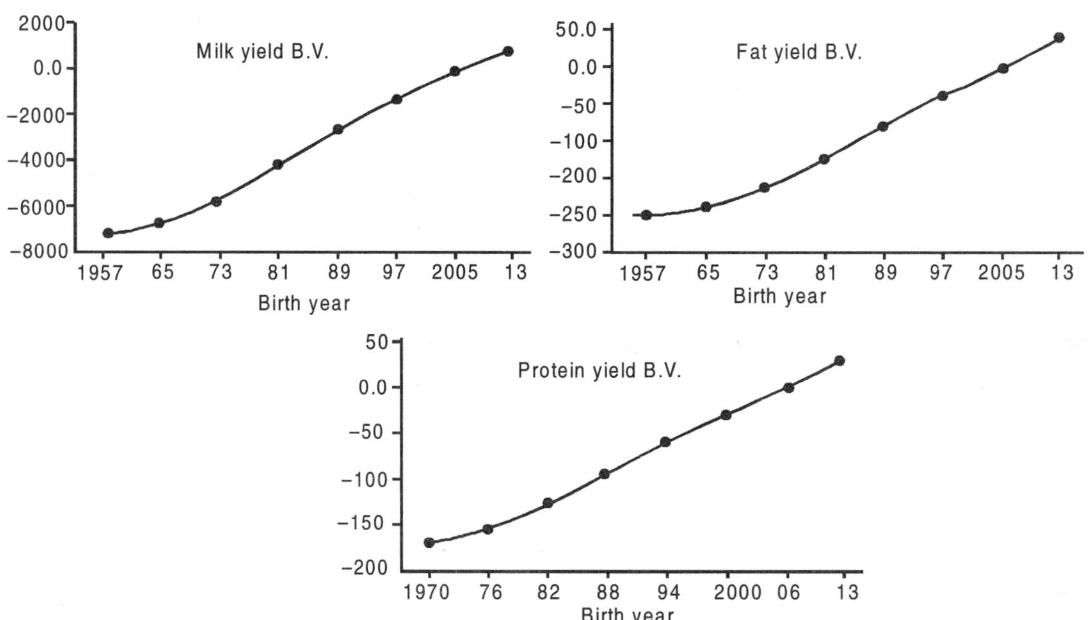

Fig. 18.5 Genetic improvement in US Holstein or Red and White population for milk yield and yield of fat and protein (breeding values have been computed using different base years).

Peak yield: Immediately after parturition, daily milk yield shows a pattern of increase, stability and decline during the lactation period. This pattern is expressed through a curve, called lactation curve. A typical lactation curve shows that milk production starts increasing, it reaches a peak, remains stable and then starts declining until the animal dries off. (Fig. 18.6). In a breeding herd, milk yield is recorded from 5th day onward (if not weaned) and from day of calving onward (if weaned at birth) until the animal dries off. The highest amount of milk produced by a female on a given day in lactation is referred to as peak yield (Fig 18.6). Typically, a crossbred dairy cow attains peak yield at around 50 to 60 days after calving that is referred to as 'days to reach peak yield'. Peak yield, though not of any direct significance, is considered an indicator of milk production potential of the cow.

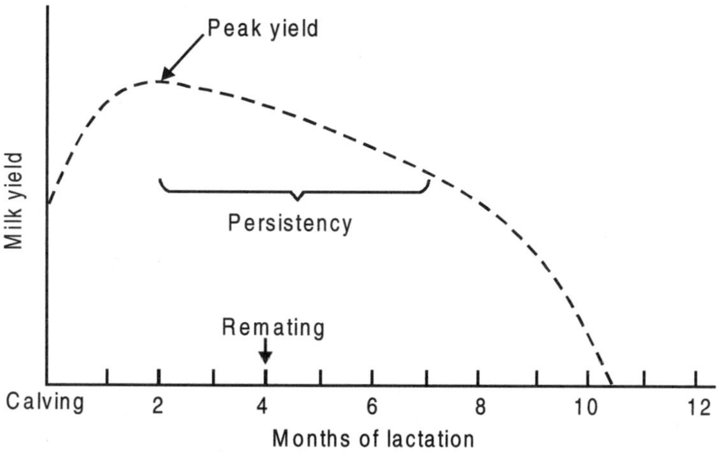

Fig. 18.6 A typical lactation curve

Persistency: After attaining peak, lactation curve starts a declining trend until the animal dries off. Persistency of milk is defined as the degree with which milk yield is maintained (after reaching peak) during declining phase of lactation (Fig 18.6). It is expressed as the average percentage of decline in milk yield each month to that of previous month. Females with greater persistency usually experience less energy deficit in early lactation and therefore less health problems.

Lactation length: It is period (the number of days) for which the female is in milk i.e. from day of calving until the animal dries off.

Peak yield, persistency, lactation length that, which low heritabilities, are not included in genetic improvement programs as some of these improve as correlated response to selection for milk yield while others can be manipulated through improved herd-management practices. Milk yield increases over lactations with the highest yield attained between 2nd and 3rd lactation.

Milk Composition Traits

Milk compositional traits (fat %, protein %, solids-not-fat % including total yield of these constituents in a lactation) are also important in dairy cattle. In Indian market, these traits had been considered relatively less important but are now becoming important as premiums are paid for milk with higher percentage of these components. These traits can be accurately measured and evaluated through automatic testing devices.

- **Fat, protein and solids-not-fat yield**: The total amount of fat produced in lactation (usually from day sixth until the animal dries off), is obtained by multiplying average fat percentage with the total milk produced in a laitation and dividing it by 100. Similarly, protein and solids-not-fat yields are calculated.
- **Fat, protein and solids-not-fat percentage**: Average percent of fat, protein and solids-not-fat in milk in a lactation are recorded through automatic measuring devices by taking milk samples at certain defined intervals in the lactation. Fat % usually increases continuously from 1st to 10th month of lactation.

Heritability of absolute yield of these components is high (about 0.35) while that of their percentages (e.g. fat percentage, protein percentage etc) is very high (0.5). Genetic correlation of milk yield with yield of fat, protein etc is positive and high but with their percentages it is generally low and negative. That the yield of these milk constituents increases due to selection for milk yield is evident from the US Holstein improvement results (Fig 18.5)

Reproductive Traits

Efficient reproduction is of considerable importance in a breeding herd for two reasons:

(i) Cows calving earlier and regularly have higher lactation as well as lifetime milk yield. Cows that calve at regular intervals spend a greater portion of their productive lives in the higher portions of lactation curve. Therefore they produce much more milk per day of productive life than cows with longer inter calving intervals.

(ii) Efficient reproduction also reduces the generation interval thus helping early genetic evaluation of breeding candidates, and it allows for higher selection accuracy.

Most of the reproductive traits can be accurately measured. Components of reproductive efficiency of a cow include a number of component traits:

- **Services or inseminations per conception:** Pregnancy rate, calculated as number of services or inseminations per conception, is a measure of fertility of female, and includes all the activities associated with fertility of the cow from heat expression to detection to conception itself.

  Heritability of fertility is quite low (0.05) which means that many non-genetic factors, in particular heat detection, health of cows, AI techniques etc, influence it.

  Genetic correlation of milk yield with fertility is low but undesirable. Continuous selection for milk yield would lead to a decline in fertility. Results of selection for milk yield in US Holstein show that direct selection for milk yield led to decline in conception rate between the years 2000 and 2007 (Fig 18.7). But once selection for fertility, alongwith milk yield, was initiated, fertility improved as is evident from the results from the year 2007 onward (Fig 18.7).

- **Age at first Calving**: It is defined as the age of female when she gives birth to her first offspring, and is recorded as the number of days or months from her birth to first calving. Age at first calving is dependent on two attributes: age at first heat, which is the number of days or months from the birth of the female till her first signs of heat are noticed, and the age at first conception which is recorded as the number of days or months from the birth till the cow first conceives or becomes pregnant. Age at first calving is important in determining the lifetime productivity of cows. A female must attain the required body weight and height (stature) for conception and calving. Herd management practices in particular nutritional regimen are important in attaining the desired weight.

- **Service period or days open**: Service period, which is period (time) from calving to conception, is the sum of two components: time from calving to first insemination and the time from first insemination to a successful conception. Service period explains a major part of variation in inter calving interval between animals in a herd/breed.

- **Dry period**: It is the period (usually expressed in days) from the date the female dries off to the next calving (freshening). Dry period is a time for the cow and her mammary glands to rejuvenate and prepare for the next lactation. Ideally, a good crossbred cow should have a dry period of 60 days.

- **Calving interval (Inter-calving interval)**: It is the period (in days or months) from one calving to the next (i.e. period between two successive calvings).It is largely dependent on how quickly a cow conceives after calving. The longer a cow is open (not pregnant), the longer the calving interval. First calving interval is usually longer than subsequent ones.

Heritability of these traits related to reproductive efficiency is low. Therefore, these are usually not included in selection program as most of these can be modified through improvements in nutrition, health practices and general herd management.

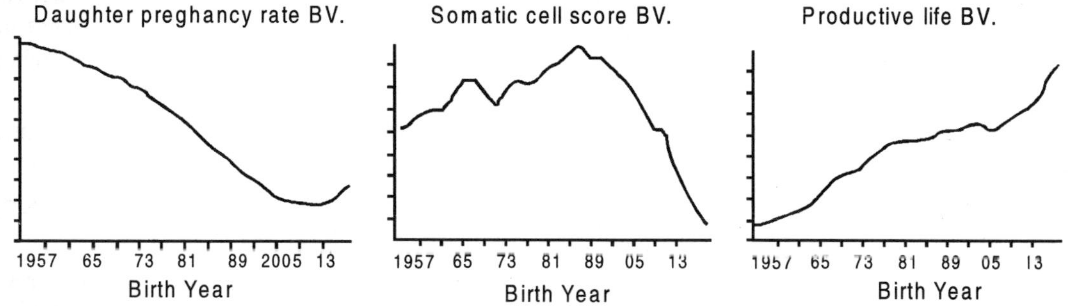

Fig. 18.7 Genetic changes (breeding values) in US Holstein (or Red and White) population for conception rate, somatic cell score and productive life(breeding values have been computed using different base years).

Other Traits

Longevity, lifetime milk production and productive life: Length of herd life is important because a high replacement rate increases replacement costs and decreases the proportion of herd in high producing age group. Longevity is expressed as the period from birth to death or disposal from herd. Sometimes it is also expressed as the period from first parturition to death or disposal from the herd. Similarly, the total milk produced by a female in all her lactations, called lifetime milk production, is a measure of cow's profitability. These traits are lowly heritable (about 0.10). Direct selection for longevity, productive life or lifetime production is possible but is made difficult because these traits are not known until the cows are culled or die (late expressed) and their heritabilities are low. The genetic correlation of milk yield with longevity or productive life is positive, though low, therefore selection for milk yield largely improves productive life. Results of selection for milk production in US Holstein show that productive life has improved as a correlated response (Fig 18.7).

Susceptibility to mastitis: Mastitis usually occurs later in life of the cow after she has produced one or more lactation. It is expressed on individual cow-basis as whether or not a cow experiences a bout of mastitis during the lactation. Indirect measure of mastitis includes somatic cell count or score (SCC or SCS) in the milk averaged over lactation. These counts range from a few thousand to over 10 million somatic cells per milliliter of milk. Heritability of somatic cell count is low (0.05 to 0.10). The genetic correlation of milk yield with mastitis (measured as somatic cell count) is positive and moderate (about 0.2) which is undesirable. Therefore, selection for milk yield slowly increases the susceptibility to mastitis. Selection for milk yield in US Holstein from 1984, when measurements on somatic cell counts were initiated, started increasing somatic cell count till around 2000-2001 (Fig. 18.7). Subsequently when concomitant selection for SCC through genetic evaluation of sires based on their daughters was started, somatic cell count started declining though milk yield continued to improve (Fig 18.7).

Type and Conformation Traits

Type and conformation traits are currently receiving considerable emphasis in dairy cattle breeding in advanced countries like USA, Canada and Europe. Procedures have now been developed for objective evaluation of type and conformation traits (linear-type scores). Each cow in US breed association programs is evaluated for several linear type traits at one time by a trained evaluator from the respective breed associations. These traits are scored from one biological extreme to another. Some of the type traits like those of udder relate to milk production (like fore-udder attachment, rear-udder height, rear-udder width, udder depth, front teat placement etc), while others (like those related to feet, legs, size etc) relate to overall productivity of the cow. Most of the linear type traits have heritability in the range of 0.15 to 0.20. Therefore, these are amenable to selection. Their genetic correlations with milk yield are low but if ignored they tend to deteriorate due to selection for lactation milk yield.

A schematic illustration of some traits related to reproductive cycle of a cow is shown in Figure 18.8

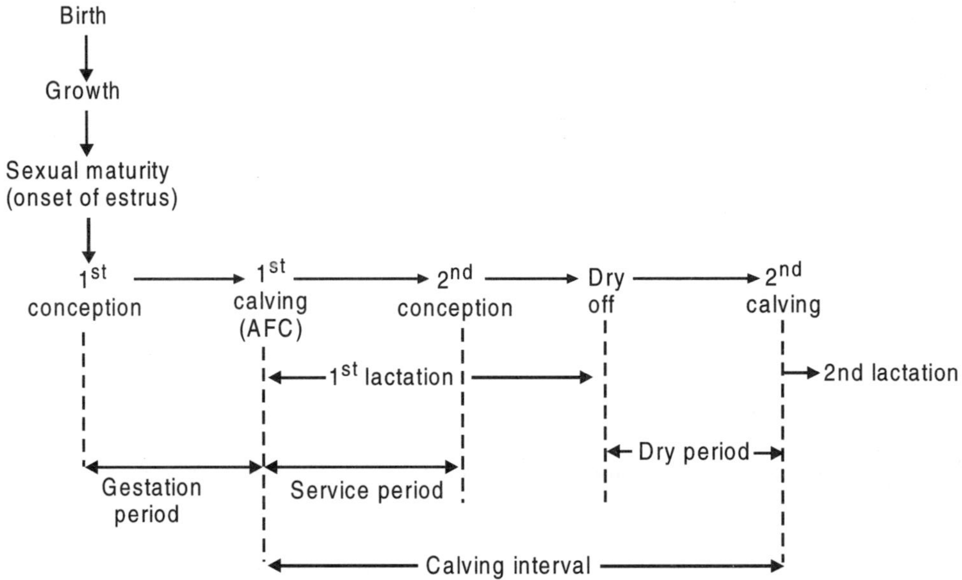

Fig. 18.8 Illustration of and measurement of some important traits of reproductive life of dairy cow.

Overview of Genetic Parameters

Heritabilities:

- Milk yield and yield of milk constituents (fat, protein and solids-not-fat yield) are moderately heritable traits (h^2 of about 0.25). Moreover, these can be measured accurately. Therefore, they are easily amenable to selection.
- Percentages of fat, protein and solids-not-fat have high heritabilities (about 0.5).
- Reproductive traits are lowly heritable therefore they frequently receive less importance in genetic improvement programs. However there is some automatic selection for improved reproductive efficiency occurs since cows that do not reproduce are usually culled from the breeding herd, therefore, they contribute lesser progeny to next generation.
- Lifetime traits like lifetime yield of milk, fat, protein, days of productive have lower heritability than milk yield (about 0.15).
- Health related traits like incidence of mastitis, incidence of ketosis, incidence of retained placenta etc have very low heritabilities.

- Linear type and conformation traits like fore udder attachment, rear udder height, rear udder width, udder depth, front teat placement etc are moderately heritable ($h^2 = 0.25$)

Genetic correlations:

- Yield of fat, protein and solids-not-fat in lactation are positively and highly correlated with lactation milk yield at the genetic level. Therefore, they automatically improve through selection for milk yield..
- Percentages of fat, protein and solids-not-fat have negative genetic correlations with lactation milk yield (about 0.20).
- Milk yield has undesirable genetic association with mastitis and fertility.

Genetic Improvement of Cattle in India

There are two to three broad approaches for genetic improvement of cattle in India:

- Selection within the indigenous breeds
- Crossing (crossbreeding) of indigenous with exotic milch breeds including up grading
- Development of new breeds/strains from the crossbred foundation

Selection within Indigenous Breeds

The Indigenous breeds of cattle are late maturing and their genetic potential for milk production is low, as they have not undergone selection for milk production traits. These breeds show wide variation in performance due to large differences in managemental practices, the geographical and agro-ecological conditions existing in their home tracts etc. Milk yield also shows wide variation among breeds, and among herds within the same breed.

Selection programs for improvement of some of the indigenous breeds were attempted in institutional herds. These programs were confined to small-sized single herds of well-defined dairy and dual cattle breeds. In most organized herds, which have been closed for varying periods with some kind of selection pressure on males and females without any systematic progeny testing, the annual genetic trends for first lactation milk yield and other important traits have been very small, and sometimes negative as well.

Reasons for low progress: The reasons for little, no or negative genetic improvement has been:

- Small herd-size and therefore absence of rigorous selection of males and females.
- Poor replacement rate and more involuntary culling of cows on the basis of traits other than low milk production.
- Unplanned breeding program including use of genetically inferior animals.

Crossbreeding with Exotic Breeds : Rationale

Crossbreeding over selection within indigenous breeds: There were some critical limitations of selection within indigenous breeds for genetic improvement of productivity, particularly in comparison to crossbreeding viz.

- Milk yield of most breeds of indigenous cattle, barring a few ones, was very low.
- There was lack of performance recording with farmer's animals, and the spread of artificial insemination was very poor.

On the other hand, crossbreeding offered a tremendous jump in productivity as compared to selection within indigenous populations. Improvement of indigenous breeds through selection has been attempted from time to time, but the rate of genetic progress has been very low or negligible. Considering an annual genetic improvement rate of 1% of the mean value through selection, the doubling of milk production of indigenous cattle would have taken over 100 years (i.e. from 1000 kg to 2000 kg per lactation). Compare this with one generation of crossbreeding assuming milk yield of indigenous as 1000 kg and of exotic as 5000 kg :

Indigenous breed (females) X Exotic breed (bulls/semen)
(Avg. yield = 1000 kg) ↓ (Avg. yield= 5000 kg)

Crossbred cows
(Avg. yield = 3000 kg)

The average yield of crossbreds is simple average of indigenous and exotic and hence excludes the additional benefit of heterosis.

Therefore to meet the demand of growing Indian population, the best option appeared introduction of superior germplasm (for milk production traits) through crossing with indigenous breeds, up grading of indigenous breeds with exotic, and finally the development of a suitable new breeds/strains.

Crossbreeding Programs

In view of the limitations of selection within indigenous breeds of cattle, introduction of exotic cattle breeds (temperate breeds of *Bos taurus* type) for crossing with local cattle or upgrading of local cattle with dairy breeds of temperate countries was initiated in the earlier part of 20th century. The primary purpose of crossbreeding was to introduce genes for higher milk productivity resulting in much faster improvement in productivity than could be achieved through selection within indigenous pure breeds. The added advantage of crossbreeding was the capitalization of heterosis resulting from crossing of widely divergent breeds (i.e. crossing of *taurus* with *indicus*).

Initial crossbreeding experiments had set backs due to outbreaks of rinderpest disease and other tropical diseases to which exotic (European) breeds are particularly susceptible. Subsequent control of these diseases with prophylactic vaccines allowed planned crossbreeding experiments to be taken up.

Crossbreeding work in India was initiated through a number of agencies at various places viz; Military Dairy Farms ; Naini Agriculture Institute, Allahabad(UP);Bilateral projects ;National Dairy Research Institute, Bangalore and Karnal (Haryana) and All India Coordinated Research Project (AICRP) on cattle, and at some state agriculture universities.

- **Military Dairy Farms:** The most extensive crossbreeding work has been carried out at Military Dairy Farms which were established by the British for supplying milk to the troops and their families. The work was started in 1909 at Allahabad and later on extended to other Military Farms. The exotic European cattle breeds used were Shorthorn, Ayrshire and Friesian (1907-1925) which were crossed with native breeds (Sahiwal, Tharparkar, Red Sindhi and Gir). The crossbreds resulting from Holstein were better to crossbreds produced from other exotic breeds; therefore, crossbreeding with Friesian was continued.
- **Naini Agriculture Institute, Allahabad:** Crossbreeding of Kankrej, Gir, Hariana, Sahiwal and Red Sindhi with exotic breeds Holstein Friesian, Brown Swiss, Jersey and Guernsey was started in 1924.On the basis the results obtained from different crosses, it was found that Jersey-Red Sindhi crosses were suitable due to their better adaptability. The breed/strain that resulted from this crossing and selection was named 'Jersind'.

- **Bilateral Projects:** Many bilateral projects were started in different states in early 1960's.These included Indo-Danish project at Hessarghata (Karnataka) in 1961, Indo-Swiss project in Kerala (in 1963) and Punjab(in 1971), Indo-German project in Himachal Pradesh and Utter Pradesh (in 1963).
- **National Dairy Research Institute, Karnal (Haryana):** Crossing of exotic breed Brown Swiss with Sahiwal and Red Sindhi was initiated during early 1960's. Over a period of time, selection and inter se mating was practiced among the crosses and a new strain carrying about 50% Brown Swiss inheritance,named 'Karan Swiss', was evolved.
- **All India Co-ordinated Research Project on Cattle (AICRP cattle):** Systematic crossbreeding work in the country was initiated with launching of the All India Coordinated Research Project (AICRP) by the Indian Council of Agricultural Research (ICAR) in 1969. Long term objective of the project, which involved five centres in the country, was synthesis of new dairy breeds/strains adaptable to different agro-climatic conditions of the country. The exotic breeds used were Friesian, Jersey and Brown Swiss. The indigenous breeds used were as follow:
  (*i*) *Hariana* at Indian Veterinary Research Institute, Bareilly(UP) and at Haryana Agricultural University(Haryana)
  (*ii*) *Gir* at Mahatma Phule University of Agricultural Sciences, Rahuri (Maharashtra) and Jawaharlal Krishi Vishawa Vidyalaya, Jabalpur (Madhya Pradesh)
  (*iii*) *Ongole* at the Cattle Breeding Farm attached to Andhra Pradesh Agricultural University at Lam (Andhra Pradesh).

These crossbreeding experiments clearly demonstrated that the crossbreeds had significantly lower age at first calving and calving interval, and about two- to three-fold higher milk yield, depending upon the level of exotic inheritance and the availability of inputs and climatic conditions, than the indigenous breeds.

Considerations in Crossbreeding Programs

Crossbreeding involving European dairy cattle breeds (HF, BS and Jersey) and Indian dairy, dual and draft type breeds has been carried both at organized farms as well as under farmer's management conditions. The voluminous data collected on performance traits (like milk yield, age at first calving, calving interval, lactation length) of different categories of crossbred cows generated from different organized herds in the country, mainly at the Military Dairy Farms, had been subjected to statistical analysis. Conclusions with regard to the breeds to be used in crossing, the level of exotic inheritance, system of maintaining exotic inheritance at a particular level (viz. through forward and backward crossing) and making further genetic improvement have been drawn from the results of analysis. The main considerations,and conclusions drawn from these exploratory crossbreeding are discussed below.

1. **Choice of Indigenous breeds:** The indigenous breed to be used for crossbreeding should be the one that is well adapted to the geographic or agro climatic conditions of the state or region in which the program is to be started. Nonetheless, the level of milk yield of the indigenous breed should also be taken into consideration as it would permit a higher average level of milk production to be obtained in the F_1 generation (crossbreds). For example in Punjab, the most suited high-producing indigenous breed was Sahiwal and was thus used in crossing. It is also important that the animals of indigenous breed to be used in crossing should be properly characterized

before starting the crossbreeding program, and only the ones having suitable traits be used rather than use of or a random set of animals.

Although the non-descript locally available animals are usually better adapted to local low-input conditions but their use in systematic crossing is not desirable because:

(i) It will also give rise to large amount of undesirable type of animals in the subsequent (segregating) generations.

(ii) The milk yield of F_1 will be less as compared to crossbreds resulting from defined indigenous breeds.

The local non-descript cows which constitute a large proportion of indigenous cattle can be up graded with suitable indigenous breeds or mated to crossbred bulls.

2. **Exotic breeds to be used:** The choice of the exotic breed(s) to be used for crossbreeding should ideally be made by trying out different exotic breeds with indigenous breeds, making the possible types of crosses, and then choosing the best combination of breeds. Such a program is however prohibitive. In the absence of such a program, information obtained from past crossbreeding experiments involving such breeds can be used as the basis for choosing the breed. A number of exotic breeds were tried for crossing with indigenous breed (s) in the crossbreeding projects undertaken by different institutes and universities. The general inferences from these were:

   (i) Performance of crossbreds resulting from crossing of Holstein, Brown Swiss, and Jersey with the important indigenous breeds ranked the exotic breeds in the following order for milk production traits: HF, Brown Swiss and Jersey.

   (ii) Holstein crosses were specifically superior to other temperate breed-crosses for growth and milk production while Jersey crosses had better reproductive efficiency.

   (iii) Red Dane crosses which were produced at some universities/institutes had pendulous udder, therefore, use of this breed was discontinued from later crossbreeding work.

   In view of the results obtained from crossing, the following general recommendations were made:

   (i) Heavy breeds (HF and Brown Swiss) were recommended for areas where physical environment was favourable and feed and fodder supply was abundant.

   (ii) Jersey was recommended for areas of low input conditions (i.e. areas of shortage of feed and fodder) like Hilly and Coastal regions.

   Although different breeds have been recommended for different areas or managemental conditions but the farmers choose the exotic breed depending upon their liking and availability of semen. For example, the crossbreds of Brown Swiss provide a good combination of milk yield and fat percentage but the farmers prefer HF. And with the development of irrigation facilities for fodder production and farm mechanization in many parts of the country, the HF breed has virtually replaced other heavy exotic breeds.

3. **Level of exotic inheritance***:* Crossing of exotic bulls (semen) with indigenous cows to produce F_1 (half-breds), and the subsequent crossing in various combinations of exotic and indigenous led to the production of animals carrying variable levels (up to 90%) of exotic inheritance. The results from the analysis of comprehensive study gave the results, which are plotted in Figure 18.9 for lactation milk yield, age at 1st calving, calving interval and lactation length.

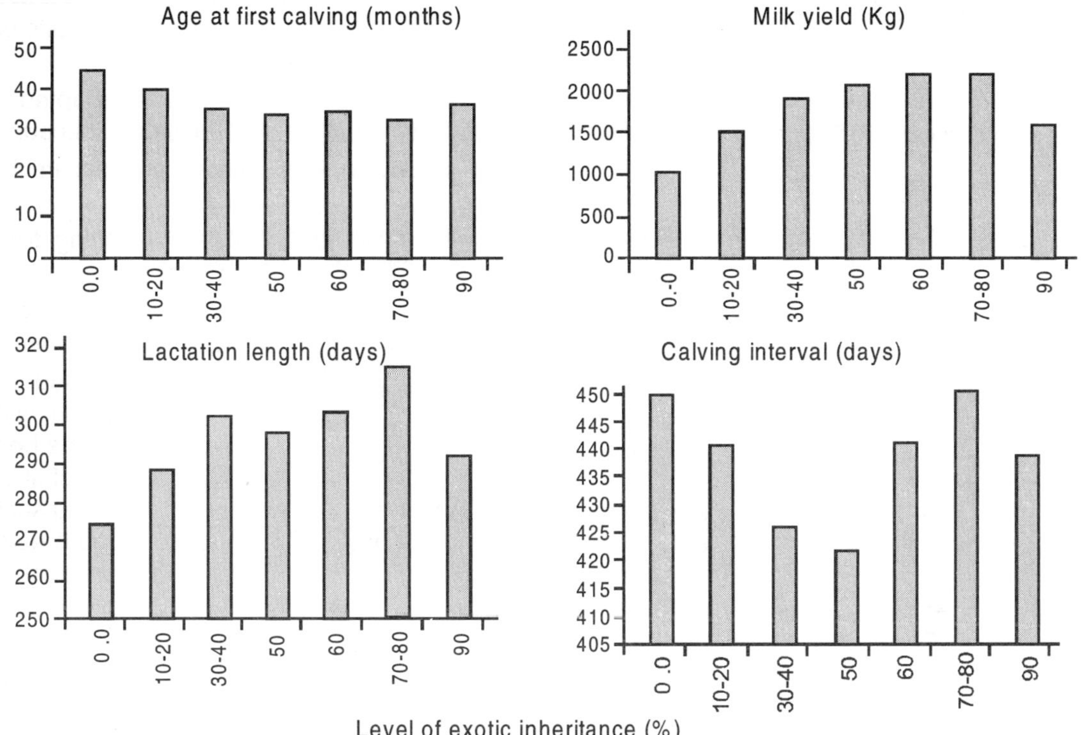

Fig. 18.9 Performance values for different levels of exotic inheritance.

General conclusions: The general conclusions from this extensive analysis with respect to level of exotic inheritance have been:

(i) In general, there has been consistent improvement in milk production and reproduction up to 50 to 60% level of exotic inheritance.

(ii) Beyond 60% exotic inheritance, there has been little improvement in overall productivity.The animals carrying higher proportion of exotic inheritance had poorer adaptability and such animals required special care and feeding. Therefore the ideal level of exotic inheritance for improvement of overall productivity was recommended to be between 50 %(half-breds) and 60% (about 5/8). In states like Punjab, Haryana etc. where good number of progressive farmers are able to provide high input conditions, animals carrying higher levels of exotic inheritance are performing very well.

Contribution of additive vs. non-additive effects: The two genetic effects that contribute to improvement through crossing are: increase resulting from additive genetic merit of exotic breed and the increase contributed by heterosis (non-additive effects). Analysis of the crossbreeding data gave the following results:

- Average additive genetic differences between exotic (HF, BS and Jersey) and indigenous breeds (measured as difference in performance of exotic breeds and indigenous breeds) were as follow:

Milk yield (kg)	Age at 1st calving (months)	Calving interval (days)	Lactation length (days)
1100	–12.9	3	52

- Heterosis computed as the superiority of crossbreds (F_1) over the average of parental breeds was estimated to be of the following order:

Milk yield (kg)	Age at 1st calving (months)	Calving interval (days)	Lactation length (days)
453	–5.6	–28	7

Therefore, collectively crossbreeding gave an immediate jump to milk production traits viz; increased milk yield, early age at calving, shorter calving interval and longer lactation length.

How to produce '5 / 8'-bred Animals ?

Based on the findings of the exploratory crossbreeding experiments conducted in the country, about 5/8 (about 60%) level of exotic inheritance was found suitable in Indian conditions. These types of animals are produced as per the following breeding plan:

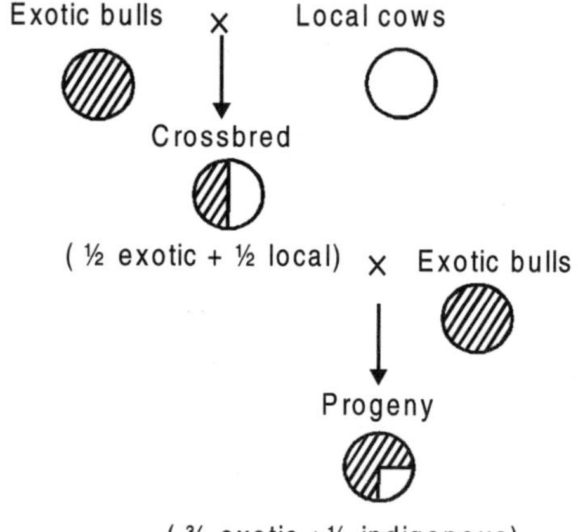

The mating between half-breds and ¾ crossbreds would produce animals carrying 5/8 exotic inheritance, as shown below:

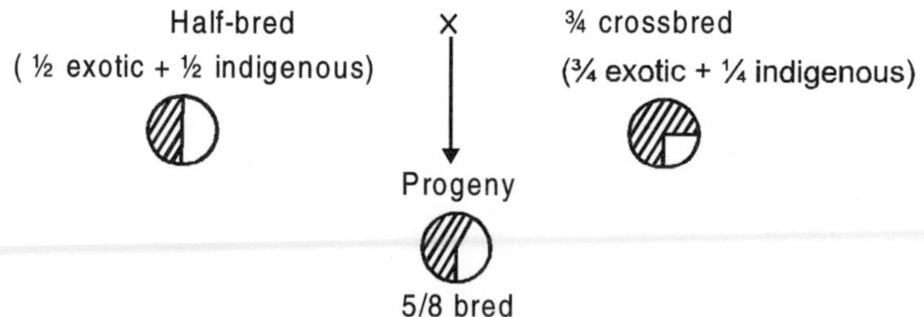

It is apparent from the procedure used for production of cows carrying '5/8' exotic inheritance that such animals are difficult to be produced, particularly at farmers field. On the other hand, it is much easier to produce halfbreds provided purebred animals of indigenous breeds are available.

However with the crossbreeding going on in the country for the past several decades, purebred animals of recognized breeds are not available as majority of the animals now are crossbreds. Therefore, the level of exotic inheritance may be restricted to 50 to 75%, in particular, for farmers who do not have large-sized organised farms. This level of exotic inheritance can be maintained by using semen from progeny tested proven crossbred bulls on crossbred females. However, in states where large-sized organized dairy farms with high input resources are available, cows with higher levels of exotic inheritance are proving successful. For example in Punjab, progressive dairy farmers having large-sized farms with modern housing and management practices are maintaining crossbred cattle carrying more than 90% exotic inheritance. Moreover, to enhance the genetic productivity of their herds, such farmers, who have formed some associations (one such being called progressive dairy farmers associations PDFA), are importing pure HF sexed semen of high genetic merit bulls from dairy developed countries.

New Strains (Breeds) Synthesized from Crossbreds

Development of a new strain/breed, as a rule, involves crossing of two or more breeds, interbreeding among the crossbreds followed by selection for the desired traits (breeding goals). In cattle interbreeding among crossbreds has mostly been done at 50% level (in case of 2-breed cross).In a few cases interbreeding was done at 5/8 exotic (2-breed cross) and ¾ exotic (3-breed cross) inheritance. Since the start of earliest crossbreeding project in the country, a few strains/breeds have been developed:

- **Taylor:** Taylor breed of cattle is said to have been evolved near Patna using crosses of Shorthorn, Channel Island and Ayrshire bulls (from UK) with local cows by Taylor (an Englishman, the then Commissioner of Patna division) in 1856. Neither published performance records nor the animals of this breed are now available.
- **Jersind:** The exotic breeds HF, Brown Swiss, Jersey and Guernsey were crossed with Sahiwal, Kankrej, Gir, Hariana and Red Sindhi at Agriculture Institute, Naini, Allahabad (UP) in a crossbreeding project launched early in 1920's. Red Sindhi and Jersey crosses were found to have desirable attributes suited for local conditions. The Jersey crossbreds between 3/8 and 5/8 exotic inheritance were interbred and named as 'Jersind'. Jersind cows yielded between 1557 and 1861 kg in first lactation. Over years, the breed has deteriorated due to small population size, and is currently confined to the institute farm.
- **Karan Swiss**: It has been evolved at National Dairy Research Institute, Karnal, by crossing of American Brown Swiss bulls with Sahiwal and Red Sindhi cows. The Brown-Swiss inheritance in Karan Swiss is around 50 per cent. The color of breed is red dun and it resembles Sahiwal in body size and general appearance. The hump is almost non-existent, the barrel is long and deep, the naval flap is from tight to slightly loose. The average age at first calving was 32 months, the first lactation yield about 2600 kg with 4.2 to 4.4%. Average yield based on several lactations has been reported to be 3257 kg with overall calving interval of 396 days.
- **Karan Fries**: The breed was developed by crossing of Holstein Friesian with indigenous breeds Tharparkar, Sahiwal and Red Sindhi at National Dairy Research Institute. The breed which has 3/8 to ½ Friesian inheritance has black patched coat. The animals are extremely docile.
- **Sunandini**: This breed is the outcome of the crossbreeding program conducted under Indo-Swiss project in Kerala.In the beginning of project, the local non-descript cows were crossed with Brown Swiss bulls, and subsequently crosses with 50%, 75% and 62.5% Brown Swiss inheritance were produced. The crossbreds with 62.5% Brown-Swiss inheritance were mated inter-se followed by selection to synthesize the breed/strain named 'Sunandini'. In later generations, some inheritance of HF, Sahiwal, Gir, and Kankrej breeds was also incorporated. For over forty years, the male and female of this mixed breed had been mating among themselves (genetic isolation) under a well-planned breeding program with moderate to intense selection pressure. Sunandini bulls are being progeny tested using performance recording under field conditions. In

B

the current development program of Sunandini, there is no further infusion of Brown Swiss while there are increasing infusions from HF and Jersey according to changing preferences of dairy farmers.

- **Frieswal**: The crossbreds produced by crossing of Holstein Friesian with Sahiwal and having Friesian inheritance between 3/8 and 5/8 at the Military Farms (in collaboration with Project Directorate of Cattle, Meerut, of Indian Council of Agricultural Research) were interbred with 5/8 Friesian inheritance for the formation of a new breed/strain. The Frieswal bulls are being progeny tested for further genetic improvement of this breed. More than 10000 cows of this breed/strain are under testing program. The Frieswal project targets to evolve a national milch breed yielding 4000 kg of milk with 4% butterfat in a mature lactation of 300 days. This work is continuing utilizing the existing crossbred herds available at Military Farms of the country.
- **Vrindavani**: This strain/breed of cattle was developed as synthetic crossbred with exotic inheritance of Holstein Friesian, Brown Swiss, Jersey and indigenous inheritance of Hariana cattle. It carries exotic inheritance of 50 to 75% with the remaining (25 to 50%) from the indigenous Hariana. The synthetic has been stabilized with inter-se-mating. Further selection is continuing. Vrindavani being a composite exhibits almost all possible coat colors though predominant being brown. Head is clean cut, well proportionate with prominent poll and concave forehead. Ears are of medium size, latterly oriented with round edge. The animals are docile to moderate in temperament. The strain has been found to be popular among dairy farmers of UP, Uttarakhand, Rajasthan and some adjoining areas.

Most of the strains/breeds developed from crossbred base have existed at Institutional farms, and that too in small numbers because the facilities in terms of housing and land resources at most of the farms were limited. The lack of extensive farm and land facilities for maintenance of large sized population of such strains for applying effective selection has also been a major constraint. Consequently, many of the strains/breeds either have deteriorated or have not been able to make significant impact on genetic enhancement of cattle productivity at the national level. Currently Frieswal, Sunandini, Vrindavani and Karan Swiss are being genetically improved. The mean values of various traits of Frieswal, Karan Swiss, Sunandini and Vrindavani cattle are listed in Table 18.2.

Table 18.2 Mean values for some traits of Frieswal, Karan Fries, Sunandini and Vrindavani cattle.

Economic traits	Frieswal	Karan Fries	Sunandini	Vrindavani	Overall ranges
Age at 1st calving(days)	976	978	966	1012	970 – 1012
300 or 305 day milk yield (kg)	3240	2864	-	3047	2900 – 3200
Total lactation milk yield (kg)	3310	3330	2435	3220	3200 – 3300
Peak yield (kg)	15	18	—	17	15 – 18
Dry period(days)	123	—	—	100	
Service period(days)	171	182	148	150	150 – 180
Calving interval(days)	445	440	425	425	425 – 445

Economic traits	Frieswal	Karan Fries	Sunandini	Vrindavani	Overall ranges
Remarks:	Data based on 27871 records of Frieswal cows that were progeny of 135 bulls at 40 Military farms located in various agro-climatic regions from 1991 to 2009.	Data based on 2602 lactation reco-rds of 1037 KF cows.	Data based on females maintained in the nucleus herd at Mattupatty from 1973-1991 (averages are based on observations varying from 580 to 1320.	Data on about 1000 observations-	–

Source: PDC Meerut (UP)

The averages given in Table 18.2 are based on data spanning a long period but the current level of performance of these strains in particular of Frieswal is higher. For example Frieswal cows at Mhow (Madhya Pradesh) and Kanpur(UP) farms recorded 300 day milk yield of over 4000 kg as compared to overall average of 3240 for all the farms. The strain is being further improved.

Box 18.1: International evaluation of cattle bulls.

Global trade of germplasm, mainly facilitated by development of techniques of artificial insemination, semen freezing and recently by embryo storage, had been expanding since 1970's. Breeders or farmers interested in purchase of genetically superior germplasm from other countries would like to have some objective and sound information on breeding values of bulls for milk production across countries.

A frame work for generating valid genetic evaluation procedures has been developed under the guidance of an organization called INTERBULL which is a subcommittee of the International Committee for Animal Recording(ICAR), and is based at the Swedish University of Agricultural Sciences in Uppsala. INTERBULL was developed in 1983 as a joint venture between ICAR, the European Association for Animal Production (EAAP) and the International Dairy Federation (IDF).

The first attempt was to run a trial comparison of germplasm of different countries in Poland. Bulls from 14 countries were nominated for the trial and semen was sent 'free' to Poland to produce about 100 daughters per country using Polish Friesian cows. The resulting daughters were managed as contemporaries on large state-owned farms. Daughters of US, Israeli and Canadian bulls ranked first, second and third respectively for milk yield. A number of other research trials that compared daughters sired by bulls from two or more countries also confirmed the high milk yield of North American Holsteins. However, by the end of the trials, genetic trends had taken each country further ahead at different rates, and so the comparisons were out of date. A less costly and less time-consuming method was needed, and it was beginning of INTERBULL.

INTERBULL decided to take EBVs of bulls from different countries and to develop a way to make accurate comparisons between animals, primarily bulls, performing both within and across countries. However, these comparisons were made difficult due to a number of differences across countries viz.

- differences in genetic evaluation methods.
- differences in breeding objectives.
- differences in genetic levels.
- differences in farming environment.

Thus, efforts were made to 'standardize' by creating minimum standards for milk recording programs. A survey of genetic evaluation models was conducted and the results published. Tests were developed to determine the quality of EBVs which different countries provided. This has benefitted all countries in that genetic evaluation methods have improved immensely over the years and all countries use very similar models now.

In 1993, a statistical model, called 'multi-trait across-country evaluation' (MACE) was developed for making comparison of bull EBVs from different countries. MACE allows the EBV of a bull in the Netherlands, for example, to be expressed on the same scale as an EBV in Canada, and it combines the daughter information of that bull from every country in which it has daughters. Thus, young bulls could be progeny tested with daughters in several countries rather than just one. Some of Canada's young bulls are simultaneously test mated in more than one country.

Today INTERBULL evaluations include data from 30 different countries, six breeds and 40 traits.

Future Breeding Plan for Genetic Improvement of Crossbred Cattle

India is a vast country characterized by varied agro-climatic conditions including farming systems. The major constraints which are hindering progress of genetic improvement of crossbred cattle, that are very large in numbers, include:

- Inadequate supply of superior germplasm (e.g. frozen semen).
- Lack of proper AI network under field conditions including high incidence of reproductive disorders (e.g. anoestrous, repeat breeding etc).
- Inadequate availability of feed and fodder in most parts of the country to sustain high yielding animals, and lack of proper health and management practices.

Nonetheless, crossing in one or the other form will continue to play significant role in improving milk productivity of cattle. The salient points to be considered in formulation of future genetic improvement of crossbred cattle are germplasm introduction and selection within existing crossbred population.

Germplasm Introduction and its Use

Genetic base of existing crossbred populations in the country over years has narrowed down.Therefore new exotic germplasm needs to be regularly introduced to broaden the genetic base of crossbreds. Many different strains/populations of the same exotic breed are available in different countries as well as with different sources/agencies within the same country. Most of the recognized breeds have their national associations (e.g. HF breed association of USA, Canada, and Israel etc). These associations publish or provide extensive information on registered animals for a wide variety of traits. International Bull Evaluation Services, INTERBULL (see Box 18.1) provide information on ranking of bulls across countries by combining genetic information provided by individual participating countries. Such information can be utilized for import of suitable germplasm.

There are wide differences in agro climatic conditions, herd sizes and management levels (including feed and fodder availability) in India. Presence of genotype-environment interaction in cattle is also well documented. Therefore, different types of exotic germplasm may have to be used or introduced for different regions taking into consideration the suitability of germplasm to particular environment

Selection within Existing Crossbred Population

- **Selection of cows:** Cows are usually selected within a herd on the basis of their own milk yield as heritability of milk yield is moderate (about 0.25). When milk yield of more than one lactation is available, cows can be selected based on average milk yield of several lactations.

  Selection of cows for more than one important trait at a time using a selection index can theoretically help in improving the overall economic value of cattle. However, in practice such a selection program is not feasible because of unavailability of sufficient number of females in organized/institutional herds (i.e. small herd size) and lack of proper performance recording at farmer's level. This can be partially overcome through herd networking (for organized herds) and through group breeding schemes involving farmers.

- **Bull selection:** Much of the genetic progress in dairy animals is achieved through selection of males because high selection intensity can be applied through this path of genetic improvement. Genetic evaluation of dairy bulls is most accurate (i.e. high accuracy of selection) when it is based on progeny performance. Progeny testing of bulls involves :

  - Use of young bulls (preliminarily selected on their dam's milk yield) on a part of cattle population,
  - Recording the performance of their progeny and evaluating genetic worth (breeding value) of bulls.

In India, like in most developing countries, proper infrastructural facilities do not exist for field-performance recording and progeny testing of young bulls. Progeny testing therefore is restricted to only Governmental and Institutional herds where herd sizes are small. Under institutional herds there are two constraints in conducting progeny testing of bulls:

(i) Accuracy of breeding value evaluation under such conditions is low because sufficient number of daughters per bull cannot be recorded.

(ii) Only a limited number of bulls can be evaluated through progeny testing in Institutional herds and hence high selection intensity cannot be applied.

In view of these constraints, systematic field progeny testing program for bull selection is necessary. Such a program would require producing large number of crossbreds under field conditions, and identification, registration and performance recording of these animals. The organized herds need to be net worked under field progeny testing programs. Bull mother farms of crossbred cattle have be developed for multiplication of superior germplasm and dissemination of genetic improvement. Some of these aspects are covered in a later section of this chapter.

Box 18.2: Project Directorate on Cattle (PDC).

The All India Co-ordinated Research Project (AICRP) on Cattle, started in early 1970's, was upgraded to Project Directorate on Cattle(PDC) in 1987 based at Meerut (UP). This institute/directorate was conceived to take advantage of the achievements made in AICRP and other crossbreeding experiments at organized farms and under field conditions for developing a national milch breed (Frieswal Project) from a reasonably large crossbred base (Sahiwal x Holstein-Friesian), which was a limiting factor under AICRP on Cattle. It was mandated to act as a nodal institution to monitor, coordinate and support all research and development projects for cattle improvement at Military Farms.

Subsequently, keeping in view the importance of indigenous breeds for their adaptability and related attributes, a project on con-servation of these breeds was initiated in collaboration with the SAUs, State Government and Non Government Organizations. Similarly, Field frogeny testing (FPT) project on crossbred cattle was undertaken to bring about improvement in crossbred cattle at farmers' herds. The aim of this project is to progeny test Frieswal and other Holstein Friesian crossbred bulls under field conditions at four different agro-climatic locations of the country, viz. Ludhiana (GADVASU), Mannuthy (KAU), Uruli-Kanchan (BAIF) and Pantnagar (GBPUAT).

The herd strength of Frieswal females is currently over 17000. The number of elite cows is over one thousand. The long-term objective of the project on Frieswal cattle is to develop a 'Frieswal strain" which produces 4000 litres of milk with 4% butterfat in a mature lactation of 300 days. This project is being carried out in collaboration with Indian Army where Frieswal cattle are maintained at 37 Military Dairy Farms located in different agro-climatic zones of the country.

The PDC has a number of sub- centres located in different states (e.g. state agriculture/ veterinary universities, BAIF, Military dairy farms, Gaushalas, Central breeding farms) for undertaking different types of activities like germplasm conservation units, field progeny testing units, bull rearing centres, data recording units etc.

Development of a New Breed (Strain) from Crossbreds: A Breeding Plan

A breeding plan, using Holstein Friesian as exotic and Sahiwal as indigenous breeds (keeping 75 % HF inheritance) which would lead to the synthesis of a new breed/strain is depicted below in Fig 18.10.

Once a new cattle breed/strain has been developed, the emphasis has be laid on producing sufficient number of progeny tested bulls, and then using them extensively for dissemination of newly developed germplasm to the farmers or developmental agencies. Further a breed society that can implement the necessary genetic improvement program also needs be formed.

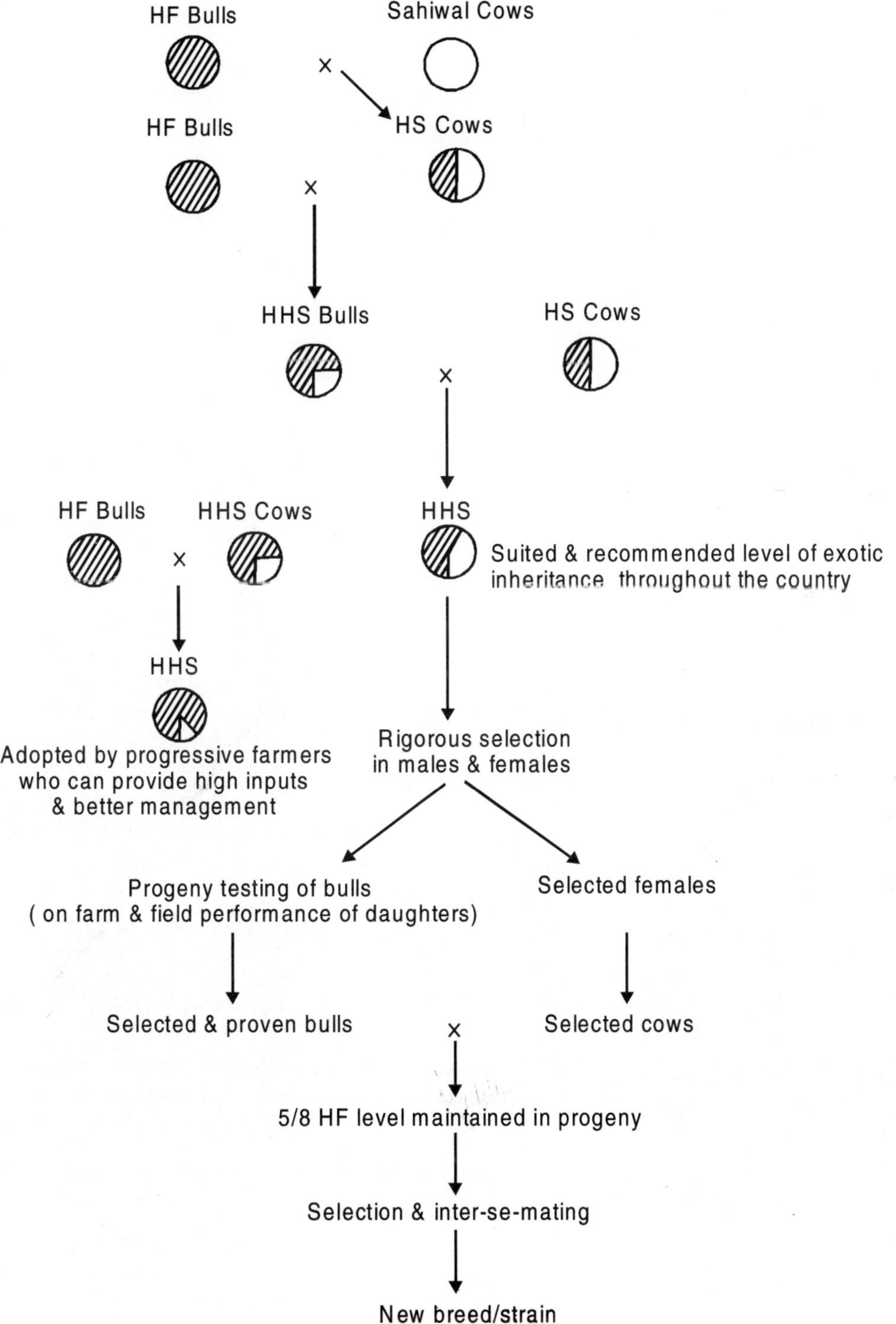

Fig. 18.10 Schematic illustration of a breeding plan for development of a new strain/breed.

Box 18.3: Crossbreeding in Punjab.

Punjab is an agrarian state and has made tremendous strides in agriculture including dairy. Currently more than 90% of cattle population in Punjab is crossbred carrying varying levels of exotic inheritance.

Crossbreeding in the state was started at the then Punjab Agricultural University(now Guru Angad Dev Veterinary and Animal Sciences University, Ludhiana) in 1967-68 with purchase of two Red Dane bulls from Hessarghata (Bangalore). Twenty cows and two more bulls of Red Dane were also imported from Denmark in 1970-71 for strengthening the crossbreeding work. Sahiwal cows were inseminated with RD bulls to produce crossbreds carrying 50% inheritance from each of the two breeds. The use of Red Dane was discontinued in 1974 due to its pendulous udder, and imported semen of progeny tested bulls of HF was introduced for producing three-breed crossbreds.

Improvement in crossbred performance: The crossbred herd of the university has been consistently improved due to combined effects of increasing level of exotic inheritance, selection especially of bulls, better nutritional, managemental and health practices :

- Average 305-day milk yield of crossbred cows has increased from 2835 kg (1974-75) to about 5200 kg (2009-10)
- Considerable improvement in other productivity and growth traits viz. 1st lactation milk yield, body weight at 18 months of age, wet average yield per day has also been achieved.
- The milk yield of crossbred cows with 50% and 75% exotic inheritance increased between 1995 and 2006 from 3168 to 3759 kg and from 2932 to 4499 kg for the two groups, respectively.

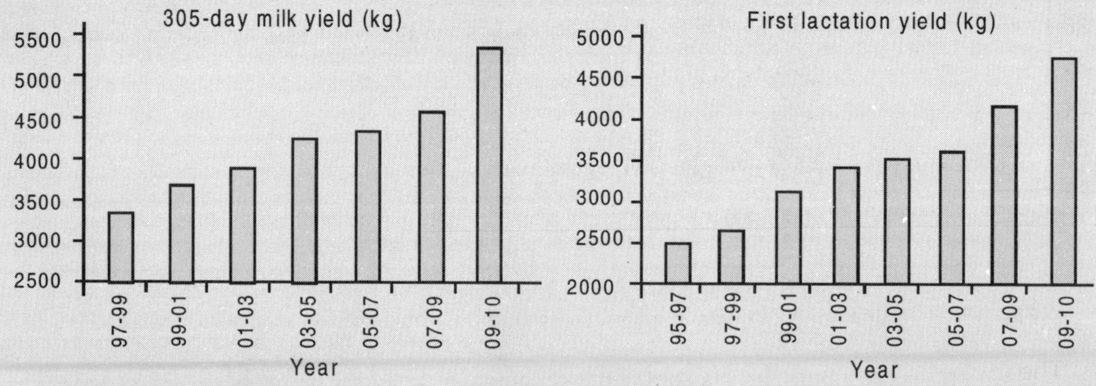

Fig. 18.11 Improvement in productivity over the years (1997 to 2010)

A number of crossbred cows have recorded peak yield of more than 50 kg a day. An elite cow with 93.75% exotic inheritances (87.5 % Holstein Friesian, 6.25 % Red Dane) recorded 305-day lactation yield of 10493 kg. Its peak yield was 47.2 kg and milk fat content 3.5%. It completed its third lactation by producing 12260 kg milk. The excellent performance of crossbreds carrying high level of exotic inheritance showed that crossbred cows could be reared economically in Punjab provided they are optimally fed and managed, especially in summer and rainy season.

Elite herd: Since a larger increase in genetic improvement comes from sire side, an elite herd of crossbred cows has been produced for the production of future bulls. Average 305-day milk yield and average peak yield of elite herd have been recorded to be over 7000 kg and 33 kg, respectively. Crossbred cows with 305-day milk yield of 10493 kg and complete lactation milk yield of 12260 kg, with peak yield of 52.5 kg per day, have been produced. The elite herd has shown consistent improvement over years, and it has shown superiority over the general herd to the tune of 16-22 % (Fig 18.12).

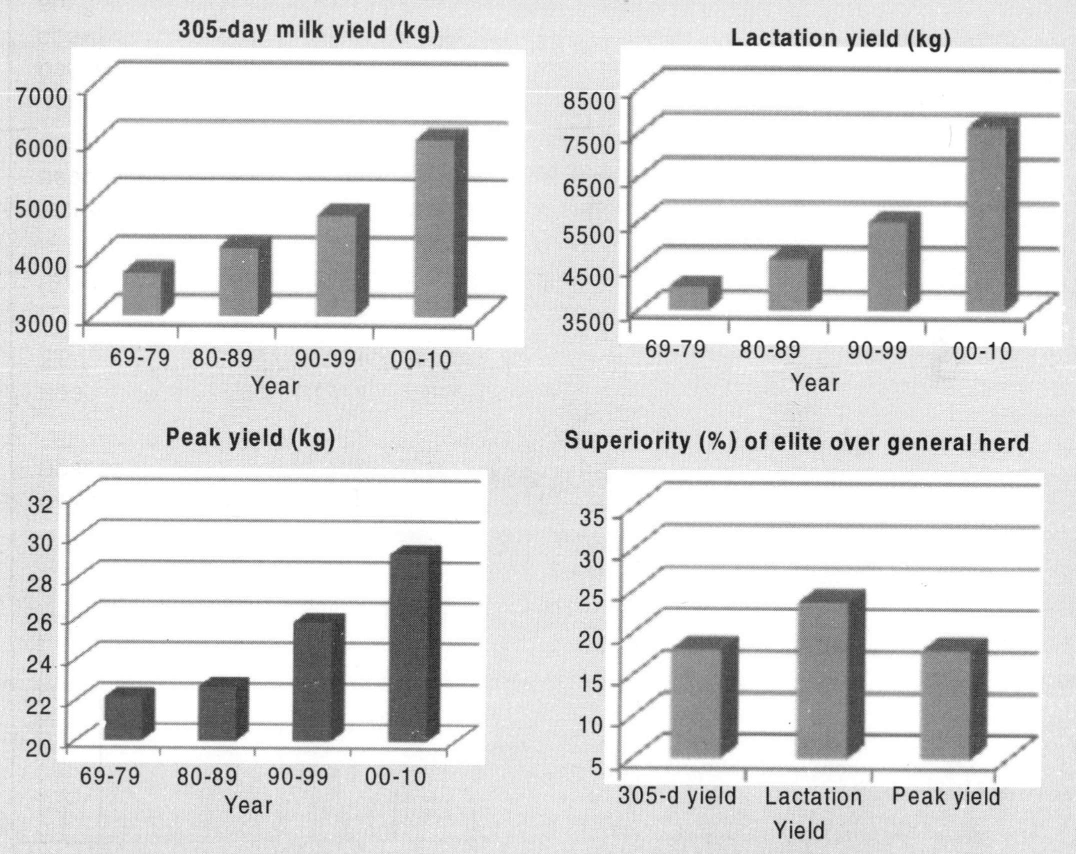

Fig. 18.12 Performance of elite herd of Cattle and its superiority over general herd.

The university has been supplying improved germplasm to farmers and developmental agencies since the start of crossbreeding program. It has adopted certain villages under its field progeny-testing program. Average at first lactation 305-day milk yield of crossbred progenies in the adopted villages 18.12 was 1698 kg in 1979. It as a result of supply of high quality semen of test bulls has increased to over 3000 kg (2009). During the past 16 years, an improvement of 18.6% in milk production has been achieved. This has been achieved by providing technical guidance to farmers about breeding, feeding and management of crossbred animals. A number of farmers have now established large dairy units to undertake commercial dairy farming, and having excellent herds with high level of exotic inheritance, and are serving as models for commercial dairy farming.

BUFFALO BREEDING

Buffaloes contribute to the Indian agricultural economy by way of milk, meat and draught. They contribute about 51% of the total milk production of the country and also make 19% contribution to the country's total meat produce. About 97% of worlds buffaloes, which are found in 42 countries, are found in Asia, and around 57% of world's buffalo population is found in India. Buffaloes thrive well under low inputs as they are efficient converter of roughages. The population of buffaloes has over the years been increasing: percentage increase in India is about 1.5% as compared to 1.45% in Asia. In India, buffaloes are spread over almost all parts of the country with varying density. Majority of the population (72%), however, is located in the Northern and Western states.

Types and Distribution

The world's buffaloes are classified into two groups – the African and the Asian.

- The African buffalo *Syncerus* which is wild has a single species (I *caffer*) and a small number of sub-species.
- The Asian buffalo (*Bubalus*) is domesticated, consists of three species: 'anoa' of Celebes, 'tomaro' of Mindori and 'arni' or Indian wild buffalo.

Phylum: Chordata (back bone),
Class: Mammalia (suckle their young ones)
Order: Artiodactyla (hooved, even footed)
Suborder: Ruminantie
Family: Bovidae (hollow-horned ruminants)
Subfamily: Bovinae
Genus: *Bubalus*
Species: *B. bubalis*

Buffaloes were domesticated in India during Harrappan period (2200 to 1600 BC). The domesticated Indian buffalo (of Indian sub-continent) is called *Bubalus bubalis* (previously called *Bubalus arnee*). It is widely distributed in Asia where about 97% of buffalo's population is found. It has also been introduced into Europe, Near East, South America, the former Soviet Union and the Caribbean.

The domesticated Indian buffalo is broadly grouped into two categories: the Riverine buffalo, with chromosome number of 50 and Swamp buffalo with chromosome number of 48. The amount of genetic material carried by both types is same, and they interbreed to produce fertile progeny with 49 chromosomes.

Riverine Buffaloes

The Riverine buffalo is the most common type found in India, Pakistan, Bulgaria, Hungary, Turkey, Italy and Egypt. They are also found in Brazil and Caucasia. The Riverine buffalo prefers to wallow in clean water of rivers, irrigation canals and ponds thereby the name, Riverine. It is found predominantly in Indian subcontinent. Riverine buffaloes have primarily been selected and used for milk production. Most of the buffaloes found in India are Riverine type.

Swamp Buffaloes

The name Swamp buffalo arises from its natural habitat in swamp or marshland areas where it wallows in mud and spends the hottest part of the day. The Swamp buffaloes belong to the same species as Riverine but resemble the wild 'arnee' in morphological traits. This type is most common in Southeast Asia (especially China, Indonesia, Philippines, Thailand and Vietnam). These are multipurpose animals: provide draught power and meat in rice growing areas and milk in other regions. Their milk yield is very low (400 to 500 kg). Their horns grow out laterally and horizontally in young animals and curve round in a semi-circle in older ones. In

India, the swamp type buffaloes are found in Eastern parts of India.Crossbreds between Riverine and swamp buffaloes have been produced in China, Thailand, Vietnam and Philippines. These are good for meat and draft, and produce relatively more milk than swamp.

Indian Breeds and their Classification

Buffaloes originated in India (the Indus valley- India and Pakistan); therefore, this region is considered as home tract for some of the best dairy breeds of this species. Buffalo population per unit area varies in different regions of India: northern region having highest concentration (47%) followed by western (22.4%), southern (19.8%) and eastern region (10.9%). The Indian Council of Agricultural Research handbook lists 15 breeds of buffaloes but 13 are registered with National Bureau of Animal Genetic Resources. The distribution of number of animals of the 13 registered breed is given in Figure 18.13.

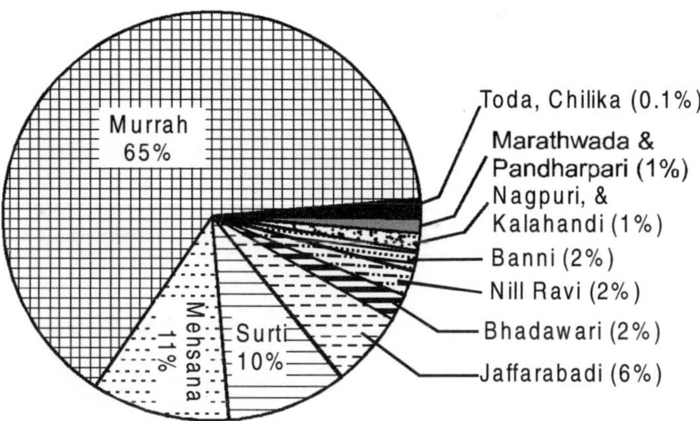

Fig. 18.13 Distribution (%) of the total population of the 13 registered breeds of buffaloes.

Murrah breed is most predominant in numbers, followed by Mehsana, Surti and Jaffarabadi. Chilika and Toda are negligible, collectively making about 0.2% of the total population of the 13 registered breeds (Fig 18.13). Of all the Indian breeds, seven are well-defined milch breeds viz. Murrah, Nili Ravi, Bhadawari, Mehsana, Surti, Jaffarabadi and Nagpuri. The defined breeds are recognized on the basis of their habitat, genetic make-up, morphological characters, body size, production and eco-geographical distribution. About 40% of the domestic buffaloes are classified into recognized breeds whereas the remaining 60% are non-descript. The Indian Riverine buffaloes have also been grouped into five distinct groups mainly on the basis of the region:

- *Murrah group*: It comprises Murrah and Nili-Ravi which originated in Punjab and Haryana.
- *Gujarat group* :It comprises Jaffarabadi, Surti and Mehsana.
- *Uttar Pradesh group:* It consists of Bhadawari and Terai.
- *Central India group:* It consists of Nagpuri(Maharashtra), Pandharpuri (Maharashtra, Andhra Pradesh, Karnataka), Kalahandi (Andhra Pradesh and Orissa), Sambalpuri (Madhya Pradesh), and Jerangi and Manda(Orissa).
- *South India group:* It comprises of Toda and South Kanara.

The swamp type buffaloes are mainly found in Assam (North-East group).

A brief description of some important breeds used for milk production is provided below. Average values for some traits recorded on institutional herds of some of the breeds are given in Table 18.3.

Murrah

This breed originated in Punjab (of India and Pakistan). In India, its breeding tract is Rohtak, Hissar and Jind of Haryana and Nabha and Patiala districts of Punjab. It is the most important milk producing breed and has spread to some other areas of the country. It is the most predominant of all the registered breeds of the country and its impact on India buffalo population is increasing. Some breeds or types of buffaloes have been developed from Murrah's crossing with local/non-descript (e.g. Godawari breed of Godavari deltaic areas). Several other countries of the world have also imported animals of this breed.

The breed's characteristics are massive body, comparatively long neck and head and short and tightly curled horns. Females have well developed udder, broad hips, and the fore- and hind-quarters are drooping. The limbs are short, straight and strong with black hooves. The tail is long with a white switch. The body color is jet black but a few brown or fawn grey animals are also seen. Adult body weight on the average is 550 kg for males and 450 kg for females. Milk yield per lactation (of 270 to 305 days) in good herds is 1800-2000 kg with 7% fat. A good number of herds with 305-day lactation yield of over 2300 to 2400 kg exist in its breeding tract. Elite herds of this breed which are maintained for producing young bulls of high genetic merit produce over 2600 to 2800 kg in 305-day lactation. Age at first calving in such herds is about 38 to 42 months.

Nili -Ravi

The Nili Ravi breed is found in Satluj valley of Ferozepur district of Punjab (India) and in Sahiwal district of Pakistan. It is believed to have developed from Murrah through geographical isolation. Previously the Nili and Ravi were described as two breeds; Nili found in Sutlej valley (Punjab) and Ravi in Sandal Bar but they are now considered a single breed. The animals are black colored, with white markings on forehead, face, muzzle and legs/ hooves and switch of tail (breed is vernacularly referred to as 'Panj Kalyani'). Peculiarity of breed is the wall eyes. Head is elongate, bulging at top and depressed between eyes. Frame is medium sized, neck is long, thin and fine. Horns are small and tightly curled. Udder is well developed with squarely placed long teats, and prominent milk veins. Average weight of male and female is 600 and 450 kg respectively. Milk yield ranges between 1,800 and 2,000 kg in 250-300 days. Pure specimens of this breed had been declining; therefore, conservations programs for the breed are in operation.

Surti

The breeding tract of Surti is Kaira and Baroda districts of Gujarat State. The color is black but brown animals are also found. The tuft of the tail is white. Horns are of medium length and sickle shaped. The peculiarity of the breed is two white collars, one round the jaw and the other at brisket. Head is long and broad, and the eyes are very prominent. Milk yield varies from 1,600 to 1900 kg with high butter content. Age at first calving is 40-50 months with inter calving interval of 400 to 500 days.

Jaffarabadi

Its breeding tract is Kutch and Jamnagar districts of Gujarat. They have long but not a compact body. Dewlap in females is somewhat loose. Head and neck are massive with a very prominent forehead. Horns are heavy and incline to droop on either side of the neck and less tightly curved than in Murrah. Udder is well developed. Animals of this breed are mainly maintained by traditional breeders called Maldhari who are nomadic.

Mehsana

Its breeding tract is Mehsana, Sabarkanda and Banaskantha districts of Gujarat. The breed is supposed to have evolved from crossing between Murrah and Surti. The color is usually black to grey with white markings often on face, legs or tail tips. Horns are less curved at end as compared to Murrah but are longer and could be of irregular shapes. Body is longer than in Murrah and the limbs lighter with longer and heavier head. Females have a well developed udder. The breed is considered to have good persistency.

Bhadawari

This breed is found in Agra (Bhadawari tehsil) and Etawah districts of Uttar Pradesh and Gwalior district of Madhya Pradesh. The animals are medium sized, and possess wedge-shaped head, short and stout legs and black hooves. Hindquarters are heavier and elevated than the forequarters. The animals have scanty hair on the body. The tail is long and thin with black and white or white switch reaching to hocks.

Nagpuri

The breeding tract of this breed is Nagpur, Akola and Amravati districts of Maharashtra. They possess long, flat and curved horns, long and thin face; somewhat long neck and lighter limbs. The tail is comparatively short, reaching below hocks.

Table 18.3 Average values* for production and reproduction characteristics of some breeds of buffaloes (2003 – 04).

Breed	Age at 1st calving (months)	Service period (days)	Dry period (days)	305 day or less milk yield(kg)	Fat (%)	Calving interval (days)
Murrah	44.9	135	167	2033	7.9	456
Nili Ravi	41.9	132	145	1941	6.9	441
Jaffarabadi	47.0	155	180	1794	8.6	466
Surti	49.8	146	177	1646	7.4	454
Pandharpuri	45.3	141	117	1385	8.1	445
Bhadawari	48.6	138	220	1029	8.1	445
Swamp**	55.4	187	193	500	7.7	511

* Weighted average from the participating centers under Network project; **Field animals

Source: National seminar 2005.Fat percentage based 4219 observations over all the centres.

There are other breeds/types but they are not important for milk production, and their numbers are negligible. Information on the performance of recognized breeds of Riverine buffaloes from different institutional herds indicate that the average first lactation milk yield in different breeds usually varies between 1000 and 2100 kg in 305 day lactation (Table 18.3). The averages age at first calving usually ranges from three to four years, and calving interval varies between 430 and 466 days. Swamp buffaloes, being not bred for milk, are poor in milk production, sexually late maturing and with longer service period, dry period, and calving interval.

Recently, **Banni** buffaloes of Kach area of Gujarat have also been registered as a breed. This breed originated in Banni area of Kach which is a part of Kach district of Gujarat. This is famous natural grassland covering around 4600 sq. k.m. The inherently saline soil, deposited by long lost river systems of this area is naturally suited for nutritious grasses. Muslim pastoral community called, Maldhari, live in this region with their animals, mainly Banni buffaloes. They maintain Banni buffaloes under locally adapted extensive production system in its breeding tract, and they are the only source of their livelihood. Purebred Banni animals are heavy with typical double and vertical coiling of the horn.They are very hardy and well adapted to harsh climatic conditions and thrive on grasses available in the area through night grazing. The performance of this breed recorded under field conditions shows its potentiality, as regular in breeding and high milk production. Banni buffaloes are trained to typical grazing on grassland during night and come to villages in the morning for milking. This traditional system of buffalo rearing has been adapted to avoid the heat stress and high temperature of the day. Maldhari have played important role in conservation and improvement of Banni Buffalo, and they have now come together to form 'Banni Breeders' Association' with membership of more than 970 breeders. Members of Breeders' Association are already participating in *in-situ* breed conservation program.

Fig. 18.14 Home tract of some important breeds of Indian buffaloes (Add Nili Ravi)

Economic Traits and Genetic Parameters

Traits of economic importance in buffaloes are the same as those of cattle since buffaloes are being bred and reared for milk. Buffaloes have not undergone long and intense selection for traits related to milk production. Therefore genetic parameter estimates of production, reproduction and milk composition traits as reported for cattle are strictly not applicable to buffaloes though they follow the same general trend. A number of estimates of genetic parameters of Indian buffalo populations have been obtained. Average values of heritabilities and genetic correlation have been as follow:

Heritabilities:

- Milk yield (per-lactation or in 305-day lactation), peak yield, fat yield have heritabilities in the range of 0.25 to 0.35.

- Lactation length, calving interval, dry period, service period have heritabilities of about 0.15 to 0.20. Number of services per conception has very low h^2 (0.05).

Genetic correlations;

- Genetic correlation of lactation milk yield with lactation length, 305-day milk yield, peak yield and milk per day of lactation length and total fat yield is positive and high.
- Genetic correlation of age at first calving with various measures of milk and fat is generally very low but positive.
- Genetic correlation of weight at first calving with lactation milk yield, 305-day milk yield, peak yield, milk yield per day of lactation length, fat yield and fat percentage is positive; usually in the range of 0.2 to 0.3.

Genetic Improvement of Buffaloes in India

Much of the best germplasm of milch breeds of buffaloes is found in India. Therefore, genetic improvement of buffaloes through introduction of superior germplasm from other countries or to capitalize the benefit of heterosis, like those in cattle, is not possible. The only breeding procedure available to genetically improve the existing breeds of buffaloes is through selection.

Genetic improvement of buffaloes through selection would be successful as the traits related to milk production are sufficiently heritable. However, there are a number of constraints in effective implementation of selection programs. These include:

(i) Non-availability of large sized herds and lack performance recording of animal's in the field.

(ii) Poor spread of AI which has been a hindrance in dissemination of high genetic merit germplasm to the farmers.

(iii) Some of the high yielding animals move to peri-uraban areas and these usually end up in slaughter houses after one or two lactations without contributing progeny

Further purebred animals of recognized breeds are declining as they are being indiscriminately bred with non-descript animals. Consequently, limited selection, mainly due to culling of less productive animals and their slaughter, has been applied for its genetic improvement.

Military dairy farms followed the policy of culling and selection since beginning.Initially those females with less than 1350 kg first lactation yield were culled. The level was then revised to 1660 kg in 1966 and to 1800 kg in 1968. Some selection in males on the basis of their dam's yield was also applied. As a result of selection coupled with improvement in nutritional, general managemental and health practices, lactation yield made some improvement but it was far less than expected. Some small-scale attempts at genetic improvement were also made by some universities. However, impetus to genetic improvement of buffaloes was through launching of the national level 'All India Coordinated Research Project' and 'Network Project', which was launched by the Indian council of Agricultural Research through its institutions and in collaboration with state agriculture/veterinary universities.

All India Coordinated Research and NetWork Project on Buffaloes

A systematic buffalo-breeding program was initiated in 1970-71 by ICAR in the name of All India Coordinated Research Project (AICRP) on buffaloes involving a number of institutes and universities. It was initiated with four centres namely PAU (Ludhiana) and NDRI (Karnal) each for large sized buffaloes, comprising mainly Murrah breed and two centres for small/medium sized buffaloes comprising mainly Surti and Mehsana breeds. Main objectives of the project were:

- To evaluate the breeding value (genetic merit) of bulls through progeny testing for identifying the bulls of high genetic merit, and
- To distribute young bulls (semen doses) produced from mating of elite buffaloes with proven sires to various farms in the country.

The AICRP was regrouped into a network mode (named 'Net Work Project' on Buffaloes) in 1993 by collaborating with various Institutions engaged in buffalo research at organized farms to meet the need of applying high selection intensity and for increasing the number of daughters per bull for increased selection accuracy. The herds of different centres were linked to for associated-herd progeny testing. Field progeny testing was initiated to strengthen the ongoing sire evaluation program of associated herds progeny testing at institutional herds by involving performance recording on farmers animals using semen of selected bulls. The participating units and breeds are :

- Murrah at CIRB (Hissar), GADVASU(Ludhiana), NDRI(Karnal), IVRI(Izatnagar), SVVU (Andhra Pradesh), KVASU(Kerala).
- Nili Ravi at CIRB Sub-campus, Nabha, (Punjab)
- Jaffarabadi at Junagarh (Gujarat)
- Pandharpuri at MPKV, Kolhapur (Maharashtra)
- Surti at RAJUVAS, Vallabhnagar (Rajasthan)
- Bhadawari IGFRI, Jhansi (U.P.)
- Swamp at AAU, Khanapara (Assam)

Additionally there are filed units of Murrah at CIRB, GADVASU and NDRI (Karnal).

The genetic improvement program under Net Work/AICRP consists of following components:

(i) *Identification of proven sires:* Progeny tested proven sires available at various institutional herds, which are part of the associated-herd progeny-testing program, are identified, and their semen is procured.

(ii) *Selection of elite buffaloes:* Elite she buffaloes are selected based on either of the following criteria:

Best lactation yield = 2500 kg

1st lactation yield = 2300 Kg

Estimated breeding value = 1950 kg

These criteria are regularly monitored and upgraded.

(iii) *Nominated matings:* The elite buffaloes are mated with proven bulls.

(iv) *Selection of test bulls:* The males born out of nominated matings in various herds are kept for subsequent selection for test mating. The criteria for inclusion of males in test mating are :

Dam's best lactation yield = 2500 Kg

Estimated breeding value = 1950 Kg

(v) *Freezing and storage of semen of test bulls:* A large number of quality semen doses (usually in thousands) from each test bull are frozen for test matings and stored till the genetic worth of bulls is determined through multiple- herd progeny testing.

(vi) *Test mating program:* Breedable buffaloes in multiple herds as well as in the field are identified as test herds, and inseminated with frozen semen of at least 15 test bulls over a test cycle of 18 months.

(vii) *Data recording:* Performance of buffaloes (progeny) produced is recorded with respect to growth, production and reproduction, age-wise mortality, bull-wise semen production and daughters born.

(viii) *Genetic evaluation of bulls :* Sires are evaluated on the basis of their daughters first 305-day milk yield across herds (multiple herds) using the following sire index:

Sire index = M + Weighting coefficient (Daughter average – contemporary daughter average)

Where:

M = overall average of daughter's 305-day 1st lactation milk yield.

n = number of daughters per sire.

D = daughter's average 305-day 1st lactation milk yield.

CD = contemporary daughter's average 305-day 1st lactation milk yield (over all the herds)

$$\text{Weighting coefficient} = \frac{2n}{n+12}$$

All the sires are ranked on the basis of above progeny test index. Out of these, the top 20 to 30 bulls are selected for future use as proven bulls for production of superior germplasm. Semen of these sires is stored for distribution to farmers and development agencies.

The program has made success as is evident from the improvement achieved in institutional as well as farmers herds (field).For example the performance of Murrah herd of GADVASU (erstwhile PAU), Ludhiana, has shown consistent performance as evident from the improvement observed from the period of 1973-77 to 2008-12 :

- 305-day milk yield has been improved from 1770 to about 2100 kg in the herd while complete lactation yield improved from 1950 to 2300 kg during the same period.
- Milk yield per day of lactation length improved from 5.2 to about 7 kg; peak yield increased from 9.3 to 11.6 kg and days to reach peak yield decreased from about 60 to about 40.
- Fat yield per lactation improved from 140 to 170 kg. Average fat, protein and solids-not- fat percentages in milk of this herd are 4.3, 4.8 and 9.8 respectively.

These improvements are not purely due to selection but also due to improved husbandry practices but a large part is definitely attributable to selection.

Genetic improvement of Non-descript Buffaloes

Nondescript buffaloes which constitute about 60 % of the Indian buffaloes are usually reared under low input production systems where feed and fodder resources are limited. Herd sizes are very low. Milk marketing facilities in such areas are not developed.

Productivity of such low producing nondescript buffaloes can be improved through mating with genetically superior bulls of improved recognized breeds (grading up).Such an up grading program would increase milk production two to three times in early generations of up grading. Within four to five generations, the low producing nondescript buffaloes can be replaced with high producing buffaloes conforming to recognized breeds.

Box 18.4: Central Institute for Research on Buffaloes (CIRB).

Central Institute for Research on Buffaloes (CIRB), based at Hissar (Haryana) has its roots in the Progeny Testing Bull Farm Hissar of the Haryana State Government. This farm was transferred to the Indian Council of Agricultural Research in 1985.All facilities including the buffalo herd of the Progeny Testing Bull Farm were transferred to the CIRB.The institute also has a sub-campus which was established in 1987 at Bir Dosanjh, Nabha, District Patiala (Punjab) with the transfer of Nili Ravi Buffalo Farm from the Punjab State Government.

The main mandate of the Institute is to undertake research on all aspects of buffalo production. The institute carries out research on various aspects of buffalo improvement including conservation of germplasm, development of optimum diets and feeding systems, enhancement of reproductive efficiency and health management practices for augmenting milk, meat and draught performance of the species

Progeny Testing of Buffaloes and Cattle

Progeny testing is the best procedure of selection of bulls for milk production. An ideal progeny testing program, integrated with reproductive technologies (e.g. AI, semen storage etc) can bring about large genetic improvement in milk yield.

Types of Progeny Testing Programs

Progeny testing of bulls for milk yield implies evaluating their genetic worth (breeding value) on the basis of milk yield of their daughters. It can be conducted in a single organized herd or across many such herds. It can also be extended to field level.

Progeny Testing under Organized Herds

(a) **Single-herd progeny testing:** Sires are evaluated on the basis of milk yield of their daughters, which have been raised in a single organised herd. The greatest lacuna of such a program is availability of less number of daughters per bull (usually 2 to 4) which lowers the accuracy of genetic evaluation.

(b) **Associated-herd progeny testing:** Most of the organised herds, which are usually maintained by institutions, are of small sizes (150-300). Under such single herd programs only a limited number of sires (4 to 8) can be progeny tested, and that too with few daughters per bull. To overcome this dual problem of low selection intensity and accuracy, the concept of associated progeny testing programs was proposed, and such associated herd progeny testing programs were initiated.

An associated- herd progeny- testing program involves use of each sire in many associated herds through exchange of semen. The sires are then evaluated on the basis of performance of their daughters that were born and reared in more than one herd and thus exposed to different environmental and managemental conditions. An associated-herd progeny-testing program involving two herds is shown in Figure 18.15.

Such associated (multi-herd) progeny testing programs serve three purposes:

(i) Higher accuracy of selection is possible than that in a single herd program because genetic evaluation of sires is based on more number of progeny, and also due to elimination of C effects (common environment as source of resemblance between progeny of same sire)

(ii) Increased selection intensity through availability of more number of progeny tested bulls.

(iii) Since the sires in such a scheme are evaluated under varied herd-management (environmental) conditions, therefore, genotype-environment interactions are accounted for in sire evaluation. The progeny of sires selected through multi-herd testing would thus be suited to a wider range of environment.

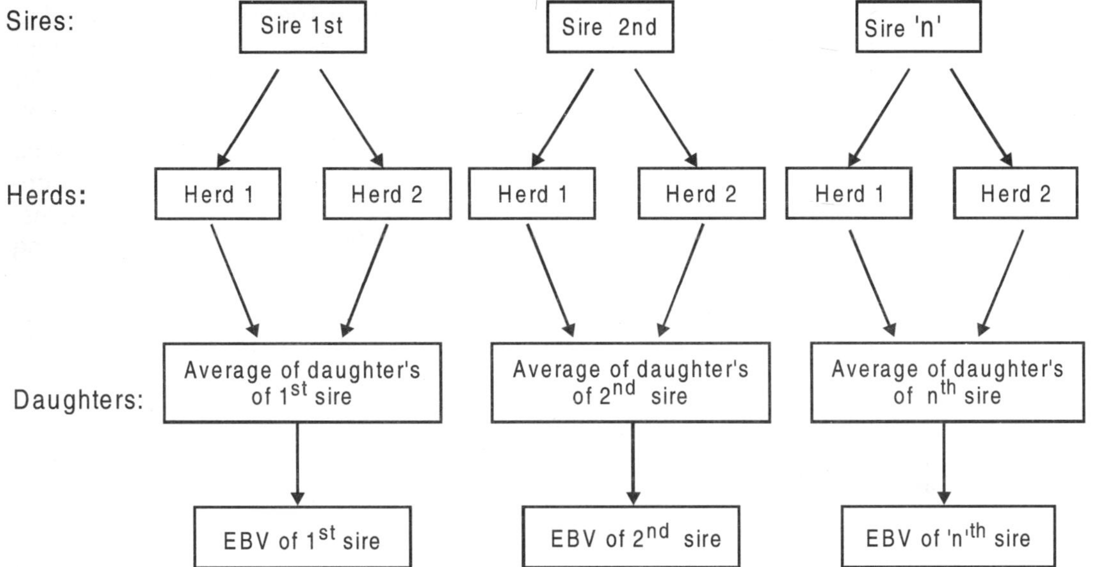

Fig. 18.15. Schematic illustration of an associated-herd progeny-testing program.

Progeny Testing under Field Conditions (Field Progeny Testing)

The buffalo and cattle progeny testing programs undertaken in India even through associated-herd programs (multi-herds evaluation in organized herds) still could not produce sufficient number of progeny tested bulls to cover large area of the bovine population. Selection accuracy was still low due to limited number of daughters as ideally a bull should have daughters in hundreds.

To overcome these problems, progeny testing programs were extended to field (farmers) conditions. Such a progeny testing program which involves evaluating the breeding values of bulls on the basis of performance of their daughters born and reared in farmer's field is called **field progeny testing.**

It is important to note that daughters of the bulls undergoing field progeny testing program are reared in farmer's herds where managemental and environmental conditions are different, usually poorer, than in organized herds. In essence, daughters of the sires being progeny tested are reared and evaluated under conditions where future daughters of the selected bulls are likely to perform. So the genotypes (sires) being selected are the ones which will be well adapted to the field environment.

Field progeny testing involves huge set up, therefore, such programs are undertaken in collaboration of different institutes and developmental organizations (e.g. state animal husbandry departments, co-operatives, farmer's associations, state livestock development boards etc.).

Features: Salient features of field progeny testing program involve/include:

(a) Selection of operational area: The area chosen for a field progeny testing be the one where other developmental schemes are operational so that sound infrastructural facilities are readily available. Areas (cluster of villages) with high density of high yielding dairy herds are best suited for implementing a field progeny-testing program. The villages/block chosen be the one where:

- no or few breeding bulls are available for natural service,
- a well-equipped artificial insemination centre easily approachable by road exists,
- sufficient number a number of trained AI workers are available for covering large livestock population,
- animal health services are available.

Farmers may have to be provided some incentives so that they retain the daughters born to bulls under progeny testing until their lactation recordings are complete.

(b) **Animal identification and performance recording:** The animals included in the program (the dam and offspring) are ear-tagged for maintaining both pedigree and individual-animal identification. Accurate recording of performance (e.g. milk yield and related traits) is extremely important. Designated milk recorders who are not associated with related agencies (AI centre, co-operatives etc) are assigned and provided with recording performa (usually, one is maintained by the owner also). Since daily recording is not feasible under field conditions, it is recorded at some regular intervals. The interval or frequency of recording of milk yield during lactation depends on a number of factors. Various procedures of weekly, fortnightly or monthly recording (both morning and evening) having been evaluated and advocated.

(c) **Data storage, analysis and breeding value evaluation:** The performance data including pedigree identity are transferred to a central storage and processing unit. Breeding values of the sire's under-test are computed using appropriate statistical procedures. The top proven bulls are used for nominated matings (with elite females which have been selected) to produce future sires (bull mother farms) for serving a large fraction of females.

Progeny testing is an accepted method for accurately selecting dairy breeding bulls and producing future bulls in India. The National Dairy Development Board (NDDB) has published a manual for implementing a field-progeny testing program. A schematic representation of various activities associated with of a bull progeny testing program (adapted from NDDB document) is given in Figure 18.16

There are two important decisions to be made before starting the program:

(i) **Number of bulls to be put under test:** Theoretically, higher the number of bulls put under progeny test, higher will be the probability of getting bulls of high genetic merit (high selection intensity would be possible). Given the dairy scenario in the country, ideally we should put at least 50 bulls under the test annually. However, recognizing the realities of field conditions, the program can be started with 20 bulls in the first year, and the number be raised to a minimum of 40 bulls per-year by 5th year.

(ii) **Number of villages to be included in the program:** The number of villages to be included in the program depends on the number of bulls put under the test and the availability of breedable animals in the area of operation. In order to obtain 30 complete first lactation records of daughters per bull under Indian conditions, a minimum of 2000 inseminations would have to be carried out per bull. Therefore depending on the number of breedable females available and the number of inseminations that can be done in the area of operation, we can decide the number of villages to be included in the program.

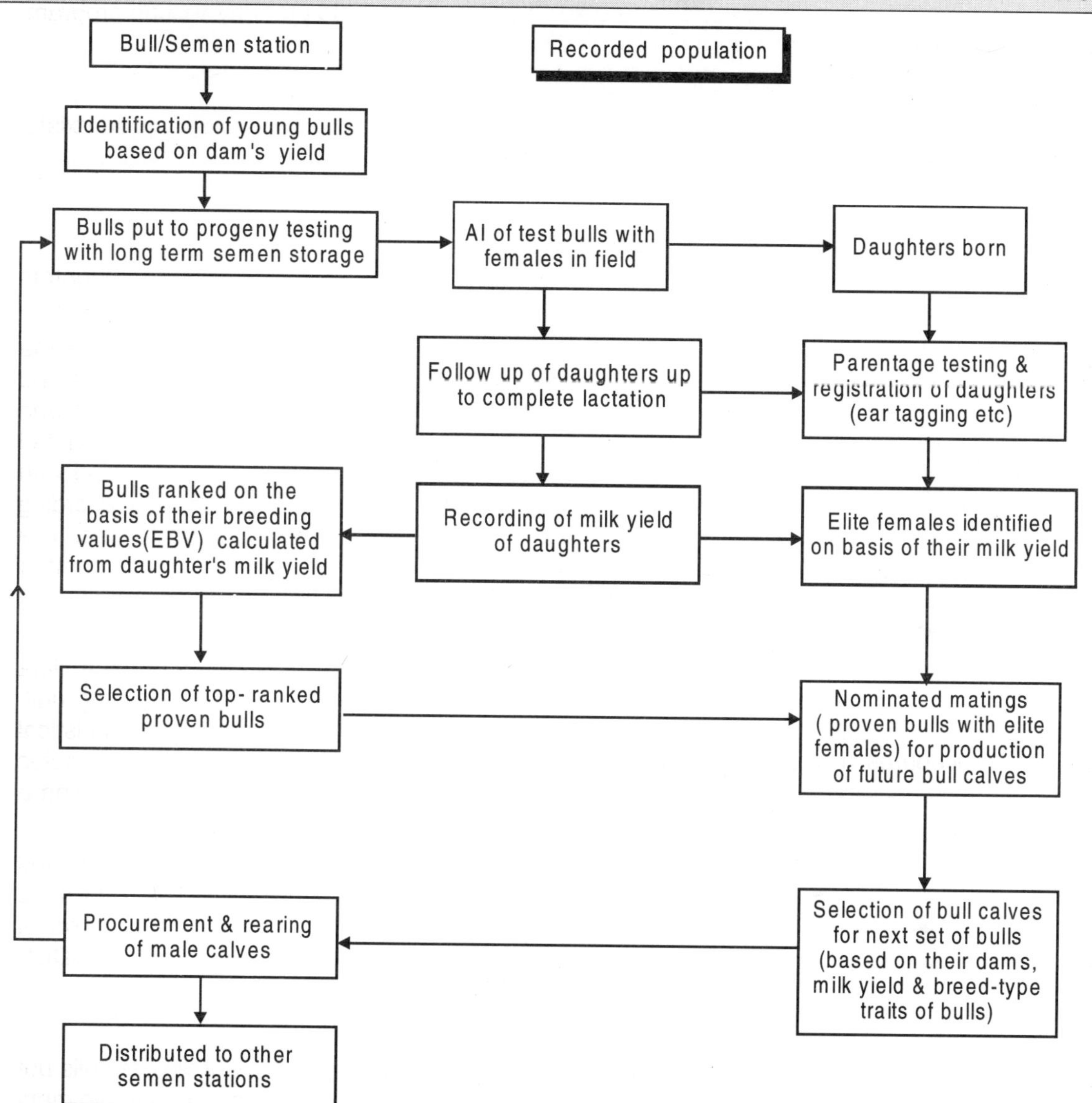

Fig. 18.16 Schematic representation of a field progeny-testing program.

Bovine Improvement Programs of Central Government and Other Agencies

The Government of India has been running a number of programs for the improvement of buffaloes and cattle in the country.

Central Cattle Breeding Farms (CCBF)

These have been established to provide support to farmers engaged in production of indigenous and exotic bull calves of Tharparkar, Red Sindhi, Jersey, Holstein Friesian and crossbred cattle and Surti and Murrah buffaloes. There are seven Central Cattle Breeding Farms (CCBFs) located at different places, and cover different breeds and crossbred cattle:

Farm location	Breed/Germplasm
Andeshnagar (U.P)	HF-Tharparkar cross
Chiplima (Orissa)	Red Sindhi(RS), and RS-Jersey cross
Sunabeda (Orissa)	Jersey
Hessarghata (Karnataka	Holstein Friesian
Suratgarh (Rajasthan)	Tharparkar
Alamadhi (Tamil Nadu)	Murrah
Dhamrod (Gujarat)	Surti

These farms are producing high genetic merit pedigreed bull calves of indigenous and exotic breeds of cattle, and important buffalo breeds for distribution to states for use in the cattle and buffalo development programs. Central Cattle Breeding Farm, Alamadhi is also participating in associated-herd progeny testing program of Central Institute for Research on Buffaloes, Hissar. The buffaloes available at the farm are used for test mating of Murrah bulls, and semen of proven pedigree bulls is used for further genetic improvement

Herd Books and Central Herd Registration Schemes

With the objective of identifying superior breeding animals and to encourage, formation of breed societies, the ICAR started herd books in 1941.This was in pursuance of participation of Government of India in International Convention which was established at Rome in 1936. As per this convention, each breed was to have a herd book in each country which will contain complete information about the breed.

At the onset, it included Red Sindhi and Sahiwal breeds. Later on, it was extended to several other breeds of cattle viz. Gir, Kankrej, Tharparkar, Ongole, and Kangayam. Certain minimum milk yield criteria have been fixed for inclusion in the herd books for each breed. Some breed characteristics were also defined for entry into the Central Herd Registers. Additionally some other requirements like maintenance of records, pedigree etc including inspection of records were also imposed on animal owners.

In 1962-63, the Government of India also started herd registration centres for registration of breeds in different states. The objective of the scheme was to locate superior germplasm in the breeding tract through milk recording and registration. About 11,000 elite cows and buffaloes have been registered during 2004-05.

The central herd registration encourages breeders to form Breeder's Associations for undertaking breed improvement programs using superior bulls. Currently a number of breeder's associations exist in a number of states.

The primary aim of central herd registration scheme is to identify elite germplasm for further improvement. Indigenous breeds covered at present are Gir, Kankrej, Hariana, Ongole of cattle and Murrah, Jaffarabadi, Surti and Mehsana of buffaloes. Under this scheme incentives for rearing of elite cows, buffaloes and their male calves are given to owners of elite animals.

National Project for Cattle and Buffalo Breeding (NPCBB)

Genetic improvement of bovines being a long-term activity, the Government of India has initiated a major program 'National Project for Cattle and Buffalo Breeding' (NPCBB).It was initiated in 2000 with the aim of genetic up gradation and conservation of important indigenous breeds. Important components of the program are:

- To arrange delivery of improved artificial insemination service at farmer's doorstep so that they are brought under organized breeding through artificial insemination or natural service by high genetic merit bulls. The objective is to cover all breedable females of cattle and buffaloes within a period of 10 years.
- To undertake breed improvement program for indigenous cattle and buffalo breeds, and to provide quality breeding inputs in breeding tracts of important indigenous breeds to prevent the breeds from deterioration and extinction.

Programs Operated by Other Agencies

There are some other organizations which are running some improvement programs. One such organization that is playing important role is Bhartya Agro Industries Foundation **(BAIF).** BAIF Development Research Foundation is a non-profit Public Trust, established by Late Dr. Manibhai Desai, in 1967(at Uruli-kanchan, Near Pune) to promote sustainable livelihood in rural India.

Cattle development was the initial preference of this trust to serve small farmers effectively. Decentralized operations in 16 states through nine associated organizations, Central Research Station at Uruli-kanchan and Regional Stations provide research and technology support. The Central Cattle Breeding Farm of BAIF maintains exotic females and bull and bull calves of various breeds of cattle and buffaloes. Under the bull procurement program, calves of varying ages are being maintained and potential bull mothers have been identified. Field performance recording is an ongoing activity. Embryo transfer technology is being integrated in genetic improvement programs.

Under the Field Progeny Testing program of Holstein Friesian, crossbred bulls of ICAR coordinated project, a number of farmers have been registered. Semen of bulls is distributed from four different institutions across the country for evaluation. Progeny performance recording of the first five-bull batches showed average milk yield of 2931.7 kg from 2324 progenies born out of 103 sires.

Review Questions

1. Differentiate between:
   (a) Zebu and European cattle
   (b) Sahiwal and Hariana breeds
   (c) Swamp and Riverine buffaloes
   (d) Murrah and Nili Ravi breeds
   (e) Single- herd and associated-herd progeny testing
2. Write short notes on:
   (a) Milch breeds of indigenous cattle
   (b) AICRP on cattle
   (c) Frieswal cattle
   (d) Sunandini cattle
   (e) Selection of cattle bulls
   (f) Milch breeds of buffaloes
   (g) AICRP on buffaloes
   (h) Filed progeny testing
   (i) Associated-herd progeny testing
3. Why are zebu cattle more adapted to tropical climatic conditions? Or What features of Zebu cattle make them more suited to tropical conditions?
4. What is the difference in cytogenetic profile of Zebu and European cattle?
5. What morphological features distinguish Zebu from European cattle ?
6. Which breeds of indigenous cattle have been exported to other countries? How have these been used in those countries ?

7. Which is the most widely used breed of exotic cattle for improvement of milk production? Describe its important traits.
8. Which is the best exotic breed of cattle for : (a) milk production (b) fat percentage?
9. Which breeds have been used for development of Australian Zebu cattle?
10. Red Dane was originally used for crossbreeding in India but its use was later on discontinued. Explain.
11. Why was crossbreeding of exotic with indigenous cattle considered a better option over selection within the indigenous breeds for enhancement of milk production in India
12. Which of the following was the primary genetic reason for crossbreeding of indigenous with exotic cattle (a) breed complementarities (b) introduction of exotic inheritance (c) use of heterosis.
13. How was crossbreeding started at Military dairy farms ?
14. Name the bilateral projects through which crossbreeding were started in different states.
15. Which three indigenous breeds were used when AICRP on cattle was launched ?
16. What factors determine the choice of indigenous breeds to be used in a crossbreeding program ?
17. Which exotic breed would you recommend for crossbreeding under : (a) low input conditions (b) abundance of feed and fodder? Explain
18. What is the ideal level of exotic inheritance under usual Indian conditions? Why higher levels are not suited?
19. Outline the procedure for producing 5/8-bred crossbreds.
20. Why is it difficult to produce 5/8-bred cattle under field conditions?
21. Outline the procedure for development of a new strain/breed of cattle using Holstein Friesian and Sahiwal cattle.
22. Name the breeds/strains of cattle developed from crossbred base.
23. Which breed/strain of cattle has been produced by the joint efforts of Military dairy farms and ICAR? Explain its status in terms of productivity and census numbers.
24. Why can't the productivity of buffaloes in India be increased through crossbreeding?
25. What have been the main objectives of AICRP and Net Work Project on buffaloes?
26. List the important elements/components of genetic improvement program of buffaloes under the net work/AICRP.
27. Which is the most predominant breed of buffaloes? What are its important attributes?
28. Which buffalo breed of Gujarat region is important for milk production ?
29. Which trait is being used for selection of buffaloes? What is the level of its heritability? What criterion is used for selection of females and males?
30. How can genetic improvement in non-descript buffaloes be brought about?
31. Give a schematic representation of various activities associated with of a bull progeny-testing program (applicable to both cattle and buffaloes).
32. Explain briefly the herd books and central herd registration schemes.
33. What are the important components of national project for cattle and buffalo breeding?
34. List the range of values for the following traits in Murrah buffaloes:
    (a) Age at first calving
    (b) Service period
    (c) Dry period
    (d) Milk yield
    (e) Peak yield
    (f) Fat percentage
    (g) Inter-calving interval

19

Poultry Breeding

The, term 'Poultry', very often used as a synonymous to chicken, means birds which are raised to serve some economic purpose to humans (i.e. meat and eggs). They include chickens, turkeys, ducks, guinea fowl, quails and some other avian species.

Phylum: Chordata
Class: Aves
Order: Galliform
Family: Phasianidae
Genus : *Gallus*
Species : *G. gallus*
Subspecies : *G.g. domesticus*

Of all these species, chicken is commercially the most widely used to provide meat and eggs irrespective of culture and religion. Chickens are taxonomically classified into order 'Galliforms', family 'Phasianidae', and genus 'Gallus' which is distinguished from all other Phasianidae species in having the comb and wattles.

This chapter deals with the breeding of chickens, as among all poultry species, it is the main species contributing to egg and meat production due to its high efficiency of feed conversion into eggs and meat. Nonetheless, the principles of breeding for other species of poultry are same as those of chickens, and in fact development in the breeding of other species has followed the same path as breeding of chickens.

Chickens have played an important role in both fundamental and applied aspects of genetics:

- Application of Mendelian principles to animals was first demonstrated by Bateson and Punnett for plumage color of chickens.
- First case of sex-linked inheritance involving barring gene was demonstrated by Bateson.
- Plumage color mutants of chickens were used by RA Fisher to support his theory of evolution of dominance.

Among farm animal species, poultry breeding was the first to successfully use the principles of quantitative genetics for genetic improvement. Chickens were also the first farm animal species to commercially exploit 'inbreeding and hybridization' technique that was originally used in breeding of maize. Naturally, poultry was the first farm animal class to take the shape of organized breeding industry. It was also the first animal industry to employ professionally trained breeders and geneticists (PhD's) for handling breeding operations including application of computers and other technologies.

Breeding of chickens, being one of the most organized breeding industries, provides a better appreciation of application of breeding principles for genetic enhancement. The coverage in this chapter of breeding of poultry (chicken) is slightly different and is more extensive than of other farm animal species. High reproductive rate, short generation interval, less cost of maintenance thus allowing large-sized populations, and better control of environment (housing, feeding and health management) of this species through confinement has facilitated its fast genetic improvement even at institutional farms in developing countries also.

Red Jungle Fowl

Classification of Chickens: Breeds, Varieties and Strains

Most breeds of domestic chicken (*Gallus gallus domesticus*) are regarded to have their origin from the 'Jungle Fowl' specifically the 'Red Jungle Fowl' (*Gallus gallus*) of India, Burma, Ceylon and other South East Asian countries. Historically chickens were domesticated for cock fighting until the mid-1800's when cock fighting became illegal in many countries. Thereafter chickens were bred by fanciers for exhibition which required uniformity in external features (e.g. for plumage color, body shape, comb type etc).These efforts resulted in development of standard breeds and varieties which were later on used for development of strains for meat and eggs.

> - *"It is a far cry from the time that man first heard the crow of the wild cock of the bamboo jungles of India to the cackle of the highly domestic hen upon celebrating her production of 1000 or more eggs"- Jull, an eminent poultry breeder in his classic paper of 1927 in National Geographic.*
> - *"To the human race this is the most important wild bird living on earth, for it represents the ancestor of all varieties of the domestic fowl. It ranges from the border of Kashmir to Singapore, and is found in the wildest regions, as well as close to native villages—" Beebe, 1926*

Standard breeds of chickens are divided into four major classes according to the area of their origin viz; Asiatic (including Indian), Mediterranean, English and American. Within each class, there are several breeds. **Breeds** within a class are classified based on similarity of body shape and size (form). Some of the important breeds of the four classes are :

- Asiatic: Brahmas, Cochins, Langshans
- Mediterranean : Leghorn, Minorca
- English: Cornish, Australorps
- American: Rhode Island Red, Plymouth Rock, New Hampshire

A breed is further divided into several varieties. A **variety** is distinguished from another by its plumage color (pattern) and comb shape. For example, the Leghorn breed has 13 standard varieties such as White Leghorn with single comb (Single Comb White Leghorn), White Leghorn with rose comb, etc. Within a variety, there may be many strains/lines. A **strain** is developed through selection by a breeder for some economic characters. Usually five generations of selection in a population is taken as the minimum for it to be recognized as a strain. Each strain is given a specific name or code by the breeders (for example, PB1 and PL1 are the strains of broiler and layer chickens respectively maintained at GADVASU, Ludhiana Punjab). Different strains of the same variety often cannot be distinguished from one another by their physical appearance. Breeders have developed a large number of strains in each of the important varieties, particularly in White Leghorn, Rhode Island Red and Plymouth Rock.

Box 19.1: Poultry terminology.

- **Broiler** (alternative term 'meat chicken')**:** A young bird of either sex that is bred and grown specifically for highly efficient meat production. Broilers are usually slaughtered at 5 to 7 weeks of age when their meat is tender and juicy.
- **Brooder:** The equipment used to provide supplementary warmth (the energy used for heating may come from electricity, gas, oil or from other sources) during the early stages of chickens' life when their thermoregulatory system is not well developed. This period of the first weeks of a chicken's life when it requires a very high standard of care including provision of special diets and supplementary warmth is called brooding period (stage), and the process is called brooding.
- **Broody:** The instinct controlled by maternal hormones that causes the female bird to want to sit on eggs for hatching and to care of baby chicks is called broodiness, and the hen that does so is called broody. In modern high laying strains, broodiness has largely been eliminated.
- **Chick:** The term used to describe chickens from day- old to end of brooding period, which can extend up to 6 to 8 weeks depending on environmental conditions.
- **Cock:** A male that has finished one season as a breeder, usually over one year of age.
- **Cockerel:** A young male from day-old to end of its first few months of breeding is called a cockerel. Often the term is used to refer to young males up to 6-8 months of age.
- **Deep litter:** The system of housing where a suitable bedding material called litter is provided in the poultry house floor for the birds to sit.
- **Fowl:** The general term used to describe all members of domestic fowl irrespective of age, sex or breed.
- **Growers:** Term used to describe birds from the end of brooding until they reach sexual maturity.
- **Hen:** The term is used to describe mature females who have laid eggs for a certain period, usually one laying cycle.
- **Incubation:** The process by which fertile eggs are subjected to conditions suitable for initiation and sustaining of embryonic development and hatching of chicks. The machine used to incubate fertile eggs is called an incubator.
- **Layer:** A female in lay; the term is usually used to refer to females kept solely for egg production.
- **Laying cycle:** The period from onset of laying until the natural moulting causes cessation of production constitutes the laying cycle. Since the modern laying stocks do not normally undergo natural moulting, therefore, laying cycle is used to describe the period during which an economic level of production is being maintained by the flock. The traditional laying cycle under intensive management is considered up to 72 weeks of age though now laying cycles are being prolonged through management and breeding.
- **Moult:** The process whereby the bird sheds its feathers and ceases egg production. It is usually initiated by hormonal influences. Birds can be induced to undergo moulting through feed and water restriction (stress) which is called forced moulting.
- **Peck(ing) order:** The social organisation of a flock leading to formation of a hierarchy from the most dominant to the subordinate member of the flock.
- **Primaries:** The ten long, stiff flight feathers at the outer extremity of trailing edge of wing. They are separated from the inner group or secondary by the axial feather.
- **Pullet:** A female in her first laying season, usually below one year of age.
- **Vent:** The common external opening from the cloaca.

Segments of Chicken industry

Chicken industry has specialized into two distinct segments viz. eggs and broilers (meat). Accordingly stocks have been specifically bred to cater to these two segments: egg-type and meat-type stocks.

Egg-type stocks: The stocks which have been specifically bred for production of eggs are referred to as egg-type or layer stocks. Single Comb White Leghorn is the most popular variety used for egg production throughout the world. It has high egg-laying capacity and a relatively small body which is desirable for economy in feeding. Hundreds of strains of this variety have been produced by different breeders. Other breeds that have been bred for egg production are Rhode Island Red and Australorps. The egg-type stocks are further divided into two types based on egg shell color:

- *White egg layers :* Virtually all the strains or strain-crosses producing white shelled eggs come White Leghorns.
- *Brown egg layers :* The strains/lines or their crosses which produce brown eggs come from a broader genetic base utilizing breeds like Rhode Island Red, Australorps, Barred Plymouth Rock and New Hampshire among others. Some breeders also incorporate White Leghorn strains in crossing to produce pale or tinted egg shell.

Meat-type stocks: The stocks which have been bred for production of meat are referred to as meat-type or broiler stocks.The major breeds used for developing broiler stocks are White Plymouth Rock, Cornish and New Hampshire. These are heavy breeds and characterized by rapid growth.

Description of Some Important Breeds

There are a very large number of breeds of chicken but only a few have been used for production of commercial birds for meat and eggs.

- **Leghorn:** The Leghorn birds have small, compact and light body with rose or single comb. There are 13 standard varieties of Leghorn out of which Single Comb White Leghorn (SCWLh) is the most popular due to its excellent laying performance, and is extensively used all over the world. The plumage color is white with white or creamy earlobes. It produces white shelled eggs.
- **Rhode Island Red:** Rhode Island Red (RIR) is a dual-purpose breed possessing both meat- and egg-laying capability. This breed originated from the cross between Brahma or Langshan (of Asiatic class) and the common fowls of Rhode Island State of U.S.A. Single and Rose Comb are the two common varieties of RIR. The birds of this breed have a long and broad body, with a deep breast carried forward. The male is dark red with black tail while the female is deep red throughout. It lays brown-shelled eggs.
- **Australorps:** This breed was developed in Australia as a utility breed in the 1920s from the Black Orpington breed primarily for egg production rather than for its meat. They lay tinted eggs. There are two plumage colors; the black, which has feathers with a lustrous green sheen, and the blue whose feathers are slate blue with darker lacing. Their comb is single and evenly serrated while their earlobes are red.

Fig. 19.1 Most commonly used breeds for eggs and meat. Note the body shape of different layer, dual and broiler breeds.

- **New Hampshire:** This breed was developed from the crosses of Rhode Island Red with some other breeds. The birds of this breed are known for rapid feathering with good quality meat and are also good egg producers. The eggs are brown shelled. The plumage color is chestnut red. Body shape is less rectangular than Rhode Island Red (Fig. 19.1) It has single comb.
- **Plymouth Rock:** The most popular varieties of this breed are Barred Plymouth Rock (BPR) and White Plymouth Rock (WPR). The birds are of good size with excellent quality of flesh and good laying capability. Body is large, deep and compact with fairly broad and well-rounded breast (Fig.19.1). This breed has a single comb. The Plymouth Rocks (often called simply Rocks or Barred Rocks, after their most popular color),) have been widely used for developing broiler stocks. The Barred Plymouth Rock was one of the foundation breeds for broiler industry in the 1920s, and the White Rock continues to be used on female side of the commercial broiler cross.
- **Cornish:** This breed was developed in England from crossing of Asiatic fighting cocks (the Aseel of India and Malay of Sumatra) and the English Game birds (hence called as Indian Game also). It has well-fleshed body. The birds are known for their especially deep and broad-breast that makes it a prized meat-type breed. It has pea comb and white skin. Cornish has been extensively used as sire line (male parent) for producing meat-type crosses for broiler production. It comes in a number of plumage colors but all lay brown eggs.

- **Asian breeds:** The chicken breeds, which originated in Asia, are no more popular as such but a number of them have been used in developing the modern commercial poultry stocks. The Aseel, also known as Indian Game Bird, is well known for its conspicuous breast and meat qualities (Fig 19.1). It has been used in developing the Cornish breed. Brahma and Langshan breeds of Asiatic class have probably contributed to the ancestry of Rhode Island Red.

LAYER BREEDING

Chickens on worldwide basis contribute about 92% of the eggs produced by all poultry species; the remaining 8% being contributed by turkeys, ducks or species other than chickens. Chicken eggs, unlike duck eggs, are acceptable to all consumers. Chickens are also better converter of feed into eggs. The chickens that are kept for production of eggs are called **layer or egg-type chickens**. Breeders of egg-type chickens concentrate on four breeding goals:

- Maximum number of saleable eggs per hen
- Low mortality
- Low feed cost per unit of egg produced (either per egg or per kg of egg mass)
- Optimum egg quality

Thus, breeders of egg type chickens are concerned with improvement of traits impacting egg production in the laying cycle, which include number of eggs that are of saleable size (egg weight), good viability, small body size to minimize maintenance cost, and optimal quality of eggs for marketing. In order to accomplish these goals, breeders of layer chicken select for a number of characters.

Economic Traits: Description, Measurement and Inheritance

The important traits of economic importance in layer chicken are described below:

Egg Production Traits

The main product of raising layer chickens is production of maximum number of eggs of saleable size during the laying cycle per hen-housed at the start. The total number of marketable eggs produced in the laying cycle is dependent on early age at sexual maturity, high laying intensity, optimal egg size that is suitable for sale, and low mortality.

Egg number (production): The number of eggs produced by a stock in a given period (*i.e.* the laying cycle), interchangeably referred to as egg number or egg production, is the most important trait contributing to economic returns. Egg production of each individual hen is measured, through trap nesting or individual cage housing as the number of eggs laid by it from onset of laying until the end of laying period. Laying period is terminated when the level of egg production drops below economic level, usually up to 72 weeks of age. Egg number up to 72 weeks of age is referred to as annual egg production (number). Though annual egg production is the most important trait for commercial purpose but poultry breeders usually record egg production for a shorter period (usually up to 40 weeks of age) which is called part-year egg number or production.

The heritability of egg number is low (about 0.15), therefore, individual selection is not effective for its improvement. This trait is improved through family selection or combined selection utilizing performance of the individual hen, and performance of its half sister's and full-sisters by assigning weighting factors ('b' values) to different sources as follow:

Combined index value of hen = b_1 *Hen's performance* + b_2 *Full sister's performance* + b_3 *Half-sister's performance*

Egg production being sex-limited, the cocks are selected using combination of their full- and half-sister's egg production:

Combined index value of cock = b_1 *Full sisters average* + b_2 *Half sisters average.*

Age at sexual maturity (ASM): A pullet is considered sexually mature when she lays her first egg. Age at sexual maturity of a pullet is, therefore, measured as the number of days taken by her from hatching to laying of first egg (hence, it is also called 'age at first egg'). The earlier a female starts laying, the longer will be the period available to her for laying, and therefore more will be the number of eggs laid by her in the laying cycle. Age at first egg thus directly contributes to economics of egg production. Modern improved strains of White Leghorn chicken become sexually mature between 20 and 22 weeks of age. Other layer breeds like Rhode Island Red etc reach sexual maturity about two or more weeks later. Strains of the same breed also differ in age at sexual maturity depending on their genetic history of selection. On a flock basis, the age when the flock attains 50% egg production, called age at 50% egg production, is often used as measure of flock's average age at sexual maturity.

Heritability of sexual maturity is moderate (about 0.25 to 0.30). Its genetic correlation with egg number is negative but desirable which means that the birds having genetic capacity to start laying at an early age also have the genetic capacity to produce more number of eggs. Selection for the improvement of egg number to a fixed age automatically leads to reduction in age at sexual maturity as a correlated response. However too early sexual maturity is also not desirable because genetic correlation between age at sexual maturity and egg weight is positive which is undesirable. Therefore, hens that start laying too early produce eggs that are not suitable for marketing.

Age at onset of laying, besides genes, is markedly influenced by photoperiod. Therefore, light management (duration and intensity of artificial lighting) is an important factor via which sexual maturity can be modified. Modern strains of layer chickens have attained sufficient early age at sexual maturity such that further genetic reduction in age at sexual maturity is not desirable.

Rate of lay (also called 'laying intensity) : It is measured as the number of eggs laid by a hen up to a fixed age divided by the number of days available from sexual maturity to the specified (fixed) age (expressed in percentage) :

$$\text{Rate of lay (\%)} = \frac{\text{Number of eggs laid upto a specified age}}{\text{Number of days in available for laying}} \times 100$$

Rate of lay per se is thus not affected by age at start of lay. Rate of lay is actually the egg number (production) adjusted for the variable age at first egg of different birds, and expressed as percentage. Rate of lay (intensity of egg production) is one of the important traits determining profitability of layer chickens.

Heritability of rate of lay is low and its genetic correlation with egg production is positive though of low magnitude. Rate of lay generally improves as correlated response to selection for egg number. But once a strain has reached a high level of egg production, selection for rate of lay (laying intensity) *per se* may have to be practiced for its further improvement.

From the definition of the age sexual maturity and rate of lay given above, it can be seen that the number of eggs produced by a bird in a fixed period is primarily made up of two components[1] : age at sexual maturity and rate of lay. A hen that lays more number of eggs up to 40 or 72 weeks of age will generally be the one that starts laying at an early age (as the potential number of days available for it will be more) and shall also have higher laying intensity

Illustrative example: A female was hatched on 1st March 2015. She laid her first egg on 27th July, 2015. The total number of eggs laid by her from first egg to 72 weeks of age was 304. Age at first egg of the bird is calculated by counting the number of days for different months as shown below:

March: 31 days + April: 30 days + May : 31 days + June : 30 days + July : 26 days (the day on which the first egg is laid excluded) = 148 days.

The calculation of rate of lay requires the number of days available for the hen to lay until 72 weeks of age (504 days of age), usually called 'potential/functional number of days. The potential number of days available for the hen from age at sexual maturity to 72 weeks of age would be 504 – 148 = 356 days.

$$\text{Rate of lay (\%)} = \frac{304}{356} \times 100 = 85.4$$

Similarly rate of lay of a flock/strain can be calculated on daily, weekly or monthly basis which can then be used to plot the laying curve over the entire laying cycle. Such a curve, called egg production/laying curve, is useful in evaluating whether a flock is performing optimally or not.

A typical egg laying curve of a modern commercial strain is given in Figure 19.2.The rate of egg laying follows a typical pattern : the flock reaches peak egg production by about 6 weeks after coming into lay, then production starts declining until termination of laying cycle, usually at 72 weeks of age. Modern laying stocks, as compared to their older counterparts, start laying earlier, reach early peak production and maintain high laying rate for a much longer duration.

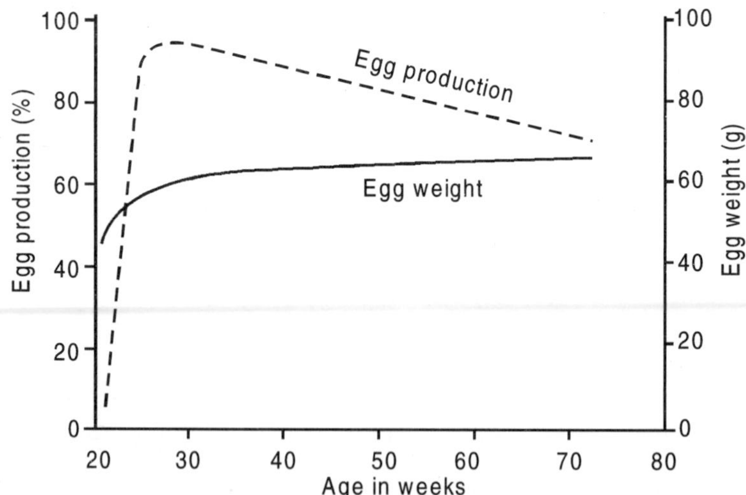

Fig. 19.2 Typical egg production and egg weight curve of layer chickens

[1]Goodale (1919) listed five components of egg production to a fixed period viz; age at sexual maturity, laying intensity,persistency of egg production, number of pauses(clutch size) and number of broody period a hen undergoes during laying cycle(broodiness). But in modern layer strains broodiness has been eliminated, and number of pauses are too few..

Measurement of egg production: Egg production can be expressed in three different ways:

Individual hen-basis:

- *Survivor's egg production*: Number of eggs laid by a hen to a fixed age which is recorded only for the hens that survive the recording period of 72 or 40 weeks of age (hence also called 'survivor's egg number/production').It can be measured on individual hen basis either through trap nesting or through housing of hens in individual cages.

Group basis (e.g. family or flock):

- *Hen-day egg production*: It is the number of eggs laid by a group of birds divided by the number of days for which the hens were alive in that period (called 'hen-days') and expressed as percentage:

$$\text{Hen-day production (\%)} = \frac{\text{Number of eggs laid upto a specified age}}{\text{Number of hen-days}} \times 100$$

- *Hen-housed egg production*: It is defined as the total number of eggs laid by a group of housed birds divided by the number of birds that were alive at the start of recording period. Thus, the hens that die after placement in the laying house, and hens that never lay any egg during the recording period are included in calculating hen-housed egg production. Since the smallest unit for measuring the hen-housed egg production in a strain/line is the dam family, hen-housed egg production of each dam family is calculated and used as the criterion of selection. This measure of egg production includes a component of mortality (livability); therefore, selection based on hen-housed egg production also helps to select families for their survivability in addition to their egg production capacity. It is calculated as follow:

$$\text{Hen-housed production (\%)} = \frac{\text{Number of eggs laid upto a specified age}}{\text{Number of birds housed} \times \text{No. of days}} \times 100$$

Illustrative example: Suppose 1000 hens were housed in a shed (fitted with trap nests or in individual cages) at start of laying beginning 1st September 2013. The number of birds alive and the number of eggs laid during different months were recorded and are listed below:

Month (days)	No. of birds	Total number of hen-days	No. of eggs laid
September (30)	1000	1000×30 = 30000	20000
October (31)	990	990×31 = 30690	23000
November (30)	970	970×30 = 29100	23500
December (31)	960	960×31 = 29760	24500
Total (122)	3920	119550	91000

The calculations of hen-housed and hen-day egg production are shown below:

$$\text{Hen-day production (\%)} = \frac{91000}{119550} \times 100 = 76.1$$

$$\text{Hen-housed production (\%)} = \frac{91000}{1000 \times 122} \times 100 = 74.6$$

It is apparent from the example that hen-day egg production is higher than hen-housed egg production because hen-house egg production includes the component of mortality while hen-day production excludes the hens which died during the period.

Egg weight (also called egg size): Weight of eggs is also an important trait that contributes to income: small-sized eggs fetch lesser price than larger eggs. Egg weight is lowest at start of laying, increases sharply, the subsequent increase in egg weight is very small (Fig 19.2). Since it is not feasible to weigh all the eggs laid by a bird, egg weight of a bird is usually measured as average weight of its 6 to 8 consecutively laid eggs at mature age (usually between 36 and 40 weeks of age). Repeatability of egg weight over the laying cycle is very high therefore, egg weight measured at mature age by sampling 6 to 8 eggs provides a useful measure of hen's genetic potential for egg weight.

Genetic correlation of egg weight with egg number is negative which means that selection directed at improvement of egg production decreasess egg weight. Since eggs are marketed on the basis of egg weight categories (e.g. large, medium and small) rather than absolute egg weight, moderate selection, either to increase it or to maintain it's current level, is practiced. Heritability of egg weight is high (more than 0.5), therefore, it responds sharply to individual selection. Usually certain independent culling levels, following preliminary selection for egg production, are used to select for egg weight.

Simultaneous selection for improvement of egg production and egg weight, the two antagonistically correlated traits, has been effective as shown by a number of selection experiments (Fig. 19.6). The effectiveness of selection for simultaneous improvement of these two traits is also evident from the profile of modern layer chicken stocks as compared to those available a few decades earlier (see Table 19.3). Through continued selection, egg weight of the modern egg-type stocks at maturity has been improved to satisfactory levels. It is however only the early egg weight (egg weight during early stages of laying) which is being primarily targeted for further improvement, and that too without causing any further increase in egg weight during late stages of laying cycle.

Both egg number and egg weight are economically important traits. These two traits are sometimes combined together to derive the trait 'egg mass'. Egg mass is obtained as product of number of eggs laid by a hen and its average egg weight.

$$\text{Egg mass} = \text{Egg number} \times \text{Average egg weight}$$

Selection for egg mass has been proposed as a way of genetically improving these two antagonistically related traits.

Mortality or viability (measure of disease resistance): Diseases cause mortality as well as morbidity leading to economic losses. The most commonly used measure of 'general disease resistance', which is the ability of birds to resist all types of diseases including stress, is mortality. Since mortality on an individual-bird basis is an 'all or none' trait, it is expressed on family basis (full-sib or half-sib family) for selection purposes. It is calculated as the number of birds which died during a specified period divided by the number of birds at the start of that recording period, and expressed as percentage:

$$\text{Mortality (\%)} = \frac{\text{No. of birds which died in a specific age period}}{\text{Number of birds at the start of a period}} \times 100$$

Mortality is sometimes expressed separately for different periods viz. brooding, growing and laying periods:

- *Brooding-house mortality:* Mortality during early period of chick's life when they are provided extra heat and special care(usually 6 to 8 weeks of age) is much more prone to environmental factors, and is expressed as follow:

$$\text{Brooding mortality (\%)} = \frac{\text{No. of birds which died in brooding period}}{\text{Number of birds (healthy chicks) put in brooder house}} \times 100$$

- *Growing-house mortality*: Mortality during growing period, which is usually from 8 to 20 weeks (i.e. from end of brooding to housing in the laying house/cages) :

$$\text{Grower mortality (\%)} = \frac{\text{No. of birds which died in growing period}}{\text{Number of birds put in grower house}} \times 100$$

- *Laying-house mortality*: Mortality in the laying house (i.e. from putting the birds in laying house to end of laying cycle).

$$\text{Laying-house mortality (\%)} = \frac{\text{No. of birds dead from housing to end of laying}}{\text{Number of birds housed}} \times 100$$

The genetic correlation of egg production with mortality is generally undesirable but low. Consistent long-term selection for egg production is thus to increase mortality percentage. The heritability of mortality (livability) is very low (about 0.05); therefore, family selection is employed to improve this trait. Mortality percentages for all the families are calculated, and the families with high mortality, usually using independent culling levels, are rejected from producing the next generation.

Feed Efficiency

Feed efficiency, also called feed conversion ratio (FCR), in layer chickens is defined in terms of feed consumed for producing a unit of egg or egg mass. Efficiency of conversion of feed into eggs is of great importance as feed accounts for more than 60% of the cost in layer chickens. Feed conversion efficiency, though primarily a function of number of eggs laid by a hen, is also influenced by its body weight and egg weight. The genetic correlation of feed efficiency with egg number is desirable. Consequently, efficiency of feed conversion improves as a correlated response to selection for improvement of egg production. For example, feed conversion in USA and Canada has improved from 2.95 gram of feed required to produce one gram of egg(1960) to less than 2.0 g feed per gram of egg (2001). The most commonly used measures of feed efficiency are: feed intake per egg, feed intake per unit of egg mass. Feed efficiency (per egg) is calculated as follow:

$$\text{Feed efficiency} = \frac{\text{Amount of feed consumed in a specific period}}{\text{Number of eggs laid by the bird in that period}} \times 100$$

Both these measures of feed efficiency require recording of feed intake of individual hens, which is not only laborious but cost prohibitive as well.

Much of the variation in feed consumption between hens in a flock is accounted for by differences in egg number, egg weight and body weight of the birds. If the differences in feed consumption between hens are corrected/ adjusted for these traits, the remaining differences for feed consumption between hens are due to differences in what is commonly called 'residual feed consumption'. Thus, residual feed consumption is a measure of feed consumption which is not accounted for by these three traits.

Egg Quality and Composition Traits

Egg quality consists of both external and internal attributes of the egg:

- Exterior quality traits include egg shape which is measured as ratio of egg width to egg length, shell-quality(i.e. shell strength which determines the susceptibility of egg to breakage from laying through transportation), shell color, shell texture (i.e. smoothness)
- Interior quality traits include albumen quality and presence of blood and meat spots (called 'egg inclusions'), and egg composition viz; percentage/proportion of yolk, albumen, and shell.

Of these traits, shell quality (shell strength) and albumen quality attain highest attention in selection programs.

Shell quality: Shell quality affects not only egg breakage (from laying of egg till it reaches the consumer which includes its transportation) but shelf-life as eggs of good shell quality can be kept longer without loss of their interior quality. The most commonly used measures of shell quality include :

- Shell thickness which is recorded by breaking-open the egg and measuring thickness of the shell at three places on its equatorial region.
- Egg specific gravity which is recorded by floatation method where eggs are immersed in saline solutions of ascending specific gravity,
- Egg shell deformation which is a measure of egg shell strength with respect to its breakage and is recorded by an instrument.

Egg shell quality declines with age. Of the various measures of shell quality, egg specific gravity, being a non-destructive is most commonly used for improvement through selection. Heritability of shell quality is moderate.

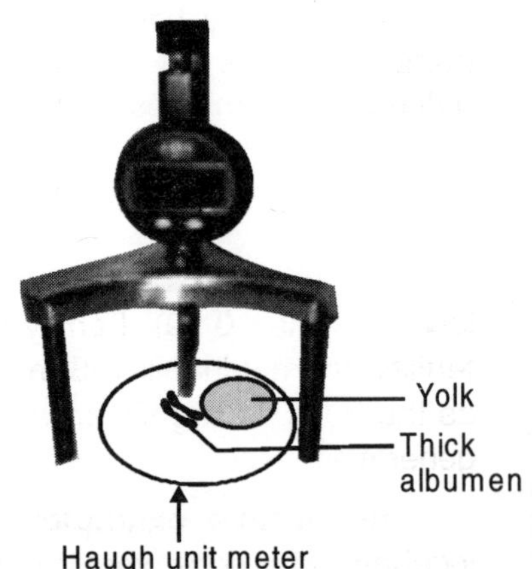

Albumen quality: Albumen quality which is measured as height of thick albumen, determines storage quality of eggs as consumers prefer eggs with firm or thick albumen. Height of thick albumen is positively correlated with egg weight; therefore, absolute height of thick albumen is not a suitable measure of albumen quality. The albumen height adjusted for egg weight, which is called Haugh Unit Score (HUS), is used as a measure of albumen quality. This measure of albumen quality was introduced by Raymond Haugh in 1937 and uses the following formula:

$$\text{HUS} = 100 \times \log\text{(height of albumen in mm)} - 1.7\ \text{(Egg weight)}^{0.37} + 7.6$$

Albumen quality of individual hens is determined by breaking a sample of eggs of each hen and measuring its HUS by Haugh unit meter which gives a direct reading (Photo above). Haugh unit score declines with age of hens. Albumen quality can be improved through individual selection as its heritability is moderate. In Western countries, standards have been set for minimum acceptable HUS. For example, the US department of agriculture (USDA) has set HUS value of 72 or more for 'AA' quality eggs.

The relative attention paid to egg quality traits varies from market to market but is usually low. Heritabilities of these egg quality traits are moderate, therefore, they respond well to individual selection. In general, the egg quality traits have undesirable genetic correlation (negative) with egg production; therefore, breeders make some selection for these traits.

Reproductive Traits

Reproductive traits per se are not important in birds reared for commercial production of eggs. But they are important for pure line breeders as higher reproductive ability of parental lines increases the number of chicks hatched and sold.

Fertility and hatchability: Fertility and hatchability are the most commonly used measures of reproduction. Fertility is calculated as the number of fertile eggs divided by the number of eggs set for incubation and expressed as percentage:

$$\text{Fertility (\%)} = \frac{\text{No. of fertile eggs}}{\text{Number of eggs set in incubator}} \times 100$$

Fertility of eggs is observed by 'candling' (i.e. passing bright light from below the eggs) during the first week of setting in the incubator. The presence of embryo is seen as a dark shadow when bright light is passed from below the eggs while eggs which do not contain embryo (infertile) are seen clear.

Hatchability is a measure of ability of embryos to complete the incubation period and hatch successfully. It is calculated as number chicks hatched divided by the number of eggs fertile, and expressed as percentage (this is specifically referred to as 'hatchability of fertile eggs'):

$$\text{Hatchability of fertile eggs (\%)} = \frac{\text{No. of chicks hatched}}{\text{Number of fertile eggs}} \times 100$$

Hatchability can also be calculated as the number of chicks hatched divided by the number of eggs set for incubation (referred to as 'hatchability of eggs set'). This measure of hatchability combines both fertility and hatchability of fertile eggs.

$$\text{Hatchability of eggs set (\%)} = \frac{\text{No. of chicks hatched}}{\text{Number of eggs set in incubator}} \times 100$$

Heritability of both these measures of reproduction (fertility and hatchability) is low to very low (less than 0.10). Family selection is usually practiced to discard the families having low fertility and hatchability. Some degree of natural selection for these traits automatically occurs as the hens having lower fertility and hatchability contribute lesser number of progeny to next generation.

Illustrative example: Suppose 1000 eggs of a strain of chicken were set in the incubator.The eggs were candled on 5th day, and infertile were discarded while the fertile eggs were incubated. The hatch was taken out on 22nd day, the chicks counted, and those which died during incubation period, called 'dead-in-shell', discarded. Suppose 900 eggs were found fertile on candling and 810 chicks hatched. The calculations of fertility and hatchability are shown below:

$$\text{Fertility (\%)} = \frac{900}{1000} \times 100 = 90.0$$

$$\text{Hatchability of fertile eggs (\%)} = \frac{810}{900} \times 100 = 90.0$$

$$\text{Hatchability of eggs set (\%)} = \frac{810}{1000} \times 100 = 81.0$$

Behavior Traits

Considerable genetic improvement in economically important traits of layer chickens has been brought about during the last five-six decades. The flocks undergoing selection were reared under intensive and mechanized management and farming systems. Under such controlled and intensive systems, the expression of behavior was restricted, and some management practices like debeaking etc were used to overcome the problems associated with behavioral changes. Also, the animal welfare was not considered a significant issue in commercial rearing practices.

It has been observed that consistent selection for production traits has resulted in some physiological changes that are reflected in behavior of birds such that modern layer strains have lowered adaptability. This reduced adaptability is leading to a range of behavior problems like aggression, feather pecking, cannibalism etc. With the animal wellfare issue gaining increasing attention, some aspects of behavior are becoming important for consideration in selection, at least in developing countries.

An important aspect of behavior is the aggression of laying hens (bird-to-bird aggression) when housed in-group cages. Aggression can lead to pecking (when one bird starts picking and pulling feathers of another bird) which then leads to cannibalism (extreme form of pecking leading to wounds and bleeding). Poor feather cover resulting from these acts also affects overall productivity of the flock (group). These aspects of behavior have been found to be under genetic control.

Selection of hens through housing in groups (called 'group selection') has been shown to be effective in reducing aggression as reflected in reduced mortality. Results of a selection experiment in which one line, called **'kinder gentler'** line, was selected by group housing of families. Comparison of cumulative mortality in kinder line along with a control line of same genetic background and an industrial stock (Fig 19.3) showed that cumulative mortality in kinder strain was reduced as the birds exhibited lower aggression.

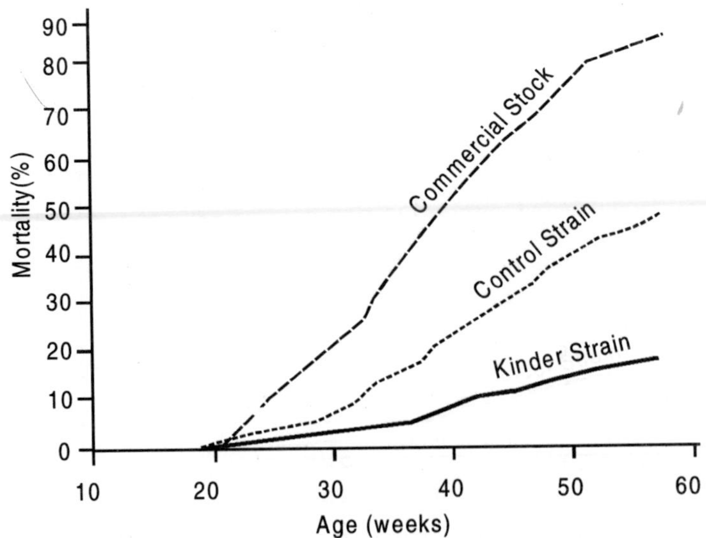

Fig. 19.3 Annual percentage mortality in 'kinder gentler' line compared with its control counterpart and a commercial stock (Source : Cheng and Muir, 2005).

Some other traits related to bird-welfare and production under intensive production system may become important for consideration in genetic improvement depending upon demands of the niche market. For example, the European Union (EU) has issued guidelines to eliminate the conventional cages for housing so that the eggs only come from free-range system. Currently the EU mandates 550 cm^2 cage area per bird. Most commercial white egg layers in the world are housed in cages at 310 to 350 cm^2 per bird.

A summary of the important traits of layer chickens is given in Table 19.1

Table 19.1 Summary of economic traits of layer chickens.

Trait	h^2	Selection criterion	r_G with egg number
Egg number :	Low	Combined selection	—
Egg weight :	High	Individual selection	–ve, undesirable
Age at first egg :	Moderate	Improved as correlated response	–ve, desirable
Rate of lay :	Low	Usually improved as correlated response	+ve, desirable
Mortality ;	Very low	Family selection	–ve, undesirable
Feed efficiency :	Moderate	Improved as correlated response	–ve, desirable
Fertility & hatchability :	Very low	Family selection	–ve, undesirable
Egg quality :	Moderate	Individual selection	–ve, undesirable

Use of Part-year Production for Genetic Improvement of Annual Egg Production

Since the laying hens are normally kept economically up to 72 weeks of age,therefore, ideally selection should be based on egg number up to this age (referred to as 'annual egg number/production'). However, selection based on egg number to 72 weeks of age (i.e. the annual record) has a few lacunae:

- It increases the generational interval and hence decreases genetic progress per annum
- It increases the cost of maintaining and recording of egg production of breeding candidates (hens)
- It also disrupts the hatching season such that hatching in successive generations occurs in different seasons which leads to management problems.

To overcome these problems, selection is based on 'part-year' egg production, (usually up to 40 weeks of age as, this period covers the peak period of egg production cycle. Since 'part-year egg production' is an integral component of annual egg production, the genetic correlation between these two is positive and very high. Therefore, selection on part-year egg production also improves annual egg production as a correlated response, which in turn enhances genetic progress per annum.

The use of part-year egg production as selection criterion was proposed by pioneer poultry breeders, Dempster and Lerner(1947),while working at the University of California(USA). However, it was later on observed from experimental results, particularly in strains that had undergone selection based on part-records for a number of generations that such a selection largely led to increased egg number during the early part of production cycle. But egg production during the latter part of the annual egg production cycle was not increasing at a corresponding rate. Egg production during the latter part of production cycle (i.e. from 40 to 72 weeks) is referred to as 'residual egg production'. Therefore, breeders initiated selection for a longer part-year records like those up to 50 to 52 weeks of age so that egg production during residual part is sustained at high levels. As a result of such selection, the modern laying strains maintain high rate of lay for a longer period (see fig. 19.8).

A schematic flow chart of the steps and activities involved in selection and reproduction of egg-type stock, with one-year generation interval, is diagrammed in Fig 19.4.

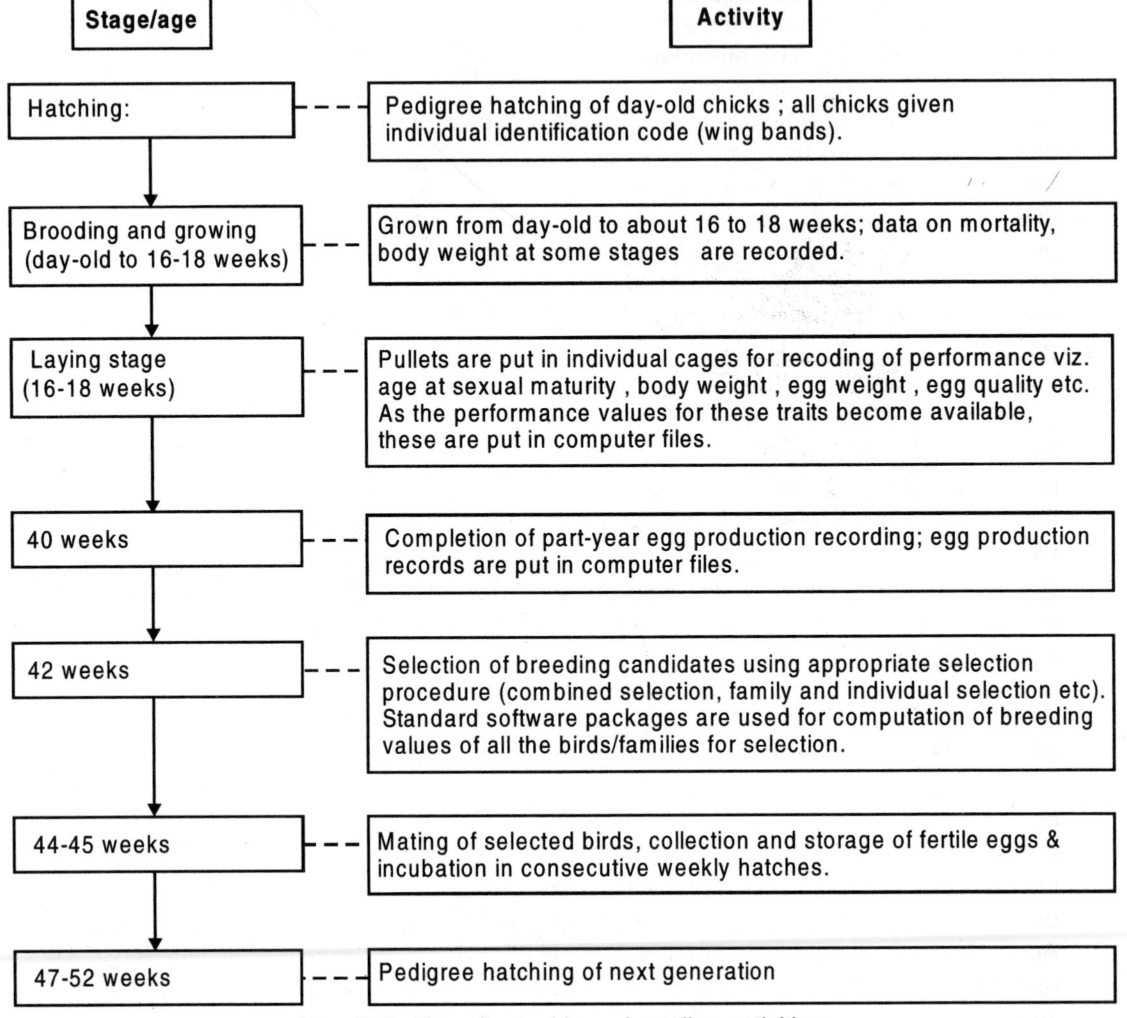

Fig. 19.4. Flow chart of layer breeding activities.

Size of the selected strains usually runs into thousands with one sire mated to 5-10 dams depending upon the mating structure.

Some Important Technological Developments in Poultry Breeding

Genetic improvement of poultry involves maintaining a number of large-sized populations recording of different traits on breeding candidates, computation of breeding values, selection of the required number of birds, making mating decisions and reproduction.

This whole process has been greatly aided by some technological developments:

- **Trap nest:** The trap nest, which is a simple mechanical device, was invented in the latter half of 19th century in Austria as a means of identifying broody hens. However, it was later employed for recording the number of eggs produced by individual hens. Interestingly the purpose of its development was different from the one for which it became popular. The trap nests are placed in laying shed. When a hen is to lay the egg, she enters the trap nest, as it prefers darkness and isolation. Once it enters the trap nest, the metallic gate closes automatically. The attendant managing the laying flock opens the nests at periodic intervals to release the hen and at the same time marks the egg with leg/wing band number of the hen who laid it (hen's identity code on egg). The mechanical trap nest has been extensively used for maintaining the egg production records of breeding hens which were reared on floor. With the development and widespread usage of cage housing and artificial insemination systems, trap nests are no more used.

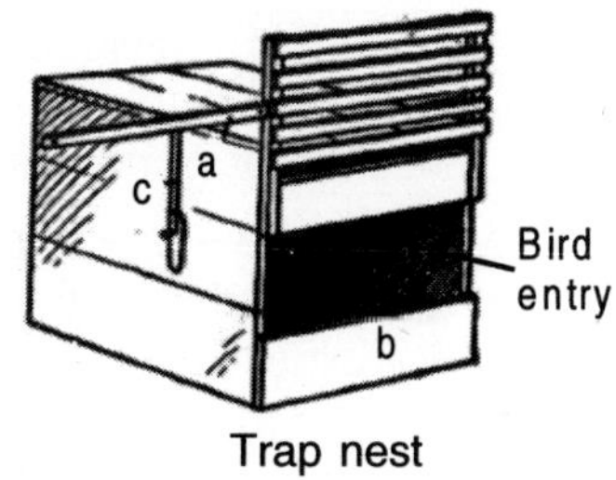

Trap nest

- **Artificial insemination:** Artificial insemination in poultry owes its origin to Burrows and Quinn (1937), who developed the method of abdominal massage and pressure to collect semen. With the ease of collecting poultry semen, and proximity of hens on large breeding farms, AI is used with freshly collected semen. It has allowed use of limited number of cocks thereby increasing selection intensity as well as overcoming the problems associated with natural mating in broiler chickens. In turkey breeding, its use is 100%.
- **Electronic methods of data recording:** Poultry breeders maintain a number of strains/lines each having a large number of breeding candidates, usually running into thousands. Performance records of all the birds of pedigreed strains, which include a number of traits, need to be recorded accurately for subsequent use in selection. A breeding program cannot succeed unless performance records are collected accurately. When the birds were reared on floor (deep-litter), the performance records were recorded manually by catching and the weighing the broilers or by trap nesting layer birds. For cage-housed birds, recording involved manual handling of birds. In the recent past, however, commercial poultry breeding companies have utilized bar- code technology, integrated with scales and hand-held computers, for accurate and efficient recording of data from hatching until completion of reproductive life of pedigreed breeders. Application of this newer technology has greatly reduced the cost of poultry breeding operations.
- **Computer processing of breeding records:** Selection of breeding candidates involves processing of large volume of data. The analysis of large volume of breeding data pertaining to several pure strains/lines, and many types of crosses needs fast processing so that timely decisions about selection and crossing can be made while breeding candidates continue to retain their reproductive ability. The new-generation high-speed computers have made the handling and processing of millions of records very easy. In fact were it not for new generation high speed computers, the application of latest breeding methodologies (e.g. Best Linear Unbiased Prediction, BLUP)which involve huge statistical computations would not have been possible.

Breeding Methods for Genetic Improvement of Layer Chickens

Like any other species of farm animals, there are two broad ways for genetic improvement of chickens: selection within strains/lines, and crossing of strains/lines.

Selection within Strains (lines)

The development of trap nest in the earlier part of 20th century made it possible to record egg production of individual hens. Various basis of selection like mass selection, progeny testing, family selection, sib selection etc were employed during the earlier part of 20th century for the selection of genetically superior birds.

By 1960's the procedure of combined selection for improvement of egg production became the standard procedure for selection, and continues to be so with some modifications/additions (like BLUP). Other traits of importance (e.g. egg weight,age at sexual maturity, laying intensity, mortality, egg quality and fertility, hatchability, etc) are selected using different criteria (like individual selection,family selection including selection index).

Table 19.2 gives a brief summary of various critical breeding and selection technologies in poultry with approximate time of their introduction.

Table 19.2. Critical technologies in poultry breeding.

Technique	Decade (approximate)
Mass selection :	1900
Trap nesting:	1930
Hybridization:	1940
Pedigreeing:	1940
Artificial insemination:	1960
Osborne index:	1960
Family feed efficiency testing:	1970
Selection index :	1980
Individual feed efficiency testing:	1980
BLUP breeding value estimation:	1990
DNA markers:	2000

Crossing of Strains (lines)

Following the commercial success of hybrid maize which resulted from crossing of inbred lines through the procedure called 'inbreeding and hybridization', poultry breeders immediately followed this line of work and developed inbred lines through close matings (brother-sister matings). All possible crosses between inbred lines were made, the superior cross-combinations identified, and then the day-old chicks of the identified crosses were sold to commercial poultry farmers.

One of the pioneers in the application of inbreeding and hybridization for producing commercial layer chickens in USA, on the lines of hybrid maize breeding, was Henry Wallace (State Secretary and subsequently Vice President of USA). His seed company, 'Pioneer Seed Company' (located in Iowa state) had been in the business of inbreeding-hybridization of maize much before it's use by other companies. His other company, named 'Hy-Line Poultry Farms' marketed its first 'hybrid layer' in 1940. Within two years, the company sold over one lakh hybrid chicks, a big achievement in those days. Shortly afterwards, another private company the 'DeKalb Association' (located in Illinois state, and later renamed as 'DeKalb AgResearch') also started producing hybrid chickens.

The main advantages/benefits of inbreeding-hybridization were:

- Highest amount of heterosis.
- Improved uniformity and predictability of commercial chicks.

However, this system of producing hybrid (crossbred) chickens was later on abandoned because:

- It required relatively large investment in developing, testing and maintenance of inbred lines as compared to that in plants.
- There was little further scope for improving the performance of hybrid chicks once a superior cross has been identified.

Though this method of inbreeding and hybridization permitted maximum exploitation of heterosis but the problems associated with it in poultry did not justify its continuation.

Strain Crossing to Overcome the Problems of Inbreeding- hybridization

As an alternative to overcome the problems associated with development of inbred lines, poultry breeders developed populations through selection for use in crossing to produce commercial hybrid chicks. Such closed populations developed through selection for at least five generations are referred to as strains, lines, or even pure lines (though they are not 'pure' in the classical genetic sense as they are segregating). The strains/lines used in developing a cross may come from the same breed, such as the White Leghorn strain crosses, or these may belong to different breeds, such as crosses between strains of White Leghorn and Australorps or between White Leghorn and Rhode Island Red. Since these strains/lines are kept as closed populations, and consistently selected, they also accumulate certain level of inbreeding (about 0.5 to 1% per generation) over time.

Though this method of strain crossing exploits lesser heterosis than inbred crosses but the lower cost of maintenance of strains compensates for the lower amount of heterosis exhibited by strain crossbreds. Both these methods of production of crossbreds/hybrids(i.e. inbreeding-hybridization and strain crossing) have proved highly successful but currently majority of the breeders are selling strain crosses.

This procedure of strain crossing also has the added advantage of making further genetic improvement both within the strains (pure lines) as well as in the crossbred chicks. Selection within strains is accomplished through selection for desired traits which also leads to improved performance of crossbreds. Procedure like reciprocal recurrent selection (RRS) can also be practiced to improves cross-line performance.

The commercial chicks (crossbreds) available in the market are produced from two to three types of crosses (Fig. 19.5):

- **Two-way crosses :** Two lines/strains, 'A' and 'B', which serve as parents are crossed to produce 'AB' crossbred chicks which are sold for commercial farming. The chicks are crossbred therefore this type of crossing exploits full individual heterosis.
- **Three-way crosses:** Males of line/strain 'A' are mated with 'BC' crossbred females to produce 'ABC' crossbred chicks. Such a system utilizes maternal heterosis

(e.g. heterosis for reproductive traits like fertility, hatchability which manifest high degree of heterosis) since the female parent of commercial chicks is crossbred ('BC').

- **Four-way crosses:** In this type of crossing, four lines/strains are used to produce what are called 4-way crosses. Two lines/strains, 'A' and 'B', are crossed to produce a cross 'AB'(called single-cross 'AB') and another two lines, 'C' and 'D', are crossed to produce the second cross, 'CD'(called single-cross 'CD'). The single crosses 'AB' and 'CD' are then crossed to give double-cross 'ABCD' chicks that are sold for commercial production. In this procedure of four-way crossing the lines/strains 'A', 'B', 'C' and 'D' are referred to as 'grand parents' and crosses 'AB' and 'CD' as 'parents' of the commercial chick, 'ABCD'. This system uses individual, maternal as well as paternal heterosis.

These different types of crosses are shown in Fig 19.5.

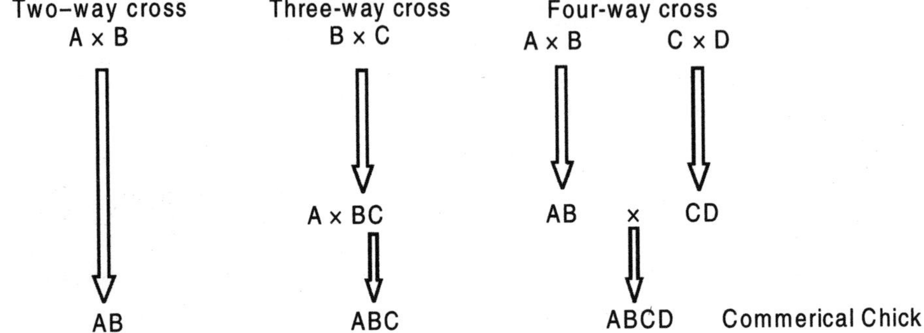

Fig. 19.5. Illustration of two-, three- and four-way crosses for production of commercial chicks

Genetic Improvement and Modern Commercial Layer's Productivity

Productivity of layer chickens, which are normally reared up to 72 weeks of age, has been tremendously improved with respect a number of economically important traits despite the existence of undesirable genetic correlations between some of the traits. Evidence for consistent genetic improvement comes from a number of sources:

Selection experiments: Specifically designed selection experiments conducted at universities and research institutes provide the first evidence for genetic improvement. The main limitation of such experiments is restricted population size of such experimental lines which yields lower rate of genetic improvement than possible in lines selected by private breeding companies. One of the most well-designed and long-term selection experiment was conducted at Animal Research Center (by Gowe and Co-workers). Canada. Selection beginning early 1950's was practiced for part-year egg production for over 30 generations which led to consistent improvement from about 65 to over 120 eggs up to 273 days of age (Fig 19.6). As a correlated response, annual egg production (up to 497 days) also improved from about 175 to over 260 eggs. Concomitantly rate of lay (%) also increased and age at sexual maturity declined. Thus, improvement in egg production resulted from improvement in both rate of lay and sexual maturity. Since simultaneous selection for egg weight was also practiced, this trait also improved.

Random Sample Tests (RST): Random sample tests are conducted to evaluate the relative performance of stocks kept by different breeders under uniform management conditions. Results of one such test conducted by North Carolina State University (USA) are illustrative.

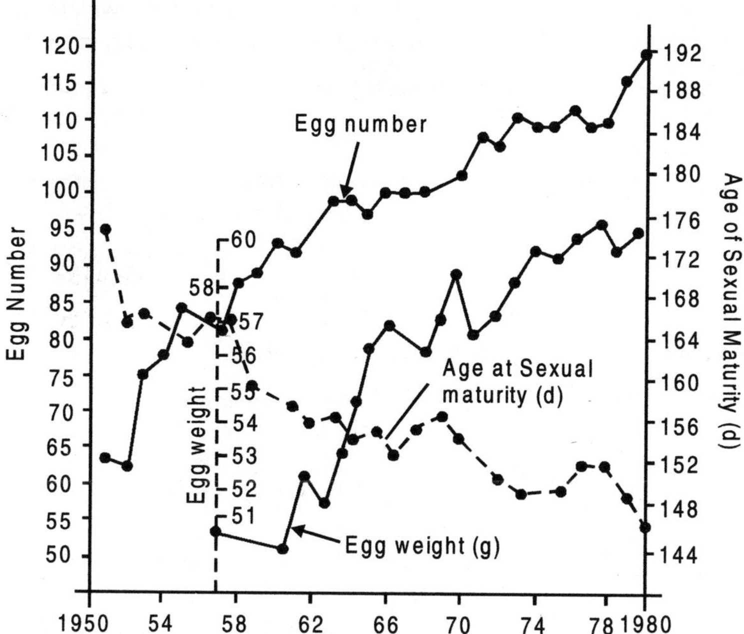

Fig. 19.6 Changes in part-year egg number and age at sexual maturity and egg weight in a multi-trait selection experiment in layer chickens.

Performance of a number of commercial layer stocks (both white and brown egg) was assessed under standard controlled environment from 1958 to 2009. Average performance values of two types of layer stocks viz; white-egg and brown-egg stocks in 1958 and 2009 are given in Table 19.3.

Table 19.3 Summary of genetic improvement in layer strains at RST in North Carolina.

Type of stock	Age at 50% production (days)	Hen-day production (%)	Egg weight (g)	Body weight (kg)	Mortality (%)	Feed consumption (g/day)
White egg :						
1958	173	70	59.5	2.05	10.9	113
2009	139	86	60.2	1.68	6.4	99
Brown egg :						
1958	166	66	60.3	2.67	16.6	113
2009	139	85	61.4	1.89	5.5	103

These results indicate that modern-day strains are early maturing, lay more eggs with higher laying intensity; are smaller in body size and have better egg size, lower mortality, lower feed consumption and better efficiency of conversion of feed into a gram of egg.

Results of German RST covering a number of commercial stock of white and brown egg layers (6 white and 6 brown stocks) over a 24 year period(1975 to 1999) are shown in Fig 19.7 for egg mass per hen-housed and feed conversion ratio, the two economically most important components of layer productivity. Large genetic changes are seen for both the traits.

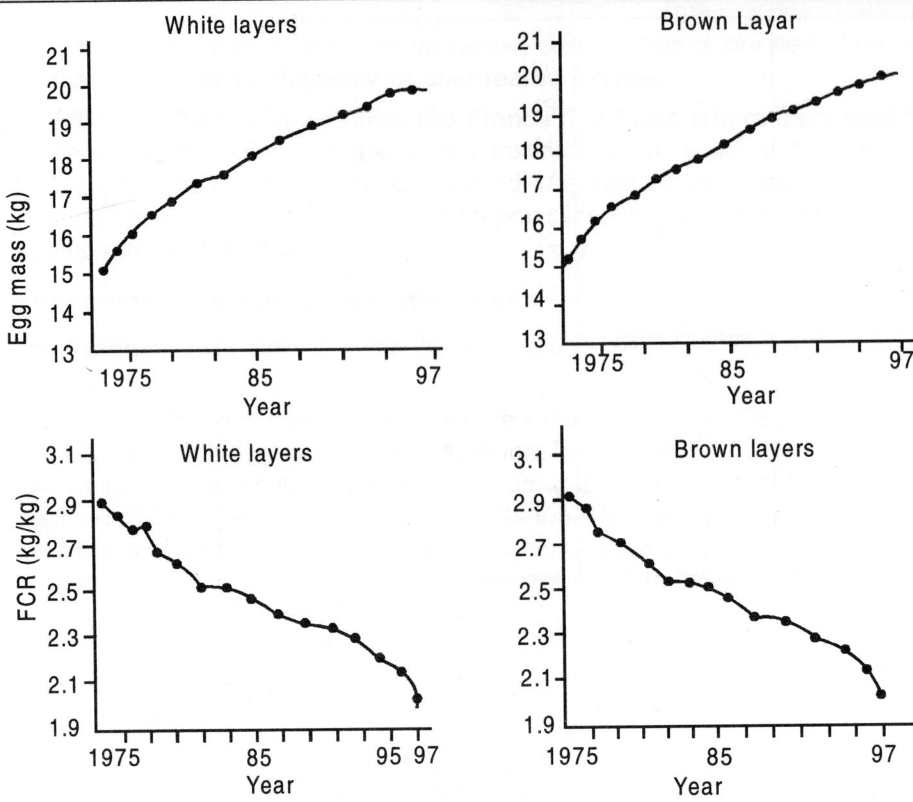

Fig. 19.7 Changes in total hen-housed egg mass(kg) and feed conversion ratio (kg of feed per kg of egg mass) in German RST (Source:Flock,2002)

Another source of information comes from a scientific publication of Jones (2001) in which performance of stocks of 1950, 1958, 1972 and 1993 under similar managemental and nutritional regime was compared (Table 19.4). Apparently, all measures of performance have improved overtime.

Table 19.4 Comparison of performance of stocks established in 1950, 1958, 1972 with a commercial stock of 1993.

Population	Age at lay (d)	Hen-day rate of lay (%)	Egg weight (g)	Egg mass per day(g)	Feed efficiency (kg feed per kg egg mass)
1950	182.9	56.9	56.5	34.2	3.13
1958	172.6	59.7	61.8	37.0	2.89
1972	166.3	64.2	61.0	41.2	2.64
1993	154.9	73.4	63.6	49.3	2.34

Industrial Data: Some private breeding companies publish performance values of their commercial stocks over years. Performance values of one commercial layer stock marketed in India are presented below for several traits for 1981 and 2007.

Trait	1981	2007
Average hen-day production (%):	73	89
Egg production to 72 weeks :	270	330
Average egg weight (g):	56	57
Feed per egg:	145	125
Mortality (%):	8.0	6.0

A large percentage of the total improvement has been due to genetic component. Annual egg number has increased at the rate of about 2.5 eggs per year. This rate of genetic improvement is consistent with some values reported in selection experiments.

Results of German random sample tests show the genetic progress achieved in hen-housed rate of lay and feed efficiency from 1980 to 2004(Fig.19.8). Modern layers (2004) as compared to their earlier counterpart (1980) not only achieve higher laying intensity but also sustain it much longer. Feed efficiency (kg of feed per kg egg mass) of modern layer strains as compared to their older counterparts is much better throughout the laying cycle.

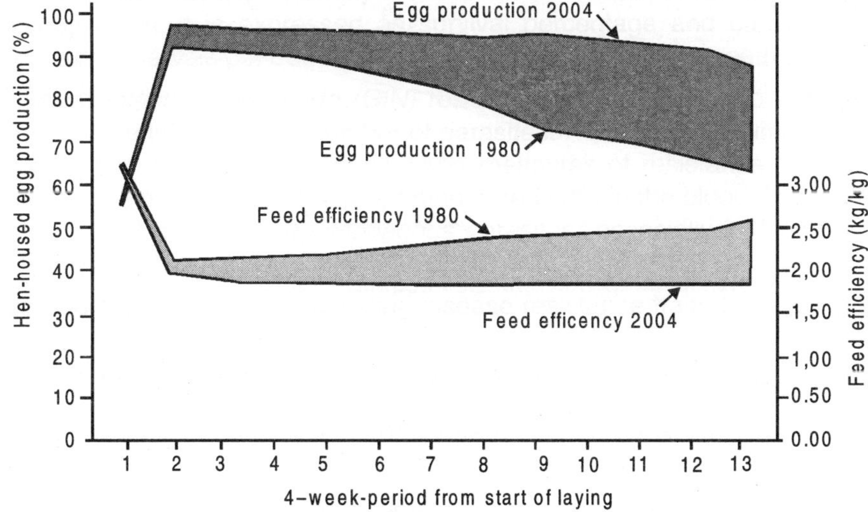

Fig. 19.8 Genetic improvement in laying intensity (rate of lay) and feed efficiency in layer chickens in German RST

A private breeding company in India indicates the following general profile of its commercial layer stock:

Age at sexual maturity(age at 50% production) :		150 days
Egg production:		
• Egg production(hen-housed)to 72 weeks:		330
• Peak production laying rate :		97%
• Number of weeks above 90% production:		40
Egg weight (average up to 72 weeks):		58 g
Feed efficiency (feed per egg):		124 g
Body weight:	20-week	1.35 kg
	72-week	1.6 kg
Mortality:	Growing	2-3%
	Laying	4-6%

The same company predicts the following magnitude of genetic improvement in performance over the next 10 years:

- Egg production: +15 eggs
- Mortality: – 1.0%
- Egg weight:
  - Early : + 1.0 g
  - Late : + 0.2 g
- Feed per egg : – 6 g.

Box 19.2: 'Double-cross' method of inbreeding-hybridization.

The method of development of inbred lines and their crossing to develop hybrid maize revolutionized maize production not only in USA but also in major maize growing countries. Though this system generated high level of heterosis but the cost of production of hybrid maize was high due to poor performance of inbred lines as they had lower vigor, and produced only a small amount of seed. Since production was low, the hybrid seed was costly, rather too expensive for the average farmer to afford. This difficulty was overcome by Dr Jones in 1919 while he was employed as a scientist at Connecticut State University's research farm following the movement of two plant geneticists, East and Hays. Initially he maintained the experiments of his predecessors. But later on, frustrated by the weakness of inbred strains and their feeble grain production but impressed by vigorous production of their single crosses, he decided to try crossing two of the single crosses to produce what later came to be known as a 'double cross'.

In this method, two inbred lines, A and B, are crossed to produce 'AB' hybrid, and another two inbred lines, C and D, are crossed to produce 'CD' hybrid. The double-cross hybrid, 'ABCD', is produced from crossing of 'AB' with 'CD". This method was then quickly applied by other maize breeders, including Henry Wallace who owned a private corn breeding company.

This method made a large impact on poultry breeding. Henery Wallace who owned a maize breeding company, also established a private poultry breeding company, 'Hy Line Poultry Farms' where crossing of inbred lines was done on the pattern of maize breeding. Thus, the breakthrough in hybrid poultry came not from conventional poultry breeders but from the success of hybrid maize. This was so because the private companies had sufficient resources, not only in terms of capital but also in terms of scientific expertise. The currently used method of four-way line/strain crossing to produce double cross hybrid, 'ABCD', as used in chickens, thus traces back to Dr Jones' work on maize.

BROILER BREEDING

Growth of poultry from small-scale backyard activity to large-scale organized farming owes its origin to the early developments initiated in North America, in particular in USA. Until 1940's most of the chickens in USA were primarily bred and reared for egg production, meat was only the by-product from commercial egg production. The breeds used were dual-purpose. This was because the males and females could not be distinguished until they were several weeks old. In this scenario, the males (cockerels) were kept until they attained marketable weight (more than 12 weeks of age) when they were sold.

The vent method of day-old chick sexing (the Japanese method) coupled with some other developments paved the way to establishment of two segments of chickens: one for meat (broiler or meat-type chicken) and the other for eggs (layer or egg-type chicken). Heavy breeds like Cornish, Plymouth Rock, New Hampshire etc. when had faster growth rate were bred for meat production. The strains when are being currently used for commercial broiler production are synthetics derived from the above mentioned heavy breeds. In fact, the concept of a pure breed in poultry has become obsolete in commercial breeding (Box 19.4).

Economic Traits: Description, Measurement and Inheritance

Broiler breeding is slightly different and complex than layer breeding due to complexity of the traits to be improved. There are two categories of traits; the ones that are important in commercial broilers for meat production and the others concern the economics of chick production (i.e. those of parents).Therefore broiler breeding programs have two distinct breeding goals:

- Commercial chicks produced from crossing need to have desirable market broiler attributes (e.g. high growth rate, low mortality, better feed efficiency, acceptable body conformation etc.)
- Parental strain/lines, especially dam lines, need to have desirable maternal traits (e.g. high egg production and reproductive fitness).

Growth rate (weight at marketing age): The primary trait of economic importance in meat-type chicken is growth rate from day-old to marketing age. Growth rate is most appropriately measured as body weight at marketable age (usually 5 to 6 weeks of age). Heritability of early body weight (juvenile weight) is moderately high (more than 0.3) and the trait can be measured in both sexes, therefore, individual selection for body weight has been highly effective in its genetic improvement.

Feed efficiency: Feed efficiency or feed conversion ratio (FCR) is another trait that directly determines the profit from broilers as feed accounts for about 70% of the cost of broiler production. The words 'feed efficiency ' and 'feed conversion ratio' are often used interchangeably, and defined as the amount of feed required for a unit gain in body weight:

$$\text{Feed conversion ratio} = \frac{\text{Amount of feed consumed}}{\text{Gain in body weight}}$$

Genetic correlation of feed efficiency with body weight is high and desirable: it is negative when feed efficiency is calculated by dividing feed consumed by gain in body weight, and is positive when feed efficiency is calculated as gain in body weight divided by feed consumed. Feed efficiency has consistently been improved as a correlated response to selection for growth rate.

Mortality or viability (up to marketing age): Heritability of mortality being very low (less than 0.10), families with highest mortality are rejected using independent culling levels.

Body conformation (indicator of breast muscle growth): Meat-type breeds, as compared to egg-type breeds, have rounded body appearance which results from their skeletal development and muscle distribution such that broilers have higher breast meat yield. Breast meat fetches higher price relative to other parts as it is tender and has low fat content.

Conformation is measured in a number of ways such as breast angle (measures breast width with a breast angle meter), thickness of breast muscle using needle catheter etc. However, experienced breeders use subjective evaluation score by examining the bird by hand. Body conformation has moderate heritability, therefore, individual selection can be effectively used for its improvement. Yield of breast meat is positively related to body conformation such that selection for conformation automatically improves yield of breast meat. Genetic correlation of body conformation with body weight is undesirable such that this trait deteriorates due to direct selection for body weight.

Abdominal fat: One of the undesirable consequences of genetic improvement of growth rate (body weight) resulting from long-term selection is the deposition of excessive fat in the carcasses as the genetic correlation between these traits is undesirable (positive). Fat deposition is most pronounced in abdominal region, hence the term 'abdominal fat'. Heritability of this trait is moderate but its measurement requires slaughter of birds. Sib selection can be used for its improvement wherein a random sample of birds from a family are slaughtered for recording the abdominal fat.On the basis of this information, selection from the remaining family members is made. Since recording of abdominal fat requires slaughter of birds, a number of indirect methods of measuring abdominal fat on live birds have been developed. These include physical palpation of abdominal region, wing web thickness, use of caliper etc.

Skeletal defects (leg and breast) : Rapid growth rate of broiler chickens exerts great stress on their skeletons (the immature cartilaginous) which leads to high incidence of leg and other skeletal abnormalities. Modern broiler processing industry discards broilers having such physical defects, the most common being defects of legs (like bowed legs, shortened/distorted long bones, angular distal tibiotarsal bone, slipped tendon etc).Exclusive selection for body weight has been shown to result in increased incidence of skeletal defects due to existence of undesirable genetic correlation. Breeding organizations practice selection for reducing the incidence of such defects in their strains/lines.

Reproductive and egg production traits: Reproductive traits, mainly fertility and hatchability, are most important for primary breeders (the breeders who maintain pure lines). Similarly, egg production of broiler mothers is an important trait as the number of eggs produced by a broiler female (dam) has direct impact on cost of broiler chick production. Therefore, selection for these traits is also practiced in parental lines as lower fertility and hatchability increase the cost of broiler chick production. Genetic correlation of growth rate with fertility, hatchability and egg production is negative (undesirable), exclusive selection for early growth rate thus leads to deterioration of these traits. The heritability of fertility and hatchability is low; therefore, family selection is used for their improvement. Egg production is improved through family or combined selection.

A summary of important traits of meat-type chickens is given in Table 19.5.

Table 19.5 Summary of economic traits of meat-type chicken.

Trait	h^2	Selection criterion	r_G with body weight:
Body weight:	High	Individual	–
Feed efficiency:	Moderate	Improved as correlated response	desirable
Mortality:	Very low	Family	–ve, undesirable
Body conformation:	Moderate	Individual	–ve, undesirable
Abdominal fat:	Moderate	Individual	+ve, undesirable
Skeletal defects:	Moderate	Individual	–ve, undesirable
Fertility & hatchability:	Very low	Family	–ve, undesirable
Egg production:	Low	Family, combined	–ve, undesirable

Breeding Methods for Genetic Improvement of Broiler Chickens

Genetic improvement of meat-type chickens follows the same general steps of selection within each line/strain followed by their crossing to produce commercial chicks which are invariably crossbreds. The commercial broiler chicks are produced from three-way or four-way crossing of purebreds (strains/ lines).

Selection within Strains (lines)

The primary trait of importance in broiler chicken, 5 or 6 week body weight, is moderate to highly heritable and is measurable in both sexes. Individual selection has been highly effective for its genetic improvement. Since this trait is measurable early in life, very high selection intensity (usually less than 1%) has ben possible which has led to dramatic improvement of body weight (50 to 100 g per generation increase in body weight and 1 to 2 day per generation reduction in age at marketing). However, this intense selection for broiler weight led to loss of reproductive traits (e.g. reduced egg production, fertility and hatchability) of the strains/lines used for crossing to produce the commercial broiler.

Crossing of Specialized Sire and Dam Lines (strains)

In order to overcome the loss of reproductive traits due to the negative genetic correlation between growth and reproduction, breeders of broiler chicken have to improve two categories of traits:

- Improvement of commercial broiler chicks for high growth rate, low mortality etc
- The parental lines used to produce commercial chicks need to have high egg production and reproduction so that large number of chicks from each parent can be produced to make the breeding operation commercially viable.

To achieve this end, there arose a need to develop specialized types of lines/strains, called **specialized sire and dam lines**. These specialized lines, which are selected for improvement of different category of traits, are then crossed in three-way or four-way combinations to produce the commercial broiler chicks. Such a system of development of specialized lines followed by their crossing helps in economizing the cost of chick production.

Sire lines: Selection within sire lines is primarily practiced for market traits (i.e. broiler body weight, mortality, feed efficiency, body conformation etc). The lines/strains used as sire lines have primarily been derived from Cornish breed which excels in growth and muscling (Fig 19.1). Characteristics of a good sire line are:

- High market weight (5 or 6 week)
- Low mortality
- Good body conformation
- Efficient in feed conversion
- Lower incidence of skeletal defects

Selection in sire lines is therefore directed towards the improvement of these traits using appropriate bases/methods of selection.

Dam lines: The lines/strains to be used as female parent for producing the crossbred broilers, referred to as dam lines/strains, need to possess high egg production, fertility and hatchability to reduce the cost of chick production. The broiler dam lines/strains have been derived from Barred Plymouth Rock, White Plymouth Rock, New Hampshire and other dual-purpose breeds. Characteristics of a good dam line, in addition to good growth rate (body weight), are:

- High number of settable eggs which includes not only high egg production but optimal egg size also.
- High fertility and hatchability(reproduction)

Sequential selection in dam lines: The traits of importance for selection in dam line are expressed at two distinct stages: 5 to 6 weeks of age and after completion of a part of laying cycle. Therefore, selection is conducted in sequential steps:

- 1st stage selection is for market traits (body weight etc) where a proportion of broilers are selected on the basis of high body weight, good conformation and related commercial broiler traits.
- 2nd stage selection is for egg production, fertility and hatchability when these traits are expressed /recorded. Since these traits are sex-limited and have low heritability, therefore, some form of family selection including combined selection is used for their improvement in dam lines.

Most of the world's commercial broilers these days are the outcome of four-way specialized crosses. This system of four-way crosses capitalizes individual, maternal and paternal heterosis. The sire lines are synthetic lines derived in part from White Cornish. The dam lines are also synthetic; they are based heavily on White Plymouth Rock.

Other breeds are used to produce chicks with colored feathers for some particular markets. Originally, market broilers came in various plumage colors. The colored broilers however faced difficulty when mechanical processing equipments were developed and commercially adopted. Melanin pigment present in pinfeathers and skin of colored broilers reduced processing speed and efficiency. To overcome this problem, breeders with large operations integrated white plumage into commercial chicks through use of appropriate parental lines.

Genetic Improvement and Modern Broiler's Productivity

Productivity of broiler chickens has improved dramatically and continues to be improved further. Like layer chickens, evidence from different sources is available:

Selection experiments: A number of selection experiments have shown continued genetic improvement in broiler body weight using individual selection. Of the various specifically designed selection experiments to improve broiler body weight, the one conducted at Virginia Tech (USA) by Prof Paul Siegel is considered a classical experiment. In this experiment, selection was conducted for high as well as low broiler weight (bi directional selection experiment). Due to limitation of facilities, the number of sires, dams and progeny in this experiment were rather small. Despite this, however, consistent genetic changes occurred in high and low body weight lines as depicted in Fig 19.9).

By 50th generation of selection, the lines selected for high and low body weight had about nine-fold difference in their body weights. The lines became so distinctly different that adult birds of low body-weight lines looked like growing chicks of the high body weight line. The two lines became distinctly different in their body weight distribution such that there was no overlapping in body weights of the two lines.

Experimental results: Another source of information comes from a scientific publication of Havenstein (2001) in which performance of broiler stocks of 1957 and 2001 was compared under similar managemental and nutritional regime as prevalent in 2001 (Table 19.6 and 19.7).

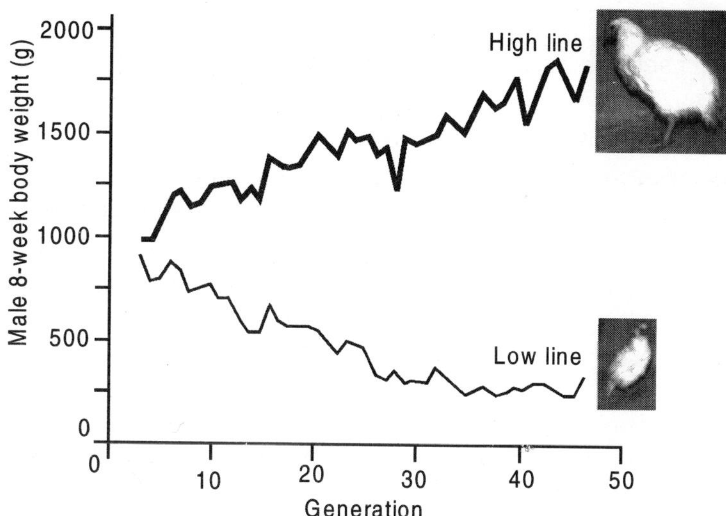

Fig. 19.9. Genetic improvement for high and low body weight lines of chickens

Table 19.6 Changes in body weight from 1957 to 2001 in broiler stocks.

Population	Body weight(g) at:				
	3 week	6 week	8 week	10 week	12 week
1957:	210	641	1009	1412	1907
2001:	791	2903	4402	5385	5958

Body weights have dramatically been improved over the 44-year period. It has been estimated that 80% of this improvement has been genetic in nature. Genetic improvement was estimated at the rate of 2% of the mean per year. Listed in Table 19.7 are the changes observed in other broiler traits.

Table 19.7 Changes in broiler body weight and body composition from 1957 to 2001 in broiler shocks.

Population	Body weight (kg)	Carcass weight (kg)	Carcass yield (%)	Breast yield (%)	Carcass fat (%)
1957:	0.81	0.48	60.8	11.4	10.6
2001:	3.95	2.81	74.4	21.3	15.9

The changes in carcass traits have also been very large. Not only body weight and carcass weight but yields of carcass and breast have also been improved. As a correlated change, carcass fat percent has increased but the rate has been slowed down due to selection directed against it.

Consistent selection over time has also greatly modified the growth curves as shown in Fig 19.10.

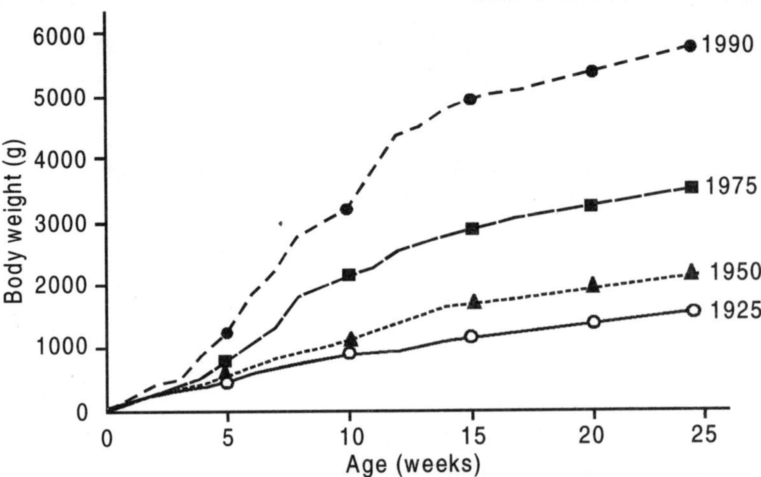

Fig. 19.10 Growth curves for the typical chicken in 1925 and meat-type chicken in 1950, 1975 and 1990 (Anthony, 1995)

As a result of genetic improvement, it is now possible to market the broilers at an early age.It has been estimated that the time required for meat-type broilers to achieve a constant market weight has been reduced by one to two day per-generation.

Industrial Data: Some private breeding companies publish performance values of their commercial stocks over years. Performance values of one commercial broiler stock marketed in India are presented below for several traits for 1981 and 2007:

Trait	1981	2007
Age to attain 1.8 kg weight(days):	55	35
Feed conversion ratio:	2.5	1.7
Mortality (%):	5.0	4.0
Eviscerated yield (%):	62.0	69.5
Number of chicks produced (per broiler mother to 68 weeks):	130	165

Desirable genetic changes have been accomplished not only in broiler traits but also in maternal trait viz. number of chicks produced by broiler dam which is an indicator of improved fertility, hatchability and egg production. Growth curve of this stock shows that its 6-week body weight has been improved from 1200 g in 1981 to about 2400 g in 2007.

The same company predicts the following magnitude of improvement in performance over the next 10 years:

- Days to reach market weight : – 7 days
- Mortality: – 0.25 %
- Feed conversion ratio: – 0.25
- Eviscerated yield: + 2–3%
- Breast meat yield: +3%
- Number of chicks per breeder: + 5

Performance Evaluation of Commercial Stocks: Random SampleTests

With the development of poultry industry and private poultry breeding companies, the poultry farmers needed to have some objective information about the relative performance of commercial stocks available in the market especially relative to the already established stocks. The first performance evaluation test, called Random Sample Test (RST), was held in Netherland in 1927. After its beginning in Europe such tests, spread to other developed countries around 1950's onward.

The main purpose of these tests was to evaluate the performance of commercial stocks developed by various organizations under similar and uniform environmental conditions. The procedure of testing consisted of:

- Collection of representative(unbiased) sample of breeder's flock in the form of eggs or day–old chicks by an independent agency/person from breeder's premises.
- Hatching, rearing and performance recording of the participating stocks (each stock being given a confidential code) under similar environmental and managemental conditions.

The tests were conducted by public agencies, and the summaries of such tests were published and made available to public and developmental organizations. At one time, these tests became very popular, in which a large number of breeders and breeding companies participated. But over years, the number of participants decreased, and eventually these tests were closed in most developed countries.

Random sample tests for layers and broiler chicken in India were started on the pattern of the American and European tests by the Department of Animal Husbandry and Dairying, Ministry of Agriculture, Government of India at its Central Poultry Breeding Farms located at Bangalore, Bombay, Bhubaneswar, and Gurgaon.

Between 1970's and 1990's, these tests were quite popular, in particular with the public sector organizations involved in poultry breeding (i.e. State Agriculture Universities, Central Research Institutes and Central Poultry Breeding Farms etc.). Testing of the new varieties or commercial stocks was carried out in several locations representing different agro climatic conditions (multi-location testing) of the country, and over several years. Testing of the stocks over years was made to reduce the contribution of chance factors (due to disease exposures etc) in a particular year. Testing of stocks at least at three to four locations for two to three years was essential prerequisite for commercial release of stock developed in the public sector by the 'Variety Release Committee' of the ICAR.

Many of the public sector stocks released during this period in India had been tested at these RST's. However, the entry of private breeding companies in India led to decline of these tests. Currently RST for layer and broiler chickens is continuing at one centre (at Central Poultry Performance Testing Centre at Gurgaon) with only ten entries of each of the layer and broiler chickens.

INDUSTRIAL POULTRY BREEDING

Poultry has developed from a small-scale backyard activity to a highly organized production system which has taken shape of a full-fledged industry in developed countries.Even in developing countries, a large proportion of chickens and eggs are contributed by organized poultry sector. It is estimated that over 75% of the world's poultry production is controlled by the industry.

A natural outcome of transformation of poultry from small-scale operation to large-scale commercial venture has been establishment of private poultry breeding companies. These companies have huge infrastructure and hire qualified poultry geneticists (PhD's who direct their breeding programs) and other scientists. Therefore, they had been quick to apply modern breeding schemes and concepts such as inbreeding and hybridization, reciprocal recurrent selection etc. for genetic improvement of the final product. Primary operations of most of these companies are based in North America and Europe but they have a worldwide net work.

Important Biological Features of Chicken for Industrial Breeding[2]

Chicken have some favorable biological features which have enabled the transition of poultry farming from backyard activity to a full-fledged industry. This transition in turn led to industrialization of poultry breeding. These features include:

- *High reproductive ability:* Tremendous natural reproductive ability of hen and short generation interval has enabled rapid multiplication of selected lines/strains, permitted high selection intensity and faster genetic progress per unit of time.
- *Artificial incubation of eggs* (instead of relying on broody hens): Elimination of broodiness of hens through continuous selection increased the number of eggs laid by a hen which, coupled with development of artificial incubators, permitted hatching of large number of chicks at a given time for commercial farming.
- *Storage of fertile eggs:* The technique of storage of fertile eggs at low temperatures (called 'pre-incubation storage') for a number of days without loss of their hatching potential provided large family- size for effective selection. It also improved flexibility of reproduction and multiplication.
- *Sexing of chicks at hatching*: The method of chick sexing (vent method) has tremendously reduced the cost of commercial egg production because males could be discarded at hatching.
- *Stimulation of egg laying and control of sexual maturity by artificial lighting* :It improved laying intensity during the season of year when day length is short (winter), and it also made possible raising pullets throughout the year.

Historical

Breeding of poultry on commercial pattern can be said to have started in early part of 20th century when some farmers in North America started making choices among the birds available for reproducing (i.e. selection) the stock. Selection in early days was almost exclusively based on appearance; physical traits were emphasized, and to some extent modified so that a certain degree of uniformity could be achieved. This helped differentiate stocks of individual farmers'.

With the invention of trap nest, it became possible to obtain individual performance records. But these records were not widely used until 1930s and '40s. Breeders emphasized progeny testing as an art, and were able to make significant improvements in egg production and other commercial traits. An early pioneer in this business was Robert Cobb who set to improve pure breeds such as Barred Plymouth Rocks, and then in 1940's, he developed a line of all-white birds.

- **1940's:** By the outbreak of Second World War, many breeders had established themselves locally as supplier of baby chicks. Most sold hatching eggs, breeding pairs or trios, to other breeders. In North America, every small town or village would have one or two hatcheries, some just using internal sources of hatching eggs, and others purchasing improved stock from outside sources. The process was labor-intensive and most of the hatcheries were small, inefficient and only operated on a seasonal basis. Customers were small with few owning more than 100 birds.

[2]Source : Adapted from Hartmann (1988)

In the late 1940s, the breeding industry began to develop more rapidly in North America. Shaver breeding company, which prior to war was in chick supply business, quickly reestablished in breeding enterprize. There were many similar breeders across North America and in most of the developed countries of world though developments in Europe were delayed due to the war. At this juncture some organized breeding companies emerged in the United States. Prof. Goodale developed the Mount Hope strain of White Leghorn which was widely sold and probably still contributes to some of today's hybrids. Kimber Farms in California employed its first professional geneticist, Dr. W.F. Lamoreux in 1943. Proximity of Kimber Farm to the University of California (Berkley) seems to have helped Lamoreux and the company to establish scientifically sound breeding programs.

- **1950's:** By the 1950s, geneticists were trying to apply scientific methods for genetic improvement of commercial traits. But the resources necessary for the useful application of this knowledge to a professional breeding operation were not cheap. These resources and expertise were beyond the capability of small town-level breeders. And, those who developed the complex breeding infrastructure did know that they required more than local markets to support large investment in their organization. This gave birth to the system of franchise, or distributor. Thus hatcheries were set up, in which the local hatchery operator obtained male and female parent breeders from a primary breeder, grew and mated them, and used the resulting hatching eggs for producing the commercial chicks.
- **1960-70's:** By late 1960's several hatchery distributors were spread across North America. Many of them originally had their own breeding programs but eventually they chose to become distributors as they could not afford huge investment. At this time, breeders were not willing to sell their stocks to each other. Rather they tried at great lengths to prevent competitors from acquiring their pure lines. When the commercial stocks consist of two-way crosses, the risk of strain piracy is high. Thus, to overcome this, the use of three-way or four-way crosses became common. This strategy thwarted the 'theft' of valuable lines by competitors. Sometimes breeders also used strategies to make acquiring of their lines by other breeders more difficult. One such practice was to place 'fake' sexing error chicks in shipment of parent and grandparent stocks.

By the early 1970s, the commercial poultry breeding in North America was dominated by approximately 20 primary breeders. These companies initially specialized in one or two products. Many primary breeders specialized only in broiler stocks, and some bred exclusively either male or female strains.

A selective list of some early breeding farms and companies operating in USA is given in Table 19.8

Table 19.8 List of some early breeding companies.

Layer chicken:	Meat-type chicken
• Mount Hope (1912)	• Arbor Acres
• Ghostley Farm (1918)	• Cobb-Vantress
• Kimber Farms (1925)	• Peterson Farms
• Hyline International (1926/39)	• Hubbard
• Babcock Poultry Farms (1938)	• Indian River
• Dekalb AgrResearch (1944)	• Avian Firms
• Shaver Poultry Farms (1946)	• Purdue Farms
• H and N Inc (1946)	

Internationalization: In 1960s, long-distance air travel became economical such that day-old chicks could be transported halfway around the world in less than 36 hours. This opened international markets to breeders based in North America. For example, Shaver Company of Canada by late 1970s was selling its product in more than 90 countries. Additionally it had company-owned (or joint venture) breeding farms in the United States, England, France, Germany, Pakistan and Barbados. Most breeding companies those days produced detailed, printed management guides of their products.

Mergers: Until 1960s, most breeding companies were privately owned. But expansion of business required greater external funding.It started with selling of some parts between companies. For example, Shaver sold a part-interest to Cargill Inc. in the early 60s and the remaining part in 1985. Similarly, other breeders started to sell out to larger companies to help facilitate their expansion. In the 1970s, a number of pharmaceutical companies like Pfizer, Merck, Upjohn and others also got into poultry breeding though these mergers were later on abandoned. Nonetheless, in this process the number of primary breeders continued to shrink, through erosion or mergers.

By the first decade of 21st century, ownership of primary breeding poultry organizations had diminished to the point at which breeding work is controlled by very few companies/groups as shown in Table 19.9.

Table 19.9 List of the main companies/group catering to worldwide commercial production.

Company (group name)	Location	Main products	Other species products
Hendrix Genetics : (Worlds 1st brown egg layers; 2nd in turkeys, 4th in broilers, 2nd in pig genetics)	Netherland Canada France Scotland	Egg layer : ISA, Babcock, Shaver, Dekalb, Bovens and Hisex brands.	Turkeys, salmon and pigs
EW group : (Worlds market leader in poultry genetics)	Germany	Egg layers: Lohman, Hy-Line, H &N brands. Aviangen Broilers:Ross, Arbor Acres, Indian River.	Turkeys and fish
Cobb-Vantress : (Owned by Tyson Foods); (Worlds largest processor of chicken & red meat)	USA	Broilers: Cobb, Avian (joint development agreement for R & D with Hendrix genetics)	-
Grimaud Freres : (2nd in avian genetics)	France	Egg layers: Novogen Broilers: Hubbard.	Pigs & ducks

Thus, in the past 100 years, poultry breeding has evolved from hundreds (or perhaps thousands) of small, independent farms that did very little in way of selection to the point at which most breeding work is is handled by a handful of multinational companies whose products are distributed and multiplied on an international basis. Also, the breeding groups are not only into poultry but in the breeding of other species as well. In this process of elimination of small breeders and mergers, significant loss of genetic diversity must have occurred. As breeders merged, many experimental lines were either discarded or combined with other breeder's gene pool.

Multiplication and distribution: Multiplication and distribution of commercial stocks has over the years also evolved. Since 1970s most integrated broiler businesses have established their own breeding farms and hatcheries. This is less so in egg industry although several of the largest companies in the United States have their own hatcheries. More common is the establishment of breeder-owned or joint venture hatcheries which have largely replaced the franchise hatcheries as the primary distribution method for egg-type chicks. Distribution system for egg stocks in North America has remained mostly as franchised hatcheries due to relatively small commercial flock sizes resulting from supply management.

Breeding methods used: The exact selection criteria and breeding methods used by various companies are closely guarded trade secrets. In early 1970's, the USDA made a survey in USA and Canada about the breeding methods used by commercial breeders or companies. The survey revealed:

- Breeders of egg-type chickens used reciprocal recurrent selection, Modified reciprocal recurrent selection, Dam family and individual selection (single trait/multiple lines), Family and individual selection (multiple traits), Family and individual selection for highly heritable traits, recurrent selection for lowly heritable traits.
- Breeders of meat-type chickens used same selection practices as used for egg-type chickens though the relative order and importance of traits varied. The listing for meat-type chicken was : Family and individual selection, Index involving family and individual performance, Profitability index function involving information on parents and progeny.

These private breeding companies have to satisfy consumer demands over a broad range of countries. Therefore, they have to continuously keep on genetically improving the product (commercial chicks) in relation to different environments and markets. Additionally, they have to provide delivery of chicks in time and provide after sale service support in nutrition, management, disease control and marketing.

Genetic improvement of their stocks is undoubtedly the major objective of breeding companies. But the market demands may vary for different countries or regions to which they cater. Therefore, to meet these special product requirements, they use different line crosses for respective markets.

Structure of Commercial Poultry Breeding Programs

Commercial poultry breeding implies breeding for commerce or business, also referred to as **industrial breeding**. Therefore, efforts in commercial poultry breeding are directed towards maximizing returns per unit of investment by poultry farmers, hatchery men and primary breeders. Naturally, the objective of breeding programs of such organizations is to develop products which farmers and hatchery owners as well as customers are willing to buy.

Breeding goals in such programs include a large number of traits as well as strains/lines for acceptance by all the stakeholders. Heterosis (individual, maternal and paternal) for most economic traits is significant particularly for reproduction, survival and production traits. Therefore, the commercial chicks that are produced through selection and multiplication are invariably crossbreds (usually called 'hybrids'). To accomplish these objectives in the competitive market, the commercial breeding programs are large-sized. Such commercial breeding programs involve several stages/components:

Pure-breeding lines/strains: A large-scale poultry program involves a number of populations, sometimes called 'basic strains' or 'pure lines'. These strains/lines, which are owned by the primary breeding company (called 'foundation breeder') are separate breeding populations, usually obtained from different genetic sources. These are maintained as pedigreed populations. The important components in pure breeding strains involve selection and testing of their crosses. These two aspects of breeding are shown in Figure 19.11.

- *Selection:* The basic strains are maintained in the whole breeding program as small pedigreed populations. These strains/lines undergo selection through different procedures and for various set of characters and with variable selection pressure. Thus some strains are subjected to family selection, some may be inbred with or without selection; others may be improved by reciprocal recurrent selection. Some lines, called 'experimental lines', which are at different stages of development are also maintained to cater to emerging markets or situations.

  Each company maintains a number of pure lines (from at least four to 10, 20 or even more) depending upon its infrastructure, breeding policy and the target market. Usually each line or strain has some distinct set of desirable attributes. Further, some of the lines are categorized into male and others into female lines, and accordingly the traits to be improved for different lines vary. For reasons of security, duplicate lines are maintained at different locations or even in different countries on high-level biosecure farms. Breeding companies thus allocate resources to both improving performance of mainlines as well as on developing new lines.

- *Testing of crosses:* The next stage in a large-scale breeding program involves testing of lines in two-, three- and four-way combinations to identify the most promising crosse (s). The initial comparison of crosses involves testing at a single farm involving small-sized groups. Based on these preliminary comparisons, the most promising crosses are remade and tested in larger numbers on several locations (farms) so that the chosen crosses are the ones that can perform well under a range of environments.

The commercial chicks in a four-way crossing are produced from only four lines (say, 'A','B', 'C' and 'D'). The remaining lines serve as back-up lines to cope with the changing market demands (Fig. 19.11). For example, line/strain 'A' could be replaced by line 'E' that might produce a superior crossbred, 'EBCD'. Requirements of some special markets may be fulfilled by using crosses of different pure lines (e.g. 'BDEG' or 'ACEF'). Once the most promising crosses are chosen, the next step (at the level of great-grandparent lines) involves multiplication.

Great-grandparent stocks: The great–grand parental lines are pure lines which have been found to be most suitable for producing the commercial chicks on basis of the crossing results (e.g. A, B, C, D). These are also under company's direct control, and subjected to limited amount of selection for desired traits. However, this generation is primarily used for multiplication of pure lines to larger numbers (may be tens of thousands) that is required to produce next generation (grand-parent stock).

Grand-parent stocks*:* The stocks which serve as grandparent stock of the commercial chick ('ABCD') are crossbreds produced from crossing of 'A' males with 'B' females, and 'C' males with 'D' females. Grandparent stocks are distributed throughout the world in at least hundreds of thousands to local or regional operators who have a network for local distribution.

Parent stocks: The parent stocks of both the broiler and layer chicken are invariably crossbreds. In case of four-way cross, the parent stocks of commercial chick are crossbred: 'AB' males and 'CD' females.

Commercial layer/broiler chicks: These are the birds (day-old) which are produced and grown for marketing (either live or slaughtered).

Structure of a commercial breeding plan is shown in Fig 19.11

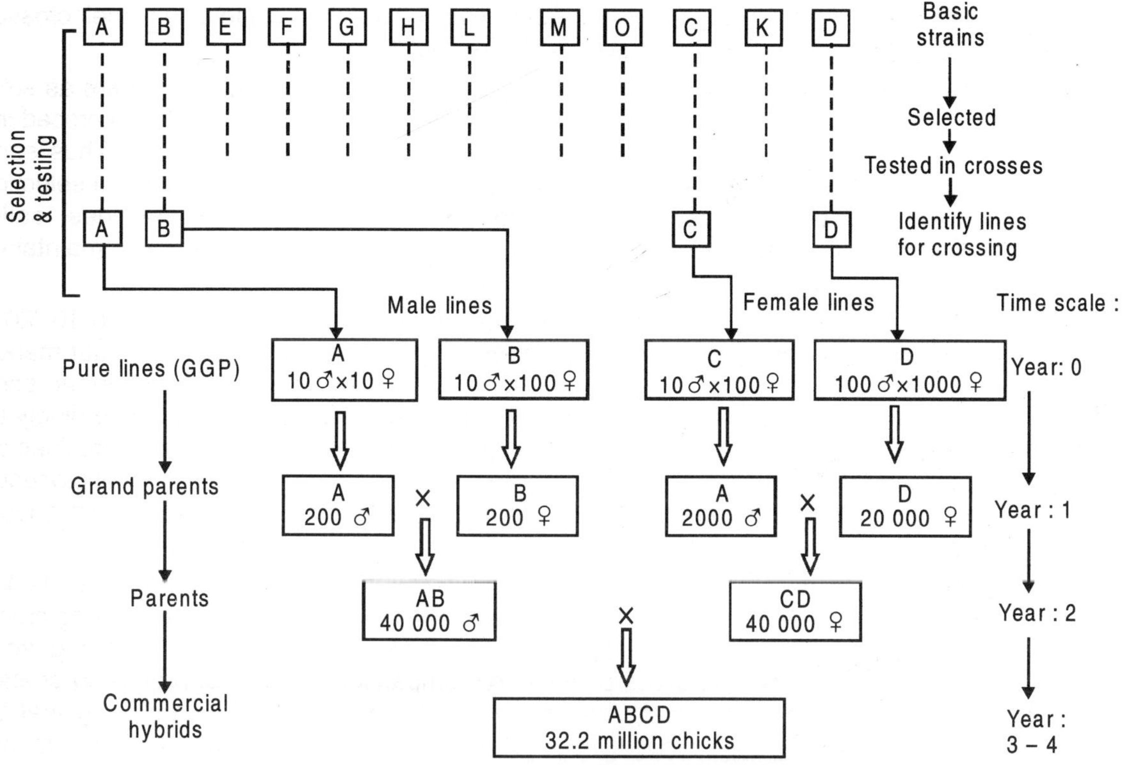

Fig. 19.11 A large scale industrial (commercial) breeding program for production of commercial hybrid chicks.

Crossing of male and female lines maximizes heterosis at grandparent and parent levels of the program. It thus allows the traits that have been genetically improved in different lines to be combined in the commercial birds optimally.

Multiplication Chart : In commercial breeding, the pure line breeder (breeding company) markets the grandparents or parents to his customers (who are called franchise hatcheries/ breeders), and they in turn produce the commercial layer or broiler for poultry farmers.

This structure of primary and franchise serves the purpose of protection of pure lines from getting into the hands of other party (i.e. protection of brand product from competitors). In such a structure whole of the genetic improvement through selection occurs at pure-line level in each generation. And, this genetic improvement is passed on to the next strata (i.e. to grandparents, parents and commercial product in that order). An important feature of this type of breeding pyramid is that pure line stocks (in relatively small numbers) are located at the apex and very large number of commercial birds at the base. It thus involves multiplication such that an extremely large volume of commercial chicks can be produced from a very small number of pure-line parents. Example of multiplication volumes for broiler and layer stocks in a typical breeding company are given in Fig 19.12.

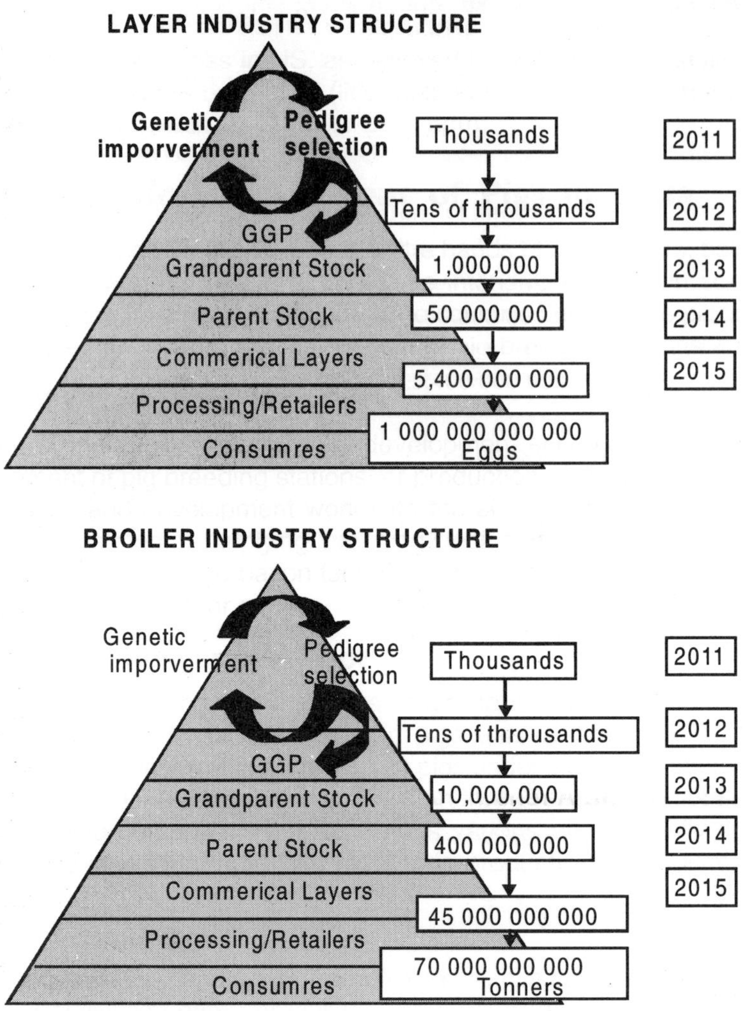

Fig. 19.12 Typical pyramidal structure of broiler and layer breeding companies

Time scale of transmission of genetic improvement: The genetic improvement made in pedigreed pure lines is passed down through multiplication process of next strata as quickly and as efficiently as possible. In case of four-way cross, it takes a minimum of four years to pass on the genetic improvement from pure line to commercial producer/product. Any inadequacy or mistake that occurs at the pure line level will be evident in commercial product after about four years. And by that time, the competitor might have taken away a larger share of market by exploiting the weakness of the product. Therefore in commercial poultry breeding, research is an integral part of production process where the small genetic improvement made each generation in pure lines is passed on to the customers. Each one of the commercial breeder lives in a competitive environment wherein primary breeding companies have to compete with every other breeding company in market. Apparently a great deal of privacy is kept by the primary breeders with respect to the selection criterion used and other aspects of breeding.

Box 19.3: Poultry geneticists and breeding companies.

Of the various farm animal species, poultry (chickens and turkeys) was the first to develop on scientific lines. The success of application of scientific breeding principles for improvement of poultry was largely due to the foresight of owners of poultry breeding companies who hired qualified (usually PhD's) poultry breeders either on full-time or as consultants.

Listed below are some of the poultry geneticists, mainly of North America (where many private breeding concerns originated) who worked or consulted for private breeding organizations. These geneticists also earned a name in academics related to genetics and breeding of poultry.

Mount Hope Farm: Mount Hope Farm was established in USA in 1912, and was one the earliest breeding farm to be set in private sector.

- **Goodale (1879-1968):** Goodale was perhaps the first qualified poultry geneticist to work with a private poultry breeding farm, the 'Mount Hope Farm', Massachusetts (USA).He originally worked for University of Massachusetts at its agricultural experimental station beginning at the turn of 20th century. His research work on genetics of fowl, in particular his paper on egg production wherein he partitioned annual egg production into five components, earned him a name among breeders of his time. He spent most of his professional life (1922 to 1962) with Mount Hope Farms.

  One of the strains of layer chickens developed by him, the Mt Hope, became popular throughout the world, was exported to many countries. The PL2 strain of chicken developed at PAU/ GADVASU, Ludhiana, India was derived from Mt Hope strain.

Dryden Farm: This breeding farm was established by the renowned poultry breeder Dryden.

- **Dryden (1863-1935):** A renowned poultry breeder, he was the founder head of the first poultry science department established in USA (at Oregon State University). His work from 1907 to 1922 at Oregon earned him worldwide acclaim. The first hen to lay 300 eggs in a year, and the first hen to lay one thousand eggs in lifetime were bred by him. In 1922 he established his own breeding farm, the 'Dryden Poultry Breeding Farm' in Modesto, California, USA. He was considered as one of the leading US poultry breeders until his premature death in a car accident.

Kimber Group: An earlier breeding concern Kimber Farm was established in 1925 as a small 800-hen farm at Niles (California, USA) by John Kimber. Kimber graduated from Stanford University (1917) with a major in agriculture, following which he went to Oregon State University to work for Master's under Prof. Dryden (though he did not complete it). After teaching in school for seven years, he established the Kimber Farm. After facing many hurdles, success came in 1932 when his farm showed a profit for the first time. A full-time geneticist (PhD) was hired, making Kimber Farm the second poultry farm to employ a qualified geneticist. The geneticist who worked as full-time or part-time with Kimber Farms included :

- **Lamoreux :** Dr Will Lamoreux was the first geneticist with a PhD to be employed by a private breeding company which was a landmark in poultry genetics. After obtaining his PhD from Cornell University in 1938, Lamoreux served on its faculty from 1939 to 1943. At Kimber's he first became the incharge of poultry breeding operations, and was later on appointed as Director of Research, a position he held until his death in 1975.

Lamoreux not only did genetic improvement work but also associated with a number of scientists working in state universities (like Lush, Hazel) for genetic analysis of performance of Kimber's strains/lines using newer breeding methodologies. His paper on genetic parameter estimation of mortality, an 'all-or-none' trait on Kimber farms' birds, is considered a pioneer piece of work. Dr Hutt, the renowned authority in poultry genetics described the 1940' to 1960' as the hay days for qualified poultry geneticists because many of the young PhD's were hired by private breeding companies at salaries higher than of the professors under whom they had trained.

- **Farnsworth:** Farnsworth joined Kimber Farms in January 1946 after leaving the US Navy. Later he completed his PhD in 1955 from Iowa State University where his PhD committee included Drs A.W. Nordskog, Jay Lush and L.N. Hazel. On return to Kimber Farms as a geneticist, he was assigned the selection programs. When the company decided to start breeding turkeys, Dr Farnsworth was put in charge of the breeding program with Dr Don Warren as his assistant. In 1967, four of the turkey breeding lines were sold to Nicholas, and Dr Farnsworth moved to Nicholas with the turkeys.
- **Warren (1890 -1994):** Dr D C Warren after obtaining his PhD in 1923 from Columbia University under Morgan (the Nobel laureate) working on Drosophila, joined Kansas State University where he remained until 1948. Subsequently, he worked at Purdue University (at the request of USDA) for implementing a poultry program until 1954.Though his MS and PhD research involved Drosophila but after joining Kansas state university, he conducted extensive studies on crossbreeding of poultry including inbreeding and hybridization which earned him great fame. He joined Kimber Farms in California as a Geneticist where he remained until his retirement in 1968. At Kimber Farms, he developed strains of chickens which were disease resistant.
- **Dickerson (1912-2000):** Dr Gordon Dickerson, a PhD from University of Wisconsin had established him in academic field, and having guided a number of PhD students at university of Missouri joined Kimber group as a geneticist. Dr Dickerson's classical paper on 'Genetic slippage in response to selection in poultry' wherein he proposed the concept of genetic platuex was published on Kimber strains. Later, he completed the development of index procedures at Kimber Farm with Lamoreux and others.

The company also utilized the professional skills of some other geneticists through summer consultancy or for short periods. These included Drs Jull (of University of Maryland),Hutt (of Cornell University) and Lush and Hazel (of Iowa State University). These geneticists used some of the latest selection techniques for improvement of pure lines and their crosses

Kimber Farm also recognized the role of the science of genetics in improving livestock productivity by establishing *'Kimber Genetics Award (1955)***.** The Kimber Medal featured Darwin, Mendel, Bateson and Morgan. Between 1955 and 1967, the medal was awarded to 14 scientists which included some eminent geneticists like Castle, Muller, Wright, Sturtevant, Dobzhansky, Beadle, Haldane, Stern, Hershey, Barbara McClintock. In 1974 DeKalb AgResearch, Inc. purchased Kimber Farms whereby it became a subsidiary of DeKalb Corporation.

Dekalb Group: DeKalb another poultry breeding company also hired a number of qualified geneticists which included:

- **Harris :** Dr Dewey Harris, after obtaining PhD in 1961 from Iowa State University, taught there until 1964 and then joined DeKalb AgResearch group as a geneticist, later becoming its Director of Research. In 1974 he moved to Canada to work on National Dairy cattle breeding project. In 1976 he returned to academics and research by joining USDA at Purdue University, and then at University of Nebraska. His research on selection index starting with his PhD work is well- recognized.

- **Briles :** Dr W E Briles, a world renowned authority on poultry immunogenetics, obtained his PhD from University of Wisconsin specializing in immunogenetics, particularly on blood group systems. This was the era of immunogenetics as it appeared to offer great opportunities for enhancing the rate of genetic improvement. After teaching from 1948 to 1957, he was hired by DeKalb for immunogenetic work where he worked until 1970. He returned to academics at Northern Illinois University where he remained professor of biological sciences until his retirement.

Shaver Group: Shaver breeding farm also distinguished itself by employing some of the best geneticists of time which included:

- **Cole (1912- 2006):** Cole graduated with a major in Poultry Husbandry from Massachusetts Agricultural College in 1934, and after doing varied works, got a position as Instructor in 1935 under Professor F.B. Hutt of Cornell University. Later he obtained MSc and PhD(Animal Genetics) under Hutt. He joined Cornell where he specialised in the role of genetics in diseases and defects in chickens. His research on genetics of avian leukosis complex (a poultry disease), for which he developed genetically resistant and susceptible lines earned him worldwide recognition. Cole provided consultancy to Shaver Poultry breeding Farms (Ontario, Canada from 1956 till his retirement in 1973). At Shaver he developed a pedigree hatching breeding program, and also used diallel crossing to identify superior crosses. This work led to development of the 'Shaver 288 layer', a stock that was most productive and enjoyed high ranking in Random Sample Tests. His zeal for work was reflected in his strenuous shuttling between Ontario (Canada) and Cornell (New York), driving from Cornell in his Volkswagen on Friday to Ontario, spend the weekend selecting breeding candidates by inspection of the manually compiled egg production records, and be back by Sunday to do research and teaching at the university.
- **Hunton** : Peter Hunton, a native of England, after obtaining his M.Sc. in genetics from University of London (1966) worked at Wye College – then the agricultural school of England's London University. In 1967 he was employed by Shaver as research co-ordinator. While working for Shaver Poultry Breeding Farms in Canada he earned a PhD from the University of London. His paper on partitioning hatchability into components viz; early and late embryo mortality is rated highly. Hunton spent his career working for Shaver Poultry Breeding Farms, Ross Poultry, and the Ontario Egg Producer's Marketing Board. Throughout his career and for many years after retirement, he wrote articles for trade journals that made science understandable to the industry. He also wrote scientific articles and contributed to major text books on poultry breeding.
- **Gowe:** Dr R S Gowe, a renowned poultry breeder, whose paper of 1959 on 'Design of poultry control strains' continues to be highly cited, worked for Animal Research Branch of Agriculture Canada at Ottawa from 1963 until 1986. There he rose to become Director of Research. His own research work focused on poultry wherein he insisted that control strains be included in all trials, and that multiple-trait selection be used to ensure overall progress in genetic merit.

Among his various research contributions was a long-term selection in layer chickens involving a number of traits. This work is considered novel as it reflected the type of selection practiced by large scale private breeding companies.

He joined Shaver in 1987 as Director of Research where his role was to increase the emphasis of selection for economic traits that enhanced the profitability of layers and broilers without diminishing their adaptability to different climates and management regimes. He developed a breeding strategy for poultry, later also applied to swine, involving pure strain selection and improvement, with crossing between strains to capture the advantages of heterosis. That approach was being used on a trial basis in Canadian dairy cattle breeding at the time of his induction into the Hall of Fame in 2009. Prior to his time at the Animal Research Centre, scientists tended to work separately on their own projects within their own areas of expertise. Gowe insisted that challenges should be tackled by teams of scientists derived from different areas.

H and N Group: Similar goes the story of 'H and N' breeding organisation which was founded in 1945 by Arthur Heisdorf who had graduated in poultry from University of Wisconsin(USA). Prior to this, he had worked for nine years with Kimber Farms which had a rich history of employing qualified poultry geneticists. In fact one of the reasons for H and N's success was that its owner understood the importance of employing trained geneticists.

The first poultry geneticist to be employed by 'H and N' was McClary (who was trained under Dr Lerner) and had worked in a university research program. This was followed by hiring PhD's from Lush-Nordskog group of Iowa State University namely Pirchner (best known for his book 'population genetics in animal breeding'), von Krosigk, Hogsett and Flock ; and Havenstein (from University of Wisconsin), and later Emsley who had been trained under Dr Dickerson. These geneticists have made significant scientific contributions to poultry breeding through their publications. Later in their career, many of them shifted to full time academics.

On similar lines, other poultry breeding companies had been hiring qualified poultry breeders in many other countries where poultry farming has developed on commercial lines.

Box 19.4: Pure breed concept vis a vis current commercial breeding programs.

Definition of a breed of poultry, within a class, has been based on the similarity of body shape and size/form. When poultry breeding programs were started in the beginning of 20th century, the concept of breed was central to such programs wherein breed crosses were used to produce commercial birds especially in broilers. However concept of breed or pure breed is now almost completely eliminated in commercial breeding programs.

Commercial breeders now deal with numbered strains that started as White Leghorn, White Plymouth or some other breed or synthesis of breeds but to which genes for traits like feathering rate, plumage feather color, dwarfing etc have been incorporated from other genetic sources. Thus these strains/lines do not descend from 'pure-bred' ancestors. Furthermore the gene pools of such populations have been selected exclusively for traits of economic merit. These populations, therefore no longer meet breed's physical standards. Currently commercial poultry breeders use White Leghorn or Rhode Island Red-like strains for egg-type stocks or White Rock-like meat dam- strains and White Cornish-like meat sire strains. But none of these populations are likely to be 'pure-bred' or meet breed standards.

Due to the virtual disappearance of pure breeds in commercial poultry, it not possible to differentiate between breed and strain crosses. Though the classical pure breeds are no more used in broiler chickens but still there are large differences between their sire and dam lines; the sire lines carry a high proportion of Cornish genes and the dam lines usually carry a high proportion of White Rock genes. Such large physical differences, however, are not seen in strains used to produce eggs (either white or brown).However, strains/lines of Leghorn (white eggs) or of Rhode Island Red (brown eggs) can have quite variable performance, and different crossing attributes.

Poultry Breeding Programs in India

India and its neighboring countries are considered the home of Red Jungle Fowl from which modern elite breeds have developed though most of the modern breeds have been developed in Western countries. The earliest attempts to genetically improve poultry in India were made by missionaries at Etah (U.P.) and Katpadi (Tamil Nadu) during the second decade of 19th century. Some important exotic breeds were imported from UK, USA and Australia. Apart from the introduction of exotic breeds, genetic improvement of indigenous stocks was also initiated by the Imperial (Now, Indian) Council of Agricultural Research through Poultry Division of its premier research institute, Indian Veterinary Research Institute (the poultry division was later upgraded to 'Central Avian Research Institute', Izatnagar, UP).

A few defined breeds of native chicken like Aseel, Chittagong and Ghagarsare were available in India. But bulk of the native germplasm consisted of some non-descript birds. A scientific report of 1950's indicates that selection for six generations in non-descript flock of *desi* fowl improved annual egg production from 116 to 140, and egg weight from 43 to 49 g. Since a genetic control line was not maintained, it is unclear as to what extent the change was genetic in nature.

The *desi* breeds were slow growing, and poor layers of small-sized eggs as compared to exotic stocks therefore, imports of high performing egg-type and meat-type chickens from developing countries were made in 1950 and 1960's. It was thought that this approach would lead to a much faster improvement in productivity than possible by selection within indigenous stocks or by upgrading of indigenous with exotic.

The main favorable features of this approach were:

- Replacement of indigenous birds with exotics was easy; the cost of each bird was low and slaughter was not a problem (unlike in cattle). The high reproductive efficiency of chicken permitted rapid multiplication of stocks.
- Control of environment for rearing of exotic poultry was not a serious problem, and genotype-environment interaction in poultry is not as important as in cattle.

However to make the country self sufficient in quality germplasm of chickens, systematic research efforts on development of suitable stocks were started in early 1970's through:

- Indian Council of Agricultural Research (ICAR) in early 1970's sponsored the 'All India Coordinated Research Projects(AICRP's) on Poultry Breeding' in collaboration with state agriculture universities, and its own institute(IVRI, later on CARI, Izatnagar, UP).
- Central Poultry Breeding Farms of the Government of India (located at Bhubaneshwar, Bombay, Bangalore and Chandigarh)
- State Government Funded Projects at respective state agriculture universities

Box 19.5: Central Avian Research Institute and Project Directorate on poultry.

Poultry research under Indian Council of Agricultural Research was carried under poultry division of IVRI. Later on realizing the increasing scope of poultry in India, two other institutes of ICAR carry the poultry research and development activities viz; CARI and DPR (previously PDP).

- **CARI:** Central Avian Research Institute (CARI) was established in 1979 at Izatnagar, Bareilly, and Uttar Pradesh under the aegis of the Indian Council of Agricultural Research (ICAR) to provide all-round support to the growth of poultry sector. Since its inception, this Institute has been playing an important role by providing need based R&D support for diversified poultry production, processing and marketing, apart from Post-graduate education, training and technology transfer activities for augmenting productivity, production and profitability of the Indian poultry sector.

  The AICRP on poultry which originally operated with a coordinating unit located at Poultry Research Division, Indian Veterinary Research Institute, Izatnagar, and U.P. until the year 1979 was subsequently at Central Avian Research Institute, Izatnagar until the year 1988.

- **DPR:** The Directorate of Poultry Research (DPR), formerly Project Directorate on Poultry (PDP) is one of the constituent research institutes of the Indian Council of Agricultural Research (ICAR). The Directorate was set up as a coordinating unit of All India Coordinated Research Project (AICRP) on Poultry in 1970 at Izatnagar, Uttar Pradesh. Subsequently the unit was elevated as a full-fledged Project Directorate during 1988 and shifted to Hyderabad. The objectives of the Directorate are to coordinate research at AICRP centres located across the country and conduct research on the development and improvement of chicken lines for commercial and rural poultry production.

  The Directorate also maintains a germplasm center to make available the improved strains and stocks to the user agencies. The Directorate has evolved synthetic strains viz. 'Vanaraja' and 'Gramapriya' free range back yard farming in rural and tribal areas and 'Krishibro' for small scale intensive broiler farming.

All India Coordinated Research Project (AICRP**):** The AICRP on poultry was started in early 1970's with a large setup to cover major agro climatic regions of the country. Most of these centeres of AICRP were located in state agriculture universities with a few in ICAR institutes. These were spread across the country to cover all the regions. There were two components: one on genetic improvement of layer chickens and the other on genetic improvement of meat-type chickens.

AICRP for egg involved testing of 13 layer strains with 3-4 strains at different centers located in major regions, as purebreds and their crosses. Improvement in purebred lines/strains was made through intra-population selection : combined selection based on part-year egg production. Similarly, in broilers, eight strains in total with 3-4 strains at different centers, were tested in pure as well as cross combinations and intra-population selection (individual selection) based on 8-week body weight was carried out.

Having accomplished its mandate of giving boost to commercial poultry farming through supply of superior germplasm, the mandate of AICRP has now been changed to breed chicken suited for low-input conditions of marginal farmers (rural poultry).

State Agriculture Universities: State agricultural universities and research institutions in India have developed a couple of stocks indigenously which have been released for commercial poultry production in the country.

Central Poultry Breeding Farms: Central Poultry breeding Farms of the Government of India (Ministry of Agriculture) had also carried out systematic genetic improvement at their breeding farms.

Research efforts of all these agencies resulted in development of high yielding layer and broiler strains. The commercial chicks, which were crossbreds, had been supplied to farmers at cheap rates. This gave a boost to commercial poultry farming which eventually developed into large-scale operations. Some of the outstanding stocks of chickens developed within the country in public sector are listed in Table 19.10.

Table 19.10 List of few commercial stocks of layer and broiler chickens developed in public sector in India.

Commercial stock	Name of Institute
Layer chicken:	
• HH 260 :	CPB Farm, Bangalore (Karnataka).
• BH 78 :	CPB Farm, Bombay (Maharashtra).
• ILI-80 :	CARI, Izatnagar (UP).
• ILM-90 :	KAU, Mannuthy (Kerala).
• ILR90 :	ANGRAU, Hyderabad (AP).
• CARI Gold :	CARI, Izatnagar (UP).
• SATLUJ layer :	PAU/GADVASU (Punjab).
Broiler chicken:	
• IBL 80 :	PAU(GADVASU), Ludhiana (Punjab).
• IBB 83 :	UAS, Bangalore (Karnataka).
• IBI 91 :	CARI, Izatnagar (UP).

Private Sector: The private sector has also played an important role in developing or making available high yielding germplasm. In 1960's, four entrepreneurs in private sector entered into foreign collaboration for commercial production of genetically superior stocks which were all hybrids:

Company	Country	Headquarter in India	Product
Hyline :	USA	Delhi	Inbred crosses
Arbor Acres :	USA	Poona	Brown crosses
Shaver :	Canada	Delhi	Single Comb WLh Hybrid
Unichix :	Czechoslovakia	Delhi	Broiler hybrid

Stocks of these companies were good performing but chick prices were high for small entrepreneurs. In 1990's the Government of India liberalized the import of great grandparent (GGP), grandparent (GP) and parent lines by private companies. This allowed the entry of International companies in collaboration with local companies into the Indian market. Subsequently, other poultry breeding farms such as Venketashwara hatcheries, Kegg farms, CSR Ross breeders, Poona Pearls etc. entered in the poultry business. Now, as a result of global liberalization, most of the leading international brand names are available in India.

Currently most of the commercial farmers which have large-scale operations buy chicks from private breeding companies and hatcheries who import grand-parental lines of world-famous stocks. A few private breeding companies have also procured pure lines and started their breeding programs in order to develop superior stocks within the country. The number of breeding companies, through mergers and acquisitions, have been declining throughout the world.

Review Questions

1. Define / explain the following (in context of poultry breeding):
   I. Poultry
   II. Breed (of poultry)
   III. Variety (of poultry)
   IV. Poultry strain

2. Write short notes on the following :
   I. Specialized sire and dam lines of chicken.
   II. Structure of a commercial poultry breeding organization.
   III. Poultry breeding programs in India.
   IV. Strain crossing.
   V. Selection in layer chickens.
   VI. Large-scale commercial (industrial) poultry breeding program.

3. Differentiate between the following :
   I. Egg-type and meat-type chickens
   II. Two-way and four-way crosses
   III. Sire and dam lines
   IV. Hen-day and hen-housed egg production

4. What are breeding goals in :
   I. Layer breeding
   II. Broiler breeding

5. Which is the most important trait in breeding of layer chickens? How is this trait measured? Which selection procedure (basis) is most widely used for its improvement?

6. List the two most important breeds each for layer and broiler chicken which have been most widely used in breeding. Describe their main features.

7. What criterion is used for selection of males in layer chicken for improvement of egg production?

8. What is meant by selection on part-year egg production? Why is it used for improvement annual egg production?

9. What is the importance of selection for early sexual maturity in layer chicken? Which selection criterion is used for its improvement ?

10. Why is reducing age at sexual maturity below a certain age not desirable?

11. What are the two most important component traits contributing to the number of eggs produced in a fixed period ?

12. How is rate of lay measured or expressed ?

13. Diagrammatically show a typical egg production curve of modern commercial layer chicken.

14. Which selection criterion is used for genetic improvement of egg weight ?

15. Which selection criterion can be used for improvement of viability (disease resistance) in chickens?

16. What are the most important traits related to reproduction in chickens? How are these measured?

17. What are the main advantages of 'inbreeding–hybridisation' (using inbreed lines) in poultry breeding? Why this procedure was later on abandoned?

18. What is meant by strain crossing? What advantages this method offered over inbreeding-hybridization ?

19. Diagrammatically show the different types of crosses used in layer breeding (viz. two-way, three-way and four-way)

20. What is modern commercial layer chicken's productivity under standard management conditions for the following traits?

I. Hen-day egg production II. Feed consumption

III. Age at 50% production IV. Average egg weight

21. List the three most important traits of economic importance in broiler chicken? Which selection criteria are used for their improvement?

22. What is the importance of body conformation in broiler breeding? How can body conformation be genetically improved?

23. Which two traits have been most unfavourably affected by long-term selection for body weight in broiler chicken?

24. Which criterion of selection can be used for reducing abdominal fat in broiler chickens?

25. List the examples of traits showing :

I. Undesirable negative genetic correlation in layer chickens

II. Desirable negative genetic correlation in layer chickens

III. Desirable genetic correlation in broiler chicken.

IV. Undesirable genetic correlation in broiler chicken.

26. List the traits of broiler chicken showing :

I. Undesirable genetic correlation with body weight.

II. Desirable genetic correlation with body weight.

27. What is meant by specialized sire and dam lines? In which type of chicken (layer or broiler) are these used and why?

28. List the three most important traits of a :

I. Broiler sire-line II. Broiler dam-line

29. Which is the most important breed that has been used to develop sire line(s) in broiler chicken? Which attribute of this breed makes it most suited to serve as sire line?

30. Which (type of) breeds are used in developing dam lines in broilers chicken.

31. How is sequential selection practised in broiler breeding ?

32. What is the current commercial broiler chicken's productivity level for the following traits?

I. Body weight (six-week) II. Feed conversion ratio

33. What is usual time scale of transmission of genetic improvement from pure lines to commercial level ? Explain.

34. List the names of at least three poultry breeding companies supplying day-old commercial chicks for each egg-type and meat-type chicken

35. List at least three commercial stocks, each for layer and broiler, developed in the public sector organization (institutes / universities) in India.

●●●

20

Sheep Breeding

Small ruminants, viz. sheep and goat, are numerically important domesticated animals in arid and semi-arid regions of Asia where they are primarily raised in traditional low input conditions. They contribute meat, milk, skin, fiber and manure, and are the sole or subsidiary source of livelihood for a large number of small and marginal fanners and landless laborers. The productivity of these species in its areas of prominence is relatively lower than that in temperate countries. Nonetheless, they contribute to the socioeconomic well being of people in developing countries in the tropics in terms of nutrition, income and intangible benefits.

Phylum: Chordata (back bone)
Class: Mammalia (suckle their young ones)
Order: Artiodactyla (hooved, even footed)
Family: Bovidae (hollow-horned ruminants)
Subfamily: Caprinae(sheep & goat)
Genus: *Ovis* (domesticated and wild sheep)
Species: *Ovis aries*

Sheep along with goats are perhaps the first ruminants to be domesticated. Reference to the role of sheep to humankind is found in the history of civilization of Mesopotamia and Mohenjo-Daro and Harappa in northern India. The domesticated sheep (*Ovis aries*) are thought to be descendents of the wild sheep: the Urial of western Asia (*Ovis vignei*) and the Mouflon of southeastern Europe (*Ovis musimomn*). Four main types of domestic sheep have been developed from this foundation:

- Fine-wooled sheep in Mediterranean area
- Medium-wooled sheep in Europe (mainly England)
- Long-wooled sheep in Europe (mainly England)
- Carpet-wool type sheep in Asia

The domestic sheep found in India and Arabia are the descendent of wild sheep (*Ovis vignei*) found in the mountains from Afghanistan to Armenia. Sheep, which constitute about 12.7 % of the total livestock population of India, are primarily raised for wool and meat, their milk is used in some places primarily for home consumption over and above lamb feeding.

The major sheep rearing states of the country, as per 2012 census, are Andhra Pradesh (40.6%), Karnataka (14.7%), Rajasthan (14.0 %), Tamil Nadu (7.4%), Jammu & Kashmir (5.2%), Maharashtra (4.0%), Gujarat (2.6%) and Orissa (2.4%).

Sheep possess the quality of utilizing uncultivated fodder growth and weeds, and has the characteristic of close grazing. They are gregarious (i.e. they like to stay close to others in a flock) which makes them easier to move together as a group. They have an excellent ability to survive over prolonged period of drought, and are less susceptible to extremes of weather. Their fiber allows ventilation and protects skin from the hot sun, rain and other insults. Sheep can also constrict or relax blood vessels of face, legs and ears for control of heat loss, and hence they are highly adaptable to dry conditions. For this reason, the North western and the Southern peninsular regions of the country have a large concentration of sheep. Sheep contribute to the agrarian Indian economy, especially in arid and semi arid areas where land is unfit for crop production. In these areas, they provide livelihood to a large number of small and marginal farmers and landless laborers.

Box 20.1: Sheep and wool terminology.

- **Apparel wool:** Wool suitable for manufacturing into apparel fabrics (fine quality wool).
- **Carpet wool:** Coarse wool used in manufacture of carpets; it usually comes from unimproved breeds.
- **Crimp:** The natural waviness of wool fiber; usually closer the crimps, the finer the wool.
- **Clip:** One or more fleeces (from a flock).
- **Clipping:** Cutting off the wool.
- **Ewe:** Female sheep capable of reproduction (adult sheep).
- **Fleece:** All the wool shorn from one sheep at one time.
- **Grease fleece:** Wool as it is shorn from the sheep before any processing/cleaning.
- **Hogget:** A British term for a young sheep of either sex from about 9 to 18 months of age (until it cuts two teeth, and has not yet been shorn); meat of hogget is also called hogget. Kemp: Opaque brittle, chalky white, weak fiber that shed periodically; such fibers are considered as an impurity in the fleece.
- **Lamb:** Young sheep in its first year; the term is also used to refer to meat of young sheep.
- **Lambing:** Act of giving birth.
- **Medullated or Medullated fiber:** Wool fibers that are hollow (have a medulla in central part of fiber) rather than solid as in true wool.
- **Mutton:** Meat of sheep, particularly of older ewe; the word is derived from an Anglo-Norman French word 'mouton'.
- **Pelt:** The skin of sheep with wool still attached to it.
- **Ram:** An uncasrated adult male sheep, capable of mating or reproduction.
- **Scouring:** Process of removing the grease and foreign matter from greasy wool. It is usually done in a lukewarm, mildly alkaline solution followed by a rinse.
- **Scoured wool:** Clean wool that has been washed to remove grease, soil, suint etc.
- **Shearing:** Removal of wool from sheep; it is done by clippers or blade shears.
- **Shrink:** The non-wool portion of greasy wool, which includes grease, vegetable matter, dirt, suint etc.

- **Staple:** A tuft of wool fibers (of shorn fleece) that form a cluster or lock as they have grown side by side.
- **Staple length:** Average length of a staple in a fleece measured from skin to tips.
- **Suint:** Water-soluble component of greasy wool produced by sweat glands in skin of sheep.
- **Tupping:** Mating in sheep
- **Whether:** Male sheep castrated before puberty
- **Wool fiber vs. hair:** Wool fibers are usually much smaller in diameter, crimped and elastic usually with little medulla in the centre; the hair is not crimped like fine wool and it is always medulated. Wool fibers grow much denser on the skin than hairs. Both hair and wool may have pigment within the cortex, but most wool is free of pigment.
- **Wool grease:** A complex mixture of esters secreted by sebaceous gland in skin of sheep.
- **Yield or clean yield (of wool):** The proportion or amount of clean wool (at standard conditions) present in a given amount of greasy wool.

Box 20.2: Wool grading and classification.

Wool grading systems: The word 'wool grade' is used to describe fineness of wool fibers. Wool has been graded by three systems:

- ***Blood system:*** Blood grade system of wool grading, an American system, began in colonial America when native coarse-wooled sheep were being bred to fine wool Merino sheep from Spain. It divides all wool, from finest to coarsest, into six market grades: fine, 1/2 blood, 3/8 blood, 1/4 blood, low 1/4 blood, and common and braid. Originally, these names referred to the fraction of Merino inheritance (blood) in sheep that produced the wool. Wool from pure merino was called "fine wool". Wool from sheep that were one-half Merino or Rambouillet and one-half some other breeds or breed-combination was called "one-half blood". As the proportion of fine wool inheritance represented in fleeces became smaller, it was described accordingly as three-eighths, or one-quarter. Wool most suitable for carpets or similar bulky products was called 'common and braid'.

  One limitation of blood grade system is that it permits only six categories with respect to fiber diameter. As greater accuracy in the description of average fiber diameter became necessary, the blood system changed to classification based on fiber diameter rather than breeding (inheritance) background. The blood system, for most useful purposes, is outdated.

- ***Spinning count system*** (also referred to as quality number): The spinning count or Bradford system, which originated in Great Britain, relates the fiber diameter to length of yarn and weight of wool top. It describes wool fiber diameter as an estimate of the number of hanks of yarn that can be spun to a minimum thickness from one lb (0.45 kg) of top. Hank is 560 yards (512 m) long. There are more fibers in a lb. of top made from fine fibers than one made from coarse fibers. Since a minimum number of fibers are required for yarn to be strong enough for weaving or knitting,, therefore more yarn can be spun from finer than coarser wools.

  Spinning count system provides a numerical designation for different grades varying from 80's (the finest) to 36's (the coarsest). Count grades commonly used are : 80s 70s,64s,62s,60s,58s,56s,50s,48s,40s and 36s. The spinning count system is more specific as it provides many more categories for describing fiber diameter as compared to only six grades used in blood system.

- ***Micron system***: It is another system of wool fineness and is considered the best method available for describing average fiber diameter of wool. It has been derived by common usage from micrometer(One micron = 10^{-3} mm or 1/25,400th of an inch).The primary advantage of micron system is that it is determined by objective measure, and it describes wool fiber diameter on a continuous scale rather than categories (as used in other two systems).

  Specifications for grades of wool for the three systems are as follow:

Type of wool	Blood grade	Spinning count	Microns
Fine	Fine	Finer than 80's to 64's	Under 17.7–22.04
Medium	½	62's–60's	22.05–24.94
	3/8 th	58–56's	24.95–27.84
	1/4th	54–50's	27.85–30.99
Coarse	Low ¼ th	48's–46's	31–34.39
Very coarse	Common & braid	40's to coarser than 36's	36.4 to over 40.2

The American blood grade system is obsolete and the British spinning count system is also likely to be abandoned in favor of microns.

Wool classification: Wool is classified for commercial purposes into five types based on relative fineness alongwith other qualities such as staple or fiber length, softness, and luster. This system is used to group wool types (rather than sheep breeds per se) :

- *Fine wool*: Wool of Merinos and its derivatives is the finest (apparel) wool. Fiber of fine wool has an average diameter of 18-24 micron and spinning count of 64's to 80's. Length ranges from 2.5 to 12.5 cm.
- *Medium wool*: It is used for knitted items, fine tweeds, blankets etc. Its spinning count is 50-58's
- *Long wool*: This type of wool is lustrous and is used for manufacture of plain clothes, tweeds, over coatings, blankets, felts etc. Spinning count is about 48's and coarser.
- *Carpet wool*: It comprises of mixture of long, hairy fibers (ranging from 30 to 40 micron or even up to 50 micron) which form the outer-coat, and fine undercoat (measuring 10 to 24 micron).Length of coarse fiber ranges from 10 to 25 cm and that of fine fiber from 5 to 24 cm. These two types of fibers make up the main part of carpet wool. A third type of fiber, kemp, is also present in carpet wool in varying amounts from 1 to 20%.
- *Crossbred wool*: It is medium type wool, which is used in knitting trade, and is longer than fine wool. Spinning count is 56-62's.

Germplasm Resources of Sheep

Sheep breeds manifest great variability in external features viz. form of horns, shape and size of ears, length of tail, arching of nasal bones etc. Variations do exist in color of face and other parts not covered with wool, quality and color of fleece. These variations have provided the basis for improving sheep for different products (e.g. wool, meat, pelt etc), and for the development of new breeds, strains or types.

Sheep Types

The number of sheep breeds in the world is perhaps more than of breeds of any other livestock species. Sheep breeds vary with respect to size, shape, color, and many other morphological features besides production differences. The breeds can be classified in several different ways: according to their primary purpose (wool, meat, or milk), the type of fibers they produce (fine, medium, long or carpet wool; or hair), the color of their faces (black, white, red etc), and/or by specific physical or production characteristics. Some common types of sheep are:

- **Fine wool sheep:** Fine wool breeds produce fibers with a very small fiber diameter, usually 20 μ or less. Their fleeces tend to be the shortest in length and contain the most lanolin (wool wax or grease). Fine wool fleeces usually yield a lower percentage of clean fiber than longer, coarser fleeces. However, fine wool is the most valuable wool in the commercial market because it is used to make the highest quality wool garments. Fine wool is less likely to itch when it is used in garments that are close to the skin.

  Fine-wool sheep tend to be hardy and long-lived. Most fine wool breeds trace their ancestry to the Spanish Merino. Fine-wool sheep have a strong flocking instinct and are well adapted to arid climates, such as South Africa, Australia, and the western United States and Canada. Most fine wool sheep belong to the Merino breed or its derivatives like Rambouillet. Worldwide, fine wool sheep comprise approximately 50% of the sheep population.

- **Long wool sheep:** Long wool sheep produce long-stapled wool with a large fiber diameter, usually greater than 30 μ. Long wool sheep are best adapted to cool, high rainfall areas with abundant forage and feed resources. Their fleeces yield more clean fiber because they contain less lanolin. The fleeces from long wooled breeds tend to be favored by hand spinners and weavers. They are commonly raised in England, Scotland, New Zealand, and the Falkland Islands. Popular long wool breeds include Border Leicester, Coopworth, Cotswold, Leicester Longwool, Lincoln, and Romney.

- **Medium wool sheep (meat-type):** Meat- or mutton-type sheep produce wool which is mostly medium or long but are raised more for their meat qualities. The length and fiber diameter of medium wool sheep is intermediate between fine and long wool breeds. Medium wool sheep account for about 15% of world's sheep population. The most popular meat breeds are Dorset, Hampshire, Southdown and Suffolk.

- **Carpet wool sheep:** The coarsest, lowest grade wool (usually over 38 microns) is used in the manufacture of carpets. Carpet wool breeds are usually double-coated, with a coarse long outer coat for protection against the elements. They are generally adapted to extreme environments.

- **Hairy sheep:** The wool sheep described above have coat consisting primarily of wool fibers. But some breeds lack wool and are covered with hair instead, like their wild ancestors. Some hair sheep have pure hair coats whereas others have coats that contain a mixture of hair and wool fibers that shed naturally. Hairy sheep do not require shearing. Generally, they are more adapted to warmer and humid regions. Hair sheep are found mostly in Africa and the Caribbean but are also raised in temperate climates, such as the U.S. and Canada. In India, most of the indigenous sheep breeds of southern and eastern region are hairy. Hair sheep account for about 10% of the world's sheep population. Interest in hair sheep is also developing in Australia, Europe, and the United Kingdom. Generally, hair sheep are more prolific and have good mothering ability.

- **Fat-tailed sheep:** Fat-tailed or fat-rumped sheep are so-named because they can store large amounts of fat in the tail and region of the rump. Fat-tailed sheep are found mostly in extremely arid regions of Africa, the Middle East and Asia. They often produce wool, but are raised primarily for meat or milk production. Fat-tailed sheep comprise 25 of the world's sheep population. The important fat-tailed sheep breeds that have been used in India are Karakul and Awassi.
- **Short- or rat-tailed breeds:** Short- or rat-tailed breeds originated primarily from Scandinavia and Northern Europe. Their tails are thin, free of wool, and do not need docking. These breeds are generally very prolific. Examples of these breeds include Finn Sheep, Romanov, East Friesian, Shetland, Icelandic, and Soay.
- **Primitive breeds:** Primitive breeds have developed with minimal human selection efforts. They typically possess lot of genetic variation but share characteristics indicative of a breed. A primitive breed typically retains survival characteristics that favor production with minimal human inputs.
- **General purpose vs. specialized sire- and dam-breeds:** For production of sheep for commercial operations, sheep breeds are sometimes grouped into general-purpose (dual) and specialized sire and dam breeds. General-purpose breeds provide a balance between wool and meat, which makes them suitable to be used as either a sire or a dam breed. General-purpose breeds are adaptable to a wide range of environmental conditions, and are usually more suited for small flocks where crossbreeding is not feasible.

Ram or 'sire' breeds excel in growth and carcass (meat) characteristics whereas ewe or 'dam' breeds excel in fitness (e.g. longevity, parasite resistance) and reproduction (early puberty, prolificacy, milk production). Sire breeds are often called 'terminal sires' because the offspring from their matings are all marketed (terminated) whereas lambs sired by 'ewe breed' ram are usually kept as flock (ewe) replacements. In USA, the most popular terminal sire breed is Suffolk. Sire breeds excel in growth and carcass traits while dam breeds excel in reproductive and mothering ability traits.

Indian Breeds and Their Distribution

The country has vast area of arid and semi-arid climate in addition to the temperate climate of Himalayan ranges. There are wide variety of sheep breeds including fine wool-type, carpet wool-type and meat cum rough wool-type. The number of native breeds is varyingly listed as 44 (in ICAR's handbook) and 39 (registered with National Bureau of Animal genetic Resources, NBAGR). A large population of sheep (about 75%) however is non-descript due to indiscriminate and large inter-mixing of breeds of adjoining areas. There are no breed societies to register animals of a particular breed and to keep purity of breed or type. Some of the sheep from adjoining areas have been named as separate breeds although they may be genetically similar. More than one strain of some breeds also exist (e.g. Deccani, Nellore etc). According to the strict definition of a breed, majority of Indian sheep breeds lack distinct traits of fixed nature.

Most of the sheep breeds in India have evolved through natural adaptation to the agro-ecological conditions coupled with some limited degree of artificial selection. These breeds have been named after their place of origin and prominent characters. Based on the agro-climatic conditions, the type of sheep and the purpose for which they are raised, the country has been divided into the following four regions: Northern Temperate region (Himalayan), North Western region (Central Arid and Semi Arid), Southern Region and Eastern region (hot and humid)

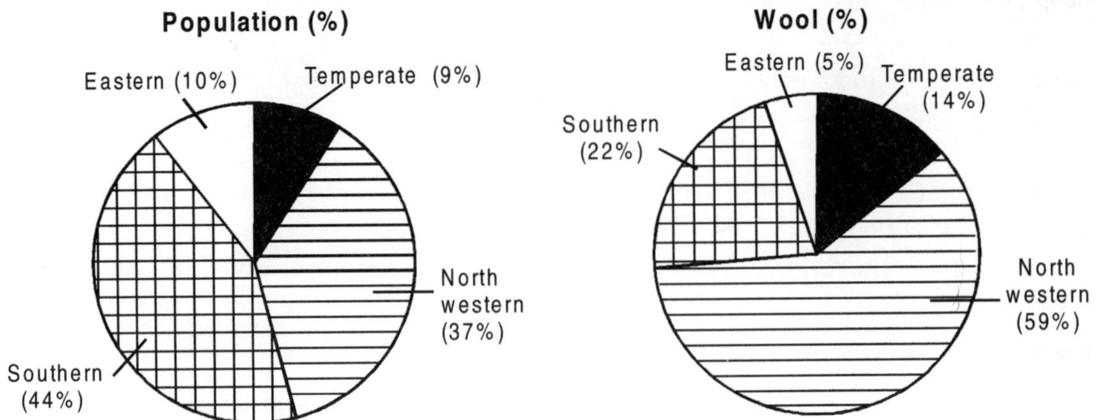

Fig. 20.1 Region-wise distribution (percentage) of total sheep population of India (a) and their contribution to total wool production (b)

Breeds of Northern Temperate Region (Himalayan)

This region comprises of Himachal Pradesh, Jammu and Kashmir, Uttaranchal. It has about 9% of country's sheep population but accounts for about 14% of the wool produced in the country (Fig. 20.1). Sheep of this region produce medium to fine quality fleece (apparel/ superior carpet type). About one-third of sheep population of this region especially in Jammu and Kashmir consists of crosses between indigenous breeds and exotic fine wool breeds. About two-third of sheep of recognized breeds of this region belong to Gaddi followed by Rampur Bushair, Changthangi, Poonchi and Gurej in that order. Karnah and Bhakarwal have very low numbers, and their numbers are further declining (Fig. 20.2).

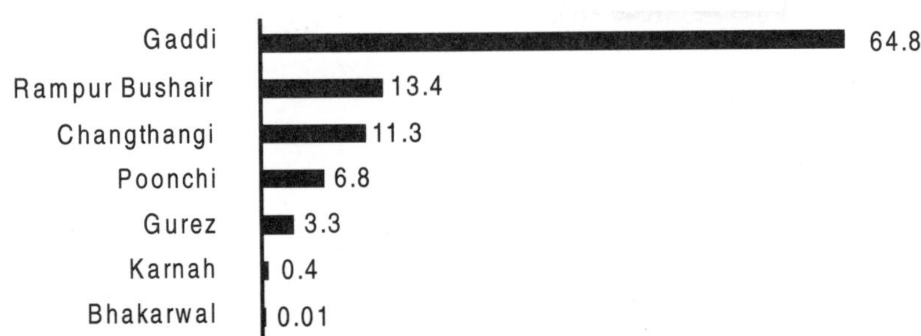

Fig. 20.2 Distribution (%) of breeds of Northern temperate region

Gaddi (also known as Bhadarwah): Animals are medium sized and usually white in color. Males are horned and so also a few females. Fleece is relatively fine and dense; it is clipped thrice a year. It is distributed in Kishtwar and Bhadarwah tehsil of Jammu and Kashmir, and in Himachal Pradesh.

Rampur Bushair. Medium sized, predominantly white colored with a convex face and Roman nose. Fleece is of medium quality and dense. It is distributed in Simla, Kinnaur areas of HP and in Uttaranchal.

Changthangi: Animals are strongly built with good fleece cover of an extraordinary long staple. Wool is of good carpet/medium apparel quality. Found in Changthangi region of Ladakh.

Poonchi: The breed derives its name from Poonch district of Jammu. Animal are similar in appearance to Gaddi except being lighter. Wool is of medium to fine quality.

Gurej: Fleece is coarse and hairy. Annual greasy fleece weight ranges from 0.5 to 1kg. Found in Gurej area of northern Kashmir

Karnah: Generally, large animals, wool is generally white. Sheep are shorn twice. Distributed in Karnah tehsil of North Kashmir.

Bhakarwal: Breed has derived its name from a nomadic tribe, which rears sheep. The breed is completely migratory; there is no home tract of this breed. Found in foothills of Kashmir.

Breeds of North Western Region (central arid and semi arid)

This region,which comprises of the states of Punjab, Haryana, Uttar Pradesh,Rajasthan, Gujarat and parts of Madhya Pradesh, has second largest sheep population (37%) of the country. Sheep of this region account for about 59% of country's wool (Fig 20.1). But the wool produced by sheep of this region is mostly coarse carpet quality except of Chokla and Patanwadi, which is medium to fine quality carpet/medium quality apparel wool.

This region is most important in the country for carpet wool production. Major sheep population of this region is concentrated in two states : Rajasthan and Gujarat. In this region, Marwari breed has the highest numbers (50%) followed by Jaisalmeri, Nali, Chokla, Patanwadi, Malpura and Magra in that order (Fig. 20.3). Pugal, Sonadi, Muzzafarnagri and Jalauni have low numbers.On the other hand, Pugal and Magra breed's numbers are declining.

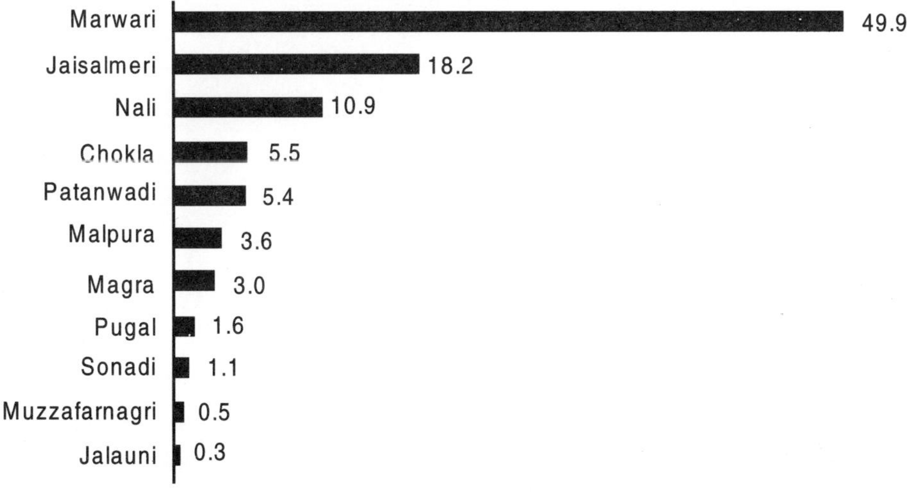

Fig. 20.3 Distribution (%) of breeds of North Western region.

Marwari and Magra breeds of this region have been identified as improver breeds and are being improved under Net Work project of ICAR for carpet wool while Mujaffarnagri is being improved as a dual-purpose breed. Sonadi has been included in the Mega Seed Sheep Project of ICAR for production of rams of superior quality for improvement of field stocks. Malpura and Chokla breeds are also undergoing genetic improvement at CSWRI for mutton and carpet wool respectively.

A brief description of breeds of this region is given below, and the average values for some important wool attributes of these breeds are given in Table 20.1.

Table 20.1 Means (alongwith SE) for some wool traits of sheep of NW region.

Breed	6 months fleece wt. (kg)	Fiber diameter (μ)	Staple length (cm)	Medullation (%)
Chokla	1.37 ± 0.01	4.7 ± 0.07	28.22 ± 0.20	24.01 ± 0.62
Nali	1.46 ± 0.16	8.31 ± 0.14	34.92 ± 0.69	30.74 ± 1.66
Marwari	0.89 ± 0.02	6.56 ± 0.05	36.93 ± 0.16	65.18 ± 3.1
Magra	1.09 ± 0.002	5.81 ± 0.02	32.45 ± 0.35	48.29 ± 0.39
Jaisalmeri	0.77 ± 0.07	6.46	39.1 ± 2.76	64.1 ± 3.1
Pugal	0.8 ± 0.02	5.71 ± 0.04	35.13 ± 1.00	61.86 ± 0.62
Malpura	0.54 ± 0.04	5.6 ± 0.0001	41.95 ± 0.37	71.84 ± 0.17
Sonadi	0.45 ± 0.01	4.58 ± 0.27	52.65 ± 1.77	88.15 ± 2.1
Patanwadi	0.628 ± 0.087	8.51 ± 0.09	31.95 ± 0.26	29.88 ± 0.95
Muzzafarnagri	0.65 ± 0.08	3.72 ± 0.05	45.17 ± 0.37	69.92 ± 0.87
Jalauni	0.90	—	41.1 ± 0.19	78.0

Chokla (syn. Chapper and Shekhawati): Sheep of this breed are light to medium sized, reddish brown in color and the coat is dense and relatively fine. The face is generally devoid of wool and both sexes are polled. Chokla perhaps grows the finest carpet wool of all the Indian breeds (ranging in quality number from 54's to 60's count).It is reared for its wool quality and suitability for migration. Among native breeds, Chokla has the overall best performance for functional characteristics of carpet. Average fiber diameter is about 28 μ, medullation percentage is around 25-30%.The breed had been undergoing selection at CSWRI. It is native to Rajasthan, and is distributed over Churu, Jhunjhunu, Sikar and border areas of Bikaner, Jaipur and Nagaur districts of the state.

Magra (formerly known as Bikaneri, also known as Bikaneri Chokla and Chakri): Medium to large sized animals. White face with light brown patches around eyes is the characteristics of this breed. Skin is pink colored. Both the sexes are polled. Fleece is of carpet quality, extremely white and lustrous but not very dense. The breed is being improved through selection under Net Work project of the Indian Council Agricultural Research (ICAR). Much crossing of this breed is occurring with other breeds of adjoining areas. It is native to Rajasthan; true to type, animals are found in eastern and southern parts of Bikaner.

Malpura: Animals are fairly well built with long legs and light brown colored face. Both sexes are polled. The fleece is white in color but extremely coarse and hairy. Belly and legs are devoid of wool. This breed is very similar to Sonadi but its wool yield and quality are better. This breed is also undergoing selection for mutton production. Ewes have good milk production to sustain the lambs. Malpura as a native breed has been used in a large number of crossbreeding experiments for improvement of mutton and wool. It is native to Rajasthan (Malpura tract)

Marwari (deriving the name from its home tract in Marwar): Sheep are medium sized. They possess black face, extremely small and tubular ears, and both sexes are polled. Fleece is white and not very dense. The breed is being improved through selection under the Net Work project of ICAR. It is found in Rajasthan(distributed in Jodhpur, Jalauni, Nagaur and Barmaur districts extending up to Ajmer and Udaipur districts) and Gujarat (Heoria region)

Muzzafarnagri (also known as Bulandshahri): Medium to large sized animals. Face and body are white and both sexes are polled. Males occasionally have rudimentary horns. Fleece is white, coarse and open. Belly and legs are devoid of wool. It is one of the heaviest breed, which is well adapted to irrigated areas. Its numbers are declining due to intensification of agriculture in its native tract of Uttar Pradesh (Mujjafarnagar, Bulandshaher, Saharanpur, Meerut, Bijnour districts) and Haryana. Pure specimens are found in Muzzafarnagri district of UP. The breed is also undergoing genetic improvement for mutton under the Net Work project of ICAR.

Sonadi. Animals are fairly well built, somewhat smaller than Malpura, with long legs. Face is light brown. Both sexes are polled. Udder is fairly well developed. They produce white fleece, which is extremely coarse and hairy. Belly and legs are devoid of wool. It is found in Rajasthan, and also extends to northern Gujarat.

Patanwadi (also called Desi, Kutchi, Vadhiyari and Charotari, Kathiawari): Animals are medium to large sized with relatively long legs, brown face and typical Roman nose. They produce white fleece of medium carpet quality, but not very dense. The breed has three distinct strains: a non-migratory which is true Patanwadi, migratory type with larger body, and a meat type with big body but low structure. It is distributed in Saurashtra, Kutch and Mehsana districts of Gujarat. By virtue of its good mothering ability traits, the breed is being used in a crossing program with Malpura and Garole for developing a prolific mutton type strain.

Jaisalmeri: Animals are tall and well built with black face and a typical Roman nose. Both sexes are polled. Fleece is white and of medium carpet quality and not very dense. Found in Jaisalmer district of Rajasthan.

Pugal (deriving its name from Pugal area of Bikaner district of Rajasthan): Animals are fairly well built with black face. It produces white but not very dense fleece of medium carpet quality. Both sexes are polled. Bikaner district of Rajasthan is its home tract.

Jalauni: Medium sized animals and both sexes are polled. Fleece is coarse, short-stapled open and generally white. Belly and legs are devoid of wool. Its body size and general appearance are close to Nali. Found in Uttar Pradesh.

Nali: Medium sized animals. Face is light brown and skin is pink. Fleece is white, coarse, dense and long stapled. Found in Rajasthan and Haryana.

Another breed/type of sheep, **Kheri** was evolved in the farmers flock under field conditions especially in the areas of Rajasthan which are important for sheep migration. The animals of this breed have established them owing to their ability to walk long distances and ability to survive on limited amount of coarse feed. The animals can sustain stress, and on return of favorable conditions, they regain faster resulting in better reproduction and growth of lambs as compared to other breeds prevalent in the area.

Breeds of Southern Region

The southern region (southern and central peninsular and hot humid coastal) comprises the states of Maharashtra, Andhra Pradesh, Karnataka, Kerala and Tamil Nadu. This region has the largest sheep population (44%) of the country (Fig 20.1) but they produce only about 22% of country's wool (low quality, usually below 36's spinning counts). About 50% sheep of this region do not produce any wool; the remaining produce coarse, hairy and colored wool. Therefore, sheep in this region are primarily reared for meat, except Nilgiri, which is restricted to Nilgiri hills of Tamil Nadu, and that too in small numbers (about 12 thousand).

Nellore is the most predominant (33% followed by Deccani, which accounts for about 23% (Fig. 20.4). Nilgiri, which has the lowest numbers and is restricted to Nilgiri hills of Tamil Nadu, is at the verge of extinction. Deccani is undergoing genetic improvement as a dual-purpose breed while Nellore and Madras Red as mutton type under ICAR funded Net Work project.

Mandya and Mecheri have been included in the Mega Sheep seed project for genetic improvement of field stocks in their native tracts.

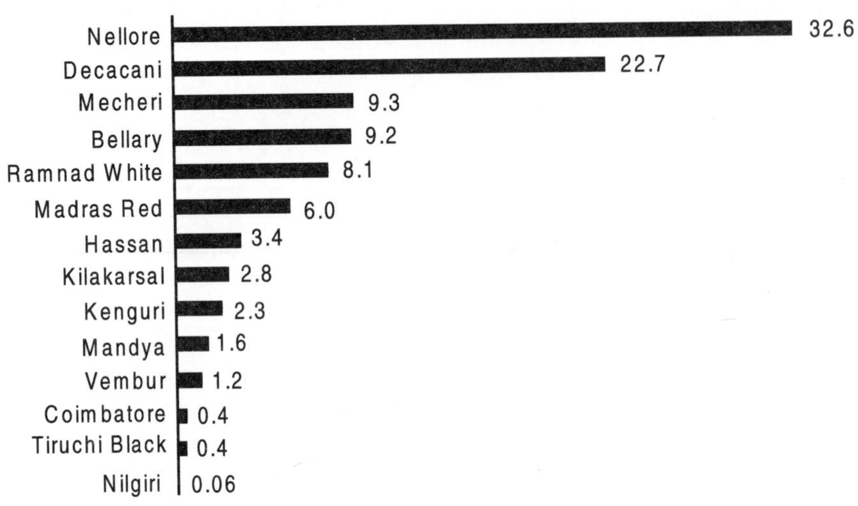

Fig. 20.4. Distribution (%) of breeds of Southern region.

Deccani: Animals of the breed are medium-sized predominantly black colored with white markings. White and brown/fawn colored animals are also seen. Ears are medium long, flat and drooping. Tail is short and thin. Fleece is coarse, hairy and open. The breed is spread over large part of central peninsular region comprising semi arid area of three states, Maharashtra, Andhra Pradesh and Karnataka. Four types (strains or within breed types) have been noted (locals call them Lonand, Sangamneri, Solapuri/ Sangola and Kolhapuri). Deccani holds potential for mutton production under intensive system of management. The breed is undergoing genetic improvement for dual purpose (meat and wool) under Net Work project of ICAR.

Nellore: Animals are tall with little hair except at brisket, withers and breech. More than 80% of the animals carry wattles. Three varieties are distinguished, primarily on the basis of color: "Palla", completely white or white with light brown spots on head, neck, back and legs; "Jodipi", white with black spots, particularly around the lips, eyes and lower jaw but also on belly and legs and "Dora", which is completely brown. Found in Andhra Pradesh (distributed in Nellore district and neighboring areas). The breed is undergoing genetic improvement for meat under the Net Work Project of ICAR.

Madras Red: Animals are medium-sized. Body color is predominantly brown with intensity varying from light tan to dark brown. Rams have strong corrugated and twisted horns while ewes are polled. This breed is very hardy and capable to travel long distances. Found in Chingalpet and Madras districts of Tamil Nadu. The breed is undergoing genetic improvement for meat under Net Work Project of ICAR.

Mandya (also known as Bannur and Bandur): Relatively small animals with a compact body and a typical U-shape conformation from the rear. Large-sized animals carry wattles. The color is generally white. The sheep are slightly Roman nosed. Coat is extremely coarse and hairy. They are mainly kept on grazing, however, a few farmers keep them on stall-feeding. Some of the farmers practice sheep tethering in their crop land, shifting from place to place and bringing back to their homestead in the evening. Mandya is recognized as one of the best meat breeds of the country by virtue of its conformation. Found in Mandya district and bordering Mysore district of Karnataka. The breed has been included in the Mega Sheep Seed project of ICAR.

Mecheri (also known as Maiylambadi and Thuvarmchambali): Medium sized and light brown colored animals. Both sexes are polled. The body is covered with very short hairs which are not shorn. Skin is of highest quality amongst all the sheep breeds of India and is highly prized. It is found in Tamil Nadu (Mecheri, Kolathoor, Nangavalli, Omalur and Tarmangalam Panchayat Union areas of Salem district and Bhavani taluk of Coimbatore district). The breed has been included in the Mega Sheep Seed project of ICAR.

Bellary: This breed is quite similar to Deccani; the sheep found to the north of Tungabhadra River (in Karnataka) are called "Deccani" and those to the south of it are called "Bellary". Medium sized animals with coat color ranging from white to black. Ears are medium long, flat and drooping. Fleece is extremely coarse, hairy and open. Belly and legs are devoid of wool. Distributed in Bellary district of Karnataka.

Hassan: Small sized animals having white body with light brown or black spots. About 40% of the males are horned. Fleece is white, extremely coarse and open. The legs and belly are generally devoid of wool. Found in Hassan district of Karnataka.

Kilakarsal (syn. Karuvai, Ramnad Karuvi and Adikarsial): Medium-sized animals. Coat is dark tan, with black spots on head (particularly on the eyelids and lower jaw), belly and legs. Most animals have wattle. Found in Tamil Nadu (Ramnathpuram, Madurai, and Thanjavur and Ramnad districts).

Vembur (also called Karnadhi): Animals are tall and white in color. Males are horned and ewes are polled. The body is covered with short hairs, which are not shorn. Found in Tamil Nadu (Tirunelveli district).

Kenguri (also known as Tenguri): Animals are medium-sized. The body coat is dark brown or coconut colored. Found in Karnataka.

Ramnad White: A medium sized breed, predominantly white with short and thin tail. Found in Ramnad district of Tamil Nadu.

Tiruchy Black (also known as Tiruchy Karungurmbai): The animals are small in size and black colored. Males are horned and ewes are polled. Fleece is extremely coarse, hairy and open. Found around Tiruchy area of Tamil Nadu.

Coimbatore (also called Kurumbai): Medium sized animals and about 40% of the males are polled. Tail is small and thin. Fleece is white, coarse, hairy and open. Animals are maintained on grazing without any supplementation. Found in Coimbatore district of Tamil Nadu.

Nilgiri: It was evolved during the 19th century from a crossbred base involving Merino, Cheviot and Southdown exotics with local hairy breeds. It contains an unknown level of inheritance of the native Coimbatore. The breed is localized to Nilgiri hills of Tamil Nadu. Animals are medium sized and white, exceptionally with brown patches on face and body. Attempts have been made to improve this breed through crossing with Rambouillet and Russian Merino. which has resulted in substantial improvement in fleece yield and quality. Fleece is fine and dense. The breed is well adapted to the conditions of Nilgiri hills where it is mainly maintained by tea planters and other flock owners. Its numbers are declining. Lack of shearing facilities in the area and disposal of wool are the contributing factors to its declining numbers.

Sandyno is another sheep type of this region, which was developed by crossing Nilgiri with Merino and Rambouillet. The breed has ¾ inheritance from Nilgiri and 5/8 inheritance from Merino/Rambouillet breeds. The breed has been field tested and found to adapt well to the local climatic and environmental conditions. A flock of Sandyno breed is maintained at Sheep Breeding Research Station, Sandynallah in Tamil Nadu.

Breeds of Eastern Region (hot and humid)

The Eastern region comprises of Orissa, Bihar, West Bengal, Assam and other eastern states (Arunanchal Pradesh, Meghalaya, Mizoram, Manipur, Tripura, and Nagaland, Sikkim).

Sheep of this region represent 10% of country's sheep population but produce only about 5% of the total wool (Fig. 20.1). The wool is extremely coarse and heavy except in Arunanchal Pradesh where small number of better quality sheep is available. Shahabadi and Chottanagpuri are important breeds of eastern region. Garole breed of this region is highly prolific, and its prolificacy inheritance has been incorporated into some improver native breeds.The Bonapala and Garole occur with high numbers while Ganjam breed is showing a declining trend in its numbers. Chottanagpuri breed of this region has been included in the Mega Sheep Seed Project of ICAR.

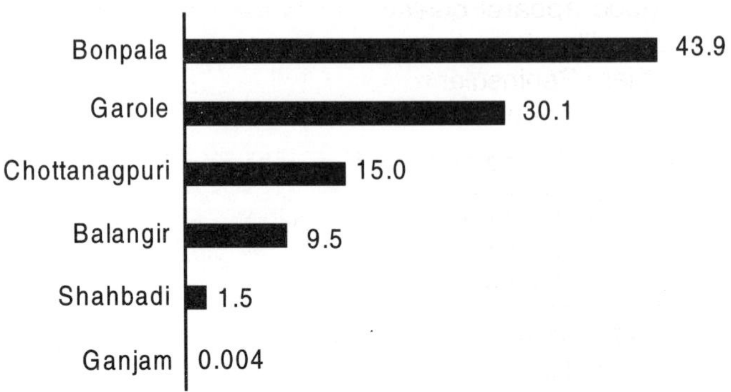

Fig. 20.5. Distribution (%) of breeds of eastern region.

Bonapala: Animals are tall, leggy and well built. Fleece is coarse, hairy and open; belly and legs are devoid of wool. Fleece colour ranges from completely white to completely black with a number of intermediate tones. Found in southern Sikkim.

Garole: This breed of sheep which is reputed for multiple births is found in the Sunderban area of West Bengal. This breed is reported to have contributed the prolificacy gene to the Booroola Merino sheep. Its breeding tract is tropical humid. Its wool is rough carpet type. Litter size at first lambing is two and at subsequent lambing, is 2 to 3. It is highly prolific: singles 25-30%, twins 55-60%, triplets 15-20% and quadruplets 1-2%.Its prolificacy gene has been incorporated into some improver breeds of India for developing prolific mutton strains.

Chottanagpuri: Animals are small, light in weight with light grey and brown color. Ears are small and parallel to the head. Tail is thin and short. Fleece is coarse, hairy and open and is generally not clipped. Found in Chottanagpur, Ranchi areas of Bihar/Jharkhand and Bankura district of West Bengal.

Balangir: Medium sized and light brown to black animals. Males are horned. Fleece is extremely coarse. The legs and belly are devoid of wool. Fleece is extremely coarse, hairy and open. Found in North western districts of Orissa

Shahabadi: It is also known as plain type sheep. The fleece is mostly grey and sometimes with black spots. The legs and belly are devoid of wool. Fleece is extremely coarse, hairy and open. Found in Shahabad, Patna and Gaya districts of Bihar

Ganjam: Animals of this breed are medium sized with coat color ranging from brown to dark tan. Fleece is hairy and short, and is not shorn. Distributed in some districts of Orissa.

Tibetan: Medium sized animals with white and black color. Both sexes are horned and have Roman nose. Fleece is relatively fine and dense. They produce excellent carpet quality wool, which was available to Indian states bordering Tibet when the border was open. The belly, legs and faces are devoid of wool. Found in Arunachal Pradesh.

Kendrapada is another sheep of this region, which has been identified as another prolific sheep type of this region after Garole. It is excellent medium statured meat type sheep. In the flocks surveyed more than 75 % ewes produce multiple births. The coat color varies from white to dark brown.

Wool Productivity in Relation to Region

The brief description of breeds of the four regions given above shows that sheep are reared under varied agro climatic conditions from temperate to arid. The type of wool (processing quality) produced likewise also differs from region to region.Generally sheep in Northern temperate region produce wool of good apparel quality ; so is the case in climatically similar areas of southern hills(Nilgiri).The quality of wool goes down as one moves from Northern temperate to North western and to Southern Peninsular region. Similarly, Eastern region of very high humidity is not suitable for raising the sheep for wool (Table 20.2).

Table 20.2 Summary of the region-wise breed/breed-groups of Indian sheep.

North Temperate (Himalayan) 7 breeds	North Western (Arid & Semi Arid) 12 breeds	Southern Peninsular 14 breeds	Eastern 6 breeds
Apparel wool:	*Carpet wool:*	*Apparel wool:*	*Medium carpet wool* :
Karnah	Chokla	Nilgiri	Balangir
Carpet wool:	Magra	*Medium carpet wool:*	Bonapala
Bhakarwal	Nali	Bellary	Chottanagpuri
Changthangi	Patanwadi	Coimbatore	Ganjam
Gaddi	*Medium carpet wool*:	Mutton:	*Carpet wool:*
Gurez	Jaisalmeri	Deccani	Tibetan
Poonchi	Jalauni	Hassan	*Mutton:*
Rampur Bushair	Kheri	Kenguri	Garole
	Malpura	Kilakarsal	
	Marwari	Madras Red	
	Muzaffarnagari	Mandya	
	Pugal	Mecheri	
	Sonadi	Nellore	
		Ramnad White	
		Tiruchy Black	
		Vembur	
Usually small sized	Medium to heavy size	Medium to heavy with exception of Deccani	Small sized
Good quality wool suitable for apparel and superior quality carpets	Superior to coarse carpet type wool-; region most important in country for carpet wool	Very little wool of low quality. Most breeds are of meat type.	Small quantity of wool, Extremely coarse, colored and hairy ; primarily meat type,
SL : 5-10 cm	SL : 4-9 cm	SL: 6-7 cm	SL: About 5
FD : 25-33 μ	FD : 28-52 μ	FD: 27-55 μ	FD: about 66 μ
Medullation : 6-18%	Medullation: 18-25%	Medullation: 12-64%	Medullation: about 90%
Wool quality counts: 36-58/80	Wool quality count: 36-54	Wool quality count: 35-36	Wool quality count: < 36

SL staple length, FD fiber diameter

Since sheep rearing is practiced by economically and socially lower strata under very low input systems, flock size in irrigated and cultivated areas is very small where it is mostly integrated with agricultural production. In dry region, where crop cultivation always depends on monsoon rain, sheep flocks are relatively large and managed on community land and migration. Sheep flocks are both stationary and migratory. Stationary flock sizes vary from less than ten to 70 while migratory flocks are larger going up to a hundred.

Of all the recognized breeds of India, Nellore, Marwari and Deccani are the most prevalent in numbers. Marwari covers the greater part of arid north western region of Rajasthan and Gujarat. The breed is highly migratory and has therefore left greatest impact on other breeds especially those with very coarse fleece like Malpura and Sonadi. Deccani occupies most of the central part of Southern peninsula, distributed over the states of Maharashtra, Karnataka and Andhra Pradesh.

Box 20.3: Central Sheep and Wool Research Institute, Avikanagar (Rajasthan).

The Central Sheep and Wool Research Institute (CSWRI) is a premier Institute of Indian Council of Agricultural Research (ICAR) engaged in research and extension activities on sheep and rabbits. It was established in 1962 at Malpura in Tonk district of Rajasthan, some 80 km from Jaipur. Now the campus is popular by the name of Avikanagar. The campus is spread over an area of over 1500 hectare. Malpura is located at 26°17'N 75°23'E26.28°N 75.38°E.[1] It has an average elevation of 132 meters (401 feet).

The institute consists of six divisions, each attending to a specific set of activities.

The CSWRI has three Regional Research Centres in different climatic zones of the country to develop region specific technologies.

- **North Temperate Regional Station (NTRS):** The North Temperate Regional Station of the Institute was started at Garsa, Kullu valley of Himachal Pradesh (in 1963) with the mutual collaboration of Ministry of Food and Agriculture, Government of India and erstwhile Punjab Government (subsequently Himachal Pradesh) as a Sheep Breeding Farm to look after the region specific research for improvement of sheep and Angora rabbits.. Later on, the station was transferred to the Indian Council of Agricultural Research (ICAR) and functioned as the Regional Station of Institute for North Temperate Region of the country.

  This station is located at latitude 31.58° N and longitude as 77.20° E with average annual rainfall of about 840 mm in Himalayan Kullu valley. This station is located at 1400-2100 meters above mean sea level on the right bank of Garsa stream. This Station was upgraded as a full-fledged division and renamed as Division of Fur Animal Breeding. Subsequently work on rabbit breeding for wool and meat purposes was also started. An elite flock of Bharat Merino is located on the station. Elite animals of this flock are supplied to farmers of the region for genetic improvement of their local animals.

- **Southern Regional Research Centre (SRRC):** The Southern Regional Research Center was established in 1965 in sub-temperate region at Mannavanur (35 Km ahead of Kodaikanal) in Tamil Nadu.The center has an area of 1347 acres including forestland. It is situated at an elevation of 2030 meters above mean sea level and located at 10°-11° N latitude and 77°-78°E longitude. The average annual rainfall is 1055 mm, well spread throughout the year. The mean monthly ambient temperature from April to June is 25°C. Maximum temperature ranges from 26 to 30°C and the minimum temperature ranges from -3°C to 5°C. Relative humidity varies from 15 to 90%. An elite flock of Bharat Merino sheep suitable to cold climate has been established at the station. Bharat Merino sheep are well in demand in southern states for improving mutton and wool. Broiler rabbits from this substation are also made available to farmers.

- **Arid Region Campus (ARC):** The Arid Region Campus (ARC) of the Institute was established in 1974 as Division of Carpet Wool and Karakul Pelt Production at Bikaner to conduct research on sheep in arid region of the country. The campus is located at 28.3° N latitude and 73.5° E longitude at 236 meters above mean sea level in the heart of Thar Desert. The average rainfall is low (250 mm) and erratic. The temperature varies between sub zero (-2°C) during winter and high (49°C) during summer. The center has 636 hectares of land. The campus is conducting research on genetic improvement of Magra and Marwari sheep breeds of arid region suitable for carpet wool production.

Exotic Breeds of Sheep

A few exotic breeds of sheep for fine wool, mutton/dual-type and pelt have been introduced in India for improvement of native sheep. Merinos, Rambouillet, Suffolk, Dorset, Corriedale, Karakul and Awassi are some of the examples.

FineWool Breeds (Apparel type wool)

The fine wool breeds produce fine and crimpy wool. Their fleece is heavy, dense[1] and of good quality, and contains a large amount of yolk. Yolk, which is secreted from skin glands, is the natural grease and suint covering on the wool fiber. It lubricates the fiber and prevents entanglement as the wool fiber grows. The main fine wool breeds which have been imported into India are Soviet Merino and Rambouillet. Both these have done well as purebreds, and their crosses with indigenous breeds have shown improvement in wool quality and production.

Merino

Merino originated in Spain. The word 'Merino' is derived from an early day Royal officers of Spain called "Merino" whose duty consisted of allotting migratory flocks of sheep to their respective grazing grounds. The breed has spread to all over the world for up grading of local breeds or for crossbreeding. The origin of most of the fine wool breeds of present day sheep can be traced to Spanish Merinos. Many countries have developed their own Marino strains (e.g. Australian, US Merinos).

Merinos are thin tailed and adapted to arid environment. They have a banding instinct, therefore, they are easy to herd. Animals are good grazers covering large areas of poor and sparse rangelands. They tend to be seasonal breeders which results in low prolificacy and lambing percentage. Average trait values of Merinos in their native countries are:

- Mature weight: Rams = 75 kg, Ewes = 65 kg.
- Annual fleece yield = 3 to 5 kg; Fiber diameter = 10-20 μ; Staple length = 5 – 10 cm.

Head and legs of Merinos are covered with wool. Merinos have been more suited in temperate or sub-temperate regions of India. Ewes of this breed have more longevity as compared to other breeds.

[1]Wool density is measured as number of fibres per square cm of skin area.

India has imported Soviet Merinos (Stavripal and Grossney strains). They have more grease in fleece, large skin folds, close face and large wool on face resulting in covering of eyes that are not desirable attributes.

Performance of Soviet Merino under semi- arid tropical environment of Rajasthan (at Central Sheep and Wool Research Institute,) was recorded from 1972 to 1979[2] as follow:

- Greasy fleece yield: 1st six monthly = 1.1 kg, 2nd six-monthly=1.5 kg.
  Fiber diameter = 16 μ
- Body weights (kg): Birth = 3.3, 3-month = 15.2, 6-month = 21.5,
  9-month = 26.2, 12-month =32.5.

Daily gain in weight from birth to weaning was 130 g and from weaning to six month was 55 g, which are indicative of their good adaptability. Males on the average were 20% percent heavier than females at 12 month. Twinning percentage was 18.4.

Rambouillet

This breed, which originated in France, is a descendant of the old Spanish Merino. There are two types: B and C which are differentiated on the basis of skin folds. The B-type has largely disappeared, and the C -type has been consistently improved both for meat and fleece.

Rambouillet has white hair around nose and ears, and produces heavy fleeces of high quality, mostly in the 62 to 64's spinning counts. They are well adaptable, and due to their hardiness and grazing ability can live and produce under extremely hot or cold conditions. The animals are large, rugged and hardy. Mature rams weigh around 100-125 kg and ewes around 60-90 kg. Ewes are good mothers and prolific. The breed is almost free of skin folds with acceptable mutton conformation, and is a good wool producer.Therefore it has good market for lamb as well as for fine wool. Majority of the Rambouillet imported into India are from USA (Texas state). Rambouillet in USA produce about 5 kg fleece per annum.

Rambouillet as purebreds and in crosses with Indian breeds has generally performed better than Russian Merinos. Performance of Rambouillet lambs at CSWRI farm was recorded over three generations. Body weight at birth and weaning (3 month) averaged to be 3.2 and 15.4 kg respectively with daily pre-weaning gain of 135 g. Twinning was 12%.

Mutton Breeds

Classification of sheep into mutton-type is mainly based on appearance of the animal. Specialized mutton breeds mature fast, have high prolificacy, higher weight gain, high feed conversion ratio (desirable), high carcass yield and good quality mutton. In developed countries mutton type animals are usually reared under relatively more intensive management with accelerated lambing, like three lambing in two years, as compared to wool breeds because the main income accrues from sale of lambs.

Most of meat-type breeds originated in England. Subsequently such breeds migrated to other countries for improvement of meat. Even the USA that possesses the best quality meat breeds has originally imported meat breeds from England, and crossed these with US breeds to develop new breeds. India has imported mutton-type breeds from UK, Australia and other countries. Brief description of some of the important exotic mutton breeds that have been imported into India is given below.

[2]Based on observations on a over 400 sheep.

Suffolk

This breed was developed in England (Suffolk, Essex and Norfolk areas) from an old strain of Norfolk and Southdown breed. The animals are characterized by straight and large legs, black face, ears and legs. Animals are very alert and active. Head and ears are free from wool and there is no wool below knees and hocks. Fleece is moderately short and fine (with 56's to 58's count). Annual greasy wool yield is 2.8 to 3.3 kg. But the breed is first and foremost a mutton breed. Carcass is full of lean meat evenly marbled, no waste fat and of excellent flavor. Mature rams weigh 100 to 135 kg and ewes 70 to 100 kg. Ewes are very prolific and excellent milkers. Because of Suffolk's activity and rustlings qualities, it is one of the best as a grazer.

The breed was imported into India and maintained at CSWRI farm where its performance was studied (under the semi-arid tropical climate) from 1976-78 (Table 20.3). Body weights in India were lower than in their home tract/country due to changes in climate (temperate vs. tropical). Six-monthly greasy fleece yield was 1.4 kg in early years but it declined in subsequent years due to stress of hot semi-arid climate.

Dorset

Dorset breed, which is native of England (Dorset and Somerset areas) has two strains: a horned and a polled. Both the strains are identical for all other traits except the presence or absence of horns. Animals of this breed produce carcass of medium weight, fine boned and high-quality mutton. Fleece weight is 2.75 to 3.25 kg and wool is short and fine (52 to 58's quality numbers). Mature rams and ewes in good condition weigh 80-100 and 50-80 kg respectively. The breed is prolific, hardy and capable of doing well under most conditions. Its growth performance at CSWRI alongwith that of Suffolk is given in Table 20.3.

Table 20.3 Body weights (kg) of Suffolk and Dorset at CSWRI (1977-78) based on few animals.

Breed	Birth	3-month	6-month	9-month	12-month
Suffolk :	4.2	24.2	28.3	35.4	38.0
Dorset :	3.8	21.7	26.8	34.2	41.0

Though the Suffolk lambs were heavier than Dorsets at early ages (up to 9 month) but at yearling stage, the Dorsets became heavier. Dorsets were found to adapt better than Suffolk to the semi-arid hot climate.

Southdown

It is one of the oldest breeds of sheep and is native to southeastern England (Southdown chalk hills). Animals are compact, wide and deep with legs set apart and broad head with excellent mutton conformation. The breed is early maturing but the ewes are not too prolific and are only average milkers. They have short, and fairly dense and fine quality fleece. Annual greasy fleece weight is around 2.2 to 3.3 kg. Mature rams weigh about 80 to 100 kg and ewes about 55 to 70 kg.

Dual Purpose Breed: Corriedale

The dual-purpose breeds combine meat production with wool production attributes. Corriedale is the only important dual-purpose breed which has been imported into India. The Corriedale (named after Corriedale estate of Otago in New Zealand) was developed in New Zealand and Australia during the late 1800's from crossing of Lincoln and Leicester breed rams with Merino females. The breed is now distributed worldwide, making up the greatest population of all sheep in South America. It thrives throughout Asia, North America and South Africa. Its popularity suggests it to be the second most significant breed in the world after Merinos. It is large-framed with good carcass quality.

The breed produces bulky, high yielding wool ranging from 25 to 30 micron fiber diameter and of 56-58's quality numbers. Annual greasy fleece yield is 4.5 to 5.5 kg with a staple length of 9-12 cm. Mature rams weigh 80 to 110 kg and ewes 55 to 85 kg. *Corriedale are excellent in efficiency as they produce more kg of lamb and wool per kg of body weight than other range breeds.* Ewes are fair in prolificacy and milking ability. In India, this breed was imported from Australia.

Pelt Breed: Karakul

Pelt sheep are reared for their lamb pelts, which are used for manufacture of garments. Mutton quality of pelt sheep is poor and the fleece from adult animals is coarse and of low value.Karakul is the chief pelt type breed. The breed is native to Bokhara (former USSR) in central Asia and derives its name after the area of Karakul (black lake).Mature Karakuls are angular and of very poor mutton type. Rams weigh about 90 kg and ewes 65 kg. The breed is very well adapted to extreme climatic conditions and sparse vegetation.

Karakuls were first imported into India in 1975 from erstwhile USSR. Their performance under hot-arid conditions (Rajasthan) and cold arid conditions (Ladakh, Jammu and Kashmir) was studied. Weights averaged to be 3.9, 14.5, 20.5 and 23 kg at birth, 3-month, 6-month and 12-month of age respectively. Six monthly greasy fleece yields in different seasons varied between 1.1 and 2.6 kg. The breed has done well as both purebreds and in crosses with indigenous breeds viz. Malpura, Sonadi and Marwari. Karakuls also had good adaptation in hot- and cold-arid conditions as reflected in their adult survival (88%) and lambing percentage (85-106%).

Mutton and Milk Breed: Awassi

Awassi is the most numerous and widespread breed of sheep in east of Mediterranean, and the main sheep breed of Iraq, Syria, Jordan, Israel, Palestine, Turkey etc. Its name is attributed to the 'El Awas' tribe between the Tigris and Euphrates rivers. In literary Arabic,"awas" is the term for the"red-and-white camel garb" or for a"white sheep". In physical and functional properties, Awassi seems to be very close to the type from which the 'fat-tailed' sheep of Asia, Africa and Europe are derived. The fat tailed sheep can store large amounts of fat in the tail and region of rump (in baggy deposits in the hind parts on both sides of tail and on first 3-5 vertebrae of tail) for subsequent use during dry seasons. For this reason, Awassi is found mostly in the extremely arid regions of Africa, the Middle East, and Asia. Awassi sheep are raised and produce under widely varied production systems. The greatest majority of Awassi in different countries in the Near East are raised extensively and/or semi-extensively (i.e. Bedouin, Steppe or Transhumant). However, specialized dairy flocks of this breed are raised intensively in farms.

In many places, Awassi has become the breed of choice, especially for dairy, due to its high performance in milk and lamb production and ability to produce under varied production environments. The breed has now spread from its place of origin to all continents including South America. Body of Awassi is covered with coarse wool of creamy white color, glossy long fibers, relatively high kemp content. The wool is classified as carpet type with spinning count of 48's-52's. Averages for some traits in their native countries have been recorded in following ranges:

- Body weight at: Birth = 3.8 – 4.5 kg, 12 – month = 35 - 42 kg.
- Annual greasy fleece yield = 1.9 – 3 kg ; Staple length =12-19 cm ;
  Fiber diameter = 31– 45 μ

Awassi also possess high potential for milk production. Milk yield in different populations of Awassi in their native countries shows wide variation. Very high yields have been recorded in Israel where for the last 40 or 50 years the breed has been genetically improved for milk.

Improved strains in Israel may yield about 500 kg in 214 days (over 2 kg per day). Litter size at birth varies from 1.1 to 1.4. Awassi mutton and lamb serve a good dish in the native region of the breed though its meat largely lacks the characteristic flavor associated with mutton in Europe. The breed was imported into India in late 1990's (at CSWRI and Nimbkar Agricultural Research Institute, NARI) for improvement of mutton and ewe productivity including milk yield.

Economic Traits : Description, Measurement and Inheritance

Sheep in India are raised for both wool production and meat, therefore, the most important traits that have direct bearing on economic returns pertain to fiber and mutton. Reproductive, survival and mothering ability traits are also economically important in breeding programs as they contribute to economics of both wool and mutton production.

Wool Traits

Wool is one of the natural fibers of animal origin which is hygroscopic (i.e. it absorbs moisture). Wool fiber has a small diameter (10^{-3} to 10^{-4} cm) as compared to its length (up to 10-12 cm) which gives it an important place among textile fibers. It grows from the follicle situated in the dermis (middle layer of skin) and has three morphological/physical components (Fig 20.6):

- **Cuticle:** The outmost layer of a fiber, which has scales, that provides cohesion to fibers when they are twisted into yarn. Cuticle cells are readily apparent at relatively low magnification (500 X).
- **Cortex:** Cortex is the major part of fiber and is made up of cortical cells (spindle shaped) that provide strength and elasticity to the fiber.
- **Medulla:** Some wool fibers have a third innermost part, medulla, which consists of a net work of hollow, air-filled pockets. Medulated fiber has a chalky white appearance. Under microscope, medulla appears black (Fig 20.2). Medulated fibers are light, and are not uniformly dyed.

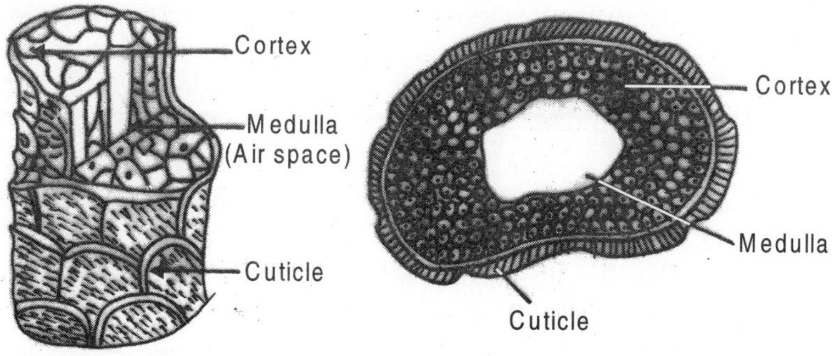

Fig. 20.6. Organization of cell types in a medulated wool fiber showing showing cuticle, cortex and medulla.

The fleece (wool) characteristics are broadly classified into two classes: yield and quality.

Fleece yield: The quantity of wool sheared can be expressed in two ways:

(a) **Greasy fleece yield:** The weight of fleece (raw) as obtained on shearing (clipping) is referred to as greasy fleece yield. Greasy fleece consists of clean wool, other skin products (wax and suint) and extraneous material (dirt, vegetable matter and moisture). The grease, which is formed in sebaceous and sweat glands, provides flexibility and durability to the wool. Greasy fleece yield is recorded either on per-shearing or on per-annum basis. The first shearing is usually done at one year of age while subsequent ones are normally done twice a year (six-monthly interval).

(b) **Clean fleece yield:** The weight of wool obtained after cleaning (i.e. removal of grease and other associated material like dust, dirt etc) is called clean fleece yield. It is the most important trait from economic standpoint but its measurement is cumbersome as compared to that of greasy fleece.

Clean fleece weight expressed as percentage of greasy fleece weight is termed **'yield'.** It is about 65% in modern Merino strains but can vary considerably depending on environmental conditions and breeds. In some of the Merino, Corriedale and Romney populations, average values have been reported to be 70 to 74%. The trait has low coefficient of variation (6-8%).

Wool quality traits: Quality of wool is assessed by the quality of individual fibers as well as by characteristics of fleece as a whole. Wool fibers are assessed on a number of attributes such as length, fineness, uniformity and waviness (crimps) together with strength and elasticity. Some of the quality traits that are important in breeding program include:

(a) **Fiber diameter:** Fineness of wool fiber is an important factor in manufacturing industry; finer fibers produce lighter weight garment which are of 'softer' finish. Softer finish is important for comfort as wool fabrics are worn close to skin. Also finer the wool, longer the yarn that can be obtained per unit of wool. Fiber diameter, which is most important dimensional measure of wool fibers, is usually recorded by a projection microscope (lanometer) in microns(1 micron = 1/10,000 of a centimeter). Projected images are measured either with a ruler or by grouping them in 2.5 micron intervals. Measurement of diameter is made at constant temperature and humidity conditions (21° C and 65 % RH). It is done under such conditions because wool is hygroscopic (i.e. fibers swell when they absorb moisture and become finer as they dry), and therefore diameter of the same fiber changes with alterations in temperature and humidity.

Average diameter of wool fiber varies widely across breed types, from 12 to 80 microns. Fibers of fine wools (e.g. wool from Merinos) fall in the range of 15 to 25 microns while modern dual-purpose breeds have average fiber diameter of 25 to 35 microns.Most of the Indian wools are of medium fine and coarse varieties, which are used for knitting wools, blankets, carpets etc.

Another useful concept in describing uniformity of fiber diameter within a given quantity of wool is its coefficient of variation. Average values for it have been reported to be 20-25% in modern Merino strains. Its heritability, like that of fiber diameter, is high. In some countries (e.g. USA), some standards have been fixed for coefficient of variability for field evaluation viz. less than 21% as excellent, 21 to 27% as average and over 27% as poor.

(b) **Staple length** (also called bundle length or lock length): Wool fibers lie arranged in tufts of about 30-40 held together by wax (grease) and special binding hairs to form a staple. Staple length, usually measured in cm, is obtained by measuring the natural staple without stretching (unstretched lock of wool) with a ruler as shown in the adjoining photo. Staple length is an important value determining attribute of wool quality as long fibers are required for worsted processing, and they normally give better yarns. Length is also important for strength and extensibility of yarn. However, staple length beyond a certain limit may be detrimental to normal combing operations. Normal values for staple length in Indian wool breeds fall in the range of 4 to 9 cm.

The fineness and length of fiber together are the major factors contributing to manufacturing quality.

(c) **Medullation:** Main body of a true fiber does not have a stiff central core in it ; it only has the cortex. But wool fibers of less developed sheep breeds (e.g. native breeds) contain a central component, medulla, which consists of a network of cell-wall remnants filled with air. Medulla may be continuous or interrupted and may occupy varying proportions (width) of centre of fibers (Figure 20.7). Those fibers with discontinuous medullas are described as heterotypes.

(a) Diagramatic representation

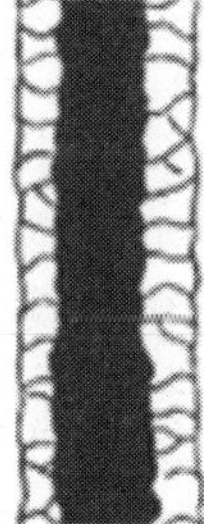

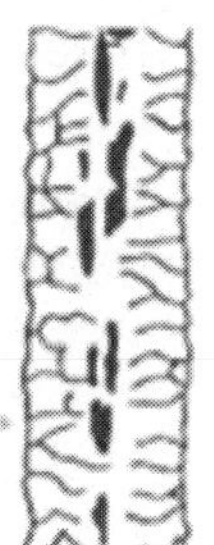

Continous Discontinous Fragmented

(b) Under microscope

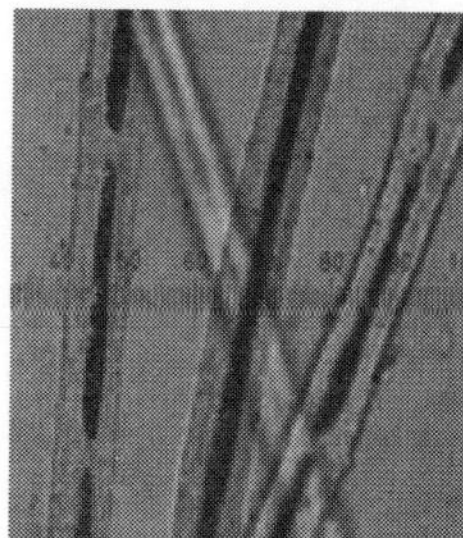

Fig 20.7 Medullated fiber : (a) diagrammatic representation of types and degree of medulation (b) a medullated wool fiber as seen under microscope (500 X magnification)

Medullated fibers, regardless of thickness or whether they are continuous or interrupted, are unfavored by processors as they are weaker and lighter on drying than non-medulated fibers. Medullation is most frequently noticed on the britch of sheep (i.e. the area just above the flank level) but medulatted fibers may be observed scattered throughout the fleece. The percent (proportion) of medullated fibers in a sample of fibers can be obtained at the time of measurement of fiber diameter by microprojection as they are easily identified (Fig 20.7 b):

$$\text{Medullation \%} = \frac{\text{No. of medullated fibres}}{\text{Total no. of fibres examined}} \times 100$$

Similarly, the degree of medullation can also be documented by a projection microscope. Medullation percentage in major wool breeds of India, (mainly of North western region, varies widely: 24 %(Chokla) to 88 %(Sonadi). But the newly developed wool-type synthetics have low medullation, from as low as zero to 20%. However, in carpet-type wool, medullation, to some extent, is desirable as it provides resiliency.

(d) **Crimps or Crimpiness:** Crimp is the natural waviness of wool fiber which is readily visible. Usually closer the crimps, the finer the wool. Crimpiness is recorded as number of waves per unit length (cm). In fine wools, these curls are larger in number and more regular in unit length of fiber than in coarse wools. A fine wool like Merino may have up to 100 crimps per inch, while the coarser wools like Karakul may have as few as one or two per inch.Though crimpiness is related to fleece elasticity but its main value lies in the fact that a distinct crimp alone indicates a well grown, sound fiber of uniform diameter and length.

(e) **Kemps:** Kemps or kemp fibers are brittle guard hairs which are generally found along with wool. They are usually similar to medulated fibers except that the degree of medullation is greater (Fig 20.8).They are also normally considered 'shedding' or discontinuously growing. They are readily visible, may vary in length from a few mm to nearly as long as the staple, and are often scattered throughout the fleece. Kemps are usually assessed as the proportion of dry lifeless fiber, and are considered undesirable.

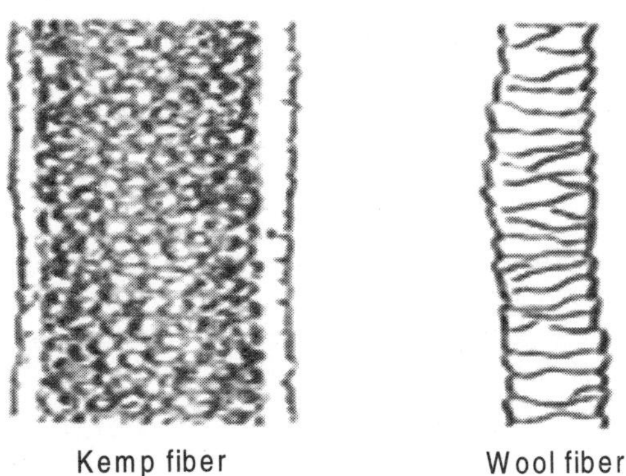

Fig. 20.8 Illustration of a kemp fiber compared to a wool fiber.

Genetic parameters of wool traits: Genetic parameters of wool traits are available in the literature.These have been obtained through a large number of investigations conducted across many populations (breeds/strains/lines):

- Heritabilities of greasy and clean fleece wool yield are medium to high (0.3 to 0.5), therefore, selection based on individual's own fleece yield is effective in bringing about genetic improvement.
- Genetic correlation between greasy and clean fleece yield is very high (more than 0.8 and favorable), hence selection for improvement of clean fleece yield can be based on greasy fleece yield as it is easy to measure.
- Heritability of 'yield' (clean fleece weight expressed as percentage of greasy fleece yield) is higher (0.5 to 0.6).The genetic correlation of 'yield' with clean fleece yield is positive but low but its genetic correlation with greasy fleece weight is close to zero or slightly negative which means that selection based on greasy fleece yield will not improve the 'yield'.
- Heritabilities of wool quality traits like fiber diameter, staple length, crimp frequency etc are high (more than 0.5). Therefore, these quality traits are also easily amenable to individual selection.

- Genetic correlations of greasy and clean fleece yield with fiber diameter are moderately positive (0.3 to 0.4) implying an undesirable genetic association. This undesirable genetic association between these two important traits determining economic returns (i.e. fleece weight and fiber fineness) has been observed both across strains and within strains. Therefore, selection for improvement of fleece yield has to be simultaneously supplemented by selection for reducing fiber diameter or maintaining it at current level (usually by a selection index or using independent culling levels).
- Genetic correlation of fleece yield with staple length is positive, therefore, selection for wool yield increases staple length.
- Genetic correlation of crimp frequency with greasy fleece weight is negative (undesirable).
- Most of the wool traits are expressed several times in the life of an individual (e.g. in different shearing's). Repeatabilities of fleece yield, fiber diameter, staple length and other wool attributes are high (0.7 to 0.8) which means that selection for improvement of these quality traits can be based on first shearing. Such a selection (based on single record) reduces the generation interval.

Meat and Carcass Traits

Several measurements like body weights at various ages (e.g. birth weight, weaning weight, yearling weight and mature weight) and gains in weight or growth rate etc. are used to denote growth.

- **Birth weight:** Weight of lamb at birth (recorded within 24 hours of birth) is an important trait as it influences subsequent growth as well as survival. Birth type viz. single, twins etc, also influences birth weight. On the average lambs born as single are heavier (up to 25%) than those born as twins. Moreover, this superiority of single-born over twins continuous up to weaning age or even later. Therefore, body weights are adjusted for birth type (i.e. singles, twins) using appropriate adjustment factors for selection of breeding candidates.

  Average birth weight across different Indian breeds ranges from 2.0 to 3.6 kg. Breeds of North western region are, in general, heavier at birth than those of Southern region. Birth weight of improver breeds under farm conditions is in the range of 3.0 to 4.0 kg.
- **Weaning weight:** Lambs can be weaned at different ages depending on the system of management viz. extensive or intensive. Under intensive system, in which lambing is accelerated (as the main income comes from lambs), lambs are weaned as early as 4 to 6 weeks of age but in extensive system they may not be weaned until 120 days (16 weeks). The usual practice is to wean the lambs at 90 days (three-month) when they are weighed individually for selection purposes. If the lambs are weaned slightly earlier or later than 90 days, their weaning weights can be adjusted to 90-day age as per the following expression:

  Adjusted 90-day weaning weight = Birth weight + (90 × Average daily gain in weight from birth to weaning)

  Average weaning weight (3-month) across Indian breeds ranges widely; from 9 to 18 kg for NW region breeds and from 8 to 14 kg for Southern region breeds. Weaning weight in improver breeds of India is in the range of 12 to 16 kg.The newly developed mutton synthetics or some improved strains weigh between 12 and 18 kg.

  Age of ewe exerts considerable influence on weaning weight of her lambs. Two-year-old ewes wean lambs that are usually 2 to 4 kg lighter than those born to mature ewes. Therefore, weaning weight is also adjusted for age of ewe. Usually two age-categories are made for age of ewes: two-year old and mature ewes.

- **Market weight:** Weight of lamb at age of marketing which can be 6-month, 9- month or even one year, depending upon the breed and market requirements, is called market weight. In Indian sheep, 9-month weight is usually called market weight. Dressing percentage at this age is high and the mutton is leaner as well.
- **Yearling weight:** Weight of ram or ewe when it is one year old, called yearling weight, is an important trait not only for meat production but also for wool production. Higher the yearling weight more will be the surface area thus giving rise to more yield of fleece. Yearling weight in Indian improver breeds falls in the range of 25 to 35 kg. Breeds of NW region are generally heavier than of the southern region.
- **Pre-weaning weight gain:** It is measured as the difference in weight from birth to weaning age, which is usually 3 month. It can also be expressed as growth rate (pre-weaning growth rate) as follow:

$$\text{Pre-weaning growth rate (\%)} = \frac{\text{Weaning weight (90th day)} - \text{Birth weight}}{90} \times 100$$

- **Post-weaning weight gain:** It is measured as difference in weight from weaning (3-month) to market age (9-month). It can also be expressed as growth rate (post-weaning growth rate) as follow:

$$\text{Pre-weaning growth rate (\%)} = \frac{\text{Weight at market age (270 day)} - \text{Weaning wt}}{\text{Age at market in days (270 day)}-\text{90 day}} \times 100$$

- **Dressing percentage:** It is weight of dressed meat expressed as percentage of live weight excluding head, tail and skin of the animal (lamb). Dressing percentage, on slaughter-weight basis, in Indian sheep breeds in feedlot experiments is about 50 %.
- **Hot carcass weight:** Total weight of the carcass immediately after slaughter (excluding the viscera) is referred to as hot carcass weight while the weight of carcass when it is frozen after slaughter is simply called carcass weight. Higher the weight of carcass, more will be the yield of mutton. Hot carcass weight in native mutton-type breeds is about 9 to 18 kg.
- **Meat-bone ratio:** The ratio of meat to those of bones in dressed carcass, referred to as meat-bone ratio (proportion), is also an important meat quality trait. Higher the ratio more will be the proportion of meat relative to those of bones and more will be the economic returns from a unit of carcass (per kg of carcass). In Indian meat and dual-purpose breeds, the meat-bone is in the range of 5:1 to 8:1.
- **Other carcass and meat traits:** There are a number of other traits that determine the quality of carcass viz; muscling quality grade, fat-depth (between muscles),eye muscle (*m. longissimus lumborum*) depth and area (generally near 12th rib),lean meat yield, meat pH, meat color, (measured by a chromometer as relative lightness, redness and yellowshiness) etc. These quality traits are not yet important considerations in developing markets (e.g. India) but attain significance in specialized mutton markets.

Genetic parameters of meat and carcass traits:

- Heritability of birth weight is moderate (around 0.25). The trait is also affected by large maternal genetic effects; heavier mothers produce heavier lambs.

- Heritability of weaning weight is higher than of birth weight (about 0.3), and it is also affected by maternal genetic effects.
- Heritability of yearling weight is higher than of early body weights (about 0.4).
- Heritability of dressing percentage is high (0.4) and of carcass yield slightly lower (0.2 to 0.3).Other carcass and meat quality traits are heritable to varying degree, hence amenable to selection. Since the carcass traits can be measured only after animal's slaughter, therefore, selection of breeding candidates for these traits can be based on sib-performance.
- Genetic correlations among body weights at various ages (from birth to yearling age) are positive (moderate to high) viz; r_G of birth weight with weaning weight is about 0.4 but with post-weaning weights it becomes weaker. Genetic correlation of weaning weight with post-weaning and yearling-weight is higher (more than 0.7 – 0.8). Improvement in marketing weight can be accomplished by sequential selection starting from birth wherein a certain proportion of lambs with low body weight can be culled at early ages.
- Correlations of body weights with wool traits: Genetic correlations among growth, carcass and wool traits have been obtained from large number of populations (wool– type, mutton type and dual– purpose), and they generally follow a similar trend across populations. Some of the conspicuous and consistent types of genetic correlations are:
    - Genetic correlations of weaning, market and yearling weight with fleece yield (greasy and clean) are positive and moderate (0.2 to 0.3). Genetic improvement in body weight is thus not antagonistic to genetic improvement in fleece yield. Rather the genetic association is favorable.
    - Body weight (yearling or adult) bears a favorable genetic correlation (about 0.15) with staple length. On the other hand, the r_G of body weights with fiber diameter is very low (varies from – ve to + ve in different populations).

Reproduction and Survival Traits

Sheep in India and other parts of tropics are bred throughout the year but in European countries they are seasonal breeders. The various components of reproduction, survival and mothering ability are important for both wool and meat type sheep. The important ones include:

- **Fertility:** Fertility is measured as the number of ewes lambing per ewe joined. Average values for fertility vary widely depending on breeds and environmental factors.
- **Litter size:** The number of lambs born per ewe-lambing, called litter size, is one of the important measures of reproductive ability/efficiency of an ewe. On an ewe-basis, it is expressed in an few classes (e.g. single, twin, triplet etc). But on group-basis (e.g. family/flock/strain etc), it can be expressed as lambing percentage.

  Lambing percentage is measured as the number of lambs born or raised per 100 ewes bred (pregnant ewes) or exposed for breeding :

$$\text{Lambing (\%)} = \frac{\text{Number of lambs born or raised}}{\text{Number of ewes pregnant or joined}} \times 100$$

  Litter size (number of lambs) can be expressed at birth (number of lambs born) or at weaning (number of lambs weaned); the latter measure includes the contribution of survivability. Litter size at birth in most of the Indian breeds is close to one. But average litter size in modern exotic breeds of US, Australia, New Zealand falls in the range of 1.3 to 2.7. Litter size on an individual-ewe basis is an 'all or none' trait, its coefficient of variation is very high (over 30%).

- **Prolificacy:** Sheep usually produce single lambs. Twinning is often desirable as an ewe which weans twins on the average produces 10-15 kg more lamb weight (at marketable age) than the ewe that produces a single. Similarly, the total output of wool from twins is higher.

  Incidence of twinning or of triplets in some production systems (e.g. intensive system) is considered a useful measure of prolificacy. It is expressed as number of multiple births (twins, triplets etc) to the total number of births (total number of ewes lambed) viz; percentage of singles, twins or triplets.

- **Tupping percentage:** The percent of ewes covered by rams in a flock, called tupping percentage, is a measure of male's reproductive ability. Larger the coverage, more will the lambing rate, and hence higher the wool yield of flock. The trait is recorded on group–basis as follow:

$$\text{Tupping \%} = \frac{\text{Total number of ewes covered}}{\text{Number of rams used}} \times 100$$

- **Scrotal circumference:** Scrotal circumference (measured as width of the testicles at the widest point) is an indicator of ram's fertility.

- **Survivability or mortality:** Survivability is important as it has direct bearing on economic returns. In a breeding program, high mortality reduces the intensity of selection, and in turn decreases the rate of genetic improvement per-unit time. On an individual-basis, survivability is an 'all-or-none' trait. Survivability is usually calculated for different periods on group- basis (e.g. for a dam, sire or flock) as percent of lambs surviving up to a fixed age as follow:

  - Pre-weaning survival (%) : It is calculated as follow:

$$\text{Pre-weaning survival (\%)} = \frac{\text{Number of lambs surviving to weaning (3 month)}}{\text{Number of lambs born alive}} \times 100$$

  - Post-weaning survival (%) : It is calculated as follow:

$$\text{Post-weaning survival (\%)} = \frac{\text{No, of lambs surviving from weaning to 9 or 12 month}}{\text{Number of lambs weaned}} \times 100$$

  Survivability can also be expressed on yearly basis.

Genetic parameters of reproduction and survival:

- Heritability of different measures of reproductive ability and survival is low. For example heritability of fertility and survival is usually in the range of 0.05 to 0.10; of litter size about 0.10; of scrotal circumference about 0.20.
- Genetic correlation of litter size with measures of ewe productivity like weight weaned per- ewe is desirable. Its genetic correlation with adult body weight is also positive (about 0.25).

Other Traits

In addition to the various reproductive and survival traits listed above, some measures of overall reproductive efficiency can also be derived. One such trait is the net reproductive rate. **Net reproductive rate** refers to the number of lambs weaned per ewe placed in breeding. It is a product of ewe viability, fertility, prolificacy, fraction born alive and fraction weaned. Average net reproductive rate in US Rambouillet strains has been reported to be about 106%.

- **Ewe productivity (efficiency or index):** The trait 'ewe productivity', commonly defined as the total weight of lambs weaned per-ewe per-lambing, is also called 'kg or lb weaned'. It is calculated for the ewes that wean at least one lamb. The weights of lambs born to the ewe but reared artificially or on a foster ewe are not included in its computation. Ewe productivity,being a composite trait, is a measure of net merit of ewe (ewe's production efficiency) as it includes the contribution of reproductive, growth and survival viz. litter size, lamb survival to weaning, growth rate of lambs- and mothering ability (mother's traits like milk yield, her behavior traits like suckling, lamb rearing ability, milking score etc). The US National Sheep Improvement Program (NSIP) considers ewe productivity a biologically optimum index for improvement of overall flock productivity where it has been identified as a target trait for genetic improvement.

  Heritability of ewe productivity is about 0.2. The trait does not bear undesirable genetic correlation with its component traits (reproductive, growth and survival), therefore, selection on ewe productivity is not expected to lead to deterioration in its components (like litter size, survival, lamb growth rate etc). Genetic improvement in ewe productivity of a population (breed/strain/line) can be achieved by direct selection for it. Such a selection program would result in biologically optimum balance among component traits. In countries where sheep industry is developed, strains/lines have been developed which are recognized for their excellence in ewe productivity

- **Resistance to parasites:** Sheep (and goat) are exposed to a number of gastro–intestinal parasites; the predominant parasite in tropics/sub– tropics is *Haemonchus contortus* (a nematode). Ability of sheep to resist such parasites is most commonly measured by fecal egg count (FEC). Heritability of FEC is low to moderate (0.2 – 0.3), therefore sheep can be genetically improved to resist its infection. Some breeds have been shown to be relatively more resistant to its infestation than others. For example, Garole sheep breed, which is a breed of Sunderban area of Bengal and is highly prolific, has been shown to offer greater resistance than Deccani and Awassi breeds to *H. contortus.*

  Fecal egg count, a measure of nematode resistance, is now included in many sheep selection programs due to increasing resistance of sheep against anthelmintic treatment in areas where such parasites are a major problem, and especially where parasite resistance to drugs is emerging as a serious issue.

  In addition to worm resistance, incidence and severity of fleece rot is also becoming an important economic attribute. Its heritability is about 0.20.

Pelt Characters

Lamb pelts are produced either through slaughtering of lambs within 24 to 48 hrs of birth or from unborn lambs recovered through slaughter or through cesarean of pregnant ewes at 13 to 140 days of pregnancy.The latter two pelts are known as broad- or fat-tailed Persian. These are twice as more valuable than the best type produced through slaughtering after birth as they are thin and more lustrous. The lightness of pelt combined with a larger size of pelt, luster combined with the richness of coloration, silky feel, beauty and grace permit it to be used for many kinds of highly priced fur garments such as overcoats, jackets and suits.

Quality of pelts is determined by the ornaments i.e. the type of curls, size and tightness of curls, luster and weight of pelt. The curls are usually formed as long as the length of the hair (fiber) is more than 20 mm. The present trend is to have light skins with short hairs. Pelts are classified into various types viz; jacket, Caucasian, ribbed etc. The relative proportion of these types is taken as a measure of pelt quality. The above mentioned attributes are dependent on the characteristics of fiber and skin follicles.

Milk Production

Milk production *per se* is not a trait of direct economic importance as under extensive system of management it is only used by sheep farmers for home consumption over and above lamb feeding. Some specialized dairy breeds of sheep have been developed in Europe and the Middle East (e.g. East Friesian; Lacaune, Awassi etc). But Indian sheep produce milk only sufficient enough to nurture young ones; average daily milk yield of important native breeds ranges from 400 to 500 ml per day during 90 to 110 days of lactation. In contrast to this Assaf and Awassi breeds of sheep in Israel and Iran produce from 1.2 to 2 kg daily inlactation extending from 100 to 200 days. Improved sheep flocks of Awassi in Israel can produce over 500 kg in a lactation of 214 days.

Heritability of milk yield in sheep is about 0.25-0.30, the trait is thus amenable to individual selection. Improvement in milk yield of sheep when they are reared for dual purpose (wool and meat) will only be beneficial for improvement of lamb survival and growth. Milk production is an important component of mothering ability; therefore, its improvement would lead to improved ewe productivity.

Genetic Improvement of Sheep: General

Like genetic improvement of any species, there are two ways of producing genetically superior sheep: selection and crossing. Selection in sheep for both wool and meat is easy for two reasons:

- Heritability of most of the traits related to wool-yield and -quality and of mutton traits is moderate to high, therefore, they are easy to be improved through selection.
- Most of the traits of economic importance are expressed early in life, and are easily measurable in both sexes.

Selection for Wool Traits

The traits concerned with wool production like greasy fleece weight, clean fleece weight, fiber diameter and staple length are moderate to highly heritable, and they respond well to individual selection. Effectiveness of selection for their improvement has been well documented through selection experiments.

A large fraction of income from fine wool breeds like Merino is determined by fleece yield and fiber diameter. There is some genetic antagonism between these two traits such that selection for either trait alone leads to slow deterioration in the other.Since the undesirable genetic correlation(negative) between these two traits is moderate, simultaneous genetic improvement in both the traits is possible. Simultaneous selection for improvement of these two traits can be effected by selection index or by independent culling method. Effectiveness of selection for simultaneous improvement of fleece yield and fiber diameter (the two antagonistically related traits) has been demonstrated through selection experiments conducted in Australia (Fig 20.9) where fleece yield increased by 0.57 kg while fiber diameter improved (decreased) by 1.46 µm.

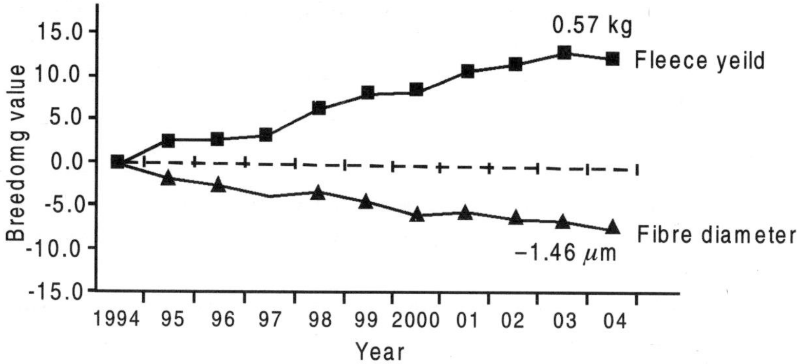

Fig. 20.9 Genetic improvement in greasy fleece yield and fiber diameter over ten generations of selection in a Merino line based on a selection index of the two traits.

Besides higher fleece yield and fine fibers, staple length is also a valuable trait. Its genetic correlation with fleece yield being positive (desirable), it improves as a correlated response to selection for fleece yield. Selection against medullation percentage by virtue of its high heritability shall be effective in reducing medullation alongwith a correlated decrease in fiber diameter as the genetic correlation between them is desirable.

Selection for Meat

Selection for mutton improvement may be directed at growth and body weight. These respond effectively to individual selection as they are moderate to highly heritable and the generation interval in meat-type sheep is short. Genetic correlation of birth weight with subsequent growth rate and body weights is positive. Therefore, initial selection may be based on lamb weight at birth because heavier lambs generally have higher vigor and weight gain up to weaning and slaughter.

Selection for weaning weight, gain in weight and feed efficiency can also be practiced. Recording of feed efficiency is laborious; it usually improves to a certain extent as correlated response to selection for growth rate. Various meat and carcass traits being moderate to highly heritable can also be improved through sib selection.

The traits concerned with growth are affected by sex, age of dam, season of birth, type of birth (born single or multiple) and age at weaning. Therefore, performance records need to be adjusted for these for making selection decisions. Appropriate adjustment factors have been developed by breed associations and other such organizations.

Selection for Reproduction, Survival and Mothering ability

Age at first lambing, interval between lambings and multiple births are the traits that determine overall reproductive efficiency of sheep flock. Heritability of the various components of reproduction is low; therefore, selection for their improvement may be based on family performance. Efforts should be made to get the first lambing early and to obtain as many lambs as possible per year (multiple births and regular lambing).

When sheep are raised for lamb production (income is derived from sale of lambs meant for slaughter), lambing is usually accelerated. Under such a situation, both growth and reproductive traits become important. Selection can then be based total weight of lambs weaned per ewe

(ewe productivity) as this trait combines growth, survival and reproduction traits in an optimal fashion.

Effectiveness of selection based on ewe productivity for improvement of overall productivity has been demonstrated through a selection experiment conducted in different breeds of sheep over 12 generations[3]. It improved ewe productivity (@ about 0.7 kg/generation). It also resulted in desirable correlated genetic changes in different component traits as follow: fertility @ 0.3%/ generation, prolificacy @1.20%/generation, percentage of lambs born alive 0.09%/generation, percentage of lambs weaned 0.54%/generation, net reproductive rate @1.95%/generation, ewe viability @0.13 %/generation, weaning weight @012 kg/generation. Milk yield of ewes (a measure of mothering ability) also improved.

Crossbreeding in Sheep

Crossbreeding, in addition to utilization of heterosis, also combines the desirable traits of two or more breeds into a single product (breed complementarity). Crossbreeding is more prevalent when sheep are raised for meat (lamb production) because both meat as well as reproductive, survival and mothering ability traits are important for improvement of lamb production efficiency. Lambing in such a scenario is accelerated. Potential heterosis values in sheep for different traits are usually in the following ranges:

- Individual heterosis for birth weight, weaning weight, gain in weight, yearling weight and survival to weaning age is 3 to 10%.
- Maternal (ewe) heterosis for fertility, prolificacy, lamb weight (birth and weaning), and lamb survival to weaning is 3 to 9%.
- Maternal heterosis for various measures of ewe productivity,like lambs surviving to weaning, lamb reared per ewe exposed, weight of lambs weaned per ewe exposed etc, is high (10 to 20%).

In order to capitalize the large cumulative benefits of individual and maternal heterosis, specialized sire (ram) and dam (ewe) breeds/strains are used in countries where sheep operations are industrially organized.

- **Specialized ram breeds:** Such breeds are raised primarily to produce rams for crossing with ewes from specialized dam breeds. Specialized sire breeds are noted for rapid early growth and desirable meat and muscling traits.
- **Specialized dam (ewe) breeds:** Such breeds are noted for reproductive efficiency, mothering ability, longevity. They are adaptable to a wide range of environments. Ewes of such breeds are of adequate size to produce lambs of desired carcass weight and produce moderate quantity of wool as well.

There are several systems of systematic crossbreeding. Of these terminal crossing makes maximum uses of both types of heterosis (individual and maternal) and is more widely used to maximize production of market lambs. In terminal crossbreeding systems, specialized ram and ewe breeds yield best results. Since flock sizes in most commercial sheep operations are usually not large, management considerations limit the choice of crossbreeding to two-breed terminal.

[3]The selection experiment was conducted at US Sheep Experimental Station, Idaho (by Ercanbrack and Knight, 1998).

Genetic Improvement of Sheep in India

Early Sheep Breeding Programs

Indian breeds of sheep have lower productivity as reflected in lower fleece yield, poor wool quality and lower growth rate as compared to improved exotic breeds of Europe. Attempts for genetic improvement of sheep had been made by the State Governments, State Agriculture Universities and by Central Government through All India Coordinated Projects involving central institutes and state universities for quite a long time.

1. Selection within Indigenous Breeds

Selection for increasing body weight or fleece traits was undertaken on some indigenous breeds in a number of Indian states:

- Bikaneri and its derivatives in Rajasthan
- Nellore and Mandya in Andhra Pradesh
- Muzzafarnagri in Uttar Pradesh
- Nali and Lohi in Haryana and Punjab(including West Punjab)

Similarly, improvement of Deccani at Poona (Maharashtra) and Bellary in Tamil Nadu for developing non-hairy sheep was undertaken. Though improvement in some traits was recorded but there was a general decline in fertility perhaps due to inbreeding resulting from small population size.

2. Crossbreeding with Exotic Breeds

Indigenous breeds differ greatly from European breeds. Nearly all the European sheep are wooled and many are multipurpose. Fleeces of European breeds range widely, from primitive carpet-wool types (such as the Scottish Blackface), with a coarse, partly medulatted outer coat and a finer inner one, to the supreme apparel-wool breed, the Merino, with its fleece of fine, relatively uniform fiber diameter. On the other hand, the indigenous tropical breeds have coarse fleeces, usually containing medulatted fibers and kemp, or a coat of short or long hair with little or no wool. They serve several purposes like meat, milk, skins. The coarse-wool or long hairy fleeces of Indigenous are used for carpets, rugs, blankets, and even tents.

Skin structure differences: There are differences in the skin structure of the wooled sheep of European origin and the coarse wooled or hairy sheep of tropics. All sheep, like other pelaged animals, produce fibers from two types of skin follicle : primary (P) and secondary (S). Large breed differences exist in the total number of follicles per unit skin area ("population density"), and in the ratios of numbers (S/P) and diameters of secondary and primary follicles (d_p/d_s) as follow:

- Merino, which has premium fleece quality has the highest total number of follicles per unit skin area i.e. population density (over 70 per square millimeter), highest secondary to primary follicular ratio (over 20), a low primary to secondary fiber diameter ratio, and the finest fleece. There are various Merino strains (fine-, medium- and strong-wool) but all show this superiority over other breeds.
- On the other hand indigenous wool breeds have low density of skin follicles (10 to 15 per square cm) for coarse carpet wool breeds), low 'S/P' ratio, high 'd_p/d_s' ratio and high incidence of medulatted fibers. In indigenous hairy breeds, fibers from secondary follicles constitute a soft short undercoat.

In view of the vast superiority of European breeds over indigenous in wool yield and quality, crossing of indigenous with European, for either crossbreeding or upgrading, appeared to be the easiest solution for improvement of wool yield and quality.

Historical Overview of Crossbreeding

- **19th century:** Sheep development programs (crossbreeding) were initiated as early as in 19th century by the East India Company through import of exotic breeds for crossbreeding with the indigenous. Crossbreeding using exotic fine wool breeds, particularly Cape Merino, was undertaken in which about 4000 to 5000 crossbred sheep were raised around Poona (Maharashtra).The crossbreds had higher wool yield as well as superior quality.
- **Early 20th century:** In the first decade of 20th century, crossbreeding of indigenous sheep with Romney Marsh was undertaken in both plains and hilly areas. Crossbreeding experiments were also undertaken in Punjab, Bengal, Mysore and Madras using Southdown and Cape Merino.
- **After establishment of ICAR:** More systematic crossbreeding programs with exotic breeds were taken up after the establishment of the Imperial (now Indian) Council of Agriculture Research (ICAR), and in particular after independence through five-year plans. The major initiatives were:
  - o An Indo-Australian Sheep Breeding Farm (now Central Sheep Breeding Farm) was set up at Hissar(Haryana) during 4th five-year plan, with Corriedale breed for production and distribution of Corriedale stud rams to different states for crossbreeding to improve wool and mutton. Corriedale did not perform well, especially with respect to reproduction and survival, and was later replaced by Rambouillet.
  - o An All-India Co-ordinated Research Project (AICRP) to improve sheep for fine wool was launched in 1971 by ICAR.
  - o Crossbreeding was also started at CSWRI, Avikanagar (Rajasthan) immediately after its establishment in 1962.

Crossbreeding for Wool (Apparel and Carpet type)

In 1971, the ICAR sponsored the AICRP for wool with the objective to develop superior fine and carpet wool strains/breeds suited to different agro climatic conditions of the country using exotic breeds. The exotic breeds used were:

- Russian Merino and Rambouillet for fine wool (apparel)
- Rambouillet, Suffolk, Dorset and Correadale for carpet wool

The native breeds used in crossbreeding were:

- Chokla, Nali, Malpura, Mujaffarnagri in North western region
- Sonadi, Coimbatore, Nellore, Bellary, Patanwadi, Nilgiri in Southern region
- Gaddi in North temperate region.

The major objectives of the program were to identify the breed combination and the level of exotic inheritance (viz. ½, 5/8 th or ¾ th) and finally to evolve a new strain/synthetic for fine wool.Salient results from crossbreeding experiments undertaken at institutional farms were:

- Crossbreeding of Merino and Rambouillet with native superior carpet wool breeds (Chokla and Nali) improved wool yield, fiber diameter (reduction) and medullation percentage (reduction). The synthetics developed from crossbreds had better performance than native breeds. The improvement was relatively lower for fleece yield but higher for fiber diameter (16 to 20%) and medullation percentage (53 to 62%). Reproduction and survival of new synthetics was similar to native breeds. The new synthetics had annual greasy fleece yield of more than 2.5 kg and other desirable attributes as shown below:

Synthetic Strain	Annual greasy woolyield (kg)	Staple length (cm)	Avg fiber diameter (μ)	Medullation (%)
Nali synthetic :	2.8	5.2	22.9	16.2
Chokla synthetic :	2.7	5.1	21.0	8.9

The new synthetics had two shortcomings: lower staple length (about 5 cm) and presence of medulatted fibers.

- Crossbreeding of Soviet Merinos and Rambouillet with Patanwadi (a carpet-wool breed native to Gujarat) showed significant improvement; the Patanwadi halfbreds(Patanwadi with Merino and Rambouillet) exhibited higher six-monthly greasy fleece yield (7- 45%), and reduction in fiber diameter (7 to 13%) and medullation (18-36%) as shown below:

Breed/Cross	6-monthly greasy fleece yield(g)	Avg fiber diameter(μ)	Medullation (%)
Patanwadi :	484	24.9	32.8
Merino × Patanwadi :	518	21.5	19.0
Rambouillet × Patanwadi :	706	23.3	26.7

- Crossbreeding of Merinos (Soviet) and Rambouillet with Nilgiri (a fine-wool breed of Nilgiri hills of Tamil Nadu) and Gaddi (a breed of North temperate region) yielded very encouraging results :
  - Gaddi half-breds showed improvement in greasy fleece weight (16 to 36%), fiber diameter (16 to 26%) and medullation (50 to 60%) as shown below:

Breed/Cross	6-monthly greasy fleece yield(g)	Avg fiber diameter (μ)	Medullation (%)
Gaddi :	499	28.3	28.0
Merino × Gaddi :	679	23.7	11.2
Rambouillet × Gaddi :	577	21.0	12.8

  - Nilgiri half-breds showed larger improvement as shown below:

Breed/Cross	Annual greasy fleece yield(g)	Staple length (cm)	Avg fiber diameter (μ)	Medullation (%)
Nilgiri :	1285	6.2	22.1	5.1
Merino × Nilgiri :	2860	7.5	21.5	Nil
Rambouillet × Nilgiri :	2508	7.2	21.8	2.1

The Merino-Nilgiri crossbreds had higher greasy fleece yield than the Rambouillet-Nilgiri cross.

- *Performance of higher grades:* Performance of higher exotic grades of Merino and Rambouillet with fine-wool breed Nilgiri viz. 5/8 and 3/4th -breds was also studied.The results are shown below:

Breed crosses	Annual greasy fleece yield(kg)	Staple length (cm)	Avg fiber diameter (μ)	Medullation (%)
Merino × Nilgiri :	2860	7.5	21.5	Nil
Merino × Nilgiri (3/4) :	3470	8.0	20.2	Nil
Merino × Nilgiri (5/8) :	3264	6.7	20.8	Nil
Rambouillet × Nilgiri	2508	7.2	21.8	2.1
Rambouillet × Nilgiri (5/8) :	2900	7.6	20.3	Nil

The higher grades viz. 5/8 and ¾ had higher greasy fleece yield than the halfbreds though there was no further improvement in wool quality traits over the halfbreds.

- Crossbreeding of Rambouillet with native breeds like Jaisalmeri and Malpura (medium carpet wool breeds) was started in 1964 at CSWRI to evolve new wool strain(s) suitable for arid and semi-arid regions. There was a general improvement in performance of F_1 (crossbreds) over the native breed for wool yield, wool quality and body weight. Beyond 50% exotic inheritance, however, there was little improvement in wool yield and body weight though there was improvement in wool quality. Higher crosses had lower survival, and lower adaptation to heat and prevalent diseases.
- Similarly crossing of other indigenous dual, carpet or meat type breeds (viz. Sonadi, Muzzafarnagri, Coimbatore, Nellore, and Bellary etc) with exotic fine— wool, and dual-purpose breeds resulted in improvement in greasy fleece yield as well as in wool quality.

Summary of farm studies: Overall, the results of crossbreeding for wool under farm conditions indicated improvement in wool yield and quality as well as in body weight mainly up to 50% exotic inheritance in North western and to some extent in Southern regions. The experience of crossbreeding in Jammu and Kashmir (North temperate) indicated that exotic inheritance up to 75% could be sustained in temperate and sub-temperate conditions. Similarly, in southern hilly regions higher exotic inheritance could do well.

Field evaluation of crossbreds: The crossbreds resulting from exotic and native breeds were distributed among farmers to obtain feedback about their suitability under farmers management conditions. Under field conditions of the North western and Southern region (under limited input conditions), the crossbreds did not perform as well as under farm conditions primarily due to their poor reproduction. In particular the crossbred rams exhibited low libido and semen production. Nonetheless, crossbreeding followed by interbreeding at various levels of exotic inheritance resulted in evolving of some new strains/synthetics suited for different regions.

Crossbreeding for Mutton

Crossbreeding for improvement of mutton production was started in early 1960's. Romney Marsh breed was imported from New Zealand for use in Kashmir. But the results were not encouraging due to poor survival and fertility. Under the AICRP, mutton type and dual purpose breeds viz. Dorset, Suffolk and Corriedale were crossed with some mutton and dual purpose

[5]Feedlot trials are the ones in which lambs are confined and fed carefully mixed, high-concentrate feed to fatten them.

breeds of Southern and North western arid regions (like Bellary, Nellore, Sonadi, Mandya, Coimbatore, Malpura, Mujaffarnagri).The results were as follow :

- Crossbreds resulting from different breed combinations showed improvement over respective native breeds. The percent improvement was conspicuous for body weights up to six months of age. However, the differences between purebreds and crossbreds narrowed down gradually from six-month onward, and became marginal at yearling age. This was indicative of the fact that crossbreds required higher plane of nutrition. There was some improvement in efficiency of feed conversion of crossbreds in feedlot trials[5] but dressing percentage was not significantly improved.
- Suffolk had a better combining ability with the natives (Malpura, Sonadi, Mandya, Nellore and Mujaffarnagri) as compared to Dorset for body weights up to yearling age as shown below:

Type of cross	Percentage superiority of crossbreds over native breeds for:		
	Weaning weight	6 month weight	Yearling weight
Dorset × Natives	18.5	22.2	7.8
Suffolk × Natives	26.7	24.6	10.0

Under field conditions, however, the crossbreds were not successful. A number of crossbreds did not survive when they had to compete with native breeds because the special care and managemental conditions, as available at experimental farms, were not available in the field. Nonetheless, some improved strains (e.g. Avimaans, and Mandya and Nellore synthetics) have been evolved from such crossbreeding.

Crossbreeding for Pelt

Russian Karakul sheep were imported into India in 1975 under the Indo-USSR cooperation scheme in the form of 200 ewes and 50 rams. The objective of import was to study the performance of purebred Karakuls as well as to explore its crossing ability with native coarse carpet wool breeds under both hot-arid (Rajasthan) and cold-arid (Ladakh in Kashmir) climatic conditions. Performance of Karakuls as purebreds for growth, survivability, greasy fleece yield and pelt quality was satisfactory, and they adapted well to the hot-arid and cold-arid climatic conditions.

The crosses of Karakul with coarse carpet wool breeds (Marwari, Sonadi and Malpura) also showed improvement in pelt quality over natives (Fig 20.10) At half-bred level, about 25-35% of the pelts (across three types of crosses viz. Karakul x Sonadi ; Karakul x Malpura and Karakul x Marwari) were of jacket type while 28-36% were Caucasian type. Further, at 3/4th exotic level, the proportion of jacket type pelts increased to about 50%. The size and weight of pelt also compared suitably with the pure Karakul pelts.

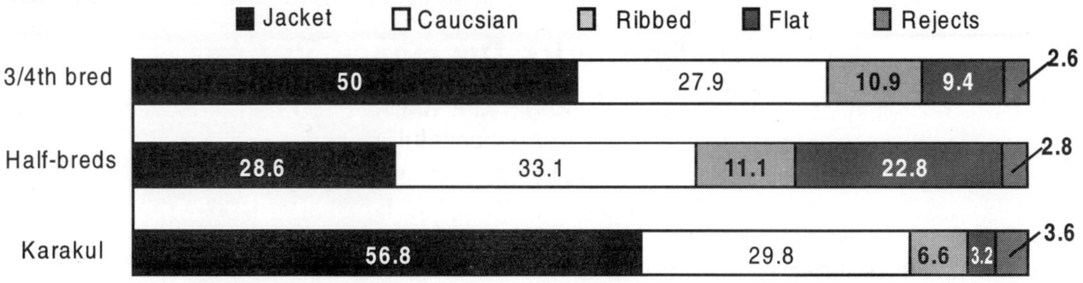

Fig. 20.10 Distribution (%) of pelt type in pure- and cross-breds

All sheep produce fibers from two types of skin follicles, primary (P) and secondary(S).The fibers growing from primary follicles are normally coarser than those from their associated secondaries, therefore, skin with greater proportion of secondary follicles (higher 'S/P' ratio) produces fibers with less variation in diameter. The characteristics of fiber (fiber diameter and length) and follicles (follicle density, follicle diameter and S/P ratio) of crossbreds were much better than those of native breeds, and compared favorably with purebred Karakuls as shown below:

Breed/Cross	Fiber diameter (µ)	Fiber length (cm)	Follicle density (no./sq cm)	Follicle diameter (µ)	S/P ratio
Karakuls :	30.7	1.03	25.4	60.4	2.2
Native breeds :	36.8	1.16	14.8	75.0	1.7
Crossbreds :	30.9	0.93	18.6	70.0	2.1

Though crossbreeding of natives with Karakuls for pelt production was a success but due to socio-political reasons, the program of pelt production was phased out.

General Conclusion from Crossbreeding with Exotics for Wool and Mutton

In most of the crossbreeding experiments, except in North temperate region (Jammu and Kashmir) where climatic and feed conditions have been favorable, crossbreds *per se* have not been very successful under field conditions. The general situation of sheep husbandry in India is such that they continue to be raised on limited input conditions. Under such a scenario, the crossbreds could not compete with natives to outdo them. Consequently, the number of crossbred sheep in the country has shown a declining trend. As per the latest national livestock census, the number of crossbred sheep has declined from 5.7 million in 2003 to about 3.8 million in 2012.

Crossbreeding among Indigenous Breeds

Crossbreeding among indigenous breeds for improving wool yield and quality was undertaken in Andhra Pradesh, Karnataka, Tamil Nadu and UP. Most of the improver breeds used were from North-western region mainly from Rajasthan like Magra, Nali etc. The crossbreds did not succeed in southern states because Rajasthani rams did not adapt to the hot humid climate (except in UP).

The crossbreds exhibited some improvement in fleece quality and quantity. But there was little improvement in body weight gain, dressing percentage or efficiency of feed conversion. In many of these experiments, population sizes were small and there was lack of consistency in the programs. A new sheep type, **Kheri,** is thought to have been evolved under field conditions from the migratory flocks by using Marwari and Jaisalmeri rams.

Strains or Synthetics Developed in India

Crossbreeding of exotic fine wool breeds with native good carpet-wool/apparel- type/ course- carpet wool breeds by itself was not successful in field conditions, except in North Temperate region. But crossbreeding followed by selection and or stabilization of exotic inheritance has been used to evolve some new synthetics, strains or types. The stocks developed in 20th century include Hissardale, Kashmir and Bharat Merino, Avivastra, Avikalin, Avimaans etc. A brief description of these is given below.

Fine Wool Strains

Hissardale : Hissardale was evolved at Government Livestock Farm(now Central Sheep Breeding Farm), Hissar, through crossbreeding of Australian Merino rams with Bikaneri (Magra) ewes, and stabilizing the exotic inheritance at about 75 per cent. Rams of this breed were earlier distributed in hilly regions of Kullu, Kangra etc. Currently, a small flock of Hissardale is being maintained at the Central Sheep breeding farm (Hissar), and for practical purposes, the breed/strain is serving little utility, if any. Its performance values for some important traits were recorded to be as follow as per ICAR Handbook:

Trait	Mean ± SE
Greasy fleece weight (kg):	
Six-monthly :	1.20 ± 0.02
Annual :	2.80 ± 0.08
Staple length (cm) :	15.6 ± 0.07
Avg. fiber diameter (μ) :	20.4 ± 0.14

Kashmir Merino: This strain has been evolved from crosses of different Merino types (initially with Australian Merino, then with Rambouillet and Russian Merino) with predominantly migratory native sheep breeds such as Gaddi, Bhakarwal and Poonchi of the North Temperate region. The level of inheritance varies from very low to almost 100 per cent of Merino, though a level from 50 to 75 per cent is predominant; majority of exotic inheritance is from Russian Merino. Due to the variable level of exotic inheritance and involvement of a number of native breeds, animals of this strain are highly variable. This strain is one of the best examples of success of introducing superior temperate inheritance into indigenous sheep. Kashmir Merino is now a well-recognized Indian sheep used for apparel wool. The present population of the Kashmir Merino type animals exceeds one million. Some flocks of the strain are maintained at state farms in Jammu and Kashmir. Annual greasy fleece yield is about 2.8 kg with fiber diameter of 20 μ.

Bharat Merino: Bharat Merino strain has been developed at CSWRI through crossbreeding of indigenous (Chokla and Nali) with Rambouillet and Merino rams, and stabilizing the exotic inheritance at 75%. Annual greasy fleece yield is in the range of 2.0 to 2.6kg with fiber diameter less than 20 μ and very low medullation. Animals maintained under hot climate have short staple length (3-5 cm) as the animals have to be clipped twice in a year. Under cold climate, however, they produce more wool with long staple length (over 9 cm) as the animals need to be clipped once in a year (Table 20.4) Peak wool yield is obtained at 4-5 years age.Wool quality of the strain is comparable with that of imported exotic Merino wool.

Performance of the strain has been evaluated at two locations viz. semi-arid (CSWRI, Avikanagar, Rajasthan) and sub-temperate (Southern Regional Research Centre, SRRC, located at Mannavanur, near Kodaikanal in Tamil Nadu) and is presented in Table 20.4.

Obviously, the strain gives better performance under sub-temperate conditions. Animals of the strain had been distributed to farmers (in Kolar district of Karnataka and in some areas of Tamil Nadu) where they have performed very well The strain holds potential as import substitute of Merino and Rambouillet for improving the performance of native breeds for carpet and fine wool.

Table 20.4 Performance of Bharat Merino under semi-arid and sub-temperate climatic conditions in India.

Trait	Semi-arid	Sub-temperate
Wool:		
Annual greasy fleece yield(kg)	2.05	2.61
Staple length(cm)	3.24*	8.11**
Fiber diameter(μ)	17.9	19.4
Medullation %	2.66	0.79
Body weights(kg):		
Birth	3.4	4.1
3 month	16.0	18.4
6 month	23.4	25.3
12 month	30.6	32.3
Reproduction (%):		
Tupping	90.0	91.5
Lambing (available basis)	71.6	82.3
Survival	92 – 96	97 – 99

*At 6 month; ** at 12 month Source: Annual report of CSWRI.

Avivastra: The Avivastra strain has been evolved from crossing of Rambouillet rams with Chokla (a native breed of Rajasthan) ewes at CSWRI. Half breds were interbred to stabilize exotic inheritance at 50%, and further improvement was brought through selection for greasy fleece yield and wool quality. Two other synthetics viz. **Chokla** and **Nali synthetics** were also produced at CSWRI in 1983 from half- and 5/8th -breds of Rambouillet and Merino with Chokla and Nali breeds.The performance of the two synthetic groups was similar to Avivastra; therefore, the Chokla and Nali synthetics were merged with 'Avivastra'

Performance of the mixed Avivastra strain in terms of greasy fleece yield (about 2.5 kg), body weights and survival was satisfactory. Under field conditions, however, the strain had lower performance and poor acceptability. It was therefore upgraded by crossing it with Bharat Merino rams. Eventually Avivastra was merged with Bharat Merino (in the year 2000).This mixed strain holds the potential of improving the indigenous coarse fiber breeds as its wool is suitable for processing on worsted[5] system.

Carpet-wool or Dual-type Strains

Avikalin (derived from 'Avi' a Sanskrit word for 'sheep' and 'kalin' a Persian word for 'carpet')**:** Avikalin strain has been produced through a crossbreeding program started in 1964-65 at CSWRI. It involved crossing of exotic fine wool breed Rambouillet and the native extremely coarse wool breed Malpura. The half-breds were interbred followed by selection for greasy fleece weight. During initial years and up to 1992, rams were selected by sequential selection(individual) based on weaning, six- and 12-month weights, and six-monthly greasy fleece yield and wool quality traits viz. staple length and fiber diameter. Once the objectives of wool quality traits were

[5]Worsted : wool yarn of long staple with fibers that have been combed prior to spinning.

achieved (in early 1990's), selection then was directed towards developing it as a dual purpose breed. In view of the change of objectives, the selection criteria were modified wherein selection was based on an index of six-month body weight and 1st six-month greasy fleece yield. Performance of Avikalin at CSWRI farm has been as follow :(2011 to 2014).

Body weights (kg): Birth : – 3.0, 3-month : 15.0 ; 6-month : 25.0, 12-month : 30.

Wool: Annual greasy yield : – 1.6 kg; fiber diameter : 30-35 μ, staple length : 5 cm; medullation: 30%.

The strain had daily weight gain of 135 g from birth to 3 month and 105 g from 6 to 9-month indicating good potential as mutton–type also. Wool yield of Avikalin is about 80% more compared to native Malpura and it also has better growth potential. Annual tupping and lambing is about 94 and 84% respectively. This strain is suitable as a dual-purpose sheep for carpet wool and mutton production.

The strain has been tested at Arid Research Centre (ARC, Bikaner, and Rajasthan) where it adapted well to hot arid conditions though its performance was lower than at CSWRI (semi-arid). Its performance has also been evaluated under sub-temperate climate (at SRRC, Mannavur, Tamil Nadu) where its growth and wool performance was better than under semi-arid climate (of CSWRI, Avikanagar, Rajasthan). Twelve-month body weight was about 38 kg and annual greasy fleece yield was 2.5 kg. Avikalin is being used in some southern parts of India as an improver strain for up grading of local sheep. It is recognized as a dual-purpose strain.

Summary of range of performance values for some important traits for some of these stocks (based on flocks reared on farms or from surveys in field) is given in Table 20.5.

Table 20.5 Summary of new fine and carpet wool strains of sheep developed in India from crossbred base.

Name	Exotic	Performance values for:			
	Inheritance (%)	GF yield (kg)	SL (cm)	FD (mm)	Medullation (%)
Nilgiri :	62.5 – 75	0.2 to 1.4	7.0 – 8.7	22 – 27	8 – 18.0
Hissardale :	75	1.4 to 2.7	5.5 – 6.2	22 – 25	0.0 –0.6
Kashmir Merino :	50 – 75	2.4 to 2.8	4.8 – 5.6	20 – 21	0.0
Bharat Merino :	75	2.6 to 2.8	2.9 – 9.4	18 – 20	1.0 –1.1
Avivastra :	50	1.9	3.5	22	7.60
Deccani Merino :	–	1.50	5.8	22	1.6
Nilgiri Synthetic :	–	1.0 to 1.5	4.6 – 5.0	21 – 22	1.0 – 3.2
Patanwadi synthetic :	50	1.2 to 1.7	4.6 – 5.2	19 – 23	9.0 – 20
Avikalin(dual) :	50	1.7	4.3	25	24.9

Source: VIIIth National Conference on Animal Genetics (2005)

Mutton-type Strains

Avimaans : This strain,also known as 'mutton synthetic', was evolved through crossbreeding of Malpura and Sonadi (indigenous) ewes with imported Dorset and Suffolk rams at CSWRI for evolving a mutton strain suitable for semi-arid conditions of the country. The level of exotic inheritance was stabilized at 50%. Avimaans lambs had potential for mutton production under intensive feeding system where they attained about 25 kg weight at 130 days of age with feed good conversion efficiency of 18%. However, Avimaans lambs had a tendency to deposit fat during terminal part of feedlot. Also, its reproductive efficiency was significantly lower than of Malpura. Since reproduction is of utmost importance especially in breeding sheep for mutton production, further propagation of this strain was stopped, and the available resources of the institute were diverted for the improvement of Malpura breed.

Mandya and Nellore synthetics: Two other synthetic mutton strains viz Mandya and Nellore synthetics have also been evolved at Palamaner in Andhra Pradesh (at Sri Venkateshwara Veterinary University) through crossing of Suffolk and Dorset with Nellore and Mandya breeds, and stabilizing the exotic inheritance at 50%. Nellore synthetic rams were supplied to farmers through field demonstration units (established in 1985-86). Performance of Nellore synthetic rams was comparable to that of Nellore rams in terms of fertility and libido.

Feedlot trials were conducted to evaluate growth and mutton traits of the newly developed mutton synthetics (viz. Avimaans, and Mandya and Nellore synthetics) alongwith important meat breeds (viz. Malpura, Sonadi, Mandya, Nellore, Deccani, Madras Red) from three to six months of age. Daily growth rate from three to six month varied between 90 and 167 g with feed efficiency of 12 to 15% (except that of Mujaffarnagri). In general, the mutton synthetics had better growth rate and feed efficiency than the mutton-type pure breeds.

Ongoing Sheep Breeding Programs in the Country

Early genetic improvement programs including those of the AICRP involved crossing of indigenous with exotic breeds (fine wool, dual purpose, mutton and pelt types), producing different levels of exotic inheritance and breed combinations, and their evaluation under farm as well as field conditions. The general lessons from these exploratory crossbreeding programs were:

- *Non-suitability of crossbreds*: Crossbreds have not been very suited in field, with the exception of North temperate region, because of the limitation of inputs where they could not compete with natives. Nonetheless, a number of new strains have been evolved from the crossbred base. Some of these strains are being further improved through selection.
- *Changes in breeding goals:* The wool scenario in the country has changed dramatically. Early research and development programs primarily targeted improvement of apparel wool production. Under the liberalized import policy, the Government of India has allowed large-scale import of cheap apparel and carpet wool to meet the need of installed capacity in worsted, woolen and carpet sectors. Therefore, efforts had to be shifted from wool (primarily from fine wool) to improvement of meat, dual-purpose or carpet wool through genetic improvement of existing well-adapted native germplasm.

In view of these changes, the genetic improvement programs were reoriented towards:

- Selection within native improver breeds including further improvement of synthetic strains.
- Genetic improvement of sheep for mutton and carpet wool including infusion of exotic germplasm for improvement of mutton of native breeds. Improvement of prolificacy has gained importance due to its contribution to increased meat (lamb) production.

The genetic improvement program supported by ICAR through AICRP was converted to Net Work Project on Sheep Improvement (NWPSI) in 1990. The CSWRI has also been improving some breeds/strains including evolving of new ones through institute funded projects. A few state agriculture/veterinary universities are also doing small scale trials on germplasm evaluation.

NetWork Project on Sheep Improvement

The AICRP on sheep for fine-wool and mutton was converted to Network Project on Sheep Improvement (in 1990) with the objective to undertake survey, evaluation and improvement of indigenous sheep breeds under native environment. This change of objective was necessitated by the experiences of crossbreeding, and from the changing world wool-trade scenario. Therefore genetic improvement programs of native stocks, which are well suited to the low-input and arid, semi-arid and hot-humid climatic conditions, for improvement of carpet wool and mutton were started.

There are six ongoing centres of the Net Work program: four are farm- based and two are field- based, located in major sheep rearing states/regions. The native breeds involved in genetic improvement program (using within-breed selection) are:

- Marwari and Magra - carpet wool type
- Mujaffarnagri and Deccani - dual type
- Nellore and Madras Red - mutton type

Institute/SAU and location	Breed	Utility
Farm based units:		
1. Arid Research Center, Bikaner(Rajasthan) :	Marwari	Carpet
2. Central Research Institute on Goats, Makhdoom(UP) :	Mujaffarnagri	Dual
3. Mahatma Phule Krishi Vidyapeeth (MPKV) Rahuri (Maharashtra) :	Deccani	Dual
4. Sri Venkateshwara Veterinary University, (SVVU), Palamner (Andhra Pradesh) :	Nellore	Mutton
Field based Units:		
1. Arid Research Center(CSWRI) Bikaner(Rajasthan) :	Magra	Carpet
2. TANUVAS, Kattupakkam(Tamil Nadu) :	Madras Red	Mutton

Farm-based Units

Each farm-based unit has about 300 breedable females and requisite number of males. The animals are reared under farm management conditions, their performance is recorded, and the requisite number of sires and dams are selected each generation for reproducing the flock. The traits included in selection vary according to the utility for which the breed is being improved (e.g. carpet wool, dual, mutton). For example:

- Selection in carpet-type wool breeds is based on six-month body weight and greasy fleece weight combined into a selection index.
- Selection in dual-purpose breeds is also based on body weight and wool yield.
- Selection in mutton-type breeds is based on body weight alone.

Body weights of the breeds included in Net Work project, as recorded under farm of the respective university/institute are tabulated below*:

Utility	Breed	Body weight at :				
		Birth	3 month	6 month	9 month	12 month
Carpet :	Marwari :	3.0	16.6	25.8	30	33
	Magra :	3.0	–	19.0	–	27.2
Dual :	Mujaffarnagri :	3.7	18	26.3	30	35
	Deccani :	3.4	16	22.3	24.0	26.0
Mutton :	Nellore :	3.1	13	19.7	23.4	27.2
	Madras Red :	2.6	10.8	15.8	19.0	21.0

* Values for different breeds are not strictly comparable due to differences in climatic and farm management conditions (Source : Annual report of AICRP).

B

Field-based Units

The field based units are using group breeding scheme in which farmers from breeding tract of the breed are made a part of breed improvement program (group breeding program). In such a breeding scheme, a nucleus flock consisting of elite animals drawn from farmer's flocks is created and maintained in home tract of the respective breed with the nucleus being open. The salient features of this open-nucleus group breeding scheme are:

- *Centres:* Each unit has four centres, three located in the field and the fourth at the institution farm, which is a ram-rearing centre. Each field unit covers a population of about 1500 ewes.
- *Farmer's involvement*: The co-operating farmers are provided incentives in the form of health cover (e.g. deworming and vaccination) and supplements (salt licks, mineral mixture etc). Emergency outbreaks in these villages are attended to by scientists of the network team. The number of farmers, which participate in the scheme varies from one unit to another (e.g. in Madras Red unit there are about 125 farmers in sixteen villages).
- *Uniformity of management*: Attempts are made to have uniform system of management across all the participating flocks in terms of pasture, grazing time, type of crops etc.The non-genetic factors considered in the program are centre (cluster of adjacent villages), season (main and off season) and sex.
- *Ram (sire) selection*: Rams are selected in the farm and field flocks based on six months body weight in mutton-type (Madras Red), and on 6-month body weight and greasy fleece yield in Magra, sheep (carpet type). Care is taken to apply effective selection differential (e.g. at least 5 kg for body weight in Madras Red). Rams purchased from one centre are usually supplied to flocks in other villages. One ram is given for 40 ewes. The wider sex ratio is used to avoid ram fighting.
- *Ram distribution*: The selected rams are purchased from farmers between 6 and 12 months and are supplied for breeding to the base flock (ram rearing centre). Breeding rams of high breeding value are produced from farm-based units every year for germplasm distribution to the farmers or development agencies (State govt, NGO's). An advantage of this scheme is the control of inbreeding through rotation of rams between farmers. A ram is kept in one flock for a maximum of 2.5 years.

Mega Sheep Seed Project (MSSP)

Realizing the need for wider distribution of superior animals of a breed in its native tract, the ICAR started Mega Sheep Seed Project in 2009. The objective was to produce of 80 breeding rams of each breed of sheep annually and to cover about 8000 breeding ewes using selected rams. The project has four cooperating units, which are located in different regions:

- Chottanagpuri breed at Birsa Agricultural University (BAU), Ranchi (Jharkhand)**:** A flock of Chottanagpuri sheep has been established for production of superior sheep seed. Survey of Chottanagpuri sheep was conducted in the breeding tract for identification of flocks, selection of flocks, their registration and collection of baseline data.
- Mandya breed at Karnataka Veterinary Animal and Fishery Science University (KVAFSU), Bidar**:** A flock of Mandya sheep has been established. Survey in the breeding tract was conducted to generate base line data. About 300 to 500 ewes per centre have been identified.

- Mecheri Sheep, TANUVAS, and Chennai: A flock of Mecheri sheep has been established for production of superior seed. Survey in the breeding tract was conducted to generate base line information. Field centres had been registered under the project.
- Sonadi Unit, RAJUVAS, and Bikaner: A flock of Sonadi sheep was established for production of superior seed. The survey for generating base line information was conducted in breeding tract of Sonadi sheep.

Genetic Improvement Program at CSWRI

Malpura and Chokla breeds, which are native to Rajasthan, are undergoing genetic improvement at central sheep and wool reserarch institute (CSWRI).

Genetic Improvement in Malpura: A flock of Malpura sheep has been undergoing selection for improvement of mutton production since 1974. The 6-month body weight of the flock through selection coupled with improved management practices has increased from 13.8 kg (1974-77) to 25.7 kg (2010-11) accounting to a cumulative improvement of over 85% Performance of Malpura animals under field conditions has been satisfactory, though lesser than under farm conditions, indicating their good adaptability to low input conditions.

Genetic improvement in Chokla: A flock of Chokla is undergoing selection for improvement of carpet wool. Selection along with improved management has increased six-month body weight from 16.5 to 24.83 kg, and the first 6-monthly greasy fleece yield increased from 0.92 kg to 1.4 kg. Fiber quality of Chokla has been stabilized around 30-35 μ fiber diameter and 30-35% medullation with staple length of more than 6.0 cm. The breed is highly suitable for good quality carpet preparation.

Infusion of New Germplasm (Awassi) for Enhancement of Mutton Productivity

Improvement in growth rate and carcass yield of native sheep in hot semi-arid environment has been attempted through crossbreeding of natives with established exotic mutton breeds (e.g. Dorset, Suffolk). The evolved crossbred lambs had been able to attain finishing weight of 35 kg at 6 months of age under intensive feeding (on 50:50 roughage and concentrate based complete feed).However the improved growth rate of crossbred lambs observed in individual feeding(on farm) was not reflected when they were reared in field. This was ascribed to poor adaptability of the crossbred to prevailing feeding and climatic conditions of hot-arid conditions as the exotic breeds belonged to European countries (temperate climate).

Therefore, it was decided to import a breed which has been evolved under hot-climatic and low-input conditions as prevalent in India. Awassi is one such breed which belongs to Middle East that has similar agro-climatic conditions and production systems as prevalent in India. This breed is characterized by high milk yield and desirable mutton traits. It was imported into India in 1994 for up grading of native sheep. It is has been used in mutton improvement program of CSWRI where it was crossed with native Malpura breed.

The meat and carcass traits of Malpura and 'Awassi X Malpura' (AM) crossbreds have been evaluated under intensive feeding/management system from weaning to 120 days of age. Weaning body weight was similar in the two genetic groups (11.1 kg) while the final body weight after 90 days of intensive feeding was higher in crossbred (22.6 kg) than in Malpura (18.2 kg) lambs. Total as well as daily gain in weight in 'AM' crossbreds was higher than in Malpura. But dressing yield and cutability standard-cuts were similar in the two genetic groups. Further, at six months of age, the 'Malpura × Awassi' lambs produced significantly more edible meat. Subsequently the crossbreds had been crossed with Patanwadi breed for improvement of mutton and maternal traits. The Awassi breed has also been used at Phalton, Pune.

Genetic Improvement of Prolificacy

Genetic improvement programs in major sheep producing countries (e.g. Australia, New Zealand, USA etc) at beginning mainly concentrated on improvement of meat and wool traits. After some time, however, it was realized that selection for improvement of wool or body weight without adjusting for litter-size at birth, which is one of the important measures of prolificacy, could lead to selection against twinning because single born have better growth and also higher wool yield. This is turn would lower the reproductive rate. This realization led to modification of genetic improvement programs with the main aim to increase the prolificacy of pure breeds, and to crossing programs with prolific breeds.

In sheep which are bred for meat, reproductive performance attains even more importance because ewes producing twins have higher meat output. For example, it has been calculated that an ewe which weans twins on the average produces 10-15 kg more lamb weight (marketable) than the ewe that produces a single. In India the relative importance of mutton over wool is increasing as farmers are fetching higher income from mutton than wool. Sheep growers are therefore opting for animals of heavier breeds with better feed efficiency for mutton production. To meet the objective of high mutton production, attempts are being made to improve the prolificacy of recognized native breeds along with improvement of meat traits.

There are two approaches to accomplish it:

(A) Selection for prolificacy: Most of sheep breeds of the world deliver one or two lambs at each lambing. The genetic correlation between prolificacy and growth rate is generally positive. Selection for both the traits can thus be practiced simultaneously without danger of antagonism. But the heritability of litter size is low (less than 0.1), therefore, response to selection for litter size will be low. Increasing prolificacy via selection within the available Indian breeds would thus be a very slow process. Maximum improvement of 1% per year in lambing can be achieved provided selection is solely for this trait. Another way of selection for improvement of prolificacy has been suggested to select the rams of high body weight from among the multiple-born lambs. This in a way adjusts body weight for litter size.

When records of several lambing are available, selection can be based on average, litter size of several lambing. Pedigree records, if available, can also be combined with ewes own records to increase the accuracy of selection. Scrotal circumference can be used as an indicator trait of prolificacy in males.

(B) Introgression of fecundity ('*FecB*') gene/allele: A single autosomal gene/locus (*FecB*) is known to exert a major effect on prolificacy in sheep. *FecB 'B'* is the allele promoting higher fecundity while *FecB '+'* is the wild-type allele. The genotype carrying two *FecB 'B'* allele is homozygous for *FecB*'B', usually designated as 'BB', the heterozygous are 'B+' and non-carrier are '++'.

The *FecB* mutation was first identified in Booroola Merino sheep from Australia, and has now spread to over 48 breeds in 19 countries, mainly through the Booroola Merino. In India, this mutation is found in Garole sheep of Sunderban area of West Bengal in which it is fixed. The mutation is co-dominant or partially dominant with one copy of gene increasing ovulation rate by 1.3–1.6, and two copies increasing it by 2.7–3.0. Similarly, litter size at birth is increased by 0.9–1.2 in ewes carrying a single copy, and by 1.1–1.7 in ewes carrying two copies of the gene.

In view of the importance of fecundity gene, introgression of this gene offers an opportunity to improve reproductive performance of sheep breeds which are otherwise fit in their native tract/

environment. This gene has been introgressed in many breeds in developed countries. In India, a program for introgression of 'B' allele from indigenous Garole breed into some non-prolific but large-sized and well-adapted breeds (e.g. Malpura of North western region and Deccani of Southern region) has been started:

- **Introgression into Malpura:** Malpura is an important breed of North western region which has been genetically improved for body weight at CSWRI since 1974. The breed has high acceptability with farmers. A breeding program for introgression of FecB 'B' allele from Garole (a prolific breed of West Bengal) into Malpura breed was started in 1997 at CSWRI. This program has involved interbreeding the first cross to establish a prolific strain comprising 50% Garole and 50% Malpura inheritance. Currently 74% of crossbred sheep carry the FecB 'B' allele. Intergression of this allele resulted in about 45% increase in prolificacy and 35% increase in weaning rate over Malpura breed. Twinning percent in 'Garole × Malpura' (GM) half-breds has been about 50% and the percentage of triplets about 7%. However, body weights of GM half-bred were lower compared to Malpura. In view of low body weight, the level of Garole inheritance was reduced to 25% by backcrossing of F_1 (GM sires) to Malpura(as dam breed) to produce 'GMM'.The growth performance (from birth to 6 month age) of backcross sheep (GMM) was 33-55% improved over GM half-breds.

  Subsequently, inheritance from Patanwadi breed, which has good maternal traits, was introduced into the prolificacy improvement program to improve milk yield and mothering ability. Crossbreds resulting from 'GMM X Patanwadi' had improved body weight (kg) over 'GMM' crosses as shown below:

Cross	Birth	3 month	6 month	12 month
GMM :	2.5	11.8	16.9	25.5
'GMM × P' :	3.6	15.9	23.4	32.1

- **Introgression into Deccani:** Deccani is raised on the Deccan plateau (in Maharashtra, Karnataka and Andhra Pradesh) by smallholder communities such as the Dhangar, the Kuruba and the Kuruma. Sheep are usually reared in flocks of 25 to 200 breeding ewes along with their progeny and one to four breeding rams. For about 8 months of the year (November to June), 80% of the flocks migrate to areas of higher rainfall or irrigated. Lambs are not artificially weaned by separating them from their mothers. More than 90% of the income from rearing Deccani sheep is earned from sale of 3 to 4 month old unweaned lambs weighing about 15 kg. Lambs are sold young because there is market for them and sheep owners consider it beneficial to have the dams conceive again once the lamb is sold.

  Deccani ewes usually produce only a single lamb with an average lambing interval of about 10 months. Lambs are reared carefully with provision of milk and small amount of supplement such as groundnut cake and wheat flour dough. Given this system of management, and the high demand and price of lambs, a moderate increase in the litter size of Deccani ewes is expected to increase the number of saleable lambs and income.

  With this background, a program on introgression of FecB 'B' allele from Garole into Deccani was started in late 1990's at the Nimb rkar Agricultural Research Institute, NARI, (Phalton, near Poona in Maharashtra). Garole ewes weigh only 15 kg compared with 28 kg for adult Deccani ewe. The FecB *'B'* mutation was introduced into a flock of the Lonand strain, which is one of four strains of Deccani breed (the other strains are Sangamneri, Sangola and Kolhapuri). The half breds produced from 'Garole × Deccani' had small- size, poor lamb-rearing ability, poor milk production, existence of horns in males, wide foreheads and other morphological features which were considered undesirable by local shepherds. Therefore, the level of Garole inheritance, through backcrossing, was maintained at 25 %(quarter-breds).

From 2001, improved Awassi strain from Israel was used for crossbreeding to improve milk yield, lamb weight and growth rate in the quarter-breds with the focus to develop two prolific strains ; named 'NARI Suwarna' and 'NARI Composite' with attributes of high fertility and prolificacy, good lamb-rearing ability and fast growth. The strains are being tested in shepherds' flocks.

Future Genetic Improvement Programs *vis a vis* Changing Sheep Scenario

Sheep husbandry faces a number of challenges. In India, it is a smallholder enterprise where marginal farmers and landless laborers raise sheep on community land under minimum input and marginal output system of management. Sheep scenario is undergoing changes. Three important issues in this changing scenario are:

- **Scarcity of grazing land:** Under the changing scenario of intensive land reclamation for conventional agriculture, the grazing land is shrinking. Therefore, sheep rearing under extensive system is likely to become limited.
- **Declining flock sizes:** Flock sizes over time had been declining. For example, the flock sizes in semi-arid region of Rajasthan were 100 to 120 (in 1990); 70 to 80 (in 2000) and 35 to 40 (in 2008). Given this trend, it is anticipated that they would become less than 25 per household by 2030. Additionally sheep numbers in the some of the important sheep rearing states (e.g. Maharashtra, Gujarat, Rajasthan, Jammu & Kashmir, Orissa, and Tamil Nadu) have declined between 2007 and 2012.
- **Changing breeding objectives vis a vis wool and mutton:** Since more income is accrued from meat, the breeding objectives for improvement of meat production shall become complex. In this situation, improvement has to be made not only in growth and mutton traits but also in reproductive efficienc so as to improve overall productivity. Reproductive efficiency is biologically complex trait as it is made up of several component traits. Therefore, breeding goals have to be clearly defined for genetic improvement of overall productivity.

In view of changing scenario, the following considerations are important in relation to genetic improvement:

- **Selection in improver breeds and group breeding:** The current breeding programs involving selection within recognized improver breeds would continue. Currently selection programs are being conducted at institutional farms with limited flock sizes where intense selection cannot be effected. Moreover, such programs can cover only a few breeds in a limited area. Future genetic enhancement can mainly come from organizing group breeding involving farmer's participation The ongoing field units which use a form of group breeding would have to be expanded by participation of more flocks/farmers, and accurate performance recording. Sheep farmers need to be educated to have their associations (Breed or Breeder's associations). Such programs would permit effective selection pressure to be applied, and help in larger dissemination of genetic improvement. Other schemes like sire referencing, which has been highly successful in UK, New Zealand and some other developing countries can also be organized.

- **Infusion of need based exotic germplasm:** Exotic breeds generally show poor adaptability and require major alterations of management and feeding systems to yield satisfactory results. Although there are many examples of failure but the possibility of utilizing some of their genes in suitable breeding programs cannot be ignored for broadening the gene pool for subsequent selection.

Review Questions

1. Write short note on :
   (a) Crossbreeding of native with exotic breeds of sheep for wool and meat
   (b) Selection for improvement of prolificacy in sheep
   (c) Selection for wool traits in sheep
   (d) Selection for improvement of meat in sheep
   (e) Crossbreeding of sheep for improvement of pelt in India
   (f) Bharat Merino sheep
   (g) Introgression of fecundity gene into native breeds of sheep
2. Define/explain the following with suitable examples wherever necessary:
   (a) Mega sheep seed project
   (b) Ewe productivity index
   (c) Use of Awassi breed for improvement of mutton
   (d) Merino breed
3. Differentiate between the following:
   (a) Wool vs. meat breeds of sheep of India
   (b) Indigenous wooly breeds of sheep and the Merino breed
   (c) Specialized sire and dam breeds of sheep
4. How wool productivity does differ among the four regions of India?
5. What are the three main economic purposes for which sheep have been raised in the country?
6. List the two most important exotic breeds of sheep that have been imported into India for improvement of wool production. Describe their main features, and their use in sheep genetic improvement program of India. Which of these has given better performance in crosses with native breeds?
7. Which dual-purpose exotic breed of sheep was imported into India in 1970's? What are its main features?
8. Which exotic breed of sheep has been imported into India for improvement of pelt? What are its main features?
9. Which exotic breed of sheep has recently been imported into India for improvement of mutton and related traits? What are its main features? How has this breed been used for genetic improvement programs in India?
10. List two important measures of wool yield of sheep. How are these traits measured? What is the type (favorable or unfavorable) and magnitude of the genetic correlation between these two? Which of the two is used for genetic improvement of wool production and why?

11. What are the main differences in skin structure of exotic fine-wool breeds and indigenous breeds of sheep?
12. Categorize the native sheep breeds of four regions into meat or wool.
13. Why is fiber diameter of wool measured at constant temperature and humidity conditions?
14. How is staple length measured?
15. What are the two most important traits contributing to income from wool-type sheep? How are these genetically related to each other (the direction and magnitude of the r_G)? Can these two traits be simultaneously improved in a breeding program? If yes, which selection procedures can be employed for their simultaneous improvement? Is there any evidence to show that these two traits can be simultaneously improved?
16. How is staple length genetically related to fleece yield? Is the r_G between these favorable or unfavorable? Explain
17. List three desirable types of genetic correlations in sheep.
18. At what age are the native sheep of India weaned?
19. Give the range of the following traits of Indian improver breeds of sheep :
    (a) birth weight (b) weaning weight
    (c) yearling weight (d) annual greasy fleece yield
20. What is the most important measure of overall productivity of ewe for the genetic improvement of meat- type sheep? What is the magnitude of its heritability?
21. List the two important components of reproductive performance in sheep. What is the magnitude of heritability of reproductive traits?
22. List three economic traits of sheep which can be measured on group basis.
23. List the selection criterion that can be used for the improvement of following traits, assuming that selection is directed for that trait only, giving the reason for choice of the criterion :

Trait	Selection criterion	Reasoning
Greasy fleece yield :	....	....
Fiber diameter :	....	....
Staple length :	....	....
Marketing weight :	....	....
Prolificacy :	....	....
Hot carcass yield	....	....

24. What were the accomplishments of the exploratory crossbreeding program of sheep in India? Why is crossbreeding of indigenous with exotic breeds no more a national breeding policy?
25. With what objectives was the AICRP on sheep improvement launched?
26. What are the objectives of the Net work project on sheep? Which breeds are included in this project? Name its different centers along with the breeds included in each.
27. List two strains of sheep developed in India for fine wool alongwith the institution where they were developed and the exotic breeds involved.

28. Name a dual-purpose strain of sheep developed in India alongwith the institution where it was developed and the exotic breeds involved in its development.
29. Name a strain of sheep developed in India for meat alongwith the institution where it was developed and the exotic breeds involved.
30. Which breed/strain of sheep was developed in Haryana? Is it being currently used?
31. Which breed/strain of sheep was developed in India in the 19th century alongwith the institution/state where it was evolved and the exotic breeds involved in its development? Is the breed being used commercially? Explain
32. Where were the Mandya and Nellore synthetics developed?
33. Trace the history of development of following strains/types of sheep: (a) Hissardale (b) Bharat Merino (c) Kashmir Merino
34. Name the indigenous breed of sheep that is known for its high prolificacy. How has this breed been used in genetic improvement program? Explain
35. Are there any changes likely to happen in the sheep breeding program *vis a vis* traits to be genetically improved? Explain
36. List two exotic breeds which were used in crossbreeding program launched by ICAR for: (a) fine wool (b) carpet wool (c) mutton
37. Name a major gene influencing prolificacy in sheep.How does this gene influence prolificacy? Has the gene been used for improvement of prolificacy of Indian breeds/germplasm? Explain
38. Why is selection for improvement of prolificacy of sheep a slow process?
39. List two features of sheep that make it easier to bring about genetic improvement.

●●●

21

Goat Breeding

Goat was the earliest ruminant to be domesticated. Archeological evidence indicates goats to have been associated with humans in a symbiotic relation for over 10,000 years. The Harappa toys contain representations of a goat. Two seals from Mohenjo-Daro show a wild goat with enormous curled horns, and a bearded domestic male goat with side-spreading horns.

Phylum: Chordata (back bone),
Class: Mammalia (suckle their young ones)
Order: Artiodactyla (hooved, even footed)
Family: Bovidae (hollow-horned ruminants)
Subfamily: Caprinae (sheep & goat)
Genus: *Capra* (domesticated and wild goat)
Species: *Capra aegagrus*
Subspecies : *Capra aegagrus hircus*

The wild goat *(Capra hircus)* is the chief ancestral stock from which the various breeds of domestic goats have been derived. The domestic goat (*Capra aegagrus hircus*) is a sub species of goat domesticated from the wild goat of Southwest Asia and Eastern Europe.

Goats are now spread all over the world because of their great adaptability to varying environmental conditions and different nutritional regimes. In developing countries, they make a very valuable contribution, especially to the poor in rural areas due to their small size, high fertility, ability to utilize a wide range of feed and fodder resources. It serves triple purposes: meat, milk, and fiber/hair besides skin. The species has an advantage of utilizing crude fiber or roughage such as bushes and thorns grown in desert and semi-desert areas.

World's total goat population is 862 million (FAOSTAT, 2008). However, there is wide variation among different parts of the world with regard to the number of goats. Maximum population of goats is found in Asia (59.7%) followed by Africa (33.8%). The Asia and Africa (tropical and sub tropical) together account for 93.5% of world's goat population (Figure 21.1). The lowest number of goats is found in Oceania accounting for 0.1% of the world's goat population.

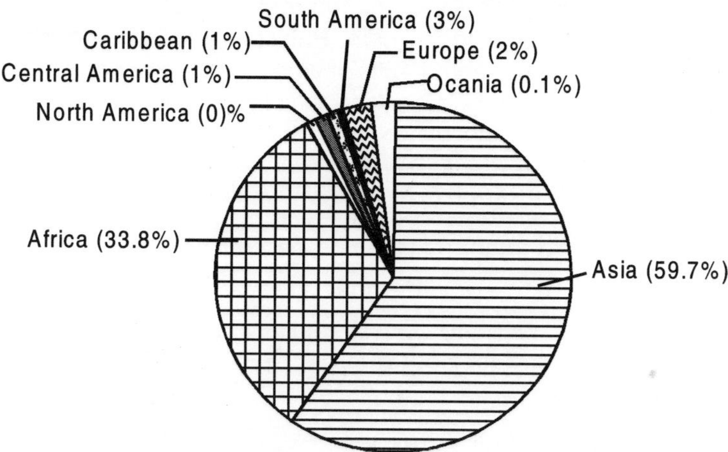

Fig. 21.1 Distribution(%) of world's goats by regions

The wide distribution of goats in tropics and sub tropics reflects their ability to adapt to a variety of environments. However the preferred environment is on lighter sandy soils in dry tropics rather than wet tropics. The inherent characteristics of goats such as resistance to dehydration, preference for browsing and wide ranging feeding habits enable them to thrive in regions that receive low rainfall (< 750 mm). For this reason, approximately 47% of all the goats of India are found in rather dry states of Rajasthan, Bihar, Utter Pradesh and Madhya Pradesh.

Worldwide the number of goats has increased by 146% over their numbers in 1990. The fact that world's goat population has been increasing by about 1 to 4% each year during this period points out to their increasing utility. Ten countries of the world viz; China, India,Pakistan, Bangladesh, Nigeria, Sudan, Iran, Ethiopia, Mongolia, Indonesia together have approximately 65.7% of the worlds total goat number as shown in Figure 21.2.

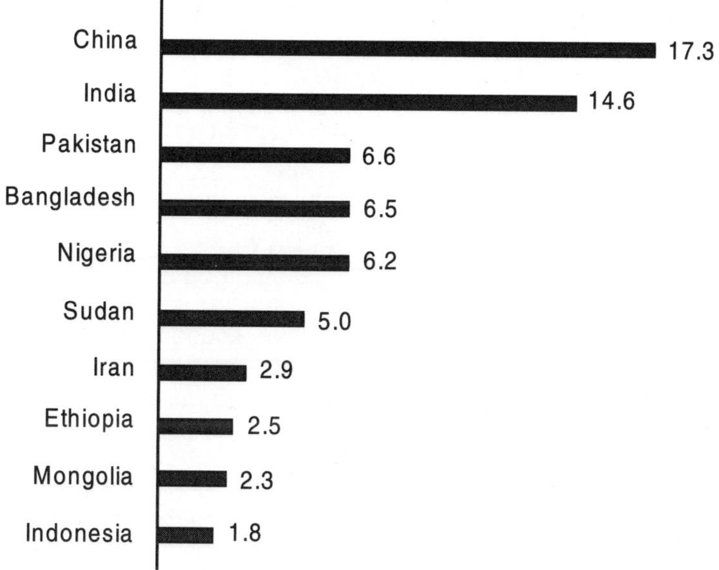

Fig. 21.2. Distribution (%) of goats in top ten countries on the basis of number of goats.

Seven countries in this list are in Asia, and the rest in Africa. The highest number of goats in the world is found in China, followed by India, Pakistan and Bangladesh, all of them are in Asia. These four countries constitute about 45% of worlds total goat population. Goat population of India (as per FAOSTAT 2008) is 125.7 million which is approximately 15.0 % of the world goat population.

Box 21.1: Goat terminology.

- **Buck:** A sexually mature intact male goat used for breeding.
- **Caprine:** The scientific name for the goat species.
- **Chevron** (French word for goat): Goats that are slaughtered near or shortly after weaning; word is also used for goat meat.
- **Doe:** A sexually mature female goat.
- **Doeling:** A young female that is not yet sexually mature.
- **Kid:** A goat less than one year old.
- **Loin:** A muscle that lies between the last rib and the hipbones of the back.
- **Mohair:** Usually a silk-like fabric or yarn made from the hair of the Angora goat. Mohair, which grows in uniform locks, is notable for its high luster and sheen. It is considered a luxury fiber and is usually more expensive than most wool that comes from sheep. Mohair fiber is approximately 25–40 microns in diameter. Shearing is done twice a year, in the spring and in the fall. One goat will produce 5–8 kg of mohair a year.
- **Pashmina** (name derived from Persian: *pašmina, syn. cashmere*): It is downy undercoat fiber derived mainly from domesticated goat indigenous to Asia. It is always mixed with coarse outer coat known as guard hair, which needs to be separated before processing. Its diameter is between 12 and 21 microns. Though the names Pashmina and Cashmere are used interchangeably but a distinct difference between Pashmina and generic Cashmere is the fiber diameter. Pashmina fibers are finer and thinner (12-14 microns) than generic cashmere fiber (15-19 microns).
- **Pelt:** The skin of a goat.

Germplasm Resources of Indian Goat

It is estimated that there are altogether about 300 breeds and types of goats. Majority of these are found in tropics and sub tropics. While many of these have been adequately described but many more yet remain to be described as they are thought to be non-descript or unimportant.

India's goat population is represented by 20 well-recognized breeds[1] with wide product diversity e.g. meat, milk and fiber (hair/ pashmina). Descript breeds, however, constitute only about 20-25% of the total goat population. The remaining goats are non-descript with mixed features. With respect to their habitat, however, there is no clear cut demarcation as some of

[1]Information on many new breeds appears every now and then but, it is not reliable enough to include them in the catalogue of goat breeds in India. The National Bureau of Animal Genetic Resources (NBAGR), is recently working on a project *"Report on the State of the World's Animal Genetic Resources"* sponsored by the Food and Agriculture Organization (FAO) for cataloguing the animal genetic resources of India.

the breeds, except Jamunapari, Jhakhrana, Surti, Beetal, Chegu, Changthangi, are spread over large areas across different states as well. Boundaries of two or three breeds also overlap which encompass migration, market and economic compulsions, besides natural existence. Similarly, the population size does not reflect the true picture as there is existence of large number of non-descript animals in the habitat of most of the breeds.This is due to intermixing which occurs due to migration during feed and fodder scarcity when indiscriminate crossing occurs leading loss of identity of recognized breeds.

Goat Production Systems in India

Three type of management systems with overlapping features are normally practiced in the country:

- ***Extensive production system:*** This is a low-input and low-output system, and is integrated with arable crop, tree, shrubs etc. Housing, feed and health care investments are minimal. This system of goat production is further sub-divided as:
  - *Nomadic*: Characterized by irregular movement of goat and their keepers in search of feed with no fixed base.
  - *Transhumant*: There is regular seasonal movement of animals which return to the fixed base in the same year. In strict sense, there is vertical movement of goats from high altitude to low altitude in search of fodder during winter.
  - *Sedentary*: Animal movement is restricted to a short radius from a permanent base.

  Nomadic and transhumant systems cover sizable goat population in different eco-zones of the country. But it is difficult to carry out improvement program in such systems due to movement of flock and their owner's. Its sustainability is decreasing due to shrinkage of common land, urbanization etc.
- ***Intensive production system:*** It is associated with high investment on housing, veterinary care, housing, feed resources, labor etc. The goats are fed in confinement with limited access to land. It is characterized by high-input and high-output and is usually practiced for high potential animals. It uses cultivated fodder with zero grazing and high plane of concentrate diet. This system is least practiced in India, mostly at institutional flocks and by a few commercial goat farmers.
- ***Semi extensive production system:*** It is a combination of extensive and intensive systems. Goats graze in pastureland (5-8 hours) while further requirements are met through concentrate supplementation. Local resources, crops by products and lopping from plants are utilized in a better way and hence it less expensive compared to intensive production system. This system is practiced to some degrees in most situations.

Breed Classification

Most of the goat breeds of India have evolved through geographical isolation and natural selection with adaptation to agro-ecological conditions. To a limited extent, there might have been artificial selection for specific needs. Most of the indigenous breeds are very well adapted to the harsh climate, long migration and lack of vegetation and drinking water. Goat breeds of India have been assigned names on the basis of morphological attributes and on the vicinity of their natural habitat. The well-defined breeds have been classified by three broad criteria:

- Ecological region (agro-climatic conditions)
- Conformational attributes (mainly body size)
- Production type (utility).

Breeds by Ecological Regions

Of the three criteria used for classifying goat breeds, the classification by ecological zones is most commonly used. By this, the country has been considered into four regions: North-Western (dry- arid and semi- arid), Southern Peninsular (hot humid), Eastern and North Temperate (Himalayan). Distribution (percentage) of total goat population of India and the descript breeds for the four ecological regions is given in Figure 21.3.

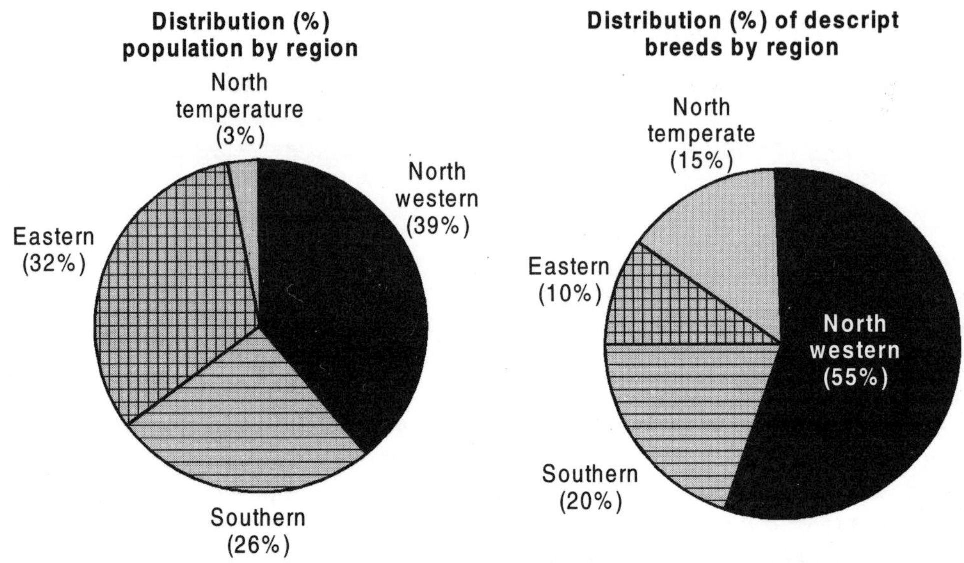

Fig. 21.3 Region-wise distribution of goat population and breeds (%).

Highest numbers of goats are found in North western region while lowest in the North temperate region. Distribution of descript breeds of different regions indicates that though the eastern region has sizeable number of goat population but it has only 10% of the descript breeds of India. Thus maximum number of non-descript breeds are found in Eastern region.

A brief description of these regions and their breeds is provided below. Salient morphological features of different breeds are given in Table 21.1.

North Western Region and Its Breeds

This region, which comprises the states of Haryana, Punjab, Rajasthan, Gujarat, Uttar Pradesh and Madhya Pradesh, is characterized by tropical and sub tropical climate (arid and semi-arid). The temperature usually varies between – 2°C (winter) and 49°C (summer). This region, which has many rivers, is highest in crop productivity. Large numbers of uplands, ravines, wastelands are available. Rainfall is scanty and erratic in arid region, which commonly encounters drought. Small ruminants are considered to possess an important mechanism for withstanding drought due to their better adaptability and mobility as compared to large ruminants. Goat density in this region is also high.

Major forage sources in arid and semi-arid regions are rangelands, community grazing land and crop residues. Average goat population in this region has shown increasing trend over the past decades since 1951. However, population trend is varying: the flock sizes are declining in high-crop intensity areas while in low-cropping intensity area the number of goat keepers is increasing coupled with large flock size. Medium and large flocks are common in Rajasthan and Gujarat whereas in Haryana, Uttar Pradesh and northern Madhya Pradesh and Punjab flock sizes are small to medium. Goat density also varies: from 0.70 (Rajasthan) to 40 per sq km (Uttar Pradesh). Goats in this region are managed under traditional extensive system where they are grazed on village community land and crop residues. Consequently, there is high competition for feed resources between small and large ruminants as population of buffaloes and cattle in this region is also quite large. Goats are mostly reared by small and marginal farmers belonging to schedule castes, tribes and backward communities.

Maximum number of breeds (11) as well as the population of goat (39%) is found in this region as the prevalent dry climatic conditions are conducive for their growth, reproduction and survival. Most breeds of this region are reared for meat and milk. An additional crop of hair is also harvested once in a year from some breeds. Most breeds are sturdy and have good resistance to climatic stress, feed, fodder and water scarcity and parasitic diseases.

The breeds found in this region are Jamunapari, Barbari, Beetal, Jhakhrana, Sirohi, Marwari, Kutchi, Mehsana, Zalawadi, Gohilwadi, and Surti (Table 21.1) In terms of numbers, Marwari has the largest numbers while Beetal has the lowest. Of the eleven breeds of this region, four breeds viz; Jamunapari, Barbari, Sirohi and Jhakhrana, are undergoing systematic genetic improvement at institutional farms. Beetal has been recognized as improver breed for medium and small size goats due to its better reproduction efficiency and adaptability. However, its numbers are declining due to intensification of crops in its native tract. A conservation program in its home tract has been started. Barbari, which is a dual-purpose breed, is one of most prolific breed amongst goat breeds of this region. It has about 60-70% multiple births, and is very well adapted for stall-feeding. However, this breed is facing the problem of genetic dilution due to lack of potential Barbari bucks, small flock size, and use of inferior non-descript bucks and crossbreeding with Sirohi goats.

Southern Peninsular Region and Its Breeds

The states of Maharashtra, Karnataka, Andhra Pradesh, Tamil Nadu and Kerala belong to this region. The climatic condition is semi-arid in the central peninsula and hot humid along the west coast. About 26% of India's goat population resides in this region. Most areas of this region are drought prone which makes livestock sustainability vulnerable to long dry periods and non-availability of feed, fodder and water. The rivers of South India are seasonal, which overflow during rainy season and become dry during summer. Rainfall distribution is highly varied and nearly 70 % area of the region gets low rainfall and remaining 30 % (coastal) gets high rainfall (1000-4200 mm). Most of goat keepers practice mixed farming system. Goat population is more intensified in western and southern plateau region.

The recognized breeds of goats found in this region are Osmanabadi, Sangamneri, Malabari and Kanni Adu which are primarily reared for meat (Table 21.1). Osmanabadi breed is well adapted for distant walking in search of feed.It also has great resistance for disease and climate stress and is known for breeding efficiency (about 70% twinning). Sangamneri breed is suited for stall feeding. Osmanabadi and Sangamneri are highly prolific (about 50% multiple births). Malabari breed occurs in southern peninsula and is spread over five districts in Kerala. It accounts for about 30% of total goat population of Kerala. Of the breeds of this region, the most populous is Kanni Adu followed by Osmanabadi, Malabari and Sangamneri in that order. Another breed/type named Attapadi black, which is medium sized and is kept by Attapadi tribe, has also been registered with NBAGR. But its numbers are very low (about eight thousand).

Eastern Region and Its Breeds

This region comprises the states of Orissa, West Bengal, Bihar, Jharkhand and the north eastern hill states. Climatic condition is warm and humid in lower Gangetic basin, coastal plains and plateau region while it is sub-temperate in the hills. High rainfall, poor drainage system, frequent floods, high human density and low land availability are common features of this region. Goat population density is moderate to high and it is the major livestock activity of this region. Goats are predominately reared by socio-economically weaker sections (schedule castes, schedule tribes and backwards) however, people of general castes of all religion also rear them as a secondary source of income. Most rural communities especially tribes attach deep socio cultural and religious sentiments to goats. Specific coat colors, sex and appendages over body are preferred for use of goats for sacrificial purposes. This liking for morphological features indirectly supports the conservation of biodiversity of small ruminants.

Goats of this region are mostly of meat type. The descript breeds found in the region are Black Bengal and Ganjam. The Black Bengal, a small sized breed, is well known for high fertility, fecundity and excellent skin quality. Black Bengal is characterized by highest incidence of multiple births (60-70%) among all the recognized breeds of the country. On the other hand Ganjam breed of this region is amongst the low prolific breeds (5-10% multiple births) amongst recognized breeds of the country.

The Assam hill goats (Khasi) are also found in this region. These goats are known for multiple births, short kidding interval and superior quality of skin. They are found in varying colour pattern; the predominant colors are white, black and brown. These goats are small and body is compact which is suitable for hilly terminus of Assam and eastern states.

NorthernTemperate Region and Its Breeds

This region includes the states of Jammu & Kashmir; Himachal Pradesh and Uttaranchal.The climate is predominantly temperate which is extremely cold with heavy downpour in monsoon and moderate to heavy snowfall in winter. The forest cover ranges from 30 to 50 % and productivity of crops is very low. Land holdings are small and fragmented. Livestock, which is predominated by goat population, play a very important role in rural economy.

Goat density is highest in Himachal Pradesh (21.0 per sq km) closely followed by Uttaranchal and is lowest in Jammu & Kashmir (8.1 per sq km). Majority (more than 80%) of the farmers depend on uncultivated forest (tree leaves, grasses and herbs), grazing land and rangelands for grazing. Agro forestry and related resources are also easily available for goat production under the extensive production system. Since cattle and buffaloes cannot graze the grass cover in upland terrains, extensive feeding system for goats at hills is relatively cheap and viable. Changpa, Gujars, Gaddi, Kumaoni, Bhakarwal, Khas, Jadhas, Marchyas, and Bhutias are pastoral tribes and traditional goat keepers of this region. Goats in Uttaranchal are mostly reared by Shilpkar and other backward community people for meat purpose. Goats at higher altitudes are also used as pack animals. Farmers practice pastorals, transhumance and nomadic type of extensive production system with large flocks of goat (50- 500) due to availability of forests and rangelands. However, semi-intensive system of management is not uncommon in small flocks (5-10).

The recognized goat breeds of this region are Changthangi, Chegu and Gaddi, which are found at a high altitude in the Himalayan/ Trans-Himalayan terrain. Two breeds of this region, Changthangi and Chegu, are of unique importance due to fine quality of Pashmina. These goats tolerate sunlight well but prefer shade and become uncomfortable under humidity. Cold arid region is more preferred by these goats. Prolificacy of Changthangi and Chegu is low (2-10%).Gaddi goat is used as beast of burden. In terms of numbers, Gaddi has the highest numbers followed by Changthangi and Chegu.

Breeds by Production Characters (Utility)

Goat is one of the species, which delivers milk, meat and fiber (hair/pashmina). Among Indian breeds, however, there is no clear-cut distinction between different types particularly between meat and dairy type breeds. In fact, there is no breed in India which is exclusively reared for milk although milk production potential of some breeds e.g. Jamunapari, Beetal, Jhakhrana and Surti is reasonably high. Goat milk is primarily used for home consumption by poorer sections of rural masses. Goat meat has wider acceptability over large population without any religious bias.

Majority of the goat breeds are used for meat. Some breeds bear luxurious body coat with long hairs while the breeds like Changthangi and Chegu found in the hilly cold desert region at high altitudes in the Himalayan ranges produce a fine fiber as under coat called Pashmina, besides hairy outer coat. Since goat farming is carried as a small scale family activity, farmers generally choose animals of the breed/type prevalent in the vicinity and use them for multiple purposes.

Despite their multiple utility, some breeds are better than others for one or more product. Goat breeds are broadly grouped into three major classes depending upon their primary production characteristics i.e. milk & meat (dual purpose), meat and fiber as follow:

Production type	Breeds
Dual (Meat and Milk) :	• Jamunapari, Beetal, Jhakhrana, Surti and Barbari (North western region)
Meat :	• Sirohi, Marwari, Zalawadi, Gohilwadi, Kutchi Mehsana (North western region)
	• Malabari, Osmanabadi, Sangamneri, Kanni adu (Southern region)
	• Black Bengal, Ganjam (Eastern region)
Fiber :	• Changthangi, Chegu, Gaddi {North temperate region}

Nonetheless certain degree of overlapping among three categories does occur. Goat skin is also an important attribute as it is a major input of leather industry. The skin produced by Black Bengal is acclaimed internationally for its fine grain quality.

Breeds by Body Size (Conformation)

Goat breeds of India have also been classified on the basis of conformational characteristics e.g. body size, ear shape, shape of horn and body coat etc. Of these, body size is of greater economic importance to farmers as it bears a positive relation with the production of milk, meat, fiber and skin. On the basis of body size, goats can be classified into three categories:

- *Large sized*: Some breeds like Jamunapari, Beetal, Jhakhrana, and Sirohi are heavier than others.. All of these belong to North-western region.
- *Medium sized*: Majority of descript breeds (more than 80%) are medium sized. All breeds of Southern peninsular region fall in this category.
- *Small sized* : Generally breeds of Eastern region, in particular Black Bengal, are small in size. Barbari breed of North western region is also a small sized breed.

Table 21.1 Morphological features including habitat of recognized Indian breeds of goats.

Breed	Coat colour	Face pattern	Horn pattern	Specific character(s)	Product utility	Habitat
North western:						
Sirohi :	Brown/dark brown patches on body	Straight/raised	Curved, backward, upward with pointed tips	Wattles in both sex	Meat & milk	Udaipur, Sirohi, Ajmer districts (Rajasthan)
Marwari :	Black, shaggy coats with long hair	Straight	Medium, backward and upward with pointed tips	Bear available in 30-40% animals	Meat & hair	Bikaner, Churu, Jodhpur (Western Rajasthan).
Jhakhrana :	Black with white patch on muzzle ear	Straight/raised	Short, thick	Long body, long conical & rounded teats	Milk & meat	Jhakhrana village in Alwar district of Rajasthan.
Beetal :	Black, brown with white patch on body	Convex with roman nose	Short & backward	Long body, bear common in male	Milk & meat	Gurdaspur, Amritsar, Ferojpur(Punjab)
Barbari :	White with brown spots	Slightly raised	Short and erect, directed upward & outward	Prominent orbit bone, protruding eyes	Meat & milk	Agra, Mathura, Etah, Aligarh,dist. (Uttar Pradesh)
Jamunapari :	White or white tan paces on head	Convex with roman nose	Short & backward	Long body, tuft hair on back of thigh, lower jaw longer	Milk & meat	Chakarnagar Block in Etawah district of U.P.
Mehsana :	Black/grayish	Convex face	Twisted, curved, backward & upward	Ears are white with black marking, short & upward tail	Milk & meat Anand	Mehsana, Kheda, Gujarat) Mehsana and Banaskantha districts in Gujarat.
Gohilwadi :	Black	Slightly convex long	Twisted, curved, backward a& upward	Developed udder with conical teats udder conical treats	Milk & meat	Bhavnagar, Amreli and Junagarh districts in Gujarat

Breed	Coat colour	Face pattern	Horn pattern	Specific character(s)	Product utility	Habitat
Zalawadi :	Black with long lustrous hairs	Convex with roman nose	Twisted screw type, backward & upward with pointed tips	Long, lustrous hairs & cork-screw type	Milk & meat	Surendranagar and Rajkot districts in Gujarat.
Kutchi :	Black with white marking on face	Slightly raised & prominent forehead	Long, backward & upward	White marking on ear	Meat & milk	Kutch district in Gujarat.
Surti :	White	Slightly raised & prominent forehead	Medium, directed backward & upward	Udder well developed with conical teats	Milk & meat	Surat and Vadodara districts in Gujarat.
Southern :						
Osmanabadi :	Predominant black	Straight	Long	Long pendulous	Meat, milk	Osmanabadi district in Maharashtra.
Sangamneri :	White	Straight	Backward & upward	Dropping ear	Meat	Ahmednagar, Nasik, Pune, Maharashtra).
Kannai Adu :	Black	Straight	Backward & upward	Legs below knee and hocks are white with white strip on face	Meat	Virudunagar ; Tiruneleveli, Tuticorin,Tamilnadu) Adu Ramanathpuram and districts in T.N.
Malabari :	White, mixed cooler	Straight face with slightly raised no.	Medium directed outward & downward	Wattles in 25%.	Milk & meat	Calicut, Cannanore

Breed	Coat colour	Face pattern	Horn pattern	Specific character(s)	Product utility	Habitat
Eastern :						
Ganjam :	Black & mixed colour	Straight face, prominent forehead	Medium strait or horizontal	Males have bear	Meat	Ganjam district, of Orissa
Black Bengal	Black & brown	Straight & slight depressed	Short, flat & horizontal	Dwarf breed, both sexes bearded	Meat & skin	West Bengal
Temperate Himalayan						
Gaddi	White	Small convex nose-line	Long, spiral directed upward and backward with pointed tips	Well built, hardy with long hairs of 20-25 cm	Fiber, meat & pack	Chamba, Kangra, Kullu, Bilaspur, Kinnanaur, Lahul-Spiti districts in HP
Changthangi	White	Straight profile	Long, spiral directed upward & backward with pointed tips	Long lustrous hair	Pashmina & Meat	Leh in Ladakh, J & K
Chegu	White	Long with straight profile	Spiral directed upward, outward & backward with pointed tips	Long lustrous hair	Pashmina & meat	Lahul Spiti valley HP and Uttarkashi, Chamoli and Pithoragarh districts of Uttaranchal

Goats for Milk

Dairy goat is often referred to as cow of the poor as it consumes little, occupies a small space and produces enough milk for average unitary family. Milk of goat has some special attributes viz; small-sized fat globules, easily digestible casein, higher calcium content, distinctly alkaline etc. Goat milk is thus well suited for babies and patients suffering from cow-milk intolerance.

Dairy goats account for about 2% of the world's total milk produced by livestock species (FAOSTAT, 2008). Of the total amount of goat milk produced in the world, developing countries account for about 83%. However, it is only In Europe where goat breeding is strongly oriented towards milk production. Europe has only 2% of the world's goat population but it produces about 15% of world's goat milk. It is mostly used for cheese production.

Of the top ten goat milk producing countries, India occupies the first position accounting for about 36 % of the total milk produced by top ten countries (Fig-21.4). It is followed by Bangladesh and Sudan. There are three European countries in the list of top ten milk producing countries viz; Spain, France and Greece, each producing about the same quantity. In France, interest in dairy goats has led to the establishment of organized programs for selection, processing and commercialization of goat milk which mainly comes from Saanen and Alpine breeds.

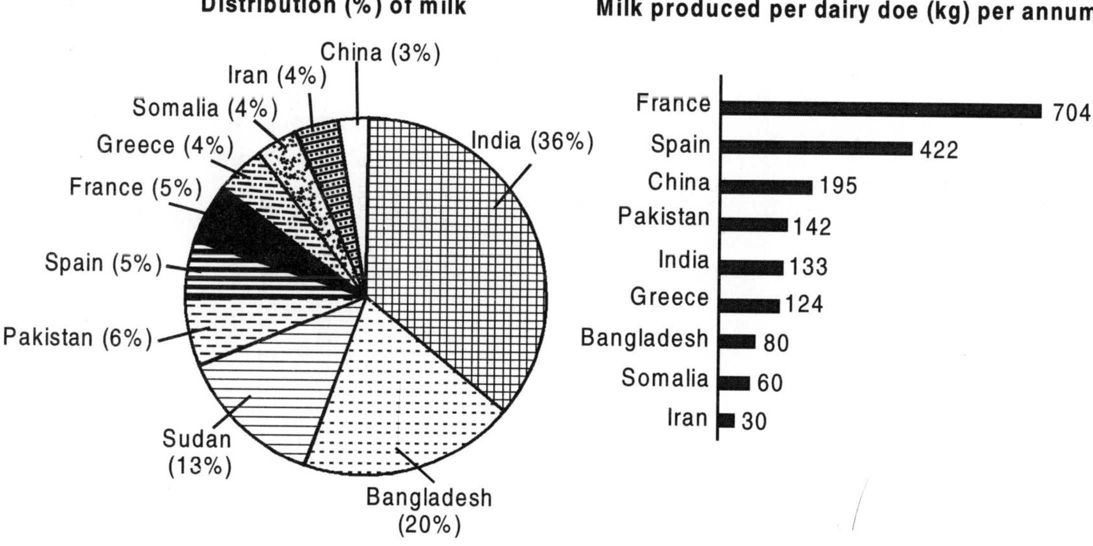

Fig. 21.4 (a) Distribution of milk produced by top ten countries (b) milk produced per dairy doe in top ten countries.

On milk produced per dairy doe basis, France leads the list (over 700 kg/year), while Iran reports the lowest milk production (30 kg/year). China has the highest total number of goats in the world, but they are mainly kept for meat. Indian dairy doe on the average produces 133 kg/year.

Since 1990, interest in dairy goats in the world has been steadily increasing as indicated by the increase in milk production from about 10 million MT (1990) to about 15.2 million MT (2008).

In developing countries, much of the milk produced by goats is used for family consumption by rural poor. Goat milk has only a very small market share and has been consumed more as a medicine than as food. In India, goat milk accounts for about 3.8 % of the total milk produced. Average annual growth rate of goat milk (2000-01 to 2010-11) was only 1.4%. The top three goat milk producing states are Uttar Pradesh, Rajasthan and Madhya Pradesh which account for 69.9% of the total goat milk produced in the country.

Though India has 20 recognized breeds of goat but there is no breed which is exclusively reared for milk. Milk production of most of the indigenous breeds of goat is low since they have not been not been subjected to selection for improvement of milk production A few breeds of North western region (e.g. Jamunapari, Beetal, Jhakhrana and Surti) have higher milk production than others. Information on the milk production and related traits of different breeds, based on small sample size collected over short period under varied conditions,has been reported (Table 21.2) but a comparison between breeds is not strictly valid.

Table 21.2 Average values for milk production and related traits in some breeds of goats.

Region/ Breed	Age at 1st kidding(days)	Lactation yield(kg)	Lactation length(days)	Kidding interval(days)
North Western :				
Jamunapari	735	202	194	350
Beetal	743	174	182	365
Jhakhrana	561	152	143	316
Surti	400 – 450	155	150	–
Sirohi	776	114	194	351
Marwari	468	102	111	507
Kutchi	778	124	195	–
Barbari	633	96	152	329
		101	124	
Mehsana	600-650	132	197	–
Gohilwadi	650-700	156	207	–
Zalawadi	500-600	165	201	–
Southern :				
Osmanabadi	364	60	100	–
Sangamneri	550	84	168	356
		65	90 – 110	
Malabari	610	90	178	290
Kanni Adu	650	–	–	–
Eastern :				
Bengal	521	35	111	263
Ganjam	—	46	120	–
North Temperate:				
Changthangi	700 – 900	150 – 250g/day	200 – 300	–
Chegu	630	69	187	–

Source: Winter school. Performance values obtained from AICRP reports (up to 1991) on institutional herds maintained under intensive or semi-intensive rearing systems. A strict comparison among breeds is not justified due to climatic and managemental differences at different centers.

Apparently, there is wide variability for the values recorded for different breeds for lactation milk yield, lactation length and age at first kidding. Lactation yield has been reported to vary from 35 kg (Black Bengal) to over 200 kg (Jamunapari) with lactation length varying from over three and half month to about 10 months. Average daily milk yield is low: varying from 250 g/day to little over 1 kg/day in the entire lactation. Some breeds like Jhakhrana, Jamunapari, have given peak yield of 4-5 kg.

Milk yield is usually low in the first lactation but increases with advance in lactation order up from 2nd to 4th. Large variability is observed between breeds with respect to peak yield as well. The reasons for low productivity of Indian goats are the low inputs (in terms of feeding, inadequate grazing resources, disease problems) and a lack of organized efforts for genetic improvement as goats are reared by poor with small flock sizes. Lately milk production of a few important breeds has been systematically recorded under field (farmer's) condition under semi-intensive and extensive managemental system. Average values for milk production and reproductive traits of Jhakhrana breed are shown below:

Traits	Extensive	Semi-extensive
Milk production(kg):		
• 90-days lactation yield	107.4	152.9
• Peak yield	2.1	2.61
• Lactation length	118.0	143.0
Reproductive :		
• Age at first conception (days)	493	426
• Age at first kidding (days)	632	561
• Service periods (days)	198	161
• Kidding interval (days)	333	288
• Twinning rate	1.32	1.54

Under semi extensive management system, milk production as well as reproductive traits contributing to efficiency of milk production show improvement over extensive system. These figures suggest the potentiality of some of the breeds to be used for further genetic improvement. A good average grade goat, if fed and kept under proper conditions, can yield about 2.25 kg milk per day over a lactation period of 7 to 10 months.

Goats for Meat

Goat meat is widely consumed in developing countries but it represents only 2% of the total meat produced in world. The developing countries produced approximately 97% of this amount, reflecting great importance of goat meat to feed millions of people in these countries (FAOSTAT 2008). The top ten countries producing goat meat are: China, Pakistan India, Nigeria, Bangladesh, Sudan, Iran, Indonesia, Ethiopia, Niger (Fig. 21.5).

It is apparent that the top ten countries producing goat meat are all from Asia and Africa indicating the importance of goat meat to people living in these areas. Among the top ten meat producing countries, India comes second to China accounting for 13% of total goat meat produced. Goat meat production has been consistently showing an increasing trend: from 2.65 million MT (1990) to 4.93 million MT (2008). But a major part of this volume is not traded like other major meats as it is usually produced and consumed locally among the poor. Meat production per goat in top ten countries (Fig 21.5) indicates it to be highest in Pakistan and lowest in Bangladesh.

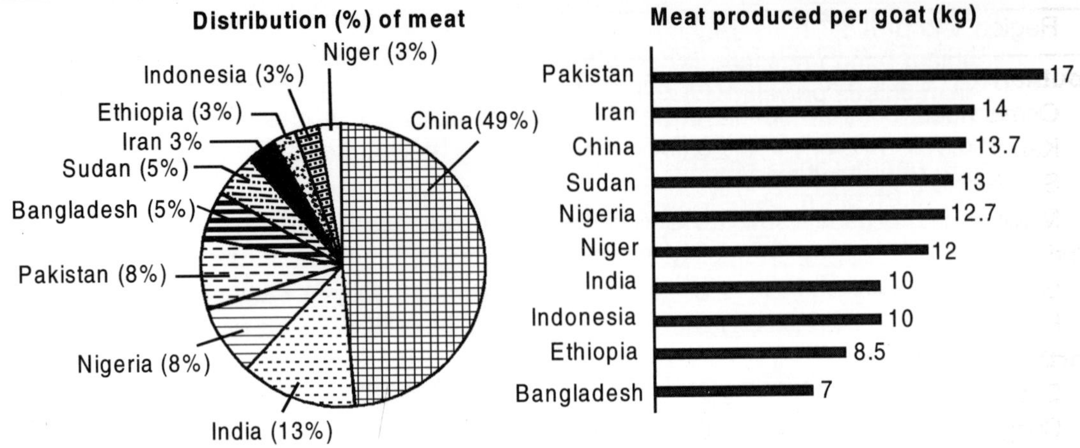

Fig. 21.5. (a) Distribution of total meat produced and (b) meat produced per goat in top ten meat producing countries.

In India, goat meat has wider acceptability over large section of population without any religious bias. Goat meat is lean and juicy, and it fetches premium over other meats. Goat meat contributes about 37% of the total meat produce of the country. However, there are no recognized meat breeds of goat in the world except the Boer breed found in South Africa. In India all, the breeds are used for meat but those found in Coastal areas and Deccan plateau are more commonly used. Such goats have rough coat of short hair having no economic value.

Traditionally definition of meat breed in India has been based on pre-slaughter body weight and dressing percentage. Since goats are reared on range, feed conversion ratio is of little value. Prolificacy contributes to total meat output but this trait relates to flock productivity. Since there are little differences in dressing percentage among different indigenous breeds usually it varies from 45 to 50%, with few exceptions, definition of a meat breed is generally restricted to body weight. Average body weights at various ages for Indigenous breeds are listed in Table 21.3.

Table 21.3 Average body weights (kg) at different ages for indigenous breeds of goats.

Region and Breed	Birth	3 month	12 month	Adult	
				Male	Female
North Western :					
• Jamunapari	3.1	10.5	26.0	59.0	38.0
• Beetal	3.0	9.3	22.0	59.0	35.0
• Sirohi	2.8	9.9	21.0	50.0	22.5
• Jhakhrana	2.6	12.3	25.3	57.8	44.5
• Marwari	2.3	6.0	21.2	33.2	25.9
• Kutchi	3.1	12.8	28.0	43.5	39.3
• Mehsana	2.8	9.9	19.2	37.0	32.4
• Gohilwadi	2.6	–	22.0	–	–
• Zalawadi	2.9	–	22.0	38.8	33.0
• Surti	2.1	10.0	23.0	33.5	32.0
• Barbari	1.8	6.7	15.0	37.9	22.6

Region and Breed	Birth	3 month	12 month	Adult	
				Male	Female
Southern :					
• Osmanabadi	2.4	7.4	21.0	33.7	32.4
• Kanni adu	2.3	–	27.0	–	–
• Sangamneri	1.9	–	20.0	–	—
• Malabari	1.6	5.8	15.0	39.0	31.1
Eastern :					
• Ganjam	2.3	–	16.0	–	–
• Bengal	1.4	5.4	14.0	–	–
North Temperate :					
• Changthangi	2.2	–	16.0	–	–
• Chegu	2.0	–	15.0	–	–
• Gaddi	2.6	–	25.0	–	–

Source: Winter school and terminal report of AICRP.

On the average, breeds of north western region, in particular Jamunapari, Beetal, Sirohi, Marwari, Kutchi and Jhakhrana, are heavier than of other regions while those of eastern region are lighter; Black Bengal being lightest. For example, breeds like Jamunapari, Beetal, Sirohi, Marwari, Kutchi and Jhakhrana, under good feeding can attain body weight of 18 to 20 kg at six months of age. Under semi-intensive management, there is significant improvement in body weights.

On the other hand, Black Bengal and Barbari can never attain similar weights even under intensive feeding due to their, small size. Most of the heavy breeds of North western region are used for both milk and meat while those of southern and eastern region are mainly used for meat. Information on carcass characteristics of a few breeds maintained at institutional farms is available.

Box 21.2: Summary of growth and related traits of Indian breeds.

Goat meat is preferred by some over sheep meat due to certain features: fat is not distributed all over the body but is mainly in viscera and therefore goat meat is more juicy and tender

- Average birth weight: Ranges from 1.4 to 3.1 kg.
  - o Heavier breeds are over 2.6 kg
  - o Smaller breeds are in range of 1.4 to 2.5 kg.
- Average weaning weight((3 month) : 5 to 12 kg
- Yearling body weight(12 month): 14-16 in small sized and 20-25 kg in large-sized breeds
- Market age: Around 9 month in heavy and 12 month in medium/small breeds.
- Average daily gain (from 3 to 12 month) : 25 to 60 g under range/semi-extensive intensive.
- Dressing percentage on pre-slaughter wt: About 45 to 50 %
  - o Dressing percentage is higher from 9 to 12 month age but lean weight is better up to 10 month (67-70%).
  - o Juiciness and tenderness is best in 8-12 month group.

Goats for Fiber (Pashmina) and Hair

Mountain goats, besides producing hair, also produce undercoat of fine silky fibers, known as Pashmina (Syn. Cashmere, pashm etc). It has unique importance due to its softness, warmth, tightness. Due to its better ability to absorb dyes, it is used to make costly fashion wears. Fiber fineness is the most important attribute giving unique value to Pashmina fiber, and differentiates it from sheep wool. It's average diameter is 12-13 micron with average fiber length of 50-60 mm though these can vary depending on source and breed/genotype. The morphological structure of Pashmina fiber is similar to fine wool fiber comprising outer cuticle and cortex but there is no medullation in the fiber. Diameter of the fiber is uniform throughout its length which property differentiates it from sheep wool.

The major Pashmina producing countries in the world are : China, Mongolia, Iran, Afghanistan, Pakistan, Nepal and India. Apart from these, Australia, Britain and New Zealand also produce the cashmere fiber. However, the quality of fiber produced in different countries is not same. China is the major producer of Pashmina with share of 70% followed by Mongolia with 20% share. Rest 10% of the total production is from other countries including India which produces less than 1% of total.

In India, two Pashmina producing goat breeds are Changthangi and Chegu which are found in hilly cold region at high altitudes. Changthangi is found in Leh and Ladakh regions of Jammu and Kashmir and Chegu in Lahul and Spiti and Kinnaur region of Himachal Pradesh and Uttarkashi, Chamoli regions of Uttarakhand. The Changathangi which produces mazority of Indian Pashmina is found at 3000 to 5000 meter above sea level. The area is is extremely arid with very low humidity and rainfall.

The Pashmina produced by Indian goats is either gray or white; the white is preferred over gray due to its luster. Its staple length is short. Average annual yield of Pashmina of the two breeds is low[2] : it varies between 100 and 150 g during 2nd to 4th year of production. Peak yield is usually obtained in 3rd year (Table 21.4).

Table 21.4 Averages for various Pashmina attributes of Changthangi and Chegu breeds.

Trait	Breeds	
	Changthangi	Chegu
Annual yield(g) :		
• 1st year :	67	86
• 2nd year	142	98
• 3rd year	154	114
• 4th year	152	121
Other traits:		
Staple length(mm)	40	55
Fiber diameter (μ)	13	12
Medullation (%)	10	24
Clean yield (%)	64	76

Gaddi breed of hill goats produces hair, which is utilized for preparation of ropes and gunny bags due to its higher tensile strength. They are clipped once a year and produce about

[2]Soviet Don breeds have been claimed to produce an avg yield of 400g

500 g to one kg coarse hair with staple length of 25 cm. The goats of North-western parts of Himalayas, parts of Rajasthan and Gujarat have long hairs which are regularly clipped. Goat skin is used by leather industry. The skin produced by Black Bengal goats is acclaimed internationally for its fine grain quality.

Exotic Breeds of Goats

There are over 300 breeds or types of goat in the world. Description of these breeds based on their physical conformation, body size with certain major production traits has been provided by Mason[3]. Some of the specialized exotic dairy breeds of temperate zone viz. Alpine, Saanen, Anglo-Nubian and Toggenburg, were imported into India and used for crossbreeding to increase milk production. Similarly, Angora which excels in mohair has been imported into India. Lately, Boer breed has also been introduced in India for meat. The origin and characteristics of some important exotic breeds are described below.

Alpine

The home tract of Alpine is France and Swiss Alps region but it has also performed well in warm climates. These hardy, adaptable animals thrive in any climate while maintaining good health and excellent production. French-alpine is also referred to as 'Alpine Dairy Goat'. The registration papers for this dairy breed use both designations, Alpine dairy goat or Alpine, as they are synonymous. Alpine possesses perky ears, slightly dished face and variable colors, such as black, tan white or combination of these. On average, buck weighs 65-80 kg and doe weighs 50-60 kg. Alpine goats are one of the top milk producers alongside Saanen. Alpine is the most common breed of France where dairy goat breeding is most organized, and breed is undergoing systematic genetic improvement with large use of artificial insemination. Its milk yield in 2007 has been reported to be 890 kg in lactation length of 290 days (i.e. average daily yield of over 3 kg). The protein and fat% average to be 3.2 and 3.7 respectively.

Anglo Nubian (Nubian)

The Anglo-Nubian breed originated in England as a cross between the Old English milch goat and the bucks imported from India, Russia, and Egypt. Coat color varies from white to black, red tan or any other combination of these. They have been exported to most countries from England. In United States and Canada, the breed is referred to as simply Nubian.

Its distinguishing characteristics include large, pendulous ears and a 'Roman' nose. Due to its Middle-Eastern background, Anglo-Nubians can thrive in very hot climates and have a longer breeding season than other dairy goats. It is considered a dairy or dual-purpose breed and is known for the high fat percentage, and often referred to as Jersey of goat breeds. The butter fat content varies from 4 to 5 percent and protein percentage is over 3.8. But on the average, the breed produces lesser milk than other dairy breeds. Anglo-Nubians are large; does weigh about 50-60 kg and bucks about 65-80 kg.

Saanen

The breed originated in the Saanen valley in Switzerland (south of Canton Berne). By late 19th century, it spread across Europe, and in the early 20th century to the US. The animals are short, sometimes with a fringe of long hairs along the back and down the hindquarters. Ears are either erect or pointed forward. They are white or cream-colored. The milk yield ranges from 2 to 5 kg per day with 3 to 4 percent butterfat. Saanen's temperament is calm and mild; breeders have often referred to them as living marshmallows as children can handle them.

[3]Mason's World Dictionary of Livestock Breeds

Angora

The home tract of Angora breed is Turkey and Asia minor; the breed is named for 'Ankara' in Turkey. The animals are small with short legs. Back is straight with no slope down to the rump. The body is square and stocky. They possess small head, flat ears that are pointed and pendant. Horns of bucks are spirally twisted. The fleece is beautiful which hangs down loosely from the body in long white spiral. Angora goats produce the lustrous fiber known as mohair (an important textile fiber). A single goat produces between four and five kilograms of hair per year. Angoras are shorn twice a year. Angoras, despite their coats, bear no relation to sheep.

Boer

The Boer breed was developed in South Africa in early 1900s for meat production. Its name is derived from the Dutch word "Boer" meaning farmer. The Boer was probably developed from the indigenous goats of the Namaqua Bushmen and the Fooku tribes with some contribution of Indian and European inheritance being possible. This breed has been selected for meat rather than milk production. Therefore, Boer goat has a fast growth rate and excellent carcass qualities, making it one of the most popular breeds of meat goat in the world.

They have a high resistance to disease and adapt well to hot, dry semi-deserts. Like Nubians, the breed possesses long, pendulous ears. They are noted for being docile, fast growing, and having high fertility rates. Does are reported to have superior mothering ability as compared to other goat breeds. Mature bucks weigh between 110–135 kg, and mature does between 90–100 kg. South Africa established its breed registry in 1959. From 1970 onward, the Boer breed has been incorporated into their National Mutton Sheep and Goat Performance Testing Scheme of South Africa making it the first goat breed involved in meat performance testing.

The Boer breed was imported into India by Maharashtra Goat and Sheep Research and Development Institute in collaboration with Nimbarkar Institute with the aim to establish a nucleus flock of Boer and promote its crossbreeding with local goats. The breed has proved to be a fast growing and hardy for Indian climatic conditions. A large number of goat keepers from Maharashtra and other States such as Tamil Nadu and Madhya Pradesh have benefited by crossbreeding their goats with the Boer. With good feeding, Boer kids attain weights of 25-30 kg in 5-6 months.

Economic Traits : Description, Measurement and Inheritance

Though goats serve triple purposes but commercially the species has acquired the status of meat production with the exception of France and a few European countries. Therefore, most of the important traits from genetic improvement standpoint pertain to growth and carcass.Reproduction is important as it influences net productivity of meat production. Traits related to milk production are also being considered in genetic improvement programs being carried out in India for som breeds.

Growth and Carcass Traits

The traits related to growth and carcass/meat in goat are the same as those of sheep.

(a) *Birth weight:* Birth weight influences viability as well as subsequent growth of kids; therefore, it indirectly influences economic returns.

(b) *Weaning weight:* Goats, which are to be sold for meat, are usually weaned at three months of age when they are individually weighed. Weaning weight varies widely among breeds and production systems. Institutional herds of important breeds of NW region undergoing genetic improvement attain about 10 to 12 kg at weaning. Under semi-extensive management system, they can attain up to average of 16 kg.

Total weight of kids weaned is used as a measure of overall efficiency as it includes contribution of survival, growth as well as mothering ability.

(c) *Post-weaning weights*: Body weights at six and nine months of age are useful measures of growth rate.Weights at six and nine months of age vary among breeds and production systems. Institutional herds of important breeds of NW region undergoing genetic improvement attain higher weights.

(d) *Yearling weight:* Weight at one year of age varies considerably among breeds but Jhakhrana and Jamunapari attain about 26 to 28 kg at yearling age.

(e) *Gain in weight and growth rate*: Pre- and post-weaning gains and growth rates are indicators of growth pattern.

(f) *Dressing percentage:* Dressing percentage at marketing age is important when goat are raised under intensive or semi-intensive system. Information on carcass characteristics including dressing percentage of different Indian breeds is limited. This limited information indicates dressing percentage to range between 39 and 52% in different breeds.

(g) *Meat and carcass quality* traits like carcass weight, intramuscular fat, carcass quality etc are important in specialized markets but have not yet attained significance in Indian market.

Genetic Parameters: Goats are usually selected based either on six- or nine- month body weights.

In a selection program, genetic parameters and the non-genetic factors influencing traits are important considerations:

- Body weight from birth to 12 month of age, in particular early body weights, is affected by birth type (single vs. twins etc) and sex. Therefore, body weights of breeding candidates be adjusted for birth- type and sex using appropriate adjustment factors of that breed.
- Heritability of body weight at various ages follows the general pattern observed for sheep: most body weights have moderate to high heritability, therefore individual selection for their improvement is effective. Body weights, particularly early body weights, are also influenced by maternal genetic effects.
- The pattern of genetic correlations among body weights at different ages is also similar to sheep; high genetic correlations among body weights at various ages, particularly among body weights at closer age groups (i.e. consecutive age groups)

Milk Production Traits

The important economic characteristics of dairy goats are same as that of dairy cattle viz;

(a) *Milk yield:* Milk yield for selection purposes can be recorded for different periods. The most important periods considered in breeding programs in India are: 90 day, 140 day or complete lactation yields.

(b) *Age at first kidding, inter-kidding period, service period, dry period* are other traits that contribute to efficiency of goat reared for milk.

(c) *Milk constituents*: In countries like France and Spain where dairy goat have good presence as dairy animals, milk constituents like yield of fat and protein are important traits, and hence considered in genetic improvement programs. But in India, goat milk is consumed by rural poor; selection is being only directed towards improvement of milk yield only.

Genetic parameters of milk production traits

- Heritability of milk yield in Indian goat is moderate as indigenous breeds have not undergone systematic selection. Heritability of fat percentage is high.
- Genetic correlation of milk yield with fat percentage is negative. Concomitant improvement in milk yield and fat percentage can therefore be effected by selection index or through use of independent culling levels for fat percentage with primary selection directed for milk yield. That these traits can be simultaneously improved despite the existence of undesirable genetic association between these, has been demonstrated through selection experiments conducted for their improvement in France (Fig 21.6)

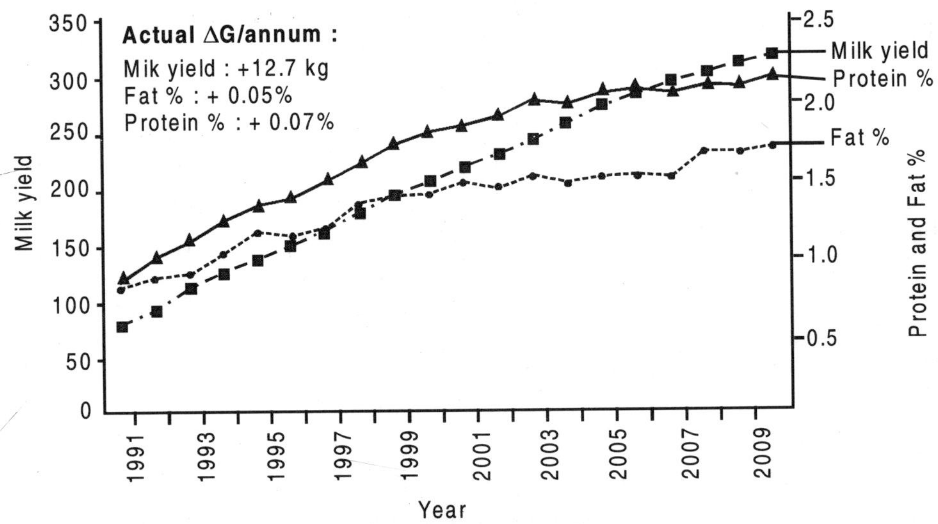

Fig. 21.6 Actual genetic improvement in milk yield, and fat and protein percentage in an Alpine strain (Source : EAAP Meeting 2011).

Reproduction and Survival Traits

(a) *Age at first kidding and inter-kidding interval*, which influence efficiency of milk production are important components of reproduction. These also contribute to efficiency of goat production for meat. Age at first kidding among Indian breeds ranges from 16 to over 25 month. Jhakhrana breed, which is undergoing genetic improvement at CIRG, has first kidding by about 14 months with first kidding interval of eight month.

(b) *Fertility:* Fertility, which is a measure of reproductive success of each female joined, is an 'all or none' trait, and is calculated as whether the animal has conceived or not.

(c) *Tupping percentage:* It is measure of fertility of a group and is expressed as the percentage of doe (she goat) covered in a year.

(d) *Litter size:* Litter size at birth is the number of kids born per doe-kidding. Average litter size in important Indian breeds ranges from 1.2 to 2.1. Among Indian breeds, small sized breeds, in particular.

Black Bengal is characterized by higher litter size.Chinese small and dwarf breeds are better than Indian breeds in terms of litter size. Litter size can also be expressed at weaning age.

(e) *Kidding percentage:* It is a measure of prolificacy in goat as twin- producing doe is more profitable than single producing except where grazing conditions are poor. On an individual-doe basis, it is an 'all or none' category trait viz. single, twins, triplets. On flock basis it is expressed, as the number of kids born per 100 doe kidding. It can be even more than hundred as the occurrence of multiple births (i.e. twinning and triplets) in goats is quite frequent.

(f) *Survivability Viability:* Viability on an individual-basis is an 'all or none' trait. On group basis, it is expressed as survival percentage and can be calculated for different intervals/periods like pre-weaning, post-weaning, yearly etc.

(g) *Gastro intestinal nematode (GIN) resistance:* Goat like sheep in tropics and sub tropics are infected by a number of parasites. Packed cell volume (PCV) and fecal cell count (FEC) are commonly used measures of resistance to GIN. The PCV, which is a measurement of proportion of red cells in the blood, is used to diagnose anemia caused by *Hematophagous* parasites, especially in infection originating from *H. contortus*.

The reproductive and survival including disease resistance traits have low heritabilities.

Genetic Improvement of Goats in India

Like sheep, selection for genetic improvement of goat particularly for meat is easy as growth traits are expressed in both sexes and their heritabilities are moderate to high.

Early Goat Breeding Programs in India

A number of schemes were initiated for the improvement of goats for milk, meat and fiber through state and Central Government sponsored projects. Some of these are listed below:

1. Selection and Development Programs

- A development program on indigenous goat breed Beetal was started at Government Livestock Farm, Hissar (Haryana) to improve milk production through selection. The program was not successful perhaps due to use of sires of low genetic merit.
- Improvement programs on Beetal, Jamunapari, Black Bengal, Sirohi breeds including their crossing with exotic breeds on small-scale were undertaken before the establishment of State agriculture universities One such crossbreeding program was initiated at Agra (at Raja Balwant Singh College) on Jamunapari, Beetal, Barbari and Black Bengal for improvement of growth under intensive management system. Performance of purebreds and their crosses based on small number of goats of each group was studied but the results were variable.

2. Crossing among Native Breeds

In view of the wide range of variability in body weights and reproduction rates among indigenous breeds, crossbreeding among them was attempted for improving meat production. The results of such experiments involving Jamunapari, Beetal, Barbari and Black Bengal showed that:

- 'Beetal x Jamunapari' and 'Barbari x Jamunapari' crosses were inferior to purebred Jamunapari.
- 'Barbari x Beetal' crosses were, however, superior to 'Jamunapari x Beetal'. The magnitude of improvement was 18% in carcass weight, 8% in dressing percentage and 18% in bone percentage over purebred Beetal.
- 'Beetal x Barbari' crosses were superior to 'Jamunapari x Beetal' and 'Barbari x Beetal' crosses. The magnitude of improvement in 'Beetal x Barbari' was 33% in carcass weight 2.5% in dressing percentage and 16% in bone percentage over purebred Barbari.
- 'Beetal x Sirohi' crossbreds were superior to Sirohi. The magnitude of improvement in carcass weight and dressing percentage being 13% and 1.4% respectively over purebred Sirohi.

Results of crosses among several indigenous breeds showed the existence of specific combining ability among various indigenous breeds but Jamunapari, the largest Indian breed, did not "nick" well with any other indigenous breeds.

3. Crossing of Indigenous with Exotic Dairy Breeds

- Saanen, Alpine, Anglo-Nubian and Toggenburg, which are recognized milk breeds of temperate zone, were imported into India for crossing with recognized Indian breeds. Crossbreeding experiments were conducted during 1970's and 80's at some institutional farms (e.g. NDRI, Karnal, Kerala agriculture university etc) for improvement of milk yield under farm conditions. The native breeds used in crossing were Beetal at NDRI and Malabari in Kerala. The Table below lists the performance of purebreds and crosses at NDRI:

Breed or Cross	Lactation yield(kg)	Lactation length(days)	Average daily yield(kg)
Beetal :	164	185	0.89
Sannen × Beetal :	306	240	1.28
Alpine × Beetal :	257	230	1.12
Barbari :	67	129	0.50
Alpine × Barbari :	177	172	1.03

In general, the milk yield of exotic breeds (as purebreds) under Indian conditions declined by 25-50% but lactation length remained the same as in their native countries. The reason for decline was attributed to low genetic merit of the imported germplasm, along with environmental and nutritional stress.

The general findings were that crossbreds (F_1) had significantly higher milk yield over the native breed used in crossing (by 57 to 147 %). Sannen was better in combining ability with native breeds than the Alpine. The performance of F_2 and F_3 born out of inter se mating deteriorated due to loss of heterosis. The crosses above 75% exotic germplasm could not adapt well and had reduced reproduction and viability. Further, the advantage of crossbreds for milk yield was neutralized by their high mortality and low viability.

- To improve the milk production In Indian goats, crossbreeding of Indian breeds with exotic dairy breeds was also started at Makhdoom (under PL-480 scheme).There was no substantial improvement over 50% exotic inheritances.

4. Crossbreeding of Angora with Native Breeds for Fiber

Mohair is a fine and non-medulatted fiber, produced from secondary skin follicles like Pashmina. There is no indigenous mohair producing breed in India. Angora breed of Turkey, which excels in mohair quality, was used for crossbreeding with local breeds, first in the Punjab in 1941, and subsequently in UP, and later in HP and Maharashtra by involving Gaddi and Sangamneri breeds respectively under the AICRP on goats.

- Angora crosses with Gaddi and Sangamneri showed an increase in staple length, and decrease in fiber diameter and medullation percentage.
- The quality of mohair fibers produced in 75% Angora cross was quite close in fineness of Pashmina. However, 87.5% and 75% Angora crosses were just able to produce fiber useable as mohair.
- Angora crosses with Gaddi showed an increase in fiber yield and staple length, and decrease in fiber diameter with increase in level of Angora inheritance. Performance of Angora crosses is shown below:

Trait	7/8 Angora X Sangamneri	7/8 Angora X Gaddi	Angora pure
Yield(kg):			
• 1st clip	0.27	0.45	0.43
• 2nd clip	0.57	0.75	0.82
• 3rd clip	0.87	1.0	1.1
Staple length(mm)	94.1	13.7	103
Mohair/Pashmina diameter (μ)	21.0	24.7	42.2
Medullation (%)	5.28	9.9	3.3
Clean yield (%)	-	-	74.9
Primary follicles(per mm^2)	3.1	-	3.6
Secondary follicles(per mm^2)	19.6	-	28.5

The fiber (mohair) production in Gaddi and Sangamneri crosses with Angora increased in direct proportion to level of exotic inheritance. Further, with 7/8 Angora inheritance the crosses produced a similar quantity of relatively fine quality mohair than that of the Angora controls.

The overall results from crossbreeding of exotic breeds with indigenous breeds conducted on Government farms/institutions showed improvement in crossbreds for milk yield, age at first kidding and body weights. But under village conditions, the crossbreds could not compete with indigenous animals. Wherever attempts have been made to introduce exotic germplasm under village conditions, the level has not exceeded 50%, and crossbreds have generally not survived as well as the natives did.

Box 21.3: Central Institute on Goat Research (CIRG), Makhdoom, UP.

The Indian Council of Agricultural Research (ICAR) established a National Goat Research Centre at Makhdoom (Mathura district of Uttar Pradesh) in 1976. This centre was upgraded to Central Institute for Research on Goats (CIRG) in 1979.The mandate of the institute is to undertake research, training and extension education for improving milk, meat and fiber production of goats including development of goat product processing technology. Research is being undertaken on all aspects of goat production and utilization. The Institute has modern laboratory facilities to conduct research on genetics and breeding, feed resource development & utilization, nutrition, physiology, reproduction, management, disease diagnosis & prevention, transfer of improved technologies, and milk and meat products technology.

The Institute has three goat breeding farms on Jamunapari, Barbari and Jhakhrana breeds and one sheep breeding farm on Muzaffarnagari breed for conducting research on genetics and breeding. The coordinating unit of the All India Coordinated Research Project(AICRP) on goat Improvement which was originally based at Central Sheep and Wool Research Institute, Avikanagar is now located at the Institute. A number of goat breeds located in different regions of the country are being studied under the AICRP. In addition, two breeds of Gujarat state are also being studied under Institute Projects.

Ongoing Goat Breeding Programs in the Country

All India Coordinated Research Project for Genetic Improvement

Systematic genetic improvement of goats was started through the All India Coordinated Research Project (started in IV Five Year Plan, 1970-71) with the main objective to develop new strains involving crossbreeding of native goat breeds with high yielding exotic breeds in order to improve the efficiency of production of milk, meat and fiber. This program was continued up to VIIth Five year Plan. There were three components:

- *Milk component*: Located at three centres viz. NDRI Karnal, Kerala Agricultural University and Assam Agricultural University. Exotic breeds used were Alpine and Saanen
- *Meat component*: Located at two centres viz. CSWRI (Rajasthan) and Rajendra Agricultural University, Ranchi (then Bihar).
- *Pashmina/mohair component*: Located at three centres viz, IVRI Mukteshwar, Leh (J & K) and Mahatma Phule Krishi Vidyapeeth at Rahuri (Maharashtra). Angora was the exotic breed used.

With the experience gained and the information generated from this project, it was considered necessary to improve the native goat breeds, which are well adapted to the local environment and feeding management systems in their natural habitat.

From the IXth Five year plan, therefore the main emphasis has been on improvement of important indigenous breeds of goats through selection. Like that of sheep, the program is based on involving the flocks maintained at the Institutional farms (nucleus herd) and village flocks maintained by the farmers.

The coordinating unit of AICRP is located at Central Institute on Research on Goats (CIRG). The project aims at improving production performance of a number of breeds (13) distributed in different regions of the country under farm and field conditions. Information about different centres is given below:

Breed	Location	Utility
Farm Units :		
Jamunapari	CIRG, Makhdoom, Uttar Pradesh	Milk & Meat
Barbari	CIRG, Makhdoom, Uttar Pradesh	Milk & Meat
Sirohi	CSWRI, Avikanagar, Rajasthan.	Milk & Meat
Field Units :		
Sirohi	RAJUVAS, Udaipur, Rajasthan	Milk & Meat
Surti	Navsari Agricultural University, Gujarat	Milk & Meat
Sangamneri	MPKV, Rahuri, Maharashtra	Milk & Meat
Malabari	Kerala Agriculture University	Milk & Meat
Marwari	RAU, Bikaner, Rajasthan	Meat
Black Bengal	WBUVS & F, Kolkata, West Bengal	Meat
Ganjam	OUA & T, Bhubaneswar, Orissa	Meat

Improvement through Farm-based Breeding Units including Institutional Unit

There are three farm based units under AICRP which involve Jamunapari, Barbari (both at CIRG) and Sirohi (at CSWRI) breeds for improvement of milk and meat, and one unit of Jhakhrana breed, which is based and maintained at CIRG (through institutional funds) for improvement of milk and meat.

One of the important objectives of these units is selection of genetically superior males and females for breed improvement. The improved germplasm is to be stored and supplied to end users (farmers and development agencies). Each unit maintains about 200-300 breedable does and 10 bucks under semi intensive system of management. The salient features of the farm-based units are:

- Selection of males and females is made on the basis of 6 -month body weight in small breeds (Black Bengal) and on 9-month body weight in medium and heavy breeds. For improvement in milk and meat production, sires are selected on the index combining 12-month body weight of bucks and 90 days milk yield of dams.
- Selected superior sires are maintained as elite germplasm and 500 doses of semen of each elite buck are stored in semen bank.
- Superior sires are distributed for improving the production of goats in farmers' flocks.

Selection has yielded positive genetic improvement for body weight at birth, three- month, six- month, nine- month and 12- month with per annum genetic improvement of 0.1, 0.4, 1.0, 2.5 and 2.4 kg respectively. Improvement in 90 days milk yield and body weight at 9 and 12 month of age has also been observed.

Genetic Improvement of Jamunapari : Jamunapari is an important breed of north western region. Its natural habitat is ravines of Jamuna, Chambal and Kawari rivers of Chakarnagar district, Etawah (UP) and adjoining area of Bhind and Morena districts of Madhya Pradesh where browse material is available in abundance. Jamunapari goats were brought to CIRG in 1982 with further introductions made in 1988, 1997 and 2008. Its genetic improvement program has been in operation since 1985. The project aims to improve the breed for milk production and body weight through use of selected bucks and production of elite germplasm for field improvement programs. Goats are being maintained under semi-intensive system of management. Because of selection, genetic improvement has been obtained for milk yield in Jamunapari goat population. Milk yield and lactation length traits of Jamunapari goats (based on records of 17 years and of over 2200 does) were as follow:

Parameter	Milk yield(kg) 90 day	140 day	Lactation	Lactation length (days)
Mean:	80.3	114.0	125.0	179.5
Range:	22 to 168	47 to 234	33 to 274	70 to 277

Under intensive management, kids had average daily weight gain (ADG) of 74, 121 and 112 g/day respectively during 3-6, 6-9 and 9-12 month age groups. Reproductive performance of Jamunapari goats in terms of breeding efficiency and kidding percent on the basis of does tupped were 86.0 and 121.1%, respectively. The kidding rate was 1.46.

Genetic improvement of Barbari: Barbari is a small-sized breed of North western region that is highly suited for rearing under restrained and stall-feeding conditions for both milk and meat, In addition to being a good milker, it is highly prolific.

The breed is being improved for meat and milk. Effective selection differential had been applied for 9- month body weight (about 5.5 kg) and milk yield (about 7.0 liter for dam's 90 days milk yield). These high selection differentials indicate further scope of improvement through selection. Being a small-sized breed, its body weights are significantly lower than of Jamunapari. Its milk yield and reproductive performance is given below (2013-14):

Traits	Mean ± SE
Milk yield (liters):	
• 90 days	57.90 ± 1.04
• 140 days	79.78 ± 1.71
• Lactation yield	68.27 ± 1.41
Lactation length (days)	131.38 ± 1.64
Age at first mating (days)	354.7 ± 6.4
Weight at first mating (kg)	15.01 ± 2.3
Age at first kidding (days)	422.3 ± 5.2
Weight at first kidding(kg)	16.01 ± 2.3
First kidding interval(days)	221.04 ± 7.2
Gestation period (days)	145.4 ± 1.4
Fertility % (does tupped)	90.7
Kidding % (does tupped)	145.3

Genetic Improvement of Jhakhrana: Jhakhrana, an important milch breed, is also used for meat due to its compact and large body size. Being hardy, it can be reared under poor resource conditions. The breed derives its name from the name of village "Jhakhrana" where it is found in most pure form.

A small unit of Jhakhrana goats is maintained at CIRG, Makhdoom (through institutional funding) for genetic improvement of milk and meat production. Selection is based on 90-day milk yield and 9-month body weight. Its milk yield (during 2013-14) has been as follow:

Period:	30 d	60 d	90 d	120 d	150 d
Milk yield:	51.2 ± 9.3	97.2 ± 3.8	141.3 ± 5.1	178.2 ± 6.3	209.2 ± 11.7

Sirohi Improvement: Selection in Sirohi goat since 1976 coupled with improved management practices has improved 12-month weight from 21.5 kg to 26.4 kg. Similarly total milk yield increased from 75 to 108 kg. Bucks have been distributed in Rajasthan as well as all over India incluidng the southern region for breed improvement program due to its sustainability in all type of environment.

Field - based Units

The purpose of field-based units is to assess production performance of goat breeds in farmer's flocks under village management system and to improve the germplasm through selection. Field units are based on the flocks owned by the farmers maintained under village management system in the native home tracts, and include Jamunapari, Marwari, Sirohi, Sangamneri, Surti, Black Bengal, Malabari, Ganjam, Osmanabadi, Gaddi, Assam Hill and Changthangi breeds.. Each unit is to maintain at least four Centres having at least 600 breedable does under the coverage area. The approach is to select the best animals, and to establish germplasm centre to cater the future need of improvement.

Review Questions

1. Write a short note on :
   (a) Early development/selection programs for genetic improvement of goat in India
   (b) Crossbreeding among native goat breeds.
   (c) Crossbreeding of indigenous with exotic dairy goats.
   (d) Crossbreeding of Angora goat with native goat breeds
2. Define/explain the following with suitable examples wherever necessary:
   (a) Angora breed (b) Boer breed
3. Differentiate between the following:
   (a) Meat vs. dairy goat breeds of India
   (b) Indigenous vs. exotic dairy goat breeds
   (c) Angora and Boer breeds of goat
4. Why majority of goat population is found in Rajsthan, UP, MP and Bihar?
5. What are the two main economic purposes for which goats have been raised in the country?
6. Is there a clear cut demarcation of habitat of Indigenous goat breeds? Explain.
7. What are the different criteria used for classification of defined goat breeds of India? Which criterion is most commonly used? Explain.
8. Why is Barbari breed of goat facing genetic dilution?
9. Why is Beetal considered an improver breed for medium- and small-sized goat breeds?
10. Which goat breed of Southern region is well adapted to distant walking
11. Which goat breed of India is well suited for stall-feeding?
12. Which breed of Indian goat is most prolific?
13. List the Indian breeds of goats classified by product utility.
14. List the Indian breeds of goats classified by body size.

15. Is there any goat breed in India, which is exclusively reared for milk? Explain.
16. Name two to three breeds of Indian goats, which are better than others for milk production.
17. What is the definition of a meat breed of goat in India?
18. Which is the lightest goat breed of north western region?
19. Which native breed of goat is known for its finest Pashmina? What is its habitat?
20. Which Indian breed of goat produces the finest skin?
21. Which goat breed is referred to as Jersey of goat?
22. Which exotic goat breed is known for its calm and mild temperament?
23. Which exotic breed of goat has recently been introduced into India for meat?
24. (a) List the two most important exotic breeds of goats that have been imported into India for improvement
    (b) What was the economic purpose or main attributes of these breeds?
    (c) Which of these breeds has given better performance in crosses with indigenous breeds?
25. At what age are indigenous goat weaned?
26. List the range of the following traits of Indian improver breeds of goats :
    (a) birth weight (b) weaning weight
    (c) nine-month weight (d) lactation length and milk yield
27. List two economic traits of goats which can be measured on group-basis.
28. What were the salient outcomes of the exploratory crossbreeding program on goats conducted in India? Why is crossbreeding of indigenous with exotic breeds no more a national breeding policy?
29. What are the objectives of the AICRP/Net work project on goat? Which breeds are included in this project? Name the different centers along with the breeds included in each.
30. Name the indigenous breed of goat that is known for its high prolificacy.
31. List two exotic breeds of goats which were used in crossbreeding program in India for
    (a) meat (b) fiber

●●●

22

Swine Breeding

Phylum: Chordata
Class: Mammalia
Order: Artiodactyla
Family: Suidae
Genus: *Sus*
Species: *S. scrofa*
Subspecies: *Sus scrofa domesticus*
Synonmous: *Sus domesticus*

The present day pigs have been evolved through gradual domestication of wild pigs. Genetic and archaeological findings suggest that pig domestication began about 10000 to 12000 BC at multiple sites across Eurasia, followed by their subsequent world-wide spread. Development of local types throughout centuries led to foundation, mostly during the nineteenth century, of current modern breeds with defined phenotypes and production abilities.

Pig, commonly referred in India as 'mortgage cleaner', is one of the most efficient feed converting animal species. It's high prolificacy and short generation interval as compared to other litter-bearing animals gives it a biological advantage for meat production. Pigs in India have traditionally been raised and managed by lower economic strata of the society under extremely low input system. Originally, pigs were raised for bristle production required for brush industry. However, after the discovery of nylon the main objective of pig rising has shifted to meat production.

Most of the pigs in India are concentrated in Eastern and North Eastern states of the country. In developed countries like Europe, America, Canada etc the pigs were developed and raised for meat. Therefore, the recognized breeds of pigs are found in these countries. There are more than 60 recognized breeds of pigs in the world.

Box 22.1: Indian wild pigs.

Indian wild boar (*Sus scrofa aristans*) is found in Himalayan forests up to a height of 4500 m. The adult animals are about 1.5 m long and 86 cm wide at shoulders and weigh over 130 kg. They have long snout, short ribs and long legs. The body color is rusty in young but with advancing age, the color changes to dark chestnut brown. They are extremely active, highly prolific and reproduce throughout the year.

Wild boars are also found in the forests of the Andaman Islands (*Sus scrofa* and *amanensis*) and in wet forests of Sikkim, Nepal, Bhutan and Eastern region (*Sus salvanilis*). They are nocturnal in habits and prefer to remain in high grasses and, therefore, rarely seen. They measure about 66 cm in length and are brownish black in color.

Germplasm Resources of Pig

Indigenous Pigs

The indigenous pigs called '*desi pigs*' have evolved from Indian wild pig. Their body color varies from black brown to rusty grey or an admixture of two colors depending on topography and climatic conditions. They have thick hair. The head and shoulders are heavier than hindquarters, the face is tapering and the ears are small and medium-sized.

Since the indigenous pigs were originally reared for bristle production, their level of productivity is very low: average litter size at birth being 5.2 and at weaning being 3.0; birth weight varies between 300 and 700 g. Adult pigs weigh around 170 kg. The indigenous pigs have early age at maturity and at first farrowing, and shorter farrowing interval.

Exotic Breeds of Pigs

Originally, two types of pig breeds were evolved in developed countries: the American breeds were all of lard-type while British and Scandinavian breeds were of bacon-type. After world war second, the objective of pig breeding switched from lard to meaty (lean) types, particularly in the USA. During the past few decades, a number of exotic breeds of pigs have been introduced into India to improve local populations:

Berkshire: It is one of the oldest improved English breed, and is typically a pork breed of medium size. Animals are black in color and have a characteristic short up turned nose. The face is dished and ears are erect but slightly inclined forward. Pigs of this breed grow to market size in about 6 months. Mature boars weigh 272-385 kg and sows weigh 204-294 kg. Average litter size at birth and weaning are 9.9 and 8.4 respectively. The breed has been found suitable in Indian conditions. It is known for its muscle quality.

Large White Yorkshire: This breed is native to UK and is of bacon-type. The skin is pink in color, free from wrinkles and is covered with long and moderately fine coat. Ears are thin and long. Mature boars weigh 300 - 400 kg and sows 240 - 320 kg. Body colour is solid white with occasional black pigmented spots with erect ears, snout of medium length and dished face. It is highly prolific with average litter size of 10.5 at birth. Sows have good milking and mothering ability traits. Carcass, growth rate and feed conversion efficiency are good.

Middle White Yorkshire: This breed has resulted from crossing between Large- and Small-Yorkshire breeds of England. It grows rapidly and has good dressing percentage but it is not as prolific as Large White Yorkshire. Males weigh 250 - 340 kg while females weigh about 180 - 270 kg

Landrace: It is a breed of Northern Europe and is a bacon type. It color is white with black skin spots, long body, large drooping ears and long snout. The breed is noted for prolific breeding and for efficiency of feed utilization. Carcass quality is close to Yorkshire. Mature boars weigh 270 - 360 kg and mature sows weigh 200 - 320 kg. It is a maternal breed that excels in litter size and milk production.

Hampshire: The Hampshire breed is possibly the oldest American breed. The breed was developed in USA from pigs which originated in southern Scotland and Northern England. It is black with a white belt around the chest. Body is elongated and well muscled.Under well-managed conditions, sows produce up to two litters per year with average litter size of eight at birth. Sows are very docile. Mature boars weigh 150 - 250 kg and sows weigh about 90 - 200 kg. The breed has adapted well under Indian conditions. It excels in leanness and muscling, and is often used as terminal sire breed.

Duroc: It is an older breed of American domestic pig that forms the basis for many mixed-breed commercial hogs. Duroc pigs are red, large-framed, medium length, and muscular, with partially drooping ears. They tend to be one of the least aggressive of all the swine breeds. They also have an excellent rate of gain. Originally, Duroc was very large pig but today it is a medium-sized breed. The color is often an orangish-brown but any color is allowed from light golden shade to a deep mahogany red. In US, this breed is used as sire line.

Other breeds used in India include Tamworth, Large Black etc.

Economic Traits : Description, Measurement and Inheritance

Pigs are reared for meat; therefore, the main economic traits pertain to growth and carcass. However, reproductive traits are equally important as they determine economic efficiency of pig production.

Growth and Feed Efficiency Traits

- **Litter weight:** The weight of all the piglets (of a litter) born alive within the first 24 hours to a sow, called litter weight, is one of the important traits. It is expressed both at birth, and at weaning, which in India is done at about 8 weeks of age. Litter weight at weaning is a measure of net merit for pre-weaning performance as it includes the contribution of traits like fertility, milking and mothering ability of sow as well as growth rate of pigs. Weight of pigs can also be expressed on per-pig basis which is useful for selection purposes.
- **Market weight:** Weight at the time of marketing is one of the most important traits. In India, pigs are marketed at around 5 - 6 months of age when they usually weigh more than 60 kg. Marketing can be done up to 9 months of age but their feed efficiency declines after 6 months of age.
- **Age for slaughter weight (days to reach market weight)**: It is the age of pigs in days when it reaches slaughter weight which in North America is about 100 to 113 kg. In USA, breeds like Yorkshire, Duroc, Hampshire, and Landrace etc take about 175 to 185 days to attain market weight of about 113 kg. Selection is usually practiced for age at marketing rather than weight *per se.*
- **Average daily gain in weight (ADG):** Average daily gain in weight (post-weaning) is a good measure of post-weaning growth rate.
- **Feed efficiency**: It is the kg of feed required to gain one kg of body weight. Lower the requirement of feed for a kg of gain, higher will be the feed efficiency, which helps in economic production of pork. Feed efficiency is seldom evaluated on an individual-pig basis as it is costly.

Body Composition and Carcass Quality Traits

- **Back fat thickness (BPT)**: It is the thickness of fat layer measured on live animals in the back region, which is after the last rib. Generally, the desirable value for BFT is less than 2.5 cm. This trait is used for improving leanness of pork in pigs : lesser the BFT, leaner the pork. It is measured (in cm or mm) by probing a scaled knife, until it encounters resistance (of the muscle).In USA the commonly used breeds currently have about 1.8 cm back fat thickness.
- **Loin eye area**: It is the measurement of attachment of *longissimus dorsi* muscle in the sternum just after last rib and is measured in sq.cm. The impression of this attachment is taken on a paper and the area is measured by planimeter. Ultrasonic devices are also now used for its measurement. This trait is used for selection of pigs for leanness, similar to that of back fat thickness. Loin eye area in US breeds is about 43 to 45 cm^2.

The percentage of different cuts such as ham (thigh), butt (shoulder) bacon (belly) etc is important in some markets. These traits which are measured as percentage of dressed carcass weight.

Reproduction, Survival and Mothering Ability

- **Litter size**: Litter size is one of the important traits contributing to overall profitability. It is measured/expressed at birth as the number of piglets born alive. It is also expressed at weaning as the number of piglets surviving to weaning age, which is generally 8 weeks in India. However, in developed countries it much earlier, at about 21 days. The reason for weaning at 21 days is that this age best reflects the females' milking potential. In western countries, litters are standardized within 24 to 36 hours of birth so that the 21-day litter size reflects pig survival and sow lactation yield.
- **Other traits :** This category includes number of farrowings/litters per year, viability of piglets, and mothering ability, which includes not only the milk yield of the sow, but its ability to take care of piglets. In western countries, a trait determining bio economic efficiency of sows is 'sow productivity index', which combines the number of pigs born alive, number weaned, and litter weight.

Genetic Parameters

Extensive information on genetic parameters of different traits of swine is available from North American and European studies where swine are reared on commercial lines. Like other litter bearing and meat-producing species the traits related to growth (e.g. gain in weight and body weights) are moderately heritable, those related to reproduction (litter size, litter weight etc) are lowly heritable while carcass and body composition traits have high heritabilities.

Genetic correlation between growth rate and back fat thickness is unfavorable, low to moderate (ranging from 0.14 to 0.42). Lean meat content is also unfavorably correlated with growth rate. Similarly days to reach slaughter age has unfavorable genetic correlation with back fat thickness. Simultaneous genetic improvement in leanness and growth rate is difficult because the genetic correlation between these traits is unfavorable.

Lean growth rate is a newly derived trait, which combines leanness (as measured by BFT and loin eye area) and growth rate (as measured body weight gain) into a single trait. Selection for this trait has been found to be successful in simultaneously improving both weight gain and leanness. Genetic correlations between growth and feed efficiency are favorable and generally high.

Genetic Improvement of Pigs : General

In Europe and North America where pigs are grown under organized and high-input systems, breeding is carried as per standard selection and crossing schemes. High reproductive ability coupled with short generation interval of the species has facilitated fast genetic improvement for a number of traits determining efficiency of pig production.

Selection

The important traits included in selection programs have been fast growth rate, increased litter size, reduced back fat thickness, desirable feed efficiency and carcass composition and quality. And these traits continue to be improved through selection though the relative emphasis on different traits has been changing as per market demands.

Growth has most commonly been expressed as days to attain a specific weight, usually 100 to 105 kg, which is called **'market weight'.** The alternative measure of growth has been average daily gain (ADG) from 40 to 105 kg. Back fat thickness is typically measured ultrasonically at the 10th or last rib. Feed efficiency has improved as a correlated response to selection for growth rate.

Effectiveness of selection for improvement of these traits has been well documented through controlled selection experiments conducted for a specific trait at universities or research stations. The organizations involved in genetic improvement of pigs for supplying to commercial producers use multi-trait selection the development of modern genetic methodologies (mainly the BLUP), it has become possible to evaluate the effectiveness of selection in registered herds of different breeds in terms of direction and amount of genetic changes. Figure 22.1 shows the results of genetic improvement for some important traits in some commonly used breeds in US registered herds (Source: Chen *et al* 2002, 03: J. Animal Science).

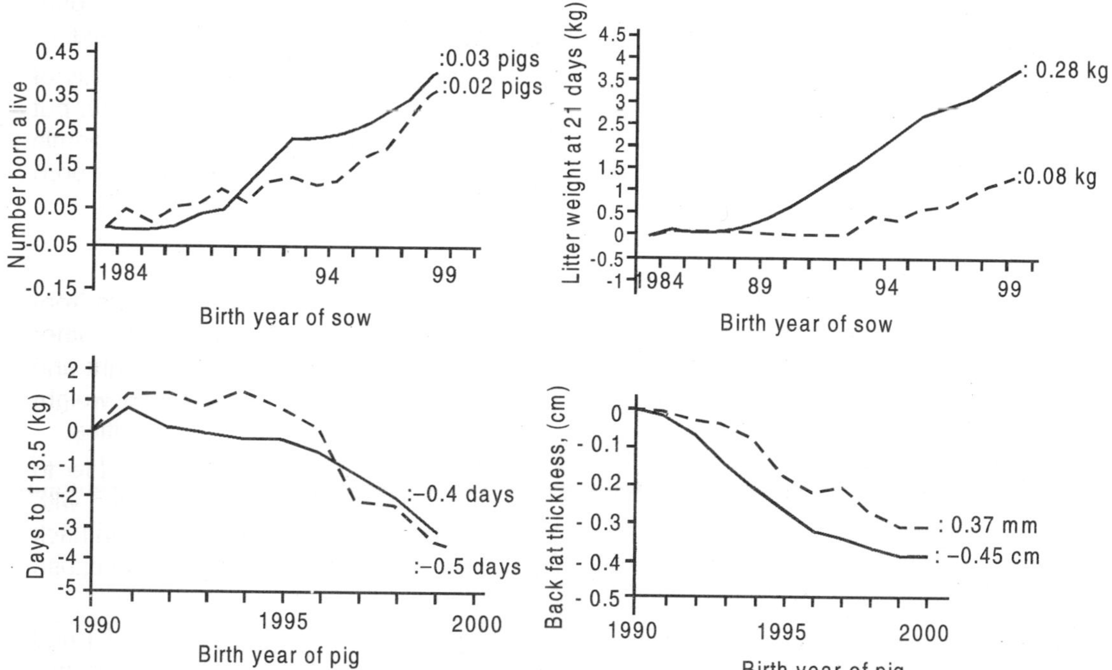

Fig. 22.1. Genetic changes (breeding values) in some traits of US pig breeds (— Yokshire, and .... Landrace). Annual genetic changes are listed along the graph lines.

The results of these selection efforts show substantial genetic improvement in growth rate(measured as days to reach market weight), carcass quality (expressed as reduced back fat thickness, increased loin eye area) and lean growth rate. At the same time, reproductive and sow productivity traits like number of pigs born alive (litter size at birth), numbers weaned (litter size at weaning) and litter weight at 21 days have been improved. Similar evidence from European countries is also available.

Box 22.2: National swine genetic evaluation and improvement programs in western countries.

Important swine breeding countries have organized some forms of collaborative genetic evaluation and improvement programs for achieving a competitive edge. One such is the National Swine Improvement Federation (NSIF) of the US, which is a quasi-governmental agency that maintains a set of guidelines for uniform swine improvement. The federation is composed of Federal and State extension geneticists, and private and corporate breeders.

Recently, public and private geneticists from Canada have also joined the federation. The guidelines cover a wide range of topics from what adjustments should be made to phenotypic measurements, what rations to feed and suggestions on methods of indexing.

Crossbreeding

Crossbreeding is widely used in swine as many important traits are influenced by non-additive genetic effects, and heterosis for overall production is substantial. In fact, nearly all commercial hogs in western countries (like the US) are crossbred. All the three types of heterosis (viz. individual, maternal and paternal) are important, therefore, the performance traits which determine efficiency of pig production are grouped into two categories:

- Production traits such as growth rate, feed efficiency, back fat thickness and carcass composition are expressed in market animals as well as animals, which are to be used in the breeding herd.
- Reproductive and sow productivity traits such as age at puberty, fertility, litter size, litter weight at weaning are expressed only in breeding females.

In order to meet the twin-objective of improvement of production (market traits) and reproduction traits, **specialized sire and dam breeds** (lines) have been identified which are being further improved through selection for the respective category of traits. The dam breeds/lines are improved for high reproductive performance while keeping desirable level of weight gain, feed conversion etc. Maternal breeds, which are typically white and excel in sow productivity traits like litter size, fertility, milk production include Yorkshire, Landrace, Chester White. The sire breeds/lines are primarily improved for gain in weight, feed efficiency and carcass and body composition traits. In US, sire breeds are typically colored and excel in leanness and muscling. The most popular are Duroc and Hampshire.

Breed complementarity of different breeds used in producing crosses has also been assessed for their appropriate use in crossing programs. To capitalize maternal and paternal heterosis, different types of crossing schemes like rotational crossing (two- or three- breed), terminal sire crossing etc are widely used.

Genetic Improvement Status in Western Countries

Estimates of rate of genetic improvement from different countries are available.

Europe: *G*enetic changes in the European pig during the last century have been evaluated. Reported below is the Dutch data for changes in performance for the main traits (based on central testing station's performance) for Dutch Landrace and Yorkshire breeds (Source: Merk, 2000, British Society of Animal Production):

Year	Weight gain (g/day)		Feed efficiency (kg of feed per kg of gain)		Back fat thickness (mm)	
	Landrace	Yorkshire	Landrace	Yorkshire	Landrace	Yorkshire
1930	500	550	3.5	3.4	45	48
1947	650	650	3.4	3.2	33	35
1972	788	788	2.6	2.5	26	22
1990	840	840	2.8	2.7	24	27

These changes are phenotypic, attributable to both improved genetic make-up and management but the traits being moderate to highly heritable; a large fraction of the change particularly during the last two periods is due to genetic improvement. From 1990-2000, most European pig breeding programs have made annual genetic progress for daily gain of +20 g/day, lean meat percentage of +0.5% and litter size of +0.2 piglet/litter. In commercial breeding programs, there are no signs of genetic limits to selection.

Canada: Canada is one of the main exporters of pork, with a well-organized industry. Canadian Center for Swine Improvement (CCSI) is a national non-profit corporation created by the Canadian swine industry to provide leadership, coordination and services for swine genetic improvement. Its activities include national standards for data recording, genetic evaluation, coordination of the national swine improvement program etc. A number of agencies (like swine breeders associations, swine testing stations, universities, agriculture Canada etc) are associated with the genetic improvement work. Listed below are average genetic changes per-year for some key traits evaluated during the period of 2006-2012:

- Days to market weight: – 1.0 days
- Feed conversion ratio: – 0.014 kg feed/kg gain
- Lean yield: + 0.04 %
- Loin eye area: + 0.21 cm^2
- Back fat thickness : – 0.06 mm
- Loin muscle depth: + 0.24 mm
- Litter size at birth : + 0.14 piglets/litter
- Piglet perinatal survival: + 0.36 (%/litter)

US: Primary selection for the past nearly 60 years in USA has been devoted to faster growth and leaner pigs. Such selection has had a clear outcome:

- Back fat thickness has been reduced from 45 mm to closer to 15 -16 mm, with growth rates increasing in some cases by more than 50% faster growth while increasing markets weights from 80 kg to 125 kg.
- Feed conversion has nearly halved with the best commercial herds having efficiency ratios of 2.5 to 2.6 kg feed consumed per kg weight gain over lifetime of the pig.

- Genetic improvement has continued even in recent years with reductions in age to market weight at the rate of – 0.4 days yr.

The level of back fatthickness in US, as believed by industry analysts, has now reached a minimum. Interms of numbers born alive (litter size at birth), genetic rate of improvement of +0.2 pigs per-litter per-generation appears to be a norm.

Genetic Improvement of Pigs in India

In India, pig keeping has traditionally been in the hands of pig keepers, belonging to lowest socio-economic stratum with no means to undertake intensive farming on scientific lines. Pigs contribute about 8% to the total meat produced in the country. The various programs which have been used for improvements include establishment of pig breeding stations and AICRP.

Establishment of Pig Breeding Stations

Government of India initiated a piggery development scheme in 1959-60. The scheme involved establishment of pig breeding stations for production of imported exotic stocks to form a base for distribution and development work with the section of population who traditionally raise pigs. The scheme aimed at supplying breeding stock and other inputs for pig development. The major thrust was for setting up bacon factories and pork processing plants for providing a ready and remunerative market for pigs.

All India Coordinated Research Project

During 1970's pig breeding research was taken up in a systematic manner and an All India Coordinated Research Project (AICRP) on pigs was initiated (1970-71) with the main objective of studying performance of purebred exotic pigs under existing managemental conditions at the following places: IVRI (Izatnagar), Jabalpur, Andhra Pradesh and Assam. Of its various objectives, the breeding component was to assess various economically important traits of existing exotic breeds of pigs in India with respect to production, reproduction and efficiency of feed utilization.

The broad achievements of the AICRP have been:

- Exotic pig breeds viz. Landrace, Large White Yorkshire and Hampshire were successfully raised and multiplied under organized farm conditions.
- All groups of crossbred had higher litter-size and weight, growth rate and better feed efficiency than indigenous pig. Large White Yorkshire crossbred (75%) and Hampshire crossbred (75%) had higher value for litter traits than their respective 50% crossbreds. The crossbreds had lower back fat thickness and higher lean cuts in carcasses than indigenous pigs.
- Genetic improvement of indigenous pig through pure-breed selection program was conducted at various centres of AICRP under different agro-climatic conditions. Litter size at birth and weaning showed continuous improvement over years. Similarly, the growth rate and body weight at 32 weeks was also increased significantly. However, genetic improvement of indigenous pigs through pure-breed selection program has been slow.

Despite the efforts of supplying superior crossbred piglets to pig keepers, the program could not succeed as the crossbreds did not perform well because traditional keepers reared them under very low input conditions. Therefore, AICRP on pig was remodeled to undertake research first on indigenous pig and then subsequently on crossbreeding between indigenous female with appropriate exotic breed.

Ghungroo : A Strain of Indigenous Pigs

Ghungroo, an indigenous strain of pig first reported from North Bengal, is popular among local people because of high prolificacy and ability to sustain in low input system. This strain produces high quality pork utilizing agricultural byproducts and kitchen wastes. Ghungroo are mostly black colored with typical 'Bull dog' face appearance, with a litter size of 6-12 piglets, individual piglet weighing about 1.0 kg at birth and 7.0–10.0 kg at weaning. Both sexes are quite docile and easy to handle. In its breeding tract, these pigs are maintained under scavenging system and mainly act as insurance to the rain fed agriculture.

At National Research Center on Pig (Guwahati), Ghungroo pigs are being maintained under intensive system of rearing with standard breeding, feeding and managemental system. Their evaluation for genetic potential for use in future breeding programs is in progress, and the strain is performing well in terms of productive and reproductive efficiency. Some of the selected sows have delivered litter size of 17 piglets at birth as compared to other indigenous strains of pigs maintained at the Institute farm.

Box 22.3: National Research Center on Pigs.

National Research Center (NRC) on Pig was established at Guwahati (Assam) towards the fag end of the IX five year plan by merging all existing AICRP centres. Its mission is performance appraisal and genetic cataloguing of indigenous pigs, development of improved pig variety together with production, health, product processing and pig based integrated farming system technologies to facilitate pig keepers of the country achieving household food, nutritional and economic security.

Swine Breeding Companies

In western countries like North America, originally small breeders supplied stocks to pig growers. The advent of corn breeding companies and the successful development of inbred lines and crosses paved the way for the development of breeding companies who began to control swine breeding.

In the beginning, a number of pig breeding and family seed-stock companies existed. Market pressures, better business approaches and changes in marketplace have reduced the number of individual family breeders to a very small share of the marketplace. It is estimated that about one-third of the stock purchased by commercial producers in these countries comes from one of the several private breeding companies and the remaining from traditional purebred producers. Currently a few companies (less than ten) actively compete worldwide. In US, the top two companies make up 65 to 70% of market share.

These companies follow a number of approaches for supplying breeding stock. Many of the companies have some lines or strains that are unique to that company. Several also sell stocks that are crosses of conventional breeds. Most of the organizations have extensive performance testing programs to meet needs of commercial producers.

The goal of genetic evaluation programs of these organizations is not simply selling boars and gilts. Rather they target to sell a 'program of organized crossbreeding' with them supplying all the needs for breeding stock. The program may be multi-breed terminal crossing system with crossbred boars and gilts supplied by the company. In other cases, it may be a rotation of several breeds of boars provided by the company.

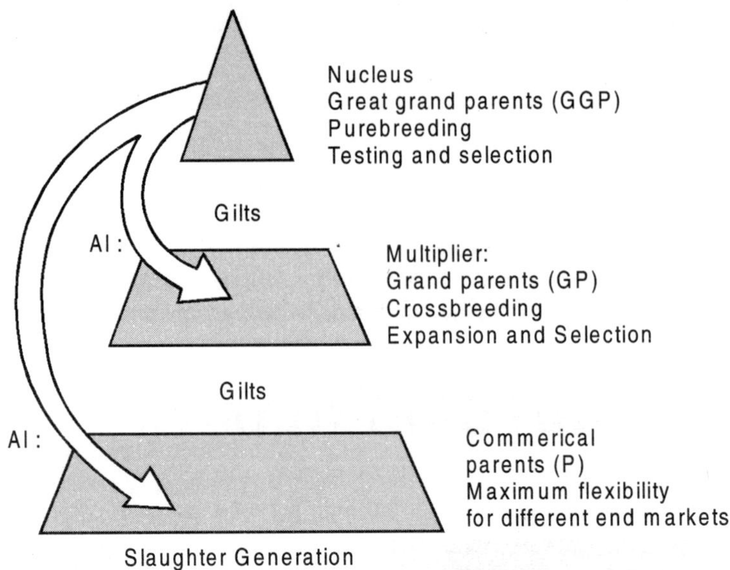

Fig. 22.2. Pyramid structure of swine breeding industry.

These organizations have a high level of integration, from the Grandparent seed stock sources through to processing sector (Fig 22.2). The forms of integration vary from complete ownership by a single company (such as Tyson's in the US) to partnerships among "family farms", to contractual arrangements between a company that owns the sows and the processing plant and small growers that contract feed pigs in the nursery or grow-finish units.

Another important change in production practice of companies is the utilization of A. I. with fresh semen. Recent estimates are that over 25% of all market pigs in North America are derived from A.I. matings and the trend is still increasing.

The Nucleus-Multiplier structure of the pig industry is now well defined. With consolidation of the industry, there is a more distinct use of genetic lines/breeds. For example in the US, the maternal crosses are typically Yorkshire-Landrace F_1s when total confinement production is used. Yorkshire and Hampshire cross females are popular among open-lot producers. Sire lines are of Hampshire and Duroc origin.

Review Questions

1. Define/explain the following with suitable examples wherever necessary:
   (a) Ghungroo strain (b) AICRP on pigs
   (c) Pig breeding stations
2. Differentiate between the following:
   (a) Indigenous and exotic pigs (b) Sire and dam breeds
3. How are back fat and loin eye area in pigs measured?
4. List the economic traits of pigs.

●●●

SECTION - 7

BIOTECHNOLOGY

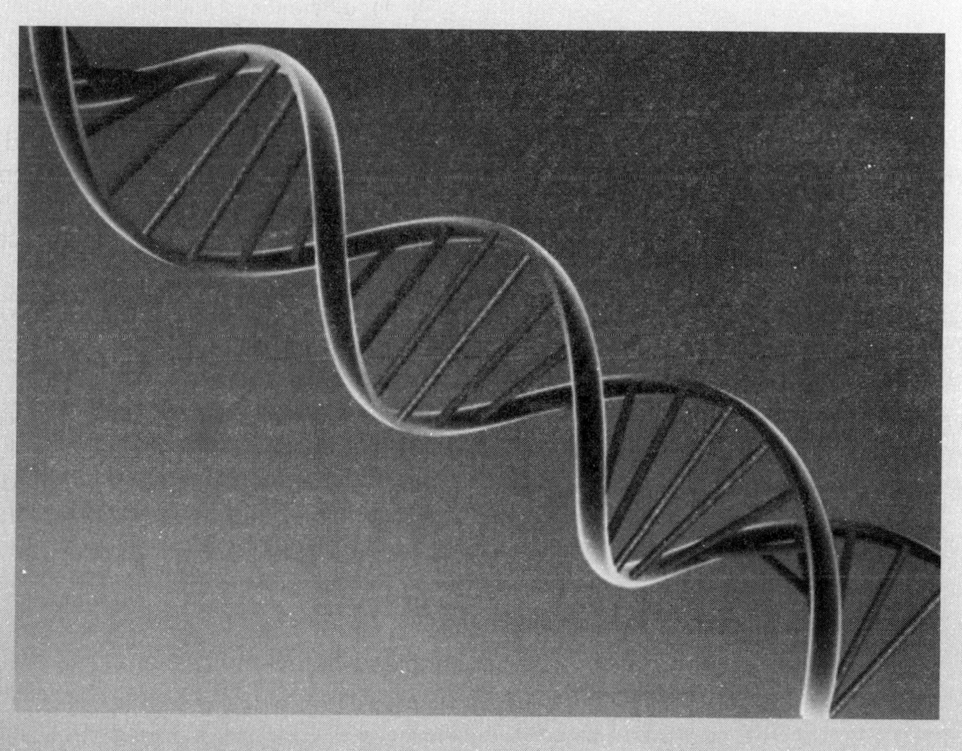

23

Biotechnology in Animal Breeding

In 1919, Karl Ereky, a Hungarian engineer, coined the term 'biotechnology' and described it as the process by which products could be synthesized from raw materials with the aid of living organisms. **Biotechnology** is broadly defined, as per FAO, as any technology (technological application) that involves the use of biological systems, living organism or their products as a way to benefit humans. In this broad context, biotechnology is not a new concept. Man has used it for thousands of years to manufacture products such as beer, wine and bread. The so called' 'traditional' plant and animal breeding that involves selection and mating of desired individuals is a good example of age-old application of biotechnology. However, breakthroughs in molecular biology and genetics since 1970's such as recombinant DNA and embryo manipulation technologies etc have provided new and improved ways to make use of the living organism for the benefit of humans. The term biotechnology has, therefore, become associated with the use of all the newer biological techniques like zygote, embryo and DNA manipulation, for the benefit of humans.

Animal breeding involves directed genetic improvement of animal populations for traits of economic importance. Selection of genetically superior individuals to become parents of next generation is one of the most important components of genetic improvement. Genetic gain attributable to selection for quantitative traits is dependent on four factors:

$$\text{Genetic gain/year} = \frac{\text{Accuracy of selection} \times \text{Selection intensity} \times \text{Genetic standard deviation}}{\text{Generation interval}}$$

From animal breeding viewpoint, therefore, any biotechnology technique that can enhance rate of genetic progress through change in any of the components of genetic gain shall be of use in genetic improvement. Broadly, the following two categories of biotechnological interventions are important in enhancing the genetic progress:

- Reproductive technologies
- Molecular genetic technologies

Reproductive Technologies

The process of reproduction serves the mechanism of transmission of genes from one generation to the next. Reproductive physiology and animal breeding are interlinked. Genetically superior animals can not contribute their superiority to next generation unless their reproductive capacity is satisfactory. Therefore, maintaining high reproductive rate is critically important in achieving high rate of genetic improvement through selection as:

- Reproductive rate determines the percentage of animals, which have to be retained to keep constant herd (flock) size, therefore, intensity of selection is directly dependent on reproductive rate.
- Accuracy of selection is also directly related to the number of offspring produced per parent (family size).
- Low reproductive rate also prolongs the generation interval.

Reproductive rate varies considerably among various species of domestic animals. Typical figures for the different components of the reproductive cycle and reproductive rate are given in Table 23.1.

Table 23.1 Typical values for different components of reproduction in farm animals.

Species	Age at first breeding (months)	Gestation length (d)	Ovulation rate	Offspring of per gestation	No. of offspring per year
Buffalo	24–30	315	1	1	1
Cattle (*B. taurus*))	13–16	283	1	1	1
Horse	24–36	336	1	1	1
Sheep	16–20	150	1-3	1-3	1-2
Goat	10–12	150	1	1	1
Swine	7–9	130	12-30	6-15	12-30
Chicken	5–6	22	1	—	150-200*

* Using artificial incubation

With the exception of chicken, and to some extent swine, reproductive rate of most species of livestock is low. Primarily because of the biological advantage of the high reproductive rate of poultry and pigs compared to ruminants, higher rate of genetic changes are possible in these species. In species like dairy cattle and buffaloes, the main traits of economic importance are sex-limited and measured late in life. These coupled with low reproductive rate puts dairy cattle and buffaloes at a disadvantage for genetic improvement. The enhancement of reproductive rate of ruminants through various reproductive technologies is expected to yield substantial benefits not only to animal breeder but to livestock producers as well.

Several reproductive technologies can enhance the rate the genetic gain. The commonest of these are:

- Artificial insemination (AI)
- Embryo transfer (ET)
- Semen and embryo sexing
- Embryo cloning

Artificial Insemination

Artificial insemination (AI) is a reproductive technology in which semen is collected from the male and is introduced into the genital tract of the female which is in heat. It consists of a number of steps viz; semen collection and its evaluation for quality; dilution of semen to increase its volume and to preserve it; semen freezing and finally delivery of the semen to the receptive female's genital tract.

Benefits: Artificial insemination has been highly valuable in enhancing the reproductive rate of males in low-prolific species of livestock. From animal breeding standpoint, it offers the following benefits:

(i) *Increased rate of genetic gain:* AI enhances genetic gain per annum through the sire path by:

- Increasing the accuracy of male selection through production of its large number of daughters and sisters across many herds (environments).
- Increasing the intensity of selection among males as a very small number of top sires can be used to serve large population of females.
- Decreasing the male generation interval by producing the required number of progeny in a shorter time compared to natural mating.

The combination of these three components of genetic gain arising from AI can lead to four-fold increase in rate of genetic improvement in dairy cattle relative to that from natural mating.

(ii) *Rapid dissemination of genetic improvement:* Artificial insemination permits extensive use of proven sires of high genetic merit across herds, regions or countries. It is thus most effective method of dissemination of genetic improvement from elite sector to the lower strata of breeding pyramid.

(iii) *Establishment of genetic links between herds*: Through AI, progeny of a sire can be produced in many different herds/flocks. This in turn makes a **'genetic connectedness'** between different herds. Such genetic connectedness makes the application of new large-scale genetic evaluation procedures, like BLUP, possible.

Constraints: The greatest genetic problem associated with AI is the increased relationship among individuals of a population leading to increased inbreeding. This can partially be overcome by exchange of semen between the cooperative herds or breeders.

Artificial insemination has now been a practical technology for over 60 years. Its use has been most widespread in dairy cattle. Perhaps no other technology in food production, except hybrid varieties and use of fertilizer, has been so widely adopted. The rapid uptake of AI in dairy cattle is due to combination of genetic, technical and economic factors. AI has made possible large scale progeny testing of bulls for identifying bulls of high genetic merit (breeding value), and also facilitated their widespread use. On the technical side, the development of semen extenders, straws, liquid nitrogen freezing have made AI successful. AI coupled with semen storage enables a single bull to be used simultaneously in several countries for upto 1,00,000 inseminations in a year.

The advantages of AI have led to the replacement of natural service with AI throuought the commercial dairy cattle populations of developed countries. However, in beef populations its usage is much less for a number of factors which include:

- Difficulty in detecting heat in large-sized beef herds kept on ranch.

- Beef production traits can be measured in both sexes, progeny testing of sires thus offers less advantage than in dairy; and
- High cost of AI as proportion of output per beef cow.

In sheep and goats, the unavailability of a simple, non-surgical insemination procedure has prevented AI's extensive use. However, technical success of laparoscopic intrauterine insemination is likely to lead to development of less invasive transcervical procedures. In breeding of turkeys and meat-type chicken, AI is perhaps the rule.

In developing countries, however, the use of AI is not widespread for a number of reasons. AI has high cost as a proportion of output value per cow which is low (1000 to 2000 liter of milk). The cost of AI is further increased by a combination of environmental and nutritional stresses together with less efficient transport. The usage of AI in developing countries will, however, be driven by economic factors as it is an established technology.

Embryo Transfer Technologies

Embryo transfer (ET) involves collection of embryos from donor females and their transfer to recipient females. There are two procedures presently available for production of embryos from donor females.

The first procedure consists of super ovulation of the donor through hormone administration,artificial insemination of donor females,flushing of uterus (after a certain interval) to collect embryos and implantation of embryos to recipients immediately after recovery or following freezing at a later time.

The other procedure consists of collection of eggs from ovaries, their maturation and fertilization in the laboratory (*in vitro).* The resulting embryos are cultured on to a certain stage at which point they either are transferred to recipients immediately or can be frozen for later transfer. This whole process is sometimes known as IVF (*in vitro* fertilization).

The main advantage of *in vitro* fertilization over traditional methods of embryo recovery is to take the benefit of thousands of eggs present in female ovaries that never develop to point of ovulation. It thus increases the number of offspring per female. The eggs can be collected from the same donor repeatedly at short intervals irrespective of the stage of estrus cycle. It has the potential of decreasing the generation interval through collection of eggs from young pre-pubertal females thus producing the required number of offspring at a younger age. In vitro production of embryos is expensive, has low success rate, and has therefore not been of practical use so far. IVF facilitates the recovery of a large number of embryos from a single female at a reduced cost thus making ET techniques economically feasible on a large scale.

Embryos transfer was originally developed as a surgical technique. Over the past 20 to 30 years, methods of recovery, storing and implanting cattle embryos have been perfected. It is now possible to super ovulate cows and to recover non-surgically upto thirty embryos at a time though average yield is about 5.5. Freezing in liquid nitrogen and thawing are now routine operations, and implantation with an insemination gun is a non-surgical routine.

Benefits: The procedure of multiple ovulation and embryo transfer, referred to as **MOET**, potentially offers a number of advantages.

- *Increased genetic gain*: The principal benefit of embryo transfer is to produce several progeny from a cow (female) just as through AI a male can produce many offspring. For example, the average lifetime production of a cow can be increased from 4 to 25 calves. This increased reproductive rate has the benefit of increasing the female selection intensity as well as the female accuracy and thus enhancing the rate of genetic gain through female path of selection upto about two-fold increase in rate of genetic improvement in cattle and sheep has been predicted to be possible through MOET.

- *International trading*: MOET makes international trading of germplasm easy by reducing the cost of international transport by shipping embryos rather than live animals, reducing risk of importing diseases, and is also beneficial from animal-welfare viewpoint. Therefore, embryo transfer is the cheapest and the safest way of import or export of germplasm.
- *Transmission of genetic improvement*: MOET could be highly effective technique for dissemination of genetic improvement from upper tier (elite sector) to the commercial sectors of livestock industries. Since embryos already contain full complement of chromosomes necessary for their development, the MOET can deliver populations of commercial animals with 100% of their genes from elite sector parents.
- *Genetic conservation*: Rapid expansion of rare or newly introduced breeds than possible with grading up (repeated) backcrossing is possible by this technology. Also, it is useful in *ex-situ* preservation of endangered breeds (species).

The greatest disadvantage of MOET is the increased level of inbreeding since fewer females make a larger contribution to genetic makeup of the next generation. This is particularly important in small and closed nucleus populations. With current MOET techniques, each donor female produces relatively few embryos at least compared to the number of animals produced by males through AI. Currently MOET is a more expensive than AI and is far less effective than AI in increasing rate of genetic improvement.

Semen and Embryo Sexing

Sexing of semen has been attempted since early 1920 has or even before. Incidentally, the PhD dissertation of Prof Lush, father of scientific animal breeding, was on semen sexing though without success. The basis of most of the methods for separation of sperms carrying X- and Y-chromosomes has been the presence of physical differences between them. These methods included using differences in mass, surface charge, antigenic properties, buoyant density of sperms carrying X and Y bearing sperms. The success rate and repeatability of these methods was low. The method of flow cytometery developed later has proved successful with higher accuracy.

The basis for differentiating between X- and Y- bearing sperms is the slight difference in their DNA content. The procedure consists of staining of sperms with fluorescent stain and using differences in fluorescence when cells pass through a laser beam to identify and sort X- and Y- bearing sperm.

Sexing of embryos is accomplished by removing a small number of cells at 16-cell or later stage of development. Sex of embryo is then determined by karyotyping to confirm the presence of sex chromosomes. Other procedures include the use of immunological test for the H-Y antigen produced only by male mammalian embryos, use of sex-linked differences in enzyme activity or newer molecular techniques to detect sequences of DNA which are specific to Y chromosome. In animal breeding, sexing of semen and embryos holds the potential of enhancing genetic gain through:

- *Increased selection intensity:* Sexing of either semen (sperms) or embryos could be useful in increasing the selection intensity applied to females by producing more of them (in species where the main traits of interest of interest are limited to females).

It will, however, be of little value when performance records are available on both sexes, and where the total number of animals tested is fixed, as an increase in the number of females implies a reduction in the number of males, and therefore, a reduction in male selection intensities. However, when the trait is sex-limited (e.g. milk and egg production), there may be benefits in creating more animals of the recorded sex. This may be beneficial in nucleus schemes if they already have access to high genetic merit males which have been accurately tested in a separate population.

- *Faster dissemination of genetic improvement:* The development of a cheap, reliable technique for sexing semen in large quantities for conventional AI could lead to major improvements in dissemination of genetic improvement.

The development of techniques for production of large numbers of cheap, sexed embryos could lead to use of high genetic-merit female's embryos to produce replacements.

Embryo Cloning

Production of groups of genetically identical embryos from a single original embryo is termed embryo cloning or embryo multiplication. At present embryo cloning can be achieved in two ways:

- *Embryo splitting:* Physical splitting of an embryo to produce twin embryos which are then transferred to recipient females, is a relatively simple technique of production of identical offspring. Splitting into more than two parts is not successful as there are few cells available for normal development.
- *Nuclear transplantation:* Nuclei from cells of early embryos (16-, 32-, or 64-cell) are transferred to unfertilized egg from which the nucleus containing single copy of chromosomes has been removed. An electric current causes fusion of introduced embryonic cell and the unfertilized egg. And the new embryo then develops as if newly fertilized.

Until mid 1990's, nuclear transfer was accomplished using early embryos. Subsequently, viable cloned embryos as well as live cloned animals had been produced by transferring cells which were originally derived from embryos but had been cultured and multiplied in the laboratory. This procedure gives the potential for production of far greater number of identical animals to be produced from a single pair of elite parents. In 1997, a viable lamb was produced following nuclear transfer from a cultured cell originating from mammary tissue of an adult sheep. These developments have the potential to increase success rates and reduce costs.

Potential benefits: Cloned embryos, in monotonous species like cattle, could be a useful for some purposes viz. evaluation of individuals via their clones; cloning of elite individuals for use in industry (e.g. beef bulls for natural mating). Cloning offers the potentiality to improve the accuracy of genetic evaluation on the female side but its impact is small.

Possible disadvantages: The potential disadvantages of embryo cloning include :

(i) The risk of producing large number of animals which may subsequently turn out to be susceptible to diseases or to be unsuitable for new production methods. A widely reported problem with animals derived from nuclear transfer embryos concerns a number of abnormalities viz; extended gestation, increased stillbirth rate and perinatal deaths.

(ii) Cloning may reduce genetic variation as a population is reduced to small numbers and inbreeding increases.

(iii) Expensive and time consuming.

Overview of Some of Developments in Molecular Biology

Advances in molecular biology beginning from 1970's have allowed the identification and flagging the location of functional genes using DNA markers. The most common of these are;

- **Discovery of restriction enzymes**: Bacteria produce a large number of enzymes, called restriction enzymes, which are capable of 'cutting up' sequences of DNA at specific recognition sites. Different restriction enzymes recognize different sequences of four to six base pairs in DNA, and make their cuts whenever they come across such sequences of bases in a strand of DNA. This property of restriction enzymes has been used in Animal and plant genomics (species characterization, cloning and in vitro gene expression, to identify and use molecular markers etc). Treating or digesting identical sequences of DNA with same enzymes produces identical sets of DNA fragments, each of a specific size. Conversely treating different sequences of DNA with the same restriction enzyme produces different sets of fragments of different sizes. Therefore, restriction enzymes are useful in detecting similarities and differences in DNA sequences of different animals.
- **Separation and detection of DNA fragments:** Fragments of DNA of different lengths, resulting from digestion with restriction enzymes, are separated by applying a sample of the fragments of mixed sizes through electrophoresis. The movement of the fragments in the gel is inversely proportional to their size; the large fragments move slowly while small fragments move quickly. The gel can be stained to display a series of bands, with each band representing DNA fragments of particular size. When samples of DNA from different animals are applied side by side to the same gel, a series of bands can be detected in the respective lanes representing the experimental sample. By comparing the patterns of bands across columns, it is possible to tell whether different samples contain DNA fragments of a similar length.
- **Multiplying sequences of DNA:** There are two methods for the rapid multiplication of a target DNA sequence viz. molecular cloning and polymerase chain reaction (PCR) of the DNA fragment. The former (i.e. cloning) is a method of multiplying DNA in living cells. In this case, the mammalian DNA fragments, which are to be multiplied, can be inserted into the DNA from a vector (such as a plasmid like, pUC, pBR322) which occurs independently in some bacterial cells). Incorporation of foreign DNA is achieved by treating the vector DNA with the same/similar restriction enzyme that was used to create fragments of mammalians DNA. The target DNA gets inserted into the vector and produces a circular recombinant plasmid. These recombinant DNA molecules can then be introduced into a rapidly reproducing host (e.g. bacterium in case of plasmid vectors). DNA fragments reproduce with the host cell DNA, and thus producing multiple copies of the original DNA fragments which can then be extracted from host cells. The other method, called polymerase chain reaction or PCR, involves DNA multiplication in a test tube rather than in living cells. PCR produces large qualities of DNA in a matter of hours, and needs only very small amounts of DNA initially e.g. from hair follicles.
- **Development of DNA sequencing technology:** DNA sequencing technology has been used to sequence a particular fragment of DNA (produced by PCR or cloning or RE digestion) or the whole genome (shotgun sequencing and next generation sequencing).. DNA sequencing is the process of determining the exact order of the bases A, T, C, and G on a strand of DNA. The DNA to be sequenced by Sanger

sequencing method is used to generate a set of fragments that differ in length from each other by one base pair. The fragments are separated by size using electrophoresis. By reading, the gel from the bottom up, the sequence of DNA can be determined. The most commonly used method of sequencing DNA was developed by Sanger in 1977. In recent years, 'next generation sequencing' has become globally the main player in the area of genomics and molecular breeding. In this approach, the whole genome is fragmented by physical or chemical shearing and the fragments are sequenced in parallel (that's why it is also called as massive parallel sequencing) by any one of the chemistry like, pyrosequencing, reversible dye terminator method etc.

Molecular Genetic Technologies

In animal breeding, two broad categories of traits are recognized viz. qualitative and quantitative. The qualitative traits are controlled by single or few genes of large effect (e.g. coat color, presence/absence of horns etc.) Such traits exhibit close and direct association between the gene and its effect (phenotype) which implies that individuals carrying such genes can be easily identified. The main problem in determining the genotype for such traits is to discriminate between homozygote dominant and heterozygote individuals. Traditional method for detection of heterozygotes, that involved test crossing, has not been very feasible in most farm animal species.

Most of the traits of economic importance in livestock (e.g. milk, meat, egg production etc) are determined by a large number of loci. Consequently, the link between the genes and phenotype (performance) for such traits is complex, and it is difficult to identify any single gene that has a significant influence on the phenotype. To accurately identify individuals with superior genes, it is necessary to measure performance of large number of their relatives (including progeny) and to compare the performance of different animals under as similar environmental conditions as possible. Such a selection does not involve knowing the number of genes influencing the trait, the effect of genes on the trait-phenotype and location of the genes on the chromosomes. Despite this limitation, tremendous genetic progress has been achieved in all species of domestic animals with the use of so-called 'traditional methods' of selection.

Within the past few decades, advances in new molecular technologies have opened up prospects of accurately identifying genetically superior individuals, and thus to supplement 'traditional' animal breeding methods. These include a number of technologies like gene transfer, DNA fingerprinting, marker assisted selection. The most important of these for enhancing the rate of genetic improvement of economic traits is marker assisted selection.

Marker Assisted Selection (MAS)

The term **'marker'** implies a 'constituent that determines the function of a construction'. Genetic markers are landmarks on chromosomes that serve the purpose of location of other genes of interest. Breeders are interested in knowing the association (linkage) of markers to genes controlling the traits of interest.

The rationale of markers is that marker is tightly linked to a trait that is difficult to be observed but is desirable for the breeder. Hence, breeders select for the trait of interest by indirectly selecting for the marker that is readily detected or observed. The basic premise of use of marker or genetic marker that it is easily detectable and is linked to the gene (trait) of interest.

The concept of marker is not new to breeding of livestock and plants. Breeders have used morphological markers for a long period. Some of the earlier examples of markers included the association of a simply inherited traits such as variation in coat color, shape of horns, body type etc with some traits of economic interest. Such markers are often detectable by eye, by simple visual inspection, hence called **morphological markers**.

The next category of markers were biochemical attributes that can be extracted and observed (e.g. isozymes, proteins etc).These involved chemical or immunological assays to detect variations in protein products (products of translation) but not the variation in DNA *per se*. One of the most well documented examples of such markers was of the association of blood-group types with genes controlling virus resistance in chickens. Their use was quite limited due to the insufficient number of assays available and their uneven distribution on the genetic map etc.

The basic premise of use of markers is based on the close linkage between the marker gene and the trait of interest (genetic trait). Therefore, the value of a marker depends on the closeness of its linkage to the trait of interest; closer the linkage lesser the chances of recombinants, and in turn more is the reliability of the marker.

Molecular or DNA Markers

DNA markers, also called molecular markers, came with the discovery of restriction fragment length polymorphisms (RFLPs) in the 1980s. A DNA-based or **molecular marker** is a unique gene (DNA sequence), occurring in proximity to the gene or locus of interest, and hence it segregates in a predictable pattern as the trait of interest. Some molecular markers are pieces of DNA that have no known function or influence on the performance of the animal. Other markers may involve the gene of interest itself.

- **Linked Markers**: Using well-designed experiments, one can find molecular markers that are located very close to major genes of interest; the molecular marker is then said to be linked to the gene of interest, and thus called **linked marker.** Linked markers are only near the gene of interest on the chromosome and are not part of the DNA of the gene of interest. As an example, suppose we want to locate a certain gene. Choosing animals randomly from a population and studying them would give us no clues about whether a marker is associated with the gene. However, if we study the progeny (offspring) of mating of sire and dams through many generations, we may be able to determine the presence of a useful molecular marker.
- **Direct Markers:** A second kind of molecular marker is one that is part of the gene of interest and thus called **direct marker.** Direct markers are easier to work with once they have been found but they often are more difficult to find than linked markers.

DNA-based markers can be identified by a range of molecular techniques. Three common technologies used for detection of molecular markers are restriction fragment length polymorphisms (RFLP), simple sequence repeats (SSR), and single nucleotide polymorphisms (SNP):

(*a*) **Restriction Fragment Length Polymorphisms (RFLPs):** Restriction fragment length polymorphisms (RFLPs) were the first molecular markers used to diagnose genetic variability in organisms. Since animals differ from one another for the number and position of sites at which a given restriction enzyme will cut the DNA molecule, this variation is called 'restriction fragment length polymorphism'. The number of DNA fragments generated by one restriction enzyme digest can be in the millions, with many being several thousand nucleotides long. This makes it difficult to determine specific DNA fragments that are associated with the trait of interest on the electrophoresis gel. To help visualize the specific DNA fragments, a technique called Southern blotting was developed.

Southern blotting uses a porous membrane containing specific radioactive DNA probes for one or more DNA fragments. **Probes** are very short pieces of DNA used to find specific sequences of "A", "C", "T", and "G" in very long pieces of DNA from a chromosome. The probe hybridizes (attaches) to the membrane at a unique DNA band on an electrophoresis gel. The membrane containing the probe is developed on X-ray film and analyzed.

RFLP's are often located in functional genes which increase their value as potential markers. The major limitation of use of this type of marker is that in many cases a given RFLP only involves two different variants or alleles[1]. Therefore, animals can only be assigned to a small number of classes or genotypes.

(b) **Simple Sequence Repeats or Microsatellites**: There are many regions throughout the genome where same sequence of base-pairs is repeated several times end to end. These repeat sequences are termed microsatellites (also called simple sequence repeats, (SSRs).These are 2 to 6 base pairs long. The sequence "ATATATAT" is one example of a dinucleotide SSR. Similarly, the sequence "GATGATGAT" is a trinucleotide SSR.

The number of times a given sequence of bases is repeated varies between animals. Because of this variation, these repeat sequences can be used as markers. SSRs or microsatellites are useful as molecular markers for selection as they are highly polymorphic (have many forms or alleles at a particular site or locus).Therefore animals can be assigned to many different classes or genotypes which makes them useful in selection (marker assisted selection).

Some important features of RFLP's and microsatellite markers which make them useful are:

(i) These are co-dominant such that homozygote and heterozygote genotypes of the marker locus can be distinguished.

(ii) Being DNA-based, they can be applied to animals of both sexes and of any age.

(iii) Those based on PCR need only a small amount of DNA.

(c) **Single Nucleotide Polymorphisms (SNPs)**: As suggested by the acronym, an **SNP** (single nucleotide polymorphism) marker is just a single base change in a DNA sequence with a usual alternative of two possible nucleotides at a given position. For such a base position with sequence alternatives in genomic DNA to be considered as an SNP, it is considered that the least frequent allele should have a frequency of 1% or greater. Thus on average, SNPs will occur in an organism's DNA more than 1% of the time. Because only about 3 to 5% of an organism's DNA codes for proteins, most SNPs are found outside the regions of genes of interest. SNPs found in a gene of interest are of particular interest to researchers because they are directly associated with a desired trait.

Recent advances have increased the role of SNPs in selection. SNP, (sometimes called epitome of molecular markers), covers a very wide range of genome such that there is a high probability for many SNPs to be in close proximity to most genes of interest. SNPs are the basis of DNA chips, which have thousands complementary

[1]The terms allele and locus are still used for markers even though the markers may not be a functional gene.

DNA fragments arranged on a small matrix and are capable of scoring large number of loci simultaneously. Sequence analysis of either specific DNA fragments or entire genomes can also be carried out.

Use of Molecular Markers in Selection

Molecular markers being DNA-based remain same throughout an individual's life and are not influenced by environment.Two broad applications of molecular markers in genetic improvement of farm animal are: gene introgression and marker assisted selection.

The traditional procedure of introgression of a desirable allele into a population that was not carrying that allele, called **gene introgression**, consisted of crossing between the donor and recipient populations followed by backcrossing to the original breed (population) for several generations. Molecular markers by accurately identifying the animals carrying the desired allele help in speeding up the process of recovering of the original genotype. Such an application of molecular markers is called **'marker assisted introgression'**. Most of the traits for which molecular-based introgression is made are single genes.

However, the major application of molecular markers is in selection of genetically superior animals. Selection based on markers is referred to as **marker assisted selection** (MAS). Majority of MAS work in the present era uses DNA-based markers. Marker assisted selection (MAS) is thus an **indirect selection** where a trait of interest is selected, not based on the trait itself, but on marker(s) which is linked to it.

MAS can be used for both single-gene and polygenic traits. A number of major genes are known to be useful in animal breeding. MAS was originally used for single-gene controlled traits where it was successful for its intended purpose. But the main problems of selection in animal breeding relates to economic traits which are controlled by a large number of genes scattered throughout the genome, and these traits are more influenced by environmental factors.

Situations for use of MAS for quantitative traits: Marker assisted selection for quantitative traits would be especially useful for enhancing the rate of genetic progress than possible through 'conventional' selection criteria. These include:

- *Low heritability traits*: Genetic gain is a function of how accurately the genetically elite animals within a population can be identified; the greater the accuracy, all else being equal, greater will be the genetic gain.
- *Traits measured late in life*: Traditional selection for such traits involves long-generation interval,and to some extent, lower selection intensity due to availability of lesser number of animals for selection at a later stage. However, being able to accurately identify genetically elite animals at a younger age it can increase annual genetic gain further and can also reduce the cost of a breeding program as breeding candidates have to be kept for a shorter time.
- *Slaughter- and sex-limited traits*: Traits requiring slaughter of animals for recording can not be selected by individual selection. Other selection procedures like sib-selection are expensive for recording and have lower accuracy as well. Similarly, traits which are expressed in one sex are difficult to be selected by conventional selection procedures. MAS can be an effective procedure for their genetic improvement.

Steps in Marker Assisted Selection

Marker assisted selection for quantitative traits is a combined product of 'traditional' selection which involves performance and pedigree of animals (individual's as well as of its relatives) and the molecular information generated through DNA-based techniques. The information from DNA testing, combined with the observed performance records for individuals, is intended to improve the accuracy of selection and increase the possibility of identifying animals carrying desirable and undesirable traits (genes) at an earlier stage of development.

Concept of QTL's : The concept of QTL (Quantitative trait loci) needs to be understood for application of MAS. As indicated earlier, quantitative traits are of a major concern to animal breeders because most of traits of economic importance are quantitative traits. These traits are controlled by polygenes or QTLs. QTLs are 'genetic factors' that are responsible for a part of the observed phenotypic variation for a quantitative trait (polygenic trait).

A QTL is similar to a gene but actually, it merely indicates a region on the genome that may comprise one or more functional genes. QTL mapping involves an analysis of the association between observed trait-values (phenotypic value) and the presence or absence of alleles of markers that have been mapped onto a linkage map. A QTL is declared to exist when it is determined that the association did not occur by some random process.

Application of MAS in a selection program involves the following:

(i) *Gene mapping :* Identification and mapping of the marker.

(ii) *Marker genotyping*: Genotyping of large number of individuals for a number of markers.

(iii) *QTL detection*: Detection of the quantitative trait loci (QTL) underlying the trait of interest in the genome.

(iv) *Establishing association:* Establishing association of genetic markers with economic trait of interest in the 'training or discovery' set of animals.

(v) *Marker validation:* Validation of the identified marker in an independent 'test population'

(vi) *Genetic evaluation*: Integration of performance (phenotypic), pedigree and marker information using appropriate statistical methodology to estimate breeding values of individuals in the population. Breeding value calculated by incorporating performance, pedigree and genomic marker information is called **molecular breeding value.**

When the markers which are spanned over the entire genome are analyzed for detecting the key SNPs (single nucleotide polymorphism) contributing to the trait of interest, it is called **genomic selection.** The basic assumption of genomic selection is that at least one marker should be in linkage disequilibrium (non-random association between the loci) with the QTL. As a result, the accuracy of selection and reliability of prediction are increased along with the genetic gain (due to decrease in generation interval).

Problems in use of MAS: The practical application of MAS for complex (quantitative) traits is still challenging partly because of the difficulty of effective detection, estimation, and utility of QTLs and their effects. The problem is more significant for economic traits, that besides being controlled by large number of genes, are also under the influence of epistasis (gene-by-gene interaction) and gene-by-environment ($G \times E$) interaction effects.

Selection using QTL is more complex than selection based on performance of the animal and or its relatives. Information generated through molecular markers by itself is not of utility in selection, as has to be integrated with traditional selection criteria. In that context, molecular/ QTL information is another source of information like an additional trait or selection criteria. For effective implementation of MAS, optimal integration of both types of information (i.e. relative weights to be given for phenotypic and marker information), with sophisticated statistical models is required.

Issues such as reduction in selection intensities resulting from multi-source information will also influence genetic gain in net merit. It has also been indicated that short-term gains due to MAS may be at the expense of medium- to long-term gains in quantitative traits.

Future practical utility of MAS in animal breading can not be forecast but it is doubtful if it will replace the existing selection procedures. MAS should therefore be regarded as a tool to supplement existing selection procedures rather them to replace them.

Gene Transfer (Transgenesis)

Gene transfer or transgenesis is a procedure, which involves stable incorporation of specific gene (s) from one species into the genome of another species in such a way that it functions in the receiving species and is passed on from one generation to the next. This entails physically inserting the DNA coding for specific gene (s) into the DNA of another species (animal). The foreign or introduced genes are referred to as 'transgenes' to distinguish them from endogenous genes. And the animal that contains the introduced genes (the foreign DNA) is called a **transgenic animal**. Transfer of gene(s) allows the manipulation of individual genes rather than the entire genomes. The premise on which transgenesis is based is the high level of conservation of genes not only within species but also between species, and potentially between different classes of organism. The gene- transfer technologies are the outcome of a number of molecular genetic techniques.

The primary features of gene transfer technology are:

- To identify, locate and multiply specific gene (s) of interest (i.e. genes with desirable effect).
- To remove the gene and its controlling elements from the DNA molecule in which it normally resides
- To insert it into a carrier DNA molecule (vector) and to have the vector transferred and integrated into the genome of the host

Two methods are predominantly used to produce transgenic animals:

- Direct microinjection of foreign DNA into the pronuclei of fertilized eggs
- Infection of preimplantation embryos with retroviral vectors. In mammals, gene insertion is mostly accomplished by the first method (e.g. microinjection of DNA into fertilized eggs nucleus at the early embryonic stage).

The first successful gene transfer in animals was carried out in mice where the gene coding for human growth hormone was introduced into the genome of a mouse. The transgenic mouse grew to about twice the size of normal mouse. Since then, over hundreds of different genes have been transferred into recipient mouse eggs or embryos and stably maintained in adult mice and their progeny. Transgenic mice are now routinely produced and studied in laboratories throughout the world.

Since then, there have been dramatic advances in gene transfer technology. The technique has now become routine in mouse and the resulting transgenic mice are able to transmit their transgenes to offspring thereby allowing a large number of transgenic animals to be produced. Transgenic sheep, pigs, rabbits, and chickens have also been produced.

The proper expression of genes from other mammals and poultry in transgenic mice indicates that many of the regulatory sequences in all species of animals are highly conserved. Consequently, species-specific regulatory sequences may not be necessary for producing transgenics.

Applications The technology for creating transgenic plants and mice has been extended to a number of domestic species of animals. Two broad uses of this technology are relevant to animal production:

(i) Improving productivity of animals: Increased growth rate of transgenic mice which overproduced bovine or human growth hormone has prompted animal breeders for the introduction of foreign genes (either extra copies of the same species growth hormone gene or copies of growth hormone genes from related species) into livestock to achieve higher growth and productivity. Genes responsible for some specific functions of the animal viz; increased growth, production and disease tolerance, if available, can be incorporated from one species into another using DNA manipulation techniques. Under natural conditions, reproductive incompatibility between species does not allow interspecific gene transfer. Some examples of interspecific gene transfer are:

- Growth hormone gene has been incorporated into pigs for producing leaner animals with improved meat quality and faster growth.
- Transgenic chickens carrying gene responsible for resistance against avian leukosis virus have been produced. The resistance gene has been transmitted to 5th generation progeny, thus indicating its stability. This example indicates that the use of transgenic chicken may provide one approach for production of virus resistant domestic animals.
- Gene responsible for cysteine synthesis has been transferred into sheep with the objective of enhancing wool production
- Gene responsible for cold- tolerance has been transferred from flounder into salmon.

In all these cases, the transferred genes fall in the category of qualitative genes which are single genes exerting major effect on trait- phenotype.

(ii) Production of valuable pharmaceutical products: There are a number of proteins used in ttreatment of human diseases. These proteins could be produced in abundant quantities in milk of livestock. Such novel pharmaceutical proteins are not a part of normal functioning of the animal but for these proteins a farm species is a convenient medium of production. Production of medically important proteins in mammary glands of livestock through introduction of cloned gene (i.e. through recombinant DNA technology) is sometimes called ***molecular pharming***.

Several human proteins have been successfully produced in milk of livestock. Compared with production of such proteins in bacteria, their production in mammals is advantageous because human protein is likely to function properly when expressed in mammals due to of their tissue-specific expression. Further milking strains of cattle, sheep and goats have some other advantages : they synthesize a range of proteins that make up about 40% of solids in milk. The relative cost of production of a transgenic protein through milk of livestock is much lesser than through recombinant bacteria. A few examples of such proteins are given in Table 23.2

Table 23.2 Examples of some proteins that can be produced in milk of domestic animals.

Protein	Host	Uses
Lactoferrin :	Cattle	Used as iron supplement in infant formula.
Tissue plasminogen activator (TPA) :	Goat	Dissolves blood clots.
L-1-Antitrypsin :	Sheep	Treatment of emphysema.
Factor IX :	Sheep	Treatment of some inherited forms of hemophilia.
Insulin-like growth factor :	Cattle	Treatment of diabetes.

Constraints: Though the transfer of single genes has been achieved in all the major livestock species but there are still a number of constraints in its wide-scale application

(i) **High cost and low success rate**: The process of gene transfer between species is slow because it involves separate steps. The success rates are often low usually one or two percent (Table 23.3).

Table 23.3 Embryo- survival and transgene- integration efficiencies.

Species	No. of injected transferred embryos	% Resulting in: Offspring	Transgenic offspring
Mice :	12314	15.0	2.6
Rabbits :	1907	11.4	1.5
Rats :	1403	25.2	4.4
Sheep :	5424	10.3	0.9
Pigs :	19397	10.0	0.9
Cattle :	1018	19.0	0.7

As a broad generalization in farm, animal species about one in ten injected and transferred embryos survives, and about one in ten of these carry the transgene. Among these transgenic animals, normally about half express the transgene. And in those which show the expression, the gene may not be activated in right tissues and at the right time in animal's development. This unpredictability of gene expression is perhaps not surprising given that it is not possible to control the site of integration into the host genome, nor the number of copies integrated. Furthermore, transgene transmission to the next generation is sometimes abnormal. The most widely accepted explanation for abnormality of transmission of the inserted gene is mosaicism where the transgene is present only in some cells of the developing parental embryo. The understanding of regulation of gene expression is far from complete. All these problems impose an enormous cost in livestock species.

(ii) **Side effects or consumers unacceptability**: One of the major problems in the production of transgenic livestock concerns its undesirable side effects and welfare of animals. For example, incorporation of growth hormone gene into pigs and sheep led to clinical diabetes, arthritis, infertility, lameness etc leading to early death. Consumers in many countries are strongly against transgenic livestock. The efficiency of transgenesis also needs to be dramatically improved, and its unacceptable side effects on health and welfare of animals need to be overcome its so that it can be come commercially viable.

Further, transgenesis for livestock production has to compete with the current technologies used for the same within a species. Genes for high productivity (meat, milk, wool, and eggs) or for reducing cost (disease resistance) are likely to be found within the species concerned. The traditional normal selection programs are quite efficient for their utilization.

DNA Fingerprinting

DNA fingerprinting (also called DNA profiling, DNA testing, or DNA typing) is a molecular technique for graphically characterizing an individual's DNA, obtained from any tissue. DNA fingerprinting is a method of isolating and identifying variable elements within the base-pair sequence of DNA (deoxyribonucleic acid). This technique was developed in 1984 by British geneticist Jeffreys, after he noticed that certain sequences of highly variable DNA (known as minisatellites), which do not contribute to functions of genes, are repeated within genes (repeated sequences in minisattelites are 10 to 60 base pairs long). Each individual has a unique pattern of minisatellites (the only exceptions being multiple individuals from a single zygote, such as identical twins).

DNA fingerprinting is widely used in parentage determination and forensic sciences. In animal breeding, DNA fingerprinting has use in parentage identification in pedigreed populations, particularly under natural mating like in beef cattle reared on multiple-sire ranches. Its other potential or academic uses include prediction of heterosis from specific crosses. By comparing DNA fingerprints of parental lines to be used in crossing, one may be able to determine the degree of genetic un-relatedness between the lines. Higher the un-relatedness, more the heterosis to expect in crossbreds of the lines.

Review Questions

1. Define/explain the following:
   (a) Biotechnology
   (b) Artificial insemination
   (c) Embryo transfer technology
   (d) Embryo cloning
   (e) Molecular marker
   (f) Restriction fragment length polymorphism
   (g) Single nucleotide polymorphism
   (h) DNA fingerprinting
   (i) Transgenesis
2. What have been the different uses of AI in animal breeding?
3. List the main genetic problem associated with-wide spread use of AI in genetic improvement programs.
4. What are the potential advantages of MOET ?
5. What are the potential advantages of semen sexing in animal breeding?
6. What potentiality does embryo cloning holds in animal breeding?
7. List the important features of RFLP and microsatellite markers
8. What are the situations for use of marker assisted selection?
9. List the steps involved in application of MAS.

●●●